Green and Sustainable Manufacturing of Advanced Materials

Green and Sustainable Manufacturing of Advanced Materials

Edited by

Mrityunjay Singh

Tatsuki Ohji

Rajiv Asthana

ELSEVIER

AMSTERDAM • BOSTON • HEIDELBERG • LONDON • NEW YORK • OXFORD
PARIS • SAN DIEGO • SAN FRANCISCO • SINGAPORE • SYDNEY • TOKYO

Elsevier
Radarweg 29, PO Box 211, 1000 AE Amsterdam, Netherlands
The Boulevard, Langford Lane, Kidlington, Oxford OX5 1GB, UK
225 Wyman Street, Waltham, MA 02451, USA

Notices

Knowledge and best practice in this field are constantly changing. As new research and experience broaden our understanding, changes in research methods, professional practices, or medical treatment may become necessary.

Practitioners and researchers must always rely on their own experience and knowledge in evaluating and using any information, methods, compounds, or experiments described herein. In using such information or methods they should be mindful of their own safety and the safety of others, including parties for whom they have a professional responsibility.

To the fullest extent of the law, neither the Publisher nor the authors, contributors, or editors, assume any liability for any injury and/or damage to persons or property as a matter of products liability, negligence or otherwise, or from any use or operation of any methods, products, instructions, or ideas contained in the material herein.

Library of Congress Cataloging-in-Publication Data
A catalog record for this book is available from the Library of Congress

British Library Cataloguing in Publication Data
A catalogue record for this book is available from the British Library

ISBN: 978-0-12-411497-5

For information on all Elsevier publications
visit our website at http://store.elsevier.com/

Printed in China

Contents

Part II
Sustainable Manufacturing—Metallic Materials

6. High-Performance Steels for Sustainable Manufacturing of Vehicles

H. Mohrbacher

7. Advanced Steel Alloys for Sustainable Power Generation

H. Mohrbacher

Part III
Sustainable Manufacturing—Ceramic Materials

8. Smart Powder Processing for Green Technologies

M. Naito and A. Kondo

14. Future Development of Lead-Free Piezoelectrics by Domain Wall Engineering

S. Wada

15. Nanostructuring of Metal Oxides in Aqueous Solutions

Yoshitake Masuda, Kazumi Kato, Tatsuki Ohji and Kunihito Koumoto

Part IV
Sustainable Manufacturing—Polymeric and Composite Materials

Contributors

D. Andersson Swerea IVF AB, Mölndal, Sweden

R. Asthana University of Wisconsin-Stout, Menomonie, WI, USA

Marsha S. Bischel Armstrong World Industries, Inc., Lancaster, PA, USA

A. Dinsdale National Physical Laboratory, Teddington, United Kingdom

J. Fu School of Materials Science and Engineering, Zhengzhou University, Zhengzhou, China

Manabu Fukushima National Institute of Advanced Industrial Science and Technology (AIST), Nagoya, Japan

S. Gupta University of North Dakota, Grand Forks, ND, USA

I. Himoto Department of Molecular Design and Engineering, Graduate School of Engineering, Nagoya University, Nagoya, Japan

Y. Hotta National Institute of Advanced Industrial Science and Technology (AIST), Nagoya, Japan

Hideki Hyuga National Institute of Advanced Industrial Science and Technology (AIST), Nagoya, Japan

H. Ipser University of Vienna, Department of Inorganic Chemistry/Materials Chemistry, Wien, Austria

Toshihiro Ishikawa Tokyo University of Science, Yamaguchi, Japan

Kazumi Kato National Institute of Advanced Industrial Science and Technology (AIST), Nagoya, Japan

Soshu Kirihara Joining and Welding Research Institute, Center of Excellence for Advanced Structural and Functional Materials Design, Osaka University, Osaka, Japan

H. Kita Department of Molecular Design and Engineering, Graduate School of Engineering, Nagoya University, Nagoya, Japan

A. Kondo Joining and Welding Research Institute, Osaka University, Ibaraki city, Osaka, Japan

Naoki Kondo National Institute of Advanced Industrial Science and Technology (AIST), Nagoya, Japan

Kunihito Koumoto Department of Applied Chemistry, Graduate School of Engineering, Nagoya University, Nagoya, Japan

A. Kroupa Institute of Physics of Materials, Brno, Czech Republic

Yin Liu Kazuo Inamori School of Engineering, New York State College of Ceramics at Alfred University, Alfred, New York, USA

Yoshitake Masuda National Institute of Advanced Industrial Science and Technology (AIST), Nagoya, Japan

H. Mohrbacher NiobelCon bvba, Schilde, Belgium

S. Mucklejohn Ceravision Limited, Milton Keynes, United Kingdom

M. Naito Joining and Welding Research Institute, Osaka University, Ibaraki city, Osaka, Japan

H. Nakano Cooperative Research Facility Center, Toyohashi University of Technology, Toyohashi, Japan

Zuoren Nie Beijing University of Technology, Beijing, China

Tatsuki Ohji National Institute of Advanced Industrial Science and Technology (AIST), Nagoya, Japan

T. Sekino ISIR, Osaka University, Osaka, Japan

Shunzo Shimai Tokyo University of Agriculture and Technology, Tokyo, Japan

Mrityunjay Singh Ohio Aerospace Institute, Cleveland OH, USA

J. Tatami Graduate School of Environment and Information Sciences, Yokohama National University, Yokohama, Japan

S. Wada Materials Science and Technology, Interdisciplinary Graduate School of Medical and Engineering, University of Yamanashi, Yamanashi, Japan

A. Watson University of Leeds, Leeds, United Kingdom

Yiquan Wu Kazuo Inamori School of Engineering, New York State College of Ceramics at Alfred University, Alfred, New York, USA

Q. Xu School of Materials Science and Engineering, Zhengzhou University, Zhengzhou, China

S. Yamashita Department of Molecular Design and Engineering, Graduate School of Engineering, Nagoya University, Nagoya, Japan

Yan Yang Kazuo Inamori School of Engineering, New York State College of Ceramics at Alfred University, Alfred, New York, USA

Yu-ichi Yoshizawa National Institute of Advanced Industrial Science and Technology (AIST), Nagoya, Japan

Preface

Over the past several years, sustainability has emerged as a recurring theme that is increasingly recognized as pervading every sphere of human activity. The interdependence of life and the ecosystem in which it resides has begun to drive the need for sustainable growth and development. Perhaps nowhere does the power of sustainability manifest itself more exquisitely than in the development of new materials and manufacturing technology. Future progress in these areas will critically depend on our engagement with sustainable practices in research and technology development.

Materials and manufacturing processes constitute a huge segment of global economy. It is thus fitting and proper that current industrial practices and new developments in materials and manufacturing technology orchestrate with the natural capacity of ecosystems. This demands global efforts to conserve energy and materials, as well as a focus on recovery, recycling, and reuse in an environmentally conscious manner. The integration of green practices is crucial to sustain long-term technological development and the economic competitiveness of modern society as well as future generations.

This book addresses green and sustainable practices by focusing on specific classes of materials. The authors are all active and recognized researchers and practitioners in their respective fields and represent universities, industry, and government and private research organizations of eight different nations. We hope that the book offers a vision for future developments and stimulates fresh thinking to integrate green and sustainable practices in diverse materials and manufacturing processes. We also hope that the book meets the educational and research needs of advanced students across multiple academic disciplines. We are grateful to all of our revered authors for their valuable contributions. We thank the publication and editorial staff of Elsevier for their excellent support during the preparation of this book.

Mrityunjay Singh,
Ohio Aerospace Institute, USA
Tatsuki Ohji,
National Institute of Advanced Industrial Science and Technology (AIST), Japan
Rajiv Asthana
University of Wisconsin-Stout, USA

Part I

Material Conservation, Recovery, Recycling and Reuse

Chapter 1

Green and Sustainable Manufacturing of Advanced Materials—Progress and Prospects

Mrityunjay Singh[1], Tatsuki Ohji[2] and R. Asthana[3]
[1]Ohio Aerospace Institute, Cleveland, OH, USA, [2]National Institute of Advanced Industrial Science and Technology (AIST), Nagoya, Japan, [3]University of Wisconsin-Stout, Menomonie, WI, USA

1 INTRODUCTION

Manufacturing is a substantial part of global economy, and manufacturing practices play a critical role in all aspects of modern life. Green and sustainable manufacturing has emerged as a globally recognized mandate. Sustainable manufacturing is defined by the U.S. Department of Commerce as "the creating of manufactured products that use processes that are nonpolluting, conserve energy and natural resources, and are economically sound and safe for employees, communities, and consumers" (http://www.nacfam.org/PolicyInitiatives/SustainableManufacturing/tabid/64/Default.aspx). It has given impetus to development of green materials and technologies that orchestrate with self-healing and replenishing capability of natural ecosystems. It has focused attention on conservation of energy and precious materials, and recovery, recycling, and reuse in virtually all industrial sectors including but not limited to transportation, agriculture, construction, aerospace, energy, nuclear power, and many others.

Historically, industry and governments have been responsive to environmental issues even before sustainability became a recognized global movement. For example, in the United States, a number of acts and Codes of Federal Regulations (CFR) have addressed key environmental issues for several decades. Examples included the Water Pollution Control Act (amended 1987 Clean Water Act), Clean Air Act (amended 1990), Resource Conservation and Recovery Act (amended 1984), Comprehensive Environmental Response, Compensation and Liability Act (1980), and many others. These regulations provided "cradle-to-grave" programs for protecting human health and the environment from the improper management of hazardous materials including toxic effluents. Other

Green and Sustainable Manufacturing of Advanced Materials. http://dx.doi.org/10.1016/B978-0-12-411497-5.00001-1

CFRs specifically addressed the health and environmental effects of specific chemicals and materials such as the known carcinogens formaldehyde (29 CFR 1910.1048) and cadmium (29 CFR 19190.1027).

Although a focus on sustainable technologies in various forms has been around for a long time in part due to government regulations and sporadic public support for isolated cases that impacted regional concerns, a paradigm shift toward and awareness of the importance of transformative green and sustainable materials and manufacturing has only recently begun to gain momentum. As a field of academic enquiry and discussion, green manufacturing is relatively young. As an emerging global movement, it has gained considerable traction as part of the broader goals of *sustainable development*. It is now being increasingly recognized that the integration of green practices is crucial to sustainable technological development and the economic competitiveness of current society as well as that of future generations.

A number of important and widely practiced industrial processes such as case hardening, plating, casting, brazing, soldering, chemical vapor deposition, organic coatings, and numerous others involve consumption or release of harmful ingredients that are injurious to both human health and the environment. All such processes and technologies are candidates for a careful reassessment of the efficiencies and structural changes that could potentially make such processes sustainable. A classic example of sustainable practices is the abolition of lead in electrical and electronic assemblies and in public utility systems owing to the possibility of water and food contamination with extremely serious consequences to human health and the environment. Major initiatives in Europe, North America, China, Korea, and elsewhere have either banned or strictly limited lead use. Major global initiatives are currently in progress to develop green substitute materials for lead and similar hazardous and/or scarce metals and materials. Critical materials including rare earths have a major economic and strategic importance, but they are limited in supply. New materials need to be developed in an environmentally conscientious manner to offset the dependence of naturally occurring critical and strategic materials.

Another focus area of sustainable development involves component weight reduction by use of light materials (foams, magnesium, and titanium) with high specific strength and other key properties. This is being vigorously pursued for reducing fuel consumption and waste emissions mainly in the transportation sector (automotive and aerospace). This also offers additional benefits of lower losses, higher operating temperatures, and higher engine efficiency. Environmentally friendly materials such as ecoceramics, ecobrass, ecosolders, and ecocoatings, as well as energy-efficient light materials such as foamed metals and ceramics and composites have gained phenomenal ascendency in research and technology. Additionally, energy and emission reduction with the aid of established and novel technology such as microwaves, lasers, and biofuels has become increasingly important.

New materials developed from natural and renewable arboreal and biological resources should continue to gain importance into the future. Ceramics such as silicon carbide developed from such resources consume less energy for their production and less waste for disposal. Environmentally conscious ceramics (ecoceramics) are produced out of renewable resources such as wood. For example, biomorphic silicon carbide is obtained by pyrolysis and infiltration of natural wood-derived preforms. It reduces energy consumption and chemical by-products of conventional ceramic production methods such as hot pressing, sintering, reaction bonding, and chemical vapor deposition (CVD). Other methods include freeze casting of ceramics and microwave sintering that are devoid of binders and fugitive chemicals. Through conscious intervention, materials and products can be designed and manufactured in a more environmentally friendly manner to facilitate assembly, recycling, and reuse with reduced waste emission and energy consumption.

A large proportion of the world's energy originates from fossil fuels while greener technologies such as nuclear, wind, and hydroelectrics generate the remaining share of total energy. The wide variety of materials used in these technologies—mainly metals, ceramics, and their composites—critically affect the performance of such technologies. Materials are enablers of advanced technology, and their properties and performance determine the system function and efficiency. Conversely, efficient development and production of current and emerging materials depends on the availability of innovative technologies for their production and fabrication. A competitive advantage in technology development can be accelerated through the development and application of new materials and processes. This complementary symbiotic relationship can help promote and advance sustainable practices with the materials producer, product designer, and manufacturer working in concert on shared concerns about environmental impact and sustainability while pushing the boundaries of the technology.

Over the next several decades, global demand for materials and energy is projected to sharply rise. This inevitably will impact the environment via increased carbon emission and energy consumption. In this context, an important goal of sustainability is the training and educational needs of a new generation of workforce that can think and act holistically about "cradle-to-grave" and "cradle-to-cradle" progression of materials and technologies. Many major industrial mishaps in the past have been linked to mistakes that could have been avoided with proper training and awareness. Examples include the mercury accumulated in fish originating from a fertilizer plant in Japan in the 1960s and leakage of toxic methyl isocyanate (MIC) gas from a former Union Carbide plant in India in the 1980s.

Nanotechnology is beginning to revolutionize modern manufacturing by offering an unprecedented range of functionalities that are possible only in the nanometer range. Novel functionalities can be achieved via atomic- and molecular-level design paradigms and methods. To achieve these functionalities, manufacturing is poised to become ever more complex and sophisticated. This will require sustainable practices to be built in production and the use or reuse of nanomaterials. Because of the growing push to increase the use of nanomaterials and nanoprocessing, it is vital that their effect on and interaction with the environment is carefully assessed lest unanticipated effects of these new materials and technology precipitate into unprecedented harm to the society and the environment.

It is thus evident that the environmental impact of materials and their manufacturing and use are vital to technological progress of modern society. In the past, performance attributes and issues related to the cost of materials and manufacturing were given precedence over environmental concerns, recyclability, and reuse. There is now a shift toward addressing in a comprehensive fashion all of the environmental attributes of materials, manufacturing, and products via green and sustainable practices and life-cycle assessment (LCA). This includes such vital issues as carbon footprint and global warming.

2 FOCUS AREAS

This book attempts to provide snapshots of selected developments and practices in green and sustainable manufacturing, focusing mainly on new material and process development. It does not purport to be a compendium on this vastly important topic but rather a contemporary resource for up-to-date information on select new developments. The collective knowledge about state-of-the-art and new developments in green and sustainable manufacturing of advanced materials from diverse fields presented in the following chapters is intended to provide further impetus to research and exploration. The following survey provides a brief overview of the main topics and issues that are discussed in different chapters.

2.1 Material Conservation, Recovery, and Recycling

In this chapter, we present a brief overview of how the development and manufacturing of advanced materials influence and are influenced by sustainable practices to serve as a backdrop for the issues and themes that subsequent chapters develop in greater depth. It is argued that future progress in developing new materials and manufacturing processes to produce them will critically depend upon purposeful engagement with sustainable practices in research and technology. In Chapter 2, Bischel from Armstrong World Industries offers insights into holistic assessment of material sustainability that relies on multiple rather than single attributes in defining the material's environmental impact. She advocates design philosophies that concurrently and seamlessly address multiple areas of material performance and sustainability and expose conflicts and trade-offs among attributes, thereby facilitating a more complete evaluation of the environmental impact of a product over its life span. The cornerstones of such a holistic approach include LCAs, environmental product declarations, and

Design for the Environment. Bischel presents several examples of new materials, processes, and products that are efficient, energy lean, economically viable, and environmentally sustainable.

Continuing the thematic focus of Bischel's chapter, Nie in Chapter 3 focuses on ecomaterials and life-cycle assessment. He considers LCA not only as a method to develop environmental profiles for materials but also as a decision-making tool to align the entire industry chain toward sustainability. He discusses ecomaterials and LCA practices to develop robust materials that are less sensitive to defects and consume less energy and raw materials, reduce emissions, and impart properties and functionalities that can be achieved via microstructure design rather than by adding precious or scarce strategic raw materials. With the Chinese experience as a case study, he reviews areas that not only focus on materials and technology but also life-cycle costing and social life-cycle assessment. He presents a number of opportunities where decision making and technical support based on LCA can be implemented in pursuit of broader goals of sustainability.

For sustainable growth, it is imperative to protect the environment, conserve precious or depleting resources, and develop technology that minimizes or eliminates environmental load. In Chapter 4, Kita et al. approach the problem of sustainable manufacturing from a thermodynamic viewpoint. The authors demonstrate how fundamental thermodynamic concepts such as energy, entropy, and exergy can be applied to environmental issues to promote sustainable design of manufacturing systems, identify types and magnitude of wastes, and develop energy-efficient systems to produce ceramics such as silicon nitride.

2.2 Sustainable Manufacturing—Metallic Materials

Many years ago, the use of lead in solders for electrical and electronic assemblies raised serious concerns about the consequences to human health and the environment. Major research and development efforts were initiated across continents to substitute lead with benign elements and to create ecosolders. In Chapter 5, Kroupa and coworkers give a comprehensive overview of scientific, technical, and regulatory issues in designing and developing lead-free solders and environmentally friendly joining approaches, and the likely impact on the industry of the phasing out of lead-bearing solders. Although they focus chiefly on European regulations and directives, their discussion and insights are universal. They highlight issues and challenges to design reliable lead-free solders to substitute the currently used high-lead, high-temperature solders.

In Chapters 6 and 7, Mohrbacher focuses on sustainability issues for the most widely used structural material—advanced ferrous alloys, especially steels—for automotive and power generation applications. Currently, the supremacy of steels in auto body structures is being challenged by lighter aluminum, plastics, and magnesium. This unprecedented challenge has led to new high-performing steels such as dual-phase steels, transformation-induced plasticity (TRIP) steels, and others. In Chapter 6, Mohrbacher describes the emerging applications of these new steels grades in frame, chassis, and other auto parts. He presents an alloy design approach that translates detailed knowledge of the manufacturing process into steel metallurgy contrary to the traditional approach that translates alloy design requirements into manufacturing process.

In the next chapter, Mohrbacher deals with the important topic of sustainable power generation in an era of continuously increasing global energy demand. In the current generation of fossil-fuel-based thermal power plants, the electrical-to-thermal energy ratio (i.e., efficiency) is about 50%. For higher efficiencies, higher operating temperatures can be sustained only by new creep and corrosion-resistance steels which satisfy performance criteria (e.g., strength, toughness, fatigue, wear resistance, weldability, etc.). Examples of advanced alloying and thermomechanical processing are offered to develop sustainable materials solutions for power generation in thermal as well as wind and hydroelectric plants. Mohrbacher notes that cost-benefit analyses establish advanced steels as irreplaceable materials in niche applications.

2.3 Sustainable Manufacturing—Ceramic Materials

Powder-based manufacturing is the most widely practiced industrial technology for ceramics. In Chapter 8, Naito and Kondo describe green technologies for powder processing that minimize energy consumption and environmental impacts. They describe an innovative direct particle bonding technology to make advanced composites via surface activation of nanoparticles but without thermal activation or fugitive binders. Their method simplifies the manufacturing process and tailor makes nano/microstructures including thin porous films that are deposited over substrates. Conversely, separation of composite materials such as glass-fiber-reinforced plastics into elemental components for recycling is also feasible with minimum expenditure of energy and resources.

In Chapter 9, Ohji and coauthors introduce two approaches for green manufacturing of silicon nitride and SiAlON ceramics: (1) sintered reaction bonding with rapid nitridation of silicon powders aided by zirconia (or rare earth oxide) additions and (2) low-temperature (<1700°C) sintering under atmospheric pressure using low-cost, low-grade powders. These processes significantly reduce the energy consumption to produce high-quality parts comparable in properties to gas-pressure sintered high-purity silicon nitride. By combining low-temperature sintering with low-cost powders saves energy and reduces production cost. Likewise, in Chapter 10, Tatami and Nakano focus on green processing of particle-dispersed advanced ceramic composite materials including but not limited to carbon nanotube (CNT)-dispersed silicon nitride composites. Excellent electrical and other properties have been achieved in many such novel composites synthesized with the aid of environmentally friendly manufacturing processes.

In Chapter 11, Fukushima and coauthors describe an environmentally friendly freeze-casting process to create macroporous ceramics for application in filtration, catalysis, thermal insulation, acoustic insulation, and other uses. Freeze casting employs water rather than fugitive organic binders as a pore former, and it can directionally align pores in ceramics with little energy consumption.

Fukushima et al. present an overview of their recent research on porous ceramics prepared by combining freeze-drying and gelcasting. They also discuss engineering properties that make macroporous ceramics promising for a variety of engineering applications. Gelcasting is further discussed by Wu et al. in Chapter 17, where they focus on the synthesis of optically transparent ceramics.

Net-shape manufacturing of engineering ceramic parts usually demands robust and reliable integration technologies. In Chapter 12, Singh and coauthors present a number of proven and emerging ceramic joining and integration technologies such as reaction bonding, diffusion, adhesive bonding, and brazing, with an emphasis on local heating methods to join materials. Conventional brazing uses fluxes that have serious health and environmental implications. These deficiencies are overcome in environmentally benign brazing technologies such as active brazing in vacuum and brazing with localized heating. The chapter describes developments and challenges in these areas and highlights opportunities in sustainable ceramic joining.

Additive manufacturing (AM) of ceramics is an emerging technology with considerable potential in diverse fields. In Chapter 13, Kirihara presents his research on a specific AM method, namely three-dimensional printing using stereolithography and nanoparticle sintering, to create patterned, complex microcomponents from functional ceramics. These components can be designed for dendritic structures having ordered microlattices that modulate the electromagnetic wave propagation and fluid flow for applications in metamaterials and microfluidics. Kirihara gives technical details of the ceramics free forming and applications of the functional dendrite structures.

The next two chapters by Wada (Chapter 14) and Masuda et al. (Chapter 15) cover a wide variety of novel multifunctional ceramics including but not limited to lead-free piezoelectrics and metal oxide nanostructures, and their processing and fabrication using sustainable practices. In Chapter 16, Ishikawa describes green manufacturing of titania, the most widely studied photocatalytic material with proven or potential applications in the degradation of volatile organic compounds, water purification and disinfection, photo-induced water splitting, photocatalytic reduction of CO_2 with H_2O, etc. He focuses on titania photocatalysts for critical water and wastewater treatments and describes their important morphologies (nanorod, nanotube, nanofiber, etc.) as well as production processes along with proven application to water treatment. In Chapter 17, Sekino continues the theme of titania, focusing chiefly on low-dimensional TiO_2-derived nanostructures such as titania nanotube (TNT) that exhibit improved photochemical properties. He reviews solution chemical processes to synthesize TNT and discusses their nanostructural characteristics, and fundamental physical, optical, and chemical properties as well as multifunctionalities. He also presents novel methodologies of tuning TNT morphology, structure, and functionalities for environmental cleaning and energy applications including high-performance oxide electrodes in dye-sensitized solar cells.

Wu and coworkers discuss the synthesis of optically transparent ceramics using environmentally friendly gelcasting in Chapter 18. They discuss the scientific theory and operational principles behind gelcasting and introduce a new water-solvable green organic gelling system that is multifunctional and serves as a dispersant, a room-temperature gelling agent, and a binder for tape casting for aqueous systems. The mechanisms behind its multifunctional properties are presented. Although it is environmentally benign, the gelling agent is neurotoxic and, therefore, research continues to focus on other gel systems which have properties similar to the new agent but are nontoxic. Progress in this line of research is presented.

A key step in all powder-based ceramic manufacturing is the sintering of a porous green body. In the pre-sintering stage, the green body experiences dehydration, organic volatilization, and decomposition, and during sintering, the ceramic densifies to the targeted microstructure and properties. In Chapter 19, Gupta offers a perspective on sustainable manufacturing of green ceramic bodies. He points out that judicious control of process parameters and proper design guidelines can usually eliminate such defects as microcracking during densification. However, where a porous ceramic structure with controlled porosity is required, stringent process control is needed to avoid microcracking of ceramic struts during pre-sintering. He presents theoretical and experimental studies as well as manufacturing practices on green body design and fabrication to produce both dense and porous ceramics in an environmentally friendly manner.

2.4 Sustainable Manufacturing—Polymeric and Composite Materials

Carbon-fiber-reinforced thermosetting and thermoplastic resins are widely used in a variety of industrial sectors including transportation. In Chapter 20, Hotta presents a microwave-assisted rapid resin-curing technology to form carbon-fiber-reinforced plastic (CFRP) with improved mechanical properties. One of the problems in many field-assisted curing technologies is damage to the carbon fiber-polymer matrix interface from uneven thermal load on the composites ingredients. The solution to the problem demands novel polymeric matrices and/or ceramic fibers with matching thermal conductivity. Recent work in microwave processing of CFRP is presented. The theme of green manufacturing of polymer composites is continued in Chapter 21 by Fu and Xu, who describe environmentally friendly manufacture and applications of high-temperature polymer—polyphosphazenes.

The collection of invited chapters that comprise this book present the state of the art on sustainable practices in research and technology of materials development. They disclose and describe emerging and innovative materials and manufacturing processes that will shape future technological developments by integrating sustainability as a key driver of progress and growth.

Chapter 2

Moving Beyond Single Attributes to Holistically Assess the Sustainability of Materials

Marsha S. Bischel
Armstrong World Industries, Inc., Lancaster, PA, USA

1 EVOLUTION OF VIEWS OF ENVIRONMENTALLY PREFERABLE MATERIALS AND PRODUCTS

Beginning in approximately 1990, views on what constitutes an environmentally preferable product or material started shifting dramatically. One reason to link the changes in perception to this particular time frame is the 1990 introduction of the United Kingdom's Building Research Establishment's (BRE's) tool to measure the sustainability of new commercial buildings, the BRE Environmental Assessment Method (BREEAM). Since then, more than 260 similar ratings systems have been developed globally [1]. The environmental performance of materials and products is typically a key component of such systems, because buildings use approximately 40% of all raw materials consumed in the world [2], while new construction generates approximately 40% of nonresidential waste. [3,4] Other ratings systems, such as the US Green Building Council's Leadership in Energy and Environmental Design (LEED) for Operations and Maintenance, have gone further and require the use of sustainable materials during the operation of buildings [5]. There are also standards for running large events that contain requirements for using sustainable materials, for example, *ASTM E2742—11, Standard Specification for Evaluation and Selection of Exhibits for Environmentally Sustainable Meetings, Events, Trade Shows, and Conferences*.

A second reason to date the beginning of the change in perception of what is "environmentally preferable" to the early 1990s was the 1990 international workshop sponsored by the Society of Environmental Toxicology and Chemistry (SETAC). This workshop formalized the term *life-cycle assessment* [6] for methods of examining the full range of environmental impacts of a material or process. This workshop is often acknowledged as the beginning of a concerted effort to look holistically at environmental impacts [6,7].

Green and Sustainable Manufacturing of Advanced Materials. http://dx.doi.org/10.1016/B978-0-12-411497-5.00002-3

During the two decades that followed these milestones, the environmental impact and performance of materials have become increasingly important to end users and specifiers. In that time, definitions of what constitutes environmentally preferred have also evolved. Historically, materials have been produced to meet a particular set of performance characteristics, such as tensile or compression strength. Achieving the desired performance was linked to the choice of raw materials, the processing parameters, and the structure of the material (the elements of the classic materials science stool). Other factors were also considered, such as linking the selection of raw materials to economic or supply concerns. Frequently it was only after the final material or product had been produced that any environmental attributes were determined, such as calculating the percent recycled content. Simultaneously, society often determined environmental preference of materials based on a narrow set of single attributes such as recycled content, recyclability, the use of bio-based components, energy efficiency, and others. The more of these attributes a product possessed, the more "environmentally preferable" it was considered to be. Ratings systems for environmental performance, such as those found in the construction sector, reflected this desire for a menu of environmental attributes, and generally awarded points based on the number of criteria met.

In the more recent past, more holistic views have gained prominence. Among these are life-cycle assessments (LCAs), multi-attribute assessments that address all the environmental impacts of a product throughout its lifetime. LCAs are detailed analyses, often either "cradle-to-grave" or "cradle-to-cradle," and include the harvesting and refining of raw materials, processing, transportation, installation, use, as well as end of life (landfilling, recycling, etc.). The impacts assessed include such things as global warming, primary energy demand, eutrophication of water, ozone depletion, acid rain generation (acidification potential), smog generation, water usage, and others. The results of an LCA can be used to generate an environmental product declaration (EPD), an environmental dossier for a product that allows the impacts of similar products to be compared. Methods known as Design for the Environment have also been developed consider a variety of environmental factors during the design and development phase of a material. When used in conjunction with one another, these three tools allow a material, product, or process to be designed in a way that optimizes traditional performance characteristics and minimizes particular environmental impacts.

LCA and EPD requirements are now incorporated into several of the leading green building ratings systems, including the US Green Building Council's LEED Version 4 [5]. In addition, the European Union has legislative requirements for the use of EPDs in certain sectors. Thus, it can be expected that the need for these multi-attribute assessments will only increase in the future, accompanied by a move away from the traditional single-attribute view of environmental preference.

These changes in perception and regulation will combine to further push the need to view materials and their environmental impacts in broad, holistic ways, rather than using traditional views that focus on a single environmental property. Thus, our view of sustainable materials will also continue to evolve.

2 EXAMINATION OF SPECIFIC SINGLE ENVIRONMENTAL ATTRIBUTES

Single environmental attributes are still widely advertised and are often more easily understood than the complex multi-attribute systems that are currently gaining in popularity. As a result, it is important to understand these single attributes, and the role they play in conversations around sustainability. It is also important to understand the drawbacks of considering only a single environmental property. The focus here will be on three of these attributes: recycled content, recyclability, and bio-based content.

2.1 Recycled Content

Recycled content is perhaps the most commonly recognized single attribute associated with sustainability, with recycled content included in items ranging from napkins (made of recycled paper), fabrics (made from recycled PET bottles), concrete (containing recycled aggregates), and fiberglass insulation (spun from recycled glass). Globally, recycled content is typically calculated using guidance provided in *ISO 14021, Environmental labels and declarations—Self-declared environmental claims (Type II environmental labeling)*. This standard provides definitions for "pre-consumer" (also known as "post-industrial") and "post-consumer" materials. Pre-consumer waste is material that is diverted from the waste stream during a manufacturing process used by another facility to create a new product, and would include materials such as fly ash and synthetic gypsum. Post-consumer waste has been generated by end users at the end of the product's useful lifetime, and would include items such as aluminum cans and newsprint.

When materials are recycled, solid waste is eliminated, and consequently so are the environmental impacts associated with landfilling. For example, when one compares the impacts associated with recycling versus landfilling 1000 kg of municipal waste, it can be demonstrated that 1,265 kg of carbon dioxide equivalents are avoided when material is diverted from landfill, significantly decreasing greenhouse gas emissions [1].

Although the inclusion of recycled content in a product is often primarily viewed as a means of diverting materials from landfills, it is important to note that the use of recycled materials also replaces natural resources that would have been used to manufacture a product, thereby decreasing the environmental burden of the product while increasing the lifespan of existing virgin reserves.

Using recycled content in lieu of virgin natural materials may also reduce the processing energy associated with extraction and harvesting. For example, less energy is needed to recycle glass into new glass than is required to make glass from sand: for each 10% of recycled material added, the energy to melt the total batch is reduced by 2-3% [8,9]. For metals, the reduction is even more dramatic: [10] there is a 95% reduction in energy use when making new aluminum ingot from recycled material as compared to using virgin bauxite [10,11].

There are also many instances when the inclusion of a recycled material has a positive impact on the environmental footprint of the new material in which it is being included beyond that of energy savings. For example, when one metric ton of reclaimed nylon carpet tiles is used to make new nylon carpet using a closed-loop system, more than 500,000 liters of water are saved, and 9265 kg of carbon dioxide equivalents are avoided because the processing of the original virgin materials is bypassed [1].

Manufacturing systems in which products are reclaimed and then remade into the same or different products are becoming increasingly important. In the building industry, the ratings systems and some construction codes now require significant amounts of construction waste be diverted from landfill, creating a large quantity of materials that must be reused in some form. For example, the state of California's Green Building Code (CALGreen) requires at least 50% of all construction waste be diverted from landfill [12]. Such mandates assume that materials can be recycled.

Globally, many items, including cars, packaging, and batteries are covered by recycling regulations known collectively as extended producer responsibility. For example, in 2000, the European Union (EU) released its End of Life Vehicles Directive (ELVA), designed to greatly reduce the estimated eight to nine million tons of waste that were being produced annually in the EU from end-of-life issues related to motor vehicles [13]. The directive included provisions for the collection of vehicles at their end of life, at the manufacturer's cost; the cars are then dismantled, and the components are put into the appropriate recycle stream.

In other instances, the types of materials that can be sent to landfill have been restricted, forcing them to be recycled. For example, in the United Kingdom, the disposal of gypsum wallboard is no longer allowed in standard landfills; due to its high sulfur content, it is now classified as hazardous waste. As a result, there is a need to recycle the gypsum wallboard that is removed during the renovation or demolition of buildings, as well as the fresh cut-offs generated from new construction. A study by the Waste and Resources Action Programmed (WRAP) stated that by 2020, 70% of waste gypsum wallboard is targeted to be recycled in the UK [14]. However, the amount of gypsum drywall available for recycling greatly exceeds the capacity of the wallboard manufacturers to use the material. As a consequence, new products and uses will need to be found for the excess material.

The use of recycled materials has many advantages: it helps to meet the goal of society to divert material from landfill; it decreases the use of raw materials and other resources; and it decreases certain environmental impacts. Coupled with the regulatory restrictions being imposed in various locales, the need to recycle materials will certainly continue, and will likely be increase in certain areas. Thus recycled content will continue to be an important attribute for materials used in all sorts of products.

2.2 Recyclability

The ability to recycle a product is known as recyclability. Some materials can be easily incorporated back into the same product and would therefore be considered to be highly recyclable: aluminum, steel, glass, and paper are classic examples. However, less traditional materials such as high-density polyethylene bottles, acoustical ceiling tiles, vinyl composition tile flooring, and gypsum can also be reused in closed-loop systems whereby they are recycled back into the original product. There are also waste materials that can be easily recycled into other products: for example, crushed concrete can be used as aggregate in road beds or in fresh concrete [15].

Other materials and products are less easily recycled. There may be any number of reasons for this difference in recyclability, including the use of multilayer or multicomponent structures in packaging, composites or layered glasses; the presence of impurities in items such as laminated glass or plastics; chemical properties associated with alloying elements or heterogeneous products; and the degradation of physical properties in materials such as polymers.

The ability to recycle a material can also be hindered by the presence of "chemicals of concerns." For example, although there are significant efforts to recycle plastics, including via curbside collection, there are also regulations and systems that prohibit the use of certain chemicals that have historically been present in many plastics. Among the chemicals considered to be hazardous by some organizations are antimony, bis-phenol A, halogenated flame retardants, and some ortho-phthalate-based plasticizers [16,17]. Their presence in many plastic products may limit the ability of those items to be recycled.

Finally, as was seen with the example of gypsum wallboard, recyclability may be limited by a lack of markets and end uses. Other barriers include the lack of economically viable recovery processes. For example, the blades of wind turbines, promoted as a source of renewable energy, are made of fiber-reinforced composite materials; [10,18] it is extremely difficult to separate the constituent components from such structures [18]. Currently, the most common way to reuse such composite materials is to shred them for use as fill; however, there are many competing materials for this end use [18]. Photovoltaics contain valuable materials that theoretically should be highly desirable, including copper, but they are not easily recycled primarily due to the presence of hazardous materials including the carcinogens cadmium and arsenic; [19] the processes for separating out the valuable materials are costly and thus recycling of photovoltaics does not generally occur [19,20].

2.3 Bio-Based Materials

Materials that are composed of rapidly renewable, plant-based ingredients have become highly marketable. Points are awarded for these materials in many of the sustainable building rating systems, and some requirements have been set for government entities to include bio-based materials in their purchases [21,22]. As a consequence of the desirability of bio-based raw materials, they are being incorporated into both established and novel products. For example, many traditional polymers are now being made fully or in part using bio-based raw material feedstocks; these include polyurethane, polyolefins, polyethylene terephthalate, nylon, and polylactic acid. In these cases, bio-based chemicals are replacing the same compounds that are traditionally made from petrochemicals, using feedstocks such as soy, caster beans, corn, and sugar cane [23]. End uses for these bio-polymers include materials for such diverse areas as construction, packaging, and automobiles.

Finding new uses for natural fibers is also a rapidly growing area of materials interest, particularly for use in the automobile industry. A 2012 report by the Center for Automotive Research on the use of bio-materials in the automobile supply chain lists specific uses of these materials, including cotton fibers in soundproofing of the car interior; wheat, flax, sisal, and hemp as reinforcing fibers in dashboards, headliners, and storage bins; flax fibers in brakes; and abaca fibers in underbody panels. In many cases, the fibers are intended to replace glass fiber reinforcements in nonstructural components, but there are also efforts to include these in structural components [23].

Other nontraditional areas for using bio-based materials include the use of sheep's wool as an insulating material for buildings, and bio-based polymers for disposable utensils. There is also a renewed interest in using natural sources of latex to replace petrochemical-based materials in automobile and medical uses [23,24].

Even ceramics and glasses may be candidates for the use of bio-based materials. Researchers at the Colorado School of Mines filed a provisional patent for a novel pyrolizing process to extract mineral content from food waste. For example, it is possible to obtain silica from rice hulls, oxides of potassium, magnesium and calcium from peanut shells, and alumina from tea. The researchers were able to make several simple soda-lime-silica glass compositions using food stock wastes, including some that were 100% bio-based materials [25]. Other researchers have investigated the use of silica from rice hulls to create precursors for ceramic glazes [26].

While the use of bio-based materials is viewed as desirable by many end users and rating systems, there are issues associated with their production that are frequently overlooked when considering the percentage of bio-based content alone. For example, a content-based view does not take into consideration the impacts of herbicides, pesticides, or fertilizers that may be applied to plants during the growing phase. It also does not consider or compare the durability of a bio-based product versus a nonbio-based product; this was noted as an important area of consideration in the automobile industry [23]. What if a bio-based product lasts only five years, but a comparable petroleum-based product lasts 15 years? Considering only the bio-based content may not provide enough information to make an informed decision regarding the overall environmental preferability of a material, in that the original bio-based product would need to be replaced twice to achieve the same overall lifetime.

3 THE USE OF LIFE-CYCLE ANALYSIS TO EVALUATE MULTIPLE PRODUCT ATTRIBUTES

As demonstrated in the previous section, the use of single attributes to evaluate the sustainability of materials and products is not always adequate if one wants to fully understand the impacts that the material has on the environment. For these more complex, holistic evaluations, additional tools and metrics are required. The most widely recognized of these is the life-cycle assessment.

3.1 The Development of LCA as a Methodology

The origins of LCA are generally traced to a resource and environmental profile analysis performed for Coca-Cola in 1969-1970 to assess the energy, material, and environmental impacts of various packaging materials [7,27,28]. Small studies continued to be performed throughout the 1970s and 1980s, largely in the area of packaging; Baumann and Tillman contain a comprehensive history of these early studies [7]. As more studies were performed, it became clear that there were no universally recognized, consistent methodologies or metrics that could be used to perform the studies; this resulted in conflicting results and distrust of the tools [7]. A 1990 symposium sponsored by SETAC was the beginning of a concerted global effort to standardize the tools and framework for life-cycle assessments [6,28], culminating in the creation of the 14000 Series of ISO standards in late 1990s and early 2000s [7]. Simultaneously, software for performing the complicated calculations and databases of impacts were developed, making it easier and more meaningful to perform these analyses [7,27].

One early driver for LCA was the 2000 Summer Olympics in Sydney, Australia, which made a commitment to being the first "green Olympics." As part of the submission process for infrastructure for the games, companies were required to submit "environmental credentials" for their products; this led to increased interest in and execution of LCA studies in Australia [28]. Since that time, the use of LCA in understanding the sustainability of construction materials has continued to drive the acceptance of the tool. Studies based on LCA are now incorporated into the BREEAM rating system [29], the latest version of the US Green Building Council's LEED program [30], the EU's Construction Product Directive [31], and others.

Other early drivers in promoting LCAs were the debates around conserving natural resources versus reducing landfill and diverting hazardous materials from landfill, areas that pitted two single environmental issues against one another. The former is epitomized by the ongoing debate between the use of cloth diapers to reduce landfill as compared to the use of disposable diapers, which conserves water [27]. The latter is demonstrated by the debate over the amount of mercury released by fluorescent lights once they are sent to the landfill versus the energy saved when they are used [27]. The amount of mercury put into landfills from traditional fluorescent bulbs is much higher than it is for incandescent bulbs; however, when the use and disposal phases are considered over the same number of service hours for the bulbs, incandescent bulbs release 4 to 10 times more mercury into the environment. In addition, the EPA estimates that fluorescent bulbs decrease electricity demand by as much as 50%, reduce carbon dioxide emissions by 232 million tons, and significantly reduce emissions of sulfur dioxide (a cause of acid rain), and nitrous oxide [27]. A recent Canadian study found that the mercury emissions and other end-of-life issues from light bulbs represent less than 1% of their total environmental impacts [32]. Only by looking at all of the impacts associated with the light bulbs could a true assessment of the environmental impacts be considered.

3.2 Use of Life-Cycle Assessment to Quantify and Reduce Environmental Impacts

Life-cycle assessment was created to be a tool for quantifying all the environmental impacts associated with the life of a product. LCAs are designed to be detailed, product oriented [7], quantitative "cradle-to-grave," or "cradle-to-cradle" analyses and include such things as the harvesting, mining, and refining of raw materials, manufacturing, transportation, installation, use, and end of life. They allow users to optimize environmental footprints, minimize particular impacts, compare products and processes, and importantly allow multiple issues to be assessed at once.

The process for conducting an LCA is described in *ISO 14044, Environmental management—Life-cycle assessment—Requirements and guidelines*. An LCA generates a list of quantified environmental impacts, which generally includes factors such as embodied energy, acidification potential, eutrophication potential, global warming potential, ozone depletion, smog creation, hazardous waste generation, and water usage, among other things. In addition, impacts are assessed during every phase of the product's life, thus making it possible to determine when an impact is at its greatest. For example, the use phase of an automobile has the greatest environmental impact, whereas for construction materials, the manufacturing phase generally has the greatest impact.

It is important to note that because the LCA is a holistic look at the entire lifetime of a product, the impact of materials in the procurement chain needs to be understood. For example, the manufacturers of aluminum window frames must consider the impacts of the aluminum refining process, as well as the impacts of their own processes and the use phases of their products. Manufacturers of fundamental materials such as alloys, glass, and the like, will therefore be under pressure to conduct their own LCAs or to provide input for their customers.

Because they look at all aspects of a product, LCAs can be used internally by manufacturers to determine the areas of greatest environmental impact, allowing them to create action plans for reducing those impacts. For example, an industry-wide LCA conducted on container glass made in North America specifies a goal of increasing the use of recycled glass from a 2007 average of 23% to 50%. This single change is predicted to reduce cradle-to-cradle carbon dioxide emissions by 6% per kilogram of finished and formed glass, or nearly 2.2 million metric tons annually. The reduction will come from a 20% decrease in furnace emissions, and a 43% reduction in carbon dioxide generated during cullet treatment [33].

These analyses can also be used to aid in the design of products that have smaller environmental impacts. An LCA of the aluminum beverage can industry found that the electrolysis processes associated with the primary aluminum production accounted for 72% to 79% of a can's environmental impacts; [34] therefore, to significantly decrease the impacts of an aluminum can, the electrolysis process would be the most logical part of the sequence to address. The study also showed that 70% of the carbon dioxide emissions associated with the electrolysis process are associated with the upstream generation of the electricity used in the process; [34] this would suggest that a move towards renewable energy sources would be of benefit if there is a desire to reduce global warming potential.

The generation of LCAs can also help identify the ramifications of specific changes in a process or material, such as increasing the recycled content, or changing alloys. A 2006 study on the LCA of automobiles showed that the switch to advanced high-strength steel alloys in car bodies has resulted in an average 9% reduction in vehicle weight. This corresponds to a 5% increase in fuel economy and a 5.7% reduction in greenhouse gas emissions during the use phase of the vehicle [35].

These analyses can also be used to determine which materials are more environmentally problematic. For example, Table 1 shows the potential impacts on global warming and water saved when four different building materials that can be recycled are used in closed-loop systems. [1] The analysis shows that different materials have different reductions; such analyses performed during the design stage of a new product could reveal raw material selections that would create the overall lowest impact on the environment.

TABLE 1 Impact of Using 1 Metric Ton of Recycled Material in Lieu of Virgin Raw Materials in a Closed-Loop System[a]

Material Recycled in Closed-Loop System	Global Warming Potential Avoided, kg CO_2 Equivalents	Water Saved (liters)
Generic vinyl composite tile	970	11,117
Generic gypsum[a]	31	11,728
Generic vinyl siding[a]	2,277	2149
Generic nylon carpet tile[a]	926	511,768

[a]*Includes the impacts associated with needed installation components; analyses based on data from NIST BEES database.*

Similar analyses can be performed using various transportation scenarios in an attempt to reduce impacts during these phases. For example, St. Gobain reports using LCA specifically to evaluate the effects of transportation on the environmental impacts associated with their wallboard products. In a trial study in Belgium, St. Gobain began using ships to deliver products and reported reductions in global warming potential [36]. In South Africa, another St. Gobain study focused on the sizes and types of trucks used and the process used to get them to wallboard distribution centers; as a result, changes were made, such as the use of larger trucks to transport more material at once. In total, the changes in South Africa resulted in a 20% decrease in fuel use, with associated reductions in global warming potential [37].

Finally, LCA can be used to identify the impacts of various materials during the use phase. A study in the United Kingdom revealed that the single most important factor in reducing the environmental impacts of a grocery bag was the number of times it was reused: one must use a paper bag four times and a nonwoven polypropylene bag 14 times to have lower impacts than those of traditional plastic bags made from high-density polyethylene [38]. By performing a series of LCAs on all life phases of the types of grocery bags, the researchers were able to determine the use phase scenarios that had the greatest reduction in environmental impact for each.

If life-cycle analyses are performed on an ongoing basis, it is possible to track reductions in environmental impacts over time. Typically, this involves obtaining baseline data for all of the impact areas, performing an LCA, and then conducting the entire analysis a second time after various improvements to the product or process have been made. For example, when Xerox Corporation switched from a process of remanufacturing entire photocopiers to one whereby the individual modules making up the photocopier are remanufactured, they more than doubled their reductions in environmental impacts [7]. Thus, it is feasible to incorporate

LCA into continuous improvement processes. It is not clear that this tool is being widely used for this purpose even though it is a stated objective of the governing ISO 14025 standard, *Environmental labels and declarations—Type III environmental declarations—Principles and procedures* [39].

3.3 The Use of Life-Cycle Analyses to Compare Products or Materials

Another important advantage of LCAs is the ability to use them to compare products for their environmental impacts, allowing end users to identify the products that best reduce the impacts of concern to them. This can be done in several different ways.

First, LCA impact data can be used on its own to compare the environmental impacts of two different products or materials. When this is done, it is important to compare similar lifetimes or outputs. For example, when comparing the impacts of incandescent and compact fluorescent light bulbs (CFLs), one must take into account the much longer useful life span of the CFLs. Once this is done, the impacts of the CFL are seen to be lower than those of the incandescent bulb. [32,39] In addition, the areas of end-use and manufacture must be consistent for a valid comparison. For example, an LCA study performed in Canada, where there is very high use of hydroelectric power, will have a much lower impact for global warming potential versus one done for the lower 48 United States, which use a much higher percentage of coal-burning electric plants, because the latter are large emitters of carbon dioxide.

LCAs have been used to compare the influence of materials choices on environmental impacts. For example, a 2011 paper from the World Steel Association compared the impacts of using steel cans as packaging for green beans versus those when the beans are frozen and put into plastic bags. The study claimed that the use of the cans resulted in a 39% reduction in global warming potential [41]. In this case, it is important to note that vastly different materials and processes were compared. When making such comparisons, it is critical, and often difficult, to ensure that all factors are comparably and fairly assessed.

In an effort to make sure that LCA data can be consistently compared across a product area, additional tools have been developed. The first of these are product category rules (PCRs), which give detailed guidance for the system boundaries and requirements for the LCA calculations. One of the most important of these is the establishment of a common lifetime for the use phase, often in the range of 30 to 50 years for durable products. If a product has an expected lifetime that is less than this number, its impacts must be normalized upward to account for this. For instance, if a bio-based composite has a lifetime that is one-third that of a standard lifetime, the impacts for the bio-based materials would need to be tripled. Another key metric outlined in the PCR is the functional unit: this allows the same volume, mass, function, or time of use to be compared. These units vary depending on the type of material and its end use, as depicted in Table 2.

TABLE 2 Functional Units for Various Types of Building Materials

Product category	Functional unit
Aerated concrete block	1 m^3 of autoclaved block [42]
Light-gauge steel profiles	1 metric ton of cold-formed steel profiles [43]
Window glass	The visible surface (in m^2) of 1 window [44]
Building insulation	The weight in kg of 1 m^2 of insulation material with a thickness that gives an average thermal resistance $R_{SI} = 1\ m^2K/W$ [45]
Softwood lumber, structural	1 m^3 of kiln dried, planed softwood lumber [46]
Wood flooring	1 m^2 of covered surface [47]

The other tool that allows for valid comparisons of products is the environmental product declaration (EPD), the formal tool for communicating the results of an LCA to end users of the impact information. An EPD can be viewed as an environmental dossier for a given product, based on a life-cycle analysis that was performed according to a recognized set of product category rules; per ISO 14025, the specific content requirements for the EPD are also outlined in the PCR. EPDs can be completed for specific products from a given manufacturer, a suite of similar products from a manufacturer, or for a representative industry-wide material. To generate an EPD, a detailed LCA must be generated and reviewed by a third party, whereas the EPD itself is generated and certified by yet another third-party provider. Although an LCA report may be 50 pages or more with detailed calculations and analysis, an EPD is a much shorter summary of the primary environmental impacts and the findings from the LCA. Simpler to use than an LCA, it should be noted that EPDs are still relatively complex documents for those who are unfamiliar with their content and structure.

3.4 Requirements for Using Life-Cycle Assessments and Environmental Product Declarations

There is a growing trend toward the inclusion of LCAs and EPDs in sustainable building systems and codes. For example, in France a specific form of EPD known as "Fiche Déclaration Environnementale et Sanitaire," (FDES), is already required for products used in green buildings certified by Haute Qualité Environnementale (HQE), and includes other environmental factors such as emissions information [48]. EPDs will be required throughout the building sector in the European Union as the requirements of Section 56 of the Construction Products Regulation of 2011 are phased in [31]. The California Green Building Code includes an option for performing a whole-building LCA that incorporates the impacts from all the materials used in the building [12]. LCA requirements were included in early drafts of the International Green Construction Code of 2012, but were removed from the final document; however, they are expected to be included in the 2015 version [1].

BREEAM has long included a requirement for EPDs in its building ratings systems [29]. LEED Version 4, released in late 2013 by the US Green Building Council, includes credits for LCAs and EPDs of building products and for whole buildings [30,49].

Other industries in which LCAs and EPDs have been gaining traction include furnishings, retail, automotive, and electronics [27]. Walmart has embarked on a program in which all of its suppliers, estimated to number at least 100,000, will disclose the environmental impacts of their products and processes [50]. As one of the world's largest retailers, this need for LCA data will greatly drive the creation of these analyses across many different product categories, and will require impact data from a broad range of materials manufacturers.

In all cases, requirements to share life-cycle impacts are being promoted as a way to transform markets by making the holistic impacts associated with a product available to consumers, thereby allowing them to make informed decisions about the sustainable attributes of the products they buy. This is much more encompassing, as well as complicated, than simply comparing a menu of single attributes. All the environmental impacts are addressed in LCA, allowing users to focus on the areas that are most meaningful to them or to their business. However, understanding the details of a life-cycle assessment can be complicated to a novice, precisely because all impacts are addressed.

Although many may view LCAs as complicated, their widespread use will ultimately enable consumers and manufacturers to make educated decisions regarding the true impacts associated with a material, product, process, or even a building. In addition to allowing end users to focus on a particular environmental attribute, LCAs and EPDs also allow them to consider potential trade-offs among different attributes and force manufacturers to more prominently integrate environmental impacts into design and manufacturing decisions.

4 DESIGN FOR THE ENVIRONMENT

The last tool that will be considered at a high level is generically known as a "Design for the Environment" (DfE). It has become clear that to use virgin and recycled materials effectively to make products that are themselves easily recycled, and which limit other environmental impacts, designers must make additional considerations of sustainability at the beginning of the design process. In the past as environmental attributes became more important to customers, many manufacturers simply modified existing products to make them more desirable, for instance, by increasing the recycled or bio-based content of the material. However, as was shown in the previous discussions, the sustainability of a material cannot be measured solely by a string of single attributes.

Life-cycle assessment can itself be used as a tool for designing materials, products, and processes with lower environmental impacts. The studies on recycled content in glass containers and the use of advanced steel alloys in automobiles are two such examples. Baumann and Tillman [7] include an entire chapter devoted to the use of LCA in product development. However, true sustainability encompasses more than just the environmental impacts of a material. Sustainability encompasses three categories: environmental, economic, and social impacts. LCA does not easily address the latter two: to address all aspects, additional tools and methods are required.

Beginning in the 1990s, there was a move toward designing products so they could be easily disassembled into their constituent parts and then be recycled [51]. In 1999, the Environmental Defense Fund published guidelines around "Design-for-Recyclability" in the auto industry, suggesting specific materials for making recycling easier, including marking parts so the material type can be easily determined; reducing the number of different materials used in the product to simplify material separation during the reclamation phase (a particular issue with laminated and composite materials); and using "compatible" materials that do not need to be separated during recycling, as mixtures of materials tend not be easily recyclable [52]. McDonough and Braungart's 2002 seminal book *Cradle to Cradle* was another early document advocating the use of environmentally sound design in the creation of products [53], but moved beyond the single attribute of recycled content to include broader sustainability concerns.

Design for the environmental approaches typically include assessing any materials of concerns (such as potential carcinogens); designing so that the product can be easily disassembled into recyclable components; including recycled and/or bio-based content; and reducing embodied energy and global warming potential. Other factors that could be included in a truly holistic sustainability model are economic issues related with the supply chain, the location of production facilities and raw materials, the use of regional materials, product durability, fair labor practices, the reduction of energy during the use phase, and the reduction of wastes.

Some DfE tools focus on specific aspects of designing for the environment. For example, the US Environmental Protection Agency administers a "Design for the Environment" labeling process that focuses on the use of safe chemicals in products [54]. The EPA and others have also promoted guidelines related to green chemistry, or the design of chemicals that reduce or eliminate the use and generation of hazardous materials. The beginnings of the green chemistry movement can also be traced to the 1990s, with requirements in the 1990 US Pollution Prevention Act to reduce the sources of hazardous materials [55].

Internally, many companies have initiated their own processes for designing for the environment. For example, furniture manufacturer Herman Miller adopted a procedure based on the cradle-to-cradle process created by McDonough Braungart Design Chemistry. Their approach addresses health and safety issues of the materials for people and the environment; the content of the material, including recycled content; end-of-life issues, such as designing for easy disassembly; and energy usage. All new products from Herman Miller must undergo this process, including the generation of an LCA [56,57]. Johnson and Johnson has also made its DfE process public. Using LCA screening and other tools, they focus on reducing impacts in seven areas: materials, packaging, energy, waste, water, social, and innovation [58].

From a materials science perspective, it is useful to understand how sustainability needs interact with traditional materials design considerations. Each of these can be viewed as a three-legged stool: the sustainability stool has economic issues, environmental issues, and social issues as legs; and the materials properties stool has raw material, process, and structure as legs. To design environmentally benign products, one must assess all six factors simultaneously. In 2009, Bischel and Costello proposed the simple model shown in Figure 1,

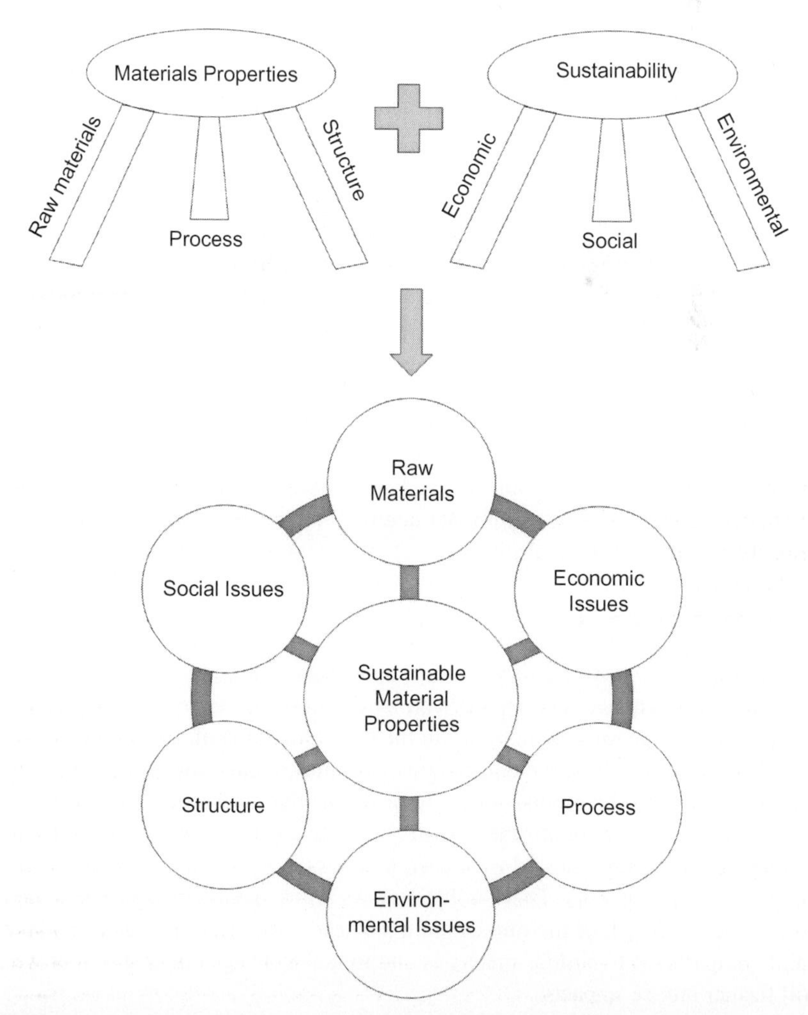

FIGURE 1 Sustainable materials model as represented by a wheel with six spokes [59].

in which each of the six components represents a spoke of the sustainable material wheel. By addressing all elements early in the design phase, a material can be optimized for both its traditional properties and its sustainable attributes [59].

The materials community needs to incorporate Design for the Environment practices into the earliest design phases for materials and products. By using LCA and other DfE tools, the potential impacts can be assessed and steps taken to minimize them, while maintaining the necessary end-use characteristics. In this way, environmentally benign materials may be thoughtfully created.

5 CONTINUAL EVOLUTION

Even as life-cycle assessment and environmental product declarations gain traction and change the discussion regarding the definition of sustainable material, portions of the sustainability community are moving even further toward environmentally neutral, or even net-positive materials and products. Among these shifts are efforts to move beyond merely reducing environmental impacts toward ultimately creating positive benefits from materials. An example of this is the Global Green Tag certification system, based in Australia. In addition to traditional LCA metrics, this system includes detailed chemical analyses, minimum requirements for design for disassembly, social responsibility requirements, and metrics for net-positive factors around reductions in carbon dioxide potential and respect for biodiversity issues [60].

As manufacturers and designers of materials embrace life-cycle assessments and other currently used tools to improve the sustainability of their products, they will need to remain aware of the continuing evolution of sustainability requirements. Continual improvements are needed in materials, products, and processes, first by reducing impacts to benign levels, and ultimately by creating materials that are at worst, impact neutral, but that preferably have positive impacts on global sustainability.

6 CONCLUSIONS

During the last two decades, views of sustainable materials have shifted dramatically. This change has been driven by a number of factors, including the rise in concerns about sustainability in the built environment, the increased use of holistic life-cycle assessment systems, and movements toward designing to reduce impacts to the environment. As a result, the assessment of a material based solely on a single attribute, such as its recycled content or its ability to be recycled, is being replaced by a need to assess it against many environmental and societal concerns. These holistic views allow specifiers, customers, and others to consider all of the impacts of a material, process, or product, to make valid comparisons to similar materials and to make changes in design that will still further reduce impacts.

As a result, the designers and manufacturers of materials will need to develop and use new tools to create materials that reduce environmental impacts, and address sustainable issues in a broader way, during the earliest stages of product development. They will need to develop new uses for the increased amount and types of materials that are being recycled, increase the recyclability of materials, convert to the use of benign raw materials, and consider the social impacts of their choices more closely. The demand by downstream users for life-cycle information and environmental product declarations will require many suppliers to conduct LCAs of their products and processes. These assessments will allow the manufacturers to identify the specific areas in which they can reduce their environmental impacts and enhance their sustainability. Ultimately, all of these together will help create more sustainable materials.

Finally, as additional requirements for sustainability evolve and are adopted in the marketplace, the materials science community will be called upon to create new materials and products that are not only environmentally benign, but which are economically viable and have a net positive contribution on global sustainability.

ACKNOWLEDGMENTS

The author would like to acknowledge the following individuals for their assistance in preparing this document: Amy A. Costello, Peter J. Oleske, and Tawnya R. Rabuck, Armstrong World Industries, Inc.; and Jeremy Sumeray, Armstrong World Industries, Ltd.

REFERENCES

[1] A.A. Costello, M.S. Bischel, Directional Drivers of Sustainable Manufacturing: The Impact of Sustainable Building Codes and Standards on the Manufacturers of Materials, Adv. Mater. Sci. Environ. Energy Technol. II: Ceram. Trans. 241 (2013).

[2] C. Cheng, S. Pouffary, N. Svenningsen, M. Callaway, The Kyoto Protocol, The Clean Development Mechanism and the Building and Construction Sector – A Report for the UNEP Sustainable Buildings and Construction Initiative, United Nations Environment Programme, Paris, France, 2008.

[3] US Green Building Council Research Committee, A National Green Building Research Agenda, (November 2007).

[4] US Environmental Protection Agency, "Lifecycle Building Challenge" (June 2009).

[5] US Green Building Council, LEED v4 for Operations & Maintenance: Existing Buildings, (2013).

[6] J.A. Fava, "SETAC and Life Cycle Assessment – Parallel Growth", SETAC Globe 12, 2011.

[7] H. Baumann, A. Tillman, The Hitch Hiker's Guide to LCA, Sweden, Lund, 2004.

[8] Glass for Europe, Recyclable waste flat glass in the context of the development of end-of-waste criteria (6/2010).

[9] St. Gobain, "Sustainable Development Report 2011" (5/2012).

[10] D.S. Ginely, D. Cahen (Eds.), Fundamentals of Materials for Energy and Environmental Sustainability, Materials Research Society, Cambridge University Press, 2012.

[11] ALCOA, "2011 Sustainability at a Glance" (2012).

[12] CALGreen, Guide to the (Non-Residential) California Green Building Standards Code (1/2012), http://www.documents.dgs.ca.gov/bsc/CALGreen/MasterCALGreenNon-ResGuide2010_2012Suppl-3rdEd_1-12.pdf.
[13] US EPA, "Recycling and Reuse: End-of Life Vehicles and Extended Producer Responsibility: European Union Directive", (11/2008).
[14] Waste and Resources Action Programme, *Waste Protocols Project: Gypsum, Partial Financial Impact Assessment of a Quality Protocol for the production and use of gypsum from waste plasterboard* (2008).
[15] Concrete Network, Recycling Concrete (2012) http://www.concretenetwork.com/concrete/demolition/recycling_concrete.htm.
[16] Cascadia Region Green Building Council, *Living Building Challenge 2.1* (5/2012).
[17] Google Healthy Materials Program.
[18] Larsen, K., Recycling wind, *Reinforced Plastics* (1/31/2009), http://www.reinforcedplastics.com/view/319/recycling-wind/.
[19] Weadock, N., "Recycling Methods for Used Photovoltaic Panels", University of Maryland Watershed Program (1/9/2011), http://2011.solarteam.org/news/recycling-methods-for-used-photovoltaic-panels.
[20] Geis, E., "Solar Panel Recycling Gears Up", The Daily Green (8/12/2010), http://www.thedailygreen.com/environmental-news/latest/solar-panel-recycling-460810.
[21] United States Department of Agriculture Biopreferred Program, "Federal Purchasing Requirement".
[22] United States General Services Administration, "Bio-Based and Bio-Preferred Products".
[23] Center for Automotive Research, "The Bio-Based Materials Automotive Value Chain", (2012).
[24] H. Mooibroek, K. Cornish, Alternative sources of natural rubber, Appl. Microbiol. Biotechnol. 53 (2000) 355.
[25] "New approach to 'mining' – extracting glassmaking raw materials from food waste", American Ceramic Society Bulletin, 92 (2013) 10.
[26] F. Bondioli, et al., Characterization of Rice Husk Ash and Its Recycling as Quartz Substitute for the Production of Ceramic Glazes, J. Am. Ceram. Soc. 93 (2010) 121.
[27] M.A. Curan (Ed.), Life Assessment Handbook: A Guide for Environmentally Sustainable Products, John Wiley and Sons, Hoboken, NJ, 2012.
[28] R. Horne, et al., Life Cycle Assessment: Principles, Practice and Prospects, CSIRO Publishing, Collingwood, Australia, 2009.
[29] BRE, BREEAM.
[30] US Green Building Council, LEED v4 for Interiors, Design and Construction (2013).
[31] "Regulation (EU) No 305/2011 of The European Parliament and of the Council of 9 March 2011laying down harmonised conditions for the marketing of construction products and repealing Council Directive 89/106/EEC," *Official Journal of the European Union* (4/4/2011).
[32] Interuniversity Research Centre for the Life Cycle of Products, Processes and Services, "Comparative life cycle assessment of compact fluorescent and incandescent light bulbs", (7/2008).
[33] Glass Packaging Institute, "Environmental Overview: Complete Life Cycle Assessment of North American Container Glass, (2010).
[34] PE Americas, "Life Cycle Impact Assessment of Aluminum Beverage Cans", (5/2010).
[35] World Steel Association, "Environmental Case Study: Automotive", (9/2008).
[36] St. Gobain, "Sustainable Development Report 2012" (5/2013).
[37] St. Gobain Gyproc SA, "Life Cycle Assessment: A Gyproc board's life cycle in 5 stages", (2012).

[38] Environment Agency, "Life Cycle Assessment of Supermarket Carrier Bags", (2/2011).
[39] International Standards Organization, *"ISO 14025: Environmental labels and declarations – Type III environmental declarations – Principles and procedures"* (2006).
[40] Ramroth, L., "Comparison of Life-Cycle Analyses of Compact Fluorescent and Incandescent Lamps Based on Rated Life of Compact Fluorescent Lamp", Rocky Mountain Institute (2/2008).
[41] World Steel Association, "Environmental Case Study: Steel Food Cans", (5/2011).
[42] Institut Bauen und Umwelt, "Environmental Product Declaration: Ytong® Autoclaved Aerated Concrete", (3/2011).
[43] Institut Bauen und Umwelt, "Environmental Product Declaration: Light Gauge Steel Profiles, Akkon Steel Structure Systems Co", (4/2012).
[44] The Swedish Environmental Management Council, "Product Category Rules: Windows" (6/2008).
[45] UL Environment, "Product Category Rules for Preparing an Environmental Product Declaration: Building Envelope Thermal Insulation", (9/2011).
[46] UL Environment, "Environmental Product Declaration: North American Softwood Lumber", (4/2013).
[47] NSF International, "Product Category Rule for Environmental Product Declarations, Flooring: Carpet, Resilient, Laminate, Ceramic, Wood", (5/2012).
[48] HQE Association, http://assohqe.org/hqe/.
[49] ASTM E2921—13-Standard Practice for Minimum Criteria for Comparing Whole Building Life Cycle Assessments for Use with Building Codes and Rating Systems (2013).
[50] Cohen, B., "Industrial Ecology at Wal-Mart", Rocky Mountain Institute (2009) http://www.rmi.org/Knowledge-Center/Library/2009-13_IndustrialEcologyAtWalmart.
[51] K. Ishii, Material Selection Issues in Design for Recyclability, in: The Second International EcoBalance Conference, Tsukuba, Japan, 1996.
[52] Environmental Defense Fund, "End-of-Life Vehicle Management: Design-for-Recyclability", (1999).
[53] W. McDonough, M. Braungart, Cradle to Cradle, North Point Press, New York, 2002.
[54] US Environmental Protection Agency, "Design for the Environment", http://www.epa.gov/dfe/.
[55] US Environmental Protection Agency, "Basics of Green Chemistry", http://www2.epa.gov/green-chemistry/basics-green-chemistry#definition.
[56] Herman Miller, Environmental Product Summary, 2012.
[57] Herman Miller, Design for the Environment Textiles, 2013.
[58] PE International, "Life Cycle Thinking: Driving Product Development, Innovation and Marketing".
[59] M.S. Bischel, A.A. Costello, How The Classic Materials Science Stool Is Being Changed By The Sustainability Stool, Adv. Mater. Sci. Environ. Nucl. Technol: Ceram. Trans. 222 (2010).
[60] Global Green Tag, http://globalgreentag.com.

Chapter 3

Eco-Materials and Life-Cycle Assessment

Zuoren Nie
Beijing University of Technology, Beijing, China

1 ECO-MATERIALS

1.1 Introduction to Eco-Materials

Materials are one of the basic founding blocks for the development of society and economy. However, for the production, manufacture, application, and disposal of materials, numerous resources and energy are consumed, and environment deterioration occurs as a result. To harmonize materials development with environmental protection, sustainable developmental eco-materials (also called environmentally conscious materials) have been actively sought in recent years throughout the world [1]. With extremely rapid industrial development, there is a danger that mistakes made in the past by the developed countries may be repeated in developing countries. Take China as an example: China is a large materials producer as well as a user. Principal Chinese raw materials output, including iron and steel, nonferrous metals, cement, glass, and so on, has led the world in recent years. At the same time, the energy and resource consumption required is sharply increasing, much higher than the world average on a gross national product (GNP) basis. This aggravates the resource and energy shortage and causes serious environmental pollution and ecological deterioration.

The traditional research and development of materials focus on the improvement of service performance, but often neglect the large amount of resource and energy consumption and pollution emission caused by materials production and related stages (e.g., mining, transportation, etc.). To solve this problem, the word *eco-materials* (the abbreviated form of "environmental conscious materials" or "ecological materials") was first put forward in October 1990 and has been widely accepted by society and the materials industry in recent years. The concept of eco-materials suggests that the development of materials should not only consider the service performance but also try to decrease any environmental impact during the whole life cycle of materials. Thus, eco-materials are defined as those materials that enhance improvement to the environmental throughout the whole life cycle, while maintaining accountable performance [2].

Green and Sustainable Manufacturing of Advanced Materials. http://dx.doi.org/10.1016/B978-0-12-411497-5.00003-5

This definition requires that environmental performance should be given equal status to service performance in the design, development, manufacture, and application of materials. Environmental performance involves the whole life cycle of materials (from mining, transportation, manufacture, until final disposal). Moreover, the social and economic issues should also be taken into consideration. Eco-materials are in demand not only for the development of advanced materials but also for the enhancement of the global environment, social development, and human existence.

In recent years, research on eco-materials has progressed within the fields of metallurgy, metals, nonmetals, polymers, wood, and minerals, among others. Functional materials for environmental protection, energy-saving materials, materials supporting low-emission systems, and materials designed by means of LCA and selected for their lower environmental impact are the most important areas on which to concentrate research when addressing global environmental problems caused by materials technology. Because all countries regard environmental measures to be part of an important strategy to protect their own industry, eco-material and eco-products developed with related technologies have not only become popular topics for basic research, but also part of the most competitive trade strategies in the market. Research on eco-materials has led to the development of new methods for the design of materials that have much lower environment loads in their life cycles and can be used to replace other, higher environmental impact materials. Additionally, eco-materials learning, to be used as a new concept of materials, is still in its initial stage of development, and its science is still in infancy. Moreover, additional efforts are needed to promote learning about eco-materials, publicity and education, law making and evaluation standards, and more.

1.2 Development of Eco-Materials Concept

As a new material concept, eco-materials have gained considerable worldwide attention over the past few decades. The concept means the attribute of environmental consciousness is now considered when evaluating materials. At first, thinking only about the material's functional characteristics was considered. In the last decade, a review of eco-materials included the concept's development, its design within the ecosystem, and an LCA emphasizing all the materials, including the construction material and every kind of functional material with an environmental attribute. Key to eco-materials include how to satisfy a function while lowering the environment load, how to improve the material's recycling, or how to use a minimal amount of material. Environmentally conscious materials are evolving from materials for end-of-pipe application to all materials designed by taking into account life-cycle thinking over the whole life cycle. Many research activities on eco-materials bring up new lessons in material science and engineering technique. Eco-materials learning involves the production, use, disposal, recovery, and each link, which is a multidisciplinary matter that crosses over into material science, environment science, biology, and more, with the complicacy of systems engineering. A guideline that has stood for more than 10 years is for the materials researcher to evaluate eco-materials in terms of coordinating the material's function with supporting good environment. According to a relevant research report covering types of materials, eco-materials should be characterized as follows:

1. Energy saving: decrease the energy consumption or increase the energy efficiency during the whole life cycle of products by improving the performance of materials, such as thermal insulation performance (e.g., low-e glass, green wall materials), lower weight (e.g., aluminum and magnesium parts in vehicles), and so on.
2. Resource saving: decrease the resource consumption during the whole life cycle of products by (a) improving the performance of materials, including abrasion resistance, intensity, heat resistance, and more; (b) increasing the resource efficiency by catalyzer; and (c) developing renewable materials.
3. Reusable: after the disposal of products, materials can be reused as the same products by a series of purification treatment, such as washing, disinfect, surface treatment, and the like.
4. Recycle: after disposal of products, the materials can be collected and used as raw materials of other products.
5. Reliability and stability: improve the performance of creep resistance, ductility, heat resistance, and so on to provide the high reliability and stability of different operational environmental conditions (strong light, moist, low temperature, etc.).
6. Nontoxic and safety: replace a poisonous substance (e.g., a carcinogen) with nontoxic materials to decrease its impact to human, animal, and plant life during the life cycle of products.
7. Comfort: provide a comfortable human living environment during the utilization phase of materials by creating and improving the performance of materials, such as deodorization performance, disinfectant performance, humidity control performance, and so on.
8. Environmental treatment: the performance of materials for purifying pollution (waste gas, wastewater, dust, etc.) also includes a pollution survey.

1.3 Research and Development of Eco-Materials and Related Technology in China

The research of eco-materials in China started in early 1990s. Universities, industries, and national laboratories have all worked on the subject, with the guidance and support of the government. The first eco-materials project financially supported by government plan was the Foundation of National Education Committee of China in 1993. Thereafter several projects for basic research and technology development were carried out by the NSFC and the National

High-Technology R&D (863) Program in the Ninth Five-Year Plan (1996-2000). In the 863 programs comprising the Tenth Five-Year Plan (2001-2005), the subject of eco-materials was set up in the new materials field as a special topic. Progress in eco-materials with related key techniques was required to meet urgent national needs. This subject launched several research projects on sustainable developmental materials and products. Goals set were to achieve eco-friendly, low-cost plastic film for farm use with complete degradation potential, ecological building materials, hazardous-substance free materials, and clean processing with lower emission and synthetically utilizable new materials and techniques for sand-fixation and vegetation materials. Also mandated were environmental evaluation techniques for materials and its application.

In the National High-Tech Demonstration Program, the eco-materials industry has been supported as a special branch. Supported by these projects, progress has been made in recent years that reformulates and upgrades widely used materials as well as eco-materials. Goals to be achieved include the following.

1. No poisonous and harmful materials used in processing

A key point for developing eco-materials is to use no poisonous and harmful components. Materials should be free of chemical and hazardous substances, provide longtime stability, and enhance biological safety. It is important not only in mass consumable structure materials but also in the rapidly developing function materials, such as electronic materials, energy materials, and biology medicine materials.

Electronics functional materials are in great demand in the information society. With rapid renewal and change of products, waste materials were increased accordingly. There are more than 700 kinds of harmful matters in electronics garbage. It is more difficult to recycle the function materials of electronics products than to recycle structure materials. Environmental issues must be taken into account at the beginning of the design and selection of materials for electronics products.

Lead poisoning has become a worldwide problem. Metal lead is a poisonous element in any alloy in which lead cannot be removed before shredding or re-melting at the end-of-life stage. The lead-free solder alloy to replace eutectic tin-lead solder in electronic product is an urgent demand for electronic material. Main problems for the lead-free solder alloy are on the eutectic point temperature, strength and elongation, thermal fatigue, surface tension, electrical conductivity, as well as industrial cost. Research efforts in lead-free solder added with rare earth and Sn-Ag-Cu-Bi alloy have reported more usable progress in some projects [3]. The melting point is near 200 °C, strength >45 MPa, elongation >20%, electricity conductivity >10% IACS, surface tension <470 mN/m. Oxygen content of the soldering powder of 25–63 μm is less than 150 ppm. For the soldering paste leaded, solid content of flux <2%, halogens contains <0.3%, expands rate >80%, insulated resistance of unwashed surface $>1.0\times10^{11}\,\Omega$. The diameters of ball grid array (BGA) soldering ball is 0.3-0.76 mm with high quantity. Also some projects are developing on Pb-free piezoelectric ceramics, Pb-free crystal and glass, and so on.

For nearly 100 years, tungsten added with thorium (ThO_2-W) has been widely applied as a cathode material to large-power transmit and medium heating tubes, TIG (argon arc welding) electrodes, and magnetrons of household-use microwave ovens, among other uses. But thorium is a radioactive element whose half-life is as long as 1.39×10^{10} years. The radioactive pollution of ThO_2-W during its production, operation, and trash is harmful to both human body and environment, causing cumulative poisoning. A Chinese researcher has developed a series of rare-earth-added tungsten cathode materials to replace ThO_2-W [4]. This not only avoids the ensuing radioactive pollution but also lowers the emission operating temperature with better start-up characteristics, with the emission stability sufficient for different applications. Especially in the argon arc welding industrial application case, the arc starting voltage and electrode consumption of rare-earth tungsten electrodes are both lower than that of ThO_2-W electrodes with the same amount added. Also some projects aimed at developing Pb-free piezoelectric ceramics, Pb-free crystal and glass, water-heating electrochemistry synthesis, and so on.

Common metal anticorrosion coatings are generally based on zinc powder; however, mass use of zinc-based coating may cause heavy metal pollution in the media. There also exists concern for the sustainable supply of zinc as a resource. Conducting polymers possess reversible redox performance, which enables them to be the new anticorrosion agents, and their anticorrosion performance has attracted intense attention since the late 1980s. Among these polymers, polyanilines are of particular interest due to their environmental stability, ease of synthesis, and relatively low cost. The prominent protection performance of polyaniline has been confirmed in numerous experiments, and some anticorrosion coating products have been developed in Germany, the United States, and China. Polyanilines are expected to be the new generation of anticorrosion materials with nontoxic and pollution-free characteristics [5].

2. Eco-cement and eco-building materials

The technology of using industrial waste as raw material for cement production has been researched since the 1950s. It has become one of the most important methods of eco-design in the cement industry, which not only decreases resource consumption but also avoids the environmental impact of the waste treatment [6]. Recently, the primary industrial waste used in the cement industry includes: (1) coal gangue from coal mining and washing; (2) tailing and slag from the metallurgy industry; (3) flay ash and desulfurized gypsum from power generation; (4) carbide slag and phosphogypsum from the chemical industry; and (5) red mud from the aluminum industry.

The wider use of eco-cement in municipal construction is a direction that prevents and cures pollution while protecting ecology and the environment. The industrial research that was carried out in the Beijing area to meet the needs of the 2008 Olympics in Beijing mostly focused on increasing the amount of waste added to the raw materials and removing impurities such as heavy metals in the incinerated ash. As a result of this research and development more attention is being given to eco-building materials with an environmental function to be used in ecosystem building. China is very active in the development of new building materials with an environmental protection function, as well as in recycling discarded and old building materials, manufacturing building materials from industry and living wastes, and more. Regarding sterilization or self-clean coating, many kinds of nonpoisonous and decontaminating water-soluble coatings, powder coatings, insoluble coating, and so on, have already been developed. Enlarging the surface contact angle to water or oil for coating can improve the self-cleaning function used on the walls of buildings. Added with TiO_2 of light activity in the coating or on the glass surface, some decontamination function may be attained.

On the other side, cement producers also have made great progress in recent years. For example, the raw meal preparation of cement production includes four stages: ore mining, pre-homogenization of raw materials, grinding, and homogenization of raw meal, which directly affects the resource efficiency, product quality, and energy consumption of the sintering system. The current raw meal preparation technology should meet the needs of processing low-quality ore and waste on the premise of maintaining the high quality of raw meal to increase its source and energy efficiency. For this purpose, computer simulation technology is widely used to analyze geological conditions and mineral composition, which can increase the use of low-quality ore and mining efficiency compared to traditional mining methods. Highly efficient pre-homogenization technology and online analytical instruments are applied to guarantee the quality and homogeneity of raw meal on the condition that a large amount of waste and low-quality ore exist. In addition, a continuous pneumatic homogenization device in raw meal storage is also helpful for thorough homogenization before entering the sintering device.

3. Degradable plastics

White pollutants caused by plastics have already been identified by countries worldwide. This is particularly true in the farmlands where plastic film was once used as a packaging material. China is the largest consumer of farmland film in the world. Research has resulted in a plastic with a controllable degradable period. One kind of low-cost degradable farmland film was made from completely degradable resin, which consisted of biodegradable polyester poly *p*-dioxanone (PPDO) and starch ($\leq$40 wt%), to which was added a degradation control agent. The farmland film with the controllable degradability can be completely degraded in a period of time, also with the similar properties as that of the traditional farmland film which is difficult to degrade (pull burden $\geq$1.6 N, break elongation rate AU: $\geq$160%, burden to right angle tear $\geq$0.6 N, flow temperature 100-130 °C). The average molecular weight of PPDO is 30,000 to 100,000, and the melting point is 100 to 120 °C.

Carbon dioxide plastics are being developed with advanced techniques and offer enormous potential for worldwide applications. The use of carbon dioxide plastics may not only solve the impact of carbon dioxide on the environment but also lead to the complete biodegradation of plastics by returning carbon dioxide into water after the incineration of plastics. The medical use of transparent thin film made from carbon dioxide plastics will resolve the polyvinyl chloride (PVC) problem that has baffled the medical profession. The use of rare earths in FCC catalysts and noumenon polymerization methods are breakthrough processes that have already been achieved in China [7]. The efficiency of catalyst is 800,000 grams polymer per mol metals; the molecular weight is more than 190,000. Now the industrial examination products are being produced to use for food packaging.

PPDO, a kind of aliphatic poly(ester-ether) with good biodegradability, biocompatibility, and bioabsorbability has been successfully used as a biomaterial. Its excellent mechanical performance and unique recyclability make it a promising candidate for replacing the existing disposable plastic products. In recent years, great progress has been made on the synthesis, crystalline structures, and properties of PPDO, especially the crystallization behavior, thermal stability and degradation, rheological behavior, and mechanical properties. As the technology improves and costs are reduced, PPDO will have a wider range of applications [8].

4. Natural resource eco-materials

Many natural minerals are widely used for decontamination, the treatment of sewage, the replacement of poisonous and harmful materials, the manufacture of eco-building materials, and catalyst support materials, among others [9]. For example, micamineral material to replace asbestos, holding water, and to enable fertilizer mineral material to prevent and cure desertification, are typical examples. Natural biological macromolecules are abundant resources. They are not only abundant raw material sources but also biodegradable ones. There are plenty of wood ceramics and bamboo ceramics made out of natural biological macromolecules [10]. By carbonization and heat treatment, the usage efficiency of these kinds of natural materials can be increased significantly, realizing a higher degree of efficiency in the use and recycling of resources.

5. Environment engineering materials

Under the guidance of the eco-material concept, environment engineering materials will not only have a decontamination function but will also coordinate with the environment. Much research on environment engineering materials supports the idea of the purification of the environment, prevention of pollution, and recycling of the waste. Specific applications may include a special kind of fire-retarding or prevention material without the use of asbestos, as well as plastics, or bromine-free flame-retardant materials, sound elimination material, smoke and dust filter material, impact wave absorb material, and more [11,12].

Humans are being increasingly affected by electromagnetic waves and have the need for electromagnetic wave defense material. To reduce radioactive contamination, researchers are developing a valid shield material [13]. The purpose would be to shield humans from electronics equipment with a protection layer that limits the outside radiation of electromagnetic waves.

Photocatalytic technology is a new form of environmental pollution control, but its application has been limited due to low quantum conversion efficiency and its requirement for UV light to excite photocatalytic activity. To solve these problems, a large number of experimental studies have been carried out with some success. A recent theoretical study has promoted the development of photocatalytic technology, including theoretical results on ion doping anatase TiO_2 photocatalyst, the influences of different ions doping or co-doping on TiO_2 crystal structure, energy band, optical and photocatalytic properties, and the methods for how to choose appropriate doping ions for effectively improving TiO_2 photocatalytic property [14].

6. Clean processing

A clean processing technique for materials, also called zero emission and zero waste processing, is a means of synthesizing every kind of process. Clean processing must be a valid synthesis technique that considers both the technique and the economic costs. It mostly reduces or avoids waste and pollutants within the material process, or realizes the material processing technology with decontamination. Biomimetic materials processing offers the starting visions of materials processing for a sustainable future such as in iron and steel metallurgy and machining, directly reducing iron craft to shorten the process and lower the environment burden. The near-last-type machining applied with the short process reduces both the energy and substance consumption in the production line. Clean production technology has also been developed for metallurgy and machining [15–17].

In the chemical industry, the economic reaction on the atom level can save consumption to lower the input and thereby reduce waste output, a process usually called Green Chemistry [18]. For example, peroxidation of alkyl aromatics is a key step in the synthesis of phenols. Compared with traditional phenol synthesis routes, the reaction features low pollution and 100% utilization of reactant atoms, which makes it an ideal green synthesis route.

2 INTRODUCTION OF LIFE-CYCLE ASSESSMENT

Since the beginning of the twenty-first century, the traditional development mode of the materials industry (production with massive resource consumption, energy consumption, and pollution) has been the main restriction of the sustainable development in materials area facing increasingly serious energy and environment problems. Therefore, energy saving and emission reduction of traditional material industry and development of eco-material are considered as one of the most important research directions of material industry. Moreover, methods for qualifying the environmental impact in the life cycle of materials are needed to support policy making for sustainable development. In this area, the LCA method is recognized as having many potential applications as a mainstream analysis tool of the impact to the environment in the life cycle of products and services.

2.1 Sustainability and Life-Cycle Thinking

Research using LCA can be dated back to the 1960s, when an analysis of Coca-Cola packaging was performed by Midwest Research Institute (MRI), which originated LCA study. The analysis intended to identify the packaging material with least resource consumption and lowest environmental impact from numerous candidates. Based on this study, the EPA published a public report, in which the preliminary LCA study frame was developed for the first time.

In the earlier 1970s, the solid waste issue became the main driver of LCA research. Many governments undertook intensive investigations on the industry energy-consumption problem after the energy crisis of 1975, which accelerated the development of the energy-consumption assessment of LCA. When the oil crisis passed toward the end of the 1970s, people learned not to focus only on energy consumption, but also to take into account other environmental problems from a more objective view; this furthered practical progress on the environmental impact analysis method. The Critical Volume Method and Environmental Priority Strategies (EPS) developed during that period are still widely used.

The global environmental issue worsened in the 1980s, which aroused the public consciousness of environmental protection and finally promoted the sustainability strategy enforcement, resulting in a rapid development of the LCA method during that decade. For example, the environment report system was built up in many developed countries, which required a standard environmental impact analysis method and dataset for industry production. Furthermore, some typical technologies of environmental impact analysis also made a remarkable progress, such as the quantitative analysis of the greenhouse effect and resource consumption.

In 1990, the Society of Environmental Toxicology and Chemistry (SETAC) held an international LCA conference to promote the standardization of worldwide LCA research, where the definition of *LCA* was proposed for the first time. Afterward, the International Organization for Standardization (ISO) initiated research in LCA and founded an environmental management committee in 1993. The establishment of ISO14001 was also achieved in the same year, when LCA was included in the fourth series standards as a tool of environmental management, which further accelerated the application and extension of LCA standardization.

In the mid 1990s, the United Nations Environment Programme (UNEP) began to participate in LCA study and published a report entitled *Towards a Global Use of LCA* to expound the acceptable and applicable level of worldwide LCA research. In 2000, UNEP cooperated with SETAC to investigate the future application of LCA. A detailed research proposal and outline was made in 2002.

2.2 Definition of Life-Cycle Assessment

The definition of *life cycle* should be established before defining *LCA*, which involves the whole process related to the production, from raw materials and energy acquisition, through the manufacture, packaging, transportation, sale, use, possible reuse/recycling, and then final disposal, which is also known as from cradle to grave. However, traditional environmental assessment focuses only on production manufacture and disposal but fails to take into account of other stages of life cycle, such as raw materials and energy acquisition, transportation and use, which are proved to greatly contribute to some important environment impact, according to certain overall environment impact assessments.

ISO14040 defines the LCA and related matters as follows: compilation and evaluation of the inputs, outputs, and the potential environmental impacts of a product system throughout its life cycle. In this definition, *product system* means the collection of materially and energetically connected unit processes that performs one or more specific functions, which includes both the manufacture phase and service phase of production.

2.3 Framework of Life-Cycle Assessment

Figure 1 shows the framework of LCA defined by ISO14040, which divides the LCA method into four parts: goal and scope definition, life-cycle inventory (LCI) analysis, life-cycle impact assessment (LCIA), and life-cycle interpretation.

1. Goal and scope definition

Goal and scope definition, as the first step of LCA, describes the purpose and sphere of the research object, which is a critical part of the method.

Goal definition explains the intention and reason for performing LCA; it covers the application area for this research as well as the intended audience for the published result. The scope of research should fulfill the goal definition, which

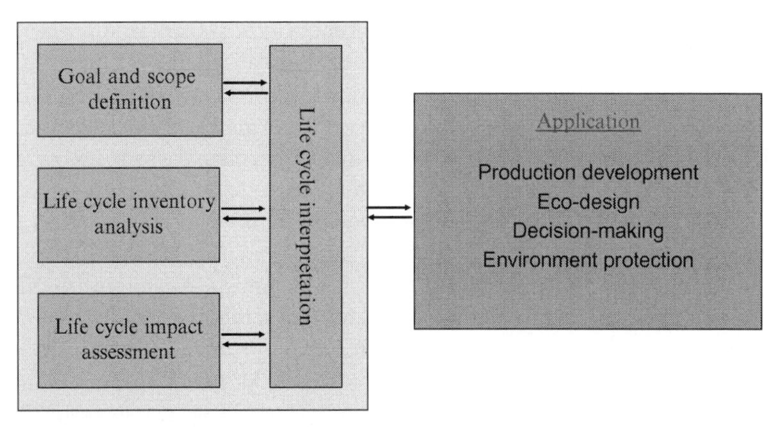

FIGURE 1 Framework of LCA.

defines the function unit and boundary of the research system, sets up allocating rules, and selects an environmental impact assessment and result interpret method.

Scope definition is intended to cover an adequate extent and depth and may be modified time and again according to the collected information and data to satisfy the goal definition. Meanwhile, the goal definition is also likely to be revised as a result of some unforeseeable condition, obstacle, or other information.

2. LCI analysis

According to the goal and scope definition, LCI analysis collects and sorts qualitative and quantitative input/output data for the system under study to offer steps to follow for an integrated inventory that covers main resource consumption, energy consumption, and pollution. The statistical expectation of input/output data in a proper production period should be used for the compliation of inventory so that to eliminate the disturbance of non-representative data. In addition, the source of data, regional and temporal limitations ought to be illustrated definitely.

3. Life-cycle impact assessment

LCIA is another method to evaluate the potential impact caused by materials and energy exchange between research system and external base on the LCI result. According to ISO14040 series standards, LCIA can be divided into mandatory and optional elements.

(a) Selection of impact categories, category indicators, and characterization models. As a preparation phase of LCIA, this step defines impact categories, category indicators, and characterization models used in classification and characterization steps, according to the goals and scope of the research. In addition, the selection of impact categories is based on natural science and satisfies the goal definition stage.

(b) Classification. LCI results are assigned to correlative impact categories in this step. If some LCI data relate to more than one impact categories, a series, parallel, or hybrid rule should be adopted to allocate the pollution into different categories.

(c) Characterization. To integrate the classification results, the data in each impact category will be converted into a uniform unit by timing a respective characterization factor and then summing to get the category indicator results. At present, the equivalency model is the most widely recognized characterization model.

(d) Optional elements. To integrate systemwide environmental impacts into a single indicator, optional elements including normalization, grouping, and weighting are performed. These are among the most controversial parts of LCIA because the elements are not natural science based and involve subjective judgment. Some LCA practitioners therefore advise that a single indicator should be used as an internal reference, whereas in the case of publishing, any limits and uncertainties should be stated specifically. The relative values of category indicator results are calculated in normalization steps. Each category indicator result will be transformed into a uniform unit by being divided by the benchmark of each impact category; this is done to compare between elements or to integrate into a single indicator. Generally, the total amount of pollution emission or resource reserves in a region or worldwide is chosen as the benchmark. The normalization result can be integrated into a single indicator by weighting factors in a weighting step. The weighting factors are usually calculated by an expert questionnaire method, which is not natural science based but reflects the social value orientation.

4. Life-cycle interpretation

Life-cycle interpretation synthetically analyzes the results of LCI and LCIA to identify the "hot spots" (i.e., parts of the life cycle that are important to the total environmental impact) of the system and search for some measures to improve the environment. This analysis thereby estimates the rationality of these measures and gives conclusions and suggestions that correspond with the goal and scope definition. The main steps of life-cycle interpretation include: (1) identifying an important environmental issue based on LCI or LCIA results; (2) evaluating LCI and LCIA results with uncertainty and sensitivity analyses; and (3) making conclusions, suggestions, and the final report.

2.4 Development and Application of Life-Cycle Assessment

1. Application of life-cycle assessment

Life-cycle assessment (LCA), as an environmental management tool, can perform effective analysis not only of environmental impact on the current phase, but also on the whole production life cycle, or cradle to grave. At present, this method is used in many areas, including the development and design of production, and environmental decision making for government and guides for the consumer. Therefore, the deployment and application are attended worldwide.

(a) Application to industry department: LCA originated from the research of some famous transnational corporations (e.g., HP, IBM, AT&T) that first applied the research in LCA methodology to the production analysis involving environment identification and diagnoses of production systems, life-cycle analysis and comparison, evaluation of production improvement, development and design of new production, management and design of recycling systems, and a cleaner production audit.

(b) Application to environment department and international organization: LCA method can be applied to assist the environmental decision making for government and international organizations, including decision support (standard of environmental productions, eco-indicator plan), improvement of government programming of energy, transportation and waste treatment, assessment for the efficiency of waste management and resource utilization, public statement of production and material information, differentiation between ordinary production, and eco-indicator production.

(c) Application to consumer organization: The LCA result can support the customer in handling a serious environmental problem (called green consumption), which has become an important driver force of the production improvement worldwide. And for a developing country, LCA can help establish the benchmark of evaluating eco-production and improving eco-consumption, extend the cleaner production theory, and promote sustainable consumption and development.

2. Development and application of LCA in China

In China, LCA and related applications became hot topics in academic circles in the 1990s, when a great deal of work was started or supported by the government, including the research of the allocation method in LCI, the calculation of Chinese characterization factors and weighting factors, and the like. Furthermore, the national standard was published in the *General Administration of Quality Supervision, Inspection and Quarantine of the People's Republic of China (AQSIQ)* in 1999, which is equivalent to the ISO14040 series standard.

Material life-cycle assessment (MLCA) is one of the most important research directions of LCA in China, and also one of the primary parts of the research of eco-material. In 1998, MLCA research was first sponsored by the National High Tech. R&D (863) program to develop the regional MLCA methods that were suitable for the Chinese situation. MLCA performed case studies on the manufacturing technologies and processes of typical materials involving steel and iron, aluminum, cement, ceramic, polymer, and construction coatings. In 2001, MLCA research was continuously supported by the 863 program to develop the MLCA database and relevant analysis software. Moreover, after 2006, the government initiative supported more and more national R&D programs including the *National High-Tech. R&D (863) Program*, the *National Basic Research Development (973) Program*, the *National Key Technology Research and Development Program,* and the *National Natural Science Foundation* to enhance MLCA research in different areas such as building materials, nonferrous metal, polymer, and so on [19].

2.5 Database and Analysis Tool of LCA

Generally, not only large numbers of environment burden data with high regional limitation but also different LCA methods and models are involved in LCA application. These data with the properties of universality, regionality, and complexity, are the basis for each LCA study and are supposed to be managed effectively. Therefore, owing to the advantage of database technology in the data management area, the development of LCA database and evaluation software has recently become one of the most important directions of LCA research.

1. Research status of international LCA database and software

For promoting the collection of LCA information, a current format for data exchange was established by the *Society for Promotion of Life-cycle Assessment*

Development (SPOLD), which performed a detailed meta-data division on each inventory record to assure the independence, handleability, and procurability of LCI. Moreover, the SPOLD format is an open source and can be embedded in different LCA software for the data exchange between these tools. Furthermore, an international standard (ISO14048) for LCA data exchange is formulated by the ISO, which put forward a normative information format including process information, model information, and management information. However, more detailed criterions for data selection and technology requirement are demanded for actual LCA study.

Several national and international public databases, such as the Swiss ecoinvent database, the European reference life-cycle database, the Japanese JEMAI database, the US NREL database, and the Australian LCI database, have been released in recent years. These databases evolved from publicly funded projects to cover a variety of inventory data on products and basic services, including raw materials, electricity generation, and transport forms as well as waste disposals and services.

On the other hand, the efficiency of LCA implement can be improved and the cost of time and worker hours is reduced by the application of LCA software, which is often divided into three groups: general software for LCA experts and consultants; professional software for the decision of engineering design, sale, or environment and waste management; and application software for specific users (mainly the enterprise users). At present, the amount of LCA software related to material and production is more than 20 worldwide, the environment database exceeds 1000, and more than 3000 commercial software packages with embedded default database are sold, in which some famous tools (such as Simapro, Gabi, Team, etc.) have been widely applied in LCI, LCIA, eco-design, and cost analysis.

2. Research status of Chinese LCA database

In recent years, LCA in China has developed rapidly due to public and government attention, even though the study started relatively late. To support the National 863 Program, initiated by the Beijing University of Technology (BJUT) and in cooperation with other colleges, research institutes, and material corporations, the environmental burden data of main material production (steel, cement, aluminum, engineering plastics, architectural coatings, ceramic, etc.) was collected and processed. Based on these data a basic MLCA database called SinoCenter and related software with independent intellectual property were also developed. As a result of the exploration and development within the past 10years, a research and consultation platform of LCA with the biggest data quantity and covered the widest range of materials in China was established in BJUT, involving six servers, firewalls and routers, 11 workstations, and some professional evaluation software (Gabi4.0, Simapro7.0, UmberTo4.0, Team3.0, etc.). Further, the website of the Center for National Materials Life Cycle Assessment (CNMLCA, www.cnmlca.com.cn) has been opened to society and the public to propagandize the origin, development, and application of LCA, introduce the latest research trend and result at home and abroad, promote the formation and development of ecological material (both in development of thought and evaluation), and most important, support the LCA and ecological design performance in China. Presently, more than 100,000 records are involved in the SinoCenter database. The structure of the SinoCenter platform (Figure 2) and main content is listed as follows [20]:

- Power supply: thermal power, hydropower, and nuclear power
- Primary energy: raw coal, crude oil, and natural gas
- Secondary energy: fuel oil, gasoline, diesel, and coal gas
- Transportation: pipeline transportation, road transportation, shipping and railway transportation
- Freshwater resources: rivers and lakes water system
- Mineral resources: ferrous metal, nonferrous metals, and inorganic nonmetals
- Materials: ferrous metal materials (steel, aluminum, magnesium, etc.), building materials (cement, glass, ceramic, concrete, plastic steel door and window, admixture, coating, carpet, floor coiled material, wallpaper, wooden furniture, adhesives, wood-based panel, etc.), chemical materials (ethylene, HDPE, PVC, PP, ABS, etc.),connecting material (solder, etc.)
- LCA methods: Eco-indicator 99, CML 2001, EDIP 2003, etc.
- Standards: steel, cement, etc.

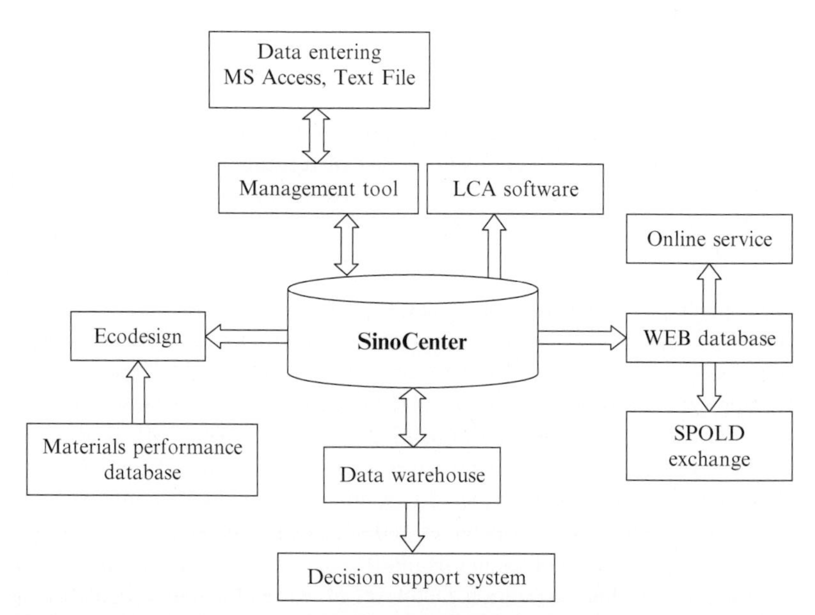

FIGURE 2 Structure and function of SinoCenter platform.

3 DEVELOPMENT OF LCIA METHODOLOGY IN CHINA

So far, the development of methodology and benchmark system of the LCIA phase is still in progress, and several models are used to determine the characteristic indicators showing the relationship between inventory data and environmental impact categories. However, a widely accepted uniform standard is not yet achieved. In current, a variety of methods have been proposed to implement impact assessment. Generally speaking, these methods could be divided into two types: midpoint methods and endpoint methods. The former focuses on the environmental impact categories and their mechanism, and uses characteristic factors to describe the relative importance of various environmental disturbance factors; the latter pays more attention to the causality of the environmental impact issue.

3.1 LCIA Model of Abiotic Resource Depletion in China

The issues of development, utilization, and depletion of mineral resources have always been an important consideration in LCA studies, and have always been given extensive attention. Nevertheless, there are still some divergences in the essence of depletion, and how to describe it scientifically has become the cardinal problem that should be solved in this field. The detailed difficulties of ADP are as follows: (1) the recognition and understanding of the essence of resource depletion; (2) the temporal and spatial assessment criteria for the mineral resource depletion; (3) socioeconomic issues related to the consumption of mineral resources; (4) the method of determining the characteristic indicators and weights of the depletion of mineral resources in the LCA. There also exists much controversy about the research on characterization methods of abiotic resource depletion impacts, and the focus of this controversy is largely centered on a number of fields, such as determination of resources function parameters, rationality of choosing characteristic factors of resource depletion, as well as the impacts caused by resource extraction, substitution, and recycling technology on resource depletion. Presently, there are mainly three kinds of models for resource depletion evaluation:

1. One model uses the ratio of resources extraction volumes to reserves to measure the level of the abiotic resources depletion. These methods use a number of characteristic factors, such as $1/R$, U/R, and U/R^2, among them: R represents the reserves of a certain resource whereas U denotes the current volumes of use or extraction of this kind of resource. The CML method developed by the group at Leiden University in Netherlands reflects this view. In related research in China, the CML method was analyzed, combined with China's characteristics of resources and statistical data, so as to modify important parameters involved in this model, and thus obtained through calculation China's characterization factor set of mineral resource depletion as well as normalization factor of resource depletion in 2004. The comparison

with the original method highlights the fact that geographical distribution differences of resources are unavoidable in the LCA study. Through case studies, the differences between the modified model and the CML model in the application were comparatively illustrated in Figure 3, and the causes for these differences were discussed, thereby a feasible basis for suggesting the modified model as the characterization method assessing Chinese mineral resource depletion provided [21].

2. Use the expected results generated by resource exploitation as a basis for characterization. This point of view suggests that humankind's current extraction of high-grade resources will cause more serious environmental and economic harm when exploiting low-grade resources in the future [22]. Such views are represented by the Eco-indicator 99 method, which uses the energy demand required for exploiting low-grade resources as the damage factor to measure resource depletion, and which believes that this kind of "additional energy" is able to interlink the functionality with technical development of the abiotic resources, rather than directly relying on estimates of hardly predictable resources reserves and annual consumption volumes in the future. The characterization model of abiotic resource depletion was modified and improved in terms of localization in China. However, the characteristic factors of the abiotic resource depletion need to be expanded in terms of both time span and resource category with the development of exploration technology, and the expansion of human demand, because some important parameters, such as resources reserves and extraction volumes, are regionally different and sensitively time-bound. And the degree of correlation between characterization factors of abiotic resource depletion and economic-social factors require further study.
3. Exergy-based model. Exergy, which is defined as the work potential of energy at a stated surrounding, is an appropriate function to express the quality of energy. The exergy model for elements was proposed by Szargut, who selected reference species at atmosphere for nine kinds of elements, hydrosphere for 23 kinds of elements, lithosphere for 53 kinds of elements, and

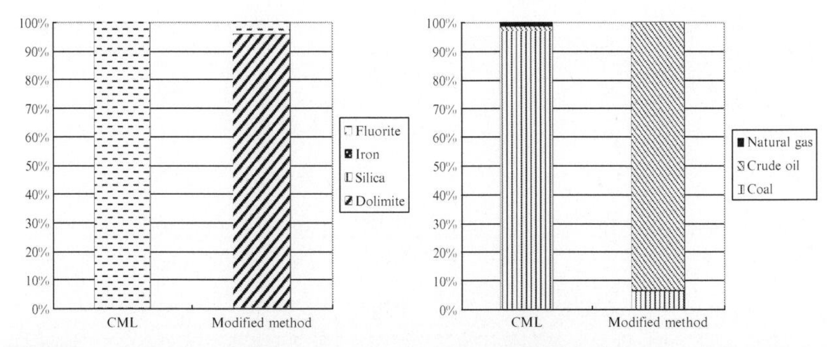

FIGURE 3 Comparison of characterization results between CML and Modified CML method.

calculated the exergy values of elements. Subsequently, methods and data based on this exergy model have been developed for natural resources in LCA by Finnveden and Östlund. To fill the gap of the lack of exergy consumption data, a series of cumulative exergy demand (CExD) indicators were set up to assess exergy scores for a large number of materials and processes. Based on the calculation model proposed by Szargut, the chemical exergy values of the natural minerals were calculated based on chemical composition data of Chinese mineral resources. The result showed that mineral resources consumed by the production system were the real minerals that could be found in nature but not specific pure chemical compounds. For example, iron ore exploited in nature is the mixture of magnetite, quartz, feldspar, etc. When iron ores are input into a system all the minerals are consumed rather than magnetite only. Exergy of the real minerals in nature should be calculated to be compatible with the research scope of LCA. The thermodynamics state of minerals can be divided into two categories: ideal solid solution (Ex1) and mechanical mixing (Ex2), and the values of Ex1 and Ex2, which represent the results of the two different models respectively, are listed in Table 1, indicating the gap between the two models [23].

3.2 LCIA Model of Land Use in China [24]

According to a 1999 report by *World Resources Institute* (WRI), almost all the ecosystem declines in the last century are related to physical changes in land use. Land occupation and the land transformation have significant impacts, whether positive or negative, in the life cycle of the production system, from the exploitation of resources and energy in the raw materials acquisition phase, the building for production and living in the manufacture phase, the construction of road and railway in the transportation phase, until the landfill in the final disposal phase. However, there is no consensus on how land use

TABLE 1 Exergy of Natural Mineral Resources in China

Minerals	Ex1	Ex2	Deviation (%)
Pencil stone	0.259	0.278	7.32
Feldspar	0242	0.265	9.51
Kaolinite	0.362	0.382	5.52
Fluorite	0.423	0.450	6.38
Wollastonite	0.251	0.274	9.16
Silicon sand	0.075	0.082	9.33

impacts should be incorporated in LCA, although the concern for the inclusion of land use impacts in LCA has led to many international publications in recent years.

In a recent study in China, an environmental analysis model for land use was established, which was suitable for conditions specific to China. The proposed model qualified the land use impact based on land quality change presented by net primary productivity (NPP) for the duration of land use. In this model, land occupation and land transformation are considered the basic land use activities that result in either damage or benefits to ecosystem quality (land transformation creates a change in ecosystem quality and land occupation delays changes to its quality). Thus, the impact of land use can be calculated by the following equations:

$$\mathrm{LUI}_{\mathrm{occ}} = \Delta Q_{\mathrm{nature},1} \cdot A_{\mathrm{occ}} \cdot t_{\mathrm{occ}} \quad (1)$$

$$\mathrm{LUI}_{\mathrm{tran}} = \frac{1}{2}\left[t_{\mathrm{res},2} \cdot \Delta Q_{\mathrm{nature},2} - t_{\mathrm{res},1} \cdot \Delta Q_{\mathrm{nature},1}\right] A_{\mathrm{tran}} \quad (2)$$

where $\mathrm{LUI}_{\mathrm{occ}}/\mathrm{LUI}_{\mathrm{tran}}$ is the impact of land occupation/transformation, ΔQ is the land quality change, which is the difference between the natural land and the occupation/transformation land, $A_{\mathrm{occ}}/A_{\mathrm{tran}}$ is the occupation/transformation area, and $t_{\mathrm{occ}}/t_{\mathrm{res}}$ is the occupation/restore duration. Based on this model, the characterization factors of both land occupation and land transformation are calculated using Chinese empirical information on NPP, which can be applied to Chinese LCA case study. The occupation factors are shown in Table 2.

To simplify, only transformation factors between primary land types are calculated, and the results are shown in Table 3. The transformation impact will increase with the difference of NPP value between two land types, and the factors will be negative when converted from low-NPP land type to high-NPP land type, which in other words, has a beneficial impact on the environment.

3.3 Methodology of LCIA Needs to Be Improved Continuously

Although the LCA methodology in China has developed and somewhat matured during the last decades, its scientific implication still needs to be continuously improved and enriched in the following several areas:

- In accordance with the requirements for consistency and comparability, the depth and breadth of the simulating environment mechanism need to be increased. The correlation between the characterization results and the environment needs to be further proved to enable the potential environmental impact assessment results to facilitate integrated decision making.

TABLE 2 Characterization Factor of Land Occupation in China

Primary Land Type	Secondary Land Type	LUF_{occ} [gC/(m^2×a)]
Forest	Subtropical evergreen coniferous forest	66
	Temperate evergreen coniferous forest	223
	Evergreen broad-leaved forest	-
	Deciduous needle-leaf forest	247
	Deciduous broad-leaved forest	185
	Temperate mixed forest	191
	Tropical/subtropical mixed forest	115
Shrub	High-density shrub	137
	Low-density shrub	441
Grassland	Meadow-herb swamp	266
	High-density grassland	218
	Low-density grassland	502
Farmland	One crop annually	312
	Two crops annually	261
	Paddy-upland rotation annually	240
	Double cropping rice	169
Desert	Semidesert	597
	Harsh desert	562
	Sand desert	608
City	City	476

- Regional LCIA methodology and related characterization factors need to be improved to support LCA case study in China; moreover, the differences of environmental impacts caused by spatial and temporal differentiation need to be identified.
- The uncertainty analysis method needs to be established to improve the application scope and result interpretation for decision making.
- Related disciplines need to be further developed to improve the method for comparing among the impact categories, such as resource depletion, human health, land use, water use, and so on, thereby providing better support for integrated decision making.

TABLE 3 Characterization Factor of Land Transformation in China [gC/(m^2×a)]

Land type	Forest	Shrub	Grassland	Farmland	Desert	City
Forest		-3.04×10^3	-8.38×10^3	-6.10×10^3	-1.62×10^4	-1.18×10^4
Shrub	3.04×10^3		-5.34×10^3	-3.06×10^3	-1.32×10^4	-8.79×10^3
Grassland	8.38×10^3	5.34×10^3		2.28×10^3	-7.83×10^3	-3.46×10^3
Farmland	6.10×10^3	3.06×10^3	-2.28×10^3		-1.01×10^4	-5.74×10^3
Desert	1.62×10^4	1.32×10^4	7.83×10^3	1.01×10^4		4.37×10^3
City	1.18×10^4	8.79×10^3	3.46×10^3	5.74×10^3	-4.37×10^3	

4 LCA PRACTICE ON MATERIALS INDUSTRY IN CHINA

With regard to case studies, there are many typical examples from industry, consulting agencies, and academia in China that use LCA methods to choose, optimize, and design technique processes, such as energy supply as well as the production and manufacturing of products. The accumulated data and case studies of LCA cover a wide range of areas in Chinese materials industry.

4.1 Case Study: LCA of Iron and Steel Production in China

Zhou *et al.* (2001) set up an LCI dataset of iron and steel materials in China by research into the production situation of more than 70 major iron and steel manufacturing plants as well as from industry statistical reports [25]. The scope of this dataset covered life-cycle stages ranging from cradle to gate, representing the environmental load of enterprises in different regions and with different levels of technology. Through the assessment of energy-saving and waste recycling and reuse technology during the iron and steel production process, a program for a large-scale integrated iron and steel enterprise to carry out the practice of recycling was put forward in China. For a large-scale integrated iron and steel enterprise with an annual production output of 10 million tons, the use of the new recycled iron and steel production process is able to absorb annually 1.2 million tons of scrap steels and 200,000 tons of waste plastics from the market, generate 9 billion kWh of electrical power, and produces 3 million tons of high-grade cement through digesting wastes produced by itself, thereby yielding huge economic and social benefits [25].

Generally, the typical processes widely used in the Chinese iron and steel industry are the BF/BOF process and the DRI/EF process. The flow diagrams of these two processes are described in Figure 4, from which we can see that BF/BOF process is more complex than DRI/EF. The former includes the sintering and coking subprocess, which is omitted in the latter. Another obvious difference between the two processes is that the scraped iron is widely used in the DRI/EF process [26].

Wang *et al.* (2005) performed a comparative analysis of the environment impacts of the BF/BOF process and DRI/EF process from cradle to gate using LCA methodology (Eco-indicator 99) [27]. The function unit was set up to 1kg ingot produced by the BF/BOF process and the DRI/EF process. The weighting and damage analysis results (including damage to mineral and fossil resources, damage to ecosystem quality, and damage to human health) are shown in Figure 5.

From the result, it can be concluded that (1) the damage to mineral and fossil resources by the DRI/EF process is only a half of the BF/BOF process; the difference is primarily attributed to the input materials, DRI/EF steels produced mainly from scrap irons, whereas the BF/BOF process mainly uses

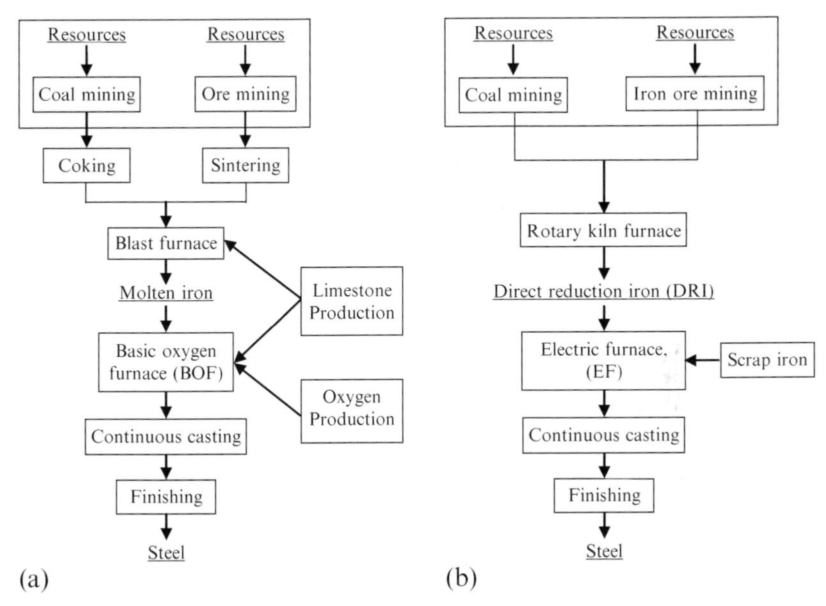

FIGURE 4 (a) The BF/BOF process. (b) The DRI/EF process.

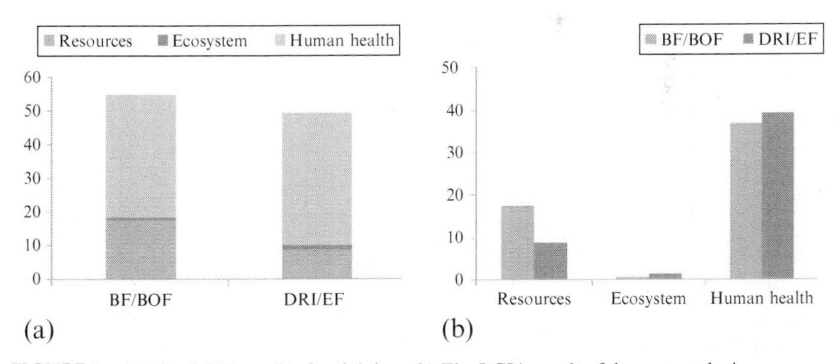

FIGURE 5 (a) The LCIA result of weighting. (b) The LCIA result of damage analysis.

molten ore as resources. (2) On the other hand, the damage to ecosystem quality and human health of the DRI/EF process are higher by 86% and 7% than that of the BF/BOF process respectively, resulting in a higher energy consumption (especially power consumption) of the DRI/EF process. (3) According to the weighting result, the environmental burden of 1kg ingot produced by the DRI/EF process is about 10% lower and that of the BF/BOF process. The results indicate that the environmental performance of the DRI/EF process is superior to that of the BF/BOF process. Great efforts should be made to develop DRI/EF process in order to ensure sustainable development of iron and steel industry in China [27].

4.2 Case Study: LCA of Magnesium Production in China [28]

Currently China is the largest primary magnesium producer and supplier in the world. However, magnesium production with the Pidgeon process is resource and energy intensive and leads to relatively severe environment pollution, which has already attracted much attention from local government and enterprises. So far, the international LCA research on both the production of primary magnesium and the magnesium products is still underway; further study is needed on the environmental impact of the extensive use of magnesium products.

Through constructing the material flow analysis (MFA) method and model of the magnesium resources and their products, Gao *et al.* [21] identified and quantified the quantity, structure, and characteristics of the substance metabolism of the magnesium metal materials in China. Moreover, these researchers followed the ISO14040 series standard, conducting an environmental assessment on the cradle-to-gate life cycle of primary magnesium production using the Pidgeon process in China. Researchers compared the accumulative environmental performance of different fuels use scenarios including abiotic depletion potential (ADP), global warming potential (GWP), acidification potential (AP), and human toxicity potential (HTP). The system boundary of this research was illustrated in Figure 6, which subdivided the Pidgeon process into four

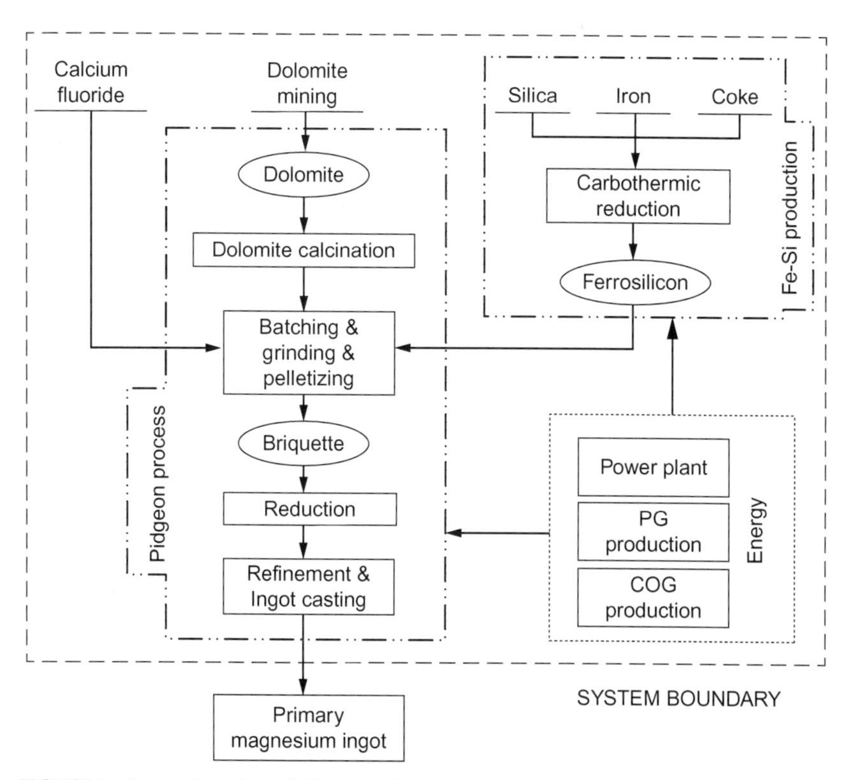

FIGURE 6 System boundary of this research.

steps: dolomite calcinations, batch pelletizing, reduction, refinement and ingot casting. Auxiliary subsystems include dolomite mining, the ferrosilicon production and transportation of involved materials such as ferrosilicon and calcium fluoride, and the gaseous fuels production and power plant supplied energy to the processes.

To identify the effects of implementing gaseous fuels, this research, which was based on the domestic practices of magnesium production, compared the environmental impacts caused by the coal and two other types of fuels: producer gas (PG) and coke oven gas (COG) used in Pidgeon process. The three scenarios of the primary magnesium production were as follows:

- Scenario 1. Factories using the Pidgeon process were located in provinces that were relatively rich in dolomite ore, and primary magnesium was basically produced by local dolomite resources. However, ferrosilicon and fluorite need to be transported over long distances to magnesium plants. Coal was the overall fuel used in the process.
- Scenario 2. PG is used in the Pidgeon process as a main energy; however, coal was still used in addition to PG, mainly because of operational issues in the calcination step. In this scenario, PG is produced in a furnace or generator in which air and steam is forced upward through a bed of burning fuel of coal. The carbon of the fuel is oxidized by the oxygen of the air from below to form carbon monoxide. The nitrogen of the air, being inert, passes through the fire without change. When steam is introduced with the air, the final gaseous product also contains hydrogen. The generator in China has advantages such as easy operating and low investment input. The gaseous sulphide, nitride, and particulates can be easily and efficiently eliminated before PG used, so it has better environmental characteristics than coal. PG, of which calorific value varies in the range of 4.6 ~ 7.5 MJ/m^3, is a mixture of approximately 26% carbon monoxide (CO), 51% nitrogen(N_2), 14% hydrogen (H_2), and 9% other gases such as CO_2, CH_4, and C_mH_n.
- Scenario 3. An associated enterprise represents a sort of production network combining the coke production, ferrosilicon production, and magnesium plants. The factories were relatively concentrated, and ferrosilicon transport distances were less than 5 km. Coke production, which is needed for the making of ferrosilicon, comes with the by-product of a considerable amount of COG, which can be used for the major fuel of magnesium production, with the merits of high calorific value, being at 17.354 MJ/m^3, and convenient combustion. For the Mg plants, most of the overall fuel consumption is achieved by implementing waste heat of coke production. After separating the ammonia and crude benzene, the main ingredients within the gas consists of hydrogen (H_2), methane (CH_4), and carbon monoxide (CO), etc.

The results show that two major components contribute primarily to the environmental effect: the Pidgeon process and ferrosilicon production. The greater or lesser impact of each of these two components varies depending on the category of impact that has been valued, while, in general, the contribution

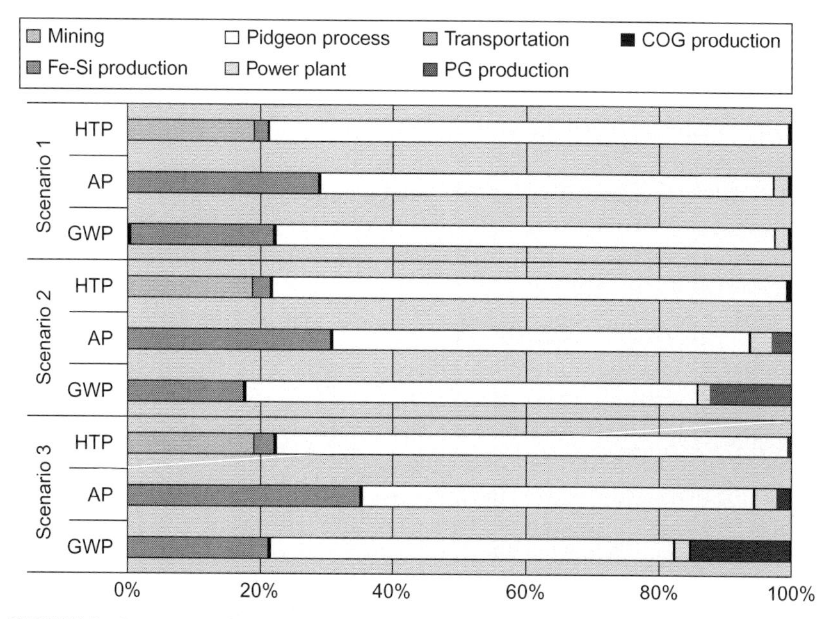

FIGURE 7 Percentage of environmental impact associated with each component category.

of them takes over 80% of the overall impact for each category. The percentage of environmental impact associated with each component category is shown in Figure 7. In the case of the Pidgeon process, the environmental impact is mainly determined by the amount of fuels used. The impact is accentuated by the amount of reducing agent and catalyst used. The emission of CO_2 from Scenario 2 was increased although the PG was used as a major fuel and aggravated, considering the burden of the PG production. In the case of gaseous fuels production, the greenhouse gases emission from COG was 2.5% more than that from PG. However, its final impact was decreased because of the reduction of COG, higher calorific value, used in the Pidgeon process. In the case of ferrosilicon production, the direct emissions of CO_2, SO_2, and PM10 were obtained from published literature of previous research, but the key element is emissions from the electricity consumed; we defined it as indirect emissions, in its manufacture. Considering the contribution of electricity consumption, the significant reduction of environmental impact in Fe-Si production is possible by converting from coal-based to gas-based electricity or hydropower.

The final single results of three scenarios were 9.40×10^{-10} year, 9.07×10^{-10} year, and 7.68×10^{-10} year, respectively. The results indicated that the accumulated environmental performance of adopting Scenario 3 to produce primary magnesium was relatively lowest, Scenario 2 comes to the next and was 18% higher than Scenario 3, and Scenario 1 showed the highest accumulative environmental load and was 22% higher than Scenario 3. Figure 8 illustrated the absolute values of each impact category of three scenarios, which showed

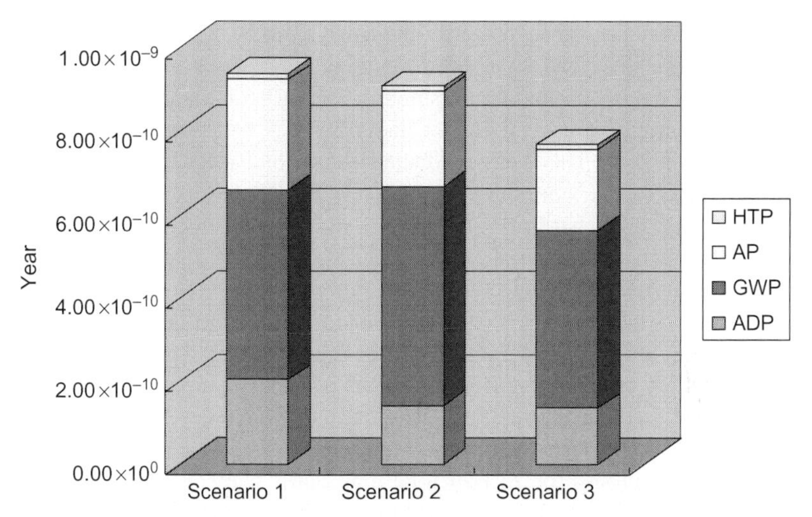

FIGURE 8 Comparative analysis of three scenarios (single indicator).

that GWP and AP were the major impacts of Mg production. Concerning the Pidgeon process, the impact of abiotic resources depletion deserves more attention although the types and the amount of mineral resources for Mg production are abundant in China.

The different fuel-use strategies in the practice of magnesium production using the Pidgeon process caused many different results on the accumulative environmental performance. The life-cycle environmental performance of Scenario 3, which adopts the local production networks of coke-ferrosilicon-magnesium, was best even if the emissions of COG production were considered. But Scenario 1, which burns coal directly as overall fuel, showed the poor environmental performance. This means that a positive improvement can be achieved by the integrated production of several commodities. Considering that the COG was limited by gas supply conditions and the location of magnesium plants, PG, in areas where coke production was not concentrated, could be used as the major fuel for primary magnesium production. Thus, this study suggested that the PG was an alternative fuel for the magnesium production rather than the coal burned directly in the areas where high-priced COG was produced. The use of "clean" energy and the reduction of greenhouse gases and acidic gases emission were the main goals of the technological improvements and cleaner production of the magnesium industry in China.

4.3 Case Study: Greenhouse Gas Analysis of Chinese Aluminum Production Based on LCA [29]

Currently China is the world's largest aluminum producer. The aluminum industry is not only an important basic raw material industry for national economic development, but also for high energy-intensive industry. The aluminum sector represents more than 75% of nonferrous metals industry's total energy consumption in China, with aluminum smelting, in particular, consuming a large amount of energy and accounting for about 5.5% of total national electricity consumption in 2006. This work conducted a study on energy consumption and greenhouse gas (GHG) emissions of primary aluminum production in China to identify the key aspects of GHG emissions reduction. The GHG emission in this research included CO_2, CH_4, and two perfluorocarbon compounds (PFCs), carbon tetrafluoride (CF_4), and carbon hexafluoride (C_2F_6). The system boundary of primary aluminum production in China is shown in Figure 9, and 1 ton primary aluminum ingot was defined as functional unit.

The result shows that the GHG emissions per ton primary aluminum ingot in China were 21.6tCO_2eq in 2003. The aluminum smelting stages contributed 72% of the overall GHG emission, and it was followed by the alumina production stages (accounting for 22% of total emission). The comparison of GHG emission between the Chinese level and the world average level (12.7tCO_2eq/t) showed that GHG emission of Chinese primary aluminum production is much higher than the world average. In particular, the GHG emission caused by aluminum smelting and alumina production in China was higher by 58% and 153% respectively than world average levels (Figure 10). The coal-dominated energy consumption structure was the major reason to account for the GHG emissions from Chinese primary aluminum production being higher than the world

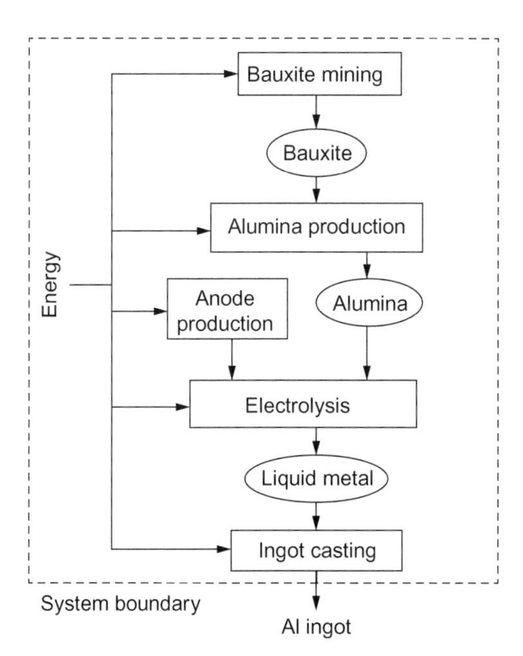

FIGURE 9 System boundary of primary aluminum production.

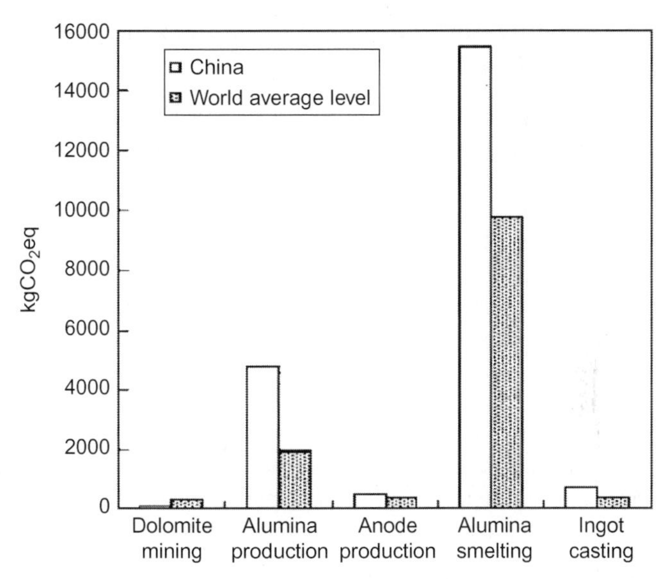

FIGURE 10 The comparison of GHG emissions between China and world average levels.

average level. Generally, secondary energy (electricity), petroleum, and natural gas were holding a large proportion (83%) of fuel supply for primary aluminum producers worldwide. However, coal and coke as well as producers of gas accounted for 79% of direct fuel consumption in the Chinese aluminum industry. Moreover, thermal power generation by coal accounted for 81% of national electricity generation capacity in China, which emitted substantially more GHG than hydropower, petroleum, and natural gas power generation. Therefore, as for the aluminum smelting process, which was a major consumer of electricity, the structure of the electricity industry also determined the major characteristics of GHG emissions.

Reducing electricity consumption at the stage of aluminum smelting is an effective way to conduct GHG emission reduction. In 2005, the Chinese aluminum industry completely eliminated Söderberg technology including 1.54 million tons of outdated production capacity, thereby further increasing the proportion of prebake technology with the features of high efficiency, low power consumption, and high capacity. At the same time, optimized control technology for the electrolyte temperature of prebake pots, the liquidus temperature, and overheating temperature of the electrolyte were developed and applied, thus the energy consumption and the PFCs emission were significantly reduced. In 2006, the electricity consumption of China aluminum smelting process is close to the international advanced level. The power consumption for aluminum ingot had already approached the target for the year 2010 (14,600kWh/t). The efforts of raising the control level of electrolytic pots and reducing the coefficient of anode effects enabled the GHG emission caused by PFCs to decrease by 75% from 2003 to 2006.

4.4 Case Study: Layout Adjustment of Cement Industry in Beijing Based on LCA [30]

As one of the most important building materials, cement is widely used. However, approximately 5% of global CO_2 emissions originate from manufacturing cement. In China, the cement industry is also one of the largest sources of carbon emissions. In addition, millions of tons of cement kiln dust and other gaseous emissions are released each year, contributing to respiratory problems and pollution health risks. Currently there is a call to reduce the environmental impact caused by cement production such as utilization of waste as raw materials and energy, waste gas treatment technology, power generation by waste heat in cement production, and so on. Eco-cement is regarded as the future of the cement industry.

For the 2008 Olympic Games, host Beijing developed a cement industry layout adjustment program that combined process LCA analysis and regional MFA. This was done to satisfy both the cement demands of Olympic construction and environmental protection requirements. Figure 11a shows the distribution of key enterprises of cement in Beijing; Figure 11b shows the distribution of production and use of cement in Beijing (a green point on the map shows where 1000 tons of cement were used, and a red point shows where 1000 tons of cement were produced).

Moreover, under the conditions that keep the cement output basically unchanged, the overall consumption of materials and energy in Beijing's cement plants, through adjustment, combination, and technological upgrading, was basically the same as in 2001, but the emissions of atmospheric pollutants, including soot, fumes, and sulfur dioxide, were decreased by 50%, 11%, and 2%, respectively, compared to 2001. This program provides an extremely important reference point for significant improvement of the quality of the atmospheric environment in Beijing and for the phased objective achievement of reducing and controlling air pollution.

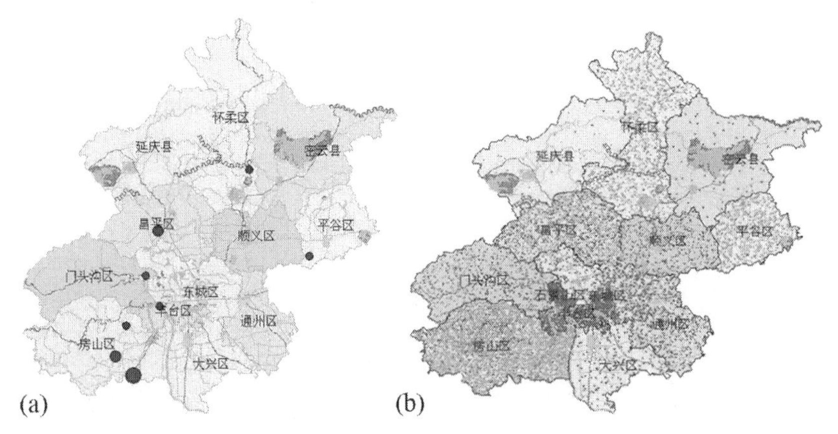

FIGURE 11 (a) Distribution of key enterprises using cement in Beijing. (b) Distribution of production and utilization of cement in Beijing.

4.5 Case Study: LCA of Flat Glass Production in China [31]

1. LCA of Chinese glass industry

Flat glass is a product of daily life and of manufacturing and industry. In recent years, the demand of flat glass has increased significantly because of the rapid development of construction, automobile, and electrical information industries in China. The goal of the study was to analyze the environmental load of the flat glass industry in China. The scope of the study covered from cradle to grave of flat glass production, including the stages of raw material and energy production, transportation, and flat glass production. For this purpose, the LCI of the Chinese flat glass industry was established, which included resources consumption and pollutants emission from 2004 to 2007 (Figure 12). The results showed that the demand of resources (e.g., dolomite, soda, limestone, etc.), especially feldspar, increased significantly from 2004 to 2007 mainly caused by the increase of glass yield. However, the emission (including slag, NO_x, SO_2, and so on) per unit glass decreased about 15% to 20% in 2007 compared to that of 2004 due to the technology improvement.

2. LCA of typical float glass production line

The most important technique of flat glass production is float technique. The float glass output accounts for about 85% in China in recent years. The goal of this study is to analyze the environmental impact status of a typical float glass factory in China, and identify the environmental hot spot and process that damages the environment most, thereby providing references for the float technique designer. A single weight case of flat glass was selected as a function unit. The weight results of different processes are shown in Figure 13, from which it can be seen that the most important impacts caused by float glass production were GWP and AP, mainly caused by the fuel combustion in the melting and mining phases.

Comparative analysis between a typical float glass production line and the average level is shown in Figure 14. The result revealed that the environmental damage of the float glass line was reduced 41.5% from the average level of the flat glass industry in China. The obvious superiority of the float glass

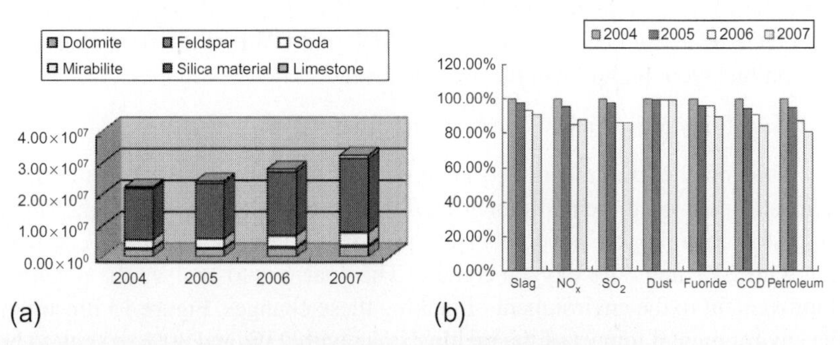

FIGURE 12 (a) Total resource consumption. (b) Emission per unit product.

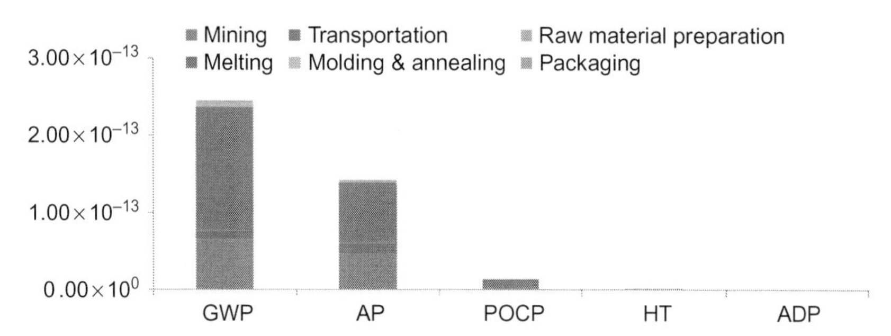

FIGURE 13 Weight results of processes in typical float glass factory (per weight case).

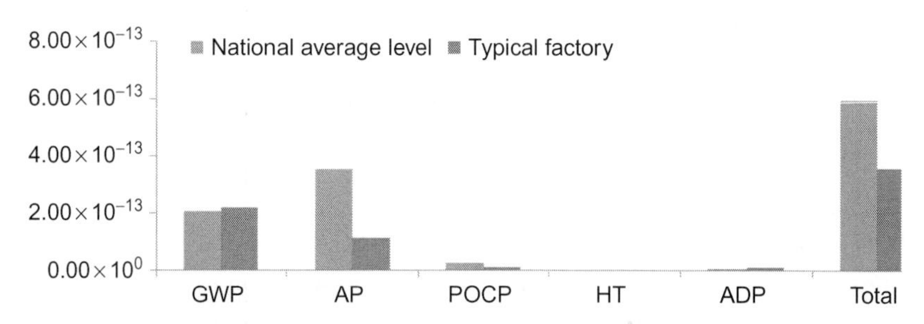

FIGURE 14 Comparative analysis between typical float glass production line and average level.

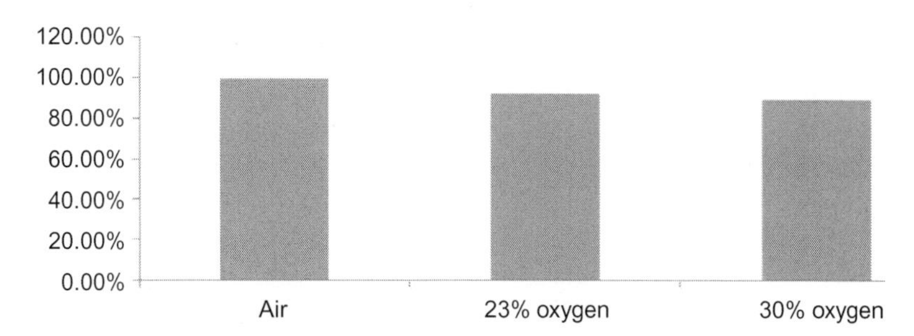

FIGURE 15 Comparative analysis among different combustion condition.

production line was decreasing in AP and POCP. There AP decreased 66.8%, and POCP decreased 62%. However, the ADP and GWP of the float glass production line were higher than the average level.

3. LCA of oxygen-enriched combustion technology

Based on the aforementioned analysis, the melting process had the most serious impact on the environment in float glass production. Thus, the study suggested changing the air combustion support into oxygen-enriched combustion with 23% and 30% oxygen content. The goal was to analyze the potential improvement to the environment of making these changes. Figure 15 illustrates the environmental impact of the melting stage with 23% and 30% oxygen to be a decrease of 8.2% and 11% respectively.

4.6 Case Study: CO_2 Emission Analysis of Calcium Carbide Sludge Clinker [32]

In recent years, China has become the largest cement producer in the world. However, cement industry is a main contributor of global carbon emissions. Substituting calcium carbide sludge for limestone is an effective method for CO_2 emission reduction in the cement industry and has developed rapidly in recent years in China. The purpose of this study is to determine the life-cycle CO_2 emission of cement clinker produced with calcium carbide sludge as secondary raw material: (1) Quantify CO_2 emission during the life cycle of calcium carbide sludge cement clinker; (2) make a comparative analysis on the CO_2 emission generated from cement clinker production between general cement clinker and calcium carbide sludge clinker.

In general, CO_2 emission mainly generates from limestone calcinations and fuel combustion in cement clinker production. Figure 16 shows CO_2 emission statistics from all main stages of the product life cycle, including mining, transportation, coal production, power generation, and cement clinker production. The phase of clinker production makes the dominant contribution of CO_2 emission for both general clinker and calcium carbide sludge clinker, which accounts for over 85% of total emission. In addition, the contribution of the other phases, ranked from high to low, is power generation, coal production, transportation, and raw materials mining.

The results of comparative analysis show that calcium carbide sludge cement clinker exhibits obvious advantages in CO_2 emission compared to general cement clinker. CO_2 emission per ton clinker is reduced by 39.1%, which is mainly achieved in mining, power generation, and clinker production. CO_2 emission of these three phases declines by 64.9%, 8%, and 42.2%, respectively. However, CO_2 emission generated from transportation and coal production was increased slightly. To analyze the reason of CO_2 emission reduction: first, as industrial waste, calcium carbide sludge doesn't need to be mined, therefore CO_2 emission declined remarkably in the mining phase. Second, the granularity of calcium carbide sludge is smaller than traditional raw material of clinker production (e.g., limestone), and therefore can decrease the power consumption for grinding. Third, to better understand and describe the CO_2 emission, Figure 17 illustrates the comparison of CO_2 emission between the two types of clinker in cement clinker production. The figure shows that CO_2 emission generated

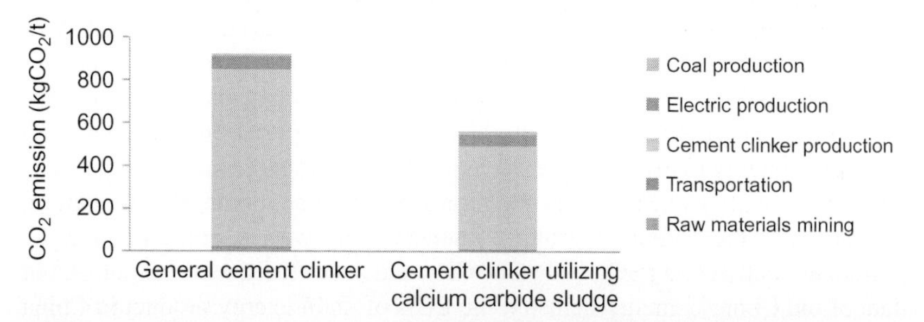

FIGURE 16 Life-cycle CO_2 emissions inventory of cement clinker production ($kgCO_2/t$).

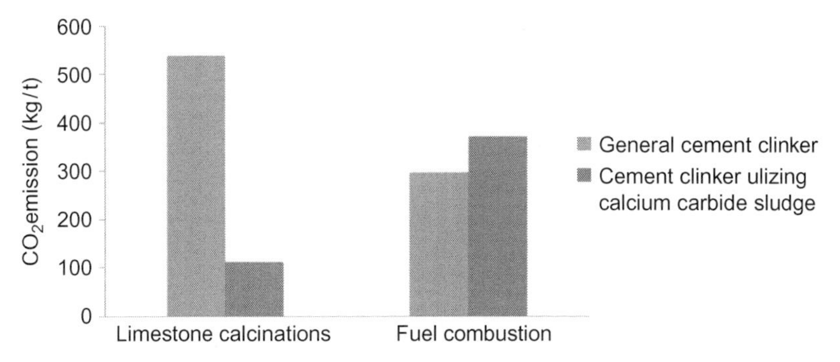

FIGURE 17 CO_2 emissions in cement clinker production (kgCO_2/t).

from limestone calcinations and fuel combustion account for 67.4% and 32.6% respectively for general cement clinker. On the other hand, for cement clinker production with calcium carbide sludge as secondary raw material, the main chemical composition of calcium carbide sludge is calcium hydrate; this therefore is able to play an important role in CO_2 emission reduction (in this case, CO_2 emission generated from limestone calcinations is reduced by approximately 79%). However, because calcium carbide sludge needs to be dried before grinding, this process consumes additional energy; thus, the CO_2 emission generated from fuel combustion is 26% higher than general clinker production.

4.7 Case Study: LCA in Chinese Energy Sector

The LCI of energy carriers is fundamental for carrying out LCA analysis on both materials industry and other industrial products. To further develop the LCA database, the inventories of primary energy are necessary, including the energy consumption and environmental emissions involved in the extraction process of coal, crude oil, and natural gas, and have compiled a full data inventory from cradle to grave of several major downstream products derived from coal and crude oil, such as cleaned coal, coke, petrol, and also electricity.

1. LCI of fossil energy product in China [33]

LCA has been an important method for systematic evaluation of the environmental impacts of chemical industry products and activities. The life-cycle inventories of fossil fuels could not only provide reference data for LCA of chemical industry and its products, but also show the environmental performance of fossil fuel production. Here, researchers investigated energy consumption and direct emissions due to production and transportation of primary and secondary energy to establish life-cycle inventories of the Chinese energy product. The inventory encompasses the inputs of raw coal, crude oil, and natural gas, and also the outputs of water pollutants, solid wastes, and gaseous state pollutants such as CO_2, SO_2, NO_x, CO, CH_4, and dust. Based on the input-output data of the Chinese energy industry, the LCIs of main energy product in China are calculated and shown in Table 4.

TABLE 4 Life-Cycle Inventory of Energy Products in China

		Primary Energy			Secondary Energy		
Consumption/Emission		Raw Coal	Crude Oil	Natural Gas	Gasoline	Coke	Refinery Gas
Energy	Raw coal	1.00×10^{0}	1.27×10^{-3}	1.18×10^{-3}	4.77×10^{-2}	1.65×10^{0}	5.10×10^{-2}
	Crude oil	8.59×10^{-4}	1.30×10^{0}	3.27×10^{-2}	1.32×10^{0}	1.43×10^{-2}	1.41×10^{0}
	Natural gas	5.13×10^{-8}	2.11×10^{-6}	1.32×10^{0}	7.95×10^{-5}	9.29×10^{-4}	8.50×10^{-5}
Pollution	CH_4	9.32×10^{-3}	7.86×10^{-6}	7.32×10^{-6}	2.17×10^{-4}	1.51×10^{-2}	2.32×10^{-4}
	SO_2	7.45×10^{-6}	2.06×10^{-4}	1.91×10^{-4}	1.17×10^{-3}	1.27×10^{-5}	1.25×10^{-3}
	NO_x	4.29×10^{-5}	2.00×10^{-4}	1.87×10^{-4}	8.48×10^{-4}	6.58×10^{-5}	9.07×10^{-4}
	CO	5.17×10^{-6}	7.77×10^{-6}	7.23×10^{-6}	1.35×10^{-4}	1.96×10^{-6}	1.44×10^{-4}

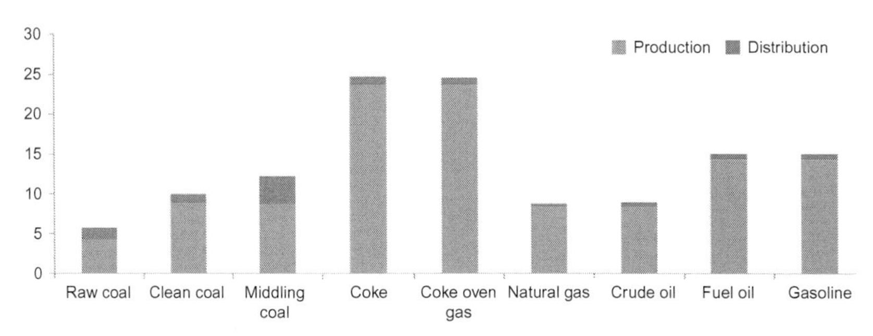

FIGURE 18 GHG emission of main energy product in China.

The GHG emission of main energy products in China is shown in Figure 18. The figure indicates that coke has the largest CO_2 emission intensity among different fuels, followed by oil refining products, middling coal, cleaned coal, natural gas, crude oil, and raw coal. For oil refining products, natural gas, crude oil, and coking products, the contribution of distribution in their life cycle CO_2 emissions is very small; in comparison, the distribution of coal-related products contribute more CO_2 emissions in life cycle than other fuels. The reasons for this are as follows: (1) Trucking on highways, which causes more CO_2 emission than other modes, is the main transport mode for coal-related products; (2) Compared with other fuels, the heat values of coal-related products are very low, and therefore, for them, more quantity should be transported for providing 1 GJ of energy than other fuels.

2. LCI of power generation in China [34]

By the end of 2002, the total national capacity of installed electricity in China had reached 356.6 GW, with a yearly power generation of 1654.2 TWh. Up to now, the electricity grid system in China has been composed of five trans-regional grids (North China, Northeast China, East China, Central China, and Northwest China), South China Electric Power joint venture network, and seven independent provincial or municipal networks. Trans-province grids and Shandong Grid have built 500 kV or 330 kV transmission lines as main trunks, and East China and Central China grids have been interconnected through 500kV DC transmission lines. These networks cover most parts of rural and urban areas in China, providing qualified, reliable electricity, and supply services. Thermal power has been the principal part of total national installed capacity and electricity generation in China. Development of hydropower is slower than that of thermal power, and nuclear power is just in its initial step. New energy resources that generate power, such as wind, solar energy, and tide, is not included in this study due to the small share of yearly electricity generation.

The benefits of public LCI data on electricity generation would be high for those who undertake LCAs and for those who draw conclusions based on LCAs. Process models of power plants were developed for the Chinese situation in the study. LCIs for the electricity industry in China were developed. The emissions of CO_2, SO_2, NO_x, CH_4, CO, nonmethane volatile organic compound (NMVOC), dust (all particulates), and heavy metals (Ni, V, As, Cd, Cr, Hg, Pb, and Zn) from thermal power plants as well as those from fuel production and transport were investigated. The emissions of CO_2 and CH_4 from hydropower plants and radioactive emissions from nuclear power plants were also calculated.

The LCI for 1 kWh of electricity generation in China is listed in Table 5. The transmission of electricity in all cases is taken to be distributed from power station via a high-voltage electricity grid to low-voltage electricity suitable for domestic use, causing a loss of 7.52% of the electricity produced at the power station. And a loss of 6.15% was caused by the electricity consumption at the power plants. The LCI for 1 kWh of electricity distributed to end users is also calculated.

Based on the LCI result, analysts compared power generation in China with that of Japan. The results showed that the emission of pollutants from power plants in China was much higher than Japan, especially in the emission of CH_4, V, Pb, dust, and Zn. The reasons of the gap between China and Japan were as follows:

(a) Compared with such developed countries as the United States, Russia, Japan, and France, the contribution of thermal power was obviously higher in the power structure of China. And general emission intensities from thermal electric plants were much more than that of hydropower plants or nuclear power plants.

(b) Coal-fired power generation was chiefly used in Chinese thermal power: In the 1990s, the proportion was more than 90%. But the amount of coal-fired power generation of Japan's electricity industry was only 30%; the remaining was oil-fired (33%) and gas-fired (33%). Compared with oil-fired and gas-fired power plants, emission intensities of coal-fired power plants are much higher.

(c) The technologies of Chinese electric power generation and distribution are still backward. The standard coal consumption related to 1 kWh of electricity distributed by Japan's electricity industry in 1997 was 324 g/kWh, while that of China's in 2002 was 383 g/kWh and was 18.2% higher. The reason was lower fuel usage efficiency and higher distribution loss of Chinese electricity industry.

(d) The treatment of stack gases in Chinese thermal power plants was much lower than the international level. The cover rate of generator sets equipped with denitrators was 80%, and the average precipitating efficiency of the precipitators was about 97% in 2002. In Japan most generator sets were equipped with denitrators, and the average precipitating efficiency of the precipitators was as high as 99.5%. The cover rate of generator sets equipped with desulfurizers in China was less than 2%, and the denitrification was just in the initial stage in Chinese thermal power plants.

TABLE 5 Life-Cycle Inventory of Electricity Generation per kWh in China

Category	Fuel Consumption				Air Pollutants		
Input/Output	Coal-fired	Oil-fired	Gas-fired	EU	CO_2	SO_2	NO_X
Unit	Kg/kWh	Kg/kWh	Kg/kWh	Kg/kWh	Kg/kWh	Kg/kWh	Kg/kWh
Per 1 kWh	3.97×10^{-1}	7.71×10^{-3}	6.90×10^{-3}	7.84×10^{-8}	7.61×10^{-1}	6.98×10^{-3}	5.50×10^{-3}
Per net 1 kWh	4.57×10^{-1}	8.88×10^{-3}	7.95×10^{-3}	9.03×10^{-8}	8.77×10^{-1}	8.04×10^{-3}	6.34×10^{-3}
Category	**Air pollutants**						
Input/Output	CO	CH4	NMVOC	Dust	As	Cd	Cr
Unit	Kg/kWh	Kg/kWh	Kg/kWh	Kg/kWh	Kg/kWh	Kg/kWh	Kg/kWh
Per 1 kWh	1.09×10^{-3}	2.30×10^{-3}	3.43×10^{-4}	1.42×10^{-2}	1.41×10^{-6}	8.94×10^{-9}	1.19×10^{-7}
Per net 1 kWh	1.25×10^{-3}	2.65×10^{-3}	3.95×10^{-4}	1.63×10^{-2}	1.62×10^{-6}	1.03×10^{-8}	1.37×10^{-7}
Category	**Air pollutants**					**Water emissions**	
Input/Output	Hg	Ni	Pb	V	Zn	Waste water	COD
Unit	Kg/kWh	Kg/kWh	Kg/kWh	Kg/kWh	Kg/kWh	Kg/kWh	Kg/kWh
Per 1 kWh	6.17×10^{-8}	1.76×10^{-7}	1.24×10^{-6}	2.02×10^{-6}	1.69×10^{-6}	1.14×10^{0}	5.23×10^{-5}
Per net 1 kWh	7.12×10^{-8}	2.03×10^{-7}	1.42×10^{-6}	2.33×10^{-6}	1.94×10^{-6}	1.31×10^{0}	6.02×10^{-5}
Category	**Solid wastes**		**Radioactive air**		**Radioactive water**		**RSW**
Input/Output	Fly ash	Slag	Inactive gas	H&G	Tritium	Non-Tritium	
Unit	Kg/kWh	Kg/kWh	Bq/kWh	Bq/kWh	Bq/kWh	Bq/kWh	m^3/Kg
Per 1 kWh	7.24×10^{-2}	1.62×10^{-2}	3.25×10^{1}	1.40×10^{-1}	3.67×10^{1}	3.53×10^{-2}	2.33×10^{-10}
Per net 1 kWh	8.34×10^{-2}	1.87×10^{-2}	3.74×10^{1}	1.61×10^{-1}	4.22×10^{1}	4.06×10^{-2}	2.68×10^{-10}

4.8 Case Study: LCA of Civilian Buildings in Beijing [35]

In the green building system currently being advocated, one is to consider, from the perspective of sustainable development, the impacts on resources, energy, and environment during the whole life cycle of buildings. China now has buildings with a total floor area of more than 40 billion square meters, and the direct energy consumption during the construction and use of buildings accounts for 30% of the total amount consumed by the whole society. To meet the urgent demand for the development of energy-saving buildings and green materials, the goal of this research is to comprehensively analyze the environmental burdens and impacts of three types of multi-unit residential buildings built in concrete framework construction (CFC), light gage steel framework construction (SFC), and wood framework construction (WFC) in their life cycle. The research objects are the three types of multi-unit residential building in Beijing with the overall floor space (5589.15 m^2) and the building lifetime of 50 years. The life cycle of buildings could be defined as the four phases of embodied materials, construction, use and disposal, and the LCI for different materials or products in each phase were collected and calculated. Furthermore, a professional LCA analysis tool, SimaPro7.1, was adopted to perform the modeling, LCI analysis, and environmental impact assessment.

The energy consumption during the building life cycle generally means the consumption of energy products (electricity, natural gas, petroleum, and coal). According to the life-cycle thinking, this can be traced back to the consumption regarding the primary fossil energy (i.e., raw coal, crude oil, and natural gas). The consumption rates of these fossil fuels are shown in Table 6.

With the aid of the calculation method of the average net calorific value, the net calorific values for raw coal, crude oil, and natural gas are determined as 20,908 kJ/kg, 41,816 kJ/kg, and 35,544 kJ/m^3, respectively. As per the corresponding energy consumption for each phase and building, the energy consumption for three buildings across their life cycles was calculated, as shown in Figure 19.

The energy consumption of the three types of buildings has similar characteristics: energy consumption is highest during the building operation phase, and the embodied materials phase occupies the second highest position. The ratios of energy consumption of WFC, SFC, and CFC to that of the life cycle in the operation phase are 87%, 76%, and 71%, respectively, and the ratios in embodied materials phase are 13%, 23%, and 27%, respectively. The energy consumption in the construction phase is higher than that of the disposal phase, and the ratio of the sum of the two phases to the life cycle for the three types of buildings is less than 2%. In the energy consumption structure of the building system, natural gas products and electricity are mainly consumed in the operation phase, and coal products are used up in the embodied materials phase. In the building operation phase, the ratio of the consumption of natural gas products to its life cycle for the three types of buildings is nearly 100%;

TABLE 6 Life-Cycle Primary Energy Consumption of the Three Buildings

Primary Energy	Life-Cycle Phase	WFC	SFC	CFC
Raw coal	Embodied material, kg	2.24×10^{5}	7.03×10^{5}	1.10×10^{6}
	Construction, kg	9.96×10^{3}	1.53×10^{4}	1.26×10^{4}
	Operation, kg	9.77×10^{5}	9.81×10^{5}	9.76×10^{5}
	Disposal, kg	9.80×10^{1}	9.22×10^{1}	4.16×10^{2}
	Life cycle, kg	1.21×10^{6}	1.70×10^{6}	2.09×10^{6}
Crude oil	Embodied material, kg	6.87×10^{4}	1.01×10^{5}	9.69×10^{4}
	Construction, kg	8.15×10^{3}	1.09×10^{4}	3.22×10^{4}
	Operation, kg	5.27×10^{4}	5.96×10^{4}	5.61×10^{4}
	Disposal, kg	3.45×10^{3}	3.24×10^{3}	1.46×10^{4}
	Life cycle, kg	1.33×10^{5}	1.75×10^{5}	2.00×10^{5}
Natural gas	Embodied material, m^3	5.76×10^{4}	1.18×10^{5}	4.05×10^{3}
	Construction, m^3	8.00×10^{0}	1.23×10^{1}	1.10×10^{1}
	Operation, m^3	1.23×10^{6}	1.50×10^{6}	1.37×10^{6}
	Disposal, m^3	2.20×10^{-1}	2.00×10^{-1}	9.20×10^{-1}
	Life cycle, m^3	1.29×10^{6}	1.62×10^{6}	1.37×10^{6}

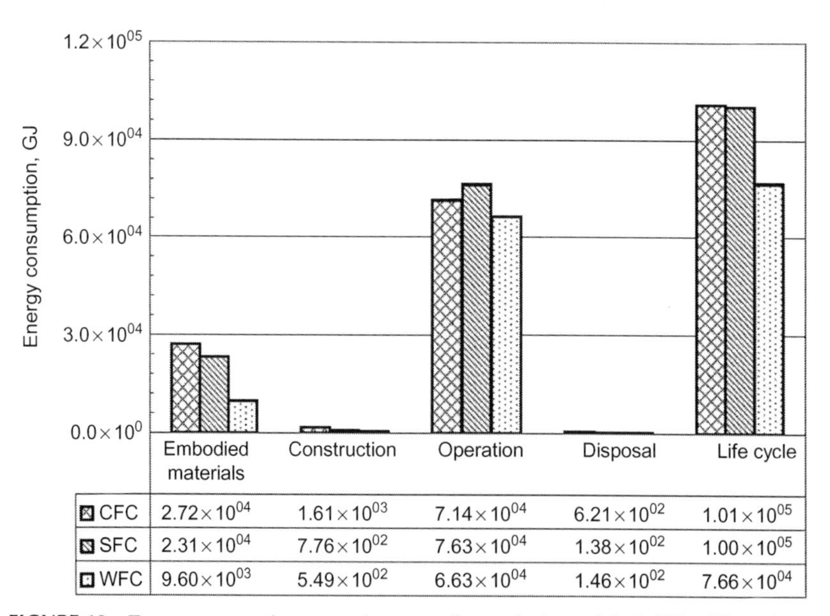

FIGURE 19 Energy consumption comparison regarding each phase of the building life cycle.

however, the SFC natural gas consumption is 10% higher than that of CFC and 23% higher than that of WFC. The ratios of the electricity consumption to its life cycle for CFC, SFC, and WFC are 87%, 92%, and 95%, respectively; however, SFC is 0.4% higher than WFC, and WFC is 0.1% higher than CFC. In the embodied materials phase, the ratios of coal product consumption to its life cycle for CFC, SFC, and WFC are 99%, 98%, and 95%, respectively; however, the coal consumption of CFC is 56% higher than that of SFC and 357% higher than that of WFC.

Greenhouse gas primarily includes carbon dioxide, methane, carbon monoxide, nitrous oxide, hydrofluorocarbons, and perfluorocarbons, to name a few. In the process of construction activities, the greenhouse gas emission covers a wide field. This study mainly analyzes the CO_2 emissions during energy utilization, material production, and transportation linkage. The material flow of the CO_2 refers to two conditions: (1) the industrial emissions in construction activities (the material productions, energy utilization, transportation, construction, and disposal, among others); and (2) the biogenic carbon dioxide of wood materials used in wood products. Apparently, from the perspective of biogenic carbon cycle, the CO_2 uptake of wood via photosynthesis from the atmosphere would be eventually emitted to the natural environment at the grave via the different ways (combustion, natural oxidation, etc.) and thus the carbon cycle of biomass materials is closed, that is, the biogenic carbon could be simply treated with carbon neutrality, which implies the CO_2 uptake equals the CO_2 emissions at the end life cycle or wood grave in this study. Therefore, using the carbon-neutrality method to calculate CO_2 emissions of natural wood materials, the net CO_2 emission is defined here as the emission in construction activities, minus the biogenic carbon during the utilization of wood products, as presented in Figure 20. The figure shows the net CO_2 emission, of which CFC has the highest, followed by SFC, and WFC has the lowest. In addition, CO_2 emission of CFC is 44% higher than that of SFC, and 49% higher than that of WFC, while that of SFC is 5% higher than that of WFC.

For the three types of buildings, the characteristics of the net CO_2 emission depicted in Figure 21 show that the net CO_2 emissions are mainly derived from energy consumption in the operation phase and have a slight difference (less than 1.5%) in this phase, but the CO_2 emissions difference of the building life cycle mostly depends on that of the embodied materials phase. At the operation phase, the CO_2 emissions for WFC, SFC, and CFC contribute nearly 67%, 64%, and 44% to the life-cycle CO_2 emission, respectively; electricity and natural gas products are primary sources of energy consumption, and their CO_2 emissions indices are 0.864 kg/kWh and 0.075 kg/m^3 respectively. That is, the index of CO_2 emissions from electricity consumption is 11.5 times larger than that from natural gas products, and the electric energy consumption of three constructions are higher than natural gas energy consumption; therefore, the CO_2 emissions from energy use arise mainly from electricity products. Furthermore, for the three types of buildings, the differences in electric consumption are less than 0.5%. The CO_2 emissions consequently have only a slight difference for the

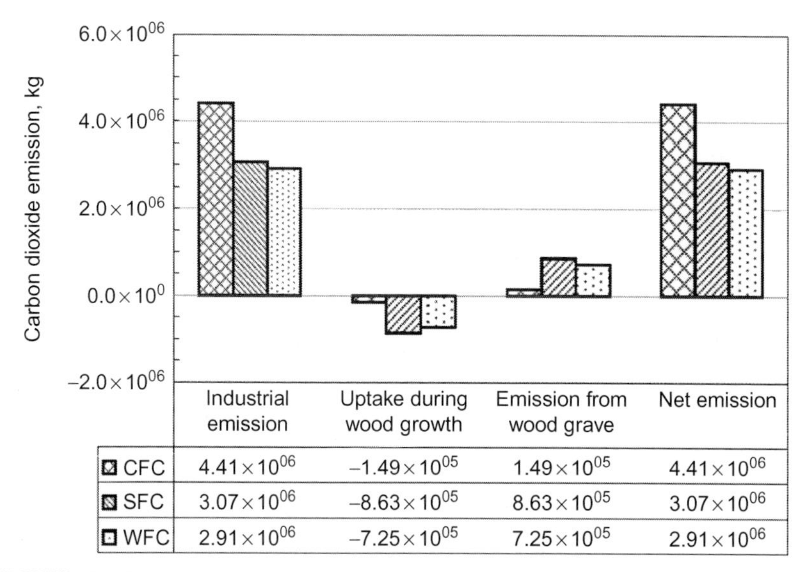

FIGURE 20 Carbon dioxide material flow analysis for the three buildings.

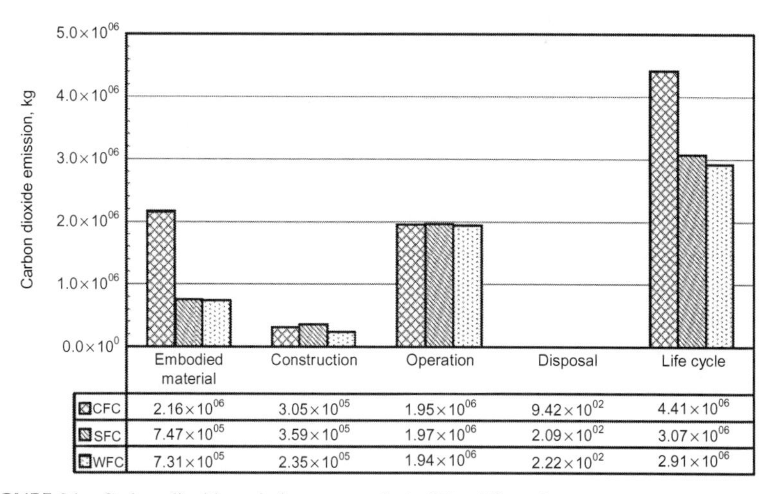

FIGURE 21 Carbon dioxide emissions across the building life cycle.

three types of buildings in the operation phase. On the other hand, for each type of building, the contribution percentage of CO_2 emissions both in the construction and disposal phase to the life cycle is less than 11%. However, at the embodied materials phase, the net CO_2 emission of CFC is approximately three times larger than that of WFC, and CFC as CFC uses a large amount of gypsum board and cement, which produce a great number of CO_2 emissions in the manufacturing process.

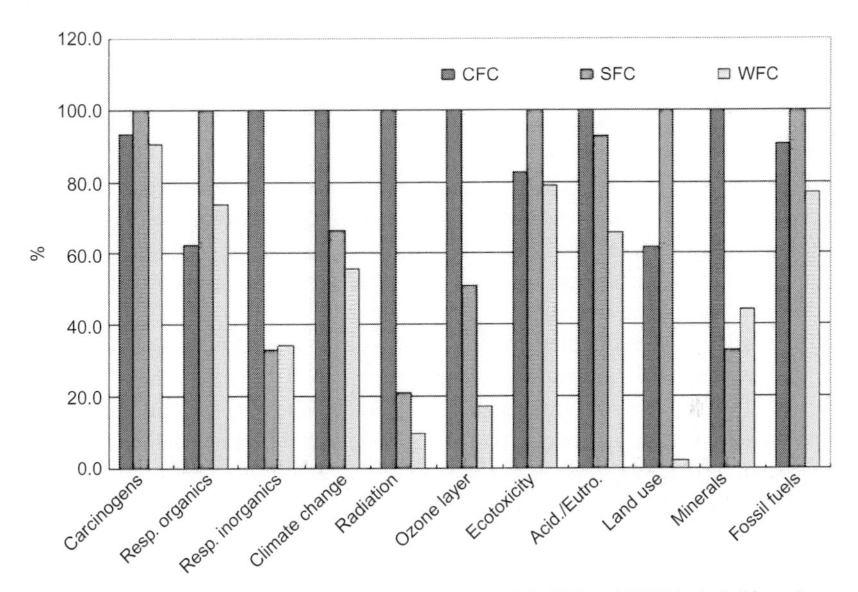

FIGURE 22 LCIA comparisons of impact categories for CFC, SFC, and WFC in their life cycles.

Based on LCI analysis and the rules of classification and characterization in Eco-indicator 99, the case study also adopted the localization characterization factors for mineral resource and heavy metal emission; the comparison of the LCIA between different phases for WFC, SFC, and CFC is illustrated in Figure 22. The results indicate that among the 11 total environmental impact categories, the nine impact indicators for WFC (e.g., carcinogens, climate change, radiation, ozone layer, eco-toxicity, acidification, eutrophication, land use, and minerals), are the lowest compared to SFC and CFC; however, for other indicators—respiratory inorganics, climate change, radiation, ozone layer, acidification, eutrophication, and minerals—the indicators of CFC are evidently larger than those for SFC and WFC. Furthermore, the environmental impact of respiratory organics, ecotoxicity, land use, and fossil fuels is relatively larger for SFC than for CFC and WFC. It is said to be a remarkable phenomenon that although the indicator of carcinogens is almost the same for CFC, SFC, and WFC, the respiratory organics indicator of WFC is 11% larger than that of CFC, and the minerals indicator of WFC is also 12% higher than that of SFC.

Consequently, the WFC design proves to be the most energy-efficient design both from a material and operating energy use perspective. Moreover, the net CO_2 emission is the lowest for the WFC design. Therefore, when considering overall environmental effect and comfort, it is strongly recommended that wood-framed residential buildings should be considered and adopted to satisfy more of China's housing needs, as these will contribute to achieving the energy-saving and decreasing emission target of Chinese government.

5 CONCLUSIONS

In conclusion, the development of eco-materials is the direction for sustainable development for the materials industry. Higher eco-efficiency, with no poisonous emissions or zero emission, and higher recycling are the basis for the eco-products processing. The ecofriendly replacement to poisonous, harmful, expensive, and difficult-to-separate elements and processing for saving energy and lowering consumption and emissions will be the mainstream for eco-materials development. For this reason, a series of new materials with related techniques need to be innovated, such that property control can be achieved not only by adding alloying elements but increasingly by microstructure control, as well as the usable insensitivity of impurities. Recycling must be taken into account for use and selection of materials. For resource protection and their sustainable utilization, research on the replacement of nonrenewable resources may be an important direction.

In addition, as a quantitative tool, LCA aims at making a comprehensive assessment of the environmental impacts of products and services and plays an important role in the eco-design of materials and products, cleaner production, decision making, and industry structure layout. This chapter reviewed several areas in which development has been active during the last several years in China. These include suitable impact assessment models for the Chinese situation, and databases for the inventory analysis, and examples of energy and materials life-cycle analysis. However, in general, the applications of LCA are still limited in several demonstration fields in China, and there is a wider gap between evaluation results and the criteria people expected. Therefore, the development of LCA study should not only extend the application range of industry and agriculture fields, but also improve LCA methodology in economic and social aspects, for example, introducing life-cycle costing and social LCA to establish indicators of sustainable development based on the three elements of environment, economy, and society. LCA research needs further improvement, primarily in the following areas:

- Lack of data is still a common problem in current LCA practice in China; thus, the database needs be continuously expanded and improved. The life-cycle inventories particularly need to be divided into regional levels because there is often a large spatial differentiation of the environmental issues in materials production and also other processes such as power supply in China.
- The database should be updated continually due to frequent technological innovations in China, and the standard of data quality analysis should be established to describe and identify the representativeness and reliability of the LCI datasets.
- Cooperation among universities, research institutions, and industry from different areas for LCA practice and LCI data sharing and exchange are very important.

- Given the lack of regional methodology and related characterization that factors in most impact categories (e.g., water, acidification, etc.), a national LCA methodology system needs to be established.
- Most of LCA practices have been performed by government and academic institutions, but there has been a lack of recognition and interest from industries.
- Most research focuses only on analysis of current environmental state of materials or products; there is a lack of methodology and related practices for predicting the environmental impact (e.g., eco-design).

Although the LCA methods and their application still needs deeper investigations, it has been widely accepted as critically important to progress in the definition of goals and scope, framework, and the challenges of implementation. It is an effort for the materials industry to perform the work of energy-saving and emission-reducing technologies, and we believe that the LCA method is potentially a great tool to provide the decision making and technical support for the achievement of the target.

REFERENCES

[1] R. Yamamoto, Ecomaterials, Chemical Industry Press, Beijing, 1997.

[2] K. Halada, R. Yamamoto, The current status of research and development on ecomaterials around the world, MRS Bull. 11 (2001) 871–879.

[3] Z.G. Chen, Y.W. Shi, Z.D. Xia, Y.F. Yan, Study on the microstructure of a novel lead-free solder alloy SnAgCu-RE and its soldered joints, J. Electron. Mater. 1 (10) (2002) 1122–1128.

[4] Z.R. Nie, T.Y. Zuo, M.L. Zhou, Y.M. Wang, J.S. Wang, J.X. Zhang, High temperature XPS/AES investigation of Mo–La_2O_3 cathodes, J. Rare Earths 18 (1) (2000) 110–114.

[5] L.I. Yingping, W.A.N.G. Xianhong, L.I. Ji, Polyaniline—a new generation of environmentally friendly anticorrosion material, Mater. China 8 (2011) 17–24.

[6] Y.M. Xi, Y.M. Cheng, S.X. Oyang, Research pulse of cement materials, Mater. Rev. 2 (2000) 8–10.

[7] US Patent, US2002/0082363 A1. A catalyst for the preparationof high molecular weight aliphatic polycarbonate and the preparation.

[8] Y.A.N.G. Keke, W.A.N.G. Yuzhong, PPDOA recyclable and biodegradable polymer: poly (P-dioxanone), Mater. China 8 (2011) 25–34.

[9] A.H. Lu, The application of environmental mineral materials to the treatment of contaminated soil, Water Air Acta Miner. 18 (4) (1999) 292–300.

[10] X.Q. Xie, D. Zhang, T.X. Fan, R.J. Wu, T. Okabe, T. Hirose, Woodceramics composites with interpenetrating network, Chinese J. Mater. Res. 16 (3) (2002) 259–262.

[11] W.P. Wang, Y. Zhang, Comprehensive utilization of chromicresidues in cleaner production of chromic salts, Modern Chem. Ind. 22 (9) (2002) 27–29.

[12] J.Y. Xu, Y. Hu, Q. Wang, W.C. Fan, L. Song, Green chemistry and technology in flame retardant industry, Polym. Mater. Sci. Eng. 18 (1) (2002) 17–21.

[13] J. Wang, Screen theory and research state of electromagneticwave screen materials, Chem. New Mater. 30 (7) (2002) 16–24.

[14] L.I.U. Qingju, Z.H.U. Liangdi, Z.H.U. Zhongqi, Research progress on theory of TiO_2 photocatalyst modified by ion doping, Mater. China 8 (2011) 42–48.

[15] H.M. Zhou, Z.R. Nie, T.Y. Zuo, Discussion of environmental problems and study of environmental impact for iron and steelmaking in China, J. Iron Steel Res. 19 (2) (2002) 39–42.
[16] J. Zhu, Basic features of metallurgical cleaner production, Nonferrous Met. 54 (2002) 173–178.
[17] G.Q. Zhang, Q.X. Zhang, A new clean production technology for tungsten hydrometallurgy, Chinese J. Rare Met. 27 (2) (2003) 254–257.
[18] J.K. Tan, X.D. Han, Present situation of researches and development into green chemistry in China, Geol. Chem. Miner. 24 (3) (2002) 157–161.
[19] Z.R. Nie, F. Gao, X.Z002.E. Gong, et al., Developments and applications of materials life cycle assessment in China, Progr. Nat. Sci.: Mater. Int. 21 (1) (2011) 1–11.
[20] X.Z. Gong, Z.R. Nie, Z.H. Wang, et al., Research and development of Chinese LCA database and LCA software, Rare Metal. 25 (Spec. Issue) (2006) 101–104.
[21] F. Gao, Z.R. Nie, Z.H. Wang, et al., Characterization and normalization factors of abiotic resource depletion for life cycle impact assessment in China, Sci. China (Ser. E: Technol. Sci.) 52 (1) (2009) 215–222.
[22] R. Müller-Wenk, Depletion of Abiotic Resources Weighted on the Base of "Virtual" Impacts of Lower Grade Deposits in Future, University of St. Gallen, Germany, 1998.
[23] B.X. Sun, Z.R. Nie, Y. Liu, et al., Exergy-based model of the depletion of mineral resources, Mater. Sci. Forum 650 (2010) 1–8.
[24] Y. Liu, Z.R. Nie, B.X. Sun, et al., Development of Chinese characterization factors for land use in life cycle impact assessment, Sci. China Ser. E: Technol. Sci. 53 (6) (2010) 1483–1488.
[25] H.M. Zhou, Life cycle assessment of iron and steel processes. Dissertation of Doctor's Degree, Beijing University of Technology, 2001.
[26] G. Li, Z. Nie, H. Zhou, et al., An accumulative model for the comparative life cycle assessment case study: iron and steel process, Int. J. Life Cycle Assess. 7 (4) (2002) 225–229.
[27] W.T. Wang, The study on life cycle assessment of advanced steel-making in China. Dissertation of Master's Degree, Beijing University of Technology, 2005.
[28] F. Gao, Z.R. Nie, Z.H. Wang, et al., Life cycle assessment of primary magnesium production using the Pidgeon process in China, Int. J. Life Cycle Assess. 14 (5) (2009) 480–489.
[29] F. GAO, Z.R. NIE, Z.H. WANG, et al., Greenhouse gas emissions and reduction potential of primary aluminum production in China, Sci. China Ser. E: Technol. Sci. 52 (8) (2009) 2161–2166.
[30] X.Z. Gong, Z.R. Nie, Z.H. Wang, Environmental burdens of Beijing cement production, J Wu Han Univ. Technol. 28 (3) (2006) 121–141.
[31] W.J. Chen, The study on life cycle assessment of flat glass production. Dissertation of Master's Degree, Beijing University of Technology, 2007.
[32] S.H.I. Feifei, W.A.N.G. Zhihong, F.A.N.G. Minghui, Analysis on the CO_2 emission of calcium carbide sludge as secondary raw material in cement clinker production, Mater. Sci. Forum 743–744 (2013) 516–522.
[33] B.R. Yuan, Measurement method for sustainable development of chemical industry and its application. Dissertation of Doctor's Degree, Beijing University of Technology, 2006.
[34] X.H. Di, Z.R. Nie, et al., Life cycle inventory for electricity industry in China, Int. J. Life Cycle Assess. 12 (4) (2007) 217–224.
[35] X.Z. Gong, Z.R. Nie, et al., Life cycle energy consumption and carbon dioxide emission of residential building designs in Beijing, J. Ind. Ecol. 16 (4) (2012) 576–587.

Chapter 4

Exergetic Aspects of Green Ceramic Processing

H. Kita, I. Himoto and S. Yamashita
Department of Molecular Design and Engineering, Graduate School of Engineering, Nagoya University, Nagoya, Japan

NOMENCLATURE

Q	amount of heat, (J)
T	temperature on thermodynamic scale, (K)
R	gas constant, 8.314 ($J\,mol^{-1}K^{-1}$)
W	mechanical work, (J)
η	energy efficiency
E_x	exergy
$E_x(X_i)$	the chemical exergy of material X_i
S	entropy, (Ons)
H	enthalpy, (J)
G	Gibbs free energy, (J)
ρ	specific density
M	molecular mass, (g/mol)
m	dry mass (kg)
H_l	lower heating value, (J/kg)
$\varphi_C, \varphi_H, \varphi_O, \varphi_N$	weight fractions of carbon, hydrogen, oxygen, and nitrogen in the organic compound
v_i	volume fraction of ingredient, i
D	thickness (mm)
v_f	volume fraction

ABBREVIATIONS

JIS	Japanese Industrial Standards
SN	silicon nitride
N-process	normal process using SN powder
LPG	liquefied petroleum gas
PVA	polyvinyl alcohol
EtOH	ethanol
LCA	life-cycle assessment
E_xLCA	life-cycle assessment based on exergy

Green and Sustainable Manufacturing of Advanced Materials. http://dx.doi.org/10.1016/B978-0-12-411497-5.00004-7

CIP	cold isostatic pressing
RBSN	reaction-bonding silicon nitride
PS	post-sintering
RBPS	reaction bonding followed by post-sintering

SUPERSCRIPTS

0	standard state at normal temperature (298 K)

SUBSCRIPTS

H	higher temperature
L	lower temperature
i	ingredient i.
0	initial

1 INTRODUCTION

Our society has nested structure in which systems with various kinds and scales exist, as schematically illustrated in Figure 1. Resources, energies, artificial products, and wastes are put into and flow through the systems transforming their shapes and quality. At the end of many years, they will return to the earth. Through these processes, many activities for production and consumption have been conducted by humans; consequently, the environment has been damaged, and resources and energy are being depleted.

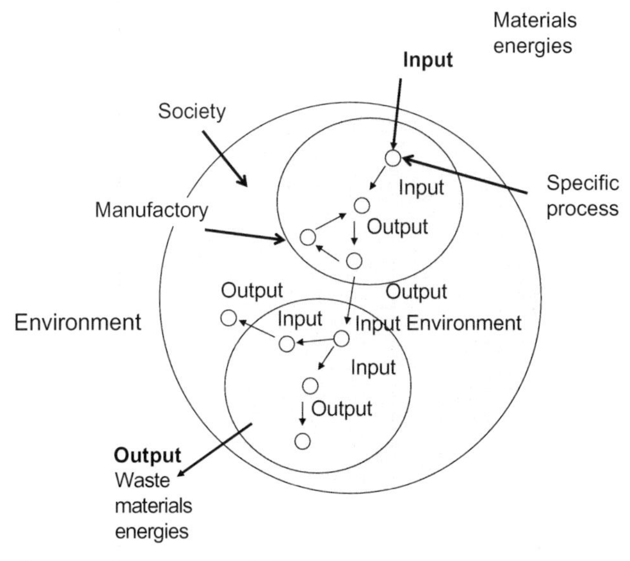

FIGURE 1 Structure of our society: nested open systems.

Toward the realization of a sustainable society, we need to proceed the research for environmental conservation while promoting advanced science and technology. The scientific principle-based method to evaluate the impact on the environment and resources consumed over a long time has been the most important issues, prior to manufacturing.

Thermodynamics is related with a much wider range of processes and applications not only in engineering, but also in environment [1–3]. A good understanding of energy, entropy, and exergy concepts derived from thermodynamics is applicable to environmental issues and is useful for those improving the design and performance of manufacturing systems, including ceramics.

According to the first law of thermodynamics, energy is conserved in an isolated system, which doesn't explain how energy can change form, but can be neither created nor consumed. Besides, it provides no information about the direction in which processes can spontaneously occur, that is, an energy balance based on the first law doesn't provide any information on the degradation of energy or resources during a process, and it does not quantify the usefulness or quality of the various energy and material streams flowing through systems.

The exergy analysis method, based on both the first and the second law of thermodynamics, may overcome the limitations of the first law of thermodynamics. The term exergy comes from the Greek words ex and ergon, meaning "from" and "work." Terms used in the literature are available energy and utilizable energy.

The exergy analysis method enables the types and magnitude of waste and losses to be identified. It has been a useful tool for determining the goal of more efficient energy resources, mainly for chemical plants and energy recovery systems [4–6]. On the other hand, applications of exergy analysis to manufacturing systems are limited, and applications specifically to ceramics have not yet been found.

In this report, application of exergy analysis to manufacturing ceramics has been attempted in several stages. First, the exergy calculations of energy and material used for manufacturing silicon nitride (SN) were conducted. Next, using these results, two types of SN process were examined from exergy viewpoints. Finally, regarding ceramic heat tubes used in aluminum casting lines, the consumption of exergy used for a 7-year period has been calculated [7–9]. Through these studies, the values of a long product life of SN ceramics, which is the heart of this report, were revealed.

2 ILLUSTRATIVE EXAMPLE FOR UNDERSTANDING EXERGY AND HEAT ENERGY [1–3]

Figure 2 shows the principle of the heat engine as an illustrative example of exergy. It delves into the use of exergy techniques as a necessary background for understanding these concepts, as well as any general definitions, basic principles, and implications.

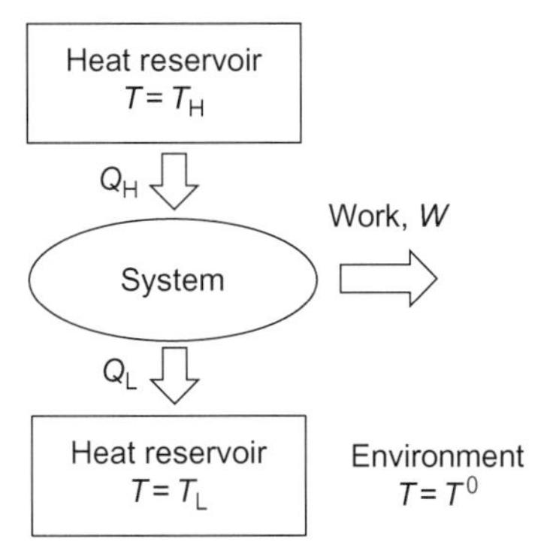

FIGURE 2 Principle of heat engine.

A heat engine generates mechanical work, W, using the temperature difference between the hot heat ($T=T_H$) and a cold heat reservoir ($T=T_L$). In terms of energy conservation, the first law of thermodynamics, the amount of W needs to be equal to the difference in the amount of heat,

$$W = Q_H - Q_L$$

Carnot discovered in 1824 that a heat engine can convert into W operating without any significant losses—only a portion of the heat (Q_H) taken from the hot reservoir. The Carnot equation is expressed as follows:

$$W = Q_H \left(1 - \frac{T_L}{T_H}\right) \tag{1}$$

Here, $\{1-(T_L/T_H)\}$ is dimensionless, and a named Carnot factor, which characterizes the "quality" of heat taken from the reservoir with a constant temperature and the factor expresses the efficiency of the Carnot cycle. It is worth stressing that the ability of heat engines cannot be characterized in terms of energy.

$$\eta = \frac{W}{Q_H}\left(= \frac{Q_H - Q_L}{Q_H} = 1 - \frac{T_L}{T_H}\right)$$

$$\eta = \frac{W}{Q_H} = 1 - \frac{T_L}{T_H}$$

At a constant pressure process,

$$dQ = dH$$

Entropy S is defined as $dS = dQ/T$. Using the differential, equation above is expressed as follows.

$$dW = dE_x = dQ\left(1 - \frac{T^0}{T_H}\right) = dQ - T^0 \frac{dQ}{T} = dH - T^0 dS$$

Integrating above equation from $T=T^0$ to $T=T$, then,

$$E_x = (H - H^0) - T^0(S - S^0) \tag{2}$$

This equation is known as the basic formula of exergy. Here, it should be noted that Equation (1) can be applied only to ideal reversible processes. According to the second law of thermodynamics, all real processes are irreversible. In actual processes, the amount of performed work is always smaller than that calculated using Equation (1). Hence, Equation (1) shows the attainable maximum amount of generated work. This quality can be expressed by means of the maximum ability to generate work using the two temperature differences between two heat reservoirs.

The ability to generate mechanical work can be understood as a measure of the quality of various kinds of energy. This quality index for energy has been termed exergy by Rant [6]. It expresses the maximum work output attainable in the natural environment, or the minimum work input necessary to realize the reverse process.

This ability is determined not only by the composition and state parameters of the considered material or energy, but also by any state environment parameters in which the considered transformation process occurs. In an irreversible process exergy is indispensably consumed. When a system is isolated, the entropy of a system increases due to irreversible processes and reaches the maximum possible value when the system attains a state of thermodynamic equilibrium. A system in complete equilibrium with its environment does not have any exergy. No difference appears in temperature, pressure, concentration, and the like, so there is no driving force for any process.

The term energy consumption needs to be corrected as "exergy consumption" or "loss of energy quality," and each of those terms means synonymously "entropy increase." From Equation (2), exergy and entropy are related with following equation:

$$\Delta E_x = -\Delta S \times T^0 \tag{2'}$$

Here, $\Delta H = 0$ is assumed.

2.1 Exergy Analysis 1: Entropy Increase due to Mixture and Exergy Calculation of N_2 and O_2 Gas

When two kinds of ideal gas A and B are mixed, the entropy increase due to mixing can be calculated by the following equation:

$$\Delta S_{mix} = n_A \cdot R \cdot \ln(1/x_A) + n_B \cdot R \cdot \ln(1/x_B)$$

Similarly, a multicomponent system, entropy due to mixing, is represented in following general forms.

TABLE 1 Ingredients in Air and Their Partial Pressure at Normal Temperature

Ingredient	Pressure (kPa)
N_2	76.57
O_2	20.61
H_2O	3.20
CO_2	0.03
Ar	0.91
Total	101.32

$$\Delta S_{mix} = R \cdot \Sigma\{(n_i) \cdot \ln(1/x_i)\} \tag{3}$$

Exergy values of pure O_2, N_2, CO_2 gas are indispensable for the exergy calculation followed by several examples that cover the materials, process, and the life cycle in industrial ceramic systems and processes. Table 1 shows the main components of air and their partial pressure at normal temperature [3].

In exergy analysis, gas with atmospheric pressure is separated from air regarded as the gas mixture of them. In other words, pure nitrogen gas has the potential of works until it becomes the state of nitrogen gas in air. The potential is understood as exergy of pure nitrogen gas with atmospheric pressure.

The exergy values of O_2, N_2, and CO_2 with atmospheric pressure are calculated using their fraction in air as follows:

$$E_x^0(O_2) = RT^0 \ln(101.3/20.61) = 3.95(kJ/mol) = 0\,12(MJ/kg)$$
$$E_x^0(N_2) = RT^0 \ln(101.3/76.57) = 0.7(kJ/mol) = 0.025(MJ/kg)$$
$$E_x^0(CO_2) = RT^0 \ln(101.3/0.03) = 20.1(kJ/mol) = 0.46(MJ/kg)$$

$E_x(X_i)$ shows the exergy for material X_i. The figures shown in parenthesis in these equations are the ratio of atmospheric pressure to the partial pressure of each gas.

2.2 Exergy Analysis 2: Chemical Exergy of Metal and Inorganic Compounds

Because the value of the exergy of a system or flow depends on the state of both the system or flow and a reference environment, the reference environment or reference material must be specified prior to the performance of an exergy analysis. Therefore the reference materials need to be clearly described.

Reference material is defined as a matter that is stable, does not make a chemical reaction alone in its geosphere, and its exergy is zero [3,4]. For example, consider SN, which is artificial material. In the exergy calculation of SN, we need to go back to SiO_2 and N_2 gas with the state in air. In case of iron (Fe), Fe_2O_3 is selected as the reference material. Reference materials corresponding to all of the elements and several kinds of compound are listed in JIS [5]; however, reference materials for specific compounds are not found in any lists. In that case, it is appropriate to select a material with the lowest free energy as the referential.

It is assumed that $X_xA_aB_b$ ... is produced by chemical reaction (4), with the change of Gibbs free energy ΔG^0. Here, X, A, and B denote the kinds of elements, and x, a, and b show atomic composition ratios. And the values of ΔG^0 for many chemical reactions are available in a commercial database [10]. The chemical exergy E_x^0 values of the inorganic compound can be calculated using Equation (6) [3]

$$xX + aA + bB + \cdots \rightarrow X_xA_aB_b \cdots \tag{4}$$

$$\Delta G^0 = E_x^0\left(X_xA_aB_b \ldots\right) - xE_x^0\left(X\right) - aE_x^0\left(A\right) - bE_x^0\left(B\right) - \cdots \tag{5}$$

$$E_x^0\left(X\right) = \frac{1}{x}\left[-\Delta G^0 - aE_x^0\left(A\right) - bE_x^0\left(B\right) - \cdots\right] \tag{6}$$

Here, $X_xA_aB_b$... is an environment reference that is defined as

$$E_x\left(X_xA_aB_b \ldots\right) = 0$$

Hereafter we describe several examples of exergy calculations. They cover materials, fuels, and energy used for SN manufacturing process, followed by exergy analysis on the life cycle of SN parts.

(a) Si

In the calculation of exergy, the chemical-based reaction is as follows:

$$Si + O_2 \rightarrow SiO_2$$

The normal standard free energy, ΔG^0 of the aforementioned reaction is −854.5 (kJ/mol). The normal standard values of the chemical exergy of the reference material, $E_x(SiO_2) = 0$ (kJ/mol), and $E_x^0(O_2) = 3.95$ (kJ/mol), as mentioned in an earlier section, Exergy Analysis 1.

According to Equation (6), the standard chemical exergy of the element E_x^0 (Si) is calculated as follows:

$$\begin{aligned} E_x^0\left(Si\right) = -\Delta G^0 + E_x^0\left(SiO_2\right) - E_x^0\left(O_2\right) &= -\left(-854.5\right) + 0 - 3.95 \\ &= 850.5\left(kJ/mol\right) \\ = 30.3\left(MJ/kg\right), \quad \text{Here, } M\left(Si\right) = 28 \end{aligned}$$

(b) Si_3N_4

$$3Si + 2N_2 = Si_3N_4$$

$$\Delta G^0 = -676.5\text{kJ/mol}, \quad E_x^0(\text{Si}) = 850.5(\text{kJ/mol}),$$
$$\text{and} \quad E_x^0(\text{N}_2) = 0.7(\text{kJ/mol}),$$

respectively. Then, standard chemical exergy, E_x^0 (Si_3N_4) is calculated as

$$E_x^0(\text{Si}_3\text{N}_4) = \Delta G^0 + 3E_x^0(\text{Si}) + 2E_x^0(\text{N}_2) = -676.5 + 3\times 850.5 + 2\times 0.7$$
$$= 1876.5(\text{kJ/mol})$$
$$= 13.4(\text{MJ/kg}), \quad \text{Here, } M(\text{Si}_3\text{N}_4) = 140$$

(c) Al_2O_3

Al_2O_3 is listed as a reference material [5] that does not make chemical reaction alone in an environment, and its exergy is E_x^0 (Al_2O_3)=0, according to the definition of exergy.

Al_2O_3 has been used as a sintering additive.

(d) Y_2O_3

Y_2O_3 has been used as a sintering additive, similar to Al_2O_3.

$$(5/3)\text{Fe}_2\text{O}_3 + \text{Y}_2\text{O}_3 \rightarrow (2/3)\text{Y}_3\text{Fe}_5\text{O}_{12}$$

$$\Delta G^0 = -47.0(\text{kJ/mol}), \quad E_x^0(\text{Fe}_2\text{O}_3) = 0(\text{kJ/mol}),$$

then

$$E_x^0(\text{Y}_2\text{O}_3) = -\Delta G^0 + (2/3)E_x^0(\text{Y}_3\text{Fe}_5\text{O}_{12}) - (5/3)E_x^0(\text{Fe}_2\text{O}_3)$$
$$= 47(\text{kJ/mol})$$
$$= 0.208(\text{MJ/kg}), \quad \text{Here, } M(\text{Y}_2\text{O}_3) = 225.8$$

(e) Fe

$$\text{Fe}_2\text{O}_3 = (3/2)\text{O}_2 + 2\text{Fe}$$

$$\Delta G^0 = 742.2(\text{kJ/mol}), \quad E_x^0(\text{Fe}_2\text{O}_3) = 0(\text{kJ/mol}),$$
$$E_x^0(\text{O}_2) = 3.95(\text{kJ/mol}),$$

then,

$$E_x^0(\text{Fe}) = (1/2)\{\Delta G^0 + E_x^0(\text{Fe}_2\text{O}_3) - (3/2)E_x^0(\text{O}_2)\}$$
$$= (1/2)\times(742.2 + 0 - 1.5\times 3.95)$$
$$= 368.1(\text{kJ/mol}) = 6.6(\text{MJ/kg}), \quad \text{Here, } M(\text{Fe}) = 55.8$$

(f) Al

$$\text{Al}_2\text{O}_3 = (3/2)\text{O}_2 + 2\text{Al}$$

$$\Delta G^0 = 1582.3(\text{kJ/mol}), \quad E_x^0(\text{Al}_2\text{O}_3) = 0(\text{kJ/mol}),$$
$$E_x^0(\text{O}_2) = 3.95(\text{kJ/mol}),$$

then,

$$\begin{aligned}E_x^0(\mathrm{Al}) &= (1/2)\times\{\Delta G^0 + E_x^0(\mathrm{Al_2O_3}) - (3/2)E_x^0(\mathrm{O_2})\} \\ &= (1/2)\times(1582.3 + 0 - 1.5\times 3.95) \\ &= 788.1(\mathrm{kJ/mol}) = 29.2(\mathrm{MJ/kg}), \quad \text{Here, } M(\mathrm{Al}) = 27\end{aligned}$$

2.3 Exergy Analysis 3: Chemical Exergy of Organic Materials

An approximate calculation method proposed by Szargut [2] is based on an analogy with the chemical exergy of pure organic substances. After calculating the chemical exergy of several organic substances, approximate formulas have been derived expressing the ratio of their chemical exergy to the lower heating value as a function of the atomic ratio of the elements C, H, O, N, S.

Although equations of Rant [6] and Szargut are known for the calculation of chemical exergy for organic material, we used the following equation derived by Nobusawa, who modified the equations for practical use [4].

$$E_x = m\cdot H_l\cdot\left(1.0064 + 0.1519\frac{\varphi_H}{\varphi_C} + 0.0616\frac{\varphi_O}{\varphi_C} + 0.0429\frac{\varphi_N}{\varphi_C}\right) \tag{8}$$

(a) PVA

PVA is used as binder. If the molecular formula of PVA is $CH_2{\cdot}CH(OH)_n$, $n = 1800$, then $M(\mathrm{PVA}) = 79{,}200$. Here, $C = 12$, $H = 1$, $O = 16$

$$\begin{aligned}&H_l(\mathrm{PVA}) = 11{,}000(\mathrm{kcal/kg}), \quad \varphi_C = 0.545, \quad \varphi_H = 0.091, \\ &\varphi_O = 0.363, \quad \varphi_N = 0.\end{aligned}$$

According to Equation (8),

$$\begin{aligned}E_x(PVA) &= 11{,}000\times 1\times(1.0064 + 0.1519\times 0.167 + 0.0616\times 0.667) \\ &= 11{,}800(\mathrm{kcal/kg}) = 49.4(\mathrm{MJ/kg})\end{aligned}$$

(b) EtOH

EtOH is used as a solvent. The molecular formula of EtOH is C_2H_5OH, $M(\mathrm{EtOH}) = 46$,

Here, C = 12, H = 1, O = 16, (EtOH) = 6400(kcal/kg), $\varphi_c = 0.545$, $\varphi_H = 0.136$, $\varphi_o = 0.363$, $\varphi_N = 0$

According to Equation (8),

$$\begin{aligned}E_x(\mathrm{EtOH}) &= 6400\times 1\times(1.0064 + 0.1519\times 0.25 + 0.0616\times 0.667) \\ &= 6947(\mathrm{kcal/kg}) \\ &= 29.1(\mathrm{MJ/kg})\end{aligned}$$

2.4 Exergy Analysis 4: Increase of Entropy during "Mixing"

During the mixing process, the different kinds of particles change their position relative to each other, resulting in the increase of entropy, *S*. *S* is determined by calculating the total volume from the density and mixture ratios:

$$\Delta S = Rln\left(\Sigma 1 / v_{\mathrm{fi}}\right)$$

It was assumed that Si_3N_4, Al_2O_3, Y_2O_3, particles were mixed with the weight ratio 0.99 kg, 0.055 kg, 0.055 kg, respectively, then, $v_f(Si_3N_4)=0.9249$, $v_f(Al_2O_3)=0.0422$, $v_f(Y_2O_3)=0.0329$, were obtained. Here, $\rho(Si_3N_4)=3.2$, $\rho(Al_2O_3)=3.9$, $\rho(Y_2O_3)=5.0$, and an interaction occurring between the solid particles is ignored, which means thus obtained value is an exergy or entropy change due to the relocation of particles.

$$\Delta S = 8.314 \times \ln\left(1/0.9249 + 1/0.0422 + 1/0.0329\right)/1000 = 0.054\left(\mathrm{kOns}\right),$$

then,

$$\Delta E_{\mathrm{x}} = 298 \times 0.054/1000 \quad 0.017\left(\mathrm{MJ/mol}\right)$$

2.5 Exergy Analysis 5: Electric Power and Gas Fuel Mixture

Electric power is energy that does not contain entropy, so it was used as value for exergy. On the other hand, exergy of mixed fuel gas was calculated using the following equation [4]. Second term corrects the entropy increase derived from mixing.

$$E_{\mathrm{x}} = \sum v_{\mathrm{i}} E_{\mathrm{x}}\left(i\right) + \mathrm{RT}_0 \sum v_{\mathrm{i}} \ln v_{\mathrm{i}} \qquad (9)$$

LPG fuel is composed of C_3H_8 and C_4H_{10} with the molar mixture ratio of 0.2 and 0.8, respectively. Here, $E_x(C_3H_8)=2{,}091{,}390$ (J/mol), $E_x(C_4H_{10})=2{,}726{,}310$ (J/mol),

According to Equation (9),

$$\begin{aligned} E_{\mathrm{x}}\left(\mathrm{LPG}\right) &= 0.2 \times \left(2091{,}390 + 8.314 \times 298 \times \ln\left(0.2\right)\right) \\ &\quad + 0.8 \times \left(2726{,}310 + 8.314 \times 298 \times \ln\left(0.8\right)\right) \\ &= 2598{,}086\left(\mathrm{J/mol}\right) = 2.60\left(\mathrm{MJ/mol}\right) \end{aligned}$$

2.6 Exergy Analysis 6: Si_3N_4 Ceramics Processing

In the previous section, the series of exergy values for the materials and fuels used for manufacturing SN ceramics were calculated. Exergy values were calculated as per unit amount, then, if the amount and type of material into and out of the system is known, exergy that flows in and out of the systems can be calculated.

In this report, system and process are defined as follows:

- System: virtual space where materials, fuels, and energies that are put into and the portion of them are fixed in the products by reaction and/or assembling into products, whereas the others are discarded as waste material and heat.
- Process: specific action to add the values using materials and energies.

Manufacturing is recognized as an assembly of systems, which corresponds to a series of processes.

In an exergy analysis of manufacturing, analysts need to know the types and quantity of all raw materials, fuels, and energies that go into and out of each subsystem.

Figure 3 shows the input/output flow of materials and energy in manufacturing process. Raw materials and energy are put into the first process; intermediate products are produced with waste materials and heats. The intermediate product becomes the raw material required for the next process, and this cycle is repeated until the last process. Finally, the process is completed and a product is generated.

Actual manufacturing is an irreversible process, which means unavoidable exergy losses (increase of entropy) are generated. According to thermodynamics, this loss is caused by degrading the quality of energy, such as temperature decrease of heat, friction loss, and transforming electric power to heat. This loss is termed "first exergy loss."

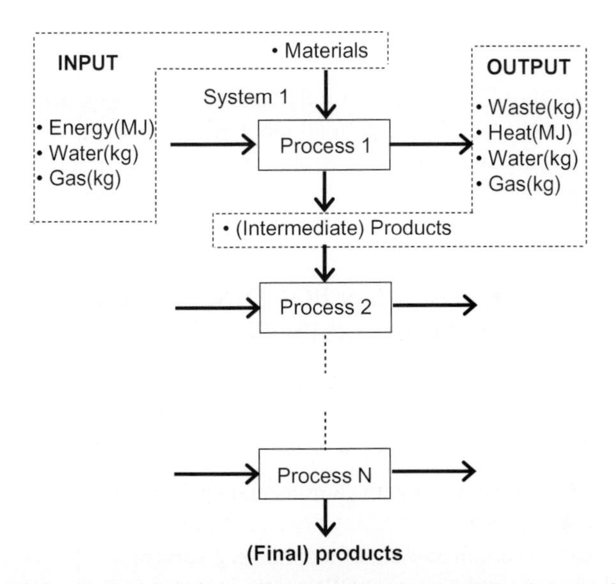

FIGURE 3 Input and output flaw of materials and energy in an assembly of systems for the manufacturing process.

In actual manufacturing, the yield can't be 100%; some portion of materials or fuels, even if they have high exergy, are wasted. The loss is termed "second exergy loss." Those two types of losses are repeated through all the processes, resulting in overall exergy efficiency: E_x(products)/E_x(input), will show a surprisingly small value.

SN was produced by two methods in the laboratory level; the necessary data for the exergy analysis related to manufacturing in each process were laboratory ones (e.g., amounts of raw fuel, wasted material, and gas as well as water). Exergy consumption for such processes were calculated followed by the exergy analysis based on the mass production data obtained from the company.

3 OVERVIEW OF NORMAL PROCESS (N-PROCESS)

In this overview, "N-process" means the process composed of the following conditions:

(1) Using SN powder (90wt%) with alumina (5wt%) and yttria (5wt%), well known as sintering additives for SN.
(2) Mixing with a certain amount of alcohol as a solvent.
(3) After drying, granulating, and CIPing, the powder mixture is sintered at 1850 °C in 0.9 MPa nitrogen atmosphere to fabricate SN sintered body.

The conditions described are quite popular in manufacturing SN. SN plates with a size of 50 × 50 × 5t (mm) were made, and the weight of consumed material to produce 1 kg SN plates was measured in each process consumed energy was measured with a wattmeter.

The loss of raw materials during the milling and granulating processes is about 10%. Then, the weights of used SN, alumina, and yttria are 0.99 kg, 0.055 kg, and 0.055 kg, respectively.

In addition, those of ethanol (EtOH) and nitrogen gas are 1.67 kg and 3.75 kg, respectively. Moreover, the total input energy is 479 kwh, in which the sintering process consumes most of it, that is, 410 kwh. On the other hand, it is understood that 5.5 kg raw materials, including nitrogen gas, are abandoned during the synthesis process of 1 kg SN sintered body.

4 OVERVIEW OF REACTION-BONDING FOLLOWED BY POST-SINTERING PROCESS (RBPS-PROCESS)

RBPS process is so called two-process sintering methods, namely reaction sintering of silicon powder and subsequent densification of sintered body. RBPS process means the process composed of the conditions as follows:

(1) Using silicon (Si) powder with alumina and yttria, well known as the sintering additives for SN.
(2) Mixing with a certain amount of alcohol as a solvent.
(3) After drying, granulating, and CIPing, the powder compacts were nitrided at 1450 °C in nitrogen gas, followed by sintering at 1850 °C in 0.9 MPa.

In RBPS process, alumina and yttria are used as sintering additives. By this mixture ratio, the weight increases by 1.67 times associated with the reaction between silicon and nitrogen to form SN in the sintering process.

The loss of raw materials during the milling and granulating processes is about 10%. Then, the weights of used silicon, alumina, and yttria are 0.6 kg, 0.055 kg, and 0.055 kg, respectively.

In addition, those of EtOH and nitrogen gas are 0.96 kg and 15.6 kg, respectively. Moreover, the total input energy is 1333 kwh, in which the nitriding and sintering process consumes most of it, that is, 1296 kwh.

5 EXERGY ANALYSIS [11]

The input exergies for the materials and the process are shown in Figures 4 and 5, respectively [11]. As for N-process, the total input exergy of raw material is

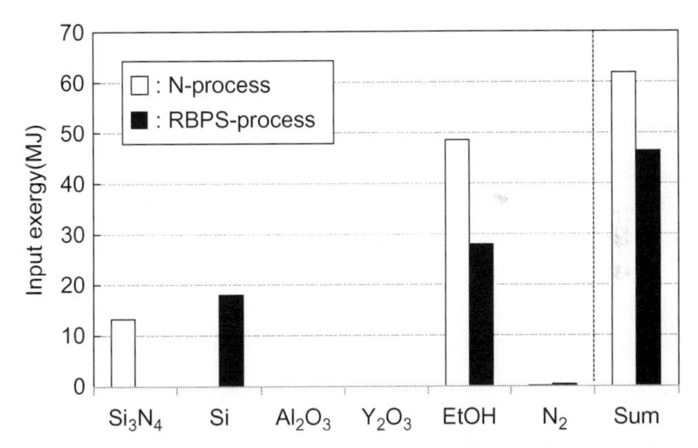

FIGURE 4 Exergy contents due to materials used in the production of two ways.

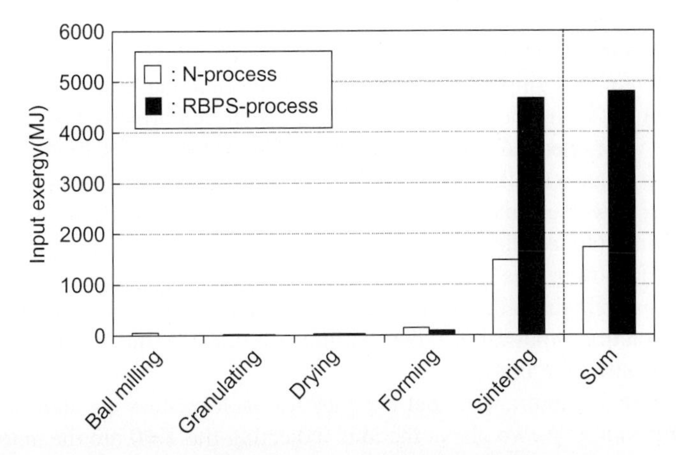

FIGURE 5 Exergy content put into each step in the production of two ways.

61.9 MJ, in which raw powders and EtOH occupy 13.3 MJ and 48.5 MJ, respectively. Input exergy due to energy is very large compared to the exergy of material. In N-process, total input exergy for the processes is 1724.3 MJ, in which the sintering process used 1476 MJ exergy. After all, input exergy for N-process in total was 1786.2 MJ.

In RBPS process, the total input exergy of raw material is 46.4 MJ, in which raw powders and EtOH occupy 18 MJ and 28 MJ, respectively. Similar to N-process, input exergy due to energy is very large compared to the exergy of material. In RBPS process, total input exergy for the processes is 4799.5 MJ, in which nitriding and sintering process used 4665.6 MJ exergy. Then, comprehensive input exergy for RBPS process was 4845.9 MJ. It was demonstrated that RBPS process, in which silicon powder as a starting raw material was used, consumed a large amount of exergy for the nitridation of silicon to form SN followed by sintering for densification.

Thus obtained SN ceramics conserved 12.1 MJ exergy; this means in N-process, only 0.7% of exergy input and in RBPS process, 0.25% of exergy input were fixed in the SN ceramics, and of the others, more than 99.3% were diffused in an environment.

In our laboratory, electric power, which is the energy with 100% exergy, was changed into heat energy. Heat energy is degraded according to the decrease of the temperature. Finally it becomes T^0, this fact vividly represents the consumption process of exergy. In addition, the exergy of EtOH used for milling and most of the nitrogen gas were also wasted completely. For improving the efficiency of the SN process, shortening the time for nitridation and sintering, substituting EtOH for water, or reducing the amount of solvent were required.

5.1 Exergy Analysis 7: Si_3N_4 Heat Tubes Manufacturing Process

Data used in this case study were based on measured values in the plant with the cooperation of the companies. The process was close to the N-process described earlier, but differs in that they use the LPG in the drying process whereas all the processes were operated using electric power in our laboratory. Figure 6 shows the exergy balance for the production of a ceramic heater protection tube [7,8]. The process involves mixing, granulation, CIPing and green machining, dewaxing, and sintering. As the input energy, LPG was used for drying in granulation, and electric power was used in dewaxing and sintering. Organic binder and water do not remain in the final product, in which the amount and type, or the influence on the output value of the post-process, consider the overall plan; their value is required for analysis.

In Figure 7, input and output exergies for each process are shown. In the figure, the values shown above the axis indicating the $Y=0$ are the amount of input exergy and the values shown below $Y=0$ axis line show output exergy.

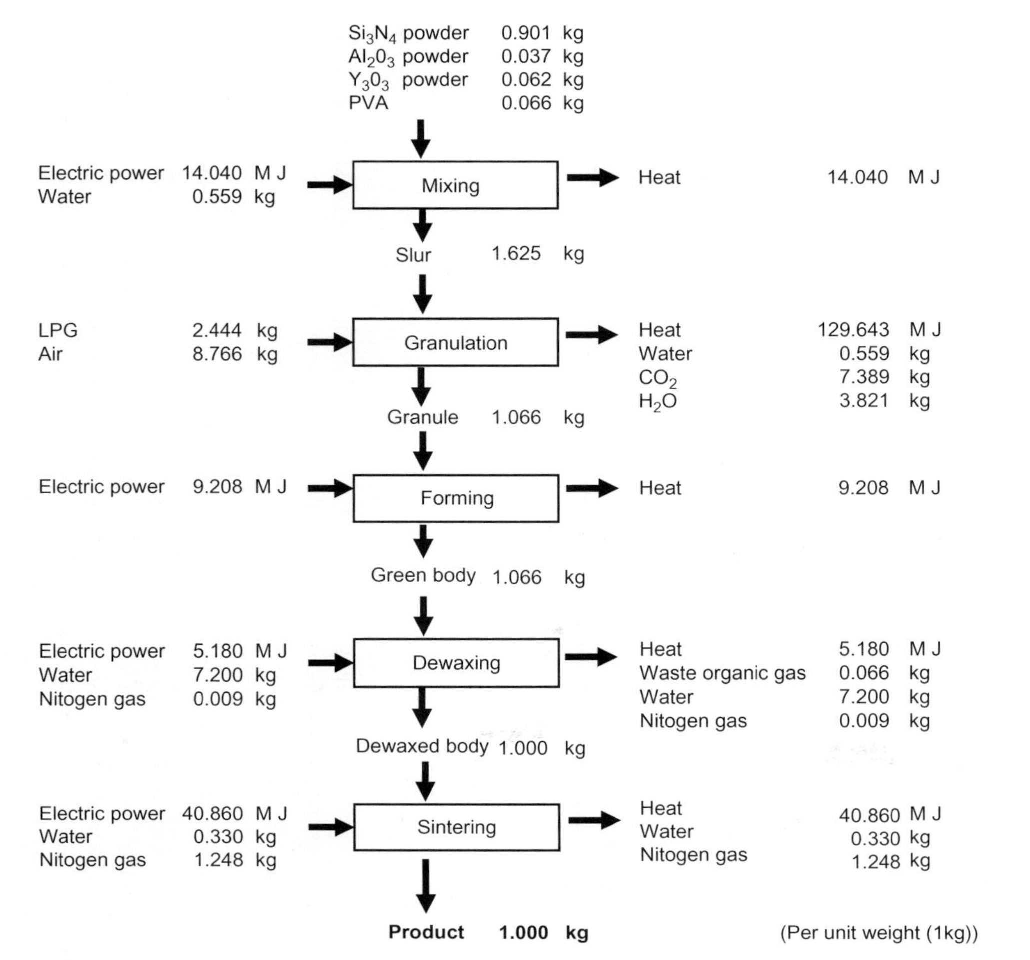

FIGURE 6 Material and energy balance for silicon nitride process 9.

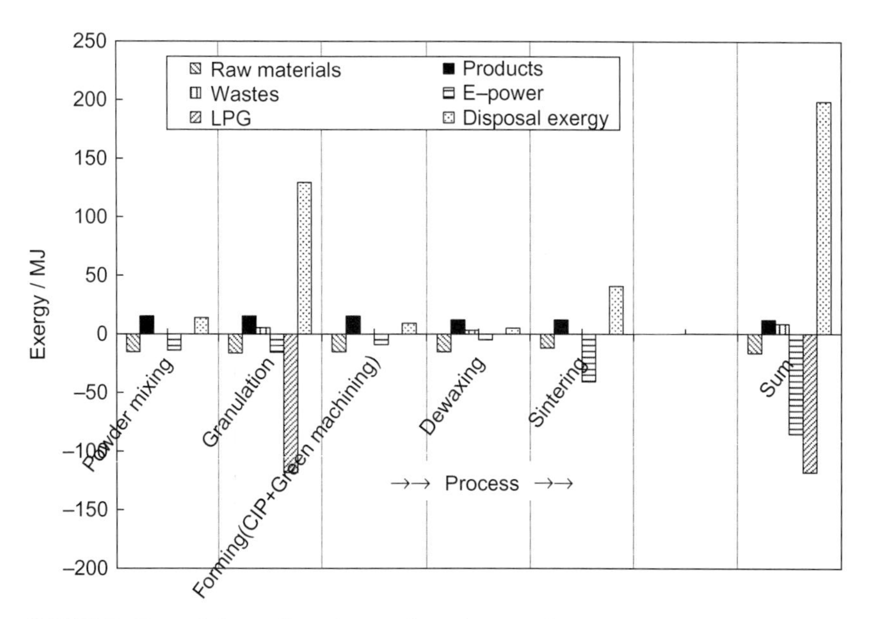

FIGURE 7 Exergy balance of in and output for each process 9.

When the same weight of SN as the aforementioned was produced, the total input exergy of 219.7 MJ, which is the sum for raw materials, electric power, and LPG (16.4 MJ, 85.1 MJ, and 118.2 MJ, respectively), was necessary. The exergy fixed in the product was 12.1 MJ, and the remaining 207.6 MJ was discharged into the environment as waste (8.8 MJ) or heat (198.8 MJ). The process efficiency of ceramics was calculated by E_x(products)/E_x(input) = 12.1 MJ/207.6 MJ, then it was approximately 5.5%, which is much higher than that obtained in our lab. For comparison, the process efficiency value of steel, calculated in the same way, was approximately 20%.

5.2 Exergy Analysis 8: Life-Cycle Assessment of Si_3N_4 Tubes

Industries and businesses have responded to this awareness by assessing how their activities affect the environment and, in many cases, providing "greener" products and using "greener" processes. LCA is a methodology for this type of assessment. In this chapter, LCA is modified and extended by considering exergy. Exergetic LCA (ExLCA) is described and, as a case study, applied to ceramics used for an aluminum casting line operation process.

6 ALUMINUM CASTING LINE OPERATION AND ROLE OF HEATER TUBE

Aluminum has excellent heat conductivity and is lightweight, and therefore is used widely in engine parts. Also, aluminum is highly recyclable, so disposed

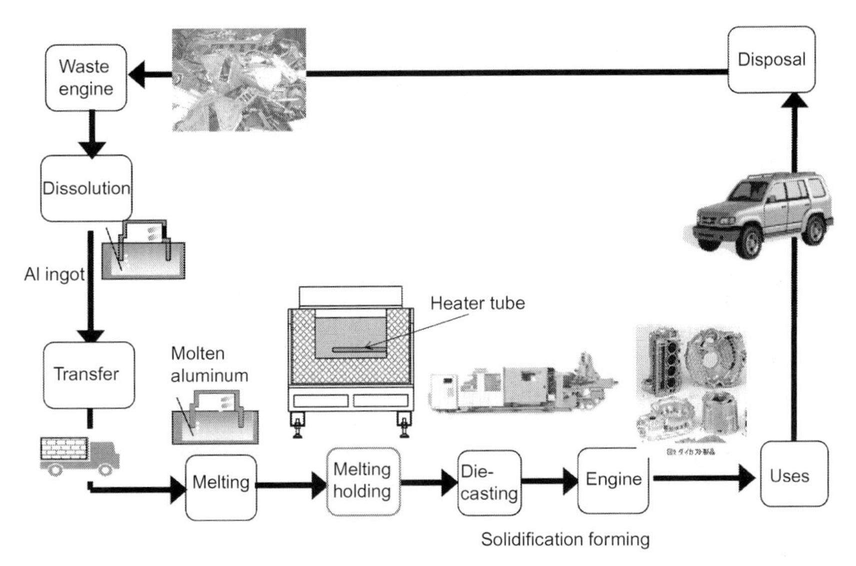

FIGURE 8 Role of heater tube in aluminum recycling system.

engines are collected as scraps and recycled as engine after undergoing some processes. Figure 8 shows the recycling system for aluminum casting line operation. First, recovered waste engine (scrap) are melted in centralized furnace.

They are made into solid ingots, delivered within the plant, melted again in centralized furnace, and then transferred to holding furnace. The molten metal is adjusted for temperature and content, distributed to the die cast machine, and formed into products. In this cycling system, there are many factors that decrease efficiency including heat loss, oxidation of molten aluminum, and inclusion of impurities. Input of energies and things from outside is unavoidable to maintain certain quality level and production volume, and reducing these inputs is expected to increase the efficiency of the cycling system.

In the melting and casting production of aluminum, SN ceramics are now being applied as many components exposed into the molten aluminum due to their superior corrosion and thermal-shock resistances. As one of measure, the use of ceramics in a production member has been attempted. The heater tube (Figure 9) used in a holding furnace is one example. It is a protective tube that envelops the heating wires and is used to maintain constant temperature of molten aluminum. Heat efficiency increases by using highly conservative SN, which allows a horizontal dip structure where the tube is fixed horizontally in the bottom of the furnace. However, ceramic tubing is much more expensive than iron tubing. We conducted exergy analysis for manufacture, use, and disposal in cases where the heater tube (weight 19 kg) was made with SN and when it was made with iron.

We conducted an exergy analyses on the production of heater protection tubes. Data for its production (quantities of raw material, fuel, waste, waste gas,

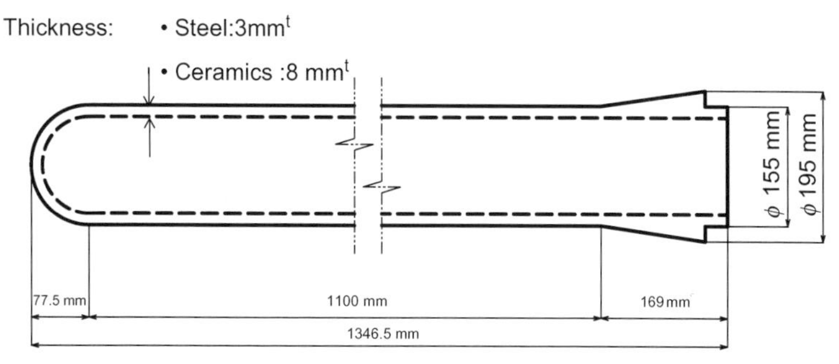

FIGURE 9 Dimension of heater protection tube 7,8.

and water, etc.), which were necessary for analyzing exergy, were obtained from companies [9]. Some data were unavailable, and they were supplemented with the author's knowledge of ceramic processing.

7 EXERGY CONSUMPTION IN EACH STAGE

7.1 Wear and Material Disposal

When steel heater tube is used in molten aluminum, it is corroded by aluminum and worn down by the passage of time t. It was assumed that wastage progressed according to the following equation.

$$D = D_o \times \left(2 - \exp\left(\mathrm{kt}\right)\right) \tag{11}$$

Here, D is the thickness of a heater tube (mm), D_0 is initial thickness (mm), k is apparent reaction speed coefficient, and D_i is thickness at time of replacement (mm). Assuming $D_0=3$ mm (from data) and $D_i=0.5$ mm, under condition of replacement every half-year, the reaction constant k was calculated to be 0.07.

Consumption exergy is expressed by the following equation.

$$E_x = E_{xo} \times \exp\left(\mathrm{kt}\right) \tag{12}$$

Exergy of steel is 6.6 MJ/kg (=368 KJ/mol) and total weight of the product is 19 kg, and when it is disposed when damage reaches D_i, consumption is 126 MJ/tube. Although steel heater tubing is exchanged once every half-year, SN is stable and does not react, and is exchanged and disposed along with a furnace that has a lifespan of seven years. Exergy consumed by disposal during this time is shown in the following equation:

- Steel: 126 (MJ/tube) × 14 (tubes) = 1764 MJ
- Silicon nitride: 229 (MJ/tube) × 1 (tube) = 229 MJ

When steel heater tube is used, damage and disposal are repeated and exergy consumption increases in a process like form. In contrast, there is hardly any consumption in seven years using ceramics, and exergy value (229 MJ) is released at the end of the furnace lifespan. Also, using ceramics, there is less of a chance of including impurities compared to steel, so clean molten metal can be obtained, and this is another advantage of ceramics.

7.2 Running

7.2.1 Melting and Holding Furnace

In vertical dip type using steel heater tube, 9.4 kW is required during run, and 4.0 kW at rest, whereas in horizontal dip type using SN, electricity consumptions at run and rest are 6.8 kW and 3.8 kW respectively, due to improved heat efficiency. Although it will be running 60% (40% rest) per day and is in operation 360 days a year, the total electricity consumed in seven years, or exergy input, will be as follows:

- Steel: $(9.4\times0.6\times24+4.0\times0.4\times24)\times360\times7\times3.6/1000=1576$ GJ
- Silicon nitride: $(6.8\times0.6\times24+3.8\times0.4\times24)\times360\times7\times3.6/1000=1219$ GJ

7.2.2 Die Cast Machine

Assuming that electricity consumption of die cast machine is 20 kW, running 60% per day for 360 days per year, the total electricity consumption, or exergy input, for seven years is as follows.

- $20\times0.6\times24\times360\times7\times3.6/1000=2612$ GJ

7.3 Manufacture, Use, and Disposal

As result of interview with companies, the total manufacture volume of cast product in seven years was estimated to be about 4300 tons. In this chapter, material loss is not considered. Therefore, the amount of molten aluminum is 4300 tons or the same as final product, and the exergy was calculated to be 126,802 GJ in molten condition (temperature 700 °C), and 125,582 GJ in solid condition.

Figure 10 shows the amount and flow of exergy input and output for manufacture using ceramics and steel heater tube, their use in a melting and holding furnace when casting was conducted for seven years. As mentioned above, when the furnace is run for seven years, 14 steel tubes are required as they are subject to damage. Therefore, energy input and output during the manufacture process is as follows.

- Input: 621 (MJ/tube) $\times$ 14 (/tubes) = 8694 MJ
- Output: 495 (MJ//tube) $\times$ 14 (/tubes) = 6930 MJ

On the other hand, only one SN tube is required during same time, and exergy for input and output will be 4175 MJ and 3946 MJ respectively according to Figure 7. Next, exergy accompanying damage and disposal during use is as follows.

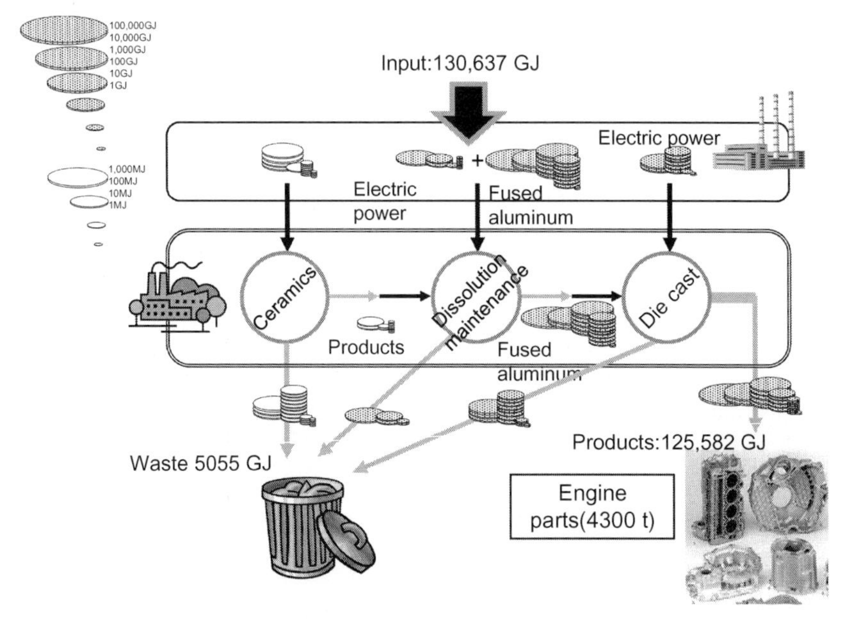

FIGURE 10 Consumed exergy for ceramic tubes through the life cycle 7.

- Steel: 126 (MJ/tube) × 14 (tubes) = 1764 MJ
- Silicon nitride: 229 (MJ/tube) × 1 (tubes) = 229 MJ

Looking over the entire process, the exergy inputs for steel and SN were 130,999 GJ and 130,637 GJ respectively, whereas exergy outputs were 5417 GJ and 5055 GJ. Using SN reduced 362 GJ of input and output exergy compared to steel. The results were summarized in Figures 10 and 11.

From the above results, it was shown that although one SN tube required seven times more exergy in manufacturing process, frequency of replacement decreased due to its high conservative property, which allowed a furnace with a highly efficient structure that reduced electricity consumption, and therefore, exergy consumption level was smaller compared to steel throughout the life cycle of manufacture, use, and disposal.

8 CONCLUSIONS

An application of exergy analysis to ceramic processing has been attempted in several steps. First, exergy calculations were conducted on the raw materials, energy, and fuels used for manufacturing SN ceramics. Next, using the values thereby obtained, exergy analysis was performed on two types of the SN process followed by the exergy life-cycle analysis of ceramic heat tubes used for aluminum casting line.

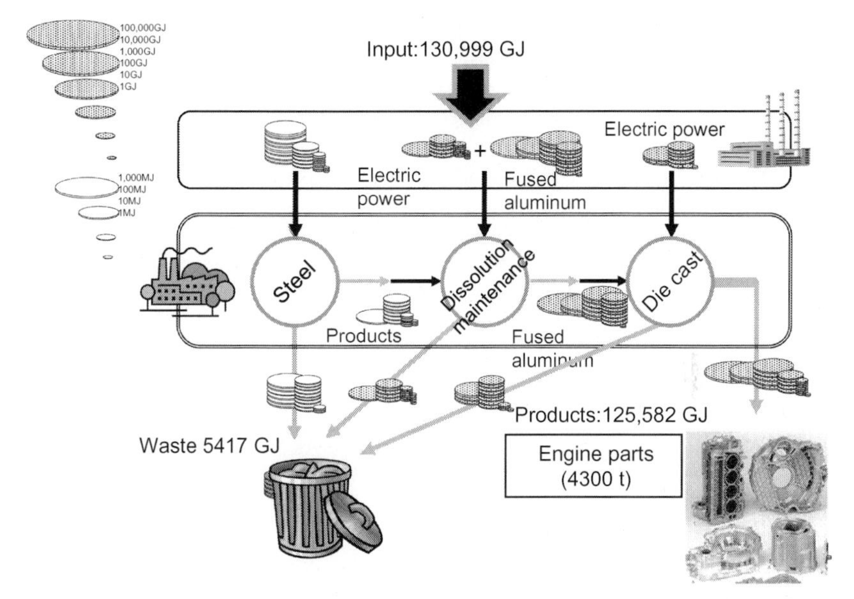

FIGURE 11 Consumed exergy for steel tubes through the life cycle 7.

1. The series of exergy values including the raw materials, energy, and fuels used for the SN process were calculated.
2. The most part of input exergy was consumed in the SN manufacturing process. In particular, the process in which the change of input electric power into low-temperature heat energy represents the loss of exergy in the process.
3. Manufacturing SN consumed significantly more exergy than steel.
4. SN heat tubing was reduced by 362 GJ exergy for seven years, comparing with steel tubes.
5. Through these studies, the values of long life with the SN were revealed.

ACKNOWLEDGMENTS

This work was supported by the New Energy and Industrial Technology Development Organization (NEDO) and METI, as part of the Innovative Development of Ceramics Production Technology for Energy Saving project.

REFERENCES

[1] I. Dincer, M.A. Rosen, Exergy, Elsevier, 2007.
[2] J. Szargut, Exergy Method, WIT Press, Southampton, 2005, 18.
[3] K. Karakida, "Ekuserugi no kiso", Ohm- sya, (2005) (in Japanese).
[4] T. Nobusawa, Nenryou oyobi Nensyou 43 (11) (1976) 49–79 (in Japanese).
[5] Nihon Kogyo Kikaku, Z 9204 (1991) (in Japanese).
[6] Z. Rant, Forsch. Exergie, ein neues Wort für "Technische Arbeitsfähigkeit" Ing-Wes 22 (1956) 36–37.

[7] H. Kita, H. Hyuga, N. Kondo, T. Ohji, Exergy Consumption through the Life Cycle of Ceramic Parts, Int. J. Appl. Ceram. Technol. 5 (4) (2008) 373–381.
[8] H. Kita, H. Hyuga, N. Kondo, A rationalization guideline for the utilization of energy and resources considering total manufacturing processes—An exergy analysis of aluminum casting processes, Synthesiology, 1 (3) (2008) 75–81.
[9] H. Kita, H. Hyuga, N. Kondo, I. Takahashi, Exergy analysis on the ceramic manufacturing process, J. Ceram. Soc. Jpn. 115 (12) (2007) 987–992.
[10] e.g. MALT(Data base), (Kagaku Gijyutu- Sya).
[11] H. Kita, H. Hyuga, T. Wubian, T. Nagaoka, N. Kondo, Exergy consumption evaluation of a silicon nitride manufacturing process, J. Ceram. Process. Res. 5 (11-6) (2010) 721–727.

Part II

Sustainable Manufacturing—Metallic Materials

Chapter 5

Lead-Free Soldering: Environmentally Friendly Electronics

A. Kroupa,[1] A. Watson,[2] S. Mucklejohn,[3] H. Ipser,[4] A. Dinsdale[5] and D. Andersson[6]

[1]Institute of Physics of Materials, Brno, Czech Republic, [2]University of Leeds, Leeds, United Kingdom, [3]Ceravision Limited, Milton Keynes, United Kingdom, [4]University of Vienna, Department of Inorganic Chemistry/Materials Chemistry, Wien, Austria, [5]National Physical Laboratory, Teddington, United Kingdom, [6]Swerea IVF AB, Mölndal, Sweden

1 INTRODUCTION

Lead has been used in a wide spectrum of applications, but in the past decades it became clear that its high toxicity could cause serious problems. Studies indicate that exposure to high concentrations of lead, particularly in young children, can result in damage to the central nervous system, and may be associated with high blood pressure in adults. Human exposure to lead typically occurs via inhalation or ingestion of lead in food, soil, water, or dust. The removal of lead from gasoline, where it had long been used as an antiknock additive, began in the 1970s, and this has dramatically reduced the lead levels in the blood of children and adults [1].

On the other hand, lead has been an important component of solders for centuries, and lead-tin alloys were used by the ancient Romans for connecting water pipes, which themselves were made from lead. With the advent of electrical appliances more than 100 years ago, and with the steep rise of the use of electronic equipment in the second half of the last century, lead-tin solders became the most important joining materials. They are robust, producing mechanically stable junctions, and by varying the alloy composition it is possible to adjust the melting temperature. Whereas the standard composition, that is, the eutectic composition with about 63 mass % Sn and 37 mass % Pb, melts at 183°C, there are alloys containing 85 and more mass % Pb, the so-called high-temperature solders, with melting temperatures of 300°C and above (all compositions in this chapter are in mass %, unless otherwise stated). Because modern advanced

Green and Sustainable Manufacturing of Advanced Materials. http://dx.doi.org/10.1016/B978-0-12-411497-5.00005-9

electronic devices are assembled in consecutive steps it is a prerequisite to have a selection of solder materials with a range of melting temperatures.

Since around 1990 there has been an increasing concern that lead from the ever-growing volume of consumer electrical and electronic appliances may end up as waste in official landfills, or worse, in unregulated dumps with the inherent danger of invading drinking water sources and consequently the human food chain. This has initiated attempts worldwide to ban, or at least strictly regulate, the use of lead (and other toxic materials) in electronics by legislation. In the European Union, the Restriction of Hazardous Substances Directive (RoHS) Directive [2] was introduced by the European Commission in January 2003. It took effect on July 1, 2006, and had to be adopted into national legislation by the member states. It restricted the use of lead in electronics in the manufacture of all types of electronic and electrical equipment. Lead-containing solders for mainstream applications were banned; however, a number of exemptions were listed, in particular for high-temperature solders with rather high contents of lead.

Several other countries followed suit with similar legislation; for example, China with its so-called China-RoHS [3] and South Korea [4] with its Act for Resource Recycling of Electrical and Electronic Equipment and Vehicles. In the United States there is no corresponding regulation on a national level, but individual states have set rather strict standards, among them the Electronic Waste Recycling Act of 2003 [5] in California.

In the search for appropriate substitute solder materials, the most important class comprises tin-rich alloys containing small amounts of silver and/or copper; the so-called SAC alloys (where SAC stands for Sn-Ag-Cu). One of the drawbacks of these solders was their considerably higher melting temperatures than the traditional Sn-Pb alloy: 221°C for eutectic Sn-Ag solders, 217°C for eutectic Sn-Ag-Cu solders, and even 227°C for eutectic Sn-Cu solders, compared to 183°C for eutectic Sn-Pb. Nevertheless, it appears that they will be used as the solder material for mainstream applications for the foreseeable future.

However, it is a different story for the high-temperature solders with high lead contents. These alloys are currently exempt from RoHS regulations as no reliable substitute is available at the moment. Even the recent RoHS 2 Directive [6] of June 2011 still lists "lead in high melting temperature type solders (i.e., lead-based alloys containing 85% or more lead)" as exempt from the regulation. However, there is still a need to find appropriate solder materials without lead, or any other toxic element, as a reliable substitute for high-lead high-temperature solders currently in use today.

This chapter aims to give an overview of the current state of the legislative regulations in the European Union and to discuss the situation in lead-free soldering at the moment, that is, the type of materials currently in industrial use. Furthermore, new routes in research and development of novel materials for soldering as well as on the type of lead-free materials will be discussed, new and promising technologies for joining in electronics will be outlined. The final part of the chapter will deal with industrial aspects of the implementation of new materials and technologies for joining in electronics.

It is hoped that the reader will gain a good overview of the attempts to provide less toxic and more benign materials for electronics application, that is, to offer more and more truly "green" electrical and electronic appliances as implied in the title of the book.

2 THE CURRENT STATE OF EU LEGISLATION IN RELATION TO THE USE OF LEAD

In most parts of the world the design, manufacture, use, and disposal of electrical and electronic equipment (EEE) is heavily regulated to safeguard the interests of the natural environment, the supplier and the end user. As mentioned above, such regulations exist in all developed countries, though in some cases not at the "federal" level. Together with some "green" U.S. states, the European Union was at the forefront of the development of such legislation. The size and importance of these markets forced other states that have major producers of electronic appliances to accept valid EU legislation as a basis for their own regulations [3,4]. Therefore, the current EU legislation will be discussed in detail in this section.

Within Europe there are two EU directives and one EU regulation that define which materials may be used in EEE and how waste (discarded) components and products (WEEE) must be treated when they come to the end of their useful life. The associated regulations may seem overbearing to producers of EEE, especially small- and medium-sized enterprises (SMEs). However, there are requirements in the formulation of EU directives and regulations to make sure they do not stifle the creativity and inventiveness of SMEs. Thus, the section below is primarily aimed at showing where small businesses can find more information and support. The most stringent requirements for producers of EEE arise from the RoHS Directive, which is therefore given most attention here.

2.1 The RoHS Directive

Directive 2002/95/EC on the restriction of the use of certain hazardous substances in electrical and electronic equipment (RoHS) [2,7] was adopted in January 2003 and came into force on July 1, 2006. The intent and aspirations of this directive are closely related to Directive 2002/96/EC on WEEE [8], which also came into effect on July 1, 2006. The RoHS Directive of 2002 has been recast as Directive 2011/65/EU of June 8, 2011 [6,9] with EU member states required to adopt and publish the laws, regulations, and administrative provisions necessary to comply with the recast Directive by January 2, 2013. Compliance with RoHS2 became a requirement for "CE" marking from this date.

The objective of the RoHS Directive is to contribute to the protection of human health and the environmentally sound recovery and disposal of waste electrical and electronic equipment. The European Commission gives first priority to *prevention* in waste legislation. Prevention is defined, *inter alia*, as measures that reduce the content of harmful substances in materials and products. Restrictions on the use of certain substances in materials and products must take into account technical and economic feasibility, including the impact on SMEs.

The recast RoHS Directive 2011/65/EU states that it should not prevent the development of renewable energy technologies that have no negative impact on health and the environment and that are sustainable and economically viable.

Exemptions from the substitution requirement are permitted if substitution is not possible from a scientific and technical point of view, taking specific account of the situation of SMEs, or if the negative environmental, health, and consumer safety impacts caused by substitution are likely to outweigh the environmental, health, and consumer safety benefits of the substitution or the reliability of substitutes is not ensured. Thus lead-containing solders are permitted for use in EEE by a number of tightly defined exemptions and are confined to high melting temperature applications.

The decision on exemptions and on the duration of possible exemptions should take into account the availability of substitutes and the socioeconomic impact of substitution. Life-cycle aspects on the overall impacts of exemptions should apply, where relevant. Substitution of the hazardous substances in EEE should also be carried out in such a way as to be compatible with the health and safety of users of EEE.

Exemptions from the restriction for certain specific materials or components are limited in their scope and duration. When the use of substances that are subject to exemptions becomes avoidable and following a detailed review, a gradual phase-out of those substances in EEE is usually implemented.

The RoHS Directive applies to EEE falling within the categories listed below.

1. Large household appliances
2. Small household appliances
3. IT and telecommunications equipment
4. Consumer equipment
5. Lighting equipment
6. Electrical and electronic tools
7. Toys, leisure, and sports equipment
8. Medical devices
9. Monitoring and control instruments including industrial monitoring and control instruments
10. Automatic dispensers
11. Other EEE not covered by any of the categories above

The categories to which the Directive does not apply are listed in 2011/65/EU [9].

The restricted substances referred to in Article 4(1) of 2011/65/EU [6] and the corresponding maximum concentration values tolerated in homogeneous materials are listed below (in mass %).

1. Lead (0.1%)
2. Mercury (0.1%)
3. Cadmium (0.01%)
4. Hexavalent chromium (0.1%)
5. Polybrominated biphenyls (PBB) (0.01%)
6. Polybrominated diphenyl ethers (PBDE) (0.1%)

The following definitions apply:

- "Electrical and electronic equipment" means equipment that is dependent on electric currents or electromagnetic fields to work properly and equipment for the generation, transfer, and measurement of such currents and fields and designed for use with a voltage rating not exceeding 1000 V for alternating current and 1500 V for direct current.
- For the purposes of the definition above, "dependent" means, with regard to EEE, needing electric currents or electromagnetic fields to fulfill at least one intended function.
- "Homogeneous material" means one material of uniform composition throughout or a material, consisting of a combination of materials, which cannot be disjointed or separated into different materials by mechanical actions such as unscrewing, cutting, crushing, grinding, and abrasive processes.

Exemptions to RoHS are granted to narrowly defined applications. Exemptions are temporary in nature and subject to review at least every four years, until such time when a reliable and safe substitution is available. Applications exempted from the restriction in Article 4(1) of 2011/65/EU are listed in Annex III of that document [6].

Reviews of existing exemptions, based on the criteria for exemptions in Article 5 of the RoHS Directive, are conducted with the involvement and consultation of stakeholders (*inter alia* producers of electrical and electronic materials, components and equipment, recyclers, treatment operators, environmental organizations, employee and consumer associations) by a panel of experts. These reviews are published as detailed reports together with a recommendation to the European Commission. Recommendations have to provide clear and unambiguous wording for any changes. Requests for new exemptions are assessed to see if they meet the criteria set out in Article 5 of the RoHS Directive.

The report of May 2006 [10] recommended that a request for a new exemption for a lead-containing solder was covered by the existing exemption 7a. Exemption 33 for the use of lead in solders for the soldering of thin copper wires in power transformers was granted as a result of an exemption request in 2007 [11].

The comprehensive review published in 2009 [12] considered all of the existing exemptions together with five new requests. All of the existing exemptions related to lead-containing solders were recommended to be allowed to continue with the exception of exemption 27 which was given an expiry date of September 24, 2010. The report recommended that one new request for a lead-containing solder was refused. The report did not make any recommendations for two further requests for exemptions for lead-containing solders. The review published in May 2011 [13] studied three existing exemptions and four new exemption requests. One recommendation in this report was that an exemption was deleted without a transition period as the manufacturer no longer wished to produce the devices.

This Directive places several obligations on the manufacturers of EEE to maintain detailed technical records of their products. These include:

- Drawing up an EU declaration of conformity and affixing the "CE" marking on the finished product.
- Keeping the technical documentation and the EU declaration of conformity for 10 years after the EEE has been placed on the market.
- Keeping the EU declaration of conformity and the technical documentation at the disposal of national surveillance authorities for 10 years following the placing on the market of the EEE.
- Providing a competent national authority with all the information and documentation necessary to demonstrate the conformity of an EEE with the Directive.
- Cooperating with the competent national authorities, to ensure compliance with the Directive.

2.2 The REACH Regulation

Regulation EC 1907/2006 concerning the registration, evaluation, authorization and restriction of chemicals and establishing a European Chemicals Agency (REACH) was published on December 18, 2006 [14] and subject to the corrigenda of May 25, 2007 [15]. The REACH regulation has seen subsequent amendments and corrigenda, and these are recorded on the legislation page of the European Chemicals Agency (ECHA) website [16].

The aim of the REACH regulation is to improve the protection of human health and the environment through the better and earlier identification of the intrinsic properties of chemical substances. At the same time, innovative capability and competitiveness of the EU chemicals industry should be enhanced. The benefits of the REACH system are expected to come gradually, as more and more substances are phased into REACH.

The REACH regulation gives greater responsibility to the industry to manage the risks from chemicals and to provide safety information relating to the substances they use. Manufacturers and importers are required to gather information on the properties of their chemical substances, which will allow their safe handling, and to register the information in a central database run by the ECHA in Helsinki.

The Agency acts as the central point in the REACH system: it manages the databases necessary to operate the system, and coordinates the in-depth evaluation of suspicious chemicals and runs a public database in which consumers and professionals can find hazard information.

A major part of REACH is the requirement for manufacturers or importers of substances to register them with the ECHA. There are thresholds for registration: substances of very high concern (SVHC), for example, have a much lower threshold value than substances regarded as nontoxic.

The principle of REACH is: "No data, no market," meaning substances cannot be placed on the market once the threshold has been passed unless the substance has been registered and authorized by ECHA.

2.3 The WEEE Directive

The aims and objectives of the WEEE Directive are closely related to those of the RoHS Directive. The WEEE Directive sets out a framework for the collection, recycling, and safe disposal of electrical and electronic equipment thereby limiting the amount of toxic waste that is put into landfill. Producers of EEE are responsible for financing the disposal of their products at end of life and hence are legally responsible for compliance with national WEEE legislation. Most producers meet their obligations by joining a WEEE producer compliance scheme.

Since the introduction of WEEE legislation there has been growing concern over the differing legal requirements in Member States and over the collection and recycling rates. The WEEE Directive was therefore recast in 2012 [17] with particular attention being paid to:

- Clarification of the scope of the Directive.
- Harmonization of requirements in all Member States.
- Setting collection and recycling targets.
- Promotion of individual producer responsibility.
- Accounting for the global nature of EEE production and markets.

3 THE CURRENT SITUATION IN LEAD-FREE SOLDERING

There are two main groups of lead-free solders, used currently in the electronics industry; solders for mainstream applications (substitution of eutectic Sn-Pb solders) and solders for high-temperature soldering with melting temperatures above 250°C (substitution of high-lead solders), where different values of melting temperatures are necessary.

The requirements that are associated with the expected transfer from lead-containing to lead-free materials have been summarized, for example, by Chidambaram *et al.* [18] for high-temperature lead-free solders, but they can be easily generalized for any lead-free solders. The conditions for manufacturing are as follows:

- Proper melting temperatures.
- A small temperature difference between solidus and liquidus.
- Good wetting (low surface tension, high wetting force, low wetting time).
- Cost-effective materials.
- Availability and manufacturability.
- Environmentally friendly materials and good recyclability.
- Possibility of miniaturization (low natural radius of curvature, easily electroplatable).

For reliability and performance, the following are necessary:

- Good thermal and electrical conductivity.
- Small variations in thermal expansion coefficient.
- Good shear and tensile strength.
- Fatigue and creep resistance.
- Oxidation and corrosion resistance.
- Intermetallic compound (IMC) formation (during wetting reaction and solid-state aging).

When changing from one material or technology to another one, the ideal solution for industry involves as few changes as possible, especially from the point of view of the complexity of materials, changes in technologies, etc. (see Section 6 and Ref. [19]).

The search for substitution is limited not only by the materials properties and processing requirements but also by economic and other less-explicit requirements. Therefore, the development of lead-free solutions can be split into two categories: the search for new materials and the search for new technologies. New materials should contain as few major elements as possible (ideally three, maximum four elements), and they should not be used in extremely low amounts (micro-alloying). There should be acceptable tolerances with respect to the nominal composition. New technologies should use existing equipment if possible, and avoid higher complexity. Otherwise, the new approach will be hindered by increased costs associated with changing the production lines.

A considerable amount of research has already been conducted on the formulation of new lead-free soldering materials in Europe, Japan, and in the United States in the new millennium. A number of promising materials, for example, Sn-Ag-Cu (SAC) or Sn-Zn based alloys were studied to replace the (near-) eutectic Sn-Pb solders for mainstream applications. Problems associated with the reactivity of Zn (corrosion, reliability, fluxing) have prevented the adoption of Sn-Zn alloys as viable lead-free solders in practice, and the SAC alloys are now generally accepted as substitutes for eutectic Sn-Pb alloys. However, open research, aimed at replacements of high-temperature, high-lead containing alloys, where the lead levels can be above 85%, has been seriously lacking until recently. Some of the problems associated with the replacement of these alloys were identified already at early stages by both Japanese and American authors [20–23]. Recently, research programs addressing the issue of high-temperature lead-free soldering have been initiated in Japan and Europe; nevertheless, a universal replacement of high-lead solders is not available, and no promising candidate exists up to now. Specific solders for special niches in the market are already available, but the existence of the exemption from the RoHS legislation [6] is still necessary.

According to a recent internal market study, 65% of the global market had converted to lead-free soldering in 2011.

3.1 The Lead-Free Solders for Mainstream Application

The main problems associated with the lead-free alternatives for lead-containing solders are related to their higher melting temperatures and lower reliability. On the other hand, some mechanical properties of these substitutions are quite acceptable, sometimes even better than for lead-containing solders. A schematic comparison of several different types of lead-free solders for mainstream application is shown in Figure 1, where the main property categories are compared for the eutectic Sn-Pb and several SAC solders. The higher the number on *y*-axis, the better the solder performs in a given property category. The results are ambiguous in the case of fatigue resistance; the lead-free solders perform better than lead-containing ones under less severe cycling, but worse under more extreme temperature cycling conditions.

A number of materials under consideration for replacing the eutectic Sn-Pb solders for mainstream applications were reviewed recently by Zhang *et al.* [25].

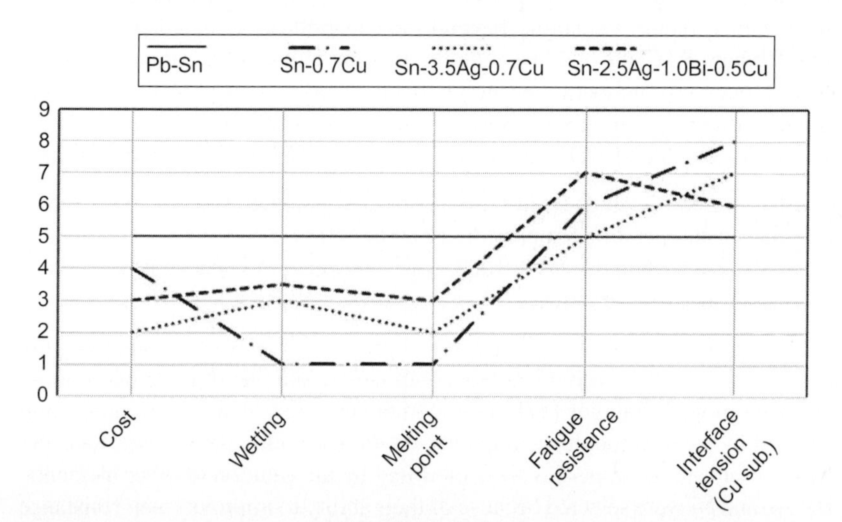

FIGURE 1 Schematic comparison of the quality of solders in selected property categories [24].

The eutectic or near eutectic Sn-Ag-Cu alloys are currently used for the majority of electronic applications, main compositions being, for example, SAC0307 (Sn-0.3Ag-0.7Cu), SAC387 (Sn-3.8Ag-0.7Cu). The melting temperature of the latter alloy is close to the eutectic temperature of the system, that is, 217.5°C [26]. The peak temperature of the reflow profile is around 225 to 235°C, significantly higher than the interval 205 to 215°C used in the case of lead-containing solders, which means that the processing window for the printed circuit boards (PCBs) assembly is much narrower, since the maximum acceptable temperature is around 250°C for standard PCBs. A higher processing temperature, usually longer processing time, and smaller margin for errors imply higher requirements on the controlling of the reflow parameters and higher energy requirements. Most producers were forced to invest in new reflow ovens to meet these more stringent conditions that influence the cost-effectiveness.

There is a desire to reduce alloy cost, particularly for wave soldering applications, and therefore the low Ag SAC solders such as SAC0307 are being considered for the assembly of mobile devices, for example, as these types of material have better drop test performance than the eutectic SAC alloys, though poorer performance with regard to temperature cycling and thermal shock. On the other hand, reducing the Ag content raises the liquidus temperature to about 225°C.

The push for cost-effectiveness has ultimately led to the use of alloys without any Ag, such as Sn-0.7Cu and Sn100C® (®Nihon Superior Co. Ltd., Japan, Sn-0.65Cu-0.05Ni + Ge), which are used in wave soldering due to their lower cost, but temperature cycling and thermal shock performance are further reduced.

Although the electronics industry at large has accepted and commonly uses the SAC alloys for an overwhelming proportion of production, the applications that involve working at elevated temperatures (e.g., under-the-hood automotive applications) require reliability beyond the capabilities of "classical" SAC alloys. Though SAC alloys do have higher melting temperatures, their reliability in high operating temperature environments is inferior even to that of traditional lead-containing eutectic solders with lower melting temperature. Highly demanding conditions were defined for solders used in the automotive industry:

- Meet RoHS standards (lead-free) and be cost-competitive.
- Work in an operating temperature of up to 150°C.
- Solder joints should survive 1000 cycles between −55 and +150°C.
- Reflow at 230°C or below.

The quest for materials with improved properties has led to abandoning the criterion for an alloy with just a few components, and one of the most promising new alloys is Innolot [27], a six-component alloy that is currently being adopted by some automotive suppliers. As the base alloy for Innolot, SAC387 was chosen and its properties were modified by the addition of other elements: Bi, Sb, and Ni were selected because of their ability to improve creep resistance and maintain an acceptable melting temperature.

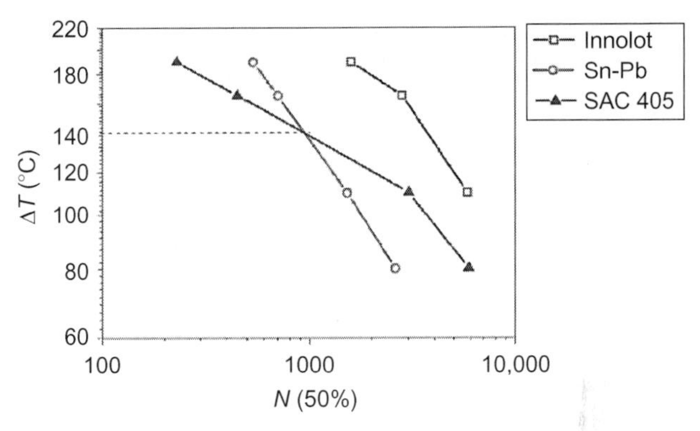

FIGURE 2 Cycles to failure vs. temperature range in thermal shock [28].

The improved mechanical properties of Innolot, and in particular the thermal cycling performance, are the most important benefits with this alloy, with superior thermal cycling characteristics as compared to SAC387 or even to traditional SnPb37 (Figure 2).

3.2 Lead-Free Solders for High-Temperature Application

The main applications for high-temperature ($T_m > 250°C$) solders within the electronics industry are the advanced packaging technologies, for example, die-attach and ball grid array (BGA) solder spheres, chip-scale package (CSP), and multichip module (MCM).

The die-attach material should be able to withstand its operational service temperature, the thermal loading during soldering and should also transfer heat away from the device. In the case of MCM technology, the so-called step soldering approach is employed. This method is used to solder various levels of the package with solders of different melting temperatures. Currently, a number of high-lead solders are used in this technology and new lead-free high-temperature solders should replace them for a broad range of melting temperatures. The upper limit of the process temperatures in the MCM technology is around 350°C, which is defined by the polymer materials used as the substrate. Each subsequent soldering temperature is defined by the melting temperature of the previous solder.

A soldered device has to withstand large mechanical or continuing fatigue stresses, too. The joints have to be able to operate in corrosive environment (high humidity at elevated temperatures) and be resistant to air pollutants (e.g., NO_2, H_2S). The key issue from the point of service is the reliability of solder joints. This is influenced by the morphological evolution inside the interconnects during fabrication and service. In addition, the strong tendency to miniaturization in the electronics industry has increased the demand on the reliability of the solder, because size effects play a significant role in the life span of solder joints; the joints become considerably more brittle with a decrease in the size of the solder gap, and high-Sn solders are more prone to such behavior.

A second reliability issue is connected with the use of barrier metals in complicated assemblies based on Si, such as flip-chip technology. The use of solders with a high-Sn content in combination with multilayered metallization (e.g., Ag, Ti, W, Cu, Ni, V) gives rise to multicomponent alloy systems during manufacturing and service and resulting complex processes (galvanic corrosion, destruction of interconnections) have been reported to take place inside the assembly [29].

The ongoing research in the field of high-temperature solders indicates that there is no viable high temperature lead-free solder to replace the high-lead solders, melting at about 300°C, for common applications. It is necessary to abandon traditional Sn-based soldering technology, as all attempts to adjust the melting temperature of Sn-based solders by the use of appropriate alloying elements have not been successful. Apart from Sb, no viable additions to Sn can be made that will raise the liquidus temperature high enough, leading to a solidus temperature above 232°C. Sn-Sb alloys can give a solidus around 240°C, but this is still not high enough. Therefore, attention has turned toward the use of a different set of base elements, and at the present time several combinations are under consideration. Hypoeutectic Bi-Ag alloys are very good from the point of view of the liquidus temperature, they exhibit mechanical properties close to those of the lead-based solders, and are affordable as well. Other interesting materials are Zn-Al eutectic alloys alloyed with Mg, Ge, Ga, Sn, or Bi. Also of interest are still Sn-Sb and Au-Sn-Sb alloys [21,22]. Some materials are already in use, for example, Au-Sn-based solders, but replacements are being sought as these alloys are quite expensive, and therefore the materials are used only in niche markets, specifically, in equipment, where costs are not one of the key issues.

In the continuing search for new alloys, much effort is being spent on the theoretical modeling of processes in complex multicomponent systems, that is, the description of their thermodynamic and thermo-physical properties and responses to mechanical loading. These methods are currently being applied both to model systems, and to real materials and joint/interfaces, saving the time and cost incurred for complex experimental studies [23,30–33]. Section 4 of this chapter will describe such novel approaches for the development of new materials.

3.3 Possible new Materials for Lead-Free Soldering

Recent progress in the search for alternatives for lead-containing solders can be illustrated by the literature search and the current situation in new patent applications. As open literature sources related to the high-temperature lead-free solders are sparse, more can be deduced from a patent overview, shown in Figures 3 and 4.

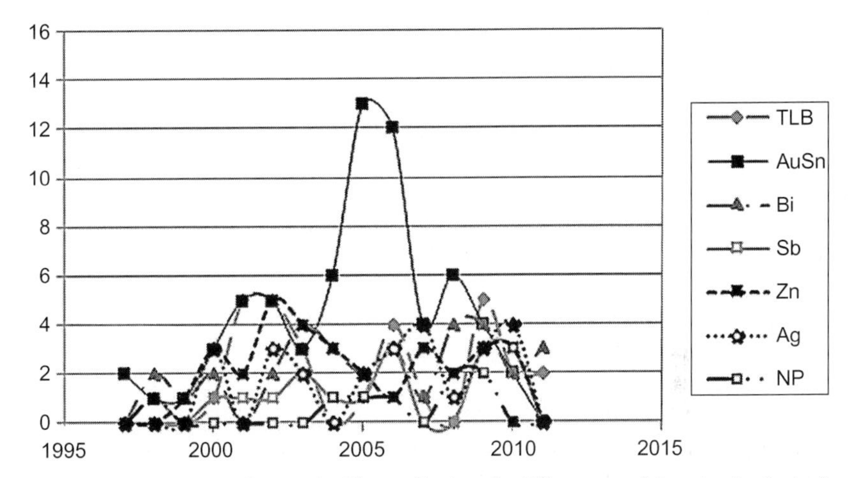

FIGURE 3 The number of patent families applications for different materials and technologies in the past 15 years, (TLB—Transient Liquid Bonding, AuSn—Au-Sn-X alloys, Bi—Bi based alloys, Sb—Sn-Sb-X alloys, Zn—Zn-(Al, Sn) based alloys, Ag—all type of Ag based alloys and technologies, NP—nanopaste technologies).

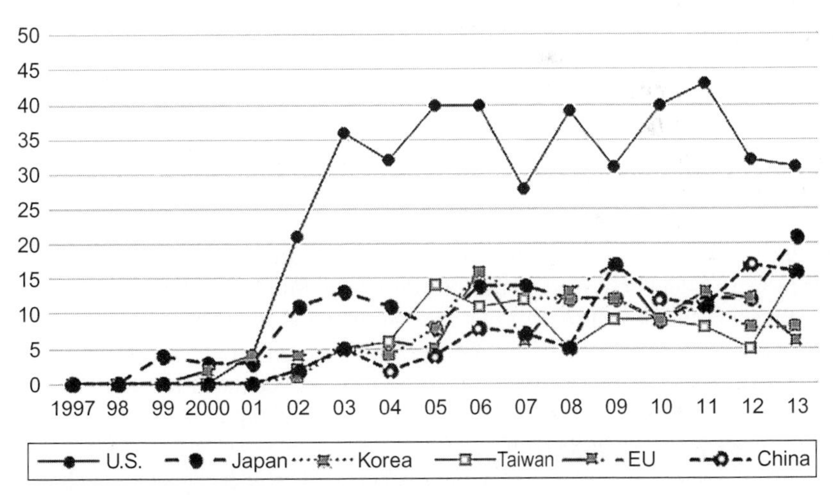

FIGURE 4 The cumulative number of patent applications in all categories for selected countries.

The patent search was carried out with respect to promising new materials and technologies. The patents related to the "classical" SAC alloys were omitted in this case. Most of them are related to the substitution of high-lead solders for high-temperature application. The results shown in Figures 3 and 4 indicate a steep increase in the number of patents starting in the new millennium and a significant number of them are associated with new technologies rather than with new materials. Aside from the transient liquid bonding and processes involving nanopaste, new technologies are also "hidden" under the Ag-based alloys (Ag sintering, Ag foils, Ag paste, or Ag glass systems, and Ag epoxy technologies). Most of the patent applications were submitted by U.S.-based companies, but a significant share of applications belong to Japanese companies, reflecting the significant attention and financial support given to this topic by Japanese manufacturers. The number of patent applications submitted in Europe is slightly influenced by admittance of new countries into the European Union; the patents submitted by new EU states are counted in the total only after the admission of the state into the EU. Generally, the sum of patent applications of all other communities except of the United States are comparable, with China currently in third position, despite a several years delay.

3.3.1 New Approaches for Mainstream Application

A large number of papers has been published in the open literature, (e.g., J. Electron. Mater, J. Alloys Comp.), devoted to the SAC type of solders (or their substitutes) and therefore only a few examples will be mentioned here.

The research papers are oriented toward the search for both improved materials properties and to lowering the melting point of alloys based on the current SAC formula. The investigators are testing new alloying elements or more complex compositelike structures. A relatively small number of papers is oriented toward the investigation of different materials.

In the search for better properties, new alloying elements are added to the standard SAC alloys, and the simplicity of the alloy is sacrificed to meet stricter requirements by the industry. A typical example of this is the above-mentioned Innolot alloy [27], which has already found practical application. From a more scientific perspective, El-Daly *et al.* [34], for example, studied recently the influence of the addition of small amounts of Ni on the mechanical and materials properties and described its positive influence. Similarly Kantarcıoğlu *et al.* [35] studied the influence of Al and Fe on the mechanical properties of SAC solders. Fe was found to have a positive influence on properties as compared with SAC or SAC+Al alloys. Kotadia *et al.* [36] studied the influence of Zn additions (0.0-1.5 mass %) on SAC alloys properties and reported the suppression of the growth of the Cu_6Sn_5 IMC.

The properties of complex structures, where SAC solders are combined with nanoparticles or nanotubes, have been intensively studied by many groups; for example, the influence of TiO_2 nanoparticles was studied by Tang *et al.* [37]. SiC particulate-reinforced SAC105 composite solders have been prepared by El-Daly *et al.* [38], and they reported significantly higher creep resistance and fracture lifetime in comparison with nonreinforced solder. Also, the effect of the addition of carbon nanotubes to the SAC solders was recently reported in Refs. [39,40].

Attempts to modify the melting temperatures of the alloys and to achieve a value suitable for application were described by El-Daly *et al.* [41,42]. They studied Sn-5Sb alloys with a minor addition of Ag, Au, and Cu. They found a significant decrease of melting temperatures to 216°C (Ag) and 203.5°C (Au), and at the same time, the alloys showed very good creep resistance. Gao *et al.* prepared SAC305 nanoalloys and achieved a melting temperature of 179°C [43]. They analyzed the structure and morphology of the nanoparticles by transmission electron microscopy (TEM).

3.3.2 New Materials for High-Temperature Application

A review of possible new materials and technologies for high-temperature lead-free soldering has been published by Suganuma *et al.* [44]. They proposed several alloy systems that have been under intensive scrutiny with respect to their application at higher temperatures; however, none of them fulfill all requirements. Three candidate alloy types have been identified that cover most of the requirements for high-temperature applications: the Zn-Sn-X (X=Al, Cu,...), Au-Sn-X (X=Ag, Cu,...), and Bi-based alloys. In a review published by Chidambaram *et al.* [45], Zn-Al alloys and Bi-based alloys were identified as possible more cost-effective alternatives and Au-Sn (both Au rich and Sn rich) and Au-Ge alloys as expensive replacement for high-lead solders.

The number of papers dealing with high-temperature lead-free solders is much lower than for mainstream solders. The properties of Zn-Sn and Zn-Al-Cu based alloys have been studied by several authors (e.g., Refs. [46–51]). Attention is also paid to Bi based alloys; for example Bi-xAg-0.4Ni-0.2Cu-0.1Ge (x=2, 5, 8, 11, 14) [52], Bi-Ag solders doped by Ce [53], Bi-5Sb-(0.5-5.0) Cu [54], and Bi-(0-11)Ag [55]. The only alloy systems that have been brought to practical application are the Au based alloys, mainly the Au-Sn, and Au-Ge based systems. The main problem with Au based solders is one of cost, but good materials properties make them suitable for special applications (e.g., space research), where the costs are not an issue [33,56–59]. Recently Wang *et al.* studied Bi-Sn alloys with 2.5 and 10% Sn [60], although only the interfacial reactions.

Another overview of materials for high-temperature soldering was published by Takaku *et al.* [61] (see Section 4.2). Their study is an excellent example of the application of theoretical modeling to the development of new materials. This approach results from the development of complex software packages for the modeling of phase diagrams, thermodynamic properties, kinetics, and materials properties of multicomponent systems. More about this will be described in the following Section 4.

4 NEW WAYS IN THE DEVELOPMENT OF NEW MATERIALS

The "classical" approach to the development of new materials is based primarily on experimental studies, where experience and existing knowledge are used to propose new materials compositions followed by rigorous testing to verify the phase stability and related reliability of products, and the required materials properties. A knowledge of the phase diagrams of the systems involved is a very important part of such an approach and, especially in the case of complex systems, this knowledge can be very limited making it quite difficult to predict the behavior of such designed materials. However, new approaches employ the theoretical modeling of thermodynamic properties and phase diagrams, and the consequent use of such data in theoretical or semi-empirical models to predict phase stabilities and materials properties. An example of such an approach, which is becoming increasingly popular and important in materials design, is Calphad-type thermodynamic modeling (CALculation of PHAse Diagrams), and its application in the development of lead-free solders will be shown here. This approach exploits parameterization of models incorporated in software packages allowing the calculation of phase diagrams and thermodynamic properties of multicomponent systems. The existence of a self-consistent thermodynamic database is crucial for such an approach, and therefore attention will be paid to the progress in the development of thermodynamic databases that is a consequence of the parameterization process. A more detailed description of the application of the *Calphad* method can be found elsewhere, for example in Refs. [62–64].

4.1 Computational Thermodynamics as a Research Tool

The *Calphad* technique is based on the sequential modeling of the thermodynamic properties of multicomponent systems, starting from the simplest—the Gibbs energy of the phases of the pure elements—followed by its variation as further components (elements) are added. The Gibbs energy parameters of phases take the form of polynomial coefficients and are fundamental for the calculation of phase diagrams. Using appropriate software [65–68], the thermodynamic properties of a binary system can be calculated using Gibbs energy expressions comprising contributions for both elements in all of the phases existing in the system, and their mutual interaction. By combining robust thermodynamic descriptions of binary alloy systems, it is possible to calculate phase equilibria in ternary and higher order systems through a constrained minimization of the total Gibbs energy for a given temperature, pressure, and overall composition. The deviation of the real behavior of the system from the prediction can be minimized by adding appropriate higher-order parameters, describing additional interactions between the elements, as long as these extra parameters do not affect the modeling of the lower-order systems.

It is possible to model such a system with high precision if reliable experimental data are available that describe both the thermodynamic properties of the individual phases (e.g., heat capacity, enthalpies of formation or mixing, activities) and the phase equilibria (e.g., invariant temperatures, compositions, and amounts of phases). A set of consistent experimental data, selected from all of the diverse data available, forms the basis for the assessment that is the derivation of the parameters giving the best fit to these properties. This is generally carried out with a thermodynamic software package (e.g., PARROT in

Thermo-Calc [65] or the Assessment Module in MTDATA [66]). The parameters for the Gibbs energy thus derived are then collected in databases, which are usually oriented toward specific groups of materials (steels, Ni-based alloys, lead-free solders, etc.).

A number of databases are available for the development of new lead-free solder materials including one developed at the National Institute of Standards and Technology (NIST) [69], the Alloy Database for Micro-Solders (ADAMIS database) [70,71], and the TCSLD1 database from Thermo-Calc AB [72]. The SOLDERS thermodynamic database [73] was developed within the scope of the COST Actions 531 and MP0602 [74,75]. The choice of elements required for a lead-free solders database is clear-cut [26,76,77]: Sn and Cu along with Ag have long been identified as important component elements for lead-free soldering materials. Cu may also be used as a substrate material. The properties of solders based on these elements are not completely satisfactory, and thus it was deemed necessary to include other elements in the database to provide a tool suitable for optimizing their properties. The low-melting-temperature elements In and Bi were included, along with termination materials such as Ni, Au, and Pd; Pb was added to the database, as it was important to be able to model the interaction between old and new solders, for example, during the repair of electronic devices. The element P also plays an important role in soldering, as it is inseparable from the Ni used in the electronics industry.

The scope of the thermodynamic database SOLDERS was broadened later [75,78] to include elements that may be useful for the development of materials for high-temperature soldering, such as Al and Ge.

4.2 Application of the *CALPHAD* Method for the Development of new Solder Materials

Most of the current research effort associated with solders has been concerned with attempts to find replacements for traditional Pb-Sn based materials. The ideal lead-free solder would be one with identical or very similar properties to the original; a drop-in replacement. Computational thermochemistry is the ideal tool to use to search for a new alloy with a eutectic temperature in the region of 183°C. Of course, it has to be recognized that the range of temperatures associated with solidification is only one of a number of properties, which are desirable for a new solder material. Among other criteria are cost, wettability, good thermal resistance, mechanical strength, creep resistance, and the ability to form an effective bond with substrates and termination materials. On the other hand, all properties besides cost can also be tested by theoretical modeling.

Figure 5 shows the phase diagram for the Cu-Sn binary system calculated from the SOLDERS database [73]. Of particular interest here is the eutectic reaction close to pure Sn, which is calculated to be at 227.2°C (just under 5°C below the melting temperature of pure Sn). This binary eutectic composition corresponds to one of the common lead-free solders currently in use commercially.

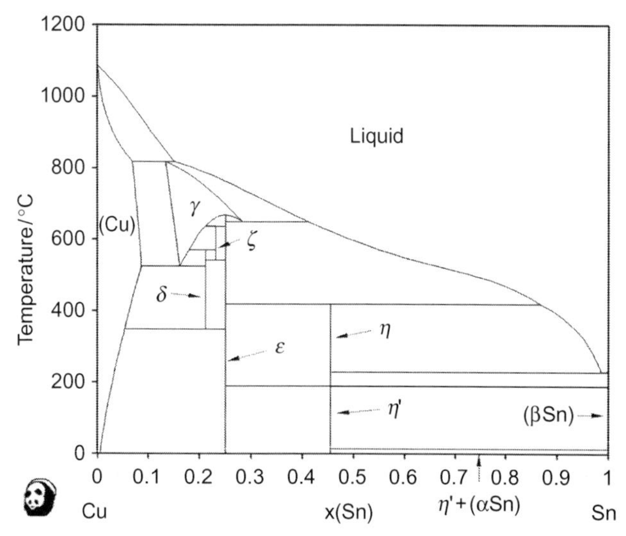

FIGURE 5 The calculated Cu-Sn phase diagram. *(Adapted with permission from Ref. [79], © (2012) John Wiley and Sons Ltd.)*

The addition of elements to this Cu-Sn system will generally reduce the eutectic temperature further by stabilizing the liquid phase at the expense of the crystalline phases. The most common lead-free solder is formed by the addition of Ag to the Cu-Sn system. This ternary system has been the subject of many experimental studies to fully understand the nature of its phase equilibria. During the course of the COST Action 531 the thermodynamic model parameters of the system were optimized to be consistent with all the experimental thermodynamic and phase diagram data. In order to explore composition ranges where the liquid can solidify at even lower temperatures, the most useful diagram is the liquidus projection. The appropriate diagram for the Ag-Cu-Sn system is shown in Figure 6, and it has the appearance of a map with temperature contours and monovariant lines, which can look like valleys. The monovariant lines extend from invariant temperatures in the boundary binary systems, and the temperature contour lines show the values where the liquid starts to freeze for specific compositions. Only at compositions where different eutectic valleys intersect will liquid solidify at a single temperature.

While Figure 6 shows very easily the compositions along the so-called eutectic valley, it is less easy to identify the actual freezing temperature. A better way of doing this is shown in Figure 7, which represents a projection of the eutectic valley from Figure 6 with the temperature as the ordinate. This allows a very clear graphical description of the temperatures and the character of any particular invariant reaction.

Computational thermochemistry also provides an easy way to trace the composition of a non-eutectic liquid phase as it solidifies. This is of particular

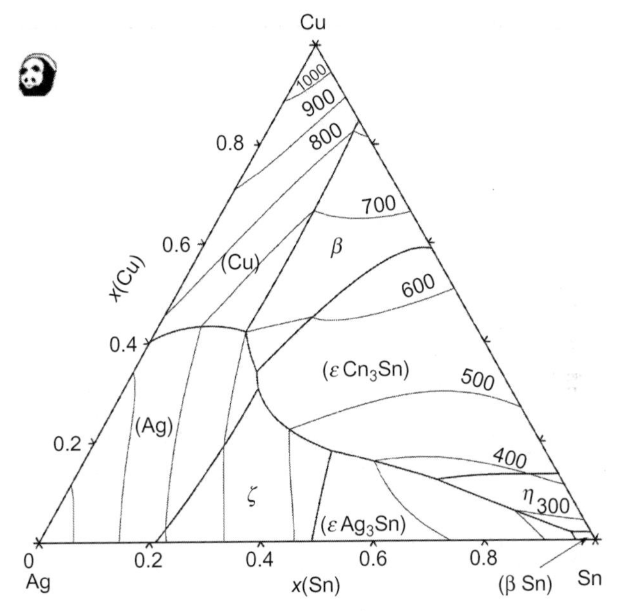

FIGURE 6 Calculated liquidus projection for the Ag-Cu-Sn system. *(Adapted with permission from Ref. [79], © (2012) John Wiley and Sons Ltd.)*

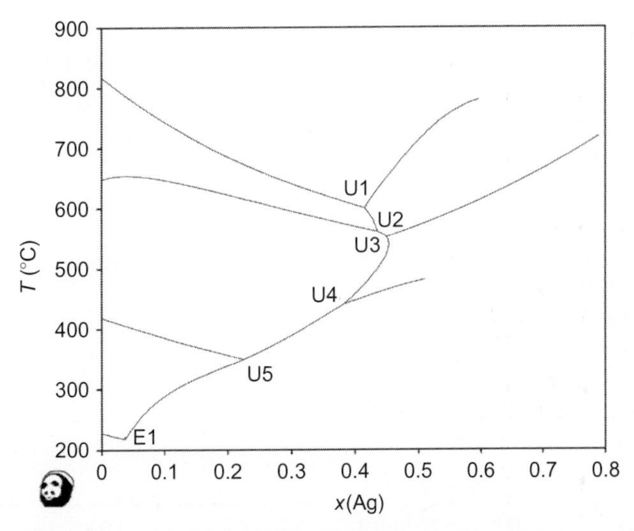

FIGURE 7 Calculated liquidus projection for the Ag-Cu-Sn system displayed on a temperature—composition axis. *(Adapted with permission from Ref. [79], © (2012) John Wiley and Sons Ltd.)*

interest where lead-free solders replace or are used to repair conventional Pb-Sn solders. There is a significant possibility of contamination of the newer solders by relatively small amounts of Pb from existing solder joints and component terminations. Calculations can be used to investigate the phases formed in a lead-contaminated solder and the microstructure, which could be expected as such a contaminated solder solidifies [80,81].

One way to explore the liquidus temperature, the phases that might form and their range of stability, is by calculating a temperature-composition section (isopleth) through a multicomponent phase diagram.

Figure 8 shows such an isopleth between the composition of the classical "electrician's solder" and the standard Sn-Ag-Cu lead-free solder. It shows that as the lead-free solder becomes more and more contaminated by the lead-containing solder, the liquidus temperature drops and also the range over which the liquid phase is stable, either on its own or in combination with other phases, becomes very much larger. This could lead to unreliability and increased porosity of the solder joint.

Computational thermochemistry is inherently predictive and can be used to predict the compositions and temperatures of eutectics in systems with more

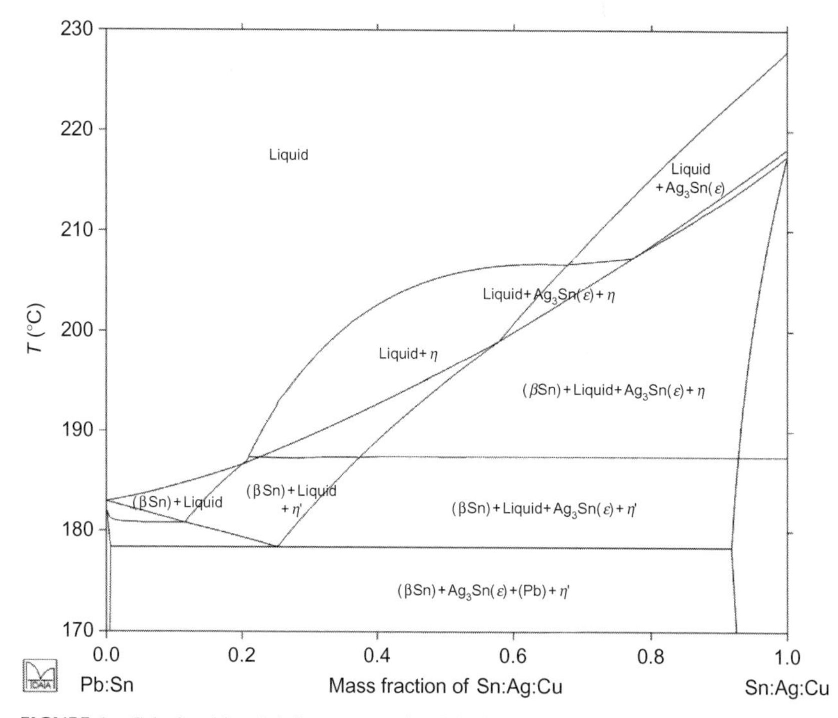

FIGURE 8 Calculated isopleth between an electrician's solder and the standard Sn-Ag-Cu lead free solder. *(Adapted with permission from Ref. [79], © (2012) John Wiley and Sons Ltd.)*

and more components, which may have not been studied experimentally. It is also possible to explore, using computational thermochemistry, the interaction between the solder and conductor metals, that is, component metallization, board surface finishes, and underlying conductors.

The study of phase diagrams and phase equilibria is not new, but the information gained is fundamental to materials science. The advent of computational thermodynamics has unleashed the "power" of the phase diagram resulting in a tool that can be used to give an insight into anything from multicomponent, multiphase equilibria to surface tension and microstructural simulation. The creation of the robust self-consistent thermodynamic databases that the technique requires needs a fair amount of effort, but the savings in both time and cost in relation to expensive experimental programs can be substantial [70,73,77,79,82].

In the field of materials properties prediction, the results of thermodynamic calculations (offering stable or metastable equilibria) in the form of expected phase compositions are used in other models, both empirical and theoretical, which predict the mechanical properties (e.g., hardness, tensile stress), thermophysical and physical properties (cooling curves, etc.) and phase transformations. One example of such a software package, which uses the thermodynamic calculations as one of the inputs, is JMatPro [83].

Processes on the interface, including the nucleation, can be also modeled by Phase Field methods. The knowledge of interface behavior is crucial in soldering processes, and this method can be applied for the study of precipitation sequences on solder/substrate interfaces [84,85]. Here, the optimized thermodynamic parameters obtained in the thermodynamic assessment of the Cu-Ni-Sn system were utilized and the 2D and 3D simulations of the growth of the IMC layers in the presence of grain boundary diffusion in the Cu-Sn system has been carried out.

As mentioned in Section 3.3.2, Takaku *et al.* [61] used the Calphad technique and the ADAMIS database [70,71] to develop promising alloy compositions. Consequent material testing allowed them the verification of the assumptions used in their original design. A number of groups of alloys were proposed that were suitable for high-temperature soldering. The first group was a Bi-based composite solder. Here, the modeling predicted the existence of a metastable liquid miscibility gap in the Bi-Cu binary system, which can be stabilized by the addition of Al. They also proposed the addition of Mn to the alloy to promote the precipitation of suitable particles. The resulting structure after solidification consisted of a Bi matrix reinforced by fine Cu-Al-Mn particles allowing for the relaxation of thermal stresses and inhibition of crack growth. The other proposed alloy was a Zn-Al based solder with the addition of Cu. This alloy can withstand thermal stresses and temperature hysteresis between 40 and 250°C, which is necessary for the packaging of power semiconductor devices. The authors also mentioned Sn-Zn alloys as possible lead-free soldering materials. These alloys have quite a wide solid/liquid region (199°C to approximately 360°C), and it is therefore important to avoid liquid formation at reflow temperatures at around 250°C during the soldering process.

TABLE 1 The Solidus and Liquidus Temperatures and the Extent of Mushy Zone Calculated using the *Calphad* Technique [18].

Alloy No.	Alloy Compositions (mass %)	Solidus, *T* (°C)	Liquidus, *T* (°C)	Temperature Difference, Δ*T* (°C)
1	Sn-35.5Au-17.5Sb	273	290	17
2	Sn-40.3Au-6.9Cu	305	322	17
3	Au-7.8Ge-6.9In	338.7	340	1.3
4	Au-11.7Sb-15.1In	285.7	288	2.3
5	Au-23.2Sn-17.9In	280.5	290	9.5
6	Au-21.5Sn-0.4Cu	278	292	14
7	Au-11.5Ge-2.5Bi	332	352	20

Another example of the application of Calphad-type theoretical modeling for the development of new materials was described in Refs. [33,56,58,59,86,87], where phase diagram modeling was used for the development of candidate alloys using the extended SOLDERS database [73,78]. Two main candidate systems were considered in these studies. The first is a ternary system Au-Sn-X (X = Ag, Cu), where compositions close to the binary eutectic Au-Sn composition [58,59] were exploited. Thermodynamic calculations were used to predict the alloying conditions necessary to suppress simultaneously the precipitation of the brittle Au_5Sn phase and the increase of the melting temperature of the alloy. Thermodynamic calculations were also used for the design of the low-Au, Sn-rich solder composition Sn-28Au-8Ag (at.%). The same approach was used for the study of the second candidate system, Au-Ge-X (X = In, Sb, Sn). Some examples of predicted compositions are shown in Table 1 [18].

The Au-Ge-X system was also studied intensively by Wang *et al.* and Leinenbach *et al.* [33,56,86,87]. They brought theoretical calculations to practical application, as the developed Au-Ge solder alloys have been used in the construction of heated conversion surface assemblies for the ESA/JAXA Mission Bepi Colombo to the planet Mercury (Launch 2014) [56].

5 NEW TECHNOLOGIES FOR MATERIALS JOINING

As the search for drop-in replacement of lead-containing solders (especially for the high-temperature applications) does not offer any cost-effective solution, great attention is paid to the development of new technologies for joining. The number of papers dealing with new technologies is very high and significantly exceeds the publications describing possible new materials.

The reviews of Suganuma *et al.* [44], and Chidambaram *et al.* [45] that were mentioned in Section 3.3.2 described not only the new prospective materials for high-temperature soldering but also several different technologies offering promising results for application in the electronics industry. They identify adhesive technologies, sintering, and transient liquid phase bonding (TLB) as the most promising. Besides these two papers, several other reviews oriented toward more specific applications have also been published. Dietrich [88] reviewed recent progress in automotive power packaging. Their conclusion was that the expected increase in operating temperatures and the advent of new devices with higher current density will require the use of new joining technologies such as sintering in the medium term. Kim and Kim [89] discussed bonding using anisotropic conductive film (ACF) and nonconductive adhesive (NCA) detailing the principles of both methods. The review by Mir and Kumar [90] deals with the recent advances in isotropic conductive adhesives (ICA). The chemistry of adhesives used in electronics (without technological details) was described in the work of Rabilloud [91].

Conductive adhesive technology can be used for bonding many different material surfaces, including those that cannot be joined by soldering. The technology is based on combining two types of materials; a polymer binder providing the mechanical bond and a filler (usually Ag) that provides the electrical conductivity. The technology offers limited shrinkage with resistance to thermal and mechanical loading [92,93]. Despite their generally good properties, there are still problems to be solved, especially related to the reliability with respect to impact strength, unstable contact resistance, and lower electrical and thermal conductivity [90,94,95]. Suganuma *et al.* [48] studied the mechanism of degradation under humid conditions at the interface between the Ag epoxy conductive adhesives (CAs) and Sn plating. Two different types of oxide phases, SnO and SnO_2, were found to be formed at the interface and are the root cause of the interfacial degradation. The power of the theoretical approach to process modeling has been demonstrated even for such a complex problem as adhesive technology, in the paper of Erinc *et al.* [96]. They presented a multiscale approach to predict the residual stresses that appear during the assembly process when using ICAs with Ag nanoparticles.

Lorenz *et al.* [97] described a method where heating during a device manufacture can be kept to a minimum by combining laser joining (localized heating technique) and adhesive wafer-level bonding. This approach was used for bonding Si to glass and enables the use of temperature-sensitive materials within the package.

A nonconductive adhesive technology described in [89] is based on an adhesive polymer resin and a curing agent without any conductive particles. During the bonding process, the bumps on the chip are in direct contact with the corresponding pads on substrate. After curing the adhesive e.g. by heat, the mechanical and conductive contact between the bumps and the corresponding pads under compression is maintained.

The next candidate technology is based on sintering technology using a paste or foils with metal (mostly Ag) micro- or nanoparticles [98–104]. For example, Ag foil was chosen to bond Si chips to Cu substrates by Dupont *et al.* [102]. Prior to bonding, the Si chips were coated with thin Cr and Au layers. The Si chip, Ag foil, and Cu substrate were bonded together in one step at 250°C under a reduced pressure of approximately 7 Pa (about 0.1 mbar), Au/Ag and Ag/Cu bonds forming at the interfaces by short-range interdiffusion. The resulting joints exhibit nearly perfect quality. No voids are observed at the Si/Ag and Ag/Cu bonding interfaces. Quintero *et al.* [105] described the thermo-mechanical reliability of novel Ag nano-colloid die attach for high-temperature applications. It was concluded that this material was not suitable for extreme conditions. Sabbah *et al.* compared the performance of Ag sintering with standard soldering methods using Au-Ge solder [106]. They concluded after thermal cycling testing that the performance of the soldered joints was better than the sintered ones.

The third technology reviewed involves joining using intermetallic phases. This approach includes evaporating films onto the interface and TLB technology. For example, a lead-based solder has been replaced by an intermetallic compound (IMC) to join an Si chip to an Ag-plated Cu substrate. This was achieved by evaporating Ag, Cu, and Sn films on the substrate at 250°C for 10 s. Heat treatment at 300°C for 30 s resulted in the evaporated films being completely transformed into intermetallic compounds (Ag_3Sn, $(Ag,Cu)_3Sn$, Cu_6Sn_5, or Cu_3Sn layers depending on the complexity of films) forming the bond. Die shear tests at 270°C showed that the IMC joint exhibits a joining strength higher than that of a lead-based solder joint [107,108].

TLB is based on creating interlayers between joined interfaces that are designed to form a thin or partial layer of a transient liquid phase, creating bonds through a brazing-like process in which the liquid phase disappears isothermally. In contrast to conventional brazing, the liquid is substituted by an intermetallic phase with a higher melting temperature than that of the parent liquid. Various alloys can be used in this technology; any system, where the liquid phase disappears by means of diffusion or other processes is suitable for this approach.

Numerous examples of TLB exist in the literature and cover a variety of filler materials, substrates, and processes. Wierzbicka-Miernik *et al.* [109] used Sn layers between Cu and Cu(Ni) substrates, where the growth of the Cu_6Sn_5 IMC layer was studied. A significant influence of even small amounts of Ni was found, for example, on the soldering time. Bosco *et al.* studied the influence of layer thickness on pore formation [110] and the strength and toughness of joints [111] produced by the TLB method in the Cu-Sn system. The TLB process temperatures of around 400°C were employed by Weyrich *et al.* [112] to bond Al_2O_3 to other ceramics using Au-12Ge and Au-3Si solder alloys. Sound joints were obtained with reasonable shear strength and remelting temperatures of more than 550°C above the melting temperature of the filler material. Lugscheider *et al.* [113] used magnetron sputtering to deposit a TLB Cu-In multilayer and showed that the joint quality was significantly dependent on the temperature of the substrate. The Cu and Ag layers were used in TLB in other studies [114,115]. The theoretical background of the method has been developed by Park *et al.* [116] who used phase-field simulations to model the growth of IMC in the Cu/Sn/Cu system under TLB conditions.

High-temperature lead-free solder joints can also be prepared by means of the high-energy droplet deposition technique. The theoretical background of this method, together with some application examples have been described by Conway *et al.* [117].

Other technologies that obviate the need for any specific materials to be used for joining are also under development. For example, direct Cu bonding [118] is based on electroless copper deposition producing all-copper chip-to-substrate connections. Li *et al.* [119] applied ultrasonic vibration to a Cu/Sn-foil/Cu interconnection system at room temperature to form homogeneous Cu_6Sn_5 and Cu_3Sn joints. The PCB process [120] is another alternative method for electronic assembly that completely bypasses the traditional high-temperature soldering process. The new process reverses the traditional approach to electronic assembly by placing components first and then making the electrical interconnections using traditional PCB manufacturing processes.

A much broader review of the field (outside the scope of this chapter) was published by Marques *et al.* [121]. They presented a literature review of PCBs, providing information about their structure and materials, the environmental problems, their recycling, and some solutions that are being studied to reduce and/or replace the solder, to minimize the impact of solder on the PCB.

Similarly, as in the case of new materials for the substitution of lead-containing solders, new technologies have not yet offered a simple and universal solution to this problem. Therefore, many new studies are published in a variety of journals, mostly oriented to specific application in the electronics industry. Currently, the sintering technologies seem to be the most useful and most ready for practical use.

6 IMPLEMENTATION OF NEW MATERIALS INTO AN INDUSTRIAL ENVIRONMENT

As this book targets researchers as well as practitioners in industry, this part of the chapter shows the difficulties that the industry faces when implementing new materials into production. It summarizes the likely impacts of phasing out of lead-containing solders for industry, especially in the case of SMEs. Any changes to an established product will need to be validated and subjected to cost-benefit analysis. Large industrial companies have well-established systems for the design, development, and validation of new products, and for material substitution into existing products, but this is not always the case for SMEs. This section is intended to provide an outline to guide businesses through the product design and validation, production validation, and material substitution processes. A fuller account, including detailed protocols, is given in Ref. [19].

6.1 A Summary of Protocols for Validating Material Substitutions in Electronic and Electrical Equipment

Proposed changes, such as the substitution of one material with another to satisfy the requirements of new regulations, to an established product will need to be rigorously examined to confirm that the substitution does not have any adverse impact. Numerous aspects of any proposed change need to be reviewed carefully, and such steps should follow well-established business processes and be recorded in detail. For example, the sequence to determine if a proposed change will have any adverse impacts on a product should include design validation, production validation, supply chain setup; waste management, and cost analysis. New products will be required to comply with all relevant regulations and applicable international standards.

In an ideal case of material substitution, a fully compatible replacement material will be found for the existing material. This would result in minimal changes to the product design and preferably little or no change to the manufacturing process. This means:

- Existing machines can be used.
- No new technology steps need to be introduced.
- No surface cleaning of parts is necessary.
- No potentially hazardous materials during processing are used.

Figure 9 outlines a typical product life cycle from manufacturing to disposal after use. A material substitution, such as the replacement of a lead-containing high melting temperature solder with a lead-free alternative, must maintain the ability of the product to meet all of the requirements of this product life cycle.

The first priority in material substitution is to ensure the proposed change will result in a product that complies fully with the appropriate regulations and international standards. The details must be documented in the product technical file. Although the product design, product technical file, and detailed assessments are all documents confidential to the company and/or institution, the associated declarations of conformity are publicly available documents.

Any proposed material substitution to an existing product must be accompanied by validation to confirm that the change does not adversely impact the product's performance and reliability. This often results in the need for new testing regimes that examine specific aspects of product performance. Such an example might be how a high melting temperature lead-free solder ages in service compared to a lead-containing solder. The results of such an examination should be recorded in the failure mode and effect analysis (FMEA) for that product.

There are, fortunately, some possibilities to shorten the testing period and thereby reduce the associated costs. One such approach is the Accelerated

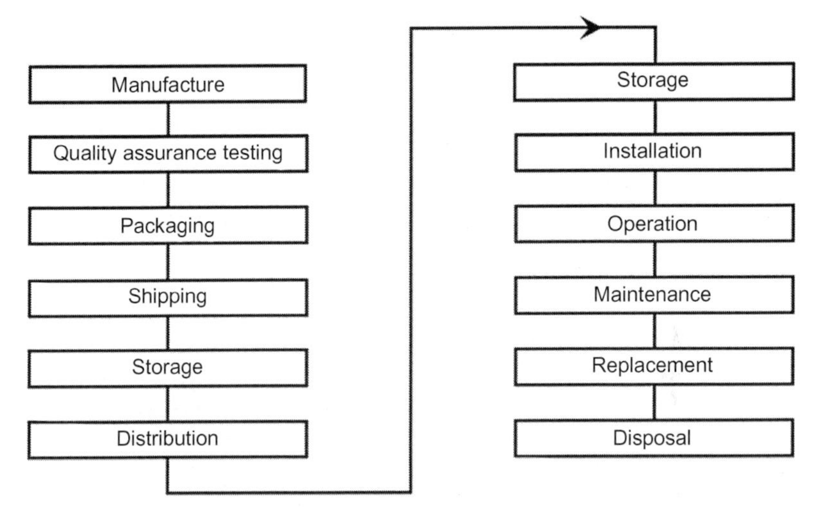

FIGURE 9 Product lifecycle from manufacture to disposal [19].

LifeTest (ALT). If the root causes of the failure modes are known, then by making the appropriate conditions more severe, failure rates can be accelerated and the life test can be shortened. In solder joints, for example, an important factor is the temperature. It is well documented in the scientific literature that an increase in temperature will elevate the failure rate.

Usually it is the mechanical characteristics that must be examined and validated. Measuring these mechanical properties at a certain level of acceleration (temperature, starting cycles, environmental conditions, etc.) and extrapolation back to the normal operation condition is another way to reduce the validation time requirement. Note, if more than one of these acceleration factors are taking place, interaction between them can influence the mechanical properties and hence the reliability. The combination of temperature and a corrosive atmosphere is a good example. Design of experiments (DOE) should be adopted in these cases.

The aim of the material substitution is that the new product should have the same performance and reliability as the original one. If the life tests do not prove that, then customers must be informed of the necessity of the change and its impact on the properties of the product.

It is vital to consider the impacts of any material changes on the ability of the product to survive transportation and storage. Degradation processes during transportation include exposure to salt atmospheres during sea transportation and vibration effects on roads.

The validations outlined above have been confined to the performance of the product; the manufacturer, however, has to quantify the cost impact of material substitutions. If the new process has a significantly different yield or a reduced throughput than it had previously, then these effects must also be taken into account.

The protocols and process flow diagrams in Ref. [19] are based on the premise that the end user will expect to see no degradation in performance and reliability with respect to the original product design. Furthermore, customers and end users will expect the availability of the product to be unchanged, that the end-of-life mechanisms will be unchanged, and the failure modes will be unchanged. There will also be an expectation that there will be no detriment to visual appearance and that the cost will not increase as a result of the material substitution.

7 CONCLUSIONS

In this chapter we have briefly described the current situation and possible further development in the field of lead-free soldering and environmentally friendly electronics joining in the electronic industry. The information covered should be useful both for scientists and people with an industrial background and should help in the transition from the traditional lead-consuming industry to a "green" one, as the pressure from the society, consumers, and accompanying legislation can be expected to increase in the future.

It is quite clear that despite investing money, time, and human effort, some problems are very difficult to solve and reasonable compromises have to be agreed on among all groups involved in the process.

ACKNOWLEDGMENT

The authors are grateful to Dr. Hector Steen for useful comments and fruitful discussion.

REFERENCES

[1] E.B. Smith, Health and environmental effects of lead and other commonly used elements in microelectronics, in: K.J. Puttlitz, K.A. Salter (Eds.), Handbook of Lead-free solder technology for Microelectronic Assemblies, Marcel Dekker Inc., New York, 2004, pp. 49–81.
[2] Directive 2002/95/EC of the European Parliament and of the Council of 27 January 2003 on the restriction of the use of certain hazardous substances in electrical and electronic equipment, Official Journal of the European Union L 037/19, 2003.
[3] Administration on the Control of Pollution Caused by Electronic Information Products (http://www.chinarohs.com/).
[4] Act for Resource Recycling of Electrical and Electronic Equipment and Vehicles—Korea RoHS, WEEE & ELV (http://uk.farnell.com/images/en/ede/pdf/rohs_korea_oct08.pdf).
[5] Electronic Waste Recycling Act of 2003, California, U.S.A., (http://www.calrecycle.ca.gov/electronics/act2003/).
[6] Directive 2011/65/EU of the European Parliament and of the Council of 8 June 2011 on the restriction of the use of certain hazardous substances in electrical and electronic equipment (recast), Official Journal of the European Union L 174/88, 2011.
[7] http://eur-lex.europa.eu/LexUriServ/LexUriServ.do?uri=OJ:L:2003:037:0019:0023:en:PDF.
[8] http://eur-lex.europa.eu/LexUriServ/LexUriServ.do?uri=OJ:L:2003:037:0024:0038:en:PDF.
[9] http://eur-lex.europa.eu/LexUriServ/LexUriServ.do?uri=OJ:L:2011:174:0088:0110:EN:PDF.
[10] http://ec.europa.eu/environment/waste/pdf/rohs_report.pdf.

[11] http://rohs.exemptions.oeko.info/fileadmin/user_upload/rohs_final_report_Oeko_Institut__22-Oct-2007_01.pdf.
[12] http://ec.europa.eu/environment/waste/weee/pdf/report_2009.pdf.
[13] http://rohs.exemptions.oeko.info/fileadmin/user_upload/RoHS_IV/RoHS_final_report_May_2011_final.pdf.
[14] http://eur-lex.europa.eu/LexUriServ/LexUriServ.do?uri=oj:l:2006:396:0001:0849:en:PDF.
[15] http://eur-lex.europa.eu/LexUriServ/LexUriServ.do?uri=OJ:L:2007:136:0003:0280:EN:PDF.
[16] http://echa.europa.eu/web/guest/regulations/reach/legislation.
[17] http://eur-lex.europa.eu/LexUriServ/LexUriServ.do?uri=OJ:L:2012:197:0038:0071:EN:PDF.
[18] V. Chidambaram, J. Hald, J. Hattel, Development of high melting point environmentally friendly solders using the CALPHAD approach, Arch. Metall. Mater. 53 (2008/4) 1111–1118.
[19] S.A. Mucklejohn, Z. Toth, D.R. Andersson, A. Watson, N. Hoo, A. Kroupa, A.A. Kodentsov, I. Plotog, GP1—Protocol for the evaluation of alloys to be used as replacement for high temperature Pb-containing soders in products & industrial processes, in: A. Kroupa (Ed.), Handbook of High-Temperature Lead-Free Solders: Group Project Reports, COST Office, Brussels, 2012, pp. 5–24.
[20] F.W. Gayle, G. Becka, J. Badgett, G. Whitten, T.Y. Pan, A. Grusd, B. Bauer, R. Lathrop, J. Slattery, I. Anderson, J. Foley, A. Gickler, D. Napp, J. Mather, C. Olson, High-temperature lead-free solders, J. Manag. 53 (6) (2001) 17–21.
[21] T. Shimizu, H. Ishikawa, I. Ohnuma, K. Ishida, Zn-Al-Mg-Ga alloys as Pb-free solder for die-attaching use, J. Electron. Mater. 28 (11) (1999) 1172–1175.
[22] M. Rettenmayr, P. Lambracht, B. Kempf, C. Tschudin, Zn-Al based alloys as Pb-free solders for die attach, J. Electron. Mater. 31 (4) (2002) 278–285.
[23] J.H. Kim, S.W. Jeong, H.M. Lee, Thermodynamics-aided alloy design and evaluation of Pb-free solders for high-temperature applications, Mater. Trans. 23 (2002) 1873–1878.
[24] I. Szendiuch, Trendy v pouzdření a bezolovnaté pájky, SMT-info konsorcium 43 (22. 10. 2003), 3-6, ISSN 1211-6947 (in Czech).
[25] L. Zhang, S.-B. Xue, L.-L. Gao, Z. Sheng, H. Ye, Z.-X. Xiao, G. Zeng, Y. Chen, S.-L. Yu, Development of Sn–Zn lead-free solders bearing alloying elements, J. Mater. Sci. Mater. Electron. 21 (2010) 1–15.
[26] A. Dinsdale, A. Watson, A. Kroupa, J. Vrestal, A. Zemanova, J. Vizdal, Atlas of Phase Diagrams for Lead-Free Soldering, COST 531 Lead-Free Solders Vol. 1, COST Office, Brussels, ISBN: 978-80-86292-28-1, 2008.
[27] H. Steen, B. Toleno, New lead-free alloy that takes under-the-hood heat in stride, innovative formulation provides high reliability for high-temp applications, private communication.
[28] R. Ratchev, Lebensdauervorhersage von Pb-freien Lotstellen unter Berucksichtigung der Feldbelastung, in: Presentation in LIVE Project Seminar 'Material Verhalten von Loten in Mikrobereichen', Berlin, 2008.
[29] Z.H. Huang, P.P. Conway, E. Jung, R.C. Thomson, C.Q. Liu, T. Loeher, M. Minkus, A reliability issue for pb-free solder joint miniaturisation, J. Electron. Mater. 35 (9) (2006) 1761–1772.
[30] V. Chidambaram, J. Hald, J. Hattel, Development of gold based solder candidates for flip chip assembly, Microelectron. Reliab. 49 (2009) 323–330.
[31] M. Maleki, J. Cugnoni, J. Botsis, On the mutual effect of viscoplasticity and interfacial damage progression in interfacial fracture of lead-free solder joints, J. Electron. Mater. 40 (10) (2011) 2081–2092.
[32] N. Moelans, A quantitative and thermodynamically consistent phase-field interpolation function for multi-phase systems, Acta Mater. 59 (2011) 1077–1086.
[33] J. Wang, C. Leinenbach, M. Roth, Thermodynamic modeling of the Au-Ge-Sn ternary system, J. Alloys Compd. 481 (2009) 830–836.

[34] A.A. El-Daly, A.M. El-Taher, T.R. Dalloul, Enhanced ductility and mechanical strength of Ni-doped Sn–3.0Ag–0.5Cu lead-free solders, Mater. Des. 55 (2014) 309–318.
[35] A. Kantarcıoğlu, Y.E. Kalay, Effects of Al and Fe additions on microstructure and mechanical properties of SnAgCu eutectic lead-free solders, Mater. Sci. Eng. A 593 (2014) 79–84.
[36] H.R. Kotadia, O. Mokhtari, M.P. Clode, M.A. Green, S.H. Mannan, Intermetallic compound growth suppression at high temperature in SAC solders with Zn addition on Cu and Ni–P substrates, J. Alloys Compd. 511 (2012) 176–188.
[37] Y. Tang, G.Y. Li, Y.C. Pan, Effects of TiO_2 nanoparticles addition on microstructure, microhardness and tensile properties of Sn–3.0Ag–0.5Cu–xTiO2 composite solder, Mater. Des. 55 (2014) 574–582.
[38] A.A. El-Daly, G.S. Al-Ganainy, A. Fawzy, M.J. Younis, Structural characterization and creep resistance of nano-silicon carbide reinforced Sn–1.0Ag–0.5Cu lead-free solder alloy, Mater. Des. 55 (2014) 837–845.
[39] Y.-K. Ko, S.-H. Kwon, Y.-K. Lee, J.-K. Kim, Ch-W. Lee, S. Yoo, Fabrication and interfacial reaction of carbon nanotube-embedded Sn–3.5Ag solder balls for ball grid arrays, J. Alloys Compd. 583 (2014) 155.
[40] Z. Yang, W. Zhou, P. Wu, Effects of Ni-coated carbon nanotubes addition on the microstructure and mechanical properties of Sn–Ag–Cu solder alloys, Mater. Sci. Eng. A 590 (2014) 295–300.
[41] A.A. El-Daly, Y. Swilem, A.E. Hammad, Creep properties of Sn–Sb based lead-free solder alloys, J. Alloys Compd. 471 (2009) 98–104.
[42] A.A. El-Daly, A.Z. Mohamad, A. Fawzy, A.M. El-Taher, Creep behavior of near-peritectic Sn–5Sb solders containing small amount of Ag and Cu, Mater. Sci. Eng. A 528 (2011) 1055–1062.
[43] Y. Gao, Ch. Zou, B. Yang, Q. Zhai, J. Liu, E. Zhuravlev, Ch. Schick, Nanoparticles of SnAgCu lead-free solder alloy with an equivalent melting temperature of SnPb solder alloy, J. Alloys Compd. 484 (2009) 777–781.
[44] K. Suganuma, S.-J. Kim, K.-S. Kim, High temperature lead-free solders: properties and possibilities, J. Manag. 61 (1) (2009) 64–71.
[45] V. Chidambaram, J. Hattel, J. Hald, High-temperature lead-free solder alternatives, Microelectron. Eng. 88 (2011) 981–989.
[46] Y. Takaku, K. Makino, K. Watanabe, I. Ohnuma, R. Kainuma, Y. Yamada, Y. Yagi, I. Nakagawa, T. Atsumi, K. Ishida, Interfacial reaction between Zn-Al-based high temperature solders and Ni substrate, J. Electron. Mater. 38 (1) (2009) 54–60.
[47] N. Kang, H.S. Na, S.J. Kim, C.Y. Kang, Alloy design of Zn-Al-Cu solder for ultra high temperatures, J. Alloys Compd. 467 (1-2) (2009) 246–250.
[48] K. Suganuma, K.-S. Kim, S.-S. Kim, D.-S. Kim, M. Kang, S.-J. Kim, Joining characteristics of various high temperature lead-free interconnection materials, in: Proceedings of the Electronic Components and Technology Conference, 2009, pp. 1764–1768, art. no. 5074255.
[49] S. Kim, K.-S. Kim, S.-S. Kim, K. Suganuma, G. Izuta, Improving the reliability of Si die attachment with Zn-Sn-based high-temperature Pb-free solder using a TiN diffusion barrier, J. Electron. Mater. 38 (12) (2009) 2668–2675.
[50] S. Kim, K.-S. Kim, S.-S. Kim, K. Suganuma, Interfacial reaction and die attach properties of Zn-Sn high-temperature solders, J. Electron. Mater. 38 (2) (2009) 266–272.
[51] G. Wnuk, M. Zielińska, Microstructural and thermal analysis of Cu-Ni-Sn-Zn alloys by means of SEM and DSC techniques, Arch. Mater. Sci. Eng. 40 (1) (2009) 27–32.
[52] G. Meng, Z. Li, Shear strength and fracture surface analysis of BiAgNiCuGe/Cu joint, Trans. China Weld. Inst. 30 (10) (2009) 45–48 (in Chinese).

[53] Y. Shi, W. Fang, Z. Xia, Y. Lei, F. Guo, X. Li, Investigation of rare earth-doped BiAg high-temperature solders, J. Mater. Sci. Mater. Electron. 21 (9) (2009) 875–881.

[54] Y. Yan, L. Feng, X. Guo, K. Tang, K. Song, Effect of the content of Cu on solderability and mechanical properties of Bi5Sb solder alloy, Mater. Sci. Forum 610–613 (2009) 537–541.

[55] J.-M. Chuang, H.-Y. Song, Faceting behavior of primary Ag in Bi-Ag alloys for high temperature soldering applications, Mater. Trans. 50 (7) (2009) 1902–1904.

[56] C. Leinenbach, N. Weyrich, H.R. Elsener, G. Gamez, Al_2O_3-Al_2O_3 and Al_2O_3-Ti solder joints—influence of ceramic metallization and thermal pretreatment on joint properties, Int. J. Appl. Ceram. Technol. 9 (4) (2012) 751–763.

[57] C. Leinenbach, F. Valenza, D. Giuranno, H.R. Elsener, S. Jin, R. Novakovic, Wetting and soldering behaviour of eutectic Au-Ge alloy on Cu and Ni substrates, J. Electron. Mater. 40 (7) (2011) 1533–1541.

[58] V. Chidambaram, J. Hattel, J. Hald, Design of lead-free candidate alloys for high-temperature soldering based on Au-Sn alloys, Mater. Des. 31 (2010) 4638–4645.

[59] V. Chidambaram, J. Hald, J. Hattel, Development of Au-Ge based candidate alloys as an alternative to high-lead content solders, J. Alloys Compd. 490 (2010) 170–179.

[60] J.-Y. Wang, Ch-M. Chen, Y.-W. Yen, Interfacial reactions of high-Bi alloys on various substrates, J. Electron. Mater. 43 (2014) 155–165.

[61] Y. Takaku, I. Ohnuma, Y. Yamada, Y. Yagi, I. Nakagawa, T. Atsumi, M. Shirai, K. Ishida, A review of high temperature solders for power-semiconductor devices: Bi-base composite solder and Zn-Al base solder, J. ASTM Int. 8 (1) (2011), JAI103042.

[62] N. Saunders, A.P. Miodownik, CALPHAD (A Comprehensive Guide), Pergamon Press, Oxford, ISBN: 0-08-0421296, 1998.

[63] H.L. Lukas, S.G. Fries, B. Sundman, Computational Thermodynamics—The Calphad Method, Cambridge University Press, Cambridge, ISBN: 978-0-521-86811-2, 2007.

[64] L. Kaufman, H. Bernstein, Computer Calculation of Phase Diagrams, Academic press, New York, 1970.

[65] J.-O. Andersson, T. Helander, L. Höglund, P. Shi, B. Sundman, Thermo-Calc & DICTRA, computational tools for materials science, CALPHAD 26 (2002) 273–312.

[66] R.H. Davies, A.T. Dinsdale, J.A. Gisby, J.A.J. Robinson, S.M. Martin, MTDATA—thermodynamics and phase equilibrium software from the national physical laboratory, CALPHAD 26 (2002) 229–271.

[67] S.-L. Chen, S. Daniel, F. Zhang, Y.A. Chang, X.-Y. Yan, F.-Y. Xie, R. Schmid-Fetzer, W.A. Oates, The PANDAT software package and its applications, CALPHAD 26 (2002) 175–188.

[68] FactSage, CRCT—Centre de Recherche en Calcul Thermochimique/Centre for Research in Computational Thermochemistry and GTTTechnologies, (http://www.factsage.com/).

[69] National Institute of Standard and Technology, Maryland, USA. (http://www.metallurgy.nist.gov/phase/solder/solder.html).

[70] ADAMIS, Alloy Database for Micro-Solders, (http://www.materialsdesign.co.jp/adamis/adamisE.pdf).

[71] I. Ohnuma, K. Ishida, Z. Moser, W. Gasior, K. Bukat, J. Pstrus, R. Kisiel, J. Sitek, Pb-free solders: part II. Application of ADAMIS database in modeling of Sn-Ag-Cu alloys with Bi additions, J. Phase Equilib. 27 (2006) 245–254.

[72] TCSLD1, Thermo-Calc lead-free soders database, (http://www.thermocalc.com/media/5990/dbd_tcsld1.pdf).

[73] A.T. Dinsdale, A. Watson, A. Kroupa, J. Vrestal, A. Zemanova, J. Vizdal, Version 3.0 of the SOLDERS Database for Lead Free Solders, (http://resource.npl.co.uk/mtdata/soldersdatabase.htm).

[74] H. Ipser, COST Action 531—Lead-free Solder Materials, March (2007, http://www.cost.esf.org/domains_actions/mpns/Actions/Lead-free_Solder_Materials).
[75] A. Kroupa, COST Action MP0602—Advanced Solder Materials for High Temperature Application (HISOLD). (http://w3.cost.esf.org/index.php?id=248&action_number=MP0602).
[76] A. Kroupa, A.T. Dinsdale, A. Watson, J. Vrestal, J. Vizdal, A. Zemanova, The development of the COST 531 lead-free solders thermodynamic database, JOM 59 (2007) 20–25.
[77] A. Kroupa, Modelling of phase diagrams and thermodynamic properties using Calphad method—development of thermodynamic databases, Comput. Mater. Sci. 66 (2013) 3–13, http://dx.doi.org/10.1016/j.commatsci.2012.02.003.
[78] A. Dinsdale, A. Kroupa, A. Watson, J. Vrestal, A. Zemanova, P. Broz, Handbook of High-Temperature Lead-Free Solders: Atlas of Phase Diagrams, COST Office, Brussels, ISBN: 978-80-905363-1-9, 2012.
[79] A. Dinsdale, A. Watson, A. Kroupa, J. Vrestal, A. Zemanova, P. Broz, Phase diagrams and alloy development, in: K.N. Subramanian (Ed.), Lead-free Solders: Materials Reliability for Electronics, John Wiley & Sons Ltd., Chichester, UK, ISBN: 978-0-470-97182-6, 2012.
[80] C.P. Hunt, J. Nottay, A. Brewin, A.T. Dinsdale, NPL Report MATC(A) 83, Teddington, UK, April 2002.
[81] Z. Huang, P.P. Conway, Ch. Liu and R.C. Thomson, 2003 IEEE/CPMT/SEMI Int'l Electronics Manufacturing Technology Symposium.
[82] R. Schmid-Fetzer, D. Anderson, P.Y. Chevalier, L. Eleno, O. Fabrichnaya, U.R. Kattner, B. Sundman, C. Wang, A. Watson, L. Zabdyr, M. Zinkewich, Assessment techniques, database design and software facilities for thermodynamics and diffusion, CALPHAD 31 (2007) 38–52.
[83] JMatPro, Sente Software Ltd., Surrey Technology Centre, United Kingdom (http://www.sentesoftware.co.uk/jmatpro.aspx).
[84] N. Moelans, A phase-field model for multi-component and multi-phase systems, Arch. Metall. Mater. 53 (2008) 1149–1156.
[85] N. Moelans, Phase-field simulations of growth and coarsening in lead-free solder joints, in: A. Kroupa (Ed.), Handbook of High-Temperature Lead-Free Solders: Group Project Reports, COST Office, Brussels, 2012, pp. 154–160.
[86] J. Wang, S. Jin, C. Leinenbach, A. Jacob, Thermodynamic assessment of the Cu-Ge system, J. Alloys Compd. 504 (2010) 159–165.
[87] J. Wang, C. Leinenbach, M. Roth, Thermodynamic assessment of the Au-Ge-Sb system, J. Alloys Compd. 485 (2009) 577–582.
[88] P. Dietrich, Trends in automotive power semiconductor packaging, Microelectron. Reliab. 53 (2013) 1681–1686.
[89] S-Ch. Kim, Y.-H. Kim, Review paper: flip chip bonding with anisotropic conductive film (ACF) and nonconductive adhesive (NCA), Curr. Appl. Phys. 13 (2013) S14–S25.
[90] I. Mir, D. Kumar, Recent advances in isotropic conductive adhesives for electronics packaging applications, Int. J. Adhes. Adhes. 28 (2008) 362–371.
[91] G. Rabilloud, Adhesives for electronics, in: S. Ebnesajjad (Ed.), in: Handbook of Adhesives and Surface preparation, vol. 1, Elsevier Ltd., Oxford, UK/Burlington, USA, 2005, pp. 259–299.
[92] D. Wojceichowski, J. Vanfleteren, E. Reese, H.-W. Hagedorn, Electro-conductive adhesives for high density package and flip-chip interconnections, Microelectron. Reliab. 40 (7) (2000) 1215–1226.
[93] F. Tan, X. Qiao, J. Chen, H. Wang, Effects of coupling agents on the properties of epoxy-based electrically conductive adhesives, Int. J. Adhes. Adhes. 26 (6) (2006) 406–413.

[94] J.E. Morris, Isotropic conductive adhesives: future trends, possibilities and risks, Microelectron. Reliab. 47 (2-3) (2007) 328–330.
[95] M. Yamashita, K. Suganuma, Nano-mechanical electro-thermal probe array used for high-density storage based on NEMS technology, Microelectron. Reliab. 46 (5-6) (2006) 850–858.
[96] M. Erinc, M. van Dijk, V.G. Kouznetsova, Multiscale modeling of residual stresses in isotropic conductive adhesives with nano-particles, Comput. Mater. Sci. 66 (2013) 50–64.
[97] N. Lorenz, M.D. Smith, D.P. Hand, Wafer-level packaging of silicon to glass with a BCB intermediate layer using localised laser heating, Microelectron. Reliab. 51 (2011) 2257–2262.
[98] T.G. Lei, J.N. Calata, G.-Q. Lu, X. Chen, S. Luo, Low-temperature sintering of nanoscale silver paste for attaching large-area ($>100\,mm^2$) chips, IEEE Trans. Compon. Packag. Technol. 33 (1) (2010) 98–104.
[99] A. Hirose, H. Tatsumi, N. Takeda, Y. Akada, T. Ogura, E. Ide, T. Morita, A novel metal-to-metal bonding process through in-situ formation of Ag nanoparticles using Ag_2O microparticles, J. Phys. Conf. Ser. 165 (2009), art. no. 012045.
[100] J. Janczak-Rusch, Research approach towards knowledge based design of lead free solder joints, in: 3rd WUT-NIMS—Empa Workshop shop: New trends in Nanomaterials Design and Engineering, September 9-10, 2010, Dübendorf, Switzerland, 2010.
[101] J.N. Calata, T.G. Lei, G.-Q. Lu, Sintered nanosilver paste for high-temperature power semiconductor device attachment, Int. J. Mater. Prod. Technol. 34 (1-2) (2009) 95–110.
[102] L. Dupont, G. Coquery, K. Kriegel, A. Melkonyan, Accelerated active ageing test on SiC JFETs power module with silver joining technology for high temperature application, Microelectron. Reliab. 49 (9-11) (2009) 1375–1380.
[103] P.J. Wang, C.C. Lee, Silver joints between silicon chips and copper substrates made by direct bonding at low-temperature, IEEE Trans. Compon. Packag. Technol. 33 (1) (2010) 10–15.
[104] J.N. Calata, G.-Q. Lu, K. Ngo, L. Nguyen, Electromigration in sintered nanoscale silver films at elevated, J. Electron. Mater. 43 (1) (2014) 109–116.
[105] P. Quintero, P. McCluskey, B. Koene, Thermomechanical reliability of a silver nano-colloid die attach for high temperature applications, Microelectron. Reliab. 54 (1) (2014) 220–225.
[106] W. Sabbah, S. Azzopardi, C. Buttay, R. Meuret, E. Woirgard, Study of die attach technologies for high temperature power electronics: silver sintering and gold–germanium alloy, Microelectron. Reliab. 53 (2013) 1617–1621.
[107] T. Takahashi, S. Komatsu, T. Kono, Development of Ag_3Sn intermetallic compound joint for power semiconductor devices, Electrochem. Solid-State Lett. 12 (7) (2009) H263–H265.
[108] T. Takahashi, S. Komatsu, H. Nishikawa, T. Takemoto, Thin film joining for high-temperature performance of power semi-conductor devices, Microelectron. Reliab. 50 (2010) 220–227.
[109] A. Wierzbicka-Miernik, J. Wojewoda-Budka, L. Litynska-Dobrzynska, A. Kodentsov, P. Zieba, Morphology and chemical composition of Cu/Sn/Cu and Cu(5 at-%Ni)/Sn/Cu(5 at-%Ni) interconnections, Sci. Technol. Weld. Join. 17 (2012) 32–35.
[110] N.S. Bosco, F.W. Zok, Critical interlayer thickness for transient liquid phase bonding in the Cu–Sn system, Acta Mater. 52 (2004) 2965–2972.
[111] N.S. Bosco, F.W. Zok, Strength of joints produced by transient liquid phase bonding in the Cu–Sn system, Acta Mater. 53 (2005) 2019–2027.
[112] N. Weyrich, Ch. Leinenbach, Low temperature TLP bonding of Al_2O_3–ceramics using eutectic Au–(Ge, Si) alloys, J. Mater. Sci. 48 (2013) 7115–7124.
[113] E. Lugscheider, K. Bobzin, A. Erdle, Solder deposition for transient liquid phase (TLP)-bonding by MSIP-PVD-process, Surf. Coat. Technol. 174–175 (2003) 704–707.

[114] S.S. Sayyedain, H.R. Salimijazi, M.R. Toroghinejad, F. Karimzadeh, Microstructure and mechanical properties of transient liquid phase bonding of Al_2O_3 nanocomposite using copper interlayer, Mater. Des. 53 (2014) 275–282.

[115] A. Sharif, Ch.L. Gan, Z. Chen, Transient liquid phase Ag-based solder technology for high-temperature packaging applications, J. Alloys Compd. 550 (2013) 57–62, http://dx.doi.org/10.1016/j.jallcom.2013.10.204.

[116] M.S. Park, S.L. Gibbons, R. Arróyave, Phase-field simulations of intermetallic compound growth in Cu/Sn/Cu sandwich structure under transient liquid phase bonding conditions, Acta Mater. 60 (2012) 6278–6287.

[117] P.P. Conway, E.K.Y. Fu, K. Willaims, Precision high temperature lead-free solders interconnections by means of high-energy droplet depositiontechnique, CIRP Ann. Manuf. Technol. 51 (2002) 177–180.

[118] T. Osborn, N. Galiba, P.A. Kohl, Electroless copper deposition with PEG suppression for all-copper flip-chip connections, J. Electrochem. Soc. 156 (2009) D226–D230.

[119] Z. Li, M. Li, Y. Xiao, Ch. Wang, Ultrarapid formation of homogeneous Cu 1 6Sn5 and Cu_3Sn intermetallic compound joints at room temperature using ultrasonic waves, Ultrason. Sonochem. 1 (2013), http://dx.doi.org/10.1016/j.ultsonch.2013.09.020.

[120] J. Fjelstad, Solderless assembly of electronic products—a more reliable and more cost effective approach to electronics manufacturing? in: IEEE Vehicle Power and Propulsion Conference, VPPC'09, 2009, pp. 11–16, art. no. 5289876.

[121] A.C. Marques, J.-M. Cabrera, C. de Fraga Malfatti, Printed circuit boards: a review on the perspective of sustainability, J. Environ. Manag. 131 (2013) 298–306.

[122] W.-M. Liu, M.-X. Chen, S. Liu, Ceramic packaging by localized induction heating, Nanotechnol. Precis. Eng. 7 (2009) 365–369.

Chapter 6

High-Performance Steels for Sustainable Manufacturing of Vehicles

H. Mohrbacher
NiobelCon bvba, Schilde, Belgium

1 INTRODUCTION

No other industrial sector has pursued weight reduction as vigorously as the automotive industry. This has been motivated by the need to reduce fuel consumption and emissions as well as the requirement to improve crash safety, and these continue to be major technical targets. The total vehicle weight has an important impact in this respect as indicated in Figure 1. Indisputably, the fuel consumption and thus the CO_2 emissions decrease with reducing vehicle weight. Several studies have indicated that a weight reduction of 100kg can lead to savings in fuel consumption of 0.15-0.5 liters per 100km [1]. This corresponds to a reduction in CO_2 emissions of between 4 and 12g/km. The intensive use of lighter materials is effectively offering a significant weight reduction potential. The weight of state-of-the-art car bodies consists of up to 80% high-strength steel (HSS), reducing the weight by 40-100kg compared to a traditional car body, made primarily of mild steel [2–4]. Lighter vehicles not only reduce fuel consumption but also facilitate faster acceleration, shorter braking distance, and an overall better agility [5]. Weight reduction in commercial vehicles has the additional advantage of considerably increasing the payload while still meeting the legally specified maximum total vehicle weight limit. Consequently freight cost per unit weight is being reduced.

Much attention has been paid to the economic advantages that high-strength low-alloy steels have to offer. These advantages include lower structural weight, better economies during construction and transportation as a result of lower cost in handling lighter sections, fewer worker hours of welding, and lower electrode consumption as a result of reduced gauges. However, light weighting is not only a question of the strength level of the steel in question: The steel selection is determined or limited by the ease of forming, welding, and other fabrication procedures. Service conditions require that the candidate steel should possess

Green and Sustainable Manufacturing of Advanced Materials. http://dx.doi.org/10.1016/B978-0-12-411497-5.00006-0

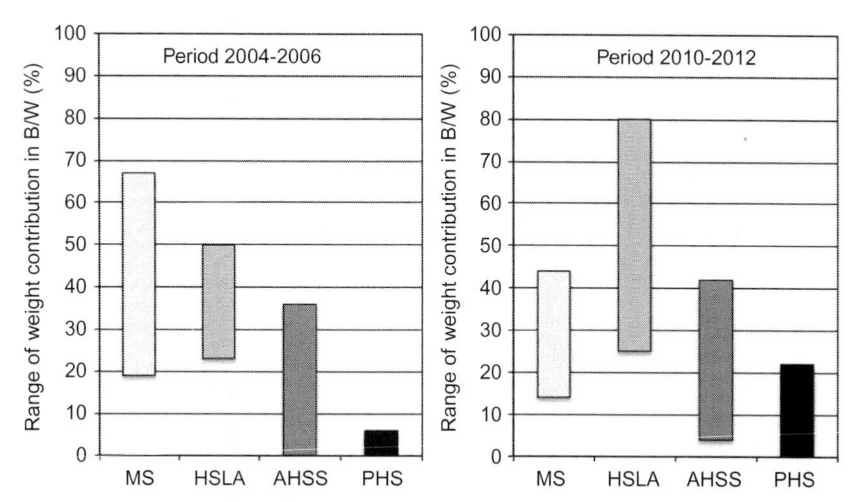

FIGURE 1 Evolution of steel utilization in passenger car bodies (min.-max. range of mild steel, high-strength low-alloy steel, advanced high-strength steel, press-hardening steel).

adequate durability under loading, impact resistance at the lowest temperatures anticipated in service, stiffness, and corrosion resistance.

Besides the pure technical benefits coming from material upgrading, the related cost is always a key concern severely influencing decision making in the automotive industry. This explains why HSS remains the preferred material for vehicle construction because its extensive use not only reduces the vehicle weight and increases the technical performance, but it also lowers the total fabrication cost. Table 1 compares current technologies for reducing fuel consumption and emissions in passenger vehicles. It is evident that intensive HSS utilization is the only technology that reduces weight and cost at the same time. The use of aluminum as a vehicle construction material severely drives up cost and is, despite its better weight-saving potential, often not affordable for car manufacturers.

The trend away from traditional mild steel (defined as having a yield strength of max. 200 MPa) toward HSS is clearly reflected when analyzing the evolution of car body design over the years (Figure 1). Within the period of the years 2004-2012 the share of mild steel in the body-in-white was reduced to a level of approximately 30%, coming down from values that were in the order of 70% earlier. The weight reduction achieved by this shift is in the range of 15-25%. In recent years the trend continues toward using progressively more ultra-high-strength steel with a strength level of up to 1800 MPa. Increasing the share of such ultra-high-strength steel to around 40% is estimated to bring down the weight of a steel car body to a similar level as could be achieved with a full aluminum car body, yet at much lower cost.

The efficiency of down gauging steel by increasing its strength depends to a significant degree on the part geometry as well as the acting load collective.

TABLE 1 Weight and Cost Effects of State-of-the-art Fuel-Saving Technologies in Passenger Vehicles

Technological solution	Weight effect (kg)	Fuel economy (%)	Cost impact
Light weighting (HSS)	−140	5	-€100
Light weighting (Al/Mg)	−280	10	€750-1250
Turbo-charged gasoline (down-sized)	−20	10	€200
Turbo-charged diesel engine (vs. gasoline)	+50	30	€1000
Advanced stop-start system	+5	5%	€200

Figure 2 indicates the weight-saving potential for various higher strength levels as compared to mild steel for numerous load cases and part geometries. It is evident that the highest efficiency is achieved for tensile loads acting within the sheet plane whereas for bending loads normal to the sheet plane the efficiency is lowest. Consequently down gauging on outer skin panels can achieve only relatively limited weight saving. This is one of the reasons why outer panels are rather likely to be manufactured from aluminum, particularly in premium class vehicles. On the contrary, open and closed profiles allow substantial down gauging and thus weight-saving potential by increasing the material strength as well as optimizing the force flow.

In commercial vehicle construction HSS has been established for a relatively long time, especially in the frame structure of trucks and trailers. The use of hot rolled steel with yield strength ranging from 235-355 MPa can be considered as being a standard solution. However, the trend in such frame applications now is to use a much higher strength steel with a yield strength reaching up to 700 MPa in advanced cases. Depending on the strength upgrade scenario, a weight reduction of 20-30% is feasible by reducing the gauge of the material correspondingly (Figure 2). This evidently enables several hundred kilograms of extra payload or reduced fuel consumption when the truck has to run unloaded. As in passenger cars, the cost surplus caused by the steel upgrade is overcompensated by the material saving so that the weight-optimized structure is effectively cheaper. Besides, the thinner gauged material brings about significant savings in the welding cost during vehicle manufacturing (Figure 3). Accordingly, upgrading a frame structure from 355 MPa to 700 MPa steel can reduce the total manufacturing cost by around 40%.

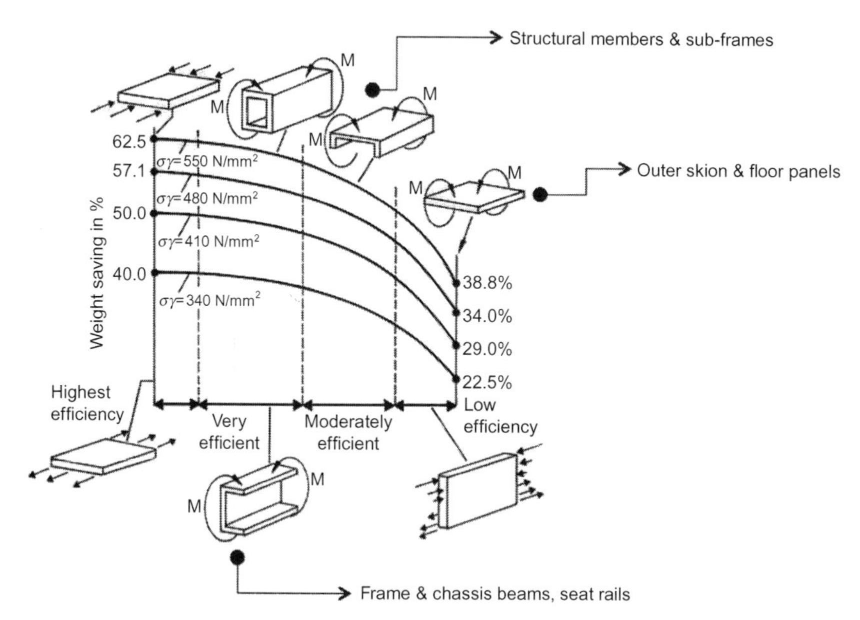

FIGURE 2 Weight-saving potential by substituting 200 MPa (YS) steel with high-strength steels for various loading conditions.

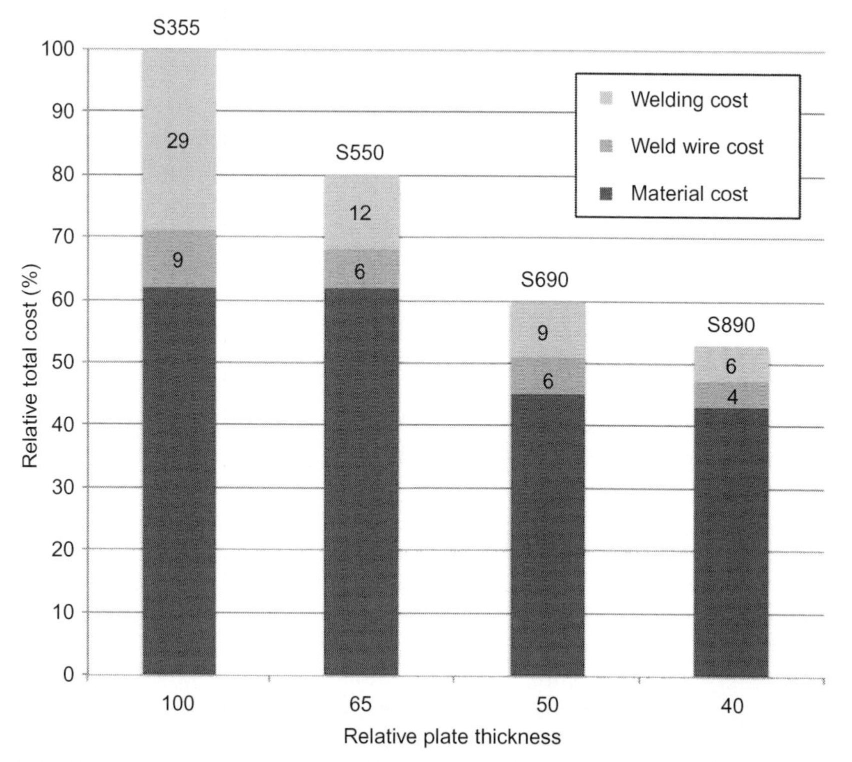

FIGURE 3 Relative cost savings of material and welding for replacing S355 with stronger HSLA grades.

2 MANUFACTURING IMPLICATIONS BY USING HSS

One of the major advantages of using HSS in vehicle manufacturing is that existing manufacturing technology and equipment can be further used to a large degree. This is in sharp contrast to alternative materials that usually require a completely new manufacturing approach and cause significant capital investments into novel manufacturing equipment. Nevertheless, also for HSS some specific upgrades have to be made and specific processing knowledge has to be gained. Typically a larger press force is required due to the higher strength, tooling materials have to be upgraded, and welding procedures have to be adapted. More recently the so-called hot-stamping technology has been widely introduced to the production of car body panels. This technology provides a solution to the classic conflict of high strength and good formability. With hot-stamping technology it became possible to produce steel components of the highest strength level (i.e., from 1500 MPa to 2000 MPa), and still maintaining a complex part shape. On a lower strength level, a large variety of steels with improved cold formability has been developed for use in conventional press stamping lines. Often total elongation of the steel being derived from a uniaxial tensile test is taken as a measure for its cold formability. However, the reality in press metal forming is more complex. Depending on the actual forming method and part shape four major forming modes can be distinguished as defined in Figure 4. The success of these forming modes depends on different material characteristics. The various material characteristics are related to the microstructure of the material, which again is determined by the alloy composition and the processing schedule in the mill. Typically it is difficult to have all forming relevant properties on a high level within one and the same steel grade.

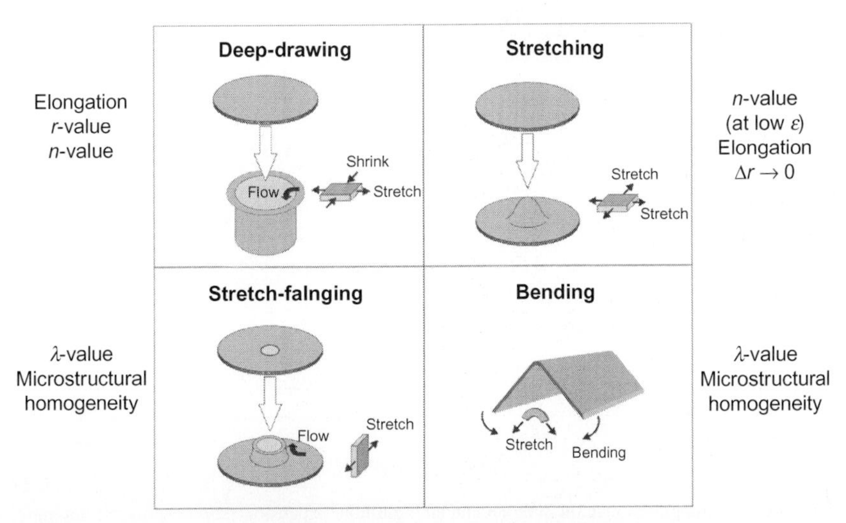

FIGURE 4 Major forming modes and relevant material characteristics.

Particularly higher strength steels having good elongation due to a multiphase microstructure tend to have lower hole expansion ratio (λ-value) (Figure 5). Part designers and stamping engineers have to consider this intrinsic contradiction when selecting a steel grade for a part. After they are formed, the individual stamped parts have to be assembled by suitable joining methods. In the majority of cases this is accomplished by welding techniques such as resistance spot welding, laser welding, and MAG welding. Accordingly, weldability of the material needs to be considered in terms of hardenability, cold cracking sensitivity, and heat-affected zone softening. Typical low-carbon steels ($C < 0.1$ wt%)

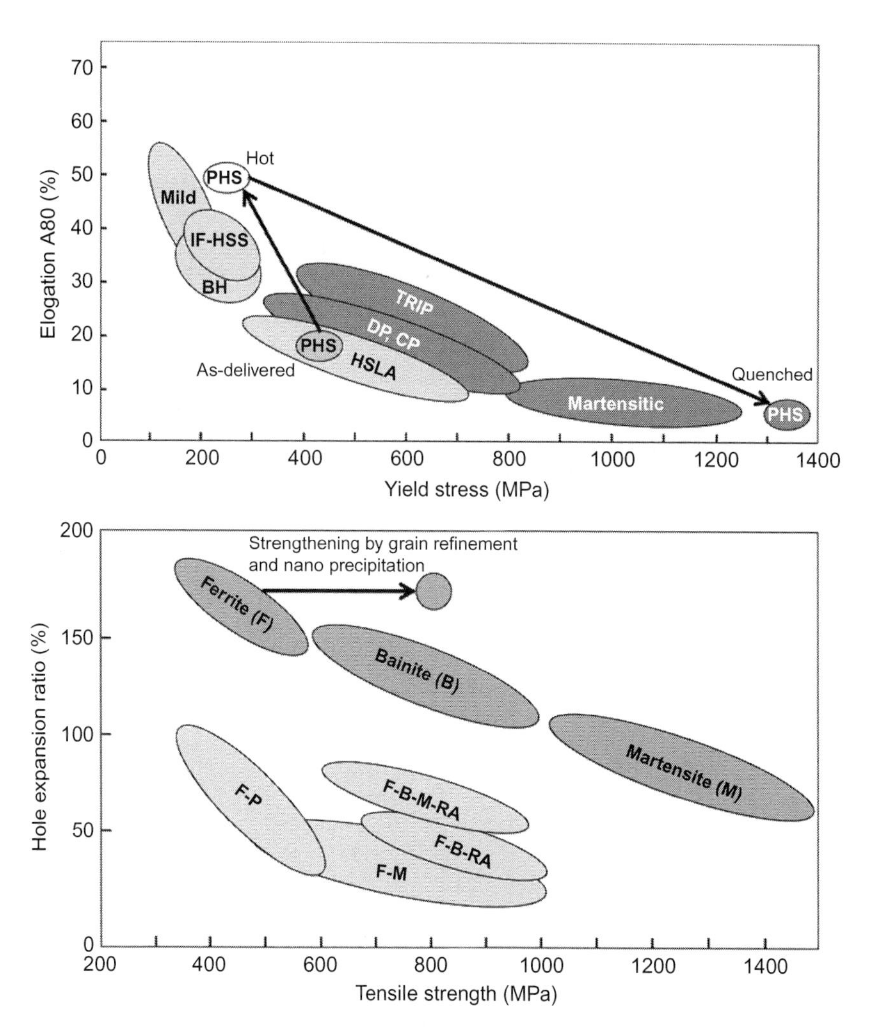

FIGURE 5 Strength-elongation diagram and strength-hole expansion ratio diagram for automotive steel grades (F: ferrite, P: pearlite, B: bainite, M: martensite, RA: retained austenite).

generally do not create a problem with regard to automotive welding procedures. However, higher carbon steels, especially steels with a strength level above 800 MPa, need to be qualified in this respect.

3 METALLURGICAL OPTIMIZATION TOWARD IMPROVED PROPERTIES OF AUTOMOTIVE STEEL

Detailed understanding of the manufacturing process in vehicle making allows optimizing steel grades for better performance. It has to be noted that these improvements are beyond the minimum-specified properties, and often they concern even nonspecified properties. Forthcoming improvements will increase the efficiency along the manufacturing chain resulting in less scrap as well as avoiding downtime and reworking efforts.

The metallurgical optimization typically involves microstructural fine-tuning that can be achieved by modifying the chemical composition and/or the thermomechanical processing in the steel mill. In this respect two specific alloying elements have proven to be particularly beneficial: niobium and molybdenum. These elements are added to steel in very small amounts, typically up to 1 kilogram of Nb (0.1 wt%) and up to 3 kilograms of Mo (0.3 wt%) per metric ton of steel, respectively. In the following, several examples will demonstrate this approach and highlight the achieved benefits.

3.1. Example 1 Extra High-Strength Hot-Rolled Steels for Chassis and Frame Parts

Chassis and frame components are typically subjected to high static and dynamic loads. Often such components involve intensive welding operations. Hence, a low carbon equivalent of such steel is mandatory for good weldability. Accordingly, high-strength low alloy (HSLA) steel is usually the preferred material. Specifications of HSLA steels typically limit the maximum carbon content to 0.1%.

HSLA steel for automotive applications is currently applied with yield strength of up to 700 MPa. Such steels typically have a ferritic-pearlitic microstructure for a yield strength level below 550 MPa. Above that level ferritic-bainitic of fully bainitic microstructures are becoming more applicable. Frame and chassis components often have only a moderate shape complexity. Hence the elongation of HSLA steel is usually sufficient to form such components. Yet, manufacturing of frame and chassis components typically involves forming methods such as profiling, press-brake bending, stretch flanging, as well as hole cutting and expansion. Consequently, the hole expansion ratio in such steels should be sufficiently high. From Figure 5 it is obvious that single-phase microstructures such as ferrite, bainite, and martensite are performing much better than microstructures containing phases of very different hardness such as ferritic-pearlitic or dual-phase steels. Furthermore, the condition of the sheet edge after mechanical cutting is of high importance. It is essential to avoid pre-damage at the edge after mechanical cutting [6]. The likelihood of edge damage is increased by the presence of hard phases such as martensite or pearlite in the microstructure. Highly localized deformation during mechanical cutting in combination with the largely different plasto-mechanical properties between the hard and soft phases leads to micro-crack formation along the edge. This initial damage can cause macroscopic cracking upon stretching the cut edge, for example, during a flanging operation. Laser cutting is a suitable alternative method to avoid damage.

With respect to avoiding or minimizing edge damage and related failure upon peripheral stretching, a ferritic or bainitic single-phase microstructure is preferable. In metallurgical terms the specific task is to adjust alloy composition and processing in the rolling mill to obtain the desired microstructure and mechanical properties. Strengthening in a polygonal ferritic microstructure has to be achieved by grain refinement and precipitation hardening. Both methods have a potential of increasing the yield strength by up to 300 MPa each on top of the base yield strength of around 200 MPa. Thus, by using both mechanisms to their full 800 MPa strength appears to be the feasible limit with this microstructure. The initial preference goes to grain refinement because it is the only strengthening mechanism that simultaneously increases toughness. Bainitic steel is made by combining a refined microstructure with dislocation strengthening, whereas precipitation strengthening is less prominent due to the specific temperature conditions after the rolling is finished. Particular alloying elements are crucial in the metallurgical concept of such optimized steels. Niobium is the strongest grain refining element known and is applied in the range of 0.03-0.1 mass percent [7]. It also has the potential for precipitation hardening by forming nano-sized NbC particles dispersed in the ferrite matrix [8]. Titanium can be used as an additional microalloying element for further increased precipitation strengthening. Molybdenum is very effective in avoiding the formation of pearlite and promotes the transformation into bainite [9]. Furthermore, molybdenum limits the coarsening of carbide precipitates formed by Nb or Ti [10]. The strengthening effect of such precipitates is best when their size is in the lower nanometer range [11].

The so-called nano Hi-Ten steel is a specific example for reverse metallurgical engineering with the target of designing steel that has very high strength and simultaneously an extremely high hole expansion ratio [10,11]. Such high hole expansion ratio necessitates a polygonal ferritic single-phase microstructure, which, however, has intrinsically a relatively low strength. To reach a target minimum tensile strength of 780 MPa, the ferritic microstructure must be severely grain refined and precipitation hardened. For this purpose Nb and Ti are being alloyed. Furthermore, any formation of pearlite must be avoided and precipitate size must be limited. Therefore low carbon content is chosen and molybdenum is added. The processing concept relies on the following steps, according to Figure 6:

- Austenite pancaking by Nb microalloying in combination with thermomechanical rolling (TMCP).
- Fast cooling after the last rolling pass into the range of ferrite formation providing additional grain refinement supported by the delay of austenite-to-ferrite transformation caused by solute Nb, Ti, and Mo.
- Quasi-isothermal holding in the ferrite range to obtain complete transformation and precipitation of the microalloying elements Ti and Nb.

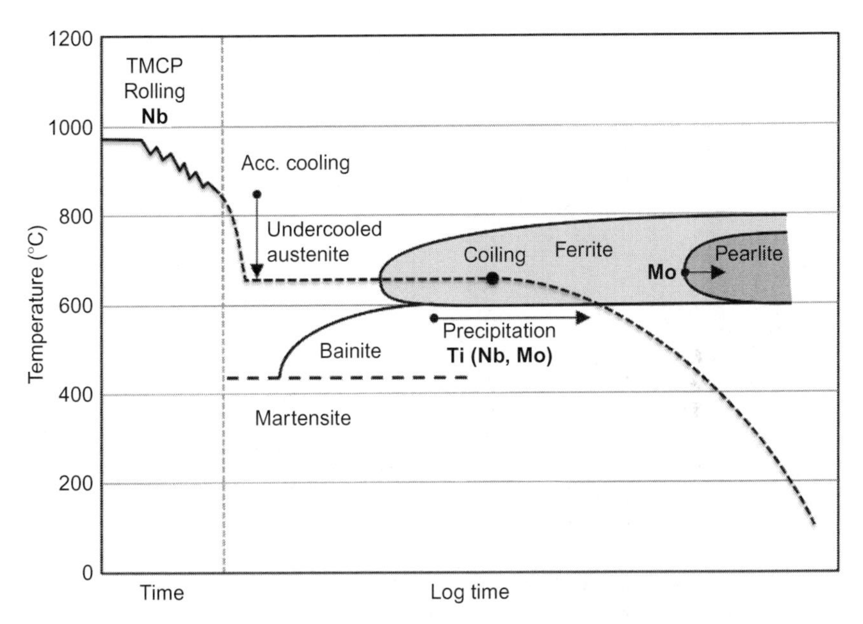

FIGURE 6 Processing concept for hot-rolled high-strength steel with grain-refined polygonal ferritic matrix and nano precipitate strengthening.

An alloy concept based only on Ti addition does not provide sufficient strength because TiC precipitates coarsen during prolonged holding time at high temperature as is the case during coiling. Combined alloying of Ti, Mo, and Nb significantly enhances the strength by grain refinement and precipitate size control even after extended holding at high temperature. Steel produced with an optimized combination of alloy and process was found to have a hole expansion ratio of more than 150% and total elongation of around 20%. However, the processing window to reach such properties is narrow and the producible sheet gauge is limited in thickness.

For the bulk of applications, being less demanding in terms of hole expansion ratio, it would be preferable to have a more robust process and metallurgical concept. A robust process design would lead to a low scattering of mechanical properties, which is beneficial with regard to avoiding springback after forming; springback can be the cause of expensive rework. The scattering of mechanical properties is related in a major way to temperature control on the runout table after rolling is finished. Temperature variations along the coil and also across the strip commonly occur. Hence a robust alloy design is needed that is insensitive to such temperature variations on the runout table. An alloy design employing competing metallurgical strengthening mechanisms is a suitable solution (Figure 7) [12]. Here the total strength is balanced by the degree of transformation and precipitation strengthening, respectively. Both effects depend on temperature, and yet they balance each other in magnitude so that the total strength being the sum of both effects remains nearly constant at sufficiently low coiling temperatures. For practical implementation of this strengthening strategy an alloy concept based

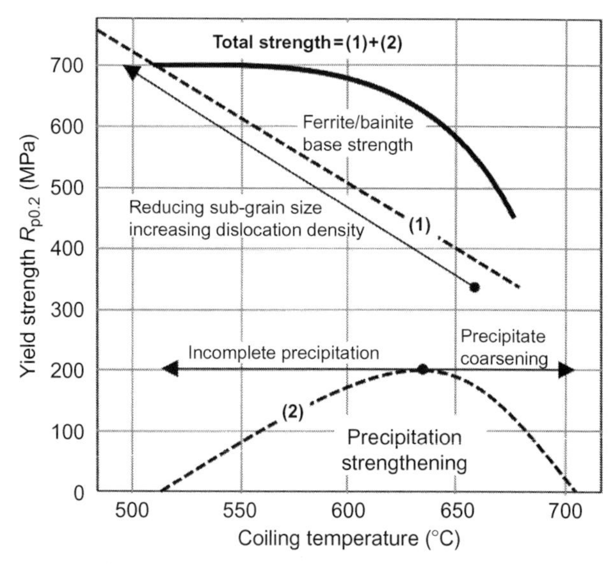

FIGURE 7 Concept of strength balancing between transformation and precipitation hardening as function of coiling temperature.

on low carbon content and a niobium addition in the range of 0.07-0.1% appears to be particularly suitable. Aided by a considerable amount of solute Nb, this alloy concept efficiently transforms into nonpolygonal ferritic or bainitic microstructure providing a high base strength. The solute amount of Nb is available for fine precipitation. Coiling temperature is preferably set to below 630 °C. Figure 8 depicts an example of such an alloy concept using a combination of 0.04%C-1.4%Mn-0.09%Nb [13]. Yield and tensile strength are very stable over a wide range of coiling temperatures from 610 °C down to 480 °C qualifying the steel for grade S550MC. Statistical evaluation of 700 coils of this material during an industrial production campaign confirmed very narrow scattering of yield and tensile strength (Figure 9). For comparison, a more traditional alloy concept based on a ferritic-pearlitic microstructure with Ti-precipitation strengthening exhibits a much wider scattering of both yield and tensile strength, indicating lower robustness against process variations. Owing to the absence of hard second-phase particles, the new alloy concept also reveals excellent bending properties as well as high-quality cutting edges. Steel processors perceived a significantly enhanced life of punching tools due to the absence of hard, abrasive particles such as pearlite. The hole expansion ratio of this steel is the range of 120% and elongation (A50) typically ranges from 20-25%.

Production of grade S700MC is possible by some modifications of the alloy concept still using the same strength-balancing strategy. This was achieved by increasing the carbon and manganese content as well as by adding molybdenum. Also this concept shows stable properties over coiling temperature, varying in the range of 580-500 °C. The microstructure is basically bainitic, containing a smaller fraction of martensite-austenite (MA) phase (Figure 8).

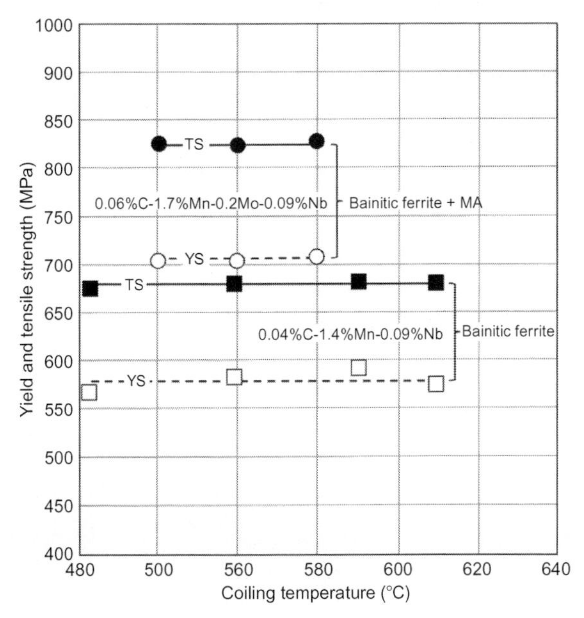

FIGURE 8 Strength properties of two alloy concepts as a function of coiling temperature and respective matrix microstructures.

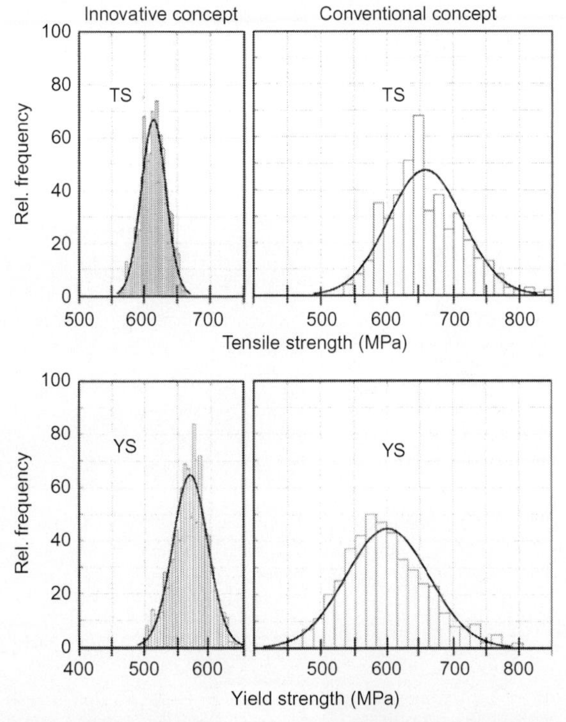

FIGURE 9 Statistical evaluation (based on 700 coils) of industrial production of grade S550MC for a conventional (0.09%C-0.7%Mn-0.13%Ti) and innovative (0.04%C-1.4%Mn-0.09%Nb) alloy concept.

3.2. Example 2 DP Steel with Improved Formability and Weldability

Dual-phase (DP) steels consist of a ferrite-martensite microstructure and have been developed to combine very high strength with increased elongation. Thus, DP steel allows forming rather complex part geometries that are not possible to manufacture with conventional HSLA steel of the same strength level. The volume fraction of hard martensite islands determines the strength of DP steel, whereas the ductile ferrite matrix provides good formability. The characteristic of as-delivered DP steel is a relatively low-yield strength and high initial work hardening resulting in a high *n*-value. The high n-value provides good protection against local thinning under the conditions of deep drawing and stretch forming.

Although DP steel allows forming of complex shapes, practical experience repeatedly reveals unexpected failure even when manufacturing quite simple

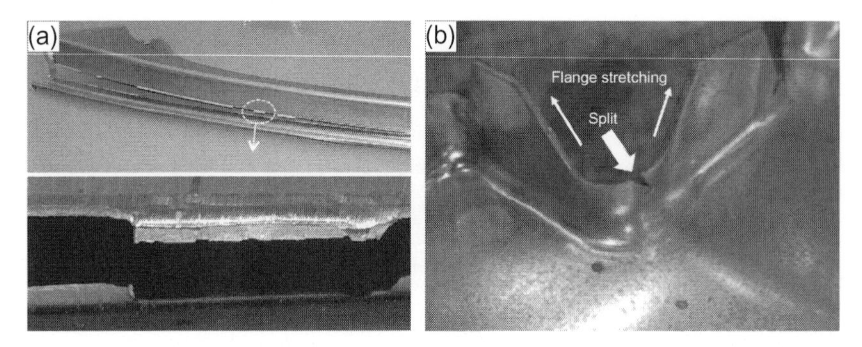

FIGURE 10 Shear fracture during die bending (a) and sheared edge fracture during stretch flanging (b).

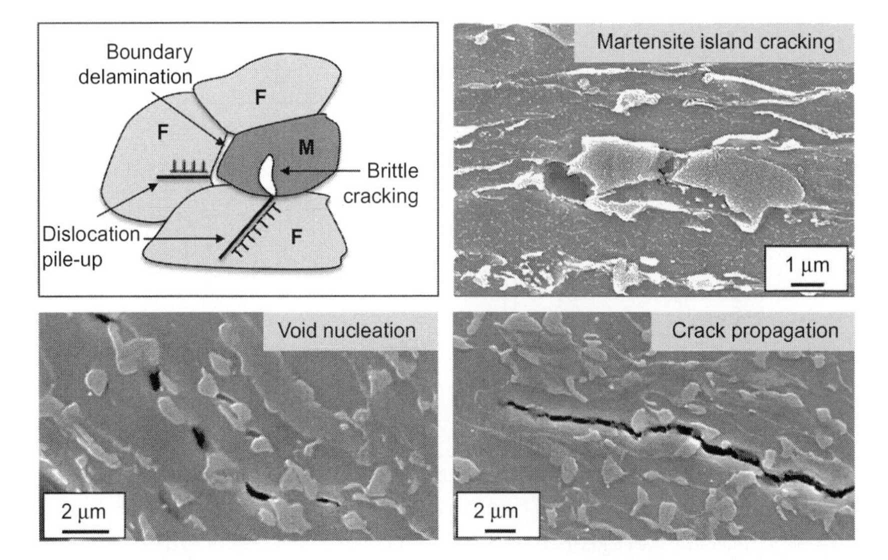

FIGURE 11 Principle and features of damage initiation and crack propagation in dual-phase steel.

geometries [14,15]. Shear fracture has regularly been observed in die bending operations (Figure 10a). Another common problem is sheared edge cracking during flanging operations (Figure 10b). Roll profiling leads to splitting in the outer bend when the bend radius is very tight causing significant downtime of production lines. All that these forming operations have in common is that strain is highly localized. Thus, severe dislocation pileup is occurring in individual ferrite grains, leading to delamination at the ferrite-martensite phase boundary or to cracking of martensite islands as demonstrated in Figure 11. Such in situ induced micro damage can grow into a propagating crack under applied stress and/or residual stress. The larger the size of an initial damage site the smaller is the critical stress required for crack propagation. A crack typically propagates along ferrite-martensite interface.

4 OPTIMIZING DUAL-PHASE MICROSTRUCTURE

Dislocation pileup causing micro damage under the condition of localized straining cannot be avoided in DP steel. Nevertheless measures can be taken to reduce the criticality of this phenomenon. Most important in this respect is refining the microstructure, that is, reducing the size of the ferrite and martensite grains. Consequently the size of initial damage is reduced due to the smaller phase boundary area raising the critical stress for crack propagation to a higher level. Furthermore, it is important to homogenize the microstructure, that is, to avoid the formation of martensite clusters as these provide an easy path for crack propagation. Both microstructural characteristics, a refined grain size as well as homogeneous phase distribution, can be achieved by Nb microalloying in combination with appropriate hot-rolling conditions. Application of this concept leads to a refined ferritic-pearlitic microstructure of the hot-rolled strip. The final cold-rolled annealed DP microstructure is in a significant way determined by the prior hot-strip grain size although some coarsening usually occurs during annealing [16]. A secondary effect of the refined hot-strip microstructure is an accelerated nucleation rate of newly formed phases during the intercritical annealing cycle since the new phases preferably nucleate on grain boundaries. In the refined hot-strip microstructure the total grain boundary area is significantly enhanced. As such, the statistical probability of having martensite clusters is reduced.

The effectiveness of microstructural refinement by Nb microalloying is demonstrated in Figure 12, taking DP780 as an example [17]. Standard and grain-refined DP steels are benchmarked in an instrumented three-point bending test. It is obvious that the grain-refined steel supports a higher bending force at an increased bending angle. The standard steel fails at a bending angle of around 90 degrees so that the production of typically hat-shaped profiles is critical. The grain-refined steel on the contrary offers a sufficient margin in critical bending angle, providing a clearly improved robustness for this forming process. Similarly positive results were obtained with DP980 material where grain refinement allows tighter bending radii in roll profiling (Figure 13). In this case

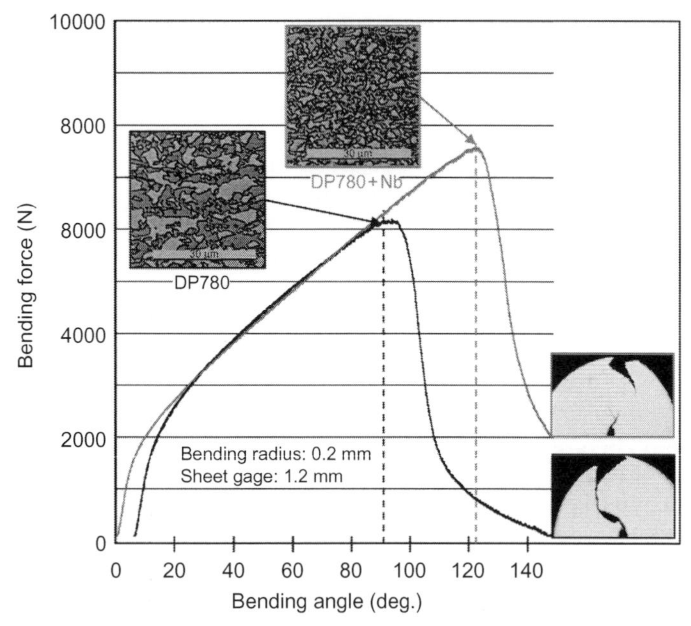

FIGURE 12 Effect of microstructural refinement (Nb addition) on the performance of DP780 steel under three-point bending conditions.

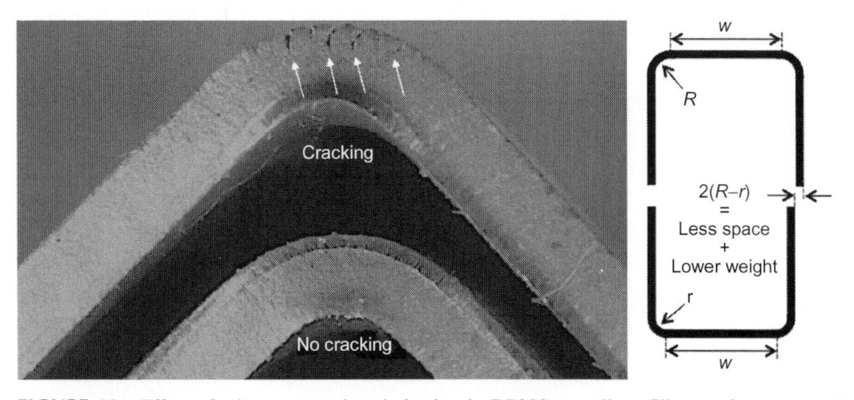

FIGURE 13 Effect of microstructural optimization in DP980 on roll-profiling performance and related technical benefits.

a profile with specified inner-width W has a more compact outer dimension facilitating packaging and reducing total weight.

Besides better bendability and hole expansion behavior, microstructural refinement also results in an increase of yield and tensile strength [18]. Thus, maintaining the targeted strength, the volume fraction of martensite phase can be reduced in grain-refined steel. The enhanced amount of ferrite results in an increased elongation and n-value, improving the steel's overall forming characteristic.

5 LOW-CARBON DP STEEL

Most of the currently produced standard DP steels use over-peritectic alloy designs, that is, the carbon content is higher than 0.14 wt%. The increased carbon content facilitates the formation of martensite needed for the DP microstructure; on the other hand, it is detrimental with regard to weldability [19]. Typical low-heat input welding processes used in automotive assembly processes such as resistance spot welding and laser-welding lead to substantial hardening in the heat-affected zone. Such welds can be brittle and sensitive to cold cracking if they are not post-weld heat treated. For that reason there is growing interest on the manufacturing side to reduce the carbon content in DP steel to below 0.1 mass percent.

However, producing DP steel of increased strength level such as DP780 or DP980 with reduced carbon content is challenging, particularly when using a hot-dip galvanizing process. The enrichment of austenite with carbon is relatively low in such an alloy concept and, hence, the risk of partial bainite formation during the holding period at around zinc bath temperature is increased (Figure 14). The presence of bainite in the microstructure reduces the tensile strength and raises the yield strength as compared to ideal ferritic-martensitic DP steel, thus deteriorating the key characteristics of low yield-to-tensile ratio and high strain hardening coefficient (n-value). Avoiding bainite formation during the isothermal holding phase requires delaying sufficiently the bainite phase field. This can be practically achieved by increasing the amount of alloying elements such as Mn, Cr, and Mo.

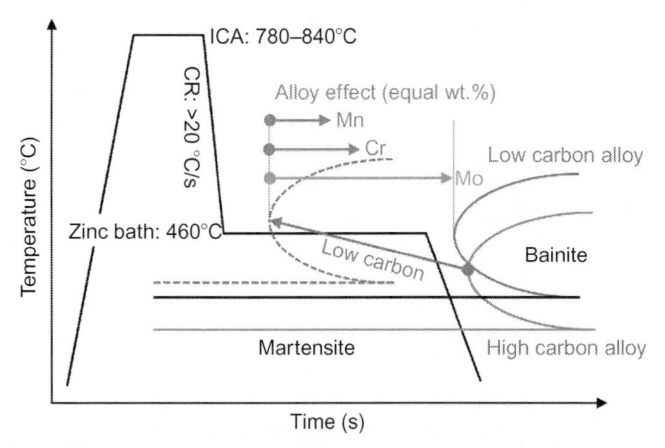

FIGURE 14 Alloying strategy for avoiding bainite formation in low-carbon DP steel during a hot dip galvanizing cycle.

The addition of molybdenum to low-carbon DP steel most efficiently delays the formation of bainite (Figure 14) because its effect is approximately 2.6 and 1.3 times larger than that of Mn and Cr, respectively [20]. Furthermore, the solid solution hardening potential of Mo is lower than that of Mn. Accordingly, ferrite remains relatively softer resulting in a lower yield-to-tensile ratio. Mo-based alloy concepts are therefore most appropriate for producing low-carbon DP grades of 780 MPa strength or higher. Producing low-carbon grades of 980 MPa strength boron microalloying in addition to Mo alloying may become necessary.

Reducing the overall carbon content in DP steel to below 0.1 wt% has a consequence that the average carbon enrichment in austenite during intercritical annealing is lower. The resulting hardness of the martensite formed after fast cooling is thereby reduced. In that case, a higher martensite share is needed to achieve target strength so that total elongation will be reduced due to less ferrite phase. On the other hand, the reduced hardness difference between ferrite and martensite has a very beneficial effect on the hole expansion ratio as shown in Figure 15. It has further been found that DP980 based on a low-carbon chemistry shows a reduced sensitivity to notch damage relative to the higher carbon versions [21]. The low-carbon DP980 fracture strain could tolerate a higher local fracture strain despite its lower elongation as compared to the higher carbon versions.

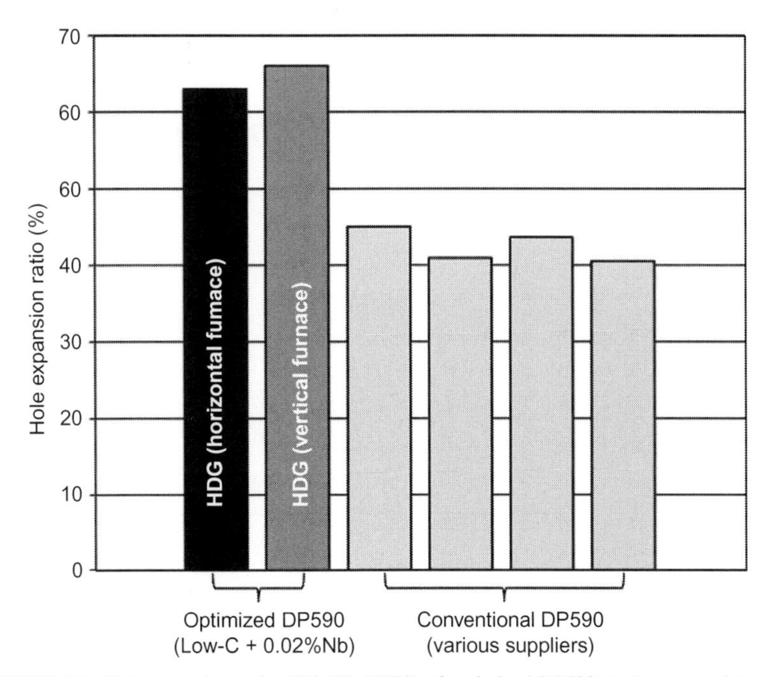

FIGURE 15 Hole expansion ratio (ISO-TS 16630) of optimized DP590 steel compared to conventional DP590 grades.

Example 3 Press-Hardening Steel for Highly Loaded Crash Parts

Press-hardening steel, often also designated as "hot-stamping" steel, has seen a remarkably increasing market in automotive body construction over the recent years. For the majority of applications steel with a tensile strength of 1500 MPa (22MnB5) is used saving significant weight in the car body. Contrary to cold-forming steels, press-hardening steel is heated to a temperature of around 950 °C before press forming. Accordingly, the microstructure is austenitic providing low yield strength and high elongation during the forming stage. The quick removal of heat when closing the forming die results in a quenching effect, transforming the austenitic microstructure into a martensitic one. Martensite is the strongest microstructure of steel but also the most brittle. By applying press hardening technology a number of conflicting issues inherent to sheet metal forming of other HSSs could be solved. Hot stamping provides the following advantages:

- Good formability at low stamping force.
- Low wear of forming tools.
- Reduced elastic springback resulting in high shape accuracy.

Extensive use of press-hardening steel—carmakers consider 45% of the total structural weight to be an upper limit—could bring the weight of a steel body structure to the same low level as would be achieved with a full aluminum body.

The alloy concept of 22MnB5 was originally designed for providing a cost-efficient abrasion-resistant steel but was not used specifically for automotive applications. Hot-stamping technology was first applied in the 1980s using grade 22MnB5 for producing automotive safety components such as door anti-intrusion beams and bumper beams with a tensile strength level of 1500 MPa. Contrary to the rapid development of a press hardening process technology, the typically applied steel grade 22MnB5 (especially since the year 2000) has remained unchanged since the 1980s. On the materials side, efforts were mainly focused on suitable surface coatings that prevent oxidation in the furnace and provide corrosion protection during service. The suitability of the metallurgical concept of the base steel with respect to the load cases prevailing in automotive applications has never been questioned in principle. A first conceptual approach to microstructural optimization of press-hardening steel based on analogies with martensitic steels used outside the automotive industry was defined in [22].

Press hardening technology as currently being practiced in the majority of applications produces full hard (i.e., nontempered, martensitic microstructure). This microstructure provides the highest possible strength for a given carbon content and simultaneously the lowest elongation. Although this low elongation is not relevant for the forming process some carmakers have been demanding an increased minimum elongation in the press-hardened component. This was perceived to be beneficial with regard to energy dissipation during a crash impact. An increased elongation can be achieved by tempering the press-hardened component (known as "tailored tempering") or by adjusting a small fraction of softer phases such as ferrite, bainite, or austenite in the martensitic matrix (known as "tailored quenching"). However, achieving a significant increase in elongation by either approach inevitably results in a severe loss of strength.

Yet, it is questionable whether elongation, being a property derived from a low-speed uniaxial tensile test, is really relevant with regard to the cracking behavior under a high-speed impact load. In this context toughness is a much more appropriate material property. Test procedures measuring toughness reveal two relevant characteristics: the upper shelf energy (USE) and the ductile-to-brittle transition temperature (DBTT). USE indicates the potential of the microstructure to arrest a propagating crack and should be as high as possible.

In full hard martensitic steel improvements to toughness can be made by minimizing the effects that lead to embrittlement. This particularly addresses the occupation of grain boundaries by impurities or precipitates as well as hydrogen embrittlement. With regard to the actual performance of a press-hardened component, specific details of the microstructure are important. USE in martensitic steel achieved under ductile failure mode is naturally lower than in softer ferritic or bainitic steels. However, it has been established over decades that microstructural refinement is always beneficial to toughness in any type of steel. More precisely, microstructural refinement increases the USE and simultaneously lowers DBTT.

6 REQUIREMENTS TO IMPROVED PRESS-HARDENING STEEL

Components made from press-hardening steel are typically located in areas where high-impact loads occur during a crash, such as A-pillar, B-pillar, bumper beam, door anti-intrusion beam, and the like. The primary requirement is that these components minimize intrusion and redistribute the crash load into areas where the kinetic energy can be dissipated by plastic deformation without threatening vehicle occupants. Such a passive safety concept demands that press-hardening steel should have the following general features:

- Sufficient toughness to avoid cracking or even rupture during crash over the entire operating temperature range.
- Ability to allow certain amount of localized deformation (bendability) during crash in order to absorb crash energy.
- High resistance to hydrogen-induced cracking.

These specific characteristics were not considered in the initial alloy design of 22MnB5. Therefore it has to be investigated how these specific properties can be optimized by metallurgical modification and particularly how toughness of the present alloying concept can be improved. The target of the improvement is primarily to decrease DBTT to lower temperature for the safe crash performance in cold climate regions and to increase the USE as schematically shown in Figure 16. Lowering the carbon content would be very helpful in this respect. However, in press-hardening steel this is not an option because strength in fully martensitic steel is almost entirely determined by the carbon content. On the other hand, further strength increase from what is currently 1500MPa toward 1800 or 2000MPa in new press-hardening steel grades is only possible by raising the carbon content from currently 0.22% towards 0.28 and 0.34%, respectively. This means that toughness properties become necessarily worse in such stronger grades. Consequently, countermeasures regaining toughness become even more relevant in such steels. As an example, a recently introduced 1800MPa press-hardening grade used by Mazda

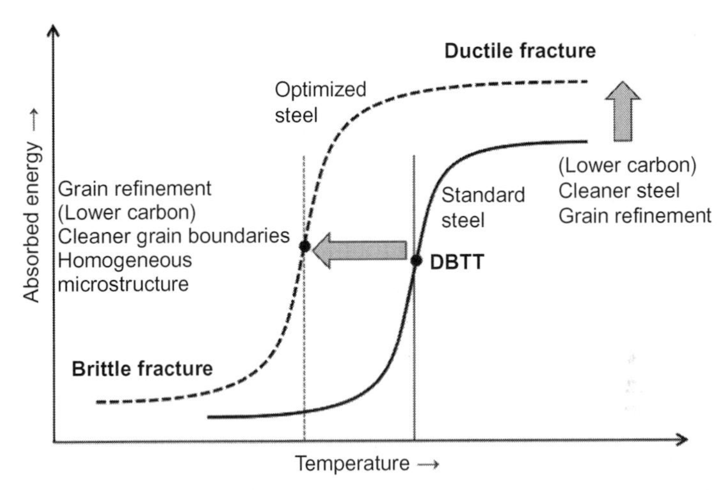

FIGURE 16 Concept of optimizing toughness by lowering the ductile-to-brittle transition temperature (DBTT) and influencing factors to be considered.

was required to have the same toughness performance as the traditional 1500 MPa grade. This was successfully achieved by grain refining the new 1800 MPa grade using Nb microalloying at an addition of >0.06% [23].

Bending is a form of highly localized deformation providing micro-cracks, which can grow into macroscopic fracture if the prevailing stress level is above a critical value. Due to the very high-yield strength of martensitic steel the stress level can indeed attain such high values. Fracture mechanics theory describes the relationship between critical stress level and the initial crack size to be inversely proportional (similar to the Hall-Petch relationship). Assuming that the initial defect size is related to the microstructural scale ("effective grain size"), microstructural refinement is beneficial for raising crack propagation stress and lowering DBTT according to Equations (1) and (2), respectively [24]. The constant K in these equations reflects relevant material properties.

$$\sigma_{\text{fracture}} = K_{\text{fracture}} \frac{1}{\sqrt{d_{\text{eff}}}} \tag{1}$$

$$T_{\text{DBT}} = T_0 + K_{\text{DBT}} \frac{1}{\sqrt{d_{\text{eff}}}} \tag{2}$$

Accordingly, the means of optimizing press-hardening steel for obtaining better crash relevant performance are refining the relevant effective grain size and reducing the chance of micro-crack initiation.

Car components subjected to crash impact often experience sharp bending or buckling deformation. Accordingly, despite the thin material gauge the outer fiber in the bending zone is subjected to highly localized deformation and simultaneously very high tensile stress. A specific three-point bending test has been designed to evaluate the maximum bending angle of the material without obvious cracking as specified by VDA238-100. European OEMs require press-hardening steel to achieve a minimum bending angle that is typically in the order of 60 degrees. A low critical bending angle indicates that the steel easily forms micro-cracks, which readily propagate. A high bending angle indicates that either crack initiation is retarded and/or the critical stress level for crack propagation is enhanced. Which of the two phenomena is more relevant can be derived from a detailed analysis of the force-displacement curve as well as from acoustic emission measurements made during the bending test.

To improve the bending performance, tempering treatment after press hardening was sometimes considered as a possible solution. Considerable research work on tempering press-hardened components revealed that both toughness and bending angle could be improved to some extent. For instance, by tempering quenched 22MnB5 at 500 °C the bending displacement in the three-point bending test could be increased by 53%. However, this treatment compromised the maximum bending force by 24% due to a decrease of tensile strength from 1500 to 1020 MPa (Figure 17) [25]. An additional problem related to tempering is the so-called tempering embrittlement typically occurring in the temperature range of 300-500 °C. Besides, integrating a tempering treatment into the press hardening process flow would increase the cost for the press-hardened components and potentially reduce productivity. Accordingly, car makers wanted the new alloying concepts for press-hardening steels to be developed having the full strength level but inherently better toughness and bendability without additional tempering treatment [25].

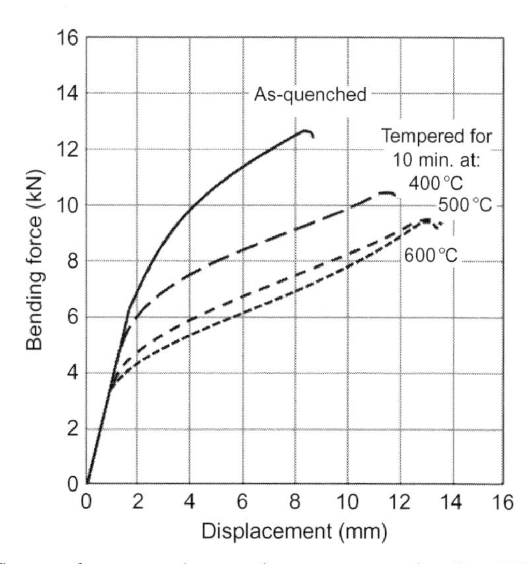

FIGURE 17 Influence of post-quench tempering treatment on bending (VDA238-100) performance of press-hardening steel 22MnB5.

7 IMPROVED ALLOY DESIGN FOR PRESS-HARDENING STEEL

Considering the fracture mechanisms in martensitic steels with regard to crack initiation and propagation three strategies have been worked out to optimize conventional alloy design based on 22MnB5:

1. Grain refinement by Nb microalloying.
2. Removal of B and Ti to avoid hard TiN particles and complex grain boundary precipitates.
3. Strengthening austenite grain boundary by Mo addition to impede intergranular fracture.

Alloying concepts related to these strategies have been explained in detail in [26–28]. In the following, the metallurgical effects of Nb microalloying on the improvement of toughness and bendability (three-point bending test) in press-hardening steel will be demonstrated.

8 TOUGHNESS IMPROVEMENT BY Nb MICROALLOYING

The relationship between the microstructural features of martensite and its failure behavior have been extensively researched and comprehensively summarized by Morris et al. [24]. The microstructure of the different forms of martensite is quite complex. Generally, however, one can distinguish laths, blocks, and packets that develop as a substructure within the prior austenite grain at the moment of austenite-to-martensite transformation. It has been clearly established that any substructural feature in martensite cannot be larger than the prior austenite grain size (PAGS) (Figure 18) [29,30]. Therefore the PAGS is the key characteristic to be considered when aiming to refine the microstructure of martensite. The beneficial effect of microstructural refinement on the toughness of martensitic steels has been verified by several studies. Reducing the PAGS increases toughness especially when the packet size is kept below 20 μm as is shown in [31].

Considering the entire processing sequence of press-hardening steel, the following steps have an influence on the PAGS:

- Reheating of the slab—austenite grain growth.
- Roughing rolling in the hot strip mill—austenite grain homogenization and moderate refinement.
- Finish hot rolling—austenite grain refinement depending on rolling schedule (current practice gives little refinement).
- Annealing after cold rolling—recrystallization and ferrite grain growth (severity increases with temperature and duration).
- Reheating before hot stamping induces austenite grain coarsening (severity increases with temperature and duration).

In the aforementioned process, the step that finishes hot rolling is the one that allows the largest grain size reduction provided the rolling conditions are

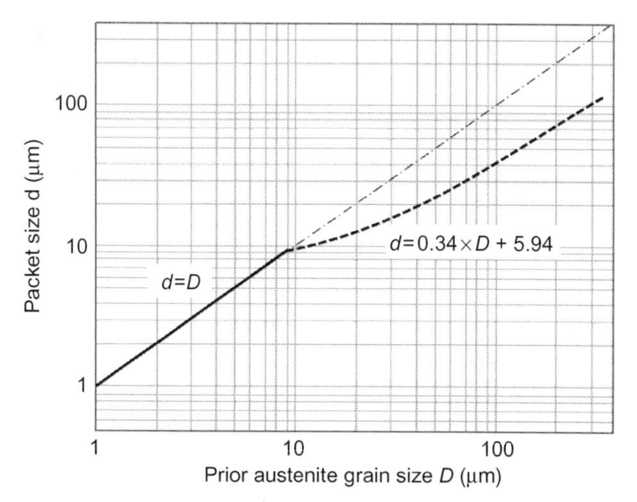

FIGURE 18 Relationship between PAGS and martensite packet size.

adequate. In all subsequent steps grain coarsening can only be obstructed, but further grain refinement can actually not be achieved. Current hot rolling practice in the production of press-hardening steel is typically applying an uncontrolled rolling schedule, which focuses on high productivity but not on austenite conditioning, resulting in grain refinement. The means of inducing grain refinement during finish rolling are well established in the production of HSLA steels and require a reduced finish entry temperature in combination with Nb microalloying that effectively prevents recrystallization of austenite [7]. Increasing the addition of Nb raises the temperature of recrystallization delay and accordingly allows higher finish entry temperature. By adding around 0.06%Nb a good compromise between productivity (finish rolling run-in temperature) and the degree of grain refinement can be achieved. Co-addition of Mo further delays recrystallization and thus supports the effect of Nb, providing a fine-grained microstructure in hot-strip, which will automatically create a finer microstructure in the cold-rolled strip provided the process conditions are the same for cold rolling, annealing, and press hardening.

9 MICROSTRUCTURAL CONTROL AND ROBUSTNESS IN PRESS-HARDENING STEEL

Nb precipitates formed during hot rolling as well as in situ formed precipitates during the reheating process before hot stamping are stable and have the capability of pinning the austenite grain boundary preventing grain coarsening. The reheating temperature of 22MnB5 for the press hardening process is usually 950 °C and the quenching temperature (after furnace transfer and forming) should be 880 ± 10 °C according to EN 10083-3. This is intended to prevent the grain size of prior austenite becoming coarser than ASTM 5. The relatively high reheating temperature will inevitably cause austenite grain growth during the reheating process and consequently lead to a coarser martensite substructure after quenching. Previous and current results demonstrate that Nb microalloying by around 0.05% can effectively obstruct grain coarsening even at reheating temperatures in the order of 1000 °C. Unfavorable conditions promoting austenite grain coarsening can particularly occur in the case of a process disturbance, for example, when the press line is stopped and material flow has to be stopped, thus residing longer in the furnace.

By increasing the Nb content in press-hardening steel the grain size of prior austenite decreases (Figure 19) and the Charpy energy (USE) increases (Figure 20). There are also indications that the DBTT can be lowered by in the order of 30 °C as a result of PAGS refinement. The toughness-enhancing effect of PAGS refinement is applicable at any strength (hardness) level.

Figures 21 and 22 demonstrate the effect of Nb microalloying on the PAGS for a wide range of furnace conditions. It is evident that Nb microalloyed press-hardening steel shows finer PAGS under any of the simulated furnace conditions. In the case of severe overheating (1000 °C), austenite grains in conventional 22MnB5 grow to nearly double the size than in the Nb-added variant. In another case of extra-long furnace residence (60 min.) conventional 22MnB5 again exhibits about double the austenite grain size as compared to the Nb-added variant. Thus, Nb microalloying clearly reduces the microstructural sensitivity of press-hardening steel to variations in the reheating process. Consequently, all

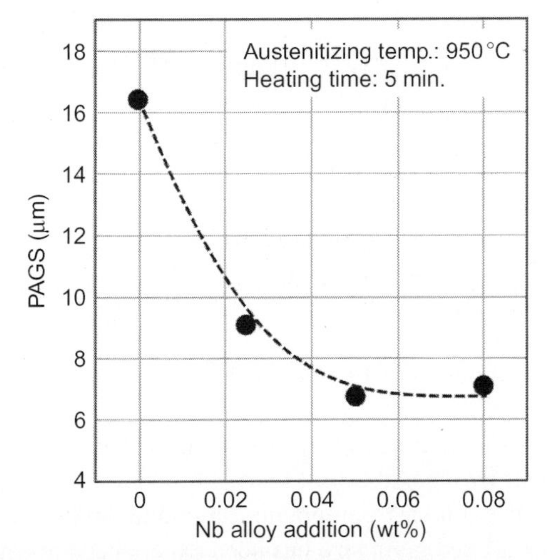

FIGURE 19 Influence of Nb content on the prior austenite grain size of press-hardening steel treated under conventional process conditions (alloy base: 22MnB5).

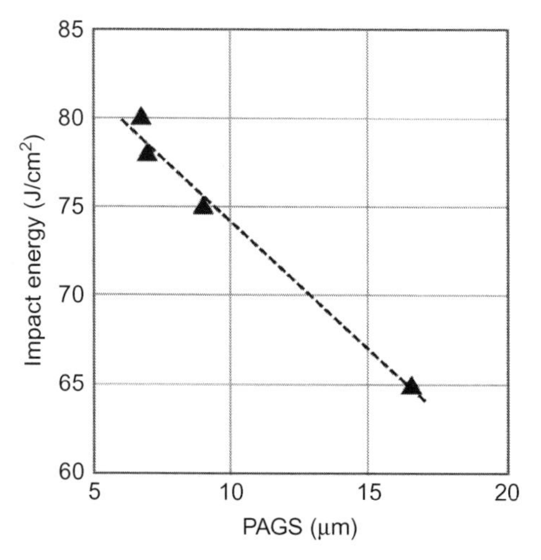

FIGURE 20 Influence of prior austenite grain size on impact energy (modified Charpy test) of press-hardened steel treated under conventional process conditions (alloy base: 22MnB5).

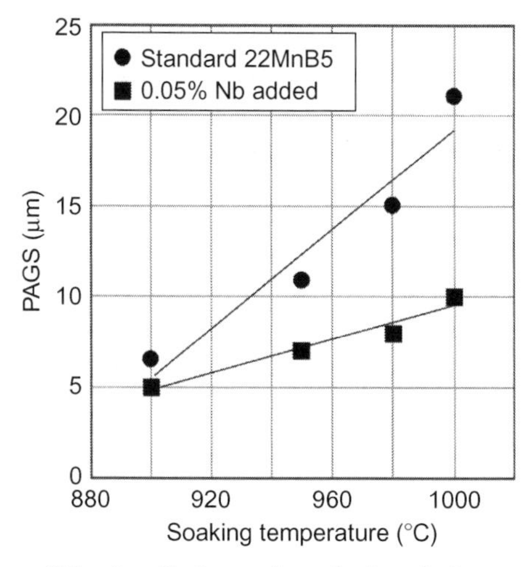

FIGURE 21 Influence of Nb microalloying on the grain size of prior austenite after reheating at different soaking temperatures.

properties depending on microstructure such as toughness become more robust by Nb microalloying.

Besides the PAGS being an average value, uniformity of the austenite grain size becomes better by Nb addition (Figure 23). Conventional 22MnB5 shows increasingly mixed grain size comprising large and small grains as the temperature increases. Mixed grain size has been experienced to cause component distortion after the quenching process due to different transformation kinetics of large and small grains. This simultaneously indicates the presence of

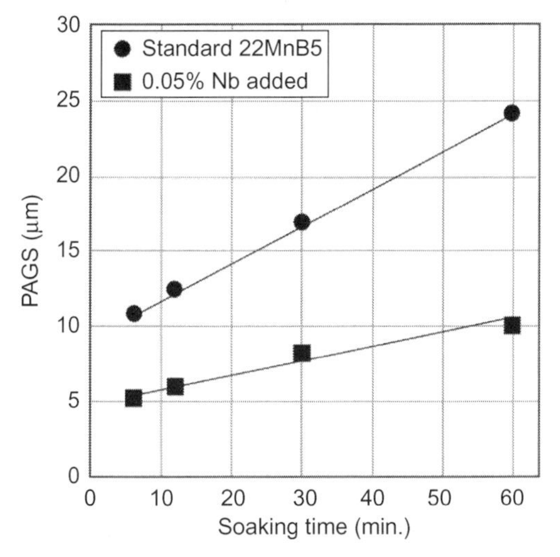

FIGURE 22 Influence of Nb microalloying on the grain size of prior austenite at different soaking times.

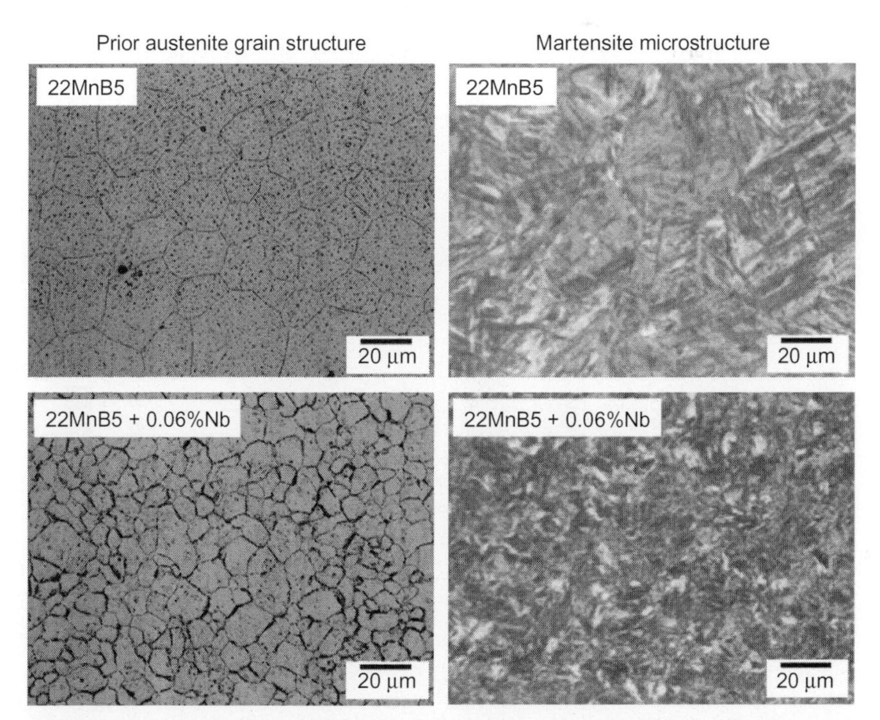

FIGURE 23 Effect of Nb microalloying on prior austenite grain structure and martensite microstructure in press-hardening steel 22MnB5.

considerable residual stresses that can later add to service stress and particularly can play a negative role with regard to delayed cracking. Mixed grain size also has a negative effect on the fatigue behavior.

Generally it was observed that the prior austenite grain size in final components made from standard press-hardening steel (22MnB5) is often in the range of ASTM 5-6. By niobium microalloying the grain size refines to typically ASTM 7-9 under conventional hot rolling conditions, that is, uncontrolled rolling. When applying a TMCP rolling schedule PAGS of ASTM 10 or finer have been shown feasible in recent trials producing press-hardened components of an Nb microalloyed variant of 22MnB5.

10 BENDABILITY IMPROVEMENT BY Nb MICROALLOYING

The positive effect of Nb microalloying and the accompanying PAGS refinement on the critical bending angle could be demonstrated for numerous press-hardening steel grades. Several improvement programs were performed at different steel producers on the 1500 MPa grade (22MnB5) and consistently revealed a continuous increase of critical bending angle. An example is shown in Figure 24a where the critical bending angle in conventional 22MnB5 grade averaged to only around 47 degrees. An addition of 0.05%Nb to that steel improved the critical bending angle to approximately 90 degrees. Higher addition of Nb did not result in a further improvement of the critical bending angle. Lower Nb addition showed a smaller improvement. In a 1900 MPa press-hardening grade (34MnB5) the improvement of the critical bending angle was on the order of 20% for an Nb addition of around

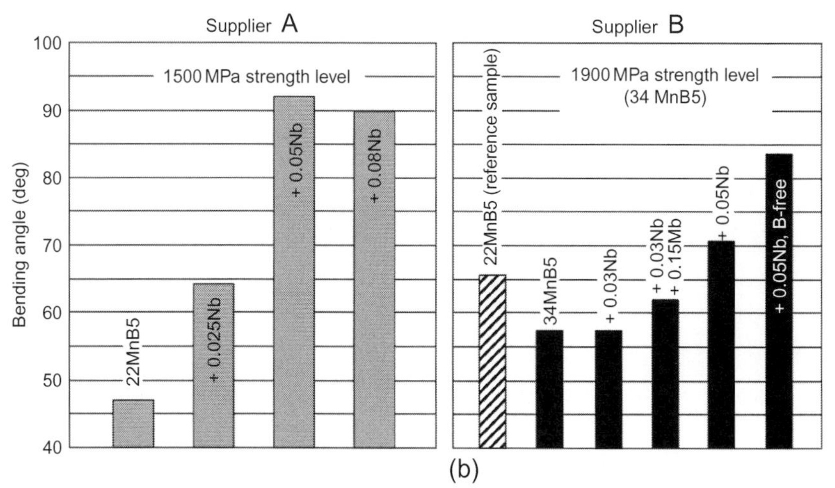

FIGURE 24 Effect of microalloy variants on the critical bending angle (VDA238-100) of 1500 MPa and 1900 MPa press-hardening steel grades.

0.05% [32]. Remarkably, the 0.05%Nb microalloyed 1900 MPa grade showed a better critical bending angle than a standard (i.e., Nb-free) 1500 MPa grade of the same steel producer (Figure 24b). Another interesting observation was that removal of boron in the 0.05%Nb added grade led to a further significant increase of the critical bending angle. Consequently, further steps in the materials improvement program will focus on the possibility of removing boron.

11 CONCLUSIONS

Weight reduction delivers an important contribution with regard to reducing the fuel consumption and thus the CO_2 emission of vehicles. Weight reduction can be achieved by using lighter components that are made from either low-density materials or stronger materials with accordingly thinner gauge. However, with generally increasing efficiency of powertrain technology and the forthcoming reduction of fuel consumption during the use-phase of vehicles, emission contributions originating form vehicle production and recycling become relatively more important.

Considering available materials for vehicle construction, steel has by far the lowest primary production CO_2 emissions on an equal part weight basis. Yet, alternative materials having significantly lower density enable weight reductions of between 40 to 60% reducing emissions during vehicle life. This has to be balanced against the higher emission set free during primary production of these materials. The intensive use of advanced HSS allows weight reduction of around 25%.

The production of steel is highly efficient, approaching the thermodynamic minimum energy consumption in state-of-the-art mills. Thus, further improvements in this respect will have marginal impact on the primary production CO_2 emissions. On the other hand, production of HSS does not significantly increase primary production CO_2 emissions. The addition of key alloying elements such as niobium and molybdenum providing high strength in steel is very small. Hence, there is little impact on steel's primary production CO_2 balance originating from mining and refining of these alloying elements.

Optimized HSSs reveal a larger processing window and higher robustness during production and manufacturing. This results in a more efficient use of the material with simultaneously lower scrap rates, thus reducing CO_2 emissions during the manufacturing phase. The alloying elements of niobium and molybdenum are added either singly or in combination to make a significant contribution to achieve this optimization.

REFERENCES

[1] Institute for Energy and Environmental Research, Heidelberg, Germany: www.ifeu.de.
[2] Automotive Circle International, Euro Car Body Conference Series (2008-2012), Vincentz Network GmbH, Hanover, Germany.
[3] B. Lüdke, M. Pfestorf, Functional Design of a Lightweight Body-in-White. in: S. Hashimoto, S. Jansto, H. Mohrbacher, F. Siciliano (Eds.), International Symposium on Niobium Microalloyed Sheet Steel for Automotive Application, TMS, Warrendale, 2006, p. 27.

[4] ULSAB AVC Study, International Iron & Steel Institute.
[5] T. Hallfeldt, Possibilities and Challenges Using Advanced High Strength Steel Sheets for Automotive Applications. in: S. Hashimoto, S. Jansto, H. Mohrbacher, F. Siciliano (Eds.), International Symposium on Niobium Microalloyed Sheet Steel for Automotive Application, TMS, Warrendale, 2006, p. 117.
[6] H. Mohrbacher, Reverse metallurgical engineering towards sustainable manufacturing of vehicles using Nb and Mo alloyed high performance steels. Adv. Manuf. 1 (1) (2013) 28–41.
[7] L. Cuddy, J. Raley, Austenite grain coarsening in microalloyed steels. J. Met. Trans. A14 (10) (1983) 1989.
[8] J.M. Gray, R.B.G. Yeo, Columbium Carbonitride Precipitation in Low-Alloy Steels with Particular Emphasis on 'Precipitation-Row' Formation. Trans. ASM 61 (1968) 255–269.
[9] S. Wang, P. Kao, The effect of alloying elements on the structure and mechanical properties of ultra low carbon bainitic steels. J. Mater. Sci. 28 (1993) 5196.
[10] Y. Funakawa, K. Seto, Stabilization In Strength Of Hot-rolled Sheet Steel. Tetsu to Hagane 93 (2007) 49.
[11] Y. Funakawa, et al., Development of High Strength Hot-rolled Sheet Steel Consisting of Ferrite and Nanometer-sized Carbides. ISIJ Int. 44 (2004) 1945.
[12] A.P. Coldren, R. Cryderman, M. Semchyshen, Strength and impact properties of low-carbon structural steels containing molybdenum. in: Proc. of Symp. on Steel Strengthening Mechanisms, Climax Molybdenum Company, Zürich, 1969, p. 17.
[13] W. Haensch, C. Klinkenberg, Low-carbon Niobium-alloyed high strength steel for automotive strip. in: Proc. 2nd Int. Conf. On Thermomechanical Rolling, Liège, 2004, p. 155.
[14] J. Fekete, International Symposium on Niobium Microalloyed Sheet Steel for Automotive Application, in: S. Hashimoto, S. Jansto, H. Mohrbacher, F. Siciliano (Eds.), Manufacturing Challenges in Stamping and Fabrication of Components from Advanced High Strength Steel. TMS, Warrendale, 2006, p. 107.
[15] T. Hebesberger, A. Pichler, H. Pauli, S. Ritsche, Dual-Phase and Complex-Phase Steels: AHSS for Wide Range of Applications. in: Proc. of Steels in Cars and Trucks, Verlag Stahleisen, 2008, p. 456.
[16] Y. Granbom, Influence of niobium and coiling temperature on the mechanical properties of a cold rolled dual phase steel. Rev. Métall. 4 (2007) 191.
[17] P. Larour, H. Pauli, T. Kurz, T. Hebesberger, "Influence of post uniform tensile and bending properties on the crash behaviour of AHSS and press-hardening steel grades", in: IDDRG 2010 Biennial Conf. (Graz, Austria), 2010.
[18] M. Calcagnotto, D. Ponge, D. Raabe, Effect of Grain Refinement to 1 μm on Deformation and Fracture Mechanisms in Ferrite/Martensite Dual-Phase Steels. in: Proc. Of Super High Strength Steel Conf., Verona, Italy, 2010, paper 44.
[19] K. Takagi, T. Yoshida, A. Sato, Material Application Development for Weight Reduction of Body-in-White. in: S. Hashimoto, S. Jansto, H. Mohrbacher, F. Siciliano (Eds.), International Symposium on Niobium Microalloyed Sheet Steel for Automotive Application, TMS, Warrendale, 2006, p. 81.
[20] T. Irie, S. Satoh, K. Hashiguchi, I. Takahashi, O. Hashimoto, Characteristics of Formable Cold Rolled High Strength Steel Sheets for Automotive Use. Kawasaki Steel Technical Report, No. 2, 14 (1981).
[21] J. Dykeman, S. Malcolm, G. Huang, H. Zhu, N. Ramisetti, B. Yan, J. Chintamani, Characterization of Edge Fracture in Various Types of Advanced High Strength Steel. SAE 2011 World Congress & Exhibition, 2011: paper 2011-01-1058.

[22] H. Mohrbacher, "Delayed Cracking in Ultra-high Strength Automotive Steels: Damage Mechanisms and Remedies by Microstructural Engineering", in: Proc. of Materials Science and Technology (MS&T), 2008, p. 1744.

[23] T. Nishibata, T. Suzuki, The Nature and Consequences of Coherent Transformations in Steel. CAMP-ISIJ 21 (2008) 598.

[24] J.W. Morris, C.S. Lee, Z. Guo, The Effect of Alloing Element on Properties of TS1.8GPa Grade Hot-Stamped Parts. ISIJ Int. 43 (3) (2003) 410.

[25] M. Glatzer, Tagungsband zum 4. Einfluss unterschiedlicher Wärmebehandlungsrouten auf die Robustheit der mechanischen Eigenschaften des Stahls 22MnB5. Erlanger Workshop Warmblechumformung (2009).

[26] J. Bian, "Progress in press hardening technology and innovative alloying designs", in: Proc. of the 1th Taiwan Symposium on Fundamentals and Applications of Mo and Nb alloying in high performance steels, Taipei, 2011, TMS in print.

[27] J. Bian, H. Mohrbacher, "Novel alloying design for press hardening steels with better crash Performance", in: Proc. of AIST International Symposium on New Developments in Advanced High Strength Sheet Steels, Colorado USA, 2013, p. 251.

[28] J. Bian, H. Mohrbacher, "Process technology and steel development in press hardening", BAC2013 Baosteel Shanghai, China.

[29] T. Maki, I. Tamura, Morphology and Substructure of Lath Martensite in Steels. Tetsu to Hagane 67 (1981) 852–866.

[30] S. Morito, H. Saito, T. Ogawa, T. Furuhara, T. Maki, Effect of Austenite Grain Size on the Morphology and Crystallography of Lath Martensite in Low Carbon Steels. ISIJ Int. 45 (1) (2005) 91–94.

[31] C. Wang, M. Wang, J. Shi, W. Hui, H. Dong, Effect of Microstructure Refinement on the Strength and Toughness of Low Alloy Martensitic Steel. J. Mater. Sci. Technol. 23 (5) (2007) 659.

[32] J. Mura, T. Gerber, S. Sikora, F.-J. Lenze, MBW1900 mit Mikrolegierung zur Optimierung der technologischen Eigenschaften nach dem Presshärten. Tagungsband zum 7. Erlanger Workshop Warmblechumformung (2012).

Chapter 7

Advanced Steel Alloys for Sustainable Power Generation

H. Mohrbacher
NiobelCon bvba, Schilde, Belgium

1 INTRODUCTION

Electrical energy production is estimated to grow by an average of 2.4% per annum. About 65% of the world's electric power currently comes from fossil fuel combustion [1], but stringent demands imposed by governments and international agreements are pushing CO_2-free sources of power. Figure 1 indicates the total life cycle of CO_2 emission per kWh of electrical energy for several primary energy sources. Nuclear, wind, and hydroelectric power have the lowest CO_2 emissions by far. Because the latter two also produce no hazardous waste, they can be considered truly "green and clean" power generation. Yet even for these sources, improvements in life-cycle efficiency, installed cost and cost of ownership, along with the associated reduction in CO_2 emissions, are possible. These improvements are inherently connected to the material performance of key components, usually made from iron and steel alloys. Performance criteria include strength, toughness, fatigue, and wear resistance, formability, and weldability. Often, niobium and molybdenum alloying, combined with appropriate thermomechanical processing, provides superior material performance. Examples of the beneficial sustainability effect by alloy optimization on components for thermal, wind, and hydropower generation equipment will be presented.

2 STEEL IN PLANTS FOR THERMAL POWER GENERATION

The efficiency of a thermal power plant represents the ratio between the net produced electrical energy and the thermal energy originally contained in the fossil fuel. This ratio can be improved by increasing steam temperature and pressure. Power plants are defined as subcritical, supercritical (SC), and ultra-supercritical (USC) depending on the temperature and pressure prevailing in the steam circuit. On a global basis, the average efficiency of coal-fired power plants is currently below 35%. State-of-the-art SC and USC power plants achieve efficiencies of approximately 45%, and future 700°C technology is

Green and Sustainable Manufacturing of Advanced Materials. http://dx.doi.org/10.1016/B978-0-12-411497-5.00007-2

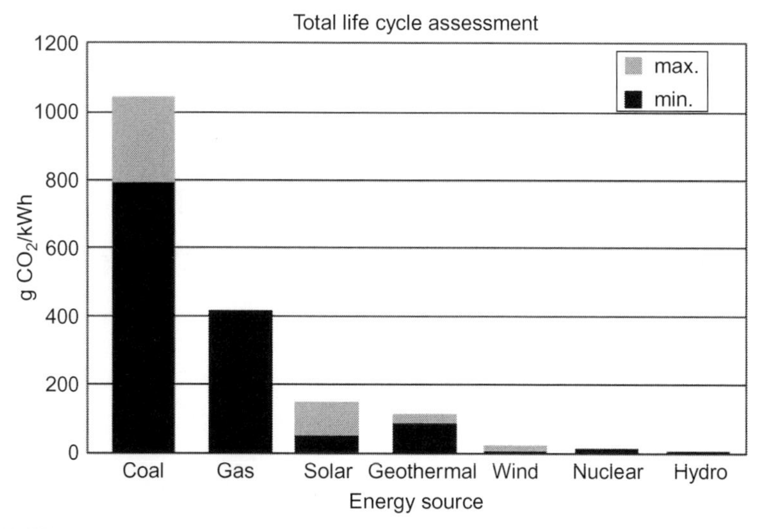

FIGURE 1 Environmental impact of electric power plants for various primary energy sources. *(World Energy Council (2007).)*

predicted to further increase efficiency to more than 50%. Efficiency improvement to the coal-fired power plants by raising steam parameters is a fast and reasonable step toward the necessary reduction of CO_2 emissions. Raising efficiency from 35% to 50% is estimated to lower CO_2 emissions by around 40% (Figure 2) [2].

Higher efficiency is also a good foundation for the upcoming application of carbon capture storage (CCS) technologies to achieve near zero emissions.

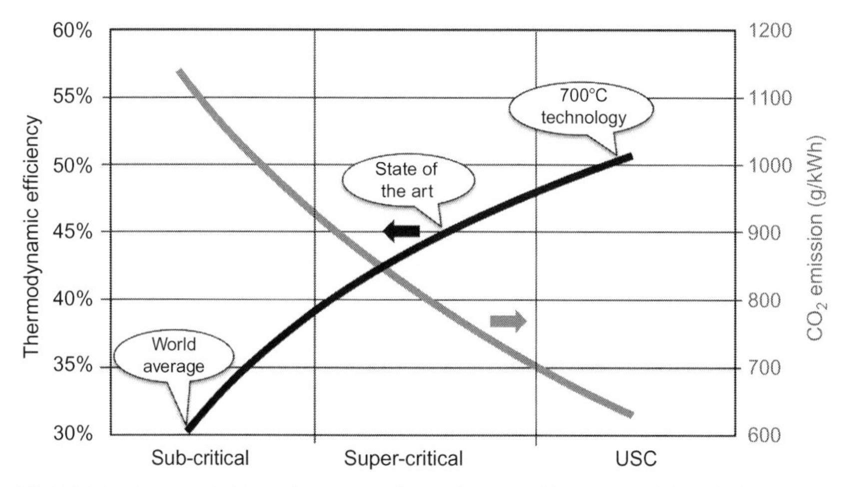

FIGURE 2 Impact of boiler technology on thermodynamic efficiency and CO_2 emission in thermal power plants. *(Vallourec.)*

The efficiency reduction by 8-12 percentage points caused by CCS technology can at least partly be compensated by the higher efficiency of a 700°C power plant. Improved materials with adequate strength and resistance to oxidation and corrosion are required to withstand the significantly higher steam parameters in such a power plant.

If a material is exposed to high temperature and pressure for extended periods, it will creep, an initially slow deformation, which accelerates over time and eventually destroys the component. A material is rated "creep resistant" at a given temperature if it can withstand a stress of 100 MPa for at least 100,000 h without fracture (Figure 3). In addition to high stress and temperature, the operating environment in the boiler is corrosive. Steam oxidation on the inside and flue gas corrosion on the outside of boiler tubes becomes more aggressive as temperature increases. Thus the high-temperature corrosion resistance of the material is also a crucial property. Therefore, the development of thermal power plant technology toward larger units and higher efficiencies directly depend on the development of suitable creep-resistant materials [3]. Creep strength has been improved successively by introducing new alloy concepts and suitable microstructures. Figure 4 indicates the typical material mix for increasing steam parameters, that is, power plant efficiency [4]. Austenitic steel grades are standardly used when steam temperature reaches between 600 and 700°C. For even higher temperatures, the creep strength of these materials is lower than 100 MPa at 100,000 h. Thus, the wall thicknesses of high-temperature components to be used in a 700°C power plant would be too high for fabrication. Therefore, the application of materials with higher creep strength such as nickel-base alloys becomes mandatory. The investment cost of a 700°C power plant is expected to be about 15 to 25% higher than those of a conventional power plant with the same power output.

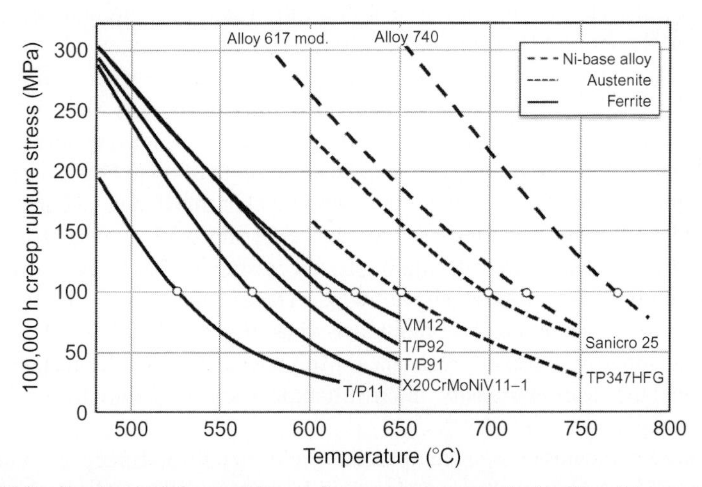

FIGURE 3 One hundred thousand hours of creep-rupture stress in function of temperature for commercially available heat-resistant alloys.

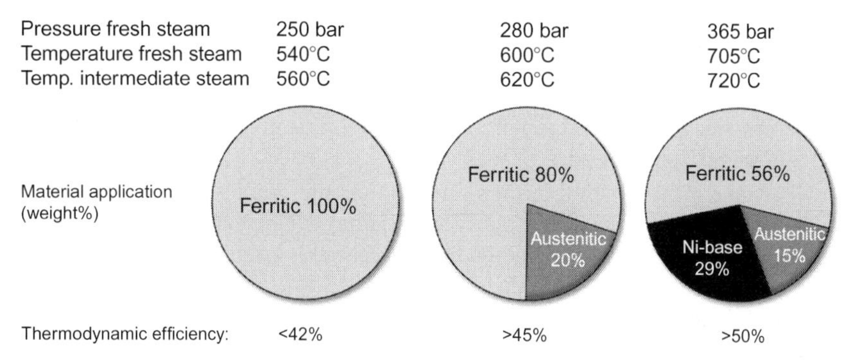

FIGURE 4 Applicable boiler material classes and achievable thermodynamic efficiency for selected steam parameters.

To reduce investment costs, the use of nickel-base alloys should be limited to the unavoidable minimum. Substitution with the martensitic steels VM12 or T92, for example, has promising potential. The price difference between these martensitic materials and a nickel-base alloy can reach several tens of thousands of dollars per ton. Considering that several hundred tons of material is required for constructing a boiler, the savings would be huge. Accordingly, considerable effort focuses on optimizing creep and corrosion resistance of ferritic and austenitic steels.

3 FERRITIC STEELS WITH HIGH CREEP RESISTANCE

Molybdenum is an established alloying element in such materials as it improves resistance to creep and high-temperature corrosion. Niobium has been applied more recently and is contained in nearly all the latest high-strength steels belonging to the group of 9-12%Cr steels. Table 1 gives an overview of customary creep-resistant ferritic steels, which are used in power plants for tubing and piping [5]. The list of steels can be subdivided into CMn-steels, Mo steels, low-alloyed CrMo steels, and 9-12%Cr steels. Ferritic steels are typically alloyed with molybdenum, providing sufficient creep strength of up to 500°C operating temperature. These steels are the most economical in this lineup and are widely used under subcritical conditions. Variants of ferritic steels having bainitic or martensitic microstructure offering higher creep strength have been developed for operation at temperatures above 500°C. These steels contain an increased amount of molybdenum and niobium as well as higher additions of chromium, providing superior corrosion resistance. Molybdenum assists the formation of the strong bainitic or martensitic microstructure. The temperature limit of steels in this group is reached at 650°C.

Figure 5 compares elevated temperature yield strength and creep resistance of standard CMn steels with those of Mo steels. It is apparent that CMn steels generally have a low creep resistance limiting their use to temperatures below 450°C.

TABLE 1 Chemical Composition of Ferritic Creep-Resistant Steels for Power Generation

EN-Designation	Comparable ASTM Grade	Chemical Composition (Mass%)													
		C	Si	Mn	Al	Cu	Cr	Ni	Mo	W	V	Nb	B	N	Other
P 235	A	max.0.16	max.0.35	0.40-0.80	min.0.02	max.0.30	max.0.30	max.0.30	max.0.08						
P 355		max.0.22	0.15-0.35	1.00-1.50	max.0.06							0.015-0.10			
16Mo3		0.12-0.20	0.15-0.35	0.40-0.80	max.0.04				0.25-0.35						
9NiCuMoNb5-6-4		max.0.17	0.25-0.50	0.80-1.20	max.0.05	0.50-0.80	max.0.30	1.00-1.30	0.25-0.50			0.015-0.045			
13CrMo4-5	T/P11	0.10-0.17	0.10-0.35	0.40-0.70	max.0.04		0.70-1.10		0.45-0.65						
11CrMo9-10	T/P22	0.08-0.15	0.15-0.40	0.30-0.70	max.0.04		2.00-2.50		0.90-1.20						
8CrMoNiNb9-10		max.0.10	0.15-0.50	0.40-0.80	max.0.05		2.00-2.50	0.30-0.80	0.90-1.10			min.10×%C			
7CrMoVTiB10-10	T/P24	0.05-0.10	0.15-0.45	0.30-0.70	max.0.02		2.20-2.60		0.90-1.10		0.20-0.30		0.0015-0.007	max.0.01	Ti: 0.05-0.10
	T/P23	0.04-0.10	max.0.50	0.10-0.60	max.0.03		1.90-2.60		0.05-0.30	1.45-1.75	0.20-0.30	0.02-0.08	0.0005-0.006	max.0.03	
X11CrMo9-1	T/P9	0.08-0.15	0.25-1.00	0.30-0.60	max.0.04		8.0-10.0		0.90-1.00						
X20CrMoNiV11-1		0.17-0.23	0.15-0.50	max.1.00	max.0.04		10.0-12.5	0.30-0.80	0.80-1.20		0.25-0.35				

Continued

TABLE 1 Chemical Composition of Ferritic Creep-Resistant Steels for Power Generation—cont'd

EN-Designation	Comparable ASTM Grade	Chemical Composition (Mass%)													
		C	Si	Mn	Al	Cu	Cr	Ni	Mo	W	V	Nb	B	N	Other
X10CrMoVNb9-1	T/P91	0.08-0.12	0.20-0.50	0.30-0.60	max.0.04		8.00-9.50	max.0.40	0.85-1.05		0.18-0.25	0.06-0.10		0.03-0.07	
X11CrMoWVNb9-1-1	T/P911	0.09-0.13	0.10-0.50	0.30-0.60	max.0.04		8.50-9.50	0.10-0.40	0.90-1.10	0.90-1.10	0.18-0.25	0.06-0.10	0.0005-0.005	0.05-0.09	
	T/P92	0.07-0.13	max.0.50	0.30-0.60	max.0.04		8.50-9.50	max.0.40	0.30-0.60	1.50-2.00	0.15-0.25	0.04-0.09	0.001-0.006	0.03-0.07	
	T/P122	0.07-0.13	max.0.50	max.0.70	max.0.04	0.30-1.70	10.0-12.5	max.0.50	0.25-0.60	1.50-2.50	0.15-0.30	0.04-0.10	max.0.005	0.04-0.10	
X12CrCoWVNb12-2-2 (VM12)		0.08-0.18	0.20-0.60	0.10-0.80	max.0.04		10.0-13.0	max.0.60	max.0.80	1.00-1.80	0.18-0.30	0.03-0.08	0.001-0.010	0.03-0.09	Co: 0.5-2.0

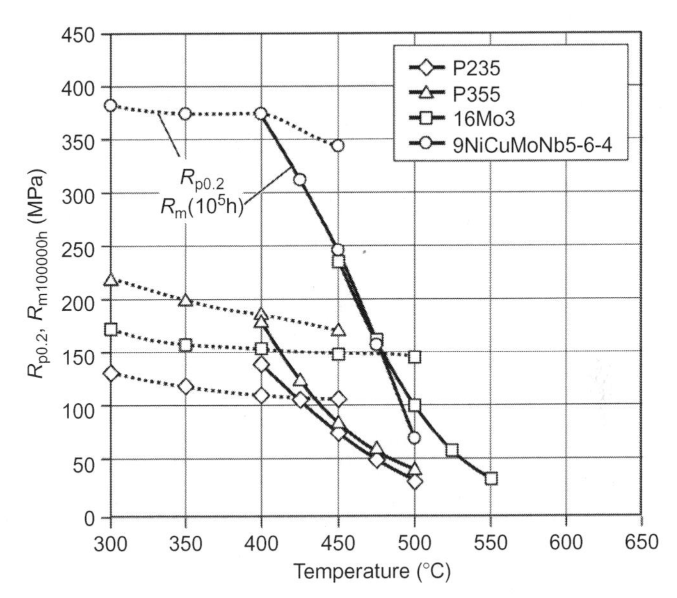

FIGURE 5 Elevated temperature strength and creep properties of CMn-steels and Mo steels [5].

The increased C and Mn content as well as the addition of Nb in P335 results in a substantial gain of yield strength over P235 due to grain refinement. The addition of around 0.3% molybdenum on the other hand results in a clear improvement of creep rupture strength as compared to P235, while the yield strength increases only moderately. The solution hardening effect of Mo is the main cause for the increase of creep rupture strength. The addition of Nb-inducing grain refinement and copper-forming precipitates dramatically enhances yield strength as seen in grade 9NiCuMoNb5-6-4. The creep rupture strength remains similar to that of 16Mo3, limiting the use of such steels to temperatures below 500°C.

Because creep ductility strongly decreases with increasing molybdenum content the apparent strengthening effect of molybdenum cannot be fully used. In addition, decomposition of iron carbides above 500°C (graphitization) further limits the application of Mo steels. However, the alloying chromium in combination with molybdenum raises the limit temperature to above 500°C.

Creep rupture strength of classic CrMo steels such as 13CrMo4-5 (T/P11) and 11CrMo9-10 (T/P22) is distinctly higher than that of Mo steels (Figure 6). This is mainly due to the higher Mo content (see Table 1). Chromium added in CrMo steels forms carbides, which are stable above 500°C. Therefore, graphitization is no longer a problem. Chromium also raises the oxidation resistance, which is an additional asset at higher operating temperatures. More sophisticated developments based on T/P22 are grades T/P23 and T/P24. In grade T/P23 a significant amount of W is added in combination with small amounts of Nb, V, B, whereas C and Mo are reduced. In grade P24, Mo and Cr additions

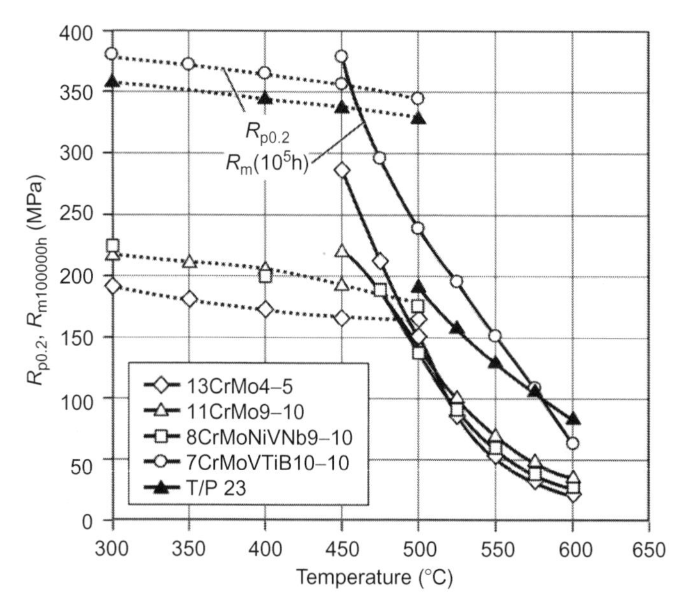

FIGURE 6 Elevated temperature strength and creep properties of CrMo steels [5].

are comparable to those in T/P22, and small amounts of Ti, V, B are added while C is reduced. Both steels comprise very attractive strength properties at high temperature.

Hot yield strength of T/P23 and T/P24 is much higher than that of T/P22, while their creep strength is approaching that of T/P91. Compared to T/P22 the pipe wall thickness can be reduced without using the expensive 9%Cr material T/P91. Due to microalloying, T/P23 and T/P24 provide a better resistance against hot hydrogen attack. Grade 8CrMoNiNb9-10 comprises an over-stoichiometric Nb addition with regard to the carbon content. Because Nb is a very strong carbide former, carbon will be completely fixed as NbC. The less carbon affine alloying elements Cr and Mo remain in a solute state. These elements provide resistance to creep and high-temperature corrosion only when they remain in a solid solution.

Increasing the chromium content to above 7% leads to a group of CrMo steels, which have a martensitic microstructure as common feature. Martensite is the strongest microstructure of steel. Its strength is predominantly controlled by the carbon content. Martensite comprises a high dislocation density and a fine lath structure which is stabilized by $M_{23}C_6$ precipitates. T/P22 differs from T/P9 primarily by the higher Cr content in the latter. The resulting change to martensitic microstructure in T/P9 raises the hot yield strength by more than 100 MPa and the creep resistance by around 20°C. The addition of alloying elements niobium, vanadium, and boron as well as tungsten further improves the creep strength, as can be seen in Figure 7. T/P91 is a good example for highly creep-resistant steel due to alloying with niobium and vanadium.

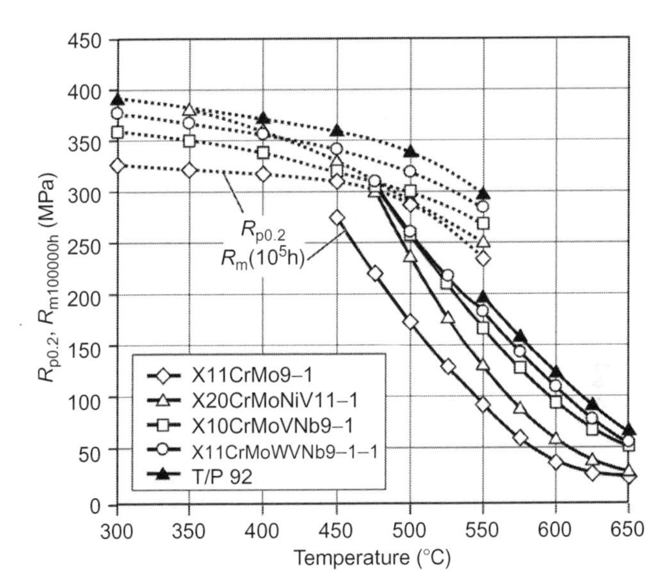

FIGURE 7 Elevated temperature strength and creep properties of 9-12% Cr steels [5].

Both elements precipitate as fine MX-type carbo-nitride particles. The grade is well established for high-pressure/high-temperature piping systems in power plants all over the world.

Other steel grades such as X11CrMoWVNb9-1-1 (T/P911), T/P92, and T/P122 have been developed on the basis of T/P91. In the state-of-the-art power plants the screen being the first convective heating surface above the furnace and the supporting tubes is built of ferritic or martensitic materials such as 7CrWMoNb 9-6 (T23), 7CrMoVTiB10-10 (T24), or X10CrMoVNb9-1 (T91). The elevated chromium addition (12%Cr) in grade VM12 combines the high-creep resistance of martensitic steel with excellent high-temperature corrosion resistance. Hence, VM12 can be applied for temperatures up to 650°C.

An example for material application for the different components in a supercritical coal-fired boiler is shown in Table 2.

4 STEEL IN HYDROELECTRIC POWER PLANTS

Hydroelectric power plants have the highest operating efficiency of all known power generation systems. They are largely automated, and their operating costs are relatively low. There are four different types of hydro power plants.

- Impoundment hydroelectric power plants:
  - Most common type
  - Used with high water heads (around 200 m)
  - Very large diameter penstock

TABLE 2 Selected Creep-Resistant Steels Used in a Super-Critical 460 MW Coal-Fired Boiler (Lagisza, Poland)

Heat Exchanger Tubes	Tube Material	Combustion Chamber Material
Economizer	15Mo3	15NiCuMoNb5
Combustion chamber walls	13CrMo4-4	15NiCuMoNb5 13CrMo4-4
Cyclones	7CrMoVTiB10-10	X10CrMoVNb9-1 X11CrMoWVNb9-1-1
Superheaters and reheaters	13CrMo4-4 7CrMoVTiB10-10	13CrMo44 X10CrMoVNb9-1
	X20CrMoV12-1 TP347HFG	X11CrMoWVNb9-1-1
Superheated steam pipe		X11CrMoWVNb9-1-1

- Diversion hydroelectric power plants:
  - Used with low water heads and strong current
  - Divert part of river with strong current (no reservoir)
  - Lower capacity than impoundment type
- Pumped storage hydroelectric power plants:
  - Dual-action water flow system
  - Used with very high water heads (typically 500 up to 1200 m)
  - Smaller diameter penstock
- Small-scale hydroelectric plants

Impoundment hydroelectric power plants store water behind a dam in a natural or artificial lake, and feed the water into the lower-lying power station through a conduit called a penstock. The head height (the difference between the elevations of the dam and the power station) tends to be moderate, typically between 50 and 200 m and the penstock tends to be very large in diameter (around 10 m) allowing high water flow rates. Table 3 summarizes information about the largest power plants of this type. As an example, each year Itaipú generates 75-77 TWh of electricity, thereby avoiding 67.5 million tons of CO_2 emission. Itaipú operates 20 penstocks, each having a 10.5 m inner diameter and weighing 883 metric tons. The recently built Karahnjukar hydroelectric power plant in Iceland, while not generating as much power as those listed in the table, operates some of the highest vertical penstock lines in the world. They reach 420 m upward and have an internal diameter of 3.4 m. The penstock lines and bifurcations are made from fine-grained steel (S355-S420MC) rolled to thicknesses up to 102 mm using controlled thermomechanical processing techniques.

TABLE 3 The World's Largest Impoundment Hydroelectric Power Plants

Name of Facility	Location	Capacity (MW)	Start of Operation
Three Gorges	China (Hubei province)	22,500	2008
Itaipú	Brazil, Paraguay	20,000	1983
Guri	Venezuela	10,000	1986
Grand Coulee	United States (Washington state)	6500	1942
Krasnoyarsk	Russia (Siberia)	6000	1968
Churchill Falls	Canada (Labrador)	5430	1971
Bratsk	Russia (Siberia)	4500	1961

Weldability was the most important selection criterion for this steel type. For components subjected to the highest loads, water-quenched and tempered steel (S690QL) in thicknesses ranging up to 150 mm was used.

Pumped storage hydroelectric plants pump water from a lower reservoir to a higher reservoir during periods of low electricity demand. The water is then released from the upper reservoir to generate electricity during periods of high demand. The higher the water head, the more compact and the more environmentally friendly hydropower facilities become. Hydropower projects with high water head require smaller dams, smaller waterway systems, smaller diameter penstock, and smaller electro-mechanical equipment than those with lower water head to generate the same energy. Designers of a pure pumped-storage plant typically seek to minimize its capital and operating costs either by making its hydro turbines more efficient or by increasing the available head to boost the overall plant's rated capacity. The typical head heights for a pumped storage plant, therefore, tend to be high compared to impoundment type plants, between 500 m and 1200 m, and the penstock tends to have a smaller diameter, typically around 3 m, with thick walls due to the high water pressure.

The parameter $H \times D$ (product of the water head and penstock diameter) is a key performance indicator (Figure 8) for hydroelectric power plants. However, increasing diameter (larger D) and pressure (larger H) to increase the performance of the power plant also increases the stress in the penstock walls. This requires increased penstock wall thickness, and often leads to a decreased reliability of the structure and to complications in manufacturing and assembly. Higher strength steels allow reduction of the penstock wall thickness, solving both these problems. The most promising steels in hydropower construction are high-strength low-alloy (HSLA) steels and, in particular, heat-treated low-alloy steels for the most highly loaded parts of penstocks.

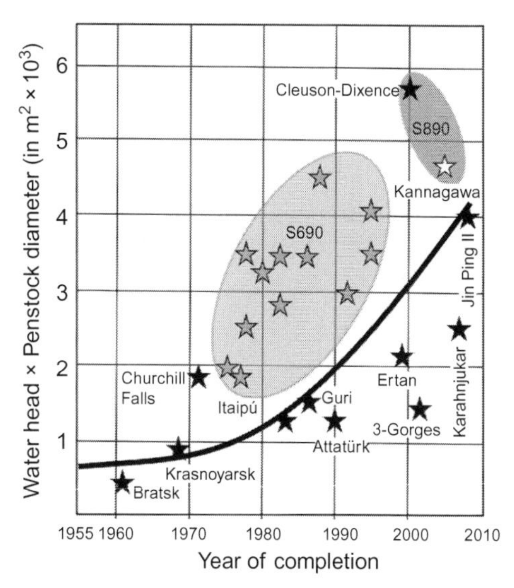

FIGURE 8 Performance indicator $H \times D$ for hydroelectric power plant of impoundment and pumped-storage type.

5 DEVELOPMENT OF PENSTOCK MATERIALS

In the 1960s, the steels used in penstocks had yield strengths in the range of 400-500 MPa. From then until the mid 1970s, steels with yield strengths of up to 600 MPa were applied. These high-tensile strength steels were used in both conventional hydropower and pumped storage power projects. The S690 steel grade (minimum yield strength of 690 MPa) has been used for penstocks since the mid 1970s, mainly for pumped storage power projects having high water head and therefore high pressure. The high-strength steel S890 (minimum yield strength of 890 MPa) was first applied in the 1200 MW Cleuson-Dixence hydropower project in Switzerland that began operation in July 1998 [6]. Depending on the internal pressure, three different grades of steels were selected. In the upper 1 km long section of the penstock grade P355 NL1 is used, while the intermediate section uses grade S690 QL. The lower section uses grade S 890 QL (Table 4) over a distance of about 3 km, with the plate gauge varying between 20 mm and 60 mm. A final short section comprises plate gauges between mm 60 and 80 mm. The maximum water pressure in the penstock during operation is approximately 200 bar (20.3 MPa).

The use of higher strength steel for penstock construction is intended to reduce the wall thickness and with it the total structure weight [7]. Figure 9 shows the wall thickness reduction achievable by increasing the yield strength of the steel from the standard high-strength S355 grade. Reducing the penstock wall thickness also increases the cross section of the pipe, raising the efficiency of a plant. More importantly, reduced penstock wall thickness and weight reduce the cost due to lower material consumption, less welding work, and reduced

TABLE 4 Chemical Composition of S890QL Plate Grades Used for the Cleuson-Dixence Penstock

	Alloy Content (Mass%, min.-max.)											
S890QL	C	Si	Mn	Cr	Mo	Ni	Cu	Nb	Ti	V	Al	B
Supplier A	0.15-0.20	0.25-0.35	0.80-1.10	0.40-0.50	0.50	1.00	–	0.015	≤0.01	0.05-0.08	0.05-0.10	≤0.002
Supplier B	0.15-0.20	0.20-0.25	1.40	0.20-0.30	0.50	–	–	0.025	≤0.01	0.04-0.08	0.05-0.08	≤0.002
Supplier C	0.10-0.15	0.15-0.20	0.80	0.70	0.45	1.30	0.20-0.25	0.015	≤0.01	–	–	≤0.001
Supplier D	0.15-0.20	0.25-0.30	0.80	0.60	0.45	1.80	–	–	≤0.01	–	–	–

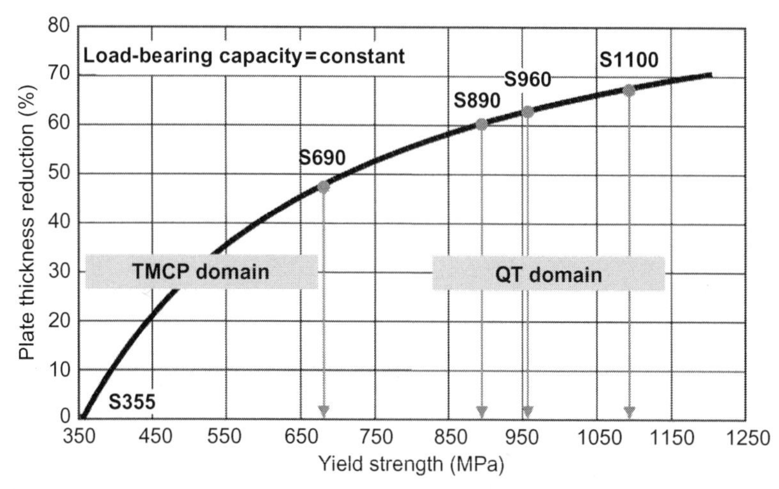

FIGURE 9 Weight-saving potential by application of ultra-high-strength steel grades (YS 690-1100 MPa) compared to a standard 355 MPa YS grade.

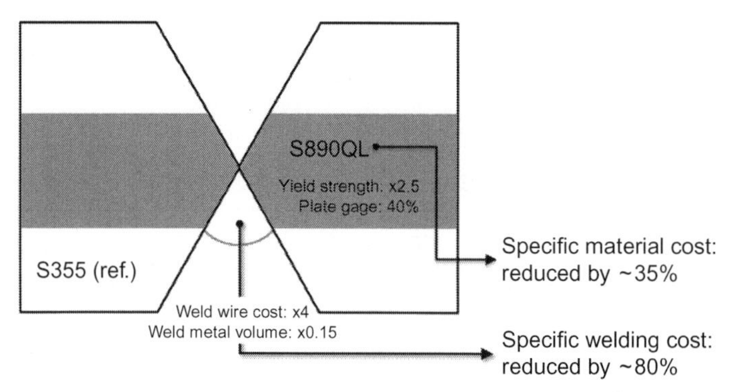

FIGURE 10 Cost benefits by upgrading penstock material from conventional S355 to S890QL grade.

transport and hoisting efforts. In fact, the higher material cost per ton for the higher strength steel is more than compensated by these cost savings (Figure 10). Although all these steel grades rely on Nb microalloying for grain refinement, molybdenum alloying becomes relevant in these steel grades for yield strengths above 500 MPa. The need for molybdenum alloying also depends on the plate gauge and the specific production facilities of the steelmaker [8,9].

6 WELDABILITY OF HIGH-STRENGTH STEELS

The real concern when shifting to ultra-high-strength steel is its weldability. The penstock segments are produced by cold rolling plate to ring segments and longitudinally welding them in the shop using submerged arc welding (SAW).

Single rings, sometimes preassembled into a section by circumferential SAW, are transported to the construction site. In front of the tunnel, these single or preassembled rings are connected using a SAW process. The preassembled parts are then brought into the tunnel and welded to the growing penstock by a coated electrode method (SMAW) or by an automated or semiautomated metal active gas (MAG) process.

The critical nature of weldability is exemplified by an accident that occurred in the Cleuson-Dixence penstock in 2000, shortly after it was taken into operation [10]. A weld seam in the S890QL section of the penstock fractured over a length of 9 m causing fatalities, major damage, and a long plant outage while repairs were made. The accident investigation revealed that SAW and SMAW welds are highly susceptible to hydrogen-induced cold cracking in the weld metal, even though the weldability of the base material is satisfactory. Further research revealed that the welds also show sufficient performance, as long as proper welding practice—including preheating—is followed. All microstructures, base metal, heat-affected zone (HAZ) SAW weld metal, as well as the SMAW weld metal show the typical tempered martensitic structure.

The most important criterion by which to judge weldability of a steel grade is the carbon equivalent (CE)ACC [8,11]. Different definitions of this parameter exist, but the most commonly used are the CE(IIW) and the Pcm value defined as:

$$\mathrm{CE} = \mathrm{C} + \mathrm{Mn}/6 + (\mathrm{Cr} + \mathrm{Mo} + \mathrm{V})/5 + (\mathrm{Cu} + \mathrm{Ni})/15$$

$$\mathrm{Pcm} = \mathrm{C} + \mathrm{Si}/30 + (\mathrm{Mn} + \mathrm{Cu} + \mathrm{Cr})/20 + \mathrm{Ni}/60 + \mathrm{Mo}/15 + \mathrm{V}/10 + 5\mathrm{B}$$

As the carbon equivalent increases, the steel's weldability decreases. The formulas show that reducing carbon content in steel is the most effective way to improve its weldability. Reducing carbon content also improves toughness, but at the same time reduces strength. Therefore, strength must be regained by using other alloying elements and thermomechanical processing during production of the plate material. To achieve the required strength without exceeding the maximum specified carbon equivalent, the alloy composition must be optimized with the help of the carbon equivalent weighing factors and combined with the proper thermomechanical treatment.

7 THERMOMECHANICAL TREATMENT AND MICROSTRUCTURES

The thermomechanical treatments used for heavy plate are rolling and normalizing (N); rolling, quenching, and tempering (QT); thermomechanical rolling (TM); and accelerated cooling (ACC). Of these processes, N-steel has the lowest strength and QT-steel the highest (up to 1160 MPa yield strength). TM-steel in combination with ACC can achieve 690 MPa, and for a specified strength always has a significantly lower carbon equivalent than N- and QT-steels, so the weldability is better. However, the thickness range for TM steels is limited, especially at higher strengths. To obtain sufficient strength and toughness even in the mid-thickness of heavy penstock plate, QT treatment is required. The chemical composition of heavy QT plate should be designed to obtain a martensitic microstructure at plate mid-thickness upon quenching. The martensite will gain toughness and ductility during subsequent tempering, showing a very favorable combination of properties at the highest strength levels. Molybdenum is one of the most effective alloying elements to achieve this goal.

Unfortunately, martensite can also be formed in any steel during welding if the cooling rate is high enough locally. This martensite remains hard and brittle if no post-weld treatment is tempering it. Untempered martensite is susceptible to cold cracking if its hardness is greater than 350 HV, a value reached at a carbon content as low as 0.05%. The steel grades applied in penstock usually are higher in carbon than that, so special welding processes are required to ensure no untempered martensite exists after welding.

Alternatively, TM + ACC in so-called ultra-low carbon bainitic (ULCB) steel makes it possible to manufacture high-strength steel plates greater than 50 mm thick, with minimum yield strength of 500 MPa [12]. The carbon content in ULCB steels is below 0.05% and Nb (0.05-0.10%), Mo (0.2-0.5%), B, Cr, Ni, and/or Cu are alloyed to achieve a fine-grained, fully bainitic microstructure that provides high strength and good toughness. The maximum hardness of the HAZ is 280 HV or less, even under welding conditions with very rapid cooling such as arc strike. Welds in these materials have excellent resistance against cold cracking.

However, the mainstream high-strength steels for penstock have carbon contents between 0.08% and 0.15%. Representative alloy concepts for 500 and 690 MPa yield strength grades are shown in Table 5. TM-steels have a lower CE at the same specified strength than QT-steels. The table also shows that TM-steels are limited in their maximum gauge. In TM-steels Mo alloying is usually necessary to obtain a yield strength of 550 MPa or more, whereas QT-steels

TABLE 5 Typical Alloying Concepts for 25 mm High Strength Plate by TM and QT Processing

	Alloy Content (Mass%)									
Grade	C	Si	Mn	Cr	Mo	Ni	Cu	Nb	V	CE(IIW)
500 TM	0.11	0.45	1.65	–	–	–	–	0.05	0.07	0.41
500 QT	0.10	0.30	1.40	0.15	0.20	0.60	0.20	0.025	0.05	0.47
690 TM	0.08	0.30	1.80	–	0.30	0.50	0.30	0.03	0.05	0.51
690 QT	0.13	0.30	0.90	0.40	0.40	1.00	0.25	0.025	0.04	0.53

always require Mo alloying. For the latter, the Mo content increases with strength level and plate gauge to a maximum of around 0.7%. Molybdenum not only provides better through hardening capability during quenching, but also provides good tempering resistance allowing the strength-toughness combination to be optimized.

8 PRACTICAL CONSIDERATION DURING FIELD WELDING

The weldability of several high-strength plate grades is compared in Figure 11 by indicating the material hardness as a function of the cooling time from 800 to 500°C after welding ($\Delta T8/5$) [8,11]. Although the carbon content controls the plateau hardness of the martensite at very short cooling time, the CE determines at which cooling time martensite forms. The critical $\Delta T8/5$, the cooling rate at which a hardness remains below 350 HV, is shifted to longer times with increasing CE, and tends to be in the typical working range of SMAW and MAG processes used for on-site welding of penstock sections. Only the 500 MPa TM-steel, with its relatively low CE of 0.40, appears to be uncritical under these cooling conditions. Grades with a higher CE demand special precautions for welding, including preheating of the weld zone. The process window becomes smaller as the CE rises (Figure 12). Welding with too high heat input degrades strength and toughness, whereas too little heat input results in excessive hardness and increased cold cracking susceptibility in the HAZ.

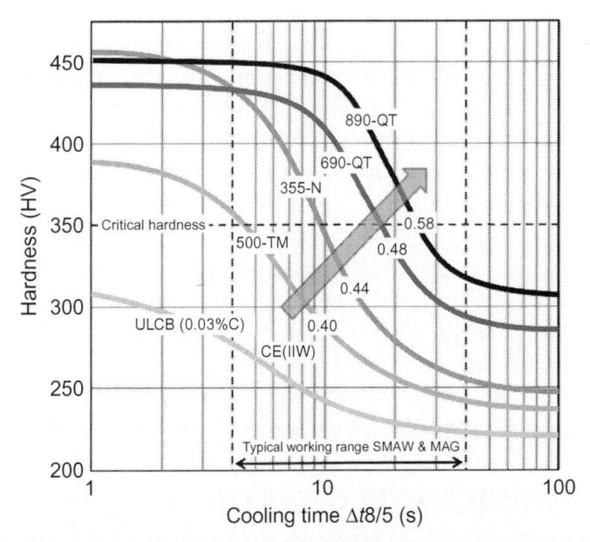

FIGURE 11 HAZ hardening of various high-strength plate grades as a function of cooling rate after welding.

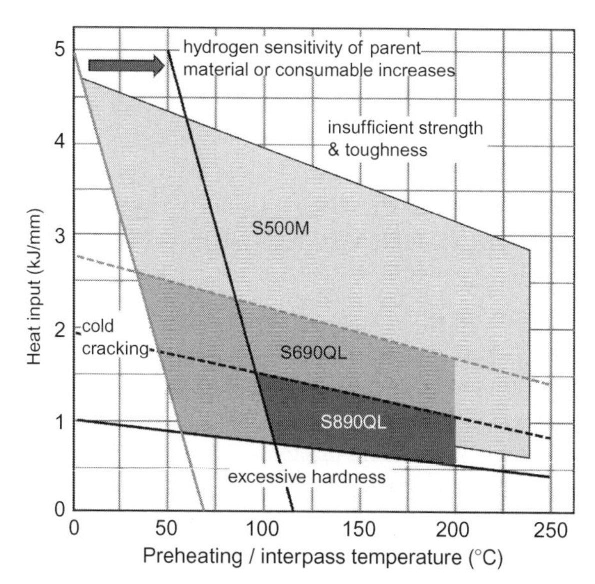

FIGURE 12 Operating window for welding of different high-strength plate grades.

The welding consumables that produce tensile properties overmatching the ones of 690 or 890 MPa QT-steel demand a high alloying content. A combination of 1.0-2.5% Ni, 0.5-1.5% Cr, and about 0.5% Mo is typical for high-strength consumables. Due to the chemical composition with high CE and the as-cast microstructure, weld metals for such ultra-high-strength steels are often more susceptible to hydrogen-induced cold cracking than the HAZ of the parent material. To avoid cracking of the weld, preheat and interpass temperature must then be adapted to the weld metal. Steel with reduced CE therefore does not allow dropping these welding precautions. In other words, high-strength steels with reduced CE, as achieved by TM rolling, need essentially the same welding precautions as the higher alloyed QT-steel. However, TM-rolled steels and particularly ULCB steels demand less stringent welding conditions with regard to preheating temperature, range of applicable cooling rates, and heat input.

Evolution in alloy design to increase the strength and weldability of steels has brought significant cost and efficiency benefits to hydroelectric plant design and construction. Molybdenum plays an important role in these alloys as moderate additions efficiently help providing high strength, especially when a large-plate gauge is required. Furthermore, Mo alloying in combination with quench and tempering treatment produces excellent toughness across the entire plate thickness.

9 STEEL IN WIND POWER GENERATION

Wind power is a fast-growing sector, having evolved from a negligible contribution in the early 1990s to a consolidated world production of around 100 GW by the end of 2007 [13]. Forecasts predict a further rapid increase toward 1000 GW by 2020. Utility-scale turbines range in size from 100 kW to as large as several megawatts; the larger turbines are more efficient and cost effective. Turbines are grouped together into wind farms that provide bulk power to the electrical grid. These wind farms are increasingly being constructed offshore because land-based sites for wind are becoming scarce, especially in densely populated regions, and wind resources are generally better offshore. Recent offshore windmills are operating in the 5 MW class, and future designs are expected to provide even higher performance. Investment, operations, and maintenance costs are roughly 70% higher for offshore windmills as compared to land-based mills. However, offshore plants produce about 50% more energy due to higher wind speeds, so the life-cycle cost per MWh becomes competitive with land-based plants.

Measured in tons of material per MW, wind power is the most iron and steel intensive of all power generation methods. From existing equipment designs, it is estimated that about 300 tons of iron and steel are required per installed MW. Table 6 lists the major components of a wind power system, together with their cost contribution. The numbers are based on a 2 MW land-based design with 100 m hub height [14]. In offshore installations, a jacket, pile or tripod foundation anchoring the tower to the sea floor is also needed. It weighs 200 to 700 tons depending on the water depth.

The wind energy business also has a big impact on other steel-intensive industries because large mobile cranes and, in the case of offshore placement, crane barges and jack-up platforms are needed to erect the turbines. Due to the hoisting heights and weights involved, crane booms made from ultra-high-strength steel are required. Applicable steel grades are in the range of S690 to S1160 [7]. They are usually made from quench and tempered plate and require Mo additions of 0.3 to 0.5%, as discussed before.

Table 7 indicates the weight of major components in a 5 MW land-based turbine [15]. The weight of the nacelle including the rotor is over 400 tons. This weight residing at a height of over 100 m is a challenge during both construction and operation, as it applies large forces to the supporting tower. With the anticipated increase of power performance, especially for offshore turbines, and the associated increase of component sizes, weight reduction becomes an important issue. In this respect, housings and support frame structures can be reduced in thickness by switching to stronger materials. For the support frames, steel grades can be upgraded according to the principle shown in Figure 9, suggesting that a weight reduction of 20 to 40% is feasible. Another considerable weight saving opportunity exists for the larger castings such as hub, hollow shaft, and gearbox housing currently manufactured from spheroidal cast iron (GJS) offering strength of up to 400 MPa [16]. Austempered ductile iron (ADI) is a material that has twice the tensile and yield strength of standard ductile irons (GJS), and 50% higher fatigue strength than these alloys [17,18].

TABLE 6 Major Components of a 2 MW Land-Based Windmill and Their Relative Impact on the Total Equipment Cost and Typical Materials Used

Component	Cost Share	Material	Function/Remarks	Alloy Design
Tower	26%	Steel plate	Range in height from 40 m up to more than 100 m	HSLA (Nb, V) plate, typically S355 or grade 50
Gearbox	13%	Steel forgings, cast iron	Gears increase the low rotational speed of the rotor shaft in several stages to the high speed needed to drive the generator	Case carburizing steel, typically 18CrNiMo7-6; Nb, Ti addition for high temperature carburization; GJS or ADI (Mo alloyed)
Transformer	3.5%	Steel sheet	Converts the electricity from the turbine to higher voltage required by the grid	Electrical sheet (high Si)
Generator	3.4%	Steel forgings	Converts mechanical energy into electrical energy	Heat-treatable CrNiMo(V) steel
Main frame	2.8%	Steel plate or cast iron	Must be strong enough to support the entire turbine drive train, but not too heavy	HSLA (Nb, V) plate, Mo alloying for extra strength; GJS or ADI (Mo alloyed)

Pitch system	2.6%	Steel forgings	Adjusts the angle of the blades to make best use of the prevailing wind	Heat-treatable CrMo steel
Main shaft	1.9%	Steel forgings or cast iron	Transfers the rotational force of the rotor to the gearbox	Heat-treatable CrMo steel; GJS or ADI (Mo alloyed)
Rotor hub	1.4%	Cast iron	Holds the blades in position as they turn	GJS or ADI (Mo alloyed)
Yaw system	1.3%	Steel forgings	Mechanism that rotates the nacelle to face the changing wind direction	Heat-treatable CrMo steel
Brake system	1.3%	Cast iron	Disc brakes bring the turbine to a halt	GJL
Rotor bearings	1.2%	Steel forgings	Withstands the varying forces and loads generated by the wind	Through hardening Cr steel (100Cr6) or CrMo steel (100CrMo7-3)
Screws, studs	1.1%	Steel bar	Hold the main components in place, must be designed for extreme loads	Heat-treatable steel, CrMo, or CrNiMo type

TABLE 7 Weight of Major Components of a 5 MW Land-Based Windmill and Potential for Material Optimization

Component	Weight (in tons)	Improvement Potential
Tower	750	Reduce plate gauge by upgrading strength from grade 50 to grade 70 or 80 (ksi) → less steel; reduced welding effort during segment fabrication; reduced transport and hoisting weight
Machine support frame	69	Reduce plate gauge by upgrading strength to grade 80 or 100 (ksi) → less steel; reduced welding effort; reduced transport and hoisting weight
Generator support frame	20	
Rotor hub	66	Replace standard ductile iron by ADI → reduce weight by up to 50%, increase toughness
Rotor shaft	27	
Gearbox	63	Housing: Replace standard ductile iron by ADI → reduce weight up to 50%, increase toughness Gear: Increase shell hardness and core toughness → reduce gear wear and failure. Increase carburizing temperature by microalloying → reduce treatment time and cost
Generator	17	Increase tempering temperature (enabled by increased Mo addition and Nb microalloying) → optimize combination of strength and toughness

10 HIGH-STRENGTH CASTING ALLOYS

ADI is a cast iron material in which carbon appears as graphite nodules and the matrix containing the nodules consists of ausferrite. Ausferrite is a fine-grained mixture of ferrite and stabilized austenite providing the high strength and ductility of ADI (similar to the TRIP effect in steel). ADI is produced by a heat-treating cycle (Figure 13) applied to cast ductile iron to which nickel (0.6-2.5%), molybdenum (0.15-0.30%) or copper (0.6-1.0%) have been added. Cooling from the austenitizing temperature to the austempering temperature must be completed rapidly enough to avoid the formation of pearlite [17]. Strength, elongation, and toughness are reduced when pearlite is formed.

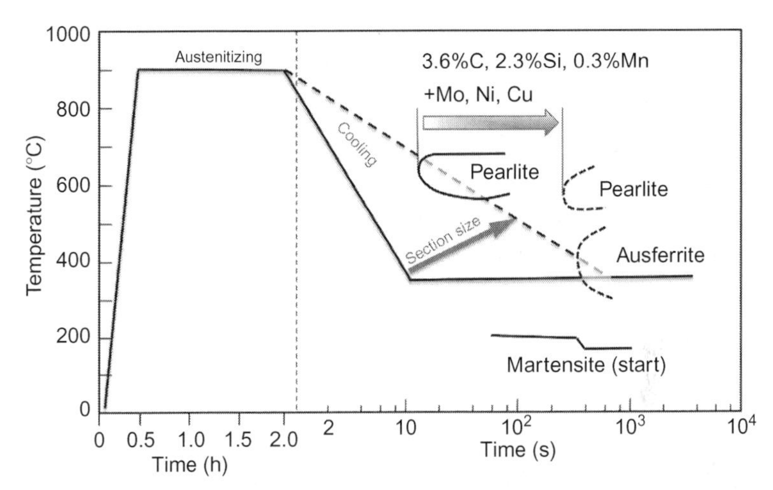

FIGURE 13 Schematic heat treatment cycle and CCT transformation diagram for the production of austempered ductile iron (ADI).

The addition of Mo, Ni, and Cu shifts the pearlite nose toward longer times and promotes hardenability. In general, section sizes greater than 19 mm require alloying. Alloying with Mn is not recommended because Mn segregates to the regions in between the graphite nodules. It delays the austempering reaction there, and can result in martensite formation and reduced toughness. Copper additions in excess of 0.80% can create diffusion barriers around the graphite nodules and inhibit carbon diffusion during austenitizing. Ni additions of up to 2% are technically and economically viable. Molybdenum is very powerful in delaying pearlite as well as the bainite formation, hence promoting the formation of metastable carbon-enriched austenite. However, it strongly segregates to intercellular/interdendritic locations between the graphite nodules where it can form hard undesirable Mo carbides [18]. These carbides are undesirable, especially if a component is to be machined after heat treatment. The Mo alloy content is usually limited to 0.3% for this reason.

11 HIGH-PERFORMANCE GEAR STEELS

Wind turbines in the higher power range usually have a gearbox converting the low-rpm rotor shaft revolution into the high-rpm rotation required by the generator. The gears in wind turbines are sometimes exposed to extremely high loads at the flanks and toes of the gear teeth, for example, during sudden changes of wind speed or hard stops. Most failures and breakdowns of wind turbines thereby occur in the gearbox, leading to significant outages and replacement costs. A hard case and a tough core are needed for a more wear-resistant gear capable of handling high-impact loads [19]. Gearboxes for wind turbines have also been specifically developed for quiet operation with reduced mechanical noise.

Gear noise increases during the turbine's lifetime due to abrasion of the gear tooth surface. Consequently, increasing the surface hardness and abrasion resistance of gears can reduce noise. The combination of a hard case and a tough core is therefore also helpful for noise reduction of gearboxes. Typical low-alloy case-hardening steels such as 16MnCr5, 20MnCr5, or 27MnCr5 cannot be used for wind turbine gear applications requiring long fatigue life and high toughness. High-performance NiCrMo carburizing steels as specified in Table 7 provide deep hardening ability and have high fatigue resistance. Currently, the grade 18CrNiMo7-6 is the standard gear steel for windmill gearboxes.

The following strategies can be used to further improve carburizing steels for large and heavily loaded gears:

- Increasing the core tensile strength and toughness.
- Increasing the fatigue strength in both core and case.
- Improving the hardenability.
- Reducing the quench distortion, with resulting improvement of reproducibility.
- Improving microstructural stability to withstand elevated temperatures during service.

Additionally, a typical surface near defect in a carburized layer, the intergranular oxidation layer has to be avoided because it acts as a fatigue fracture initiation site and reduces the fatigue strength of the gear tooth. The soft zone caused by intergranular oxidation results in surface softening in the carburized layer [20]. Elimination of surface structure anomalies is thus an important subject in the development of high fatigue strength gears. Improving toughness can effectively be achieved by raising the tempering temperature. However, this requires reinforcing the tempering resistance in order not to lose strength. A fundamental way to deal with these issues is to adjust the chemical composition of the carburizing steel. Accordingly, the chemical composition of carburizing steels can be further developed to achieve the above goals using the following guidelines:

- Prevent intergranular oxidation → reduce Si, Mn, and Cr.
- Improve hardenability → increase Mo.
- Improve toughness → increase Ni and Mo.
- Refine and homogenize grain size → balance Nb, Ti, Al, and N microalloying addition.
- Strengthen grain boundaries → reduce P and S.

Figure 14 highlights this alloy optimization starting from a standard 18NiCrMo7-6 by showing Jominy curves of noncarburized base material [19]. Steel with 0.18%C can achieve a maximum martensite hardness of 49 HRC. The standard material nearly reaches this value at the surface. In a modified concept, where the Ni content is further increased and Mo is reduced, the hardness toward the core increases due to enhanced bainite formation. However, at

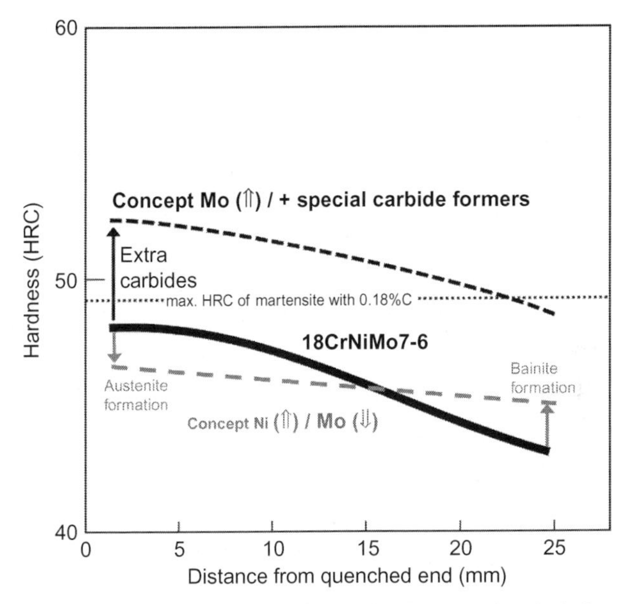

FIGURE 14 Alloying concepts for the modification of 18CrNiMo7-6 to obtain improved performance of windmill gearboxes.

the surface, where the highest hardness is needed, the material softens due to an enhanced formation of residual austenite. In contrast, the modified concept with increased Mo content shifts the entire Jominy curve up toward higher hardness [19,21]. Thus, the surface becomes harder and the core is strengthened as well. The hardness can exceed even the theoretical martensite hardness due to the formation and dispersion of ultra-hard Mo or Nb carbides. This higher hardness core of a carburized gear offers better mechanical support to the carburized case or to a potential hard surface coating. Simply raising the bulk carbon content would, of course, raise the hardenability of the steel, but the toughness would drop too much.

After quenching, the carburized component is tempered to improve its toughness. The tempering temperature has to be optimized so that the toughness of the steel improves without giving up too much of the strength. In case hardening steels such as 17NiCr6-6, 15NiCr13, 14NiCrMo13-4, the mechanical properties of the case decrease rapidly for tempering temperatures higher than 180°C. Because of that, critical applications are restricted to a maximum operating temperature of 120 to 160°C, making gearbox cooling an important issue. The tempering resistance of a steel can be strongly improved by significantly increasing the Mo content and optionally adding Nb. Figure 15 demonstrates that the addition of 2% Mo instead of the standard 0.25% provides a hardness greater than 700 HV (60 HRC) even after tempering at 300°C [22].

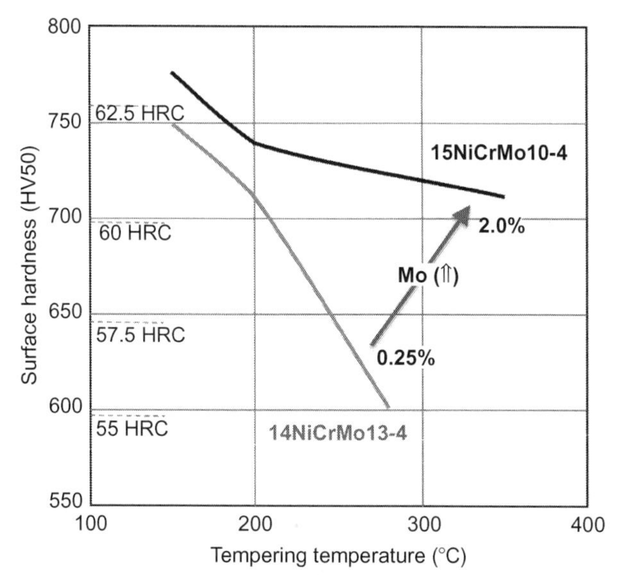

FIGURE 15 Hardness of the carburized case on quenched and tempered substrates demonstrating the effect of Mo on the tempering resistance.

Raising the carburizing temperature is an opportunity to reduce cost substantially. Because carbon diffuses more quickly into the surface layer at higher temperatures, the treatment time required to achieve a specified case depth is reduced significantly. At the standard carburizing temperature of 950°C, a carburization depth of 1.0 mm is achieved after about five hours. However, at 1050°C the same carburization depth is reached in only two hours, representing a time saving of 60% [23]. There is one obstacle to apply this increase in carburizing temperature to standard carburizing grades: excessive grain growth. Performance requirements for wind power generators require an austenitic grain size smaller than ASTM 5 after carburization, with a maximum of 10% of individual grains having size ASTM 3 and 4. Recent research shows that a balanced addition of Nb, Ti, Al, and N leads to precipitates that impede grain growth even after long times at 1050°C. The grain size distribution of standard and Nb-Ti modified 18CrNiMo7-6, treated at 1030°C for 25 h is compared in Figure 16. The modified grade (Nb, Ti microalloyed) fulfills the minimum grain size demands, and has a narrower grain size distribution. The smaller grain size is beneficial with regard to fatigue properties and a narrow size distribution avoids geometrical distortions after the heat treatment, reducing costly finish machining of the hard material.

These examples show that there is much room for advancement in the quality of materials to improve the cost and efficiency of wind power generation equipment. As the industry is moving toward larger turbines and harsher

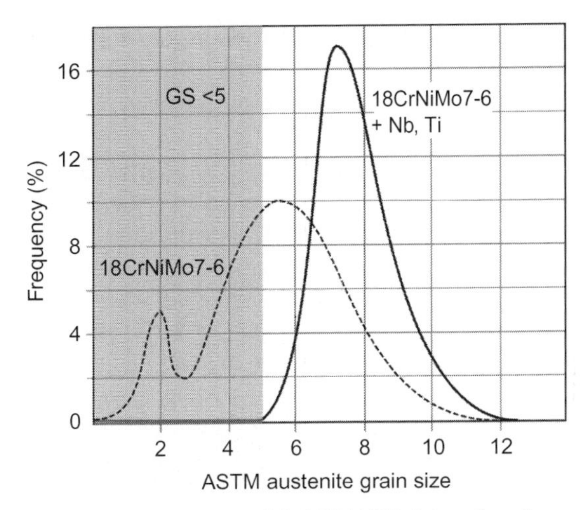

FIGURE 16 Effect of microalloying in modified 18CrNiMo7-6 steel on the austenite grain size distribution after high temperature carburizing at 1030°C for 25 h; for grain size below ASTM 5 apply restrictions.

operating conditions (e.g., offshore), demands for weight reduction, especially of the nacelle, as well as higher reliability are imminent.

12 SUMMARY

The examples given clearly indicate that upgraded steel alloys offer significant opportunities for increased efficiency, weight savings, and improved reliability of key components in facilities for sustainable power generation.

Naturally, higher alloy contents make these materials more expensive, and this additional cost must provide benefits. In all the case studies outlined here, this cost-benefit analysis reveals an unequivocally positive result.

Advanced ferritic creep-resistant steel grades allow raising the operating steam temperature and pressure in thermal boilers leading to increased thermodynamic efficiency and thus reduced CO_2 emission per kilowatt of produced electrical energy. Among others, molybdenum and niobium are key alloying elements in such alloys. The best-performing ferritic steels have the potential to limit the use of austenitic steels and nickel-base alloys resulting in major savings with regard to investment costs.

In the case of upgrading the steel of a wind tower structure from S355 to S500 or a penstock from S500 to S890, respectively, a weight saving of 30% can be achieved under constant load conditions. Niobium is the key alloying element for TMCP-produced steels while molybdenum becomes essential for QT-produced steels. The purchase cost increase for the upgraded steel grades is in the range of 20 to 25%. Thus, the balance at that stage is already positive because 30% less material needs to be purchased. Further savings accrue due to lower transport and erection costs, that is, welding and hoisting efforts. Finally, by producing a smaller amount of stronger steel, less CO_2 is generated, further improving the total CO_2 life-cycle balance of these sustainable power generation methods.

Upgrading steel for windmill gearboxes has a particularly high impact as this component represents a major investment and is critical for the reliability of the mill. Alloy concepts have been worked out using increased Mo additions in combination with Nb microalloying to achieve superior properties for gear. The additional alloying cost is, however, only a fraction of what a one-day downtime of the windmill causes in loss of income should the gearbox fail. Appropriate microalloying of such steel with Nb, Ti, and N allows reducing the carburizing treatment time by up to 60%, leading to major cost savings in processing the component and, of course, significantly reducing its CO_2 footprint.

REFERENCES

[1] International Energy Agency, IEA Energy Statistics, www.iea.org.
[2] L. Mäenpäa, F. Klauke, K.D. Tigges, Material Aspects of a 700°C—Power Plant, Hitachi Power Systems Ltd., Duisburg, 2010.
[3] F. Abe, High performance creep resistant steels for 21th Century power plants, in: Proceedings of the 1st International Conference Super-High Strength Steels, November 2005, Rome, 2005.
[4] Q. Chen, A. Helmrich, G. Stamatelopoulos, VGB Workshop, "Material and Quality Assurance", May 13-15, 2009, Copenhagen.
[5] I. von Hagen, W. Bendick, Creep resistant ferritic steels for power plants, in: Proceedings of the International Symposium Niobium 2001, TMS, December 2001, Orlando, 2001, p. 753.
[6] T. Kamo, H. Sakaibori, K. Onishi, T. Kawabata, S. Okaguchi, Metallurgical basis of 950 MPa class high tensile strength steel (HT950) plates for penstock, in: International Steel Technology Symposium, Kaohsiung (Taiwan), 2008.
[7] ThyssenKrupp Steel Europe, N-A-XTRA® and XABO® high-strength steels—processing recommendations, www.thyssenkrupp-steel.com.
[8] F. Hanus, F. Schröter, W. Schütz, State of art in the production and use of high-strength heavy plates for hydropower applications, in: Conference on High Strength Steel for Hydropower Plants, Graz (Austria), 2005.
[9] K. Hayashi, A. Nagao,Y. Matsuda, JFE TECHNICAL REPORT, No. 11 (June 2008), p. 19.
[10] N. Enzinger, H. Cerjak, Th. Böllinghaus, Th. Dorsch, K. Saarinen, E. Roos, Properties of high strength steel S890 in dry and aqueous environments, in: 1st International Conference on Super-High Strength Steels, Rome (Italy), 2005.
[11] Ming Li, Wenlin Jiang, The application and the problems of high strength steel on penstock in chinese hydroelectric stations, ISIJ Int. 42 (12) (2002) 1419.
[12] K. Nishimura, K. Matsui, N. Tsumura, JFE TECHNICAL REPORT, No. 5 (Mar. 2005), p. 30.
[13] International Energy Agency, Technology Roadmap Wind Energy, www.iea.org.
[14] NN., Supply chain: the race to meet demand, wind_directions—January/February 2007, p. 27.
[15] J. Goesswein, Operational results of world's largest wind energy converter REpower 5 M and project status of the first offshore wind farm using 5 MW turbines, in: 5th Symposium on Offshore Wind and Other Marine Renewable Energies in Mediterranean and European Seas, Civitavecchia (Italy), 2006.

[16] H. Roedter, M. Gagné, Ductile iron for heavy section wind mill castings, in: Keith Millis Symposium on Ductile Cast Iron, Hilton Head Island (USA), 2003.
[17] R.B. Gundlach, J.F. Janowak, S. Bechet, K. Röhrig, Transformation behaviour in austempering nodular iron, the physical metallurgy of cast iron, MRS Proc. 34 (1985) 399.
[18] K. Hayrynen, The production of austempered ductile iron, in: World Conference on ADI, Louisville (USA), 2002.
[19] F. Hippenstiel, K.-P. Johann, R. Caspari, Tailor made carburizing steels for use in power generation plants, in: European Conference on Heat Treatment 2009, Strasbourg (France), 2009.
[20] M. Uno, M. Hirai, F. Nakasato, Effect of alloying elements on the properties of Cr-free carburizing steels, The Sumitomo Search No. 39, September 1989, p. 33.
[21] L. Flacelière, Th. Sourmail, E. d'Eramo, P. Daguier, H. Dabas, J. Marchand, Steel solutions designed for wind mills, in: European Conference on Heat Treatment 2009, Strasbourg (France), 2009.
[22] G. Gay, Special steels for new energy sources, in: European Conference on Heat Treatment 2009, Strasbourg (France), 2009.
[23] F. Hippenstiel, Tailored solutions in microalloyed engineering steels for the power transmission industry, Mater. Sci. Forum 539-543 (2007) 4131.

Part III

Sustainable Manufacturing— Ceramic Materials

Chapter 8

Smart Powder Processing for Green Technologies

M. Naito and A. Kondo
Joining and Welding Research Institute, Osaka University, Ibaraki city, Osaka, Japan

1 INTRODUCTION

Recently, various novel powder processing techniques[1] were rapidly developed for advanced material production due to the growth of the high-tech industry, especially the fields of green and sustainable manufacturing. Smart powder processing is a green and sustainable powder processing technique that creates advanced materials with minimal energy consumption and environmental impact. Particle bonding technology is a typical smart powder processing technique to make advanced composites [1–4]. The technology has many unique features. First, it creates direct bonding between particles without any heat support or binders of any kind in the dry phase. The bonding is achieved through the enhanced particle surface activation induced by mechanical energy, in addition to the intrinsic high surface reactivity of nanoparticles. Using this feature, the composite particles desired can be successfully fabricated.

By making use of particle bonding, a new "one-pot" processing method to synthesize nanoparticles without applying extra heat was developed. Further, one-pot processing achieving both the synthesis of nanoparticles and their bonding with other kinds of particles to make nanocomposite granules was also developed. The assembling of these composite particles and granules will lead to the control of nano/microstructure of advanced materials. As a result, it can customize various kinds of nano/microstructures and can produce new materials with a simpler manufacturing process. Furthermore, the particle bonding process can also be applied to bond nanoparticles with substrate to form mechanically deposited porous films. In the first part of this chapter, we deal with the particle bonding process and its applications for making advanced materials.

By carefully controlling the bonding between different kinds of materials, separation of composite structure into elemental components is also possible, which leads to the development of novel technology for recycling composite materials and turns all of them to highly functional applications [1]. In the second part of this chapter, we introduce the development of novel recycling

Green and Sustainable Manufacturing of Advanced Materials. http://dx.doi.org/10.1016/B978-0-12-411497-5.00008-4

methods of glass fiber reinforced plastics (GFRPs). The trial to recover useful elements such as precious metals from waste composites by using particle disassembling will be also explained.

2 PARTICLE BONDING PROCESS

A grinding process produces fine particles by giving feed materials the mechanical forces strong enough to fracture them into small pieces. Alternatively, the process can be used to act as an effective force onto the surface of fine particles for their boning. Table 1 shows the machines that have been used for particle surface modification or for making composite particles [4]. In the table, group I shows the fine grinding machines used for this purpose. It means that most kinds of fine grinding machines can perform the particle bonding process. Group II shows a mixing apparatus used for particle bonding. Intensive mixing has also the capability of the particle bonding process. Based on these mechanical principles, several advanced machines for particle bonding have been

TABLE 1 Main Machines Used for Particle Bonding

Type		Advanced machines for fabricating composite particles
I	Impaction-type Pin mill, disc mill	Hybridization System (Nara Machinery Co.)
	Centrifugal classifying type	KOSMOS (Kawasaki Heavy Industry Co.)
	Frictional grinding mill	Mechanofusion System (Hosokawa Micron Corp.)
	Ball mill tumbling vibration planetary centrifugal fluidizing (CF mill)	
	Agitated ball mill Mixing vessel type	
	Jet mill	
II	Mortar	
	Cylindrical vessel type with rotating disc	Mechanomill (Okada Seiko Co.)
	Elliptical vessel type with high-speed elliptical rotor	Theta-composer (Tokuju Corp.)

developed as shown in the right column of Table 1. Although, each advanced machine works on a different principle, all of them give mechanical actions to powder materials by the rotation of rotor or vessels.

Figure 1 shows the mechanical principle of the Mechanofusion system [2,4]. The main parts are a rotating chamber and an arm head fixed with a certain clearance against the inside wall of the chamber. The powder material compacted by the centrifugal force onto the inside wall of the chamber is compressed into the clearance and receives various kinds of forces, such as compression, attrition, and rolling. The processed powder is dispersed by the scraper not shown in the figure. These actions are repeated during the chamber rotation to achieve particle bonding. The apparatus supports the option of an electric discharge part, which causes glow discharge between the arm head and the chamber during its operation [2]. The apparatus is usually operated under ambient conditions, but has an option to control the atmosphere including a vacuum condition.

Many factors affect the particle composite process. Figure 2 summarizes the main factors controlling the particle bonding process [4]. They include the conditions of starting powder materials and the processing conditions such as mechanical stress, temperature, atmosphere, and charging procedures of starting powder materials into the apparatus.

The properties of powder materials affect the structure and bonding mechanism of composite particles. For example, when the purpose is to make core/shell type composite particles, a larger size of core particle and a finer size of shell particle should be selected. The mixing ratio is also an important factor to decide the structure of composite particles. For example, the weight fraction of

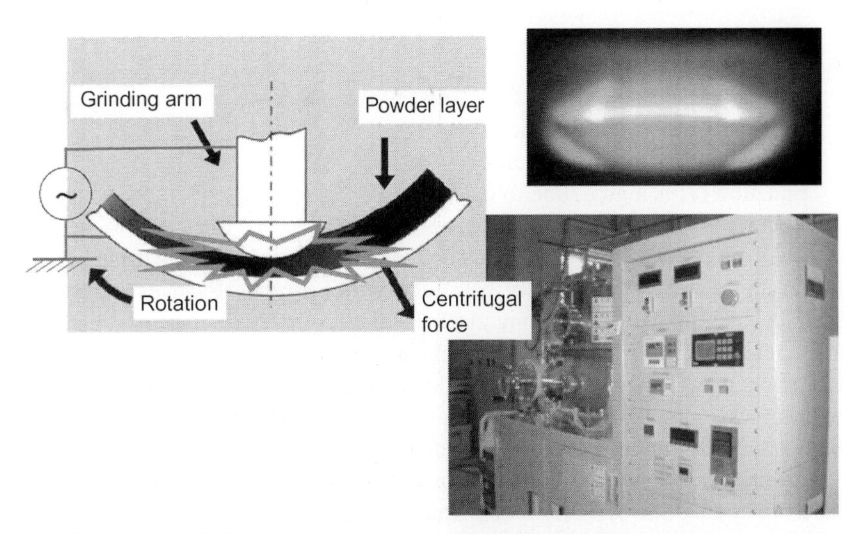

FIGURE 1 Mechanical principle of composite apparatus (Mechanofusion system).

Factors controlling the composite process

Powder materials

- Particle size (core particle, fine particle)
- Particle properties
  (shape, surface,
  Mechanical strength etc.)
- Mixing ratio
  etc

Processing conditions

- Mechanical actions
  (Mechanical stress, processing time etc.)
- Temperature
  (average, local temperature)
- Atmosphere
- Charging procedures of powder materials
  etc.

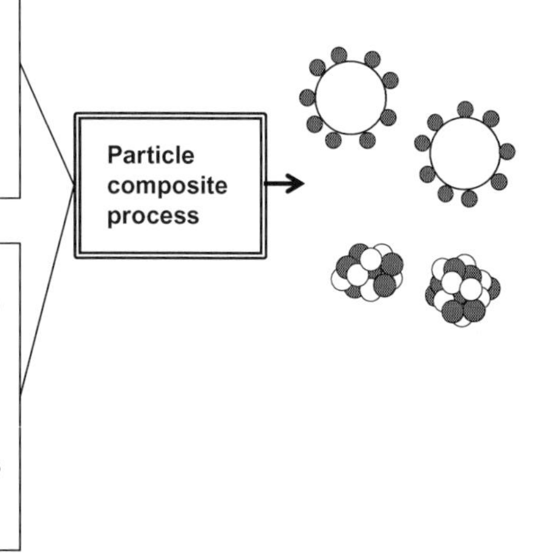

FIGURE 2 Factors controlling particle composite process.

fine particles n is calculated by the following equation [4], on the condition that the fine particles are closely allocated onto core particle surface, and $d_p \gg d_a$,

$$n = \left(1 + \frac{\sqrt{3}}{2\pi} \cdot \frac{\rho_p}{\rho_a} \cdot \frac{d_p}{d_a}\right)^{-1} \tag{1}$$

where d_p: core particle size
d_a: fine particle size
ρ_p: true density of core particles
ρ_a: true density of fine particles

Equation (1) is a guideline to determine the fine particle fraction. Actually, the fraction is selected from a wide range.

The bonding mechanism depends on the combinations of powder materials. For example, the contact surface between powder particles receives extremely high local temperature and strong mechanical stress, where mechanical actions are actually given [5]. The authors previously reported that the local temperature at the interface between particles during particle bonding processing could be 10 times higher than the apparent temperature of the processing chamber [6]. Such a local high temperature is expected to cause unique phenomena such as "micro-welding" or "chemical interaction" between fine particles and core particles, or among fine particles. For example, when coating nano titania particles on the glass beads, the peak of binding energy of Ti 2p shifted away from its original position after only 5 min of mechanical processing [4]. It suggests that there is a chemical interaction on their surface during the processing.

From the macroscopic viewpoint, processing conditions such as mechanical actions directly affect the particle composite process. As was previously mentioned, mechanical actions are given to powder materials by the rotation of rotor or vessels. For the case of a core particle coated with fine guest particles, the particle bonding process as proposed is shown in Figure 3 [4]. The rotation speed of a rotor or vessel along with the processing time decide the mechanical actions and its repeated number for powder materials. In this case, the composite process is carried out in the following two steps: First, the surfaces of fine particles and core particles are mechanically activated; as a result, fine particles adhere onto the surfaces of core particles. At this step, as fine particles are deposited on the core particle surface, the BET specific surface area of composites gradually decreases with processing time. However, at the second step, as fine particles adhered onto the surfaces of core particles are gradually compacted, BET specific surface area decreases with the processing time. It leads to a higher mechanical strength of surface layers and stronger bonding between surface layers and core particle. Therefore, by controlling the rotation speed and its processing time, the apparent density and strength of surface layers can be changed. However, when the rotation speed exceeds a certain value, the layer formed on the core particle surface is detached after a certain processing time as shown in the dotted line in Figure 3. It is because mechanical stress works for surface grinding of composite particles when it exceeds certain mechanical stress.

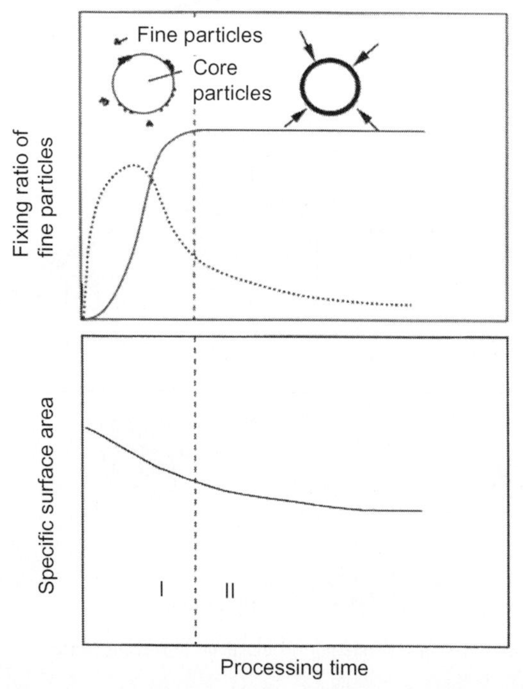

FIGURE 3 Particle composite process for core/fine particles.

During mechanical processing temperature and atmosphere also affect the particle composite process. Similar to the powder grinding process, mechanical energy is converted to heat energy; therefore, careful temperature control is needed when heat-sensitive materials are processed. Control can be achieved by cooling the apparatus. Conversely, temperature is also controlled to make the bonding strong. Atmosphere is important to process powder materials in a dry phase. It affects their properties and the particle composite process.

Charging procedures of powder materials are key issues for controlling the structure of composite particles. For example, a uniform surface layer of composite particles is created when a few fine particles are repeatedly added into the processed powder at certain time intervals [7]. Multilayered composite particles are created by adding different kinds of fine powders at certain time intervals. They have been used for purposes of drug delivery [8].

Figure 4 shows the one-pot processing methods developed thus far by applying the particle bonding principle [9]. The particle bonding process has possibilities for developing new green manufacturing techniques because it is a simple and energy-saving process without extra heat assistance. Figure 4a depicts a particle composite process that has already been introduced, and b is the process of nanoparticle synthesis from starting powder materials in a one-pot process without heat assistance. Figure 4c is the combination process of nanoparticles synthesis and their bonding to make nanocomposite granule structure. On the other hand, by applying a particle bonding principle, it can achieve particle-substrate bonding, thus leading to developing a mechanically assisted deposition of nanocomposite film on the substrate, as shown in

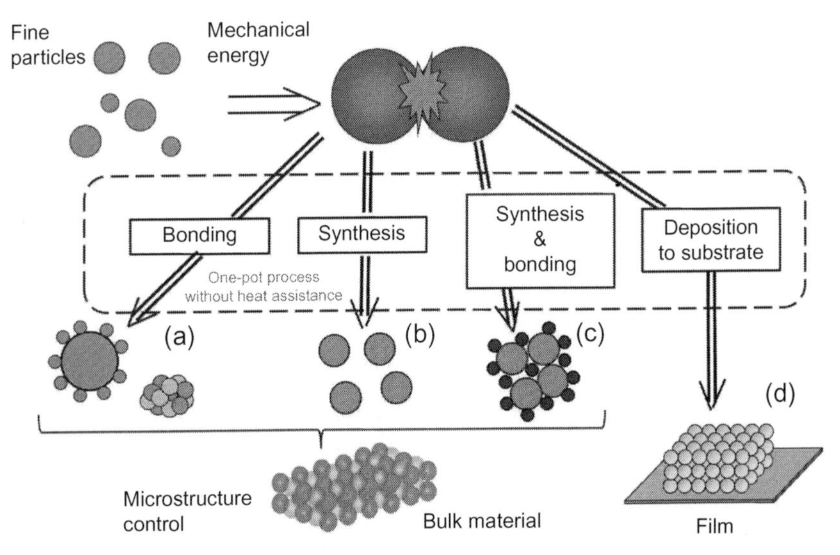

FIGURE 4 Material design by one-pot mechanical processing based on particle bonding technology.

Figure 4d. The applications of these methods to develop advanced materials are introduced in the following sections.

3 APPLICATIONS OF PARTICLE BONDING PROCESS FOR ADVANCED MATERIALS

3.1 Development of High Efficient Thermal Insulation Materials

The first example is the application for highly efficient thermal insulation materials [10–12]. Global interest in thermal insulation materials has intensified because escalating energy costs signified the importance of efficient thermal insulation. In this study, the nanoparticle bonding process was used to make composite fibers coated with a porous fumed silica layer in the dry phase. Figure 5 shows the proposed dry processing method to fabricate fumed silica compact by using composite fibers [11]. Fiberglass composites porously coated with silica nanoparticles were fabricated at the first stage and then compacted into a board by dry pressing. The composites were produced by a particle bonding process without collapsing the fiberglass and with nano-scale pores made by the fumed silica. The proposed method has the advantage of preventing contacts between fibers through a coating layer. In addition, because fumed silica was fixed on the fibers, particle segregation rarely occurred during forming. Therefore, it was easy to achieve a highly uniform dispersion of fibers in the compact.

Figure 6a shows a cross-sectional scanning electron microscope (SEM) image of a glass fiber coated with fumed silica found in the processed powder mixture. Figure 6b shows the magnified Transmission electron microscope (TEM) image of the coated fumed silica layer [10]. This layer was porously formed with a pore size of about 100 nm. The composite fibers were then successfully dry-pressed to make fiber-reinforced fumed silica porous compact.

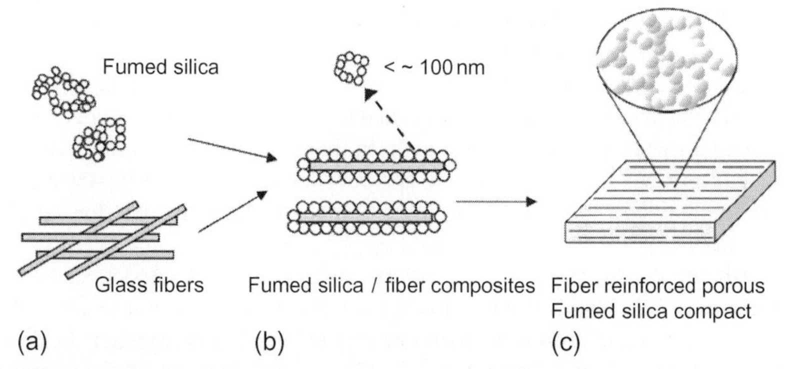

FIGURE 5 Proposed method to fabricate fibrous fumed silica compacts: (a) mixing of raw materials, (b) particle bonding to coat glass fiber with fumed silica, (c) dry pressing of the composites from (b) to produce bulk body.

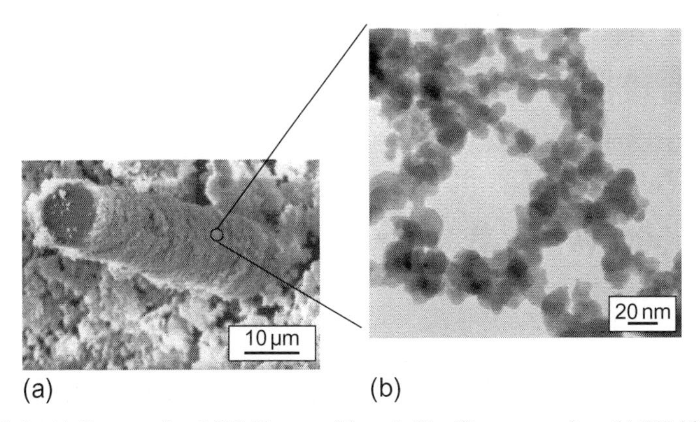

FIGURE 6 (a) Cross-sectional SEM image of fumed silica/fiber composites, (b) TEM image of the coated fumed silica surface.

TABLE 2 Thermal Conductivity and Apparent Density of the Fibrous Fumed Silica Compacts Prepared by Particle Bonding Process

Specimen	Density (kg/m^3)	Porosity (%)	Thermal Conductivity (W/(mK))	
			@100 °C	@400 °C
#1	459	81.2	0.0266	0.0269
#2	485	80.1	0.0266	0.0282

Table 2 shows the thermal conductivity of compact specimens with 80% porosity at 100 °C and 400 °C [11]. They were lower than molecular conductivity of still air (0.03 W/mK at 100 °C, 0.05 W/mK at 400 °C) and at the same level as those obtained from silica aerogel [13] and fumed silica compacts [14]. These results indicate that the obtained compacts have a nano-scale porous structure. The remarkable attribute of composite fibers is that they achieve very low thermal conductivity with a relatively large number of glass fibers. The mechanical strength of the compacts depends on their apparent density as determined by the compressive strength. In this case, fracture strength ranged from 0.4 to 1.6 MPa, corresponding to apparent densities from 400 to 480 kg/m^3, could be obtained. This made it possible to machine the compacts for various applications.

Furthermore, the thermal conductivity of fumed silica compacts at higher temperature could also be kept at lower value by adding silicon carbide (SiC) powders as an opacifier. It was found that thermal insulation compacts made of a powdered mixture consisting of fumed silica, glass fiber, and SiC at a mass ratio of 70:10:20 prepared by the particle bonding technology could achieve a thermal conductivity of 0.04 W/mK at 600 °C [15]. By changing the kind of nanoparticle additives, it is expected that thermal conductivity of the compacts can be kept lower at even higher temperatures by particle bonding in the future.

3.2 Development of Fuel Cell Electrodes

Solid oxide fuel cell (SOFC) is a promising candidate for power generation in the twenty-first century because of its high energy efficiency and clean exhaust. Current R&D efforts focus on reducing its production cost and increasing the long-term stability of cells and stacks by lowering its operation temperature without losing power density. Prefabrication of the composite particles followed by electrode forming using a particle bonding process is an ideal way to go, especially for controlling the microstructure of composite electrodes. Recently, we successfully fabricated various kinds of composite particles such as large core particles coated with nanoparticles and dispersed into the composite mixture consisting of several kinds of nanoparticles using the particle bonding technology.

Nickel-yttria stabilized zirconia (Ni-YSZ) is the most widely used SOFC anode material due to its excellent electrochemical properties at high temperatures. The electrochemical reaction (hydrogen oxidation) takes place at the triple-phase boundary (TPB) where Ni, YSZ, and fuel gas meet. The reaction rate strongly depends on the catalytic activity of anode materials and the TPB length. Because the former significantly decreases with decreasing operation temperature, the latter must be increased as much as possible in the limited effective electrode volume to keep electrochemical performance high even at lower temperatures. For the TPB enlargement, the anode microstructure such as size and arrangement of Ni and YSZ must be precisely controlled.

For example, Figure 7 shows NiO-YSZ interspersed composite (IC) particles consisting of NiO and YSZ nanoparticles processed by particle bonding

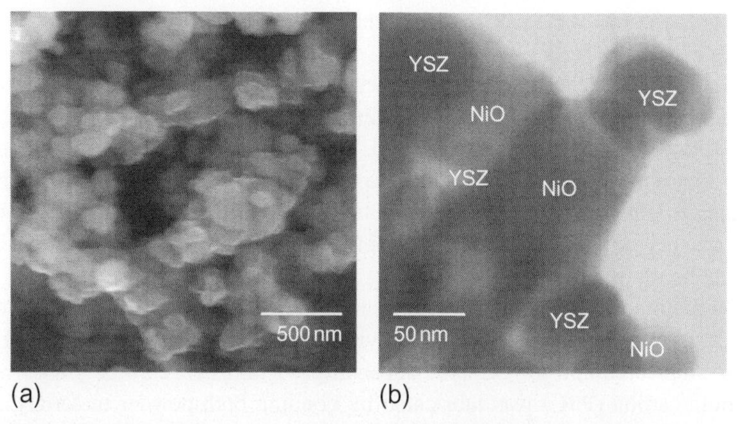

FIGURE 7 NiO-YSZ interspersed composite particles: (a) SEM image, (b) TEM image.

FIGURE 8 Cross-sectional SEM images of the anode: (a) before reduction, (b) after reduction, (c) after long-term stability test at 700 °C for 920 h.

process [16]. Figure 7a is the SEM micrograph of the composite particles. Figure 7b is the detailed structure of a composite particle observed by TEM. NiO and YSZ phases in the composite particle were identified by energy-dispersive X-ray spectroscopy (EDX) in the micrograph. The micrograph indicates successful fabrication of the IC particles. NiO and YSZ nanoparticles were well dispersed and their sizes were in good agreement with those estimated from the specific surface area of starting particles (NiO: 160 nm and YSZ: 75 nm) [16].

Figure 8 shows SEM micrographs of the anode made of IC particles before the reduction, just after the reduction, and after the long-term stability test at 700 °C for 920 h. The anode made of IC particles before the reduction shows grain sizes smaller than 1 μm without abnormally large ones. NiO shrunk when it was reduced to Ni, and porous structure evolved in the anode. Thus, the uniform porous structure after reduction suggests that NiO was uniformly distributed around the three-dimensional YSZ framework in the entire anode. Even after the long-term stability test, no significant structural change was observed in the anode. The grain size was kept at about 0.5 μm, and no abnormally large grain was observed. The insignificant micro-structural change indicates that grain growth of Ni was insignificant in the anode made of IC particles under the testing conditions [16].

Figure 9 shows anode polarization curves as a function of current density. The anode made of IC particles shows lower polarization than the anode made of coated composite particles (CC) [16]. It indicates that careful control of microstructure by the use of composite particles as starting materials is very important for improving the performance of anode. Further improvement on the electrode performance is expected by optimizing Ni-YSZ microstructure using particle bonding technology.

On the other hand, particle bonding technology has also contributed to developing low-platinum catalyst for polymer electrolyte fuel cell (PEFC) [17]. For example, composite catalyst consisting of tungsten carbide (WC) and platinum/carbon (Pt/C) was fabricated by bonding both powder materials. As a result, small Pt particles and a thin carbon layer was observed on the surface

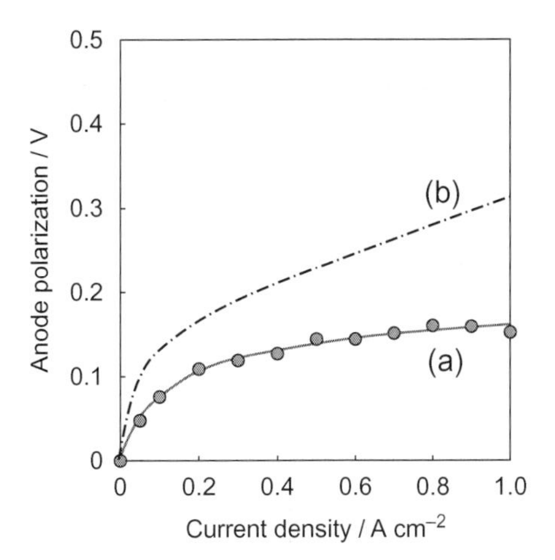

FIGURE 9 Anode polarization as a function of current density: (a) interspersed composite particles, (b) coated composite particles (coated by YSZ).

of WC particles by TEM analysis. The composite particles showed a similar catalytic activity to that obtained by pristine Pt/C catalyst for hydrogen oxidation although the amount of Pt loading was about 25% of that for the Pt/C catalyst [17]. It suggests that particle bonding is promising for developing higher performance catalysts for PEFC.

4 APPLICATIONS FOR LOW TEMPERATURE REACTION AND ONE-POT SYNTHESIS OF NANOPARTICLES

4.1 Low Temperature Reaction of Powder Materials

The first attempt to synthesize materials at low temperatures using the particle bonding principle began with the research to create MgB_2 from magnesium and boron powders [18,19]. It was found that the particle bonding process could combine the elemental powder components to form superconductive MgB_2 phase. Large magnesium particles (-330 mesh) and submicron amorphous boron particles (average particle size: 0.8 μm) were mechanically processed. As a result, boron particles were embedded into the surface of magnesium particles at a depth of about 1 μm, and MgB_2 phase was found at this embedded region after annealing at a relatively low temperature under atmospheric pressure of argon.

The second attempt at applying the particle bonding principle for low temperature reaction was to dope TiO_2 nanoparticles with nitrogen without any heat support, which usually requires annealing [20] at 500~600 °C under NH_3 flow. The high annealing temperature can lead to undesirable grain-size growth of TiO_2 nanoparticles. To overcome the problem, apparatus applying particle bonding with electric discharge was applied for this application [2]. NH_3 (10%)/Ar plasma was generated at different gas pressures in a particle bonding processing chamber, where an anatase TiO_2 powder with a BET surface area of 300 m^2/g (equivalent to about 7 nm in diameter) was uniformly irradiated. When generated plasma irradiated at 300 Pa, the TiO_2 powder had a specific surface area of 283 m^2/g, and noticeable absorption in visible light range was observed. In addition, the powder showed an improvement of the photo-catalytic oxidation activity of CH_3CHO under visible light. These results indicated that the presented plasma processing is capable of modifying TiO_2 nano-powder to improve its photo-reactivity without much reduction in specific surface area, which typically occurred when powder modification was carried out with annealing treatment [21].

4.2 One-Pot Synthesis of Nanoparticles from Raw Powder Materials

Using the features of particle bonding technology, researchers developed new one-pot processing to synthesize nanoparticles without any heat support. First, a rapid synthesis of perovskite type lanthanum manganite starting from a mixture of industrial-grade powders was demonstrated [22]. A traditional route to synthesize $LaMnO_{3+\delta}$ was through the solid-state reaction of constituent oxide powders at 1300 °C. Figure 10 shows the conventional production process of $LaMnO_3$ materials from La_2O_3 and Mn_3O_4 powders. This process needs many manufacturing steps, and the thermal reaction involved leads to particle size enlargement and limits the degree of chemical homogeneity.

On the other hand, a rapid mechano-chemical synthesis proposed by the authors is shown in Figure 11. In this method, synthesis is achieved by one-pot processing of a mixture of industrial grade La_2O_3 and Mn_3O_4. The one-pot processing is based on particle bonding technology, which applies mechanical forces such as compression and shear stresses repeatedly on the powder mixture without using media balls.

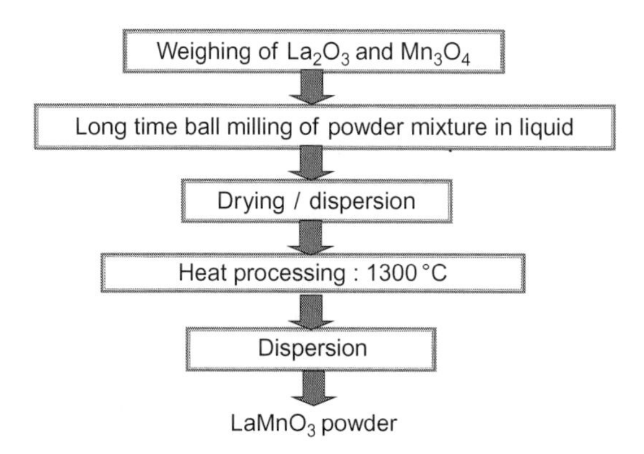

FIGURE 10 Conventional production process of $LaMnO_3$ powder from starting powder materials of La_2O_3 and Mn_3O_4.

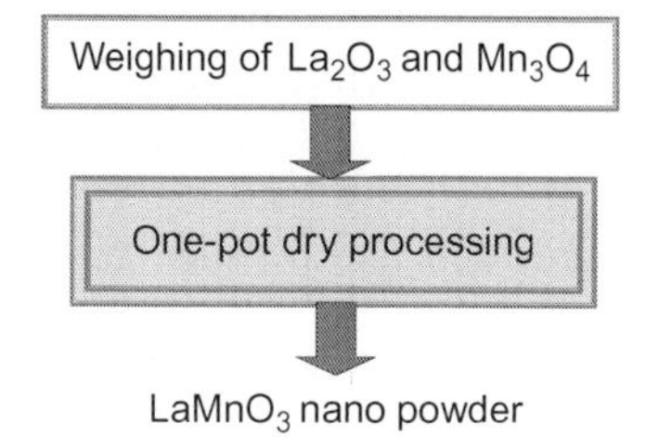

FIGURE 11 One-pot processing to synthesize $LaMnO_3$ nanosized powder.

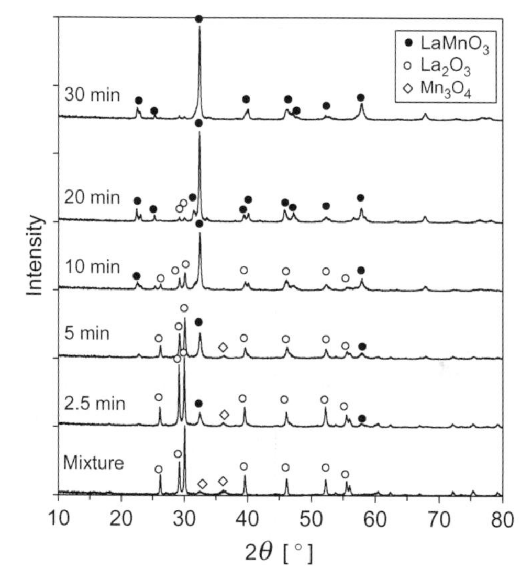

FIGURE 12 Change of XRD patterns of powders with processing time.

Figure 12 shows the phase evolution as a function of processing time examined by X-ray diffraction (XRD) with the powder mixture [23]. At the beginning, only the peaks corresponding to La_2O_3 and Mn_3O_4 were observed from the powder mixture. However, after 2.5 min, the peak of $LaMnO_3$ appears. Then, the peak intensities of raw powder materials decreased drastically as processing time increased and almost disappeared after 30 min in processing. The average particle size calculated by the BET-specific surface area was about 280 nm, and it was almost consistent with the particle size observed by SEM.

Adding to the synthesis of $LaMnO_{3+\delta}$, strontium doped materials were also synthesized by one-pot processing [24]. And $BaTiO_3$ was also rapidly synthesized by one-pot processing. In this case, an industrial grade of TiO_2 and $BaCO_3$ powders were used [25,26]. The one-pot reaction mechanism has not yet been fully understood. Further analysis would lead to the establishment of a reaction mechanism to synthesize nanoparticles in the future.

5 ONE-POT MECHANICAL PROCESS TO SYNTHESIZE NANOPARTICLES AND THEIR BONDING TO MAKE NANOCOMPOSITE GRANULES

The motivation to make nanocomposite granules came from lithium ion batteries. Nano-sized active materials are promising for improving battery properties, such as rate performance, because they contribute to the shorter diffusion path, which enhances the lithium ion diffusion [27]. In this case, nanoparticles are very difficult for higher packing density, thus leading to lower energy density. For example, $LiMnPO_4$ is expected to provide practical batteries with high voltage (4.1 V), high thermal stability, and no toxicity. However, when $LiMnPO_4$ is used, its electron conductivity must be improved. Carbon coating on its particle surface is a useful way, but it needs a very complicated manufacturing route including the fabrication of $LiMnPO_4$/carbon composites and their granulation, as shown in the left side of Figure 13.

For improving the conventional complicated manufacturing route, the authors have proposed a novel one-pot mechanical process that achieves the particle synthesis and their bonding as explained in Figure 4c. Figure 13 shows a new method to make $LiMnPO_4$/carbon nanocomposite granules by only one-pot processing [28]. It is a very simple process and does not require heating. Figure 14 shows the cross-section area of nanocomposite granules made by this method. For the experiment, three kinds of raw-powder materials and carbon nanoparticles were mixed and processed for 30 min. Near peaks of the obtained sample were identified as $LiMnPO_4$ from XRD analysis. In the figure, one can see that nanosized $LiMnPO_4$ particles make the granule structure, and the carbon is well-dispersed in the granule. This should enable a higher performance battery.

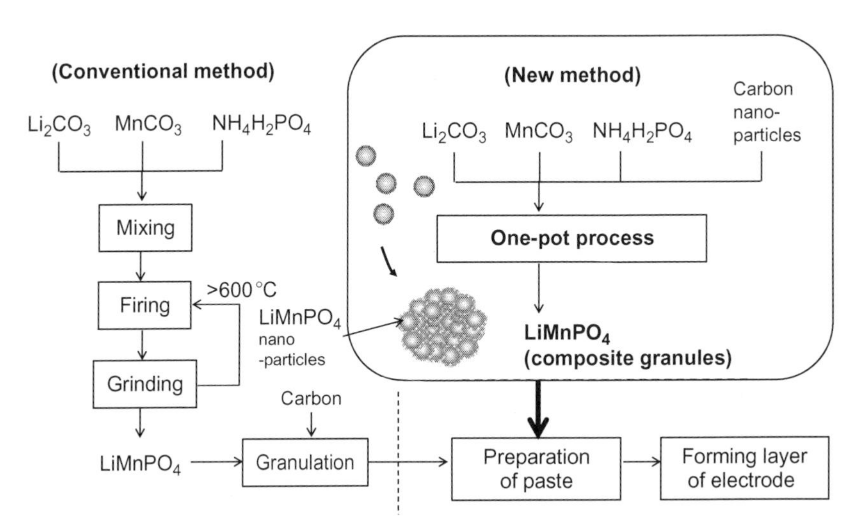

FIGURE 13 One-pot processing to synthesize nanoparticles and their bonding to make nanocomposite granules of $LiMnPO_4$/carbon.

FIGURE 14 Nanocomposite granules made by the one-pot processing as shown in Figure 13: (a) Cross-sectional SEM image of the granule, (b) $LiMnPO_4$ nanoparticles in the granule, (c) EDX mapping of carbon, (d) EDX mapping of Mn.

6 MECHANICALLY ASSISTED DEPOSITION OF NANOCOMPOSITE FILMS BY ONE-POT PROCESSING

Mechanically assisted deposition of nanoparticles on substrates was developed by applying particle bonding technology, which is explained in Figure 4d. This process aims to make porous composite films with a large surface area; the films are used for chemical sensing, energy storage, and conversion. As previously explained, various composite particles have been prepared by the particle bonding process. Such a mechanochemical process has raised concern about direct deposition of the mechanically activated nanocomposite particles on substrates in a dry phase. To investigate this possibility, the experiments have been performed using the modified bonding apparatus.

Figure 15 shows a schematic illustration of an apparatus used in this study [29]. The main components were a closed vessel and a feather-type rotor. For example, the anode of nickel-yttria stabilized zirconia (Ni-YSZ) cermet was

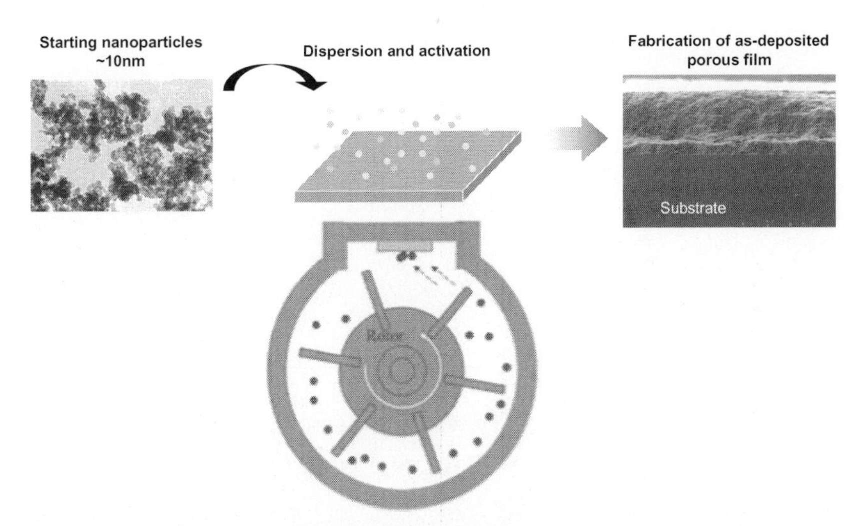

FIGURE 15 Apparatus for mechanically assisted deposition of particles on substrate.

fabricated on YSZ substrate by this apparatus. At first, NiO/YSZ composite particles were put into the vessel, and then the rotor was rotated to disperse the particles. The dispersed and activated particles were transported to the substrate and deposited on the YSZ substrate. As demonstrated by Akedo [30], dense ceramic layers can be successfully formed on a substrate caused by the consolidation of fracture or deformation of the impacted particles when submicrometer ceramic particles are accelerated up to 100-500 m/s. In this process, it is demonstrated that the mechanically impacted fine aggregates or clusters can be bonded together with the lower collision speed of about 30 m/s, producing the porous films. It should be noted that the film formation occurs efficiently for the nanoparticles with a high specific surface area.

Figure 16 shows the relationship between the processing time and the thickness of the deposited film in case of NiO-YSZ nanoparticles (120 m^2/g). The film formation occurred after 10 min. It suggests that the fine aggregates of the nanocomposite particles were produced and activated in the early stage of the processing, and then they started to be deposited with collision or impact speed of about 30 m/s. The resulting porous film was then sintered at 1200 °C. Abnormally large grains were not observed. It must be emphasized that the fine composite microstructure resulted from prevention of heterogeneous grain growth because of the homogeneously distributed NiO and YSZ nanograins in the deposited film.

Figure 17 shows the anode polarization as a function of current density at 700 and 800 °C, respectively. The polarization of the anode was 70 and 20 mV at current density of 0.5 A/cm^2 at 700 °C and 800 °C, respectively. They were significantly lower than those reported in the literatures [16,31]. Figure 18 shows the SEM image of the Ni-YSZ anode after electrochemical performance test-

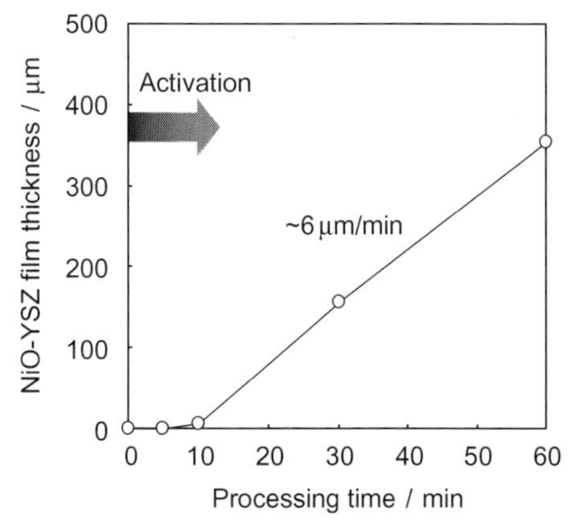

FIGURE 16 Relationship between film thickness and processing time.

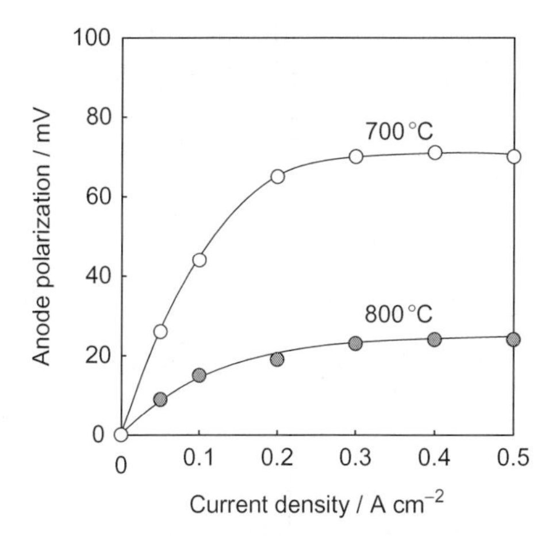

FIGURE 17 Polarization of the Ni/YSZ anode measured by current interruption method at 700 °C and 800 °C.

FIGURE 18 SEM image of the Ni/YSZ porous film after the electrochemical testing.

ing. It is obvious from the SEM image that the anode with the homogeneous porous structure consisting of dispersed Ni and YSZ grains with the size of about 200 nm was successfully fabricated, resulting in the acquisition of a large TPB region. Figure 19 shows the SEM image of a-deposited cathode film on the substrate of YSZ. It was fabricated by using the apparatus shown in Figure 15. This deposition process is very promising for making porous composite films such as fuel cell electrodes.

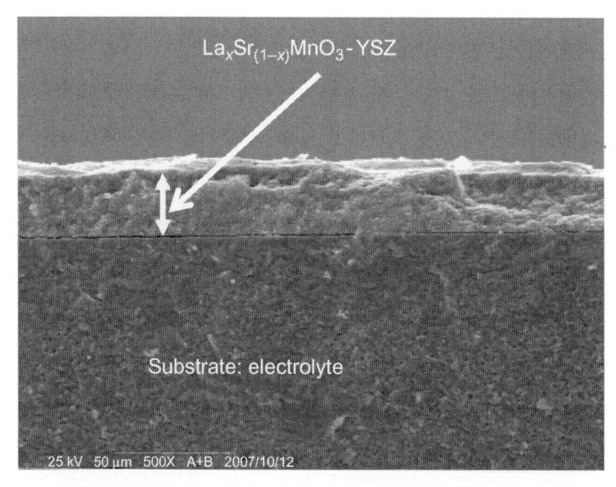

FIGURE 19 Porous as-deposited film of $La_{1-x}Sr_xMnO_3$-YSZ cathode for SOFC made by this method.

7 NOVEL RECYCLING OF COMPOSITE MATERIALS FOR SUSTAINABILITY

7.1 New Concept of Recycling

Figure 20 shows the new concept of the recycling of composite materials by applying particle bonding and disassembling between different kinds of materials [1]. From the view of conventional recycling, each element of waste composite materials must be returned back to its original state for repeat use. It is acceptable when the purpose of the recycling process is to recover only the valuable elements from the waste materials. However, the recycling costs are high and the obtained elements have a lower quality than the virgin materials. As a result, the recycling process is not practical.

On the contrary, the proposed recycling concept does not aim to obtain each original element, but to develop further advanced materials by making use of disassembled blocks of the waste composite materials. In this case, how to apply bonding and disassembling the waste materials is the key issue. As shown in Figure 20, recycling waste back to its intermediate structure and then assembling it with another material to make more advanced materials would be more energy efficient than reclaiming the original elements. This concept will be a basis for the next generation of recycling system for advanced materials.

Here, we look at an approach to recycle GFRPs. GFRP is a typical composite material having the advantages of light weight, high strength, and high weather resistance. Therefore, it has been used in various applications, including boats, bathtubs, and building materials. Its production volume reached 460,000 tons in Japan in 1996, but decreased gradually since then. However, the volume of

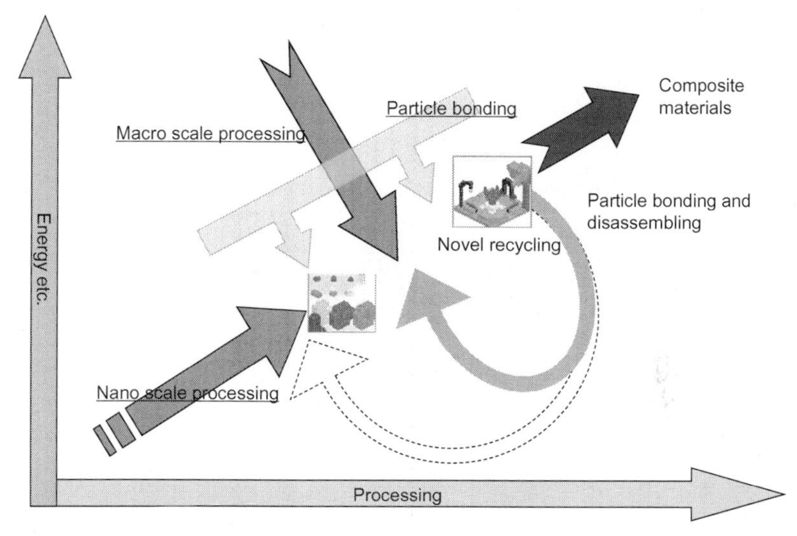

FIGURE 20 Concept of new recycling for composite materials based on particle bonding principle.

waste GFRP has increased every year [1]. So far, almost all of the waste GFRP has been incinerated or disposed of in landfill. Only 1-2% of the waste GFRP is recycled as cement raw material or additives for concrete. The Japan Reinforced Plastics Society started producing cement recycled from GFRP in 2002. The incineration of GFRP has a problem with the low calorific values on burning, and its residue needs to be disposed. GRFPs usually contain 40-50% of calcium carbonate filler and 20-30% of glass fibers [1]. These materials must be recycled through a simple and low-energy process for a profit. Therefore, we aim to develop a new recycling method to make advanced materials from the waste GFRP based on the concept of Figure 20.

7.2 Recycling of GFRP for Advanced Materials

Figure 21 illustrates an innovative recycling process for GFRP used for bathtubs proposed by the authors [1]. It consisted of two unit processes: First, the GFRP was separated into glass fibers and matrix resins, and then the surface of separated glass fibers was coated by low-cost nanoparticles. The coated composite glass fibers would be compacted to make porous materials as explained in Figure 5. High functional materials, having the properties of very low thermal conductivity, being light weight, and requiring easy machining are expected to be obtained by applying the process as shown in Figure 21.

The waste GFRP chip crushed down to about 1 cm was processed by an attrition-type mill, which applied a similar mechanical principle to that of the particle bonding process. When strong shear stress was applied to the chip layers for their surface grinding, glass fibers began to separate from matrix resins

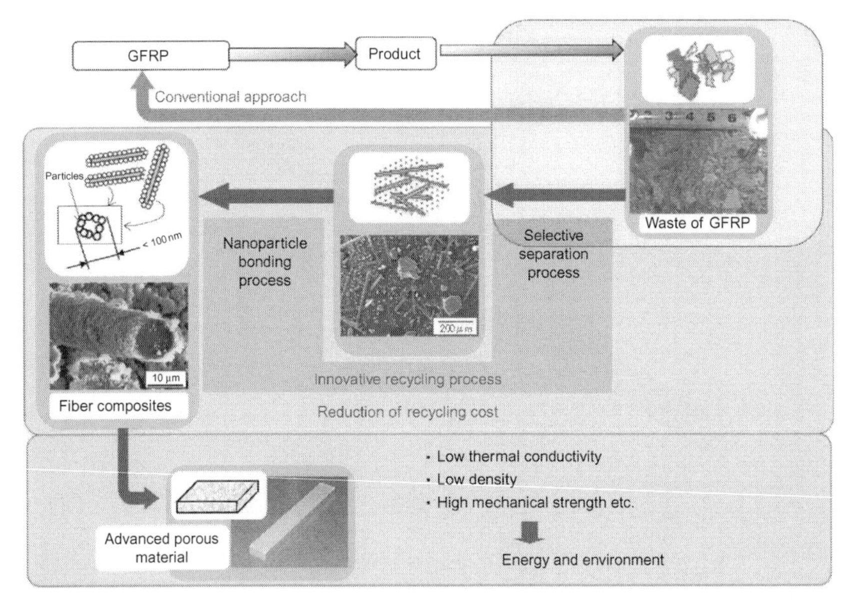

FIGURE 21 New recycling process of GFRP for advanced materials.

on the chip surfaces. As a result, all glass fibers were effectively separated from other matrix components. Then, the surface of glass fibers and that of matrix components were mechanically bonded with nanoparticles. The board was compacted with the mixture of glass fiber composites and matrix components by dry pressing. The board has a relatively high fracture strength and is therefore easy for machining into various shapes. Figure 22 shows the relationship between thermal conductivity and the bulk density of the compact [32]. When fumed silica mass percentage increases, the bulk density decreases, thereby leading to extremely lower thermal conductivity of the compact. These results suggest that the board will be used as a new material such as building material of the future.

On the other hand, GFRP has been also used together with other materials as composites. FRP mortar pipe (FRPM pipe) is a typical case, and has been widely used for many purposes. New recycling process of FRPM pipes has been proposed and applied for their recycling. The proposed process is shown in Figure 23 [33]. At first, waste FRPM pipes were ground by a rotary hammer-type crusher. As a result, the fractures were separated into mortar part and FRP part by using the different morphology between them. Optical microscope images of both products are shown in Figure 24. The mixture of mortar and resin layers shows lump blocks, but the FRP layer parts show a platelike shape. They were easily sieved into two parts. The recovered mortar part is expected to be reused for FRPM pipes.

Further, FRP parts were changed into another advanced material. At first, they were chopped by a cutter mill. The chopped FRP chips were effectively separated to glass fibers and matrix resin particles by applying strong shear

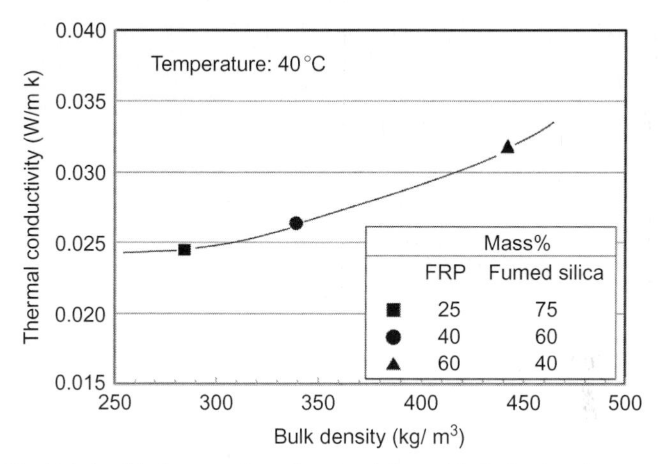

FIGURE 22 Relationship between thermal conductivity and bulk density of porous compacts made by the process as shown in Figure 21.

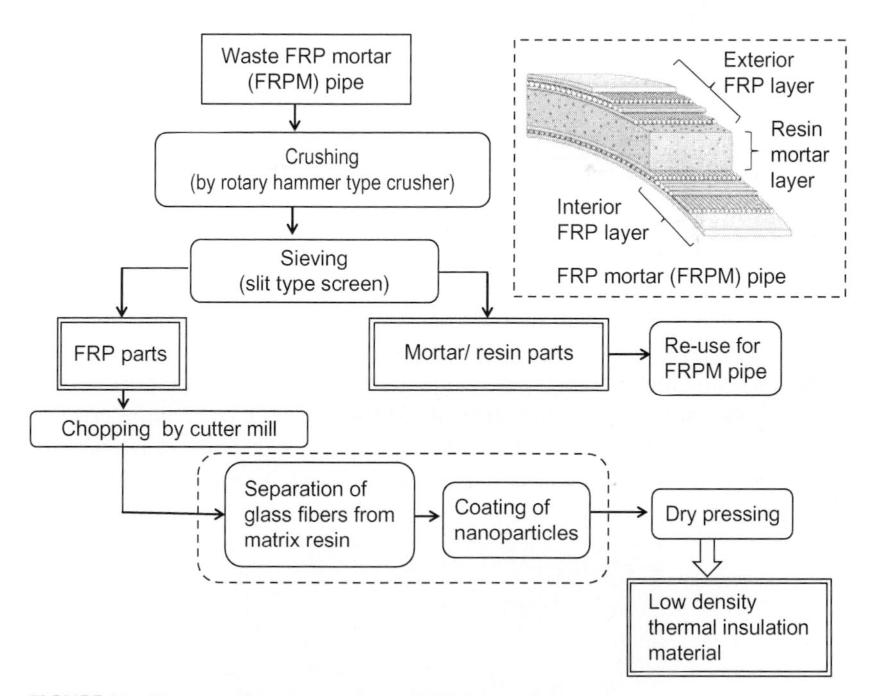

FIGURE 23 New recycling process of waste FRP mortar pipe.

stress at the interface between them as shown in Figure 21. The separated glass fibers kept their own structure without significant breakage. The separated particles and fibers were coated with a nanoparticle porous layer as shown in Figure 21. The coated particles were then pressed to make a bulk body. As a result, the bulk body was light in weight, and the thermal conductivity at

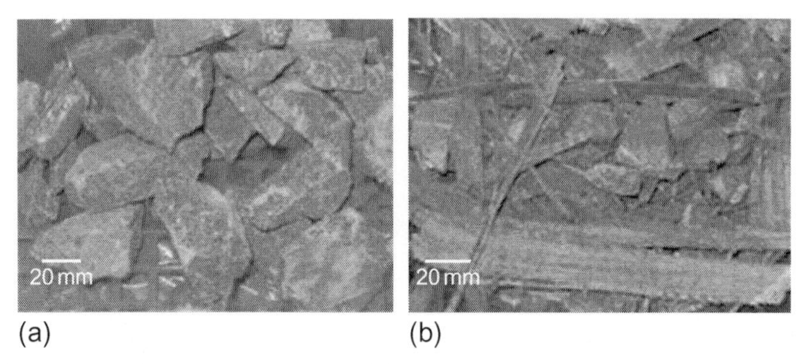

FIGURE 24 Optical microscope images of the products sieved from crushed FRPM pipe fractures: (a) Parts of mortar and resin layer, (b) FRP layer parts.

40 °C was almost the same value as those shown in Figure 22 [33]. The product is expected to be used as thermal insulation material. The proposed recycling systems shown in Figures 21 and 23 can achieve the 100% recycling of the wasted composites. It is also a simple process without any heat assistance, thereby leading to drastic energy-savings for the recycling.

7.3 Recovery of Useful Elements from Waste Composite Materials by Applying Particle Disassembling Process

The disassembling process of the composite structure by applying particle bonding is also effective for novel recycling of other composite materials. For example, flexible printed circuits (FPCs) are generally used to assemble electronics units of mobile phones and the other electronic devices. When recycling the waste FPC, separation of the layered components on metal substrates, mounted parts, and resin films covering wire is important for the effective recovery of precious metals and other valuable elements. Difficulties in the recycling process is due to their flexible features that prevent them from breaking down against impaction force loaded by conventional grinding methods. Figure 25 shows the

FIGURE 25 Photo image of the fractions of flexible printed circuits processed by cutter milling.

FIGURE 26 Photo image of the ground sample of flexible printed circuits.

chopped samples of FPC cut from the wastes [34]. They are very flexible and have various structures. Therefore, applying shearing forces for the surface of waste FPC chips by using a particle disassembling process are effective for this purpose. The experiment was actually conducted, with the result that the FPC chips were well ground and separated into flakes and powders as expected [34].

Figure 26 shows the photo image of the ground sample of FPC chips [34]. The samples are almost Cu. Actually, about 90 mass% of Cu was recovered in the flakes on 1 mm sieve by applying this method. On the other hand, this method was also effective in recovering precious metals. For example, almost all the Pd contained in the waste was condensed for the ground powder smaller than 1 mm. It was because the selective grinding for Pd existing on the surface layer of FPC was achieved, as a result Pd was dominantly obtained in finer particles [34]. It suggests that the particle disassembling process is also able to recover useful elements from waste composite materials effectively without any heat assistance. This will lead to developing new green recycling technology in future.

8 CONCLUSIONS

Particle bonding technology is very promising for the achievements of green and sustainable manufacturing. In this chapter, we introduced new one-pot processing methods developed by using this technology and explained their applications for advanced materials. We also introduced novel recycling processes by applying this technology. We saw that this recycling process can create advanced materials with minimum recycling energy. Particle bonding technology is expected to develop further advanced materials, as well as further green and sustainable manufacturing in future.

REFERENCES

[1] M. Naito, H. Abe, A. Kondo, T. Yokoyama, C.C. Huang, Smart powder processing for advanced materials, KONA Powder Part. J. 27 (2009) 130–143.

[2] M. Naito, H. Abe, Particle bonding technology for composite materials—microstructure control and its characterization, Ceram. Trans. 157 (2004) 69–76.

[3] M. Naito, H. Abe, K. Sato, Nanoparticle bonding technology for the structural control of particles and materials, J. Soc. Powder Technol., Jpn. 42 (2005) 625–631.

[4] M. Naito, A. Kondo, T. Yokoyama, Applications of comminution techniques for the surface modification of powder materials, ISIJ Int. 33 (1993) 915–924.

[5] F. Dachille, R. Roy, High pressure phase transformations in laboratory mechanical mixers and mortars, Nature 186 (1960) 34.

[6] K. Nogi, M. Naito, A. Kondo, A. Nakahira, K. Niihara, T. Yokoyama, New method for elucidation of temperature at the interface between particles under mechanical stirring, J. Jpn. Soc. Powder Powder Metall. 43 (1996) 396–401.

[7] M. Naito, T. Hotta, S. Asahi, T. Tanimoto, S. Endoh, Effect of processing conditions on particle composite process by a high-speed elliptical-rotor-type mixer, Kagaku Kogaku Ronbunshu 24 (1998) 99–103.

[8] Y. Fukumori, H. Ichikawa, T. Uemura, K. Sato, H. Abe, M. Naito, Process performance of dry powder coating for preparing controlled release microcapsules by a high speed mixer, Proceedings of 8th International Symposium on Agglomeration, Bangkok, Thailand, March 16-18, pp. 31-38.

[9] M. Naito, H. Abe, A. Kondo, T. Kozawa, Structure control of particles and powders for advanced materials, J. Soc. Powder Technol., Japan 51 (2014) 174–184.

[10] D. Tahara, Y. Itoh, T. Ohmura, H. Abe, M. Naito, Formation of nanostructure composite using advanced mechanical processing, Ceram. Trans. 146 (2004) 173–177, The Am. Ceram. Soc.

[11] H. Abe, I. Abe, K. Sato, M. Naito, Dry powder processing of fibrous fumed silica compacts for thermal insulation, J. Am. Ceram. Soc. 88 (2005) 1359–1361.

[12] I. Abe, K. Sato, H. Abe, M. Naito, Formation of porous fumed silica coating on the surface of glass fibers by a dry mechanical processing technique, Adv. Powder Technol. 19 (2008) 311–320.

[13] Y.-G. Kwon, S.-Y. Choi, E.-S. Kang, S.-S. Baek, Ambient-dried silica aerogel doped with TiO_2 powder for thermal insulation, J. Mater. Sci. 35 (2001) 6075–6079.

[14] D.R. Smith, J.G. Hust, Microporous Fumed-Silica Insulation Board as a Candidate Standard Reference Material for Thermal Resistance, NISTIR 89-3901, U.S., National Institute of Standards and Technology (1989).

[15] T. Ohmura, I. Abe, Y. Ito, K. Sato, H. Abe, M. Naito, Development and evaluation of characteristics of nanoporous materials, J. Soc. Powder Technol., Jpn. 46 (2009) 461–466.

[16] K. Sato, H. Abe, T. Misono, K. Murata, T. Fukui, M. Naito, Enhanced electrochemical activity and long-term stability of Ni-YSZ anode derived from NiO-YSZ interdispersed composite particles, J. Eur. Ceram. Soc. 29 (2009) 1119–1124.

[17] H. Munakata, T. Tashita, K. Kanamura, A. Kondo, M. Naito, Development of low-platinum catalyst for fuel cells by mechano-chemical method, J. Soc. Powder Technol., Jpn. 48 (2011) 364–369.

[18] H. Abe, M. Naito, K. Nogi, M. Matsuda, M. Miyake, S. Ohara, A. Kondo, T. Fukui, Low temperature formation of superconducting MgB_2 phase from elements by mechanical mixing, Physica C 391 (2003) 211–216.

[19] H. Abe, M. Naito, K. Nogi, A. Kondo, T. Fukui, S. Ohara, M. Matsuda, M. Miyake, Formation of MgB2 superconducting phase from mg and b composite particles produced by mechanical mixing, Ceram. Trans. 146 (2004) 277–282, The Am. Ceram. Soc.

[20] R. Asahi, T. Morikawa, T. Ohwaki, K. Aoki, Y. Taga, Visible-light photocatalysis in nitrogen-doped titanium oxides, Science 293 (2003) 269–271.

[21] H. Abe, T. Kimitani, M. Naito, Influence of NH_3/Ar plasma irradiation on physical and photocatalytic properties of TiO_2 nanopowder, J. Photochem. Photobiol. A Chem. 183 (2006) 171–175.

[22] K. Sato, J. Chaichanawong, H. Abe, M. Naito, Mechanochemical synthesis of $LaMnO_3$ fine powder assisted with water vapor, Mater. Lett. 60 (2006) 1399–1402.

[23] K. Hosokawa, A. Kondo, M. Okumiya, H. Abe, N. Naito, Effects of mechanical processing conditions on the synthesis of oxide nano-particles, J. Soc. Powder Technol., Japan 25 (2014) 1430–1434.

[24] J. Chaichanawong, K. Sato, H. Abe, K. Murata, T. Fukui, T. Charinpanitkul, W. Tanthapanichakoon, M. Naito, Formation of strontium-doped lanthanum manganite (La0.8Sr0.2MnO3) by mechanical milling without media balls, Adv. Powder Technol. 17 (2006) 613–622.

[25] A. Kondo, K. Sato, H. Abe, M. Naito, H. Shimoda, Mechanochemical synthesis of barium titanate from nanocrystalline $BaCO_3$ and TiO_2, Ceram. Trans. 198 (2007) 375–380.

[26] S. Ohara, A. Kondo, H. Shimoda, K. Sato, H. Abe, M. Naito, Rapid mechanochemical synthesis of fine barium titanate nanoparticles, Mater. Lett. 62 (2008) 2957–2959.

[27] J. Yoshida, M. Stark, J. Holzbock, N. Huesing, S. Nakanishi, H. Iba, M. Naito, Analysis of the size effect of $LiMnPO_4$ particles on the battery properties by using STEM-EELS, J. Power Sources 226 (2013) 122–126.

[28] J. Yoshida, S. Nakanishi, H. Iba, A. Kondo, H. Abe, M. Naito, One-step mechanical synthesis of nanocomposite granule of $LiMnPO_4$ nanoparticles and carbon, Advanced Powder Technology 24 (2013) 829–832.

[29] H. Abe, M. Naito, K. Sato, Mechanically assited deposition of nickel oxide-yttria stabilized zirconia nanocomposite film and its microstructural evolution for solid oxide fuel cells anode application, Int. J. Appl. Ceram. Technol. 9 (2012) 928–935.

[30] J. Akedo, Aerosol deposition of ceramic thick film at room temperature densification mechanism of ceramic layers, J. Am. Ceram. Soc. 89 (2006) 1834–1839.

[31] S.P. Jiang, A comparative study of fabrication and performance of Ni/3 mol% Y_2O_3-ZrO_2 and Ni/8 mol% Y_2O_3-ZrO_2 cermet electrodes, J. Electrochem. Soc. 150 (2003) E548–E559.

[32] A. Kondo, H. Abe, N. Isu, M. Miura, A. Mori, T. Ohmura, M. Naito, Development of lightweight materials with low thermal conductivity by making use of waste FRP, J. Soc. Powder Technol., Jpn. 47 (2010) 768–772.

[33] A. Kondo, T. Ohmura, H. Abe, J. Kano, M. Naito, Development of recycling process for waste FRP mortar pipe, J. Soc. Powder Technol., Jpn. 49 (2012) 827–831.

[34] A. Kondo, S. Koyanaka, T. Ohki, H. Abe, M. Naito, Effect of grinding method on the recycling of waste flexible printed circuits, J. Soc. Powder Technol., Jpn. 48 (2011) 750–754.

Chapter 9

Green Manufacturing of Silicon Nitride Ceramics

Hideki Hyuga, Naoki Kondo and Tatsuki Ohji
National Institute of Advanced Industrial Science and Technology (AIST), Nagoya, Japan

1 INTRODUCTION

Silicon nitride is one of the most widely used engineering ceramics for various applications because of its excellent mechanical properties including strength, fracture toughness, creep resistance, heat/corrosion resistance, wear/erosion resistance, and thermal shock resistance. However, sintering of silicon nitride is usually conducted at elevated temperatures above 1700 °C for several hours and requires substantial amount of electric energy. This is particularly true when the size of the components is large. For example, in aluminum casting industries a variety of large-scaled silicon nitride components are used including heater tubes, ladles, and stokes, and some of them exceed 20 kg in weight. Kita *et al.* [1] revealed that sintering process of a silicon nitride heater tube with ~20 kg weight requires electric power of nearly 780 megajoules (MJ), which is much higher than the total fabrication energy of a stainless-steel heater tube with the same weight. Therefore it is essentially important to reduce the energy required for manufacturing process of silicon nitride ceramic components. For examples of such efforts, this chapter deals with two topics for energy-efficient sintering of silicon nitride. One is a sintered reaction bonding process with rapid nitridation. The particular focus is placed on nitridation of silicon powder enhanced by a zirconia addition and resultant sintered-reaction-bonded silicon nitride ceramics. The other topic is low-temperature sintering, below 1700 °C, with atmospheric pressure. This approach is advantageous for saving the energy in sintering process of silicon nitride components, as well as reducing the production cost.

2 SINTERED REACTION BONDING PROCESS WITH RAPID NITRIDATION

Among the several different approaches of sintering silicon nitride, the sintered reaction bonding process has a couple of major advantages. First, raw materials of silicon powder are generally cheap compared to silicon nitride powders,

Green and Sustainable Manufacturing of Advanced Materials. http://dx.doi.org/10.1016/B978-0-12-411497-5.00009-6

leading to substantial reduction of fabrication cost. Second, shrinkage after sintering is small, making precise shape/size control of fabricated components possible [2–4]. Therefore this technique is widely used for fabricating the large-scaled components where lowering the cost of raw materials and near-net shaping are the most important issues. The sintered reaction bonding process is typically divided into two steps: in the first the nitridation of green silicon powder compacts around 1400 °C (near silicon's melting temperature) and in the other the post-sintering of nitrided compacts at 1700-1900 °C. In the nitridation process, the reaction between silicon and nitrogen is exothermic, and rapid nitridation of silicon very often leads to melting of silicon due to the self-heating reaction prior to the completion of nitridation. Therefore, in general the nitridation is conducted very slowly at temperatures where such melting of silicon can be avoided; however, this leads to a great deal of energy loss. To overcome this problem, many researchers have reported the catalytic nitridation of silicon at relatively low temperatures [5–8]. Iron is a well-known catalyst for promoting nitridation of silicon and is generally contained as an impurity in the raw material of low-cost grade silicon powders. However, it is also known that iron in a green body remains after sintering and acts as a fracture defect, reducing mechanical strength. The catalytic effects of other metals and oxides such as chromium, calcium, copper, silver, on the nitridation of silicon have also been investigated [7]. Recently, it has been reported that zirconia is a very effective catalyst for nitridation of silicon [9–14], and that, with some other appropriate oxide additives, almost full densification with good mechanical properties can be obtained after the post-sintering process [11,14–18].

Silicon nitride can be formed from silicon and nitrogen according to the following reaction.

$$3Si + 2N_2 \rightarrow Si_3N_4 \tag{1}$$

This reaction indicated that when full nitridation is obtained, the increase in mass of silicon compact is 1.665 times the original mass of silicon. Figure 1 shows nitridation behavior as a function of temperature for silicon (average particle size 1 μm) with the addition of various oxides including ZrO_2 (average particle size 0.03 μm), Y_2O_3 (average particle size 1 μm), Fe_2O_3 (average particle size 1 μm), and MgO (average particle size 0.1 μm), in comparison with silicon only. The degree of nitridation was calculated by the weight change obtained from thermogravimetric (TG) analysis of the mixed powders heated up to 1400 °C at a rate of 20 °C/min under nitrogen atmosphere. The amount of added oxides on the powder was 8.08 mass% (5 mass% of the fully nitrided silicon powder, provided that the entire initial silicon powder reacted to form silicon nitride). Even though all the oxide additions result in enhanced nitridation of silicon, the most remarkable nitridation was obtained with zirconia addition.

Table 1 shows a summary of the TG analysis results for the silicon with different amounts of zirconia, 8.08, 15.65, and 28.45 mass% (which are 5, 10, and 20 mass%, respectively, in the fully nitrided silicon powder). From these

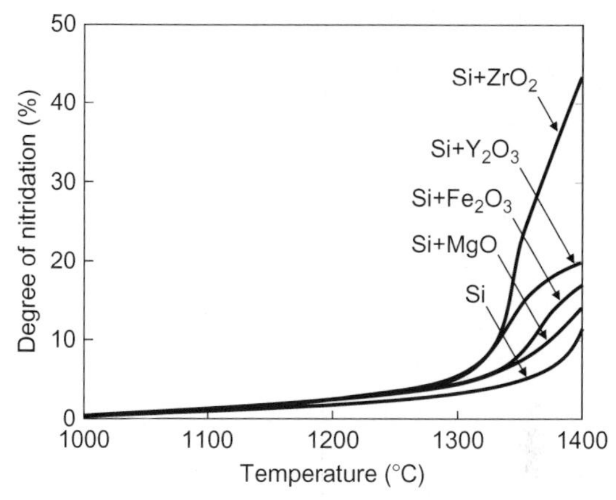

FIGURE 1 Nitridation behavior as a function of temperature for Si with the addition of various oxides including ZrO_2, Y_2O_3, Fe_2O_3, and MgO, in comparison with Si only. Nitridation of Si is remarkably enhanced with ZrO_2 addition.

results, the temperature at which the first reaction step began was similar for all samples. On the other hand, the starting temperature of the main reaction step decreased with increasing zirconia content, and at a temperature of 1400 °C, the degree of nitridation increased linearly with increasing the zirconia content. These results indicate that although additions of zirconia had little effect on the initial nitridation starting temperature or the reaction rate, remarkable effects were observed on the main nitridation reaction temperature and the amount of nitrided silicon formed at 1400 °C.

Figure 2 shows nitridation behavior as a function of heating rate for the samples of silicon with zirconia addition (average particle size 1 μm, 8.08 mass%) and silicon only, which were heated up to 1400 °C. The range of heating rate where melted silicon was observed was also indicated for each sample. Sufficient nitridation of almost 95% without melted silicon was achieved with a rapid heating of 12.5 °C/min for zirconia added silicon. The value of 95% was the maximum nitridation obtained in the series of studies on silicon nitridation, and it can be assumed to be full nitridation because of volatilization of silicon during the nitridation process and oxide layer on the surface of the silicon starting powder. On the other hand, in the case of silicon only, the heating rate was required to be reduced down to as low as 0.15 °C/min to obtain such successful nitridation; this, the former is 80 times faster than the latter. This indicates that consumed electric power in nitridation is about 80% reduced by zirconia addition, compared to the case of silicon only. Figure 3 shows samples of silicon with zirconia addition and silicon only, which were nitrided at 10 and 0.5 °C/min, respectively. Clean nitridation without melted Si was obtained for the former despite the high heating rate, while unreacted silicon seeping out from the inside was observed for the latter.

TABLE 1 TG analysis results for the silicon with different amounts of zirconia: 8.08, 15.65, and 28.45 mass% (which are 5, 10, and 20 mass%, respectively, in the fully nitrided silicon powder) [9]

ZrO_2 Content (mass%)	Weight Gain (%)		Nitridation Degree (%)	Starting Temperature (°C)	
	1200 °C	1400 °C	1400 °C	First Reaction	Main Reaction
0	1.37	9.45	14.58	1365	1365
8.08	1.62	24.1	40.28	952	1318
15.65	1.68	28.28	51.21	950	1307
28.45	1.13	33.25	71.23	949	1278

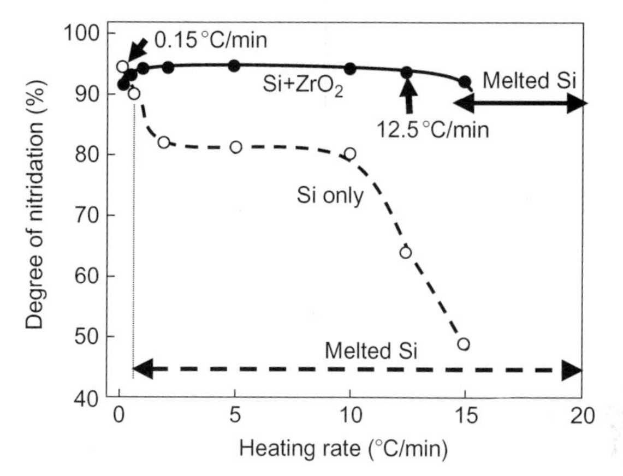

FIGURE 2 Nitridation behavior as a function of heating rate for "Si + ZrO_2," and "Si only," which were heated up to 1400 °C. The range of heating rate where melted Si was observed was also indicated for each sample. Sufficient nitridation without melted Si was obtained with a rapid heating of 12.5 °C/min for "Si + ZrO_2," but with a slow heating of 0.15 °C/min for "Si only," indicating the former is 80 times faster than the latter.

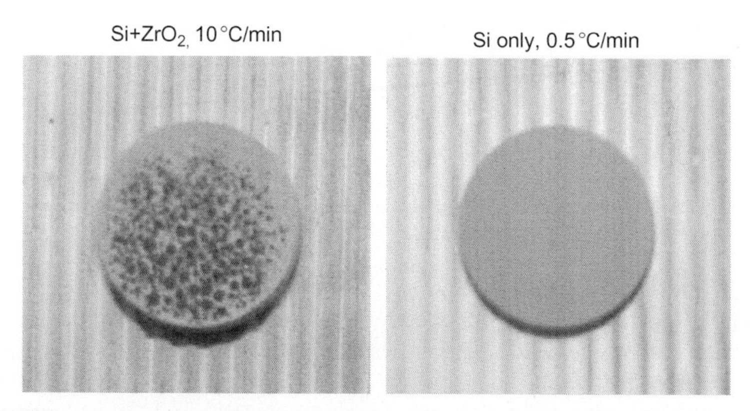

FIGURE 3 Appearance of samples of "Si + ZrO_2," and "Si only," which were nitrided at 10.5 and 0.5 °C/min, respectively. Clean nitridation without melted Si was obtained for "Si + ZrO_2," even at the high heating rate, whereas unreacted silicon seeping out from inside was observed for "Si only" even at the low heating rate.

To clarify the mechanism of nitridation enhancements by zirconia addition, high-temperature X-ray diffraction studies were performed using silicon (average particle size 9.5 μm) nitrided at 1375 and 1450 °C [10]. Figure 4a shows the X-ray diffraction spectra of silicon powder nitrided at 1375 °C both with and without zirconia. Peaks attributed to silicon were clearly identified in the silicon-only specimen demonstrating that in this specimen nitridation was not completed. None of the peaks were observed in the zirconia-added specimen. Furthermore, the monoclinic phase zirconia powder, added to the

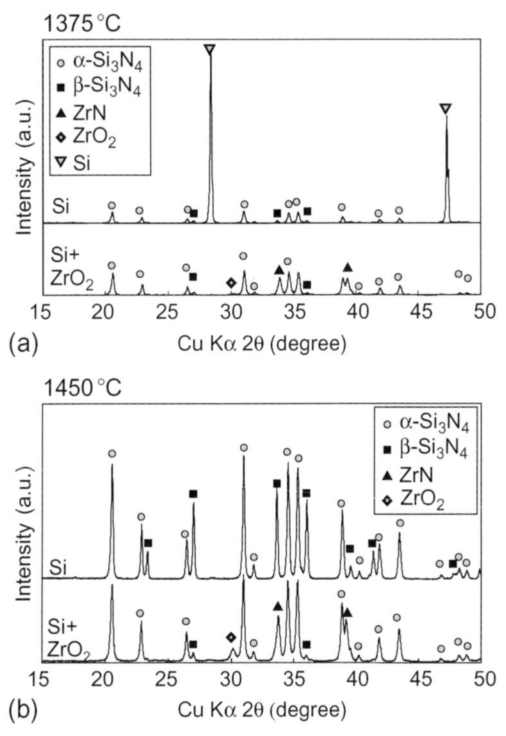

FIGURE 4 XRD spectra of samples of Si only and Si with ZrO_2 addition, nitrided at (a) 1375 °C and (b) 1450 °C. Spectrum at 1375 °C for Si shows residual Si indicating incomplete nitridation. At 1450 °C nitridation is compete for both the samples [10]. *(Reproduced with permission of the Ceramic Society of Japan. All rights reserved.)*

starting powder, was almost completely transformed to ZrN during the nitridation process at this heating condition. This transformation was observed in specimens nitrided at 1200 °C, indicative of ZrN formation below this temperature. Figure 4b shows the spectra of silicon powder nitrided at 1450 °C both with and without zirconia. In this condition, none of silicon peaks were identified in either specimen, indicating completion of nitridation. Both α and β crystalline phases of silicon nitride were identified; however, the peak intensity of the β silicon nitride phase was very low in the zirconia-added specimen indicating that the zirconia addition suppressed the formation of β-phase silicon nitride during the nitridation process. Mitomo *et al.* [5] reported that the α/β ratio of nitrided silicon with FeO addition in the reaction-bonding process was higher than that observed for nitride pure silicon. Similarly many other researchers [6,19] have reported the high α/β ratio of nitrided silicon with various added oxides, and this is most likely attributable to the high oxygen pressure and early stage nitridation caused by the oxide addition [2]. The higher α-phase content observed in the current work is also due to the zirconia acting as a catalyst for nitridation of silicon powder at lower temperature when compared with silicon powder only. The peaks attributed to ZrN were observed in the specimen with zirconia additions also in Figure 4b. The possible reactions for forming ZrN during nitridation process are as follows:

$$N_2[g] + 2ZrO_2[sl] + 2Si[sl] \rightarrow 2SiO_2[sl] + 2ZrN[sl] \quad (2)$$

$$N_2[g] + 2ZrO_2[sl] + 4Si[g] \rightarrow 4SiO[g] + 2ZrN[sl] \quad (3)$$

During heating, the original zirconia inside the powder compact forms ZrN by one or both of the above reactions. Weiss *et al.* [20] have reported on the Si-Zr-Al-O-N phase equilibria and showed that ZrN reacts with SiO_2 to form ZrO_2, SiO, and N_2 at a temperature of 1400 °C. This reaction implies that the formed ZrN can be a resource to supply nitrogen at temperatures above 1400 °C, according to the following reaction:

$$2SiO_2 + ZrN \rightarrow ZrO_2 + SiO + 0.5N_2 \quad (4)$$

In the center of reaction-bonded compacts, it is not sufficient to supply only nitrogen gas for converting silicon to silicon nitride, because the nitridation and resultant densification proceed preferentially around the surface area of the powder compact, making it difficult for nitrogen to penetrate to the center. With zirconia additions, the nitridation is promoted by forming ZrN inside the green compact at lower temperatures. The ZrN phase is produced below 1200 °C as already noted. The ZrN then reacts with oxygen and transforms back to ZrO_2, while the SiO gas produced forms the silicon nitride by reacting with nitrogen and releases the oxygen. It is thought that the transformation to ZrN (reaction (2) or (3)) and then converted back to ZrO_2 (reaction (4)) can occur reciprocally, and the nitridation is promoted by this circulating reaction mechanism. After the residual silicon has disappeared, ZrO_2 is increased with decreasing ZrN due to the termination of reaction (3). This fact agrees with the results of X-ray diffraction.

The nitridation behavior of silicon is substantially affected by the particle sizes of silicon powder as well as added zirconia powder [10,12]. Figure 5 shows the degree of conversion of both silicon-only samples and samples added with zirconia as a function of the reaction temperature for three sources of silicon powders with different average particle sizes of 1.5, 2.6, and 9.5 μm. In the case of silicon-only samples, the degree of nitridation was significantly higher for the finer silicon powder in the temperature range up to 1375 °C. At the temperature of 1450 °C, whereas the weight gain of both the samples of 1.5 and 2.6 μm indicated similar successful nitridation rate of about 95%, i.e., full nitridation, that of 9.5 μm particle size showed about 85%. As for the samples added with zirconia, the degree of nitridation showed a similar trend to that of the silicon-only powders; nitridation is more enhanced for smaller particle sizes. However, the degree of nitridation was higher when compared to the silicon-only samples at temperatures up to 1375 °C. Even for the coarse silicon powder, the level

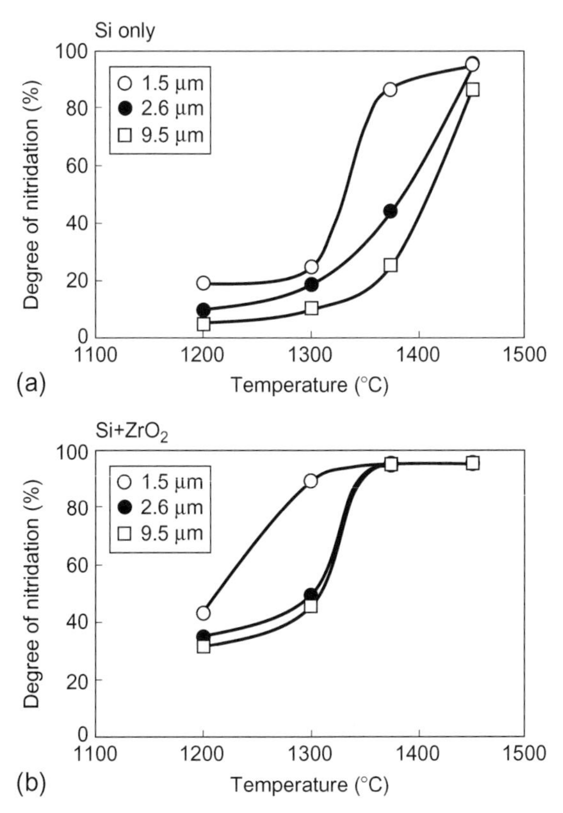

FIGURE 5 Degree of nitridation as a function of temperature for samples produced from three kinds of Si powders of different particle sizes: (a) Si only and (b) Si with ZrO_2 [10]. *(Reproduced with permission of the Ceramic Society of Japan. All rights reserved.)*

of nitridation was 96% at 1375 °C, i.e., almost full conversion was achieved at a lower temperature. Furthermore, the degree of nitridation of the samples using 2.6 and 9.5 μm powders was almost the same for all the nitridation temperatures. These results indicate that although the nitridation behavior of silicon powder is affected by the silicon particle size, the addition of zirconia so markedly enhanced the nitridation that the particle size effect diminishes and the full nitridation is obtained even for all the silicon powders at 1375 °C or higher temperatures.

The effects of zirconia particle size on nitridation behavior of silicon powder has been also investigated using three kinds of zirconia powders with different average particle sizes of 70 nm, 2.4 μm, and 6.1 μm. Figure 6 shows thermogravimetric behavior from 900 to 1400 °C for both silicon powder only and silicon powder with the three kinds of zirconia powders. In the case of silicon powder only, the increase in weight started around 955 °C, and rapid weight gain was observed around 1365 °C. The zirconia-added powders showed a

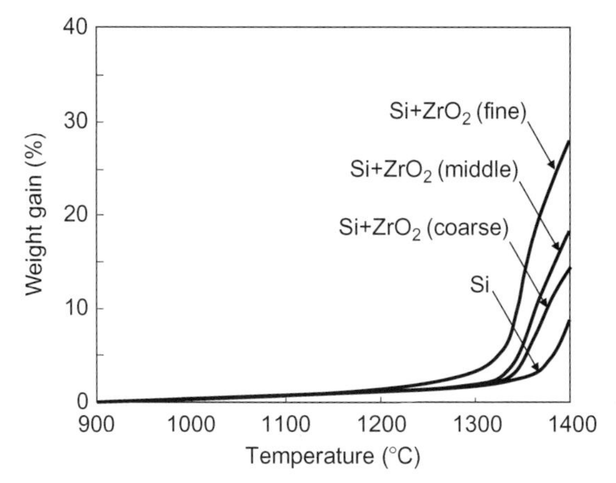

FIGURE 6 Thermogravimetric behavior from 900 to 1400 °C for both silicon powder only and silicon powder with the three kinds of zirconia powders of different particle sizes [12]. *(Reproduced with permission of the Ceramic Society of Japan. All rights reserved.)*

similar onset of weight gain onset similar to that of around 955 °C; however, rapid weight gain started at substantially low temperatures compared to that of silicon powder only, and it was lowered by decreasing the particle size. This observation was also supported by the results of differential thermal analysis for these samples [12]. The zirconia particle size dependence of nitridation behavior of silicon is most likely attributed to a contact situation between zirconia and silicon powders. With increasing the particle size of zirconia, the specific surface area of zirconia powder decreases, reducing the reaction area between these two powders and suppressing the catalysis effect of zirconia.

Densification is slowed with zirconia only as a sintering additive in gas-pressure sintering of silicon nitride. Therefore, magnesium aluminate ($MgAl_2O_4$) spinel was added together with zirconia to lower the melting point of grain boundary glassy phase and enhance the liquid phase sintering during the postsintering process after nitridation [11,21]. Figure 7a shows the degree of nitridation as a function of temperature (heated at 10 °C/h from 1100 °C to specific temperatures and cooled without soaking) in nitridation of silicon only and silicon with 7.825 mass% zirconia and 7.825 mass% magnesium aluminate (the sum of these added amounts are equivalent to 10 mass% when silicon is completely converted to silicon nitride). The degrees of nitridation were calculated from the weight change before and after the nitridation. The sample doped with zirconia and magnesium aluminate showed much enhanced nitridation from the low temperature region, compared to that of silicon only; the nitridation degree at 1375 °C was 95.2% for the former and 86% for the latter. Figure 7b shows the degree of nitridation as a function of holing time (heated at 10 °C/h from 1100 to 1300 °C and soaking in specific times) in nitridation of these samples. Although the nitridation degree for the silicon in zero soaking time was 25.1%, that for the doped sample was as high as 82.1%, and reached 96.1% in 8 h. It should

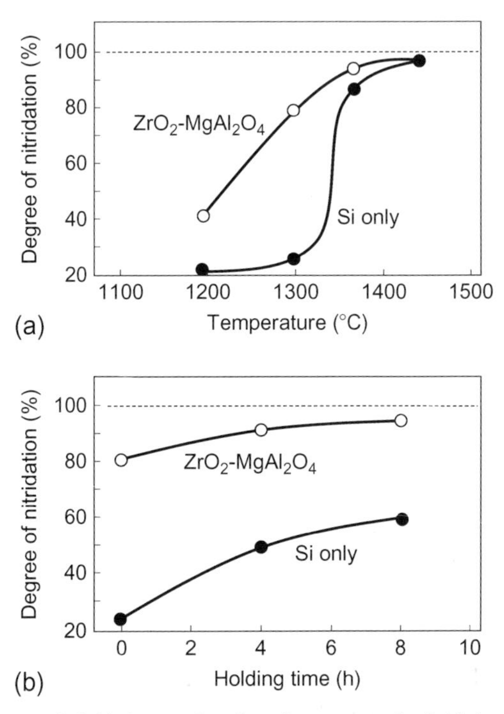

FIGURE 7 (a) Degree of nitridation as a function of temperature in nitridation of silicon only and silicon with 7.825 mass% zirconia and 7.825 mass% magnesium aluminate. (b) Degree of nitridation as a function of holing time in nitridation of the same samples [11].

be noted that, similar to the case of silicon added with zirconia, the addition of zirconia and magnesium aluminate markedly enhanced nitridation of silicon at relatively low temperatures.

Figure 8 shows the scanning electron micrographs (SEM) of the microstructures of silicon nitride postsintered at 1750 °C for 8 h under nitrogen of 0.9 MPa after nitridation of silicon with 7.825 mass% zirconia and 7.825 mass% magnesium aluminate, and silicon nitride sintered with the same sintering conditions and additives using high-purity α silicon nitride powder. Both the materials show similar microstructural feature, composed of fibrous grains that are characteristic of β silicon nitride. However, silicon nitride obtained from α powder showed some coarse fibrous grains and wide distribution of grain size, compared to silicon nitride from silicon powder, where fibrous grain size is relatively uniform. This difference is attributable to the grain growth behavior from β silicon nitride nuclei during liquid-phase sintering; it is known that when the number of the nuclei is large, grain growth is most frequently impinged each other, resulting in microstructure with uniform grain size. It can be considered that silicon nitride from silicon powder contains a larger amount of β silicon nitride nuclei during sintering than silicon nitride obtained from

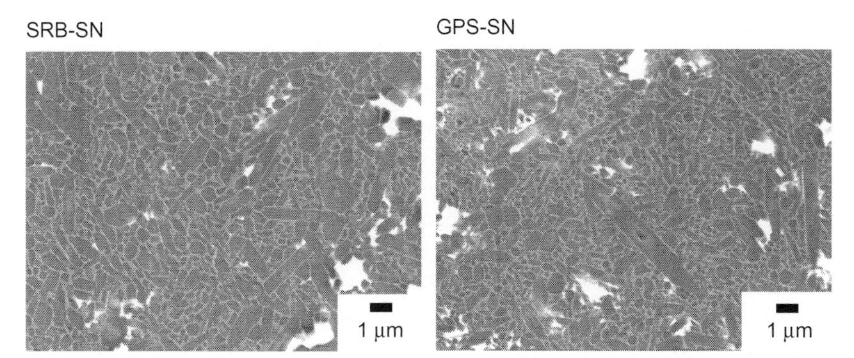

FIGURE 8 SEM images of microstructures of sintered-reaction-bonded silicon nitride (SRB-SN) using silicon powder and gas-pressure-sintered silicon nitride (GPS-SN) using α silicon nitride powder [11].

TABLE 2 Mechanical properties of sintered-reaction-bonded silicon nitride (SRB-SN) using silicon powder and gas-pressure-sintered silicon nitride (GPS-SN) using α silicon nitride powder

	SRB-SN	GPS-SN
Bending strength	851 ± 36 MPa	825 ± 20 MPa
Fracture toughness	5.3 ± 0.3 MPa m$^{1/2}$	5.9 ± 0.3 MPa m$^{1/2}$
Vickers hardness	14.0 ± 0.2 GPa	14.8 ± 0.8 GPa
Young's modulus	305 GPa	310 GPa

α powder, leading to the difference of microstructure previously noted. Table 2 shows mechanical properties of these two silicon nitrides. The fracture strength was determined by a four-point bending test, and the fracture toughness was measured by an indentation-fracture method. The Young's modulus was evaluated by ultrasonic pulse method. Vickers hardness was measured by a Vickers indentation tester applying an indentation load of 196 N. It should be noted that silicon nitride from silicon showed a fracture strength of 850 MPa, which is slightly higher than that of silicon nitride from α powder. The other properties, which include fracture toughness, Young's modulus, and Vickers hardness, are similar to each other, indicating that good mechanical properties equivalent to those of gas-pressure sintered material from high-purity silicon nitride powder can be obtained via a sintered reaction bonding technique with rapid nitridation using silicon powder, whenever the fabrication process is optimized.

The rapid nitridation with zirconia addition was also demonstrated for β SiAlON ceramics [13,14]. As it was for the case of silicon nitride ceramics, thermogravimetric analysis revealed that the addition of zirconia reduced the starting temperature of the main nitridation reaction of silicon-based mixed powders corresponding to a β SiAlON composition of $Z=1$. The fabrication of β SiAlON ceramics without zirconia addition via a rapid reaction-bonding route resulted in unreacted silicon that seeped out to the surface and substantially changed the morphology of the green compact; conversely, those with ZrO_2 were fully nitrided and no such bleeding of molten silicon was observed with keeping the original green morphology after nitridation. The rapid-nitrided and post-sintered β SiAlON samples with zirconia showed the mechanical properties and microstructure both comparable to those produced via a slower, conventional nitridation and postsintering. Only zirconia was used as an additive in this case.

Besides zirconia, rare-earth oxides, which are the most frequently used additives for sintering of silicon nitride ceramics, have also been systematically investigated as catalytic agents for nitridation enhancement, using thermogravimetric analysis [22]. Although almost all rare-earth oxide doped silicon powders exhibited a nitridation enhancing effect, the addition of CeO_2, and Eu_2O_3 to silicon powder caused the largest reduction in the temperature of the main nitridation reaction and increased the amount of silicon converted to silicon nitride at a given temperature.

3 LOW-TEMPERATURE SINTERING WITH ATMOSPHERIC PRESSURE

Silicon nitride is generally sintered at temperatures above 1700 °C as already stated; if, however, the sintering can be made at lower temperatures, around 1600 °C, it saves markedly both the consumed electric energy and production cost. In addition, in such high-temperature sintering above 1700 °C, high nitrogen gas pressure (typically up to 1 MPa) is applied for avoiding decomposition [23]; therefore, a furnace with a gas pressure chamber is needed, making the sintering process costly. On the other hand, in low-temperature sintering around 1600 °C, the use of atmospheric pressure (~0.1 MPa) nitrogen would be sufficient to avoid decomposition, making the fabrication process simple and reducing the production cost substantially.

The microstructures and properties of ceramics are significantly affected by the starting powders; they also take substantial portions of total production costs of ceramics. There are two types of silicon nitride powders, α- and β-phase powders; the former is low-temperature phase and the latter is a high-temperature one. Silicon nitride ceramics are most frequently produced by using α-phase powder as a starting material, which is transformed to β phase at 1400-1600 °C resulting in fibrous growth of grains during subsequent sintering and densification, typically above 1700 °C [24,25]. The microstructure composed of fibrous grains is advantageous in terms of good mechanical properties because crack wake toughening effects, including grain bridging and pullouts, are enhanced [26,27]. When using β powder as a raw material, higher temperatures or longer-sintering times are required to obtain such microstructures of well-developed fibrous gains, because of the impingement effects of many β-phase nuclei [24,25]. On the other hand, the β powder is generally cheaper than α powder when the purity and particle size are equal. Further, β-phase powder has relatively good sinterability compared to α-phase powder [25,28,29], indicating the potential of low-temperature sintering below 1600 °C.

This section deals with the sintering behaviors of α- and β-phase silicon nitride powders at temperatures below 1600 °C under atmospheric pressure (~0.1 MPa) nitrogen, to examine the energy and cost-saving approach to manufacture silicon nitride components via low-temperature sintering [30]. The investigation used the following three types of silicon nitride powders: an α-phase-rich powder, SN-9FWS (DENKA K.K., Tokyo, Japan), which is commonly used as a raw powder for sintering, and two β-phase-rich powders, NP-500 (DENKA K.K.) and SN-F1 (DENKA K.K.). NP-500 is specially ordered from the manufacturer, and SN-F1 is a low-cost filler-grade powder. Hereafter, SN-9FWS, NP-500, and SN-F1 are denoted as high-grade α, high-grade β, and low-grade β, respectively. Table 3 lists the prices, β-ratios, mean particle sizes, and impurities of these powders. Because the low-grade β powder had a large particle size of 2.4 μm and tended to agglomerate to ~20 μm, it was ball-milled and refined to 1.1 μm. Note that the low-grade β powder contains a relatively large amount of impurities, particularly iron. The silicon nitride raw powders were mixed with the sintering additives, 5 wt% Y_2O_3 and 3 wt% Al_2O_3

TABLE 3 Prices, β-ratios, mean particle sizes, and impurities of the used three sorts of powders

	High-Grade-α (9FWS)	High-Grade-β (NP500)	Low-Grade-β (F1)
Price	9000 JPY/kg	9000 JPY/kg (estimated)	2200 JPY/kg
β Ratio	8%	88%	95%
Mean particle size	0.8 μm	0.8 μm	1.1 μm (after ball-milling)
Impurities	O ~ 0.8%	O ~ 1.5%	O ~ 1.5%
	C ~ 0.2%	C ~ 0.3%	C ~ 0.3%
	Cl ~ 1 ppm	Cl ~ 1 ppm	Cl ~ 1 ppm
	Fe ~ 200 ppm	Fe ~ 500 ppm	Fe ~ 2000 ppm
	Al ~ 400 ppm	Al ~ 420 ppm	Al ~ 1000 ppm
	Ca ~ 200 ppm	Ca ~ 200 ppm	Ca ~ 1000 ppm

or 5 wt% Y_2O_3 and 5 wt% $MgAl_2O_4$, followed by die-pressing and cold isostatic pressing (CIP, 100 MPa).

Figure 9 shows the β-phase ratio of high-grade α powder as a function of time during sintering at 1550 and 1600 °C under 0.1 MPa N_2. At both the temperatures, the β-phase ratio increased rapidly during the first 4 h, and then gradually approached to the full transformation. Although the complete phase transformation was observed after 8 h at 1600 °C, a very small amount of α phase still remained after 16 h at 1550 °C. Both the high-grade β, and low-grade β powders, which originally contained small amounts of a phase, showed the complete transformation after 2 h at 1550 °C or 1 h at 1600 °C. Figure 10a and b shows the densification behaviors of the powder compacts of high-grade α, high-grade β, and low-grade β, during sintering at 1550 and 1600 °C (under 0.1 MPa N_2), respectively. The densities of both the high-grade β and low-grade β powder compacts gradually increased with time and reached 3.1×10^3 kg/m^3 after 16 h at 1550 °C and 8 h at 1600 °C. On the other hand, the density of the high-grade α first increased up to 4 h at 1550 °C and 2 h at 1600 °C, and then decreased with time. The highest density was 2.8×10^3 kg/m^3 at both the temperatures. This slowing of densification is probably caused by the vaporization of silicon nitride and the sintering additives through pores in the undensified bodies. It is clear that β-phase silicon nitride powder is advantageous for low-temperature sintering. Figure 11 shows microstructure of the sintered body from the high-grade α powder after 8 h at 1600 °C. The body was not densified with a density of 2.69×10^3 kg/m^3, and open pores were observed. Fibrous grains, which are supposed to suppress densification, are indicated by the arrows. Microstructures of the sintered bodies from the high-grade and low-grade β powders after 8 h at 1600 °C under 0.1 MPa N_2 are shown in Figure 12a and b, respectively. On the contrary to that from the high-grade α powder, the bodies, whose respective densities were 3.13×10^3 kg/m^3 and 3.09×10^3 kg/m^3, were

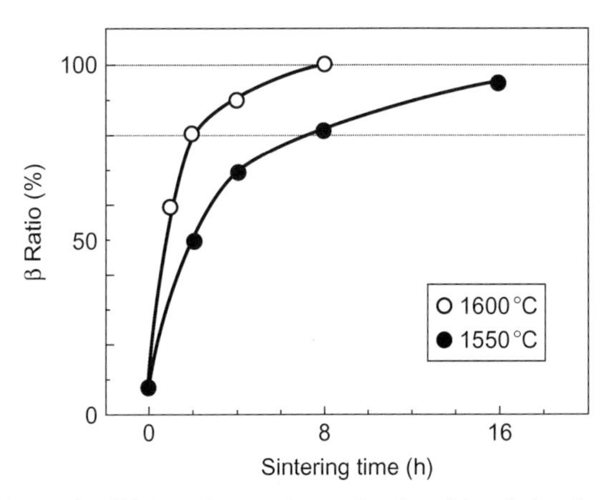

FIGURE 9 β-Phase ratio of high-grade α powder as a function of time during sintering at 1550 and 1600 °C [30]. *(Reproduced with permission of John Wiley and Sons. All rights reserved.)*

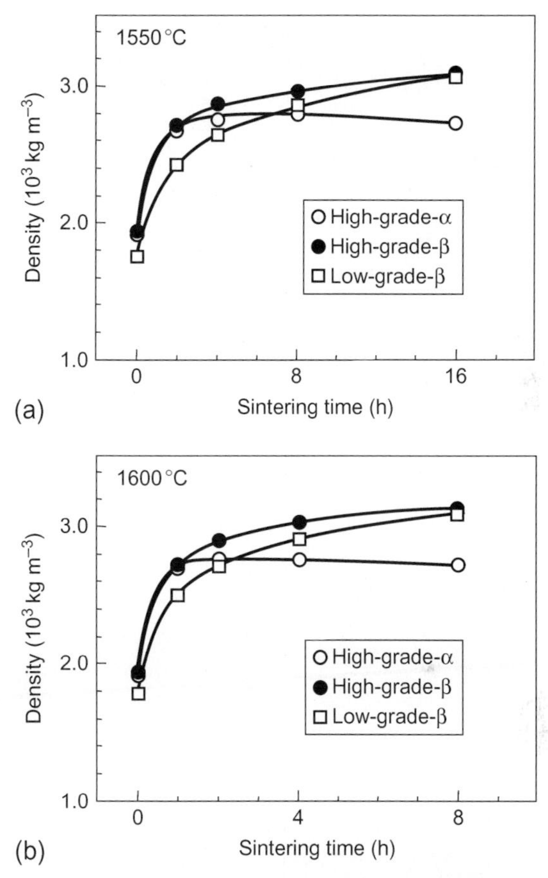

FIGURE 10 Densification behavior during sintering at (a) 1550 and (b) 1600 °C [30]. *(Reproduced with permission of John Wiley and Sons. All rights reserved.)*

almost fully densified with some residual closed pores. Note that grain sizes in the sintered bodies from the β powders were similar to those of the original starting powders. The aspect ratio of the most grains was low with a few fibrous grains. Glassy phase was observed between the grains and at the triple point junctions. The core-rim structures, where silicon nitride core from the starting powder was surrounded by SiAlON rim formed during sintering, were identified in many grains as shown in Figure 12c. Similar core-rim structures have been identified for silicon nitride ceramics sintered using β powders [31] or β seed grains [32]. However, the thickness of the rim observed in Figure 12c is substantially small compared to those of the previously reported core-rim structures due to insufficient formation and grain growth of fibrous grains at the relatively low sintering temperatures.

There are several plausible interpretations for the superior sinterability of β powder compared to α powder at 1550-1600 °C. One is the difference in

FIGURE 11 Microstructure of the sintered body from the high-grade α powder after 8 h at 1600 °C. Sintering additives are 5 wt% Y_2O_3 and 3 wt% Al_2O_3. Arrows indicate fibrous grains that prevent densification.

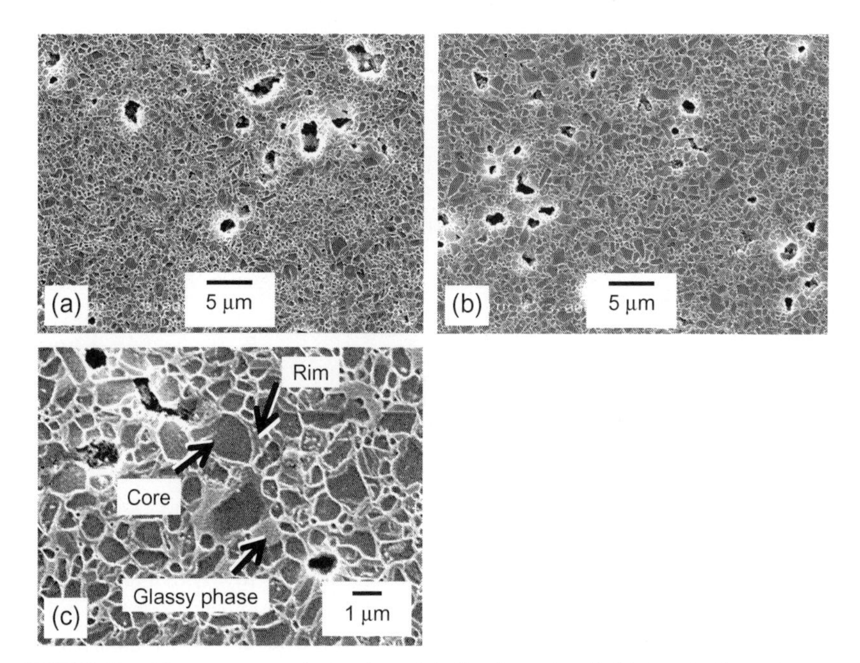

FIGURE 12 Microstructures of the sintered bodies from the (a) high-grade β powder and (b) low-grade β powder after 8 h at 1600 °C. Sintering additives are 5 wt% Y_2O_3 and 3 wt% Al_2O_3. (c) A magnified view of (b), indicating silicon nitride core surrounded by SiAlON rims.

reprecipitation behavior of ions during liquid-phase sintering between α and β grains. The phase transformation from α to β phase and grain growth in silicon nitride are generally promoted through a solution-reprecipitation mechanism, when a grain-boundary glassy phase is melted at high temperatures. Ions from α or finer β grains dissolve into the liquid phase at grain boundaries, diffuse through it and reprecipitate onto larger β grains via interfacial reaction [24,33]. Thus, the existence of α phase reduces the reprecipitation sites for interfacial reaction to that extent, when α phase does not transform well to β phase (during the first 2-4h in Figure 9), leading to the poor sinterability of α powder. However, even after most of the powder transformed to β phase, the body from a powder was still not well densified. This is plausibly attributable to the formation of fibrous β grains in the body as shown in Figure 11. The prism planes of the fibrous β grains are more stable against ion dissolution than the basal planes [34]. If fibrous grains with high-aspect ratio grow before densification, ions are less dissolved, resulting in suppressed sintering and densification. On the contrary, the body from β powder consisted of more equiaxed grains with a smaller area of the prism planes, and therefore, the densification is less suppressed than that of α powder.

The mechanical properties and densities of the sintered bodies from the high-grade and low-grade β powders after 8h at 1600°C under 0.1MPa N_2 (with 5wt% Y_2O_3 and 3wt% Al_2O_3 additives) are shown in Table 4. The strength was measured by four-point bending tests at room temperature, using outer and inner spans of 30 and 10mm, respectively, and a displacement rate of 0.5mm/min. The fracture toughness was determined by the Vickers indentation fracture method with an indentation load of 196N and a time of 15s. Because that from the high-grade α powder was not well densified, its mechanical properties are not measured. Relatively moderate values were obtained for the strength and toughness of those from the β powders; note that almost no difference was observed between the high- and low-grade powders. High strength and toughness of silicon nitride are generated by the crack wake toughening effects including bridging and pullouts of fibrous grains on the premise of grain boundary debonding [26,27]. The above moderate values of the strength and toughness are attributable to the less-fibrous grains of low-aspect ratios that are disadvantageous for the crack wake toughening effects [26,27] and the

TABLE 4 Mechanical properties and densities of sintered body from β powder

Stating Powder	High-Grade-β	Low-Grade-β		
Sintering condition	1600°C-8h-0.1MPa N_2			1750°C-8h-0.5MPa N_2
Sintering additives	5wt% Y_2O_3-3wt% Al_2O_3		5wt% Y_2O_3-5wt% $MgAl_2O_4$	
Density (10^3 kg/m^3)	3.13	3.09	3.24	3.22
Strength (MPa)	508±31	506±53	553±22	567±32
K_{IC} (MPa m$^{1/2}$)	3.9	3.8	3.5	4.0

rigid SiAlON rim structures, which suppress the grain boundary debonding, as observed in Figure 12. The mechanical properties are further deteriorated by some residual pores.

Densification and fibrous grain growth of silicon nitride can be facilitated by lowering the softening temperature of the grain boundary glassy phase and enhancing liquid-phase sintering, via selecting appropriate sintering additive systems. Figure 13a shows the microstructure of the sintered body from the low-grade β powder after 8 h at 1600 °C under 0.1 MPa N_2, with using, instead of 5 wt% Y_2O_3 and 3 wt% Al_2O_3 additives, 5 wt% Y_2O_3 and 5 wt% $MgAl_2O_4$, which are known to produce grain boundary glassy phase of lower softening temperatures. Evidently the remaining pores became fewer compared to Figure 12b, and the densification was substantially promoted, leading to a nearly full density of 3.24×10^3 kg/m^3. However, the microstructural features, including grain size, grain morphology, and core-rim structure are similar to those of Figure 12b. The strength and toughness are 553 MPa and 3.5 MPa m$^{1/2}$, respectively, as listed in Table 4; the sintering additive system of 5 wt% Y_2O_3 and 5 wt% $MgAl_2O_4$ enhanced the densification and increased the strength slightly, whereas the toughness was not improved due to the unchanged microstructural features.

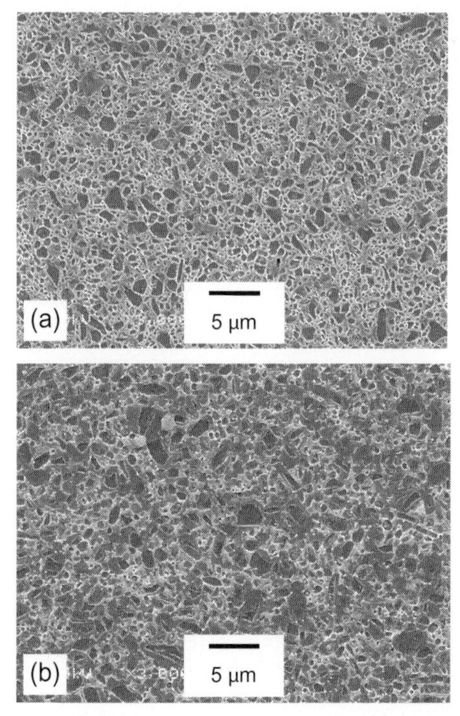

FIGURE 13 Microstructure of sintered body from low-grade β powder after 8 h at (a) 1600 °C-0.1 MPa N_2 and 8 h at (b) 1750 °C-0.5 MPa N_2. Sintering additives are 5 wt% Y_2O_3 and 5 wt% $MgAl_2O_4$.

Figure 13b shows the microstructure of the sintered body from the low-grade β powder after 8 h at higher temperature and higher gas pressure of 1750 °C and 0.5 MPa N_2, with the same additives of 5 wt% Y_2O_3 and 5 wt% $MgAl_2O_4$, and Table 4 lists the mechanical properties and density of this sintered body. Some fibrous grain growth was identified, and as a result, both the strength and toughness were slightly improved.

Silicon nitride is known as a sort of ceramics whose strength and toughness can be raised to very high values such as ~1 GPa and 6-10 MPa $m^{1/2}$, respectively, by optimizing the fabrication processes. However, there are a variety of industrial applications for engineering ceramics, and all of them do not require such superior mechanical properties. It should be noted that the relatively moderate strength (>500 MPa) and toughness (~4 MPa $m^{1/2}$) of silicon nitride prepared via the low-temperature sintering with atmospheric pressure are acceptable to some of the applications, often because of well-balanced other properties of silicon nitride including heat/corrosion resistance, wear/erosion resistance, and thermal shock resistance.

4 CONCLUSIONS

In this chapter, we gave overviews of two recent topics for a green manufacturing approach of silicon nitride ceramic components; sintered reaction bonding with rapid nitridation and low-temperature sintering with atmospheric pressure. In the former topic, the particular focus was placed on rapid nitridation of silicon powder enhanced by zirconia addition and resultant sintered-reaction-bonded silicon nitride ceramics. Adding zirconia, successful nitridation of silicon can be achieved at a rate 80 times as fast, compared with the case of silicon only, leading to a substantial reduction of consumed power. We indicated that good mechanical properties equivalent to those of gas-pressure sintered material from high-purity silicon nitride powder can be obtained via sintered reaction bonding technique with rapid nitridation using silicon powder, whenever the fabrication process is optimized. Similar rapid nitridation with zirconia addition can be applied to β-SiAlON ceramics. Besides zirconia, some rare-earth oxides, including CeO_2 and Eu_2O_3, also significantly enhance the nitridation of silicon.

As another green fabrication approach of silicon nitride components, the latter topic focused on the sintering of silicon nitride at low temperatures, below 1700 °C, under atmospheric-pressure (~0.1 MPa) nitrogen, using several types of starting powders including high-grade α phase, high-grade β phase, and low-grade β phase. Both β-phase powders showed better sinterability than α-phase powder at low temperatures around 1600 °C, and their sintered bodies showed relatively moderate mechanical properties, which are acceptable depending on sorts of applications. Thus, combining the low-temperature sintering and the use of low-cost β powder has a potential to substantially reduce the energy and cost for fabrication of silicon nitride components.

REFERENCES

[1] H. Kita, H. Hyuga, N. Kondo, T. Ohji, Exergy consumption through the life cycle of ceramic parts, Int. J. Appl. Ceram. Technol. 5 (4) (2008) 373–381.

[2] A.J. Moulson, Review; Reaction-bonded silicon nitride: its formation and properties, J. Mater. Sci. 14 (1979) 1017–1051.

[3] B.T. Lee, J.H. Yoo, H.D. Kim, Effect of sintering additives on the nitridation behavior of reaction-bonded silicon nitride, Mater. Sci. Eng. A 333 (2002) 306–313.

[4] W. Bunker, W.D. Scott, Microstructural characterization of RBSN, Am. Soc. Bull. 63 (8) (1984) 1000.

[5] M. Mitomo, Effect of Fe and Al additions on nitridation of silicon, J. Mater. Sci. 12 (1977) 273–276.

[6] V. Pavarajarn, S. Kimura, Catalytic effects metals on direct nitridation of silicon, J. Am. Ceram. Soc. 84 (8) (2001) 1669–1674.

[7] R.G. Pigeon, A. Varma, A.E. Miller, Some factors influencing the formation of reaction-bonded silicon nitride, J. Mater. Sci. 28 (1993) 1919–1936.

[8] S. Lin, Comparative studies of metal additives on the nitridation of silicon, J. Am. Ceram. Soc. 60 (1–2) (1977) 78–81.

[9] H. Hyuga, K. Yoshida, N. Kondo, H. Kita, J. Sugai, H. Okano, J. Tsuchida, Nitridation enhancing effect of ZrO_2 on silicon powder, Mater. Lett. 62 (2008) 3475–3477.

[10] H. Hyuga, K. Yoshida, N. Kondo, H. Kita, H. Okano, J. Sugai, J. Tsuchida, Influence of zirconia addition on reaction bonded silicon nitride produced from various silicon particle sizes, J. Ceram. Soc. Japan 16 (6) (2008) 688–693.

[11] H. Hyuga, K. Yoshida, N. Kondo, H. Kita, J. Sugai, H. Okano, J. Tsuchida, Fabrication and mechanical properties of Si_3N_4 ceramics with ZrO_2–$MgAl_2O_4$ additive system via reaction bonding route, Mater. Syst. 26 (2008) 29–34.

[12] H. Hyuga, K. Yoshida, N. Kondo, H. Kita, H. Okano, J. Sugai, J. Tsuchida, Nitridation behaviour of ZrO_2 added silicon powder with different ZrO_2 particle sizes, J. Ceram. Soc. Japan 117 (2) (2009) 157–161.

[13] H. Hyuga, K. Yoshida, N. Kondo, H. Kita, J. Sugai, H. Okano, J. Tsuchida, Fabrication of pressureless sintered dense beta-SiAlON via a reaction-bonding route with ZrO_2 addition, Ceram. Int. 35 (5) (2009) 1927–1932.

[14] H. Hyuga, N. Kondo, H. Kita, Fabrication of dense β-SiAlON ceramics with ZrO_2 additions via a rapid reaction-bonding and postsintering route, J. Am. Ceram. Soc. 94 (4) (2011) 1014–1018.

[15] L.K.L. Falk, T. Hermansson, J. Rundgren, Microstructures of hot-pressed $Si_3N_4/ZrO_2(+Y_2O_3)$ composites, J. Mater. Sci. Lett. 8 (1989) 1032–1034.

[16] T. Ekström, L.K.L. Falk, E.M. Knutson-Wedel, Pressureless sintered Si_3N_4-ZrO_2 composites with Al_2O_3 and Y_2O_3 additions, J. Mater. Sci. Lett. 9 (1990) 823–826.

[17] S. Dutta, B. Buzek, Microstructure, strength, and oxidation of a 10 wt-percent zyttrite-Si_3N_4 ceramic, J. Am. Ceram. Soc. 67 (2) (1984) 89–92.

[18] H.J. Kleebe, W. Braue, W. Luxem, Densification studies of SRBSN with unstabilized zirconia by means of dilatometry and electron-microscopy, J. Mater. Sci. 29 (1994) 1265–1275.

[19] C.G. Cofer, J.A. Lewis, Chromium catalysed silicon nitridation, J. Mater. Sci. 29 (22) (1994) 5880–5886.

[20] J. Weiss, L. Gauckler, H. Lukas, G. Petzow, T.Y. Tie, Determination of phase equilibria in the system Si-Al-Zr/N-O by experiment and thermodynamic calculation, J. Mater. Sci 16 (1981) 2997–3005.

[21] J.F. Yang, T. Sekino, K. Nihara, Effect of grain growth and measurement on fracture toughness of silicon nitride ceramics, J. Mater. Sci. 34 (1999) 5543–5548.

[22] H. Hyuga, Y. Zhou, D. Kusano, K. Hirao, H. Kita, Nitridation behaviors of silicon powder doped with various rare earth oxides, J. Ceram. Soc. Japan 19 (3) (2011) 251–253.

[23] M. Mitomo, Pressure sintering of Si_3N_4, J. Mater. Sci 11 (1976) 1103–1107.

[24] M. Mitomo, M. Tsutsumi, H. Tanaka, S. Uenosono, F. Saito, Grain growth during gas-pressure sintering of β-silicon nitride, J. Am. Ceram. Soc. 73 (1990) 2441–2445.

[25] M. Mitomo, S. Uenosono, Gas pressure sintering of β-silicon nitride, J. Mater. Sci. 26 (1991) 3940–3944.

[26] P.F. Becher, Microstructural design of toughened ceramics, J. Am. Ceram. Soc. 74 (1991) 255–269.

[27] P.F. Becher, E.Y. Sun, K.P. Plucknett, K.B. Alexander, C.H. Hsueh, H.T. Lin, S.B. Waters, C.G. Westmoreland, E.S. Kang, K. Hirao, M.E. Brito, Microstructural design of silicon nitride with improved fracture toughness: II, effects of yttria and alumina additives, J. Am. Ceram. Soc. 81 (1998) 2821–2830.

[28] D.-D. Lee, S.-K.L. Kang, G. Petzow, D.N. Yoon, Effect of α to β (β′) phase transition on the sintering of silicon nitride ceramics, J. Am. Ceram. Soc. 73 (1990) 767–769.

[29] N. Wangmooklang, K. Sujirote, T. Wasanapiarnpong, S. Jinawath, S. Wada, Properties of Si_3N_4 ceramics sintered in air and nitrogen atmosphere furnaces, J. Ceram. Soc. Japan 115 (2007) 974–977.

[30] N. Kondo, M. Hotta, T. Ohji, Low-cost silicon nitride from β-silicon nitride powder and by low-temperature sintering, Int. J. Appl. Ceram. Technol. 1 (6) (2013), http://dx.doi.org/10.1111/ijac.12157.

[31] N. Hirosaki, Y. Okamoto, Y. Akimune, M. Mitomo, Sintering of Y_2O_3-Al_2O_3-doped β-Si_3N_4 powder and mechanical properties of sintered materials, J. Ceram. Soc. Japan 102 (1994) 790–794.

[32] K. Hirao, T. Nagaoka, M.E. Brito, S. Kanzaki, Microstructure control of silicon nitride by seeding with rod-like β-silicon nitride particles, J. Am. Ceram. Soc. 77 (1994) 1857–1862.

[33] M. Kitayama, K. Hirao, S. Kanzaki, Effect of rare earth oxide additives on the phase transformation rates of Si_3N_4, J. Am. Ceram. Soc. 89 (2006) 2612–2618.

[34] M.J. Hoffmann, G. Petzow, Tailored microstructures of silicon nitride ceramics, Pure Appl. Chem. 66 (1994) 1807–1814.

Chapter 10

Green Processing of Particle Dispersed Composite Materials

J. Tatami[1] and H. Nakano[2]

[1]*Graduate School of Environment and Information Sciences, Yokohama National University, Yokohama, Japan,* [2]*Cooperative Research Facility Center, Toyohashi University of Technology, Toyohashi, Japan*

1 INTRODUCTION

Advanced ceramics, especially advanced ceramic composites, are significant materials that have the potential to solve present and future environmental and energy problems. The properties of these ceramic composites depend not only on the atomic bonding and crystal structure, but also on the microstructural properties such as the grain size, grain boundary, impurity, secondary phase, pores, and defects in the sintered body. In particular, the powder processing has a large influence on the latter because advanced ceramics are fabricated by shaping and sintering fine raw powders. Therefore, the advancement of ceramic powder processing techniques, including the homogeneous dispersion of raw materials, formation of a uniform green body, and analytical control of the sintering shrinkage behavior, is very important to make better ceramics. The homogeneous mixing of additives can be better achieved by using nanoparticles rather than large particles as the raw material. However, the use of the wet milling process with liquid dispersion media such as water and ethanol might lead to the reagglomeration of nanoparticles during drying. Furthermore, since the evaporation of these liquids is an endothermic process, the environmental burden of the wet milling process sometimes imposes a heavy environmental burden. The fabrication of advanced ceramics is accomplished by using a powder composite process in which nanoparticles are mechanically mixed using only a dry mixing process, which allows us to obtain nanocomposite particles. In this process, nanoparticles are bonded to submicron particles by applying an external mechanical force, specifically a shear force [1,2]. This dry process is environmentally friendly because of the lack of the need for the elimination of liquids.

Many types of powder composite processes have been commercially developed [1,2]. These nanocomposite particle processes are divided into two types: the bonding of fine particles to a core particle and the homogeneous mixing of various

Green and Sustainable Manufacturing of Advanced Materials. http://dx.doi.org/10.1016/B978-0-12-411497-5.00010-2

fine particles. The former is regarded as surface modification on the microscale and nanoscale. The powder composite process is characterized by (a) a simple dry process, (b) unlimited raw material, (c) a large amount of treatment, and (d) the easy control of the coating conditions. The particle composite behavior depends on several factors. Naito *et al.* reported the effects of the rotor shape, mechanical treatment conditions (rotation speed and time), and raw material particle sizes on the particle composite process [3–6]. Nogi *et al.* also estimated the relationship between the temperature of the container and the interface of the particles [7].

Some researchers have studied the synthesis of ceramic particles and the preparation of porous ceramics using a powder composite process [8–12]. We also developed several kinds of dense advanced ceramics and controlled their microstructures to improve their properties [13–15]. In this chapter, we will introduce the microstructure and properties of TiN-particle-dispersed Si_3N_4 ceramics and CNT-dispersed Al_2O_3 and Si_3N_4 ceramics, which are typical particle-dispersed composites, fabricated by the particle composite process.

2 TiN-NANOPARTICLE-DISPERSED Si_3N_4 CERAMICS

More than 40 years have passed since Si_3N_4 ceramics were first developed. In the meantime, Si_3N_4 ceramics with high strength and fracture toughness have been developed as a result of using newly discovered SiAlONs [16,17]; the invention of sintering aids such as Y_2O_3 [18,19]; the development of fine, pure, and highly sinterable Si_3N_4 powder [20]; the invention of a gas pressure sintering technique [21]; the advancement of the science and technology for microstructure control, and more. Si_3N_4 ceramics have been applied to automobile components such as glow plugs [22], hot chambers [23], and turbocharger rotors [24]. Around the same time, cutting tools and bearing components were also developed [25–27]. Si_3N_4 ceramic bearings were put to practical use in 1983 [25]. Although the cost of Si_3N_4 ceramics was high, they were used as bearing materials in machine tools because of properties such as high strength, high toughness, high elastic modulus, high hardness, light weight, and good corrosion resistance [26,27]. Further, Si_3N_4 ceramics fabricated from Si_3N_4-Y_2O_3-Al_2O_3-TiO_2-AlN have been widely used for bearing applications. The authors have studied the sintering behavior and microstructure control of Si_3N_4-Y_2O_3-Al_2O_3-TiO_2-AlN. We found that the densification is enhanced at lower temperatures by simultaneous additions of TiO_2 and AlN to Si_3N_4-Y_2O_3-Al_2O_3 [28]. It was also shown that TiN is formed by a reaction between TiO_2 and AlN or Si_3N_4, and that its size is almost the same as that of the added TiO_2 [29]. Also, it was established that the contact fatigue of Si_3N_4 ceramics is suppressed by the dispersed TiN particles [30].

Hybrid bearings composed of Si_3N_4 ceramic balls and metal rings are more popular than all-ceramic bearings because of their cost. In hybrid bearings, it is possible that hard and large TiN particles damage the mating metals in a manner analogous to the wear map concept [31]. The dispersion of TiN nanoparticles is expected to be effective in solving this problem. TiN nanoparticles can be formed from TiO_2 nanoparticles because the size of TiN should be almost the same as that of TiO_2.

However, it is difficult to realize TiN nanoparticle-dispersed Si_3N_4 ceramics even if TiO_2 nanoparticles are completely dispersed in the slurry [32]. This difficulty might be caused by the reagglomeration of TiO_2 nanoparticles in the drying process. We applied the powder composite process to develop TiN-nanoparticle-dispersed Si_3N_4 ceramics and lower the aggressiveness to the mating metals in a wear test.

To fabricate TiN-nanoparticle-dispersed Si_3N_4 ceramics, high-purity, fine Si_3N_4 powder (SN-E-10, Ube Co. Ltd., Japan), Y_2O_3 (RU, Shinetsu Chemical Co., Japan), Al_2O_3 (AKP-30, Sumitomo Chemical Co., Japan), AlN (F grade, Tokuyama Co., Japan), and TiO_2 (Aeroxide P 25, Nippon Aerosil Co., Ltd., Japan) were used as the raw materials. First, TiO_2 nanoparticles were dispersed in ethanol according to our previous study [32], using polyethylenimine (EPOMIN, MW1200, Nippon Shokubai Co. Ltd., Japan) by ball milling for 48 h. The Si_3N_4 powder was then mixed into the TiO_2 slurry by ball milling for 48 h, followed by the elimination of the ethanol. The premixed powder was mechanically treated using a powder composer (Nobilta NOB-130, Hosokawa Micron Co., Japan) to prepare composite particles. After the powder composite process, the other sintering aids were added by ball milling in ethanol for 48 h using β-sialon balls in a silicon nitride container. After removing the ethanol, 4 wt% paraffin (melting point: 46-48°C, Junsei Chemical Co., Japan) and 2 wt% dioctyl phthalate (DOP, Wako Junyaku Co., Japan) were added as a binder and lubricant, respectively. For reference, five types of powder mixtures were also prepared using only mechanical treatment without premixing, using conventional wet ball milling with 20-, 200-, or 540-nm TiO_2 particles, and not using TiO_2 and AlN (i.e., no TiN formation in Si_3N_4 ceramics). The notations for these powder mixtures are listed in Table 1.

TABLE 1 Composition and Mixing Process for TiN-Dispersed Si_3N_4 Ceramics

Notation	Composition (Weight Ratio)					Mixing Process	Remarks
	Si_3N_4	Y_2O_3	Al_2O_3	TiO_2	AlN		
PM-N	92	5	3	5	5	Premixing + powder composite process	TiO_2:20 nm
M-N	92	5	3	5	5	Powder composite process	TiO_2:20 nm
W-N	92	5	3	5	5	Conventional wet ball milling	TiO_2:20 nm
W-SM	92	5	3	5	5		TiO_2:200 nm
W-M	92	5	3	5	5		TiO_2:540 nm
NT	92	5	3	0	0		No TiN in Si_3N_4 ceramics

Figure 1 shows scanning electron microscope (SEM) images of the powder mixtures before and after the mechanical treatment. The TiO_2 nanoparticles formed aggregates before the mechanical treatment [Figure 1a]. On the other hand, as shown in Figure 1b, no TiO_2 nanoparticle aggregates were found in the powder mixture after the mechanical treatment, that is, the TiO_2 nanoparticles were well dispersed. Figure 2 presents TEM images of the powder mixtures before and after the mechanical treatment. Although the TiO_2 nanoparticles were dispersed in ethanol by wet mixing, as reported in our previous paper [32], they reagglomerated as a result of mixing with Si_3N_4 powder and/or drying. As shown in Figure 2b, the mechanical treatment resulted in the uniform dispersion of the TiO_2 nanoparticles, thus suggesting that they might be strongly attached to the Si_3N_4 particles. A high-resolution TEM (HRTEM) image of a composite particle is shown in Figure 3. Note that the TiO_2 nanoparticle is directly bonded onto a submicron Si_3N_4 particle. At the atomic scale, the interface between the TiO_2 and Si_3N_4 was flat. Such a direct-bonded interface should be stronger than the interface of physically adsorbed particles. Si was detected in the TiO_2 particles by an energy-dispersive X-ray spectroscopy (EDS) analysis. In addition, neck growth occurred between the TiO_2 and Si_3N_4 particles, similar to the initial stage of sintering, in spite of the mechanical treatment at ambient temperature. This phenomenon should result from the reaction between the TiO_2 and SiO_2 and/or Si_3N_4. Nogi *et al.* reported the relationship between the temperature of

FIGURE 1 SEM images of powder mixture (a) before and (b) after a powder composite process.

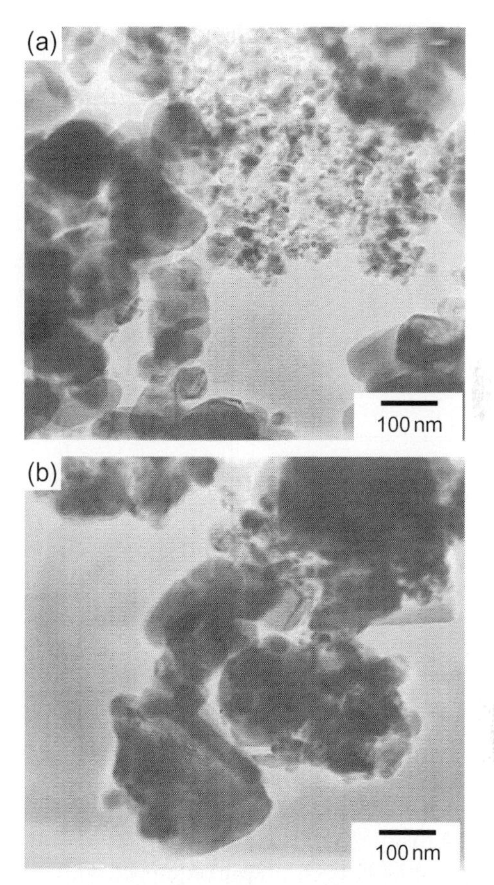

FIGURE 2 TEM images of powder mixture (a) before and (b) after a powder composite process.

the container and the particle interface [7]. In this case, because the container temperature is 50-60°C, the interfacial temperature is expected to be around 700-800°C. On the other hand, the crystalline phase of the TiO_2 was anatase, both before and after the mechanical treatment, indicating that the temperature of the particle itself was not very high. Therefore, a mechanochemical reaction occurring just at the interface resulted in diffusion and sintering, forming the flat interface and the neck between the TiO_2 and SiO_2 layer and/or Si_3N_4 particles.

The mixed powders were sieved using a #60 nylon sieve and then molded into ϕ 15×7 mm pellets by uniaxial pressing at 50 MPa followed by cold isostatic pressing at 200 MPa. After binder burnout in the air at 500°C for 3 h, the green bodies were fired at 1800°C in 0.9 MPa N_2 for 2 h using a gas pressure sintering furnace (Himulti 5000, Fujidenpa Kogyo Co., Japan). The sintered bodies were hot isostatically pressed at 1700°C for 1 h under 100 MPa N_2. The density of the fired samples was measured using the Archimedes method. The phase

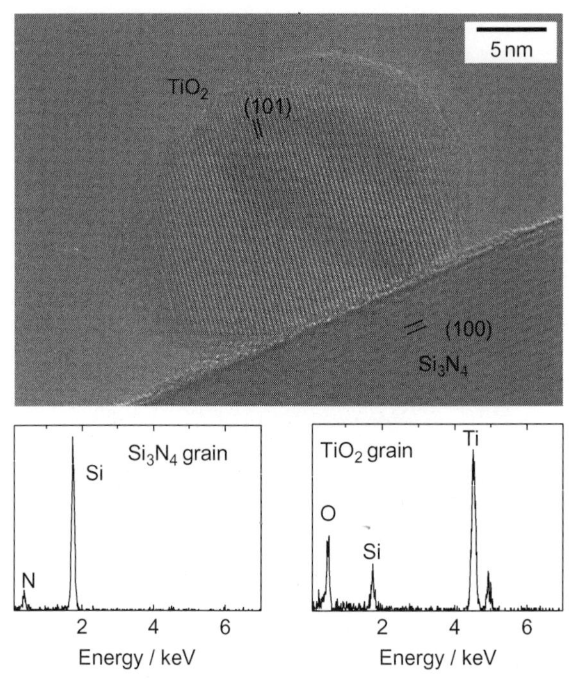

FIGURE 3 High-resolution TEM image of composite particle prepared by a powder composite process.

present in the samples was identified by X-ray diffraction (RINT2000, Rigaku Co., Japan). The microstructure was observed by a transmission electron microscope (TEM, JEM-3000F, JEOL, Japan) equipped with an energy-dispersive spectroscope (EDS, Voyager III, NORAN instruments). The Si_3N_4 ceramics were machined using a grinding machine (SG-45FIIH, Wasino Engineering Co., Japan) and then polished using diamond slurry. The wear property was estimated by a ball-on-disk test, which consisted of a polished Si_3N_4 disk and a steel ball bearing (SUJ-2). The testing conditions were relative humidity of $50 \pm 2\%$ and temperature of 22-24°C. The radius of the SUJ-2 ball was 3 mm. The rotation speed and radius were $10\,cm\,s^{-1}$ and 3 mm, respectively. The weight was 5 N, and the running distance was 250 m. The polished and worn surfaces of the specimens were observed by scanning probe microscopy (SPM, SPA-400, Seiko Nanotechnologies, Japan).

The relative density of all of the specimens was over 98%, which was sufficiently high to measure the mechanical properties. β-Si_3N_4 and TiN were also identified as the main phases of the products in the sample, in addition to TiO_2 and AlN. Figure 4 shows TEM images of the TiN-dispersed Si_3N_4 ceramics. As shown in Figure 4a, the size of the TiN particles in the Si_3N_4 ceramics fabricated using just wet mixing was 300-500 nm. On the other hand, in the case

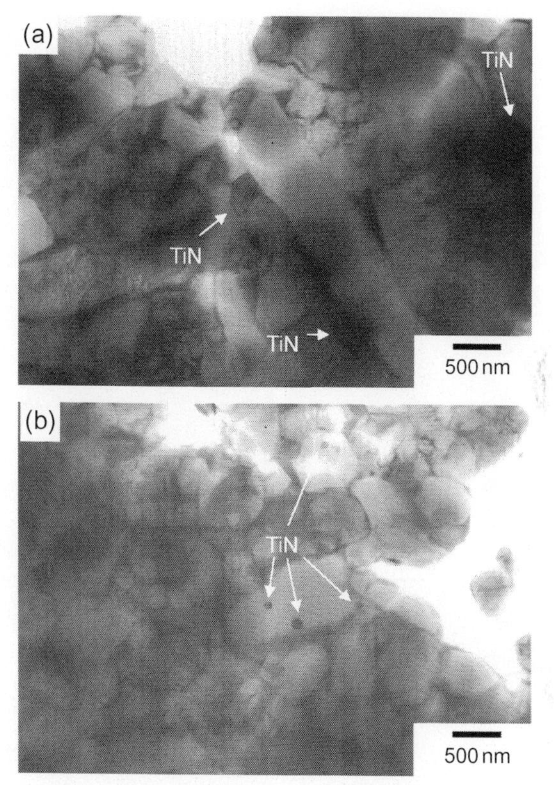

FIGURE 4 TEM images of Si_3N_4 ceramics. (a) Sample W-N and (b) sample PM-N.

of using composite particles prepared by premixing and mechanical treatment, 20-100-nm TiN nanoparticles were found in Si_3N_4 grains and in the grain boundary (Figure 4b). Thus, it was shown that TiN nanoparticle-dispersed Si_3N_4 ceramics were fabricated using composite powder prepared by premixing followed by mechanical treatment.

Figures 5 and 6 show optical micrographs of the worn surfaces of the SUJ2 balls and Si_3N_4 disks after the ball-on-disk tests. It can be observed that the areas of the SUJ2 balls worn by sample PM-N were smaller than all of the other areas, even though the wear track width of the Si_3N_4 ceramics is almost the same. The wear volume was calculated from the wear area of the ball and the worn surface profile of the Si_3N_4 disk. As listed in Table 2, the wear volume of the Si_3N_4 disk was independent of the Si_3N_4 ceramics, except for sample W-M. On the other hand, the wear volume of the SUJ2 ball depended on the mating Si_3N_4 ceramics, that is, the wear volume of the ball worn by sample PM-N was not only smaller than those worn by samples M-N and W-N but was also the same as that worn by sample NT. Furthermore, the wear volumes of the steel

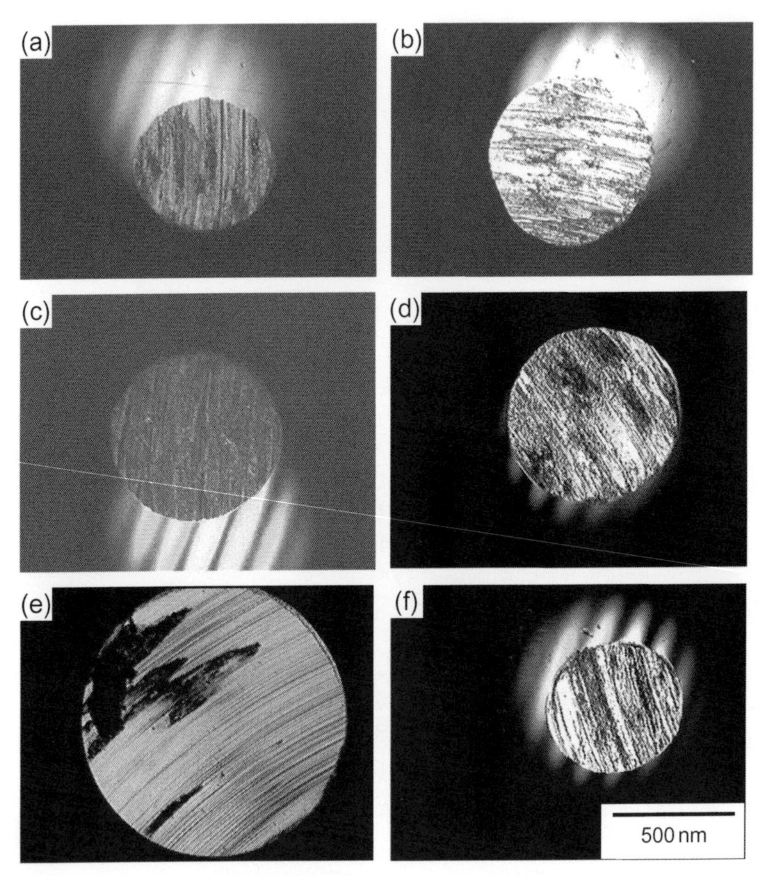

FIGURE 5 Optical micrographs of worn surfaces of the steel ball after ball-on-disk test. (a) sample PM-N, (b) M-N, (c) W-N, (d) W-SM, (e) W-M, and (f) NT.

balls worn by samples M-SM and M-M were higher than those of PM-N, thus suggesting that TiN nanoparticle dispersion should cause less damage to the mating metals.

The wear behavior was evaluated by SPM observations of the worn surfaces. Figure 7 shows SPM images of the worn surfaces of the steel balls after the ball-on-disk tests. Many grooves on the surface were worn and the distance between the grooves was comparable to the size of the TiN. Figure 8 shows the polished surface of sample W-M. Even though the surface was polished, it can be observed that a large TiN particle projects from the surface because of its hardness. This indicates that aggressive wear was caused by hard and large TiN particles during the ball-on-disk test, similar to abrasive wear [33]. In other words, because sample PM-N had much smaller TiN particles, the abrasiveness toward the mating metallic ball was considered to be comparable to Si_3N_4 ceramics without TiN particles.

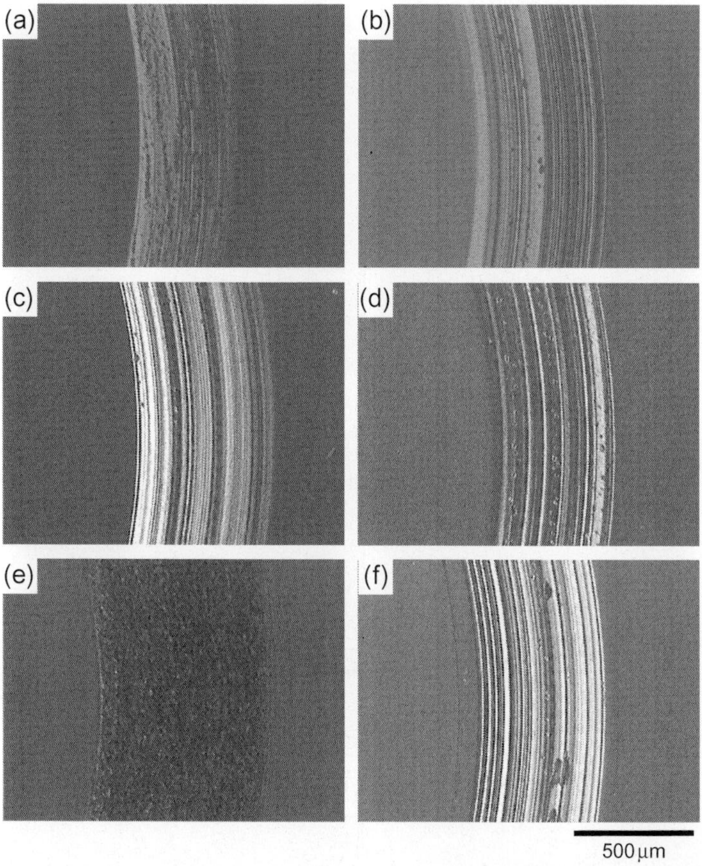

(a) (b) (c) (d) (e) (f)

500 μm

FIGURE 6 Optical micrographs of worn surfaces of the Si_3N_4 disk after ball-on-disk test. (a) sample PM-N, (b) M-N, (c) W-N, (d) W-SM, (e) W-M, and (f) NT.

TABLE 2 Wear Volume of Si_3N_4 Disk and SUJ2 Ball after Ball-on-Disk Test

	Wear Volume/$10^{-12}m^3$	
Sample	Si_3N_4 Disk	SUJ2 Ball
PM-N	9.5	1.5
M-N	9.4	2.7
W-N	9.8	3.9
W-SM	10.9	20.7
W-M	4.4	3.5
NT	10.0	1.6

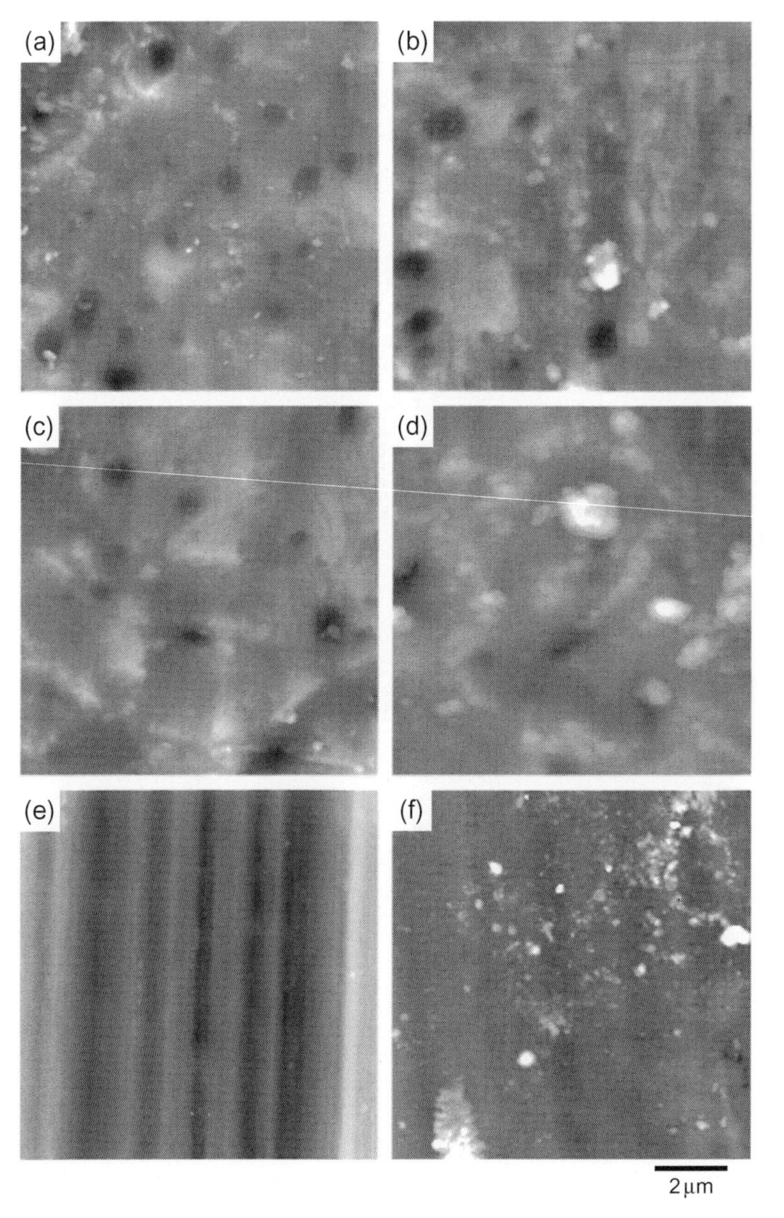

FIGURE 7 SPM images of worn surface of steel ball after ball-on-disk test. (a) sample PM-N, (b) M-N, (c) W-N, (d) W-SM, (e) W-M, and (f) NT.

Figure 9 shows SPM images of the worn surfaces of the TiN-dispersed Si_3N_4 ceramics. The agglomerates observed on the worn surface of sample W-M might consist of metallic wear particles. Because of the adhesion of metallic particles on the worn surface of Si_3N_4 ceramics, the wear volume of

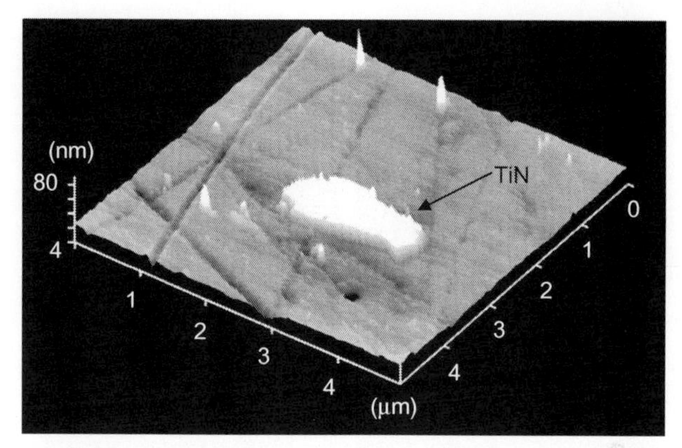

FIGURE 8 SPM images of polished surface of sample W-M.

sample W-M appears to be smaller than the other samples. Because the grain boundary phase is easily removed, the Si_3N_4 grain shape is emphasized. The Si_3N_4 ceramics in this study were composed of Si_3N_4 grains, TiN grains, and a grain boundary phase. It is well known that softer materials are more easily worn. Therefore, a softer grain boundary phase is preferred during a ball-on-disk test. Further, because the Si_3N_4 ceramics used in this study were fabricated from the same composition of sintering aids, it seems that the wear volumes of the Si_3N_4 ceramics were comparable. Thus, it was shown that TiN nanoparticle-dispersed Si_3N_4 ceramics were successfully developed using TiO_2 and Si_3N_4 composite particles prepared by premixing and mechanical treatment, and that the dispersion of TiN nanoparticles lowered the aggressiveness to the mating metals in the wear test.

The developed Si_3N_4 ceramics need high mechanical properties for bearing application. ISO 26602:2009 provides a classification defining the physical and mechanical properties of silicon nitride preprocessed ball-bearing materials. These materials are classified in three categories by the specification of their characteristics and microstructures. Figure 10 shows the Weibull plot of the bending strength of sample PM-N. The shape factor in the Weibull plot of the bending strength was 13, and the average bending strength was 1109 MPa. The fracture toughness and Vickers hardness were 6.7 $MPam^{1/2}$ and 15.4 GPa, respectively. These are higher than the Class I values in ISO 26602:2009. We prepared balls using the developed material to measure the crushing strength. The crushing strength of this material was about 31.7 N, which is 1.5 times that of the conventional Si_3N_4 ceramics used for bearings. A rolling fatigue test was carried out using 3×10 mm diameter balls under a pressure of 5.9 GPa. The rolling fatigue lifetime of the developed TiN-nanoparticle-dispersed Si_3N_4 ceramics was longer than 10^7 cycles. Consequently, it was confirmed that the developed material has sufficient mechanical properties for use in the next generation of ceramic bearings.

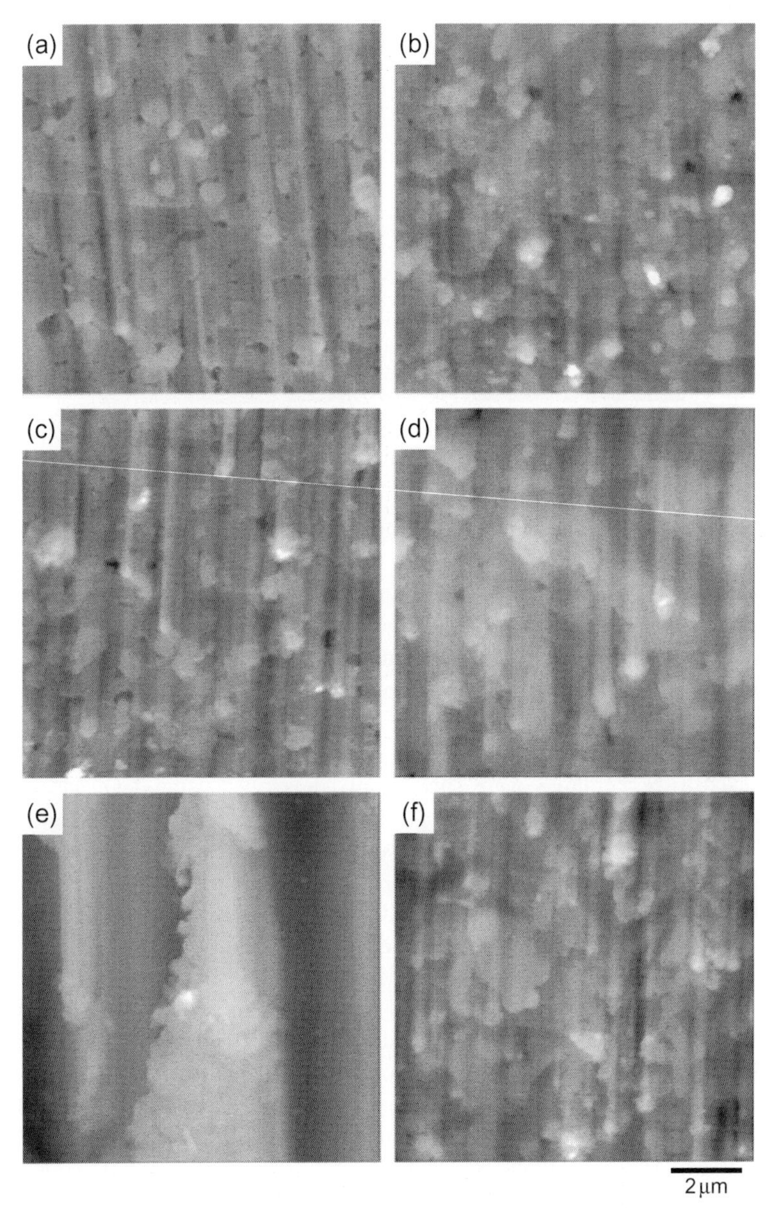

FIGURE 9 SPM images of worn surface of Si_3N_4 disk after ball-on-disk test. (a) sample PM-N, (b) M-N, (c) W-N, (d) W-SM, (e) W-M, and (f) NT.

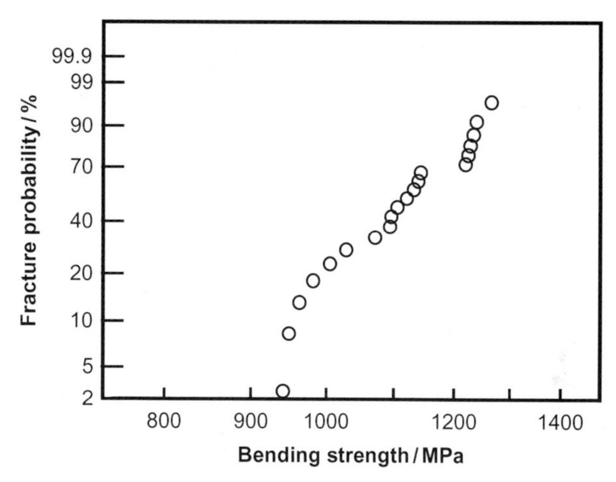

FIGURE 10 Weibull plot of bending strength of sample PM-N.

3 CNT-DISPERSED CERAMICS

Carbon nanotubes have high strength, high elastic modulus, and excellent thermal and electrical conductivities, resulting in their use as fillers for composites to improve their mechanical, electrical, and thermal properties. In our previous research, electrically conductive CNT-dispersed Si_3N_4 ceramics were developed [34–37]. In the case of the CNT composite process, one of the most important problems to be solved is the uniform dispersion of CNTs. Here, we tried to fabricate CNT-dispersed Al_2O_3 ceramics using the nanocomposite particles.

In CNT-dispersed Al_2O_3 ceramic composites, the raw materials were multiwall carbon nanotubes and an alumina powder (Taimei Chemical, TM-DAR, 0.2 μm); 2 wt% (4 vol%) of the carbon nanotubes were mixed with alumina using a powder composer. Figure 11a shows the CNT-Al_2O_3 powder mixture prepared by the powder composite process. It was found that granules composed of CNTs and Al_2O_3 particles were formed, and there were no agglomerates of CNTs. Figures 11b and c are SEM photographs of the powder mixture shown in Figure 11a after ultrasonification. The granules were easily pulverized by ultrasonic irradiation, and the CNTs were homogeneously dispersed in the Al_2O_3 powder. This indicated that the CNTs and Al_2O_3 were uniformly mixed by the particle composite process. Figure 11d is an enlarged view of the CNT-Al_2O_3 powder mixture. Fine Al_2O_3 particles were fixed on the CNTs. This phenomenon is similar to that of the SiO_2 nanoparticle-SiO_2 fiber prepared by a powder composite process, as reported by Naito *et al.* [9]

The powder mixture was fired at 1750°C for 1 h in an Ar atmosphere using a hot pressing technique. The applied pressure was 30 MPa. A 20×20×5 mm ceramic plate was obtained. After cutting the sample into 3×4×20 mm pieces, the relative density, bending strength, and electrical conductivity were evaluated. The microstructure was observed by SEM. Figure 12 shows the microstructure of the CNT-Al_2O_3 composite fabricated by hot pressing. There were

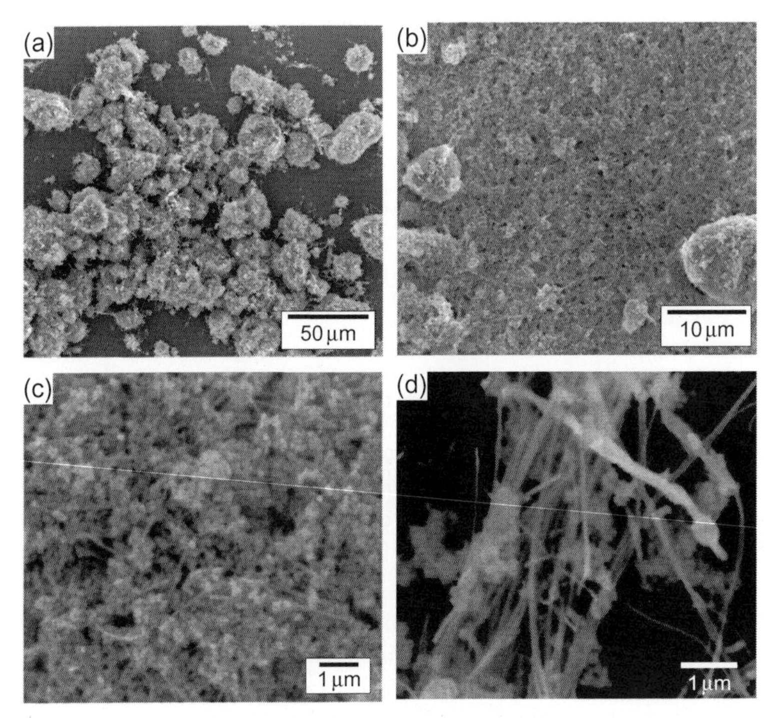

FIGURE 11 SEM images of powder mixture of CNTs and Al_2O_3 using a powder mixture prepared by a powder composite process. (a) As-received powder. (b) After ultrasonification in ethanol. (c) and (d) Enlarged view of (b).

no large pores in the sintered body in spite of the inhibition of the densification by the CNTs. The bulk density of this sample was 3.82 g/cm^3, and the relative density was 92%. The grain size of the Al_2O_3 in this composite was about 2 μm, which is much smaller than that of pure Al_2O_3 ceramics hot-pressed under the same conditions (Figure 12b). The inhibition of grain growth resulted in the lowering of grain boundary migration because of the dispersed CNTs. In the enlarged photograph (Figure 12c), many projecting CNTs and small holes can be observed in the sintered body. These is evidence of the bridging or pullout of CNTs, which is similar to fiber-reinforced ceramics.

The bending strength and electrical conductivity of the CNT-Al_2O_3 composites using the nanocomposite particles listed in Table 3. The bending strength of the CNT-Al_2O_3 composites was 423 MPa, which is higher than that of pure Al_2O_3 ceramics sintered by hot-pressing at the same temperature. This strengthening should be caused by the limitation of the grain growth and the reinforcement by the bridging and pullout of CNTs. The electrical conductivity of the developed CNT-Al_2O_3 composites was 448 S/m. This value was almost the same as the 12 wt% CNT-dispersed Si_3N_4 ceramics [34]. The achievement of such a high electrical conductivity by the addition of only 2 wt% CNTs is attributed to the homogeneous dispersion of the CNTs by the powder composite process, which formed numerous electrically conductive paths in the ceramics.

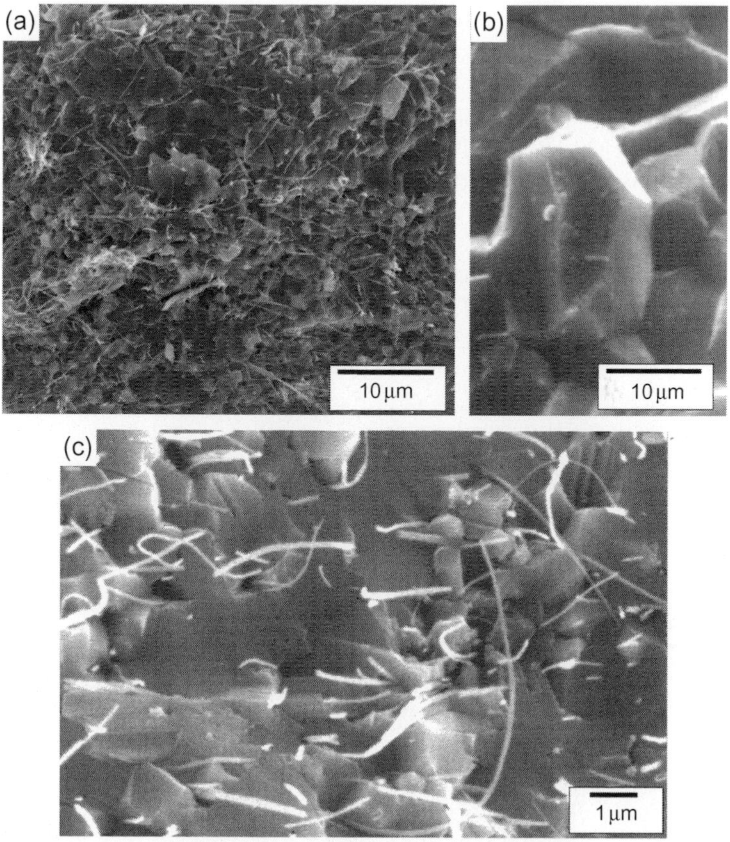

FIGURE 12 SEM images of fracture surfaces of (a) CNT-dispersed Al_2O_3 ceramics fired at 1700°C by hot pressing, (b) Al_2O_3 ceramics without CNTs fired at 1700°C by hot pressing, and (c) enlarged view of (a).

TABLE 3 Density, Bending Strength, and Electrical Conductivity of CNT-Dispersed Al_2O_3 Ceramics

CNT Content	2 wt%
Bulk density	3.82 g cm^{-3}
Relative density	97.8%
Bending strength (3-point bending test)	423 MPa
Electrical conductivity (DC 4 terminal method)	448 S/m (=0.22 Ω cm)

In CNT-dispersed Si_3N_4 ceramic composites [38], the basic chemical composition of raw materials was Si_3N_4:Y_2O_3:Al_2O_3:AlN:TiO_2 = 92:5:3:5:5. The CNT quantity was changed from 0.5 wt% to 1.0 wt%. The CNTs were dispersed by dry mechanical mixing technique with TiO_2 under 1.5 kW for 10 min. Figure 13 shows the SEM and TEM images of the microstructure after the mechanical dry mixing. It was observed that granules about 10 μm in diameter were formed,

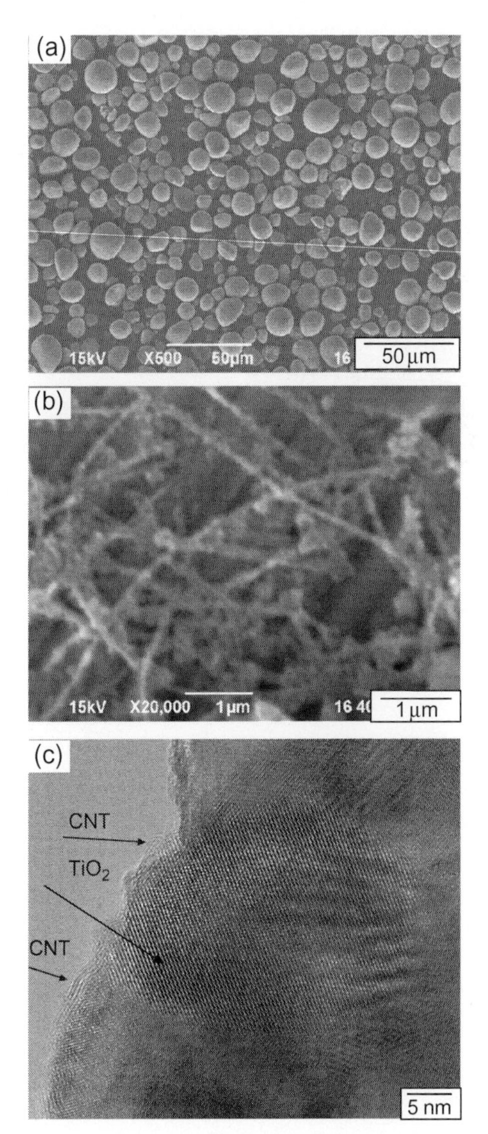

FIGURE 13 SEM images and TEM image of microstructure after mechanical dry mixing. (a) After mechanical dry mixing, (b) after ultrasonication in ethanol, and (c) TEM image.

and there were no agglomerates of CNTs (Figure 13a). Figure 13b shows the SEM image after ultrasonication in ethanol for several minutes. The granules as shown in Figure 13a were easily disintegrated in ethanol. Moreover, TiO_2 and CNTs were well-dispersed and TiO_2 nanoparticles existed on the CNTs. Figure 13c shows the TEM image of the microstructure after mechanical dry mixing. It was confirmed that TiO_2 nanoparticles were bonded on the CNTs. Therefore, we succeeded in preparing TiO_2-CNT nanocomposites.

The powder mixture of the nanocomposite particles were molded to make green bodies. After dewaxing, they were fired at 1700-1800°C in 0.9 MPa N_2 for 2 h followed by HIPping at 1700°C in 100 MPa N_2 for 1 h. Table 4 lists the values of density, and Figure 14 shows the XRD pattern. The relative density of any specimen was 95% or more; therefore, we were able to obtain highly dense sintering compacts. Only β-Si_3N_4 and TiN were confirmed to have formed as a result of the identification of the phase present. The phases present were independent of the CNT quantity and the sintering temperature.

Table 4 also lists the bending strength of CNT-dispersed Si_3N_4 ceramics. The three-point bending strength of the samples fabricated by the mechanical dry mixing technique was higher than 700 MPa. In particular, when the firing temperature was 1750°C, the bending strength of CNT-dispersed Si_3N_4 ceramics was more than 920 MPa, which is almost the same value as the Si_3N_4 ceramics without CNTs. The fracture origin was observed after the three-point bending test. The typical SEM image of the fracture origin after the three-point bending test is shown in Figure 15. The circle shown in the figure represents the fracture origin. The fracture origin was not agglomerates of CNTs. The electrical conductivity was also listed in Table 4. The specimen with 0.5 wt% CNTs did not exhibit electrical conductivity. On the other hand, the specimen with 1.0 wt% CNTs exhibited electrical conductivity. Figure 16 shows the SEM image of the fracture surface of the Si_3N_4 ceramics by adding 1.0 wt% CNTs fired at 1750°C.

TABLE 4 Relative Density, Bending Strength, and Electrical Conductivity of CNT-Dispersed Si_3N_4 Ceramics

CNT Quantity/wt%	Firing Temperature/°C	Relative Density After HIP/%	3-Point Bending Strength/MPa	Electrical Conductivity/Sm^{-1}
0.5	1700	95.6	705±39	–
	1750	94.9	945±70	–
	1800	95.0	820±87	–
1.0	1700	96.0	727±111	2.8
	1750	95.4	923±74	2.9
	1800	95.0	831±73	6.5

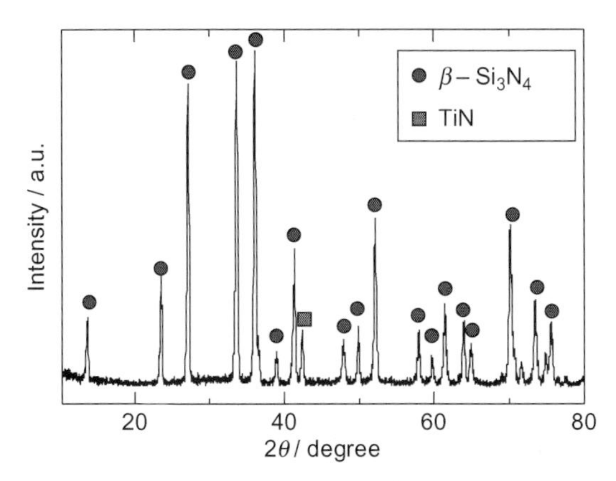

FIGURE 14 XRD pattern of CNT-dispersed Si_3N_4 ceramics.

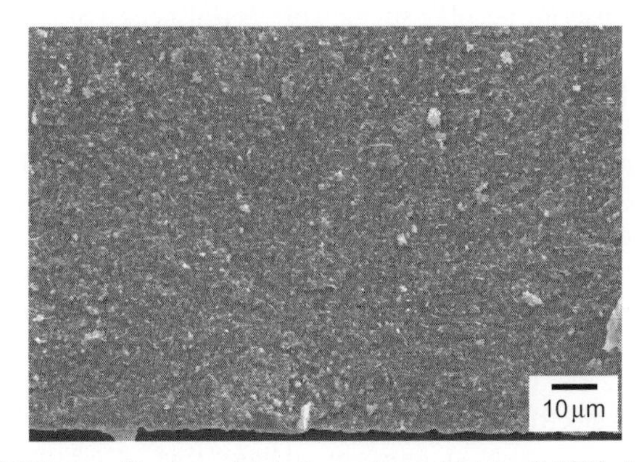

FIGURE 15 SEM image of fracture origin in the specimen with 1.0 wt% CNTs sintered at 1750°C.

FIGURE 16 SEM image of the fracture surface of the Si_3N_4 ceramics by adding 1.0 wt% CNTs fired at 1750°C.

Many CNTs remained, and there were no agglomerates of CNTs. These CNTs formed the electrical conductive path.

Consequently, CNT-dispersed Al_2O_3 and Si_3N_4 ceramic composites with high strength and high electrical conductivity were developed using nanocomposite particles prepared by the dry powder composite process.

4 CONCLUSIONS

A particle composite process was applied to develop several kinds of advanced ceramics. TiO_2 nanoparticles and Si_3N_4 particles were mixed using this powder composite process to disperse TiN nanoparticles in Si_3N_4 ceramics, which are expected to be used as novel materials for next-generation hybrid ceramic bearings. TEM observations showed that the TiO_2 nanoparticles were directly bonded to submicron Si_3N_4 particles. Si_3N_4 ceramics in which TiN nanoparticles were uniformly dispersed were fabricated using the composite particles. The amount of damage caused by the Si_3N_4 ceramics with TiN nanoparticles to the mating metals in a ball-on-disk test was comparable to the damage caused by the Si_3N_4 ceramics without TiN particles. CNT-dispersed Al_2O_3 ceramics were fabricated using a powder mixture of CNTs and a fine Al_2O_3 powder prepared by the powder composite process. It was shown that the CNTs were uniformly dispersed in the developed CNT-dispersed Al_2O_3 ceramics, and they had a high electrical conductivity and strength. CNT-dispersed Si_3N_4 ceramics were also easily fabricated by the mechanical dry mixing technique. Samples with 1.0wt% CNTs exhibited electrical conductivity. The mechanical properties improved because the CNTs were homogenously dispersed.

REFERENCES

[1] M. Naito, A. Kondo, T. Yokoyama, Applications of communication techniques for the surface modification of powder materials, ISIJ Int. 33 (1993) 915–924.

[2] M. Hosokawa, K. Nogi, M. Naito, T. Yokoyama, Nanoparticle Technology Handbook, Elsevier, Oxford, UK, 2007.

[3] M. Naito, T. Hotta, S. Asahi, T. Tanimoto, Deposition of fine particles on surface of core particles by high-speed elliptical rotor-type mixer, Kagaku Kogaku Ronbunshu 24 (1998) 52–56.

[4] M. Naito, T. Hotta, S. Asahi, T. Tanimoto, S. Endoh, Effect of processing conditions on particle composite process by a high-speed elliptical-rotor-type mixer, Kagaku Kogaku Ronbunshu 24 (1998) 99–103.

[5] M. Naito, T. Hotta, T. Tanimoto, S. Endoh, K. Nogi, Effect of fine particle size on particle composite process by a high-speed elliptical-rotor-type mixer, Kagaku Kogaku Ronbunshu 26 (2000) 62–67.

[6] T. Hotta, M. Naito, J. Szeplvolgyi, S. Endoh, K. Nogi, Effect of rotor shape on particle composite process by a high-speed elliptical-rotor-type mixer, Kagaku Kogaku Ronbunshu 27 (2001) 141–143.

[7] K. Nogi, M. Naito, A. Kondo, A. Nakahira, K. Niihara, T. Yokoyama, New method for elucidation of temperature at the interface between particles under mechanical stirring, J. Jpn. Soc. Powder Powder Metall. 43 (1996) 396–401.

[8] T. Fukui, K. Murata, S. Ohara, H. Abe, M. Naito, K. Nogi, Morphology control of Ni-YSZ cermet anode for lower temperature operation of SOFCs, J. Power Sources 125 (2004) 17–21.

[9] M. Naito, H. Abe, A. Kondo, T. Yokoyama, C.C. Huang, Smart powder processing for advanced materials, KONA Powder Part. J. 27 (2009) 130–143.

[10] H. Abe, I. Abe, K. Sato, M. Naito, Dry powder processing of fibrous fumed silica compacts for thermal insulation, J. Am. Ceram. Soc. 88 (2005) 1359–1361.

[11] K. Sato, J. Chaichanawong, H. Abe, M. Naito, Mechanochemical synthesis of $LaMnO_3$ fine powder assisted with water vapor, Mater. Lett. 60 (2006) 1399–1402.

[12] J. Tatami, E. Kodama, H. Watanabe, H. Nakano, T. Wakihara, K. Komeya, T. Meguro, A. Azushima, Fabrication and wear properties of TiN nanoparticle-dispersed Si_3N_4 ceramics, J. Ceram. Soc. Jpn. 116 (2008) 749–754.

[13] S. Tasaki, J. Tatami, H. Nakano, T. Wakihara, K. Komeya, T. Meguro, Fabrication of ZnO ceramics using ZnO/Al_2O_3 nanocomposite particles prepared by mechanical treatment, J. Ceram. Soc. Jpn. 118 (2010) 118–121.

[14] D. Hiratsuka, T. Junichi, T. Wakihara, K. Katsutoshi, T. Meguro, Fabrication of AlN ceramics using AlN and nano-Y_2O_3 composite particles prepared by mechanical treatment, Key Eng. Mater. 403 (2009) 245–248.

[15] E. Kodama, J. Tatami, T. Wakihara, T. Meguro, K. Komeya, H. Nakano, Fabrication and mechanical properties of TiN nanoparticle-dispersed Si_3N_4 ceramics from Si_3N_4-nano TiO_2 composite particles obtained by mechanical treatment, Key Eng. Mater. 403 (2009) 221–224.

[16] Y. Oyama, S. Kamigaito, Solid solubility of some oxides in Si_3N_4, Jpn. J. Appl. Phys. 10 (1971) 1637.

[17] K.H. Jack, W.I. Wilson, Ceramics based on the Si-Al-O-N and related systems, Nat. Phys. Sci. 238 (1972) 28–29.

[18] A. Tsuge, K. Nishida, M. Komatsu, Effect of crystallizing the grain-boundary glass phase on the high temperature strength of hot-pressed Si_3N_4, J. Am. Ceram. Soc. 58 (1975) 323–326.

[19] K. Komeya, Development of nitrogen ceramics, Am. Ceram. Soc. Bull. 63 (1984) 1158–1159.

[20] T. Yamada, Y. Kotoku, Commercial production of high purity silicon nitride powder by the imide thermal decomposition method, Jpn. Chem. Ind. Assoc. Mon. 42 (12) (1989) 8.

[21] M. Mitomo, Pressure sintering of Si_3N_4, J. Mater. Sci. 11 (1976) 1103–1107.

[22] H. Kawamura, S. Yamamoto, Improvement of diesel engine startability by ceramic glow plug start system, SAE Paper, No. 830580 (1983).

[23] S. Kamiya, M. Murachi, H. Kawamoto, S. Kato, S. Kawakami, Y. Suzuki, Silicon nitride swirl chambers for high power charged diesel engines, SAE, No. 850523 (1985).

[24] H. Hattori, Y. Tajima, K. Yabuta, Y. Matsuo, M. Kawamura, T. Watanabe, Gas pressure sintered silicon nitride ceramics for turbocharger application, in: Proc. 2nd International Symposium bon Ceramic Materials and Components for Engines, 1986, pp. 165.

[25] K. Komeya, H. Kotani, Development of ceramic antifriction bearing, JSAE Rev. 7 (1986) 72–79.

[26] H. Takebayashi, K. Tanimoto, T. Hattori, Performance of hybrid ceramic bearing at high speed condition (Part 1), J. Gas Turbine Soc. Jpn. 26 (1998) 55–60.

[27] H. Takebayashi, K. Tanimoto, T. Hattori, Performance of hybrid ceramic bearing at high speed condition (Prat 2), J. Gas Turbine Soc. Jpn. 26 (1998) 61–66.

[28] J. Tatami, M. Toyama, K. Noguchi, K. Komeya, T. Meguro, M. Komatsu, Effect of TiO_2 and AlN additions on the sintering behavior of the Si_3N_4-Y_2O_3-Al_2O_3 system, Ceram. Trans. 247 (2003) 83–86.

[29] T. Yano, J. Tatami, K. Komeya, T. Meguro, Microstructural observation of silicon nitride ceramics sintered with addition of Titania, J. Ceram. Soc. Jpn. 109 (2001) 396–400.

[30] J. Tatami, I.W. Chen, Y. Yamamoto, M. Komastu, K. Komeya, D.K. Kim, T. Wakihara, T. Meguro, Fracture resistance and contact damage of TiN particle reinforced Si_3N_4 ceramics, J. Ceram. Soc. Jpn. 114 (2006) 1049–1053.

[31] K. Adachi, K. Kato, N. Chen, Wear map of ceramics, Wear 203 (1997) 291–301.

[32] S. Zheng, L. Gao, H. Watanabe, J. Tatami, T. Wakihara, K. Komeya, T. Meguro, Improving the microstructure of Si_3N_4-TiN composites using various PEIs to disperse raw TiO_2 powder, Ceram. Int. 33 (2007) 355–359.

[33] R. Gahlin, S. Jacobson, The particle size effect in abrasion studied by controlled abrasive surfaces, Wear 224 (1999) 118–125.

[34] J. Tatami, T. Katashima, K. Komeya, T. Meguro, T. Wakihara, Electrically conductive CNT-dispersed silicon nitride ceramics, J. Am. Ceram. Soc. 88 (2005) 2899, 2893.

[35] S. Yoshio, J. Tatami, T. Wakihara, K. Komeya, T. Meguro, K. Aramaki, K. Yasuda, Dispersion of carbon nanotubes in ethanol by a bead milling process, Carbon 49 (2011) 4131–4137.

[36] S. Yoshio, J. Tatami, T. Wakihara, T. Yamakawa, H. Nakano, K. Komeya, T. Meguro, Effect of CNT quantity and sintering temperature on electrical and mechanical properties of CNT-dispersed Si_3N_4 ceramics, J. Ceram. Soc. Jpn. 119 (2011) 70–75.

[37] M. Matsuoka, J. Tatami, T. Wakihara, K. Komeya, T. Meguro, Improvement of strength of carbon nanotube-dispersed Si_3N_4 ceramics by bead milling and adding lower-temperature sintering aids, J. Asian Ceram. Soc. 2 (2014) 199–203.

[38] A. Hashimoto, S. Yoshio, J. Tatami, H. Nakano, T. Wakihara, K. Komeya, T. Meguro, Fabrication of CNT-dispersed Si_3N_4 ceramics by mechanical dry mixing technique, in: T. Ohji, M. Singh, S. Widjaja, D. Singh (Eds.), Advanced Processing and Manufacturing Technologies for Structural and Multifunctional Materials V: Ceramic Engineering and Science Proceedings, vol. 32, John Wiley & Sons, Inc., Hoboken, NJ, USA, 2011.

Chapter 11

Environmentally Friendly Processing of Macroporous Materials

Manabu Fukushima, Yu-ichi Yoshizawa and Tatsuki Ohji
National Institute of Advanced Industrial Science and Technology (AIST), Nagoya, Japan

1 INTRODUCTION

Macroporous ceramics have attracted plenty of research interest due to their wide variety of applications such as filtration, catalysts, catalyst supports, absorption, lightweight structural components, and thermal insulation. During the past decade, there has been a tremendous effort to research processing technologies for porous ceramics [1–5]. Freeze casting is one of the most attractive of these technologies; it can create highly porous ceramics with a directional cellular microstructure. Fukasawa et al. [6–10] first reported using the freeze-dry process to create macro-porous ceramics; Figure 1 shows an illustration of their process and an example of porous alumina obtained after employing it. They poured ceramic slurry into a mold, immersed that slurry into a freezing bath, sublimated the ice crystals, and sintered the green body. Water has been most frequently used as a solvent in this technique. This method has several advantages as an environmentally friendly and sustainable manufacturing process; it is a very simple sintering process without organic materials to be burnt. Both the pore formers and the waste materials are water, which is environmentally friendly and does not emit harmful sub-products. Another important advantage is the small amount of energy consumption required. For example, energy consumption ratios of raw powder mixing/dispersion, degreasing, purification, and sintering are 4, 43, 26, and 27%, respectively, as shown in Figure 2, when fabricating dense alumina ceramics (using an organic binder of 10 mass%, degreasing at 600 °C, and sintering at 1400 °C) [11]. It should be noted that the energy consumed for degreasing and purification accounts for about 70% of the total consumption and that, in the case of porous materials, further energy may be required to remove fugitive pore forming agents. The freeze-casting route, which is free

Green and Sustainable Manufacturing of Advanced Materials. http://dx.doi.org/10.1016/B978-0-12-411497-5.00011-4

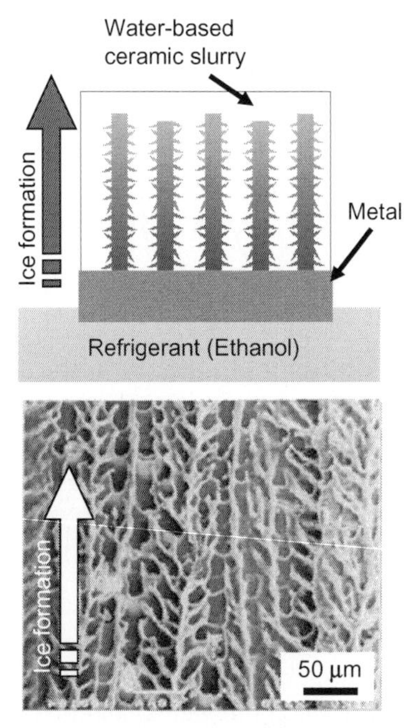

FIGURE 1 Illustration of the freeze-dry process for macroporous ceramics and an example of the resulting inside wall of a cylindrical pore with a dendritic structure in porous alumina [8]. *(Reproduced with permission of John Wiley and Sons. All rights reserved.)*

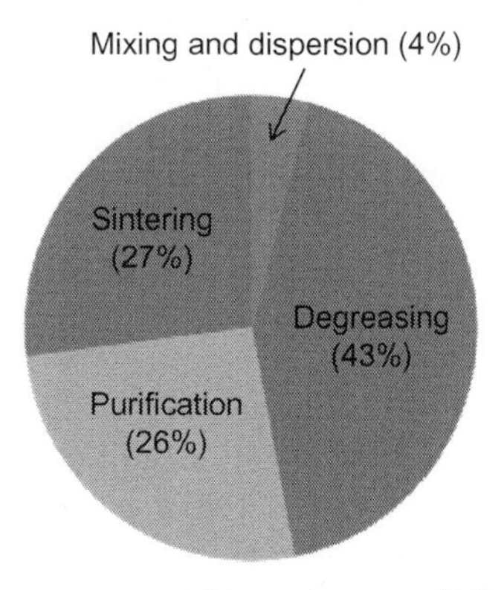

FIGURE 2 Percentage of energy consumed by each process [11] (all are laboratory level; the energy required for powder production is not included). *(Reproduced with permission of AIST. All rights reserved.)*

of such organic binders and fugitive agents, is an eco-friendly approach in terms of energy consumption. In addition, this technique has several other merits, including availability of a wide range of porosity, applicability to various types of ceramics, and cost effectiveness with simple equipment required.

A great deal of research has been devoted to freeze-casting processes for porous ceramics; some related review articles are also available [12–14]. Macroporous ceramics prepared by freeze casting are highly promising for various engineering applications such as thermal and acoustic insulators, filters, gas distributors, piezoelectric devices, as well as catalysts and their supports because of their excellent permeability, high specific strength, and great adsorption ability as well as good chemical and thermal resistances [15–24]. Deville et al. [25–30] studied the freezing kinetics and the porous structures created by freeze casting, and they revealed that the porous structures, including the dimensions, morphology, and orientation of the ice crystals, can be controlled by varying the freezing conditions of the initial slurry compositions. They also reported that bridges between neighboring lamellae have been formed due to the particle–particle interactions in the highly concentrated slurry. A great deal of effort has been devoted by many researchers, who have clarified the effects of freezing conditions, solid load, solid dispersion and composition in slurry, particle size, and so on on porous structures [19,31–35].

The freezing process under extremely cold temperatures requires a cooling system. In order to avoid the need for such a system, Araki and Halloran [36–38] used camphene ($C_{10}H_{16}$) as a vehicle in the freeze-drying process for porous ceramics. They prepared ceramic slurries in the molten camphene at 55 °C, and quickly solidified (froze) them by pouring them into polyurethane molds at room temperature. A similar camphene-based freeze-casting approach has been employed to fabricate a variety of highly porous ceramics, including Al_2O_3 [39–43], SiC [44], PZT-based ceramics [24,45], hydroxyapatite [46,47], glass ceramics [48,49], ZrO_2 [50–52], ZrB_2-SiC [53], and α-SiAlON [54]. A terpene–acrylate photopolymerizable vehicle was also adopted to fabricate porous polymer-ceramic composites and sintered porous ceramics [55]. Because the melting point of the vehicle is slightly higher than room temperature, it is relatively easy to prepare the liquid compositions and solidify them at room temperature.

It has been discovered that combining freeze-drying and gel-casting techniques can lead to macroporous ceramics with improved mechanical properties [56]. By using an organic polymer in freeze-casting methods, the ice crystal growth during freezing can be controlled, resulting in a well-tailored morphology. When polyvinyl alcohol (PVA) was used for the freeze-drying process, the pore sizes could be reduced in porous mullite [57], porous yttria-stabilized ziroconia (YSZ) [17], porous alumina [58,59], porous silicon nitride [60], and porous hydroxyapatite [61] because the growth of their ice crystals was suppressed. Among the gelation agents used for the gelation-freeze-drying process, gelatin is the most promising; it creates well-controlled porous structures because it has the ability to retain a large amount of water in a gel state and to compact ceramic particles during freezing [62–69]. Thus, the use of gelatin can lead to several advantages including

(1) the creation of cell walls that are highly packed and dense without defects because of no dendritic growth of ice crystals, leading to improved mechanical strength, and (2) cross-sections of pore channels that are honeycomb-like shapes due to the formation of columnar ice crystals, unlike ellipsoidal or lamellar morphologies obtained via conventional freeze cast. In this chapter, we give an overview of our recent work on porous ceramics prepared by a gelation-freeze-dry process using gelatin and the engineering properties of air permeability, mechanical properties, electrochemical performance, and thermal insulation.

2 PORE STRUCTURES CREATED BY THE GELATIN-GELATION-FREEZING METHOD

2.1 Overview of the Processing Strategy and Method in Gelatin-Gelation-Freezing

Lamellar or dendritic ice crystals can usually be found accompanying many defects that are grown perpendicularly to the direction of the main ice growth in wall skeletons; these defects substantially lower the strength [70,71] when an aqueous slurry is unidirectionally frozen. In order to avoid these problems, we tried to develop a combined process of gel casting and freeze casting, effectively using the advantages of gel casting [72] and eliminating the disadvantages of both castings. In conventional gel casting, poorly interconnected pores are usually obtained [73], whereas the combination of gel casting and freeze casting can create fully open porosities due to unidirectional ice growth. For a gelation agent, a gelatin gel was selected because gelatin gel suppresses the dendritic and lamellar growth of ice during freezing [62,63]. However, when gelatin is dissolved into an aqueous slurry without gelation, lamellar or dendritic growth of ice crystals can occur during freezing (see later in this chapter) [65]. Gelation processing of a slurry with a gelatin solution is essential to avoid the formation of lamellar or dendritic ice growth. A flowchart of the gelatin-gelation-freezing method is shown in Figure 3. The dried powder is mixed with a warm gelatin solution setting at around 50 °C in order to avoid the gelation of the gelatin solution. The mixing ratio of the powder/gelatin solution was in the range of 1/99-10/90 in volume where the solid load in the slurry can be varied depending on the porosity desired. The slurry is poured into a plastic mold and kept at 7 °C to achieve the smooth gelation of the gelatin solution, including the raw powder dispersed. The gelatin based gel is subsequently frozen at various freezing temperatures of approximately between −70 and −10 °C. By modulating the freezing temperature, the pore morphology can be controlled (see later in this chapter). The ice crystals in the frozen gel are sublimated by a vacuum freeze drier at 10-35 °C. The dried green bodies are finally sintered to obtain porous ceramics.

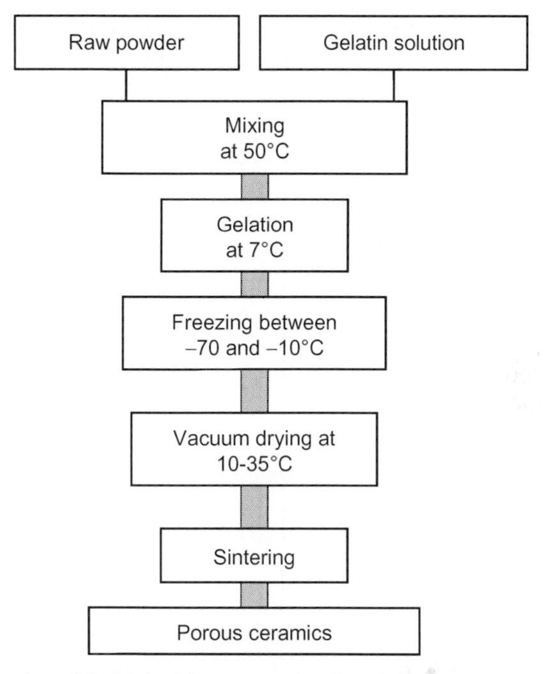

FIGURE 3 Flowchart of the fabrication process using the gelatin-gelation-freezing method.

2.2 Porous Morphology

In the freeze-casting process for porous ceramics, the porosity is controllable simply by the slurry concentration. Figure 4 shows the porosities of a typical porous silicon carbide (SiC) prepared by the gelation freezing route [66]. Regardless of the freezing temperature, the relative densities of the specimens exhibited almost constant values around 14 vol%, which was close to the solid

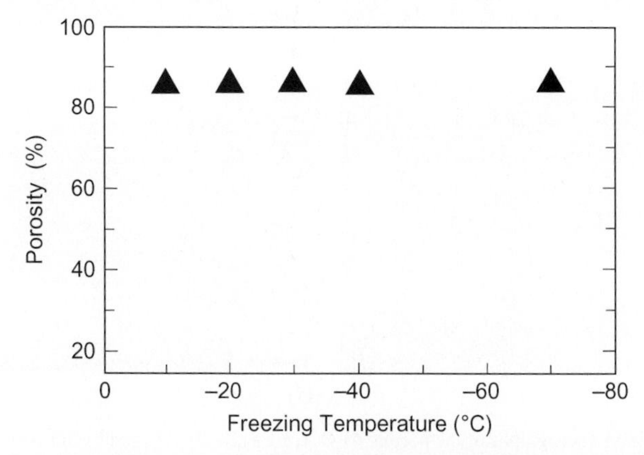

FIGURE 4 Porosities of the porous SiC obtained at various freezing temperatures [66].

load in the slurry (10%). On the other hand, the closed porosities of the samples were less than 0.5%, suggesting that almost all the water in the gelatin gel converted into interconnected open pores. Figure 5 shows typical SEM micrographs of the porous Al_2O_3 frozen at −20 °C (a-b) and −50 °C (c-d), in which the micrographs were observed in perpendicular (a and c) and parallel (b and d) to the freezing direction as indicated by the arrows. The channels running through and straight over the whole body of the sample indicate that the ice crystals have grown from the bottom (in contact with the ethanol bath) to the inside of the gel body, accompanied by the concurrent rejection of alumina particles by ice crystals grown along the temperature gradient of the gel body. The micrograph of the porous silicon carbide obtained by freezing at −10 °C and sintering at 1800 °C also clearly shows a similar cellular structure (Figure 6), in which the 3D images were collected parallel (a) and perpendicular (b) to the freezing direction. The 3D images constructed from X-ray CT scans revealed a highly porous structure with unidirectionally aligned, well-interconnected pores and homogeneous cell wall thicknesses.

The porous morphology created via the gelation-freezing method can be seen in (1) the honeycomb-like structures in the plane perpendicular to the freezing direction and (2) the unidirectional cylindrical pore morphologies parallel to the ice growth. It should be noted that this method can clearly inhibit the formation

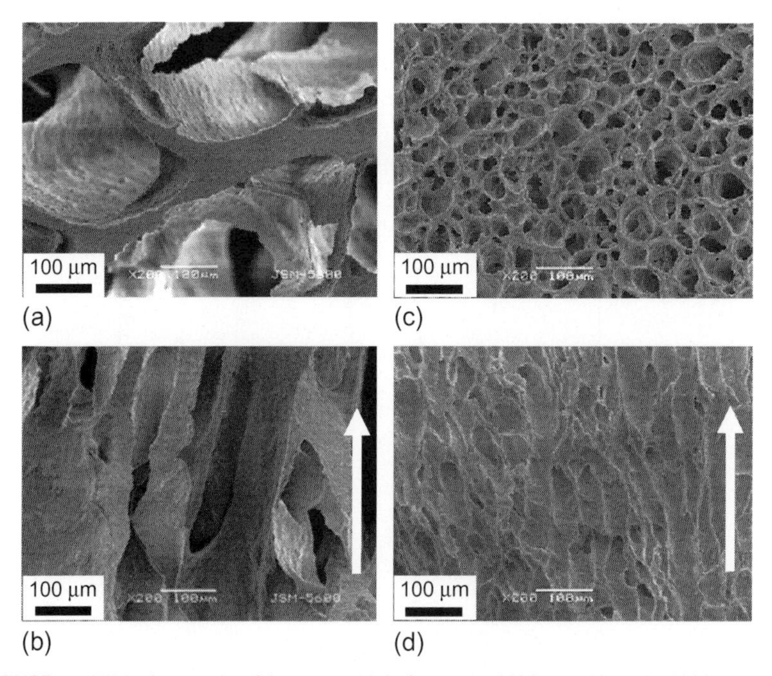

FIGURE 5 SEM micrographs of the porous Al_2O_3 frozen at −20°C (a and b) and −50 °C (c and d), both parallel (a and c) and perpendicular (b and d) to the direction of freezing.

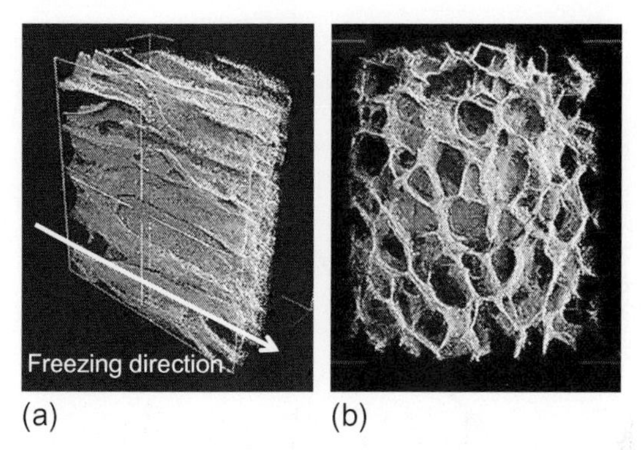

FIGURE 6 Three-dimensional (3D) images constructed from micrographs obtained from X-rays CT scans of the porous SiC obtained by freezing at −10 °C sintered at 1800 °C, in which the 3D images observed were parallel (a) and perpendicular (b) to the freezing direction.

of lamellar or dendritic structures. The typical examples of average cell size, cell wall thickness, and number of cells in the porous SiC as a function of freezing temperature are shown in Figure 7 [74]. As seen from either Figure 5a and c or Figure 7, the effect of the freezing temperature on cell size can be visibly monitored. The average cell sizes decreased with the decreasing temperature: 147 μm and 34 μm for the specimens frozen at −10°C and −70 °C, respectively. The cell wall thickness also decreased with decreasing temperature, whereas the number of cells increased with the decreasing temperature.

2.3 Pore Formation Mechanism

Here we discuss the formation mechanism of macroporous morphologies by the gelatin-gelation-freezing method, as illustrated in Figure 8. Three factors of the pore formation mechanism can affect the size of the ice crystals: (1) the ice nuclei formed, (2) the release of latent heat, and (3) the grain growth (recrystallization) of the ice. Rapid cooling can cause many ice nuclei to form, leading to numerous smaller ice crystals, while fewer ice nuclei are formed at higher freezing temperatures, resulting in the relatively large sizes of the ice crystals formed [62]. The latent heat during freezing is continuously generated by physically changing water to ice, and this heat is subsequently transferred to an ethanol bath (a cooling medium), as will be shown later. Thus, the thermal conductivity of the raw powder can also affect the size of the formed ice crystals. As we know from our previous work [66,67,75], when gelatin-based gels containing alumina or SiC were individually frozen at the same freezing temperatures, the cell sizes of the porous SiC was always smaller than those of the alumina, meaning that the raw powders play a role in the heat transfer path. Ice columns contain no gelatin or ceramic particles because growing ice crystals always reject impurities due to the very limited solubility of the crystalline lattice of ice [58,76]. Thus, it is reasonable to consider that during the

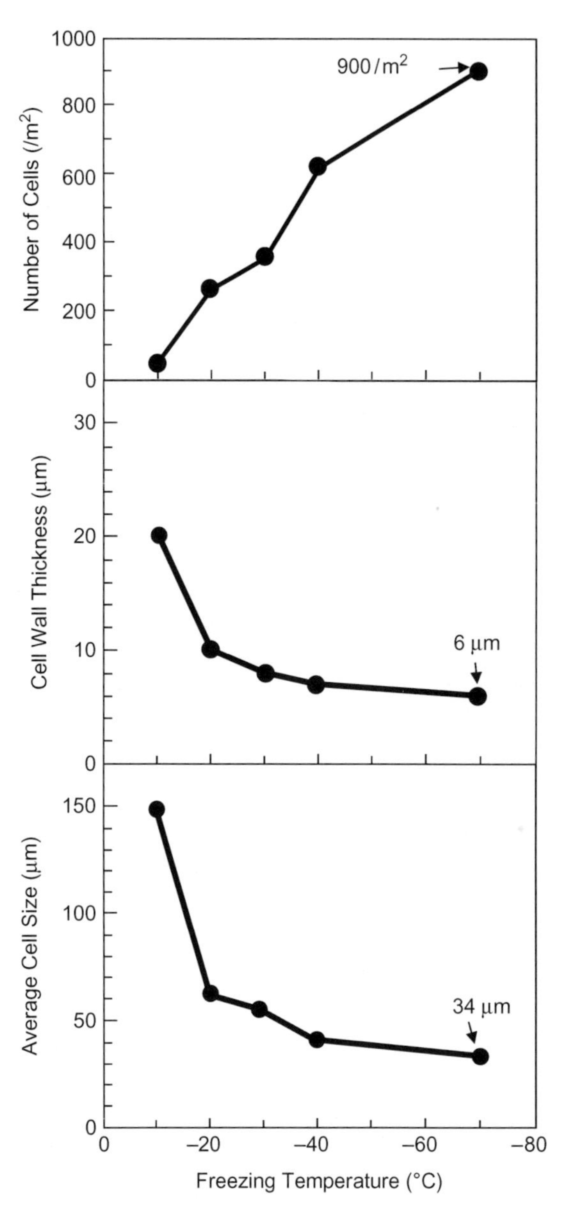

FIGURE 7 The number of cells, cell wall thickness, and average cell size as a function of the freezing temperature in porous silicon carbides prepared by the gelatin-gelation–freezing method [74]. *(Reproduced with permission of John Wiley and Sons. All rights reserved.)*

growth of the ice crystals, which result in cylindrical pores, ceramic particles are concurrently rejected by the moving freezing front. The resulting ice morphology is obviously different from dendritic or lamellar structures produced by conventional freeze-casting methods [6–10,25–27] or other fabrication

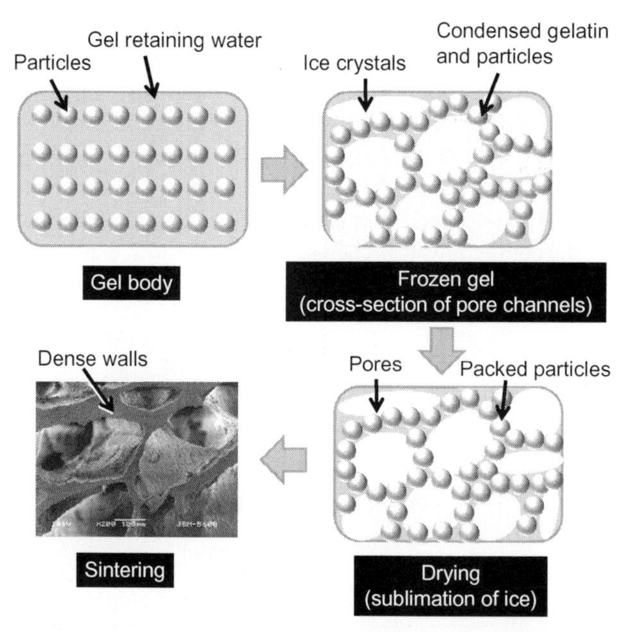

FIGURE 8 An illustration of pore formation in the gelation–freezing method. During freezing, the gel releases retained water for ice formation and rejects water to come between particles, and, in drying, the remaining gel supports the structure and packs in raw particles, leading to a firm porous structure with relatively denser cell walls after sintering.

methods such as the partial sintering technique [75,77–82], organic sacrificial pore formers [4,83,84], replica template [85–89], and chemical/physical blowing [90–94]. While the ice crystals in conventional freeze-casting are freely and preferentially grown, the gel state can substantially enhance the migration resistance of water during freezing [62]. In actuality, when gelatin is dissolved into a slurry without gelation, dendritic or lamellar structures are formed, indicating that the state of the gel is essential to avoid dendritic and/or lamellar growth of ice. The water molecules in the gel form a discontinuous network to prevent the concurrent and preferential growth of ice crystals, and then the ice nuclei formed are spatially segregated in the gelatin network [95]. Thus, it can be concluded that the honeycomb-like macroporous structures formed result from the inhibited ice growth of side branches from the main ice columns, which also leads to defect-less cell walls. In our previous work [96], the gelatin-derived carbon layer was observed on the surface of or among particles after sintering under an inert atmosphere, indicating that the ice crystals grown were wrapped by the condensed gelatin and raw particles. Presumably this wrapped structure further lowers the freezing point at the ice front, resulting in the formation of a supercooling zone. In such a way, the constitutional freeze-condensation should form a relatively higher packing density in the cell walls of the green body.

To investigate the relationship between the freezing conditions and the resulting porous structure, temperature variations of the gel body at several points were measured during freezing. Figure 9 shows an illustration of the temperature measurement setup for a gel body and temperature variations at the (a) surface, (b) center, and (c) bottom of a gel during freezing at −10, −40, and −70 °C [66]. All the freezing curves at −10 °C showed temporary and rapid temperature increases up to −1 °C at about 7 min after the freezing started due to the release of latent heat during the physical change from water to ice after supercooling.

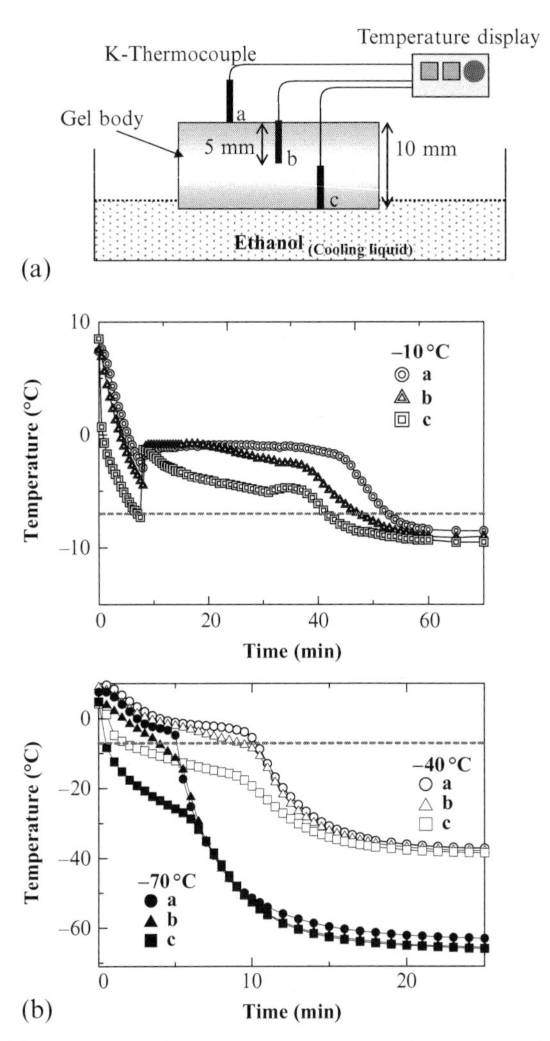

FIGURE 9 An illustration of the temperature measurement and the measured temperature of the gel body as a function of time during freezing [66]. *(Reproduced with permission of Elsevier. All rights reserved.)*

Then the curves became relatively flat until about 40 min, indicating the ongoing formation of ice crystals. This phenomenon is a well-known equilibrium between the latent heat generated and the heat flow into the cooling medium. The temperature difference among (a), (b), and (c) in this plateau area suggests that the temperature of ice crystal formation is different at the bottom of the gel than at the surface. Thus, the size of the ice crystals formed at (a) is larger than those formed at (b). At about 40 min, a temperature decrease at all measurement points was observed, indicating that the formation of ice crystals was completed in the whole body. In the plots of the specimens frozen at −40 and −70 °C, the above phenomenon was not observed due to the sufficiently lower cooling temperatures. The plateau area, which was present for the specimens frozen at −10 °C, was little observed, indicating that the ice crystals were formed in a very short period. Freezing at −40 °C resulted in gentle temperature curves during 0-10 min for (a) and (b) and during 0-9 min for (c), followed by steep slopes to the setting temperature of −40 °C. The curves of the gel frozen at −70 °C also showed relatively gentle slopes during the first 5-6 min, particularly for (a) and (b), and finally the temperatures steeply decreased to −70°C. The cell size formed at the surface of the specimen is larger than that at the bottom, which has contact with the ethanol bath, because the size of ice crystals formed at higher freezing temperatures is larger than those formed at lower ones. Ice growth in the range from 0 to −10 °C has been studied by many researchers [97–101]; the range from 0 to −7 °C is particularly known as the maximum formation zone of ice crystals, where the ice readily grows along the main ice growth direction (a-axis) [66]. When the time required to pass this zone is short (i.e., when rapid cooling is employed), the size of the ice crystals formed can be reduced.

2.4 Effects of Antifreeze Additives

As mentioned above, the size of the ice crystals depends on the distance from the cooling medium. However, using antifreeze additives, it is possible to homogenize the cell sizes. Figure 10 shows typical SEM micrographs of porous alumina prepared by (a) the conventional gelation-freezing method and by (b) doping with antifreeze glycoprotein (AFP: the mixing ratio of AFP/gelatin solution = 0.5/99.5 in weight) viewed in parallel to the freezing direction (freezing temperature at −40 °C) [67]. The microstructures visibly show that the freezing occurred from the bottom to the top, indicating that ice crystals were grown along a temperature gradient from the bottom to the top surface of the gel body and that alumina particles were concurrently pushed aside by the growing ice crystal. In the conventional gelation-freezing technique, the cell size in the region of the top surface is clearly larger than that in the bottom. The cell size gradation from bottom to top results from the temperature gradient, leading to a gradual increase of the size of the ice crystals. On the contrary, the channels of porous alumina prepared with the AFP addition were substantially homogeneous, fine, and straight over the whole body. Adsorption inhibition of AFP is a widely accepted method to prevent the grain

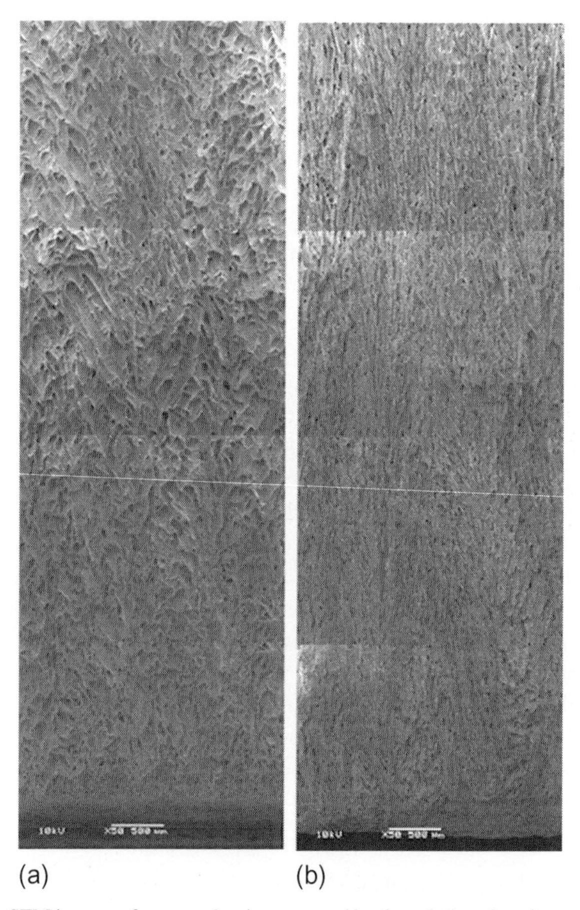

(a) (b)

FIGURE 10 SEM images of porous alumina prepared by the gelation–freezing method without (a) and with (b) anti-freeze protein (AFP) [67]. *(Reproduced with permission of John Wiley and Sons. All rights reserved.)*

growth of ice crystals [97,102]. AFP can bind to the prism planes of embryo ice crystals generated in the supercooling water including AFP, and it can inhibit the crystal growth [97,100,102–105]. This inhibition causes thermal hysteresis, that is, the temperature difference between the freezing point of a solution containing ice and its melting point [97,100,104,106]. The growing rate of ice (Ih) along the a-axis is two to three orders of magnitude higher than that of the c-axis under 1 atm [26,76]. When AFP accumulates on the prism plane of ice, convex-shaped ice can be formed from the limited interspaces among the AFPs absorbed by the ice [102]. The free energy of the convex surfaces can increase as the surface curvature increases (known as the Kelvin effect) [102,107]. Eventually, the further binding of water molecules onto the convex ice surfaces is inhibited by the increasing number of bound AFP, leading to hindrance of the ice growth along the a-axis. In addition to the adsorption of AFP by the ice crystals, grain growth is continuously inhibited.

3 ENGINEERING PROPERTIES OF MACROPOROUS CERAMICS PREPARED BY THE GELATION-FREEZING METHOD

3.1 Air Permeability

Gas permeability is one of the most important properties for porous ceramics used as gas filters, such as diesel particulate filters (DPFs), because pressure drops should be expected in such applications. Among numerous fabrication processes, the freeze-dry technique is one of the most promising processes to create unidirectionally aligned pores that can be expected to provide excellent permeability. Figure 11 shows the Darcian permeabilities as a function of pore size for macroporous ceramics prepared by freeze-dry processes [58,66] compared with those fabricated by other techniques such as organic spherical

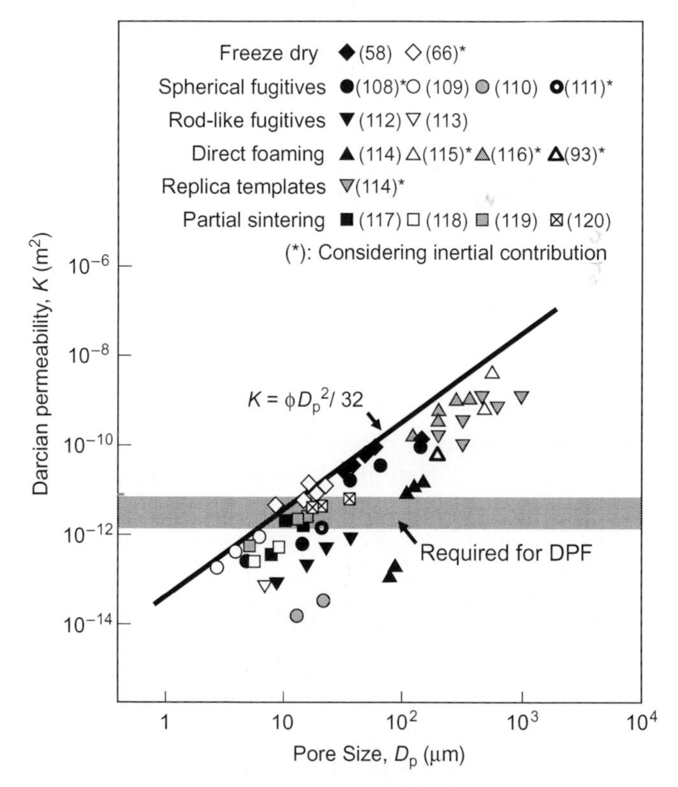

FIGURE 11 Darcian air permeability as a function of pore size for porous ceramics by the freeze-dry processes [58,66] compared with other processes such as organic pore formers [108,109], graphite pore formers [110,111], organic rod-like fugitives [112,113], pyhsical/chemical blowing [94,114–116], replica templates [114], and partial sintering [117–120]. The solid line indicates the theoretical permeability $K = \phi D_p^2 / 32$ (ϕ=0.85) for the case of unidirectional cylindrical pores penetrating in parallel.

fugitives [108,109], graphite fugitives [110,111], extruded organic fibrous fugitives [112,113], chemical/physical foaming [94,114–116], replica templates [114], and partial sintering [117–120]. The Darcian permeability, K, is calculated by the pressure drop and flow rate of air, according to the Darcy's law [121]. Based on the capillary model to compute an ideal permeability, *K* is expressed by

$$K = \phi D_p^2 / C \tag{1}$$

where ϕ is the porosity, D_p is the pore diameter, and C is a constant depending on the pore structure [121,122]. The K values of Figure 11 are adjusted from the reported values at a porosity of 0.85 using Equation (1) for comparison. The inertial contribution (non-Darcian permeability) was considered in addition to the viscous one (Darcian permeability) in Ref. [58,109], [111], [116], [114], which leads to high values of K compared with cases of neglected inertial effect [66,108,110,112,113,115,117–120]. The total ratio of viscous contribution is estimated to be 60-90% [58]. The porous ceramics fabricated by the freeze-dry processes exhibit higher Darician permeability than those created by other processes and have pore sizes ranging from 10 to 100 μm because the freeze-dry processes can provide highly aligned unidirectional cylindrical pores. The solid line of the figure means the fluid flows through the unidirectional cylindrical pores like capillary penetration in parallel ($C=32$) [122], and the permeability of the porous ceramics prepared by the freeze-dry method is very close to this solid line, indicating an ideal pore morphology for fluid flow. The permeability required for a commercially available DPF is 10 – 11 or 10 – 12 m^2 [123], and the permeability of the porous ceramics prepared by the freeze-dry technique is greater than this criterion.

3.2 Mechanical Strength

To evaluate the mechanical strength of porous ceramics, their compressive strength is usually measured. Figure 12 shows the compressive strengths of porous ceramics fabricated by a variety of freeze-dry methods as a function of porosity above 80%, in which a compressive load was applied in parallel to the orientation of the pores (the freezing direction) [35,42,51,64,66,69,125–130]. The pore morphologies are divided into cellular (shown as open circles in Figure 12) [42,51,64,66,124] and lamellar with dendritic (closed circles in Figure 12) [35,125–130] morphologies. The strength of the cellular morphologies is always higher than that of the lamellar ones because the pore size of the cellular structures was observed to be much smaller than that of the lamellar. Fewer defects can be formed in the cellular walls because the cellular structure is formed by preventing the dendritic growth of ice. The compressive strength of porous SiC with a porosity of 87% prepared by the present gelation-freezing was in the range of 5.2-16.6 MPa, depending on the cell size (147 to 34 μm). It has been reported that compressive strengths of approximately 10-16 MPa were reported in the porous SiC with a porosity of about 70-80% and a pore size of 20-100 μm [131–133], indicating that the above strength is relatively high for high porosity ceramics because of the honeycomb-like denser cell walls with fewer defects.

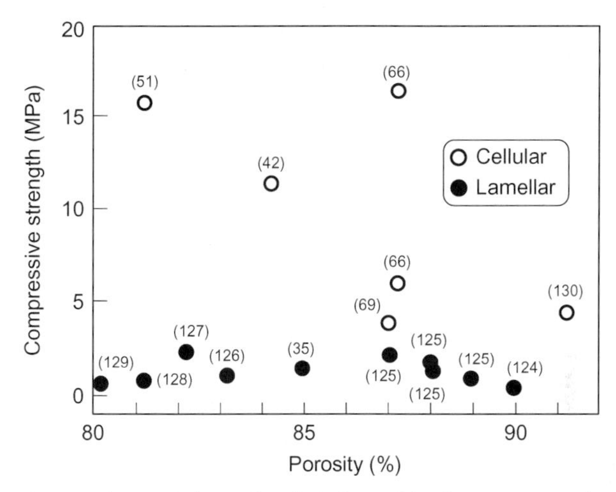

FIGURE 12 Compressive strength as a function of porosities for porous ceramics with cellular [42,51,66,69,124] (plotted by circles) and lamellar [35,124–130] (plotted by closed circles) morphologies produced by various freeze-dry processes.

3.3 Machinability

Porous ceramics, particularly with high porosities, generally are not machinable because of their fragility and weakness; however, the gelation-freezing derived porous ceramics demonstrate excellent machinability. Figure 13 shows photographs of drilled sections of porous alumina specimens with a porosity of 89 vol% (obtained by freezing at −50 °C and subsequent sintering at 1200 °C). It was possible to cut and drill the specimens to various sizes and shapes without

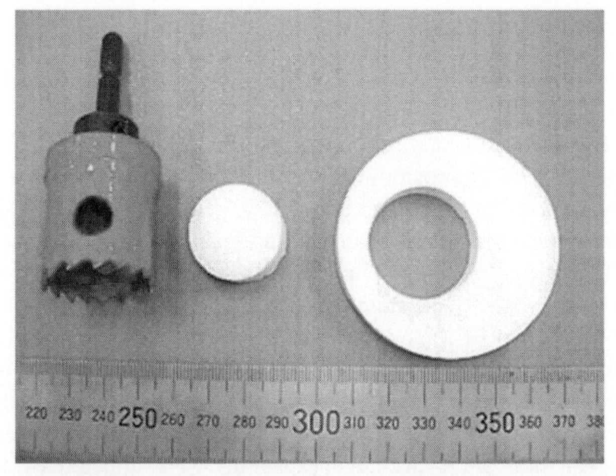

FIGURE 13 Photograph of drilled sections of porous alumina specimens with a porosity of 89 vol% (obtained by freezing at −50 °C and subsequent sintering at 1200 °C).

any chipping during either wet or dry machining processes. In addition, the specimens could be subjected to surface grinding and polishing in spite of their very high porosity of 98% [68]. In conventional freeze-casting techniques, when the solid load is very small below 5 vol%, the green bodies are critically fragile after the sublimation of ice crystals because of weak bonding between particles. Therefore, the porosity should be limited to about 70-80% at most. On the contrary, the gelatin-gelation-freezing method can give relatively high strength and sufficient rigidity to green bodies by incorporating the gelatin acting as the organic binder. Even after removal of the gelatin during degreasing, the structural stability can be kept due to the relatively dense packing of the particles derived from the gelatin-gelation-freezing process.

3.4 Electrochemical Performances

The gelation-freezing approach has been used to fabricate $LiFePO_4$/carbon composite cathodes of lithium ion batteries, and their electrochemical properties have been investigated [96]. In order to enhance electronic conductivity and the lithium diffusion constant in electrodes, a great deal of carbon has generally been used in the conventional lithium ion battery fabrication process [134,135]. The gelation-freezing technique, which uses carbonization of gelation agents, has the potential to create composite cathodes with limited carbon for lithium ion batteries that have high volumetric energy density and good rate capability. Figure 14 shows the SEM micrographs of the $LiFePO_4$/carbon composites obtained by freezing at −40 °C (a and b), −60 °C (c and d), and −80 °C (e and f) [96]. While the cellular structures varied by the freezing temperature, the open porosities of all the composite cathodes exhibited similar values of about 88% regardless of the freezing temperature. Figures 15 also show film-like carbon structures (indicated by the arrowheads) that originate from the gelatin. It is thought that these aligned carbon films work as a good electron pathway. The weight ratio of the carbon in the $LiFePO_4$/carbon composites was estimated to be 3.71 wt%, which is very small compared with that reported in other research, typically 10-15 wt% [134,135].

Figure 16 shows the charge/discharge curves at the 5th cycle for $LiFePO_4$/carbon cathodes with a porosity of 88% and a low carbon content of 3.7 wt% prepared by freezing at −40, −60 and −80 °C [96]. The measurements have been galvanostatically made at a current density of 10 mAg-1 in the range of 2.0-4.2 V. The charge/discharge curves showed the typical voltage plateau at about 3.43 V vs Li/Li+. Discharge capacities in the $LiFePO_4$/carbon composites were observed to be 152, 141, and 138 mAhg-1 for the composite cathodes prepared by freezing at −40, −60, and −80 °C, respectively. The charge/discharge capacities decreased with the decreasing temperature. Although the particle size, carbon content, and total density of all cathodes were almost similar to each other, the discharge capacities were affected by the freezing temperature because the electronic conductivity of the cell wall depends

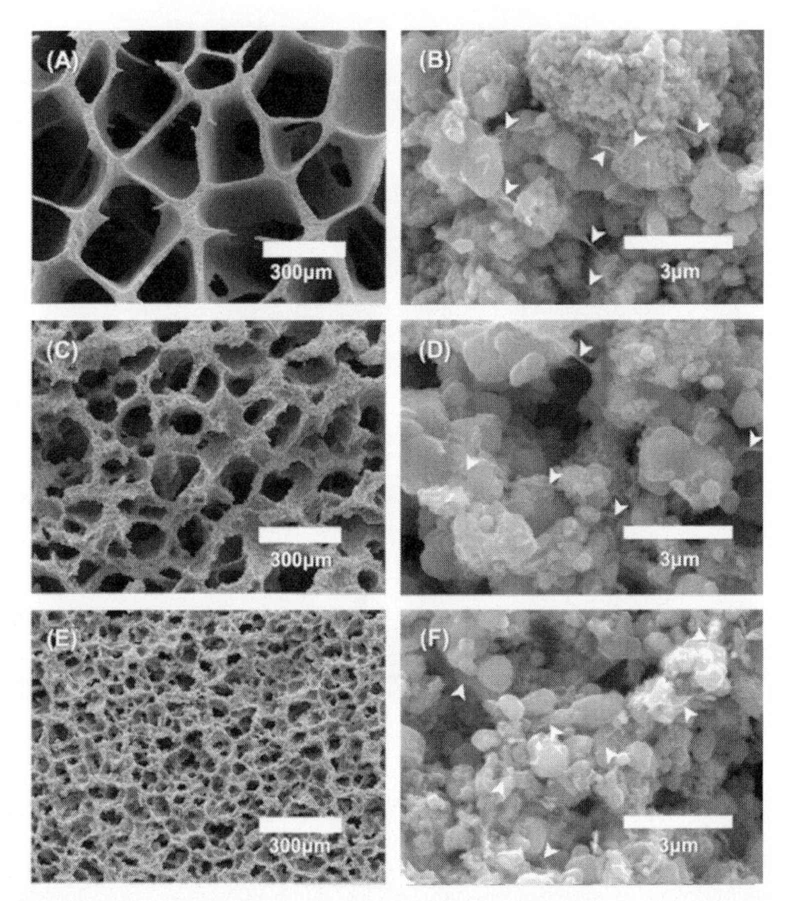

FIGURE 14 SEM micrographs of the $LiFePO_4$/carbon composites: frozen at −40 °C (a and b), −60 °C (c and d), and −80 °C (e and f), perpendicular to the direction of freezing [96]. *(Reproduced with permission of Elsevier.*

on the packing density, which varies at freezing temperatures. The present gelation-freezing technique is a promising approach to fabricate porous $LiFePO_4$/carbon composites for high-performance electrodes for lithium batteries.

3.5 Thermal Insulation Performances

Porous ceramics prepared by the present gelation-freezing method can be also used as thermal insulators [68]. Silica insulators were prepared by gelation of a gelatin solution with colloidal silica of 10, 5, and 1 vol% solid loads, followed by freezing of the gel bodies at −80 °C and heating at 500-800 °C. The porosity of the silica insulators heated at 800 °C was 88, 94, and 98% for the solid loads of 10, 5, and 1%, respectively. Figure 17 shows SEM micrographs of the silica insulators with porosities of (a) 98% and (b)

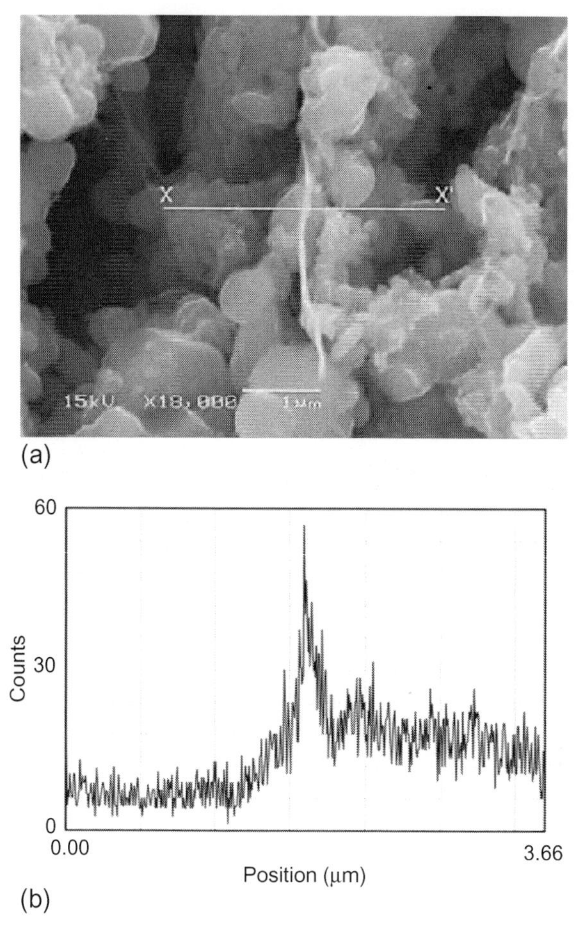

FIGURE 15 (a) An SEM image of a film-like carbon structure in the $LiFePO_4$/carbon composite. The line labeled XX' indicates positions of the EDX line. (b) The EDX line scan along line XX' of Figure 15a [96].

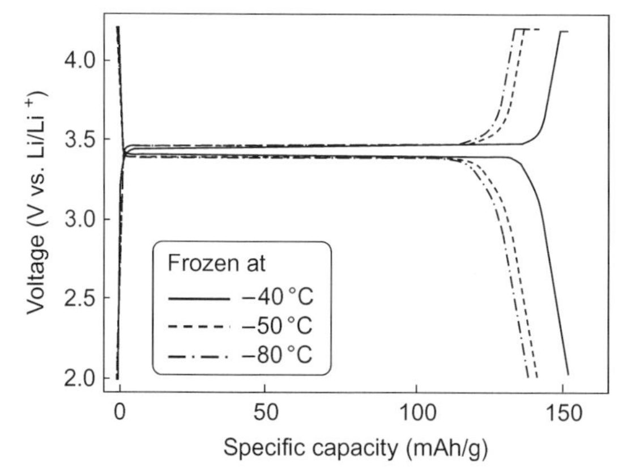

FIGURE 16 Charge/discharge curves for the $LiFePO_4$/carbon composite cathodes obtained at various freezing temperatures [96]. *(Reproduced with permission of Elsevier.)*

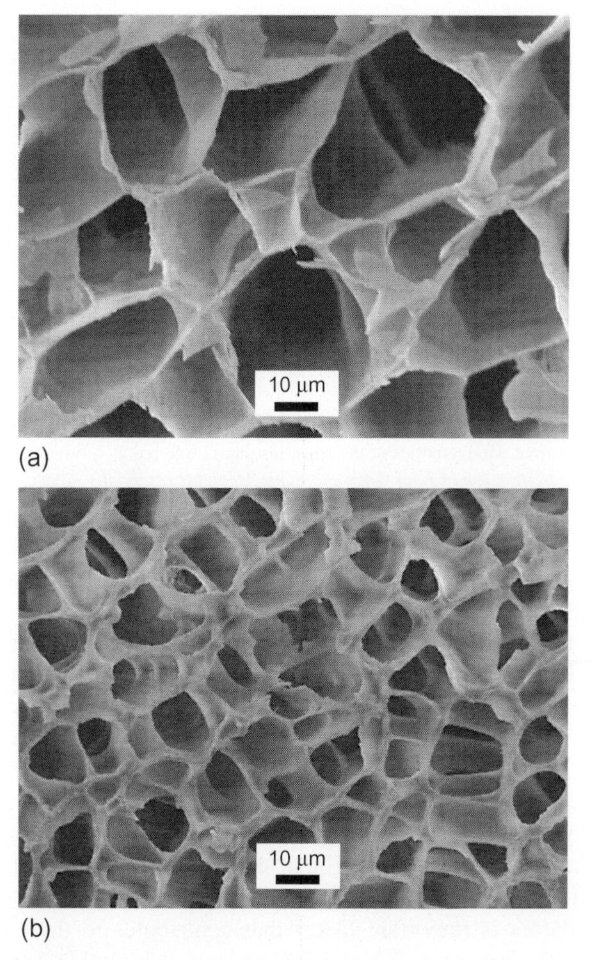

FIGURE 17 SEM micrographs of the silica insulators sintered at 800 °C with porosities of (a) 98% and (b) 88%, viewed perpendicular to the direction of freezing [68]. *Reproduced with permission of John Wiley and Sons. All rights reserved.*

88%, which again exhibit honeycomb-like cellular structures without side-branching ice dendrites. The cell diameter and wall thickness for Figure 17a were 24.5 and 0.6 μm, respectively, and those for Figure 17b were 17.0 and 2.6 μm, respectively. The cell size decreased, but the wall thickness of the skeleton increased with the increasing solid load in the slurry. Figure 18 shows the thermal conductivities of the silica insulators as a function of the sintering temperature, in which the measurements have been carried out at room temperatures. The thermal conductivities measured at room temperature were 0.054 W/mK and 0.168 W/mK, for silica contents of 1 and 10 vol%, respectively, which did not vary in this sintering temperature range. These

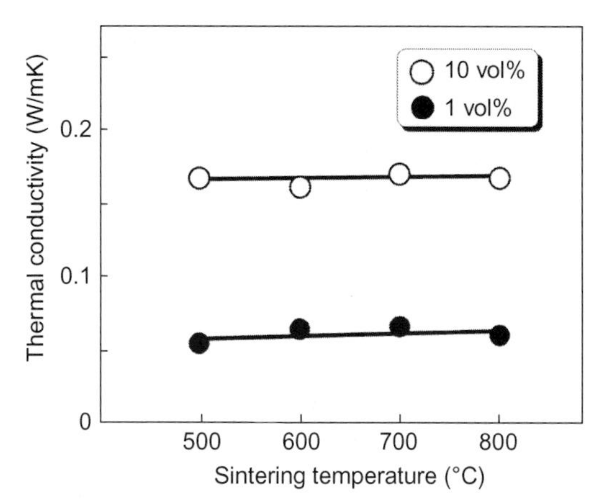

FIGURE 18 Thermal conductivities of the silica insulators sintered at various temperatures [68]. *(Reproduced with permission of John Wiley and Sons. All rights reserved.)*

values are substantially smaller than those of foam and fibrous insulators [136–138]. Heat transfer through insulators is usually affected by (1) convection flow, (2) radiation, (3) solid transfer, and (4) gas transfer [136,139–142]. The convective flow through insulators with pore sizes above 10 mm can be significantly activated throughout the entire material volume. In addition, it depends on Grashof number, which is proportional to the third power of the pore size [136,137,143]; it occurs when the Grashof number exceeds 10^3. In the present gelation-freezing method, the cell size of the derived porous ceramics is typically about 10-300 μm, which is too small to cause convective flow. In addition, solid transfers through skeletons are suppressed by the high porosity and thin cell walls of the present insulator. Gas conduction through cells in insulators is important factor that contributes to the total thermal conductivity. The estimated Knudsen number was about 0.003-0.004 for the present insulators, suggesting much larger cell sizes than the mean free path of air, that is, very limited collisions between the gas molecules and the cell walls. Therefore, the thermal conductivity of the present insulators can be reduced to 0.05 W/mK, which is close to that of air (~0.03 W/mK) at room temperature, when the porosity is as high as 98% due to the suppression of convective flow and solid conduction.

4 CONCLUSIONS

This chapter gave an overview of the advanced gelation freezing method, which is an environmentally friendly approach for fabricating macroporous ceramics. It is a very simple process that does not use organic materials that must be burnt. This means that harmful sub-products are not generated, and only a small amount of energy is consumed during the process. The macroporous ceramics prepared by this technique have several features:

(1) The walls surrounding the pore channels are relatively dense without noticeable defects, leading to excellent mechanical strength.
(2) The cross-sections of the pore channels are honeycomb-like shapes, unlike the ellipsoidal or lamellar morphologies created by conventional aqueous freeze-casting techniques, and dendritic structures are avoidable.
(3) A very high porosity and wide range of cell sizes can be produced together with unidirectional-oriented and interconnected pores.
(4) Pore morphologies, such as the controlled cell size, wall thickness, and cell interconnection in the porous monolith, can be varied by strategically modulated freezing conditions.

The fluid permeability, mechanical strength, electrochemical performance, and thermal insulation properties of the resultant porous components were examined with various microstructures and compositions, leading to promising capabilities for a variety of engineering applications.

REFERENCES

[1] P. Greil, Advanced Engineering Ceramics, Adv. Mater. 14 (2002) 709–716.

[2] P. Colombo, Conventional and novel processing methods for cellular ceramics, Philos. Trans. Roy. Soc. A-Math. Phys. Eng. Sci. 364 (2006) 109–124.

[3] A.R. Studart, U.T. Gonzenbach, E. Tervoort, L.J. Gauckler, Processing routes to macroporous ceramics: A review, J. Am. Ceram. Soc. 89 (2006) 1771–1789.

[4] P. Colombo, Engineering porosity in polymer-derived ceramics, J. Eur. Ceram. Soc. 28 (2008) 1389–1395.

[5] T. Ohji, M. Fukushima, Macro-porous ceramics: processing and properties, Int. Mater. Rev. 57 (2012) 115–131.

[6] T. Fukasawa, Z.Y. Deng, M. Ando, T. Ohji, High-surface-area alumina ceramics with aligned macroscopic pores, J. Ceram. Soc. Jpn. 109 (2001) 1035–1038.

[7] T. Fukasawa, Z.Y. Deng, M. Ando, T. Ohji, Y. Goto, Pore structure of porous ceramics synthesized from water-based slurry by freeze-dry process, J. Mater. Sci. 36 (2001) 2523–2527.

[8] T. Fukasawa, M. Ando, T. Ohji, S. Kanzaki, Synthesis of porous ceramics with complex pore structure by freeze-dry processing, J. Am. Ceram. Soc. 84 (2001) 230–232.

[9] T. Fukasawa, Z.Y. Deng, M. Ando, T. Ohji, S. Kanzaki, Synthesis of porous silicon nitride with unidirectionally aligned channels using freeze-drying process, J. Am. Ceram. Soc. 85 (2002) 2151–2155.

[10] T. Fukasawa, M. Ando, T. Ohji, Filtering properties of porous ceramics with unidirectionally aligned pores, J. Ceram. Soc. Jpn. 110 (2002) 627–631.

[11] K. Watari, T. Nagaoka, K. Sato, Y. Hotta, A strategy to reduce energy usage in ceramic fabrication, Synthesiology 2 (2009) 132–141.

[12] S. Deville, Freeze-casting of porous ceramics: A review of current achievements and issues, Adv. Eng. Mater. 10 (2008) 155–169.

[13] M.C. Gutierrez, M.L. Ferrer, F. del Monte, Ice-templated materials: Sophisticated structures exhibiting enhanced functionalities obtained after unidirectional freezing and ice-segregation-induced self-assembly, Chem. Mater. 20 (2008) 634–648.
[14] W.L. Li, K. Lu, J.Y. Walz, Freeze casting of porous materials: review of critical factors in microstructure evolution, Int. Mater. Rev. 57 (2012) 37–60.
[15] T. Moritz, H.-J. Richter, Ice-mould freeze casting of porous ceramic components, J. Eur. Ceram. Soc. 27 (2007) 4595–4601.
[16] A.-H. Lu, F. Schueth, Nanocasting: A versatile strategy for creating nanostructured porous materials, Adv. Mater. 18 (2006) 1793–1805.
[17] K.H. Zuo, Y.-P. Zeng, D. Jiang, Properties of microstructure-controllable porous yttria-stabilized ziroconia ceramics fabricated by freeze casting, Int. J. Appl. Ceram. Technol. 5 (2008) 198–203.
[18] X.W. Zhu, D.L. Jiang, S.H. Tan, Z.Q. Zhang, Improvement in the strut thickness of reticulated porous ceramics, J. Am. Ceram. Soc. 84 (2001) 1654–1656.
[19] F. Ye, J. Zhang, L. Liu, H. Zhan, Effect of solid content on pore structure and mechanical properties of porous silicon nitride ceramics produced by freeze casting, Mat. Sci. Eng. A-Struct. Mat. Prop. Microstruct. Process. 528 (2011) 1421–1424.
[20] P. Szabo-Revesz, A. Szepes, J. Ulrich, Z. Farkas, J. Kovacs, Freeze-casting technique in the development of solid drug delivery systems, Chem. Eng. Process. 46 (2007) 230–238.
[21] G. Frank, E. Christian, K. Dietmar, A Novel Production Method for Porous Sound-Absorbing Ceramic Material for High-Temperature Applications, Int. J. Appl. Ceram. Technol. 8 (2011) 646–652.
[22] T.L. Cable, J.A. Setlock, S.C. Farmer, A.J. Eckel, Regenerative Performance of the NASA Symmetrical Solid Oxide Fuel Cell Design, Int. J. Appl. Ceram. Technol. 8 (2011) 1–12.
[23] T.L. Cable, S.W. Sofie, A symmetrical, planar SOFC design for NASN' s high specific power density requirements, J. Power Sources 174 (2007) 221–227.
[24] S.-H. Lee, S.-H. Jun, H.-E. Kim, Y.-H. Koh, Fabrication of porous PZT-PZN piezoelectric ceramics with high hydrostatic figure of merits using camphene-based freeze casting, J. Am. Ceram. Soc. 90 (2007) 2807–2813.
[25] S. Deville, E. Saiz, R.K. Nalla, A.P. Tomsia, Freezing as a path to build complex composites, Science 311 (2006) 515–518.
[26] S. Deville, E. Saiz, A.P. Tomsia, Ice-templated porous alumina structures, Acta Mater. 55 (2007) 1965–1974.
[27] E. Munch, E. Saiz, A.P. Tomsia, S. Deville, Architectural Control of Freeze-Cast Ceramics Through Additives and Templating, J. Am. Ceram. Soc. 92 (2009) 1534–1539.
[28] S. Deville, E. Maire, A. Lasalle, A. Bogner, C. Gauthier, J. Leloup, C. Guizard, In Situ X-Ray Radiography and Tomography Observations of the Solidification of Aqueous Alumina Particle Suspensions. Part I: Initial Instants., J. Am. Ceram. Soc. 92 (2009) 2489–2496.
[29] S. Deville, E. Maire, A. Lasalle, A. Bogner, C. Gauthier, J. Leloup, C. Guizard, In Situ X-Ray Radiography and Tomography Observations of the Solidification of Aqueous Alumina Particles Suspensions. Part II: Steady State., J. Am. Ceram. Soc. 92 (2009) 2497–2503.
[30] A. Lasalle, C. Guizard, J. Leloup, S. Deville, E. Maire, A. Bogner, C. Gauthier, J. Adrien, L. Courtois, Ice-Templating of Alumina Suspensions: Effect of Supercooling and Crystal Growth During the Initial Freezing Regime, J. Am. Ceram. Soc. 95 (2012) 799–804.
[31] L. Hu, C.-A. Wang, Y. Huang, C. Sun, S. Lu, Z. Hu, Control of pore channel size during freeze casting of porous YSZ ceramics with unidirectionally aligned channels using different freezing temperatures, J. Eur. Ceram. Soc. 30 (2010) 3389–3396.

[32] L. Jing, K. Zuo, F. Zhang, X. Chun, Y. Fu, D. Jiang, Y.-P. Zeng, The controllable microstructure of porous Al2O3 ceramics prepared via a novel freeze casting route, Ceram. Int. 36 (2010) 2499–2503.

[33] J. Zou, Y. Zhang, R. Li, Effect of Suspension State on the Pore Structure of Freeze-Cast Ceramics, Int. J. Appl. Ceram. Technol. 8 (2011) 482–489.

[34] J.C. Li, D.C. Dunand, Mechanical properties of directionally freeze-cast titanium foams, Acta Mater. 59 (2011) 146–158.

[35] K. Zhao, Y.F. Tang, Y.S. Qin, J.Q. Wei, Porous hydroxyapatite ceramics by ice templating: Freezing characteristics and mechanical properties, Ceram. Int. 37 (2011) 635–639.

[36] K. Araki, J.W. Halloran, Room-temperature freeze casting for ceramics with nonaqueous sublimable vehicles in the naphthalene-camphor eutectic system, J. Am. Ceram. Soc. 87 (2004) 2014–2019.

[37] K. Araki, J.W. Halloran, New freeze-casting technique for ceramics with sublimable vehicles, J. Am. Ceram. Soc. 87 (2004) 1859–1863.

[38] K. Araki, J.W. Halloran, Porous ceramic bodies with interconnected pore channels by a novel freeze casting technique, J. Am. Ceram. Soc. 88 (2005) 1108–1114.

[39] Y.-H. Koh, J.-H. Song, E.-J. Lee, H.-E. Kim, Freezing Dilute Ceramic/Camphene Slurry for Ultra-High Porosity Ceramics with Completely Interconnected Pore Networks, J. Am. Ceram. Soc. 89 (2006) 3089–3093.

[40] Y.-H. Koh, E.-J. Lee, B.-H. Yoon, J.-H. Song, H.-E. Kim, H.-W. Kim, Effect of Polystyrene Addition on Freeze Casting of Ceramic/Camphene Slurry for Ultra-High Porosity Ceramics with Aligned Pore Channels, J. Am. Ceram. Soc. 89 (2006) 3646–3653.

[41] B.-H. Yoon, W.-Y. Choi, H.-E. Kim, J.-H. Kim, Y.-H. Koh, Aligned porous alumina ceramics with high compressive strengths for bone tissue engineering, Scr. Mater. 58 (2008) 537–540.

[42] Y.-W. Moon, K.-H. Shin, Y.-H. Koh, W.-Y. Choi, H.-E. Kim, Porous alumina ceramics with highly aligned pores by heat-treating extruded alumina/camphene body at temperature near its solidification point, J. Eur. Ceram. Soc. 32 (2012) 1029–1034.

[43] Y.-W. Moon, K.-H. Shin, Y.-H. Koh, S.-W. Yook, C.-M. Han, H.-E. Kim, F. Clemens, Novel Ceramic/Camphene-Based Co-Extrusion for Highly Aligned Porous Alumina Ceramic Tubes, J. Am. Ceram. Soc. 95 (2012) 1803–1806.

[44] B.-H. Yoon, C.-S. Park, H.-E. Kim, Y.-H. Koh, In Situ Synthesis of Porous Silicon Carbide (SiC) Ceramics Decorated with SiC Nanowires, J. Am. Ceram. Soc. 90 (2007) 3759–3766.

[45] S.-H. Lee, S.-H. Jun, H.-E. Kim, Y.-H. Koh, Piezoelectric Properties of PZT-Based Ceramic with Highly Aligned Pores, J. Am. Ceram. Soc. 91 (2008) 1912–1915.

[46] B.-H. Yoon, C.-S. Park, H.-E. Kim, Y.-H. Koh, In-situ fabrication of porous hydroxyapatite (HA) scaffolds with dense shells by freezing HA/camphene slurry, Mater. Lett. 62 (2008) 1700–1703.

[47] B.-H. Yoon, Y.-H. Koh, C.-S. Park, H.-E. Kim, Generation of Large Pore Channels for Bone Tissue Engineering Using Camphene-Based Freeze Casting, J. Am. Ceram. Soc. 90 (2007) 1744–1752.

[48] J.-H. Song, Y.-H. Koh, H.-E. Kim, L.-H. Li, H.-J. Bahn, Fabrication of a Porous Bioactive Glass–Ceramic Using Room-Temperature Freeze Casting, J. Am. Ceram. Soc. 89 (2006) 2649–2653.

[49] X. Liu, M.N. Rahaman, Q. Fu, A.P. Tomsia, Porous and strong bioactive glass (13-93) scaffolds prepared by unidirectional freezing of camphene-based suspensions, Acta Biomater. 8 (2012) 415–423.

[50] C. Hong, X. Zhang, J. Han, J. Du, W. Han, Ultra-high-porosity zirconia ceramics fabricated by novel room-temperature freeze-casting, Scr. Mater. 60 (2009) 563–566.

[51] J. Han, C. Hong, X. Zhang, J. Du, W. Zhang, Highly porous ZrO2 ceramics fabricated by a camphene-based freeze-casting route: Microstructure and properties, J. Eur. Ceram. Soc. 30 (2010) 53–60.

[52] C. Hong, X. Zhang, J. Han, J. Du, W. Zhang, Camphene-based freeze-cast ZrO2 foam with high compressive strength, Mater. Chem. Phys. 119 (2010) 359–362.

[53] J. Du, X. Zhang, C. Hong, W. Han, Microstructure and mechanical properties of ZrB2–SiC porous ceramic by camphene-based freeze casting, Ceram. Int. 39 (2013) 953–957.

[54] Z. Hou, F. Ye, L. Liu, Q. Liu, H. Zhang, Effects of solid content on the phase assemblages, mechanical and dielectric properties of porous α-SiAlON ceramics fabricated by freeze casting, Ceram. Int. 39 (2013) 1075–1079.

[55] V. Tomeckova, J.W. Halloran, L. Gauckler, Porous Ceramics by Photopolymerization with Terpene-Acrylate Vehicles, J. Am. Ceram. Soc. 95 (2012) 3763–3768.

[56] R. Chen, Y. Huang, C.-A. Wang, J. Qi, Ceramics with ultra-low density fabricated by gelcasting: An unconventional view, J. Am. Ceram. Soc. 90 (2007) 3424–3429.

[57] S. Ding, Y.-P. Zeng, D. Jiang, Fabrication of mullite ceramics with ultrahigh porosity by gel freeze drying, J. Am. Ceram. Soc. 90 (2007) 2276–2279.

[58] C. Pekor, B. Groth, I. Nettleship, The Effect of Polyvinyl Alcohol on the Microstructure and Permeability of Freeze-Cast Alumina, J. Am. Ceram. Soc. 93 (2010) 115–120.

[59] D. Zhang, Y. Zhang, R. Xie, K. Zhou, Freeze gelcasting of aqueous alumina suspensions for porous ceramics, Ceram. Int. 38 (2012) 6063–6066.

[60] D. Yao, Y. Xia, Y.-P. Zeng, K.-H. Zuo, D. Jiang, Fabrication porous Si3N4 ceramics via starch consolidation–freeze drying process, Mater. Lett. 68 (2012) 75–77.

[61] N. Monmaturapoj, W. Soodsawang, W. Thepsuwan, Porous hydroxyapatite scaffolds produced by the combination of the gel-casting and freeze drying techniques, J. Porous. Mater. 19 (2011) 441–447.

[62] H.W. Kang, Y. Tabata, Y. Ikada, Fabrication of porous gelatin scaffolds for tissue engineering, Biomaterials 20 (1999) 1339–1344.

[63] F. Shen, Y.L. Cui, L.F. Yang, K.D. Yao, X.H. Dong, W.Y. Jia, H.D. Shi, A study on the fabrication of porous chitosan/gelatin network scaffold for tissue engineering, Polym. Int. 49 (2000) 1596–1599.

[64] M. Fukushima, M. Nakata, Y.-i Yoshizawa, Fabrication and properties of ultra highly porous cordierite with oriented micrometer-sized cylindrical pores by gelation and freezing method, J. Ceram. Soc. Jpn. 116 (2008) 1322–1325.

[65] Y. Zhang, K. Zuo, Y.-P. Zeng, Effects of gelatin addition on the microstructure of freeze-cast porous hydroxyapatite ceramics, Ceram. Int. 35 (2009) 2151–2154.

[66] M. Fukushima, M. Nakata, Y. Zhou, T. Ohji, Y.-i. Yoshizawa, Fabrication and properties of ultra highly porous silicon carbide by the gelation-freezing method, J. Eur. Ceram. Soc. 30 (2010) 2889–2896.

[67] M. Fukushima, S. Tsuda, Y.-i. Yoshizawa, Fabrication of Highly Porous Alumina Prepared by Gelation Freezing Route with Antifreeze Protein, J. Am. Ceram. Soc. 96 (2013) 1029–1031.

[68] M. Fukushima, Y.-i. Yoshizawa, Fabrication of highly porous silica thermal insulators prepared by gelation freezing route, J. Am. Ceram. Soc. 97 (2014) 713–717.

[69] M. Fukushima, M. Nakata, Y.-i. Yoshizawa, Processing Strategy for Producing Ultra-Highly Porous Cordierite, in: Advanced Processing and Manufacturing Technologies for Structural and Multifunctional Materials III, John Wiley & Sons, Inc, Hoboken, 2010.

[70] A. Lasalle, C. Guizard, E. Maire, J. Adrien, S. Deville, Particle redistribution and structural defect development during ice templating, Acta Mater. 60 (2012) 4594–4603.

[71] S. Deville, Freeze-Casting of Porous Biomaterials: Structure, Properties and Opportunities, Materials 3 (2010) 1913–1927.

[72] O.O. Omatete, M.A. Janney, R.A. Strehlow, Gelcasting - A New Ceramic Forming Process, Am. Ceram. Soc. Bull. 70 (1991) 1641–1649.

[73] P. Sepulveda, J.G.P. Binner, S.O. Rogero, O.Z. Higa, J.C. Bressiani, Production of porous hydroxyapatite by the gel-casting of foams and cytotoxic evaluation, J. Biomed. Mater. Res. 50 (2000) 27–34.

[74] M. Fukushima, Y.-i Yoshizawa, T. Ohji, Macroporous Ceramics by Gelation Freezing Route Using Gelatin, Adv. Eng. Mater. 16 (2014) 607–620.

[75] M. Fukushima, Microstructural control of macroporous silicon carbide, J. Ceram. Soc. Jpn. 121 (2013) 162–168.

[76] V.F. Petrenko, R.W. Whitworth, Physics of Ice, Oxford University Press, Oxford, 2002.

[77] Y. Suzuki, P.E.D. Morgan, T. Ohji, New uniformly porous CaZrO3/MgO composites with three-dimensional network structure from natural dolomite, J. Am. Ceram. Soc. 83 (2000) 2091–2093.

[78] D.D. Jayaseelan, N. Kondo, M.E. Brito, T. Ohji, High-strength porous alumina ceramics by the pulse electric current sintering technique, J. Am. Ceram. Soc. 85 (2002) 267–269.

[79] J.H. She, T. Ohji, Fabrication and characterization of highly porous mullite ceramics, Mater. Chem. Phys. 80 (2003) 610–614.

[80] M. Fukushima, Y. Zhou, H. Miyazaki, Y.-i. Yoshizawa, K. Hirao, Y. Iwamoto, S. Yamazaki, T. Nagano, Microstructural Characterization of Porous Silicon Carbide Membrane Support With and Without Alumina Additive, J. Am. Ceram. Soc. 89 (2006) 1523–1529.

[81] M. Fukushima, Y. Zhou, Y.-i. Yoshizawa, Fabrication and microstructural characterization of porous silicon carbide with nano-sized powders, Mater. Sci. Eng. B148 (2008) 211–214.

[82] M. Fukushima, Y. Zhou, Y.-i. Yoshizawa, Fabrication and microstructural characterization of porous SiC membrane supports with Al2O3–Y2O3 additives, J. Membr. Sci. 339 (2009) 78–84.

[83] P. Colombo, E. Bernardo, L. Biasetto, Novel microcellular ceramics from a silicone resin, J. Am. Ceram. Soc. 87 (2004) 152–154.

[84] P. Colombo, E. Bernardo, Macro- and micro-cellular porous ceramics from preceramic polymers, Compos. Sci. Technol. 63 (2003) 2353–2359.

[85] P. Greil, T. Lifka, A. Kaindl, Biomorphic cellular silicon carbide ceramics from wood: I. Processing and microstructure., J. Eur. Ceram. Soc. 18 (1998) 1961–1973.

[86] P. Greil, E. Vogli, T. Fey, A. Bezold, N. Popovska, H. Gerhard, H. Sieber, Effect of microstructure on the fracture behavior of biomorphous silicon carbide ceramics, J. Eur. Ceram. Soc. 22 (2002) 2697–2707.

[87] M. Singh, J.A. Salem, Mechanical properties and microstructure of biomorphic silicon carbide ceramics fabricated from wood precursors, J. Eur. Ceram. Soc. 22 (2002) 2709–2717.

[88] M. Singh, B.M. Yee, Reactive processing of environmentally conscious, biomorphic ceramics from natural wood precursors, J. Eur. Ceram. Soc. 24 (2004) 209–217.

[89] M. Fukushima, Y.-i. Yoshizawa, P. Colombo, Decoration of Ceramic Foams by Ceramic Nanowires via Catalyst-Assisted Pyrolysis of Preceramic Polymers, J. Am. Ceram. Soc. 95 (2012) 3071–3077.

[90] P. Colombo, M. Modesti, Silicon oxycarbide ceramic foams from a preceramic polymer, J. Am. Ceram. Soc. 82 (1999) 573–578.

[91] P. Colombo, J.R. Hellmann, D.L. Shelleman, Mechanical properties of silicon oxycarbide ceramic foams, J. Am. Ceram. Soc. 84 (2001) 2245–2251.

[92] S. Barg, C. Soltmann, M. Andrade, D. Koch, G. Grathwohl, Cellular ceramics by direct foaming of emulsified ceramic powder suspensions, J. Am. Ceram. Soc. 91 (2008) 2823–2829.
[93] M. Fukushima, P. Colombo, Silicon carbide-based foams from direct blowing of polycarbosilane, J. Eur. Ceram. Soc. 32 (2012) 503–510.
[94] A. Idesaki, P. Colombo, Synthesis of a Ni-Containing Porous SiOC Material From Polyphenylmethylsiloxane by a Direct Foaming Technique, Adv. Eng. Mater. 14 (2012) 1116–1122.
[95] L.G. Dowell, S.W. Moline, A.P. Rinfret, A low-temperature x-ray diffraction study of ice structures formed in aqueous gelatin gels, Biochimica Et Biophysica Acta 59 (1962) 158–167.
[96] K. Hamamoto, M. Fukushima, M. Mamiya, Y. Yoshizawa, J. Akimoto, T. Suzuki, Y. Fujishiro, Morphology control and electrochemical properties of LiFePO4/C composite cathode for lithium ion batteries, Solid State Ionics 225 (2012) 560–563.
[97] Y. Yeh, R.E. Feeney, Antifreeze proteins: Structures and mechanisms of function, Chem. Rev. 96 (1996) 601–617.
[98] C.A. Knight, J. Hallett, A.L. Devries, Solute effects on ice recrystallization - an assessment technique, Cryobiology 25 (1988) 55–60.
[99] R.L. McKown, G.J. Warren, Enhanced survival of yeast expressing an antifreeze gene analog after freezing, Cryobiology 28 (1991) 474–482.
[100] C.A. Knight, A.L. Devries, L.D. Oolman, Fish antifreeze protein and the freezing and recrystallization of ice, Nature 308 (1984) 295–296.
[101] Y. Yeh, R.E. Feeney, R.L. McKown, G.J. Warren, Measurement of grain-growth in the recrystallization of rapidly frozen-solutions of antifreeze glycoproteins, Biopolymers 34 (1994) 1495–1504.
[102] J.A. Raymond, A.L. Devries, Adsorption inhibition as a mechanism of freezing resistance in polar fishes, Proc. Natl. Acad. Sci. U. S. A. 74 (1977) 2589–2593.
[103] Y. Yamashita, R. Miura, Y. Takemoto, S. Tsuda, H. Kawahara, H. Obata, Type II antifreeze protein from a mid-latitude freshwater fish, Japanese smelt (Hypomesus nipponensis), Biosci. Biotechnol. Biochem. 67 (2003) 461–466.
[104] M.M. Harding, P.I. Anderberg, A.D.J. Haymet, 'Antifreeze' glycoproteins from polar fish, Eur. J. Biochem. 270 (2003) 1381–1392.
[105] A.L. Devries, D.E. Wohlschlag, Freezing resistance in some antarctic fishes, Science 163 (1969) 1073–1075.
[106] G.L. Fletcher, M.H. Kao, R.M. Fourney, Antifreeze peptides confer freezing resistance to fish, Canadian Journal of Zoology-Revue Canadienne De Zoologie 64 (1986) 1897–1901.
[107] C.A. Knight, C.C. Cheng, A.L. Devries, Adsorption of alpha-helical antifreeze peptides on specific ice crystal-surface planes, Biophys. J. 59 (1991) 409–418.
[108] I.-H. Song, I.-M. Kwon, H.-D. Kim, Y.-W. Kim, Processing of microcellular silicon carbide ceramics with a duplex pore structure, J. Eur. Ceram. Soc. 30 (2010) 2671–2676.
[109] L. Biasetto, P. Colombo, M.D.M. Innocentini, S. Mullens, Gas permeability of microcellular ceramic foams, Ind. Eng. Chem. Res. 46 (2007) 3366–3372.
[110] B.A. Latella, L. Henkel, E.G. Mehrtens, Permeability and high temperature strength of porous mullite-alumina ceramics for hot gas filtration, J. Mater. Sci. 41 (2006) 423–430.
[111] S. Ding, Y.-P. Zeng, D. Jiang, Gas permeability behavior of mullite-bonded porous silicon carbide ceramics, J. Mater. Sci. 42 (2007) 7171–7175.
[112] T. Isobe, Y. Kameshima, A. Nakajima, K. Okada, Y. Hotta, Gas permeability and mechanical properties of porous alumina ceramics with unidirectionally aligned pores, J. Eur. Ceram. Soc. 27 (2007) 53–59.

[113] K. Okada, M. Shimizu, T. Isobe, Y. Kameshima, M. Sakai, A. Nakajima, T. Kurata, Characteristics of microbubbles generated by porous mullite ceramics prepared by an extrusion method using organic fibers as the pore former, J. Eur. Ceram. Soc. 30 (2010) 1245–1251.

[114] M.D.M. Innocentini, P. Sepulveda, V.R. Salvini, V.C. Pandolfelli, J.R. Coury, Permeability and structure of cellular ceramics: A comparison between two preparation techniques, J. Am. Ceram. Soc. 81 (1998) 3349–3352.

[115] T. Tomita, S. Kawasaki, K. Okada, Effect of viscosity on preparation of foamed silica ceramics by a rapid gelation foaming method, J. Porous. Mater. 12 (2005) 123–129.

[116] M.D.M. Innocentini, R.K. Faleiros, R. Pisani Jr., I. Thijs, J. Luyten, S. Mullens, Permeability of porous gelcast scaffolds for bone tissue engineering, J. Porous. Mater. 17 (2010) 615–627.

[117] Y.-J. Park, I.-H. Song, Si3N4 with comparable permeability to SiC, J. Eur. Ceram. Soc. 32 (2012) 471–475.

[118] Y.-J. Park, J.-W. Lee, H.-S. Yun, I.-H. Song, Porous Sintered Reaction-Bonded Silicon Nitrides with Dual-Sized Pore Channel using Presintered Si-Additive Mixture Granules, Int. J. Appl. Ceram. Technol. 9 (2012) 1104–1111.

[119] J.-H. Eom, Y.-W. Kim, I.-H. Song, Effects of the initial alpha-SiC content on the microstructure, mechanical properties, and permeability of macroporous silicon carbide ceramics, J. Eur. Ceram. Soc. 32 (2012) 1283–1290.

[120] S. In-Hyuck, H. Jang-Hoon, P. Mi-Jung, K. Hai-Doo, K. Young-Wook, Effects of silicon particle size on microstructure and permeability of silicon-bonded SiC ceramics, J. Ceram. Soc. Jpn. 120 (2012) 370–374.

[121] K. Ishizaki, S. Komarneni, K. Nanko, Porous materials; process technology and applications, Kluwer Academic Publisher, Dordrecht, 1998.

[122] F.A.L. Dullien, Porous Media Fluid Transport and Pore Structure, Academic Press, San Diego, 1991.

[123] T. Tomita, S. Kawasaki, K. Okada, A novel preparation method for foamed silica ceramics by sol-gel reaction and mechanical foaming, J. Porous. Mater. 11 (2004) 107–115.

[124] H.-J. Yoon, U.-C. Kim, J.-H. Kim, Y.-H. Koh, W.-Y. Choi, H.-E. Kim, Macroporous Alumina Ceramics with Aligned Microporous Walls by Unidirectionally Freezing Foamed Aqueous Ceramic Suspensions, J. Am. Ceram. Soc. 93 (2010) 1580–1582.

[125] M.-K. Ahn, K.-H. Shin, Y.-W. Moon, Y.-H. Koh, W.-Y. Choi, H.-E. Kim, Highly Porous Biphasic Calcium Phosphate (BCP) Ceramics with Large Interconnected Pores by Freezing Vigorously Foamed BCP Suspensions under Reduced Pressure, J. Am. Ceram. Soc. 94 (2011) 4154–4156.

[126] M.-K. Ahn, Y.-W. Moon, Y.-H. Koh, H.-E. Kim, G. Franks, Use of Glycerol as a Cryoprotectant in Vacuum-Assisted Foaming of Ceramic Suspension Technique for Improving Compressive Strength of Porous Biphasic Calcium Phosphate Ceramics, J. Am. Ceram. Soc. 95 (2012) 3360–3362.

[127] H.-J. Yoon, U.-C. Kim, J.-H. Kim, Y.-H. Koh, W.-Y. Choi, H.-E. Kim, Fabrication and characterization of highly porous calcium phosphate (CaP) ceramics by freezing foamed aqueous CaP suspensions, J. Ceram. Soc. Jpn. 119 (2011) 573–576.

[128] D. Li, M. Li, Preparation of porous alumina ceramic with ultra-high porosity and long straight pores by freeze casting, J. Porous. Mater. 19 (2011) 345–349.

[129] Y. Zhang, K. Zhou, Y. Bao, D. Zhang, Effects of rheological properties on ice-templated porous hydroxyapatite ceramics, Mater. Sci. Eng. C33 (2013) 340–346.

[130] T.Y. Yang, W.Y. Kim, S.Y. Yoon, H.C. Park, Macroporous silicate ceramics prepared by freeze casting combined with polymer sponge method, J. Phys. Chem. Solids 71 (2010) 436–439.

[131] J.-H. Eom, Y.-W. Kim, I.-H. Song, H.-D. Kim, Processing and properties of polysiloxane-derived porous silicon carbide ceramics using hollow microspheres as templates, J. Eur. Ceram. Soc. 28 (2008) 1029–1035.

[132] Y.-W. Kim, J.-H. Eom, C. Wang, C.B. Park, Processing of porous silicon carbide ceramics from carbon-filled polysiloxane by extrusion and carbothermal reduction, J. Am. Ceram. Soc. 91 (2008) 1361–1364.

[133] I. Ganesh, D.C. Jana, S. Shaik, N. Thiyagarajan, An aqueous gelcasting process for sintered silicon carbide ceramics, J. Am. Ceram. Soc. 89 (2006) 3056–3064.

[134] H. Huang, S.-C. Yin, L.F. Nazar, Approaching Theoretical Capacity of LiFePO4 at Room Temperature at High Rates, Electrochem. Solid-State Lett. 4 (2001) A170–A172.

[135] B. Kang, G. Ceder, Battery materials for ultrafast charging and discharging, Nature 458 (2009) 190–193.

[136] E. Litovsky, M. Shapiro, A. Shavit, Gas pressure and temperature dependences of thermal conductivity of porous ceramic materials: Part 2, Refractories and ceramics with porosity exceeding 30%, J. Am. Ceram. Soc. 79 (1996) 1366–1376.

[137] Z. Zivcova, E. Gregorova, W. Pabst, D.S. Smith, A. Michot, C. Poulier, Thermal conductivity of porous alumina ceramics prepared using starch as a pore-forming agent, J. Eur. Ceram. Soc. 29 (2009) 347–353.

[138] J.D. Verschoor, P. Greebler, N.J. Manville, Heat transfer by gas conduction and radiation in fibrous insulation, Trans. Am. Soc. Mech. Eng. 961–968 (1952).

[139] P.G. Klemens, R.K. Williams, Thermal conductivity of metals and alloys, Int. Met. Rev. 31 (1986) 197–215.

[140] J. Francl, W.D. Kingery, Thermal conductivity: 9, Experimental investigation of effect of porosity on thermal conductivity, J. Am. Ceram. Soc. 37 (1954) 99–107.

[141] T.J. Lu, Heat transfer efficiency of metal honeycombs, Int. J. Heat Mass Transf. 42 (1999) 2031–2040.

[142] E.Y. Litovsky, M. Shapiro, Gas-pressure and temperature dependences of thermal-conductivity of porous ceramic materials: Part 1. Refractories and ceramics with porosity below 30 percent, J. Am. Ceram. Soc. 75 (1992) 3425–3439.

[143] T.W. Clyne, I.O. Golosnoy, J.C. Tan, A.E. Markaki, Porous materials for thermal management under extreme conditions, Philos. Trans. Roy. Soc. A-Math. Phys. Eng. Sci. 364 (2006) 125–146.

Chapter 12

Manufacturing of Ceramic Components using Robust Integration Technologies

Mrityunjay Singh,[1] Naoki Kondo[2] and R. Asthana[3]
[1]Ohio Aerospace Institute, Cleveland OH, USA, [2]National Institute of Advanced Industrial Science and Technology (AIST), Nagoya, Japan, [3]University of Wisconsin-Stout, Menomonie, WI, USA

1 INTRODUCTION

Net-shape manufacture of ceramic parts is often limited because of their inherent brittleness and the related machining and forming challenges. As an alternative, advanced ceramic components could be built by joining and integrating discrete units. Joining could potentially lower cost for producing components instead of fabricating large three-dimensional parts. In addition, many ceramic components need to be integrated with other materials such as other ceramics, metals, and alloys. Development and/or adaptation of robust and efficient joining technologies thus acquires considerable importance in the manufacture of ceramic parts. Most ceramics are joined using processes such as diffusion bonding; fusion welding; adhesive bonding; active metal brazing; brazing with oxides, glasses, and oxynitrides; reaction forming; and a variety of other methods. Reaction bonding makes use of carbonaceous mixtures to form silicon carbide. Active brazing uses metallic braze fillers that contain a reactive element to form chemical bonds. Transient liquid phase (TLP) bonding uses multiple interlayers to form joints via formation of a thin transient liquid phase at temperatures that are significantly lower than those required for conventional methods. Diffusion bonding has been used to join ceramics using interlayers of nickel, molybdenum, tungsten, and titanium. In this chapter, we review selected ceramic joining methods that are environmentally benign, including active brazing, joining with rapid localized heating, diffusion bonding, and reaction bonding. We review recent research developments in these areas and highlight challenges and opportunities in joining ceramics using these technologies.

Green and Sustainable Manufacturing of Advanced Materials. http://dx.doi.org/10.1016/B978-0-12-411497-5.00012-6

2 ACTIVE METAL BRAZING

Active metal brazing is particularly versatile due to its ability to join dissimilar ceramics, create complex assemblies, and offer an efficient and inexpensive route for assembly and integration. Brazing has been applied to a wide range of bulk ceramics, ceramic-matrix composites, foams, honeycomb structures, laminates, and other forms. Ceramic brazing with metallic fillers uses either pre-metallized ceramic surfaces or braze foils, or pastes or wires that usually contain a reactive filler metal such as Ti, Zr, Cr, Nb, or Y that promotes braze wettability and flow by inducing chemical reactions with the ceramic. Titanium is one of the most commonly used active metals because it forms compounds that strongly bond to both metals and ceramics. In a manner similar to self-brazed ceramics, brazing of ceramics to metals becomes feasible and joint strength improves when either a reactive filler is used or the ceramic surface is metallized (resulting in a metal/metal bond). Pre-coating ceramics with a Ti-bearing compound that forms a Ti layer on the ceramic creates a wettable surface that strongly bonds the mating surfaces together upon cooling and solidification. Besides metallic brazes, non-metallic brazes such as glasses (or mixtures of glass with crystalline materials) are also used. Use of glass as a filler is facilitated by the formation of amorphous phases at ceramic's grain boundaries during sintering, and good wetting and bonding can usually be achieved between the glass filler and grain boundary phases.

Brazes must show good wetting and adherence to the substrates, be ductile and resistant to grain growth, resist creep and oxidation, possess closely matched coefficients of thermal expansion (CTE) with the joined materials, high thermal conductivity (for thermal management), and melting points greater than the operating temperature of the joint but lower than the joined materials' melting temperatures. A large number of metallic braze alloys have been designed and developed for use with ceramic-based materials, and are commercially available.

Many commercial brazing technologies used in large-scale production employ fluxes to protect and clean the joint during brazing. Such fluxes often contain toxic volatile compounds. Common brazing flux chemicals include hydrogen fluoride, potassium bifluoride, potassium fluoride, potassium pentaborate, and other chemicals. During joint formation, the fluxes release highly toxic and corrosive gases such as hydrogen fluoride (HF) and boron trifluoride (BF_3) that are injurious to human health and the environment. Many serious health issues have been associated with exposure to brazing fluxes. These include irritation to the eyes and respiratory system, calcium depletion (sclerosis) of the bones, mottled teeth, and other illnesses. In addition, discharge of raw flux and flux residues into plant effluents poses serious hazards to the environment.

Use of fluxes is avoided in vacuum brazing technology, which is considered far more environment-friendly. However, even though conventional flux-less vacuum brazing does not require use of fluxes, the ceramics, cermets, steels,

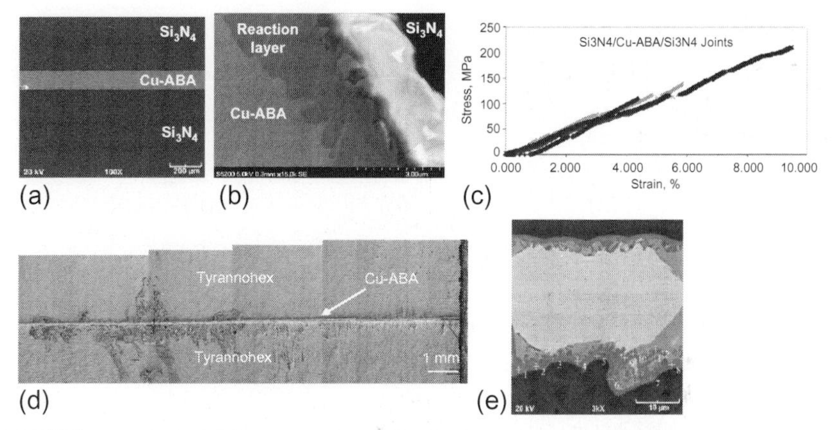

FIGURE 1 (a) & (b) SEM images of silicon nitride joints brazed using an active braze (Cu-ABA), (c) stress-strain plot showing joint strength of silicon nitride/Cu-ABA joints; (d) & (e) microstructure of a silicon carbide ceramic (SA-Tyrannohex) joint brazed using Cu-ABA.

titanium, and other alloys need to be plated to facilitate wetting and joining. Unfortunately, many electroplating processes use bath compositions that require discharge of toxic chemicals such as cyanide compounds and heavy metals via effluents. Other plating methods such as electro-less plating also involve use of toxic materials (e.g., formaldehyde, a known carcinogen).

A number of research studies have demonstrated successful joining of conventional and advanced ceramics and ceramic-matrix composites using vacuum brazing. Examples of some brazed materials such as silicon nitride, silicon carbide, and carbon foam are shown in Figures 1 and 2. Further details about brazed ceramics are presented in ref. [1–7].

During the past several years, many other environmentally benign joining technologies, such as ultrasonic brazing, nano-foil joining, rapid localized heating, and others, have emerged. Extremely clean, high-strength, flux-free joints form using these methods. Rapid localized heating that makes use of ignition and a combustion wave is an emerging method used to bond dissimilar materials. Rather than heat the entire assembly to be joined as in furnace heating, localized heating confines high temperatures to the joint seam, thus permitting rapid heat dissipation via conduction through the substrate.

3 HIGH TEMPERATURE BONDING BY LOCALIZED HEATING

3.1 Joining of Silicon Nitride Ceramics

Because of their excellent properties such as good heat-corrosion resistance, good wear resistance, high specific rigidity, and light weight, ceramics such as silicon nitride are widely used in various manufacturing industries to save

FIGURE 2 (a) A SEM images of a low-density carbon foam, (b) & (c) carbon foam bonded to Cu-clad-Mo for thermal management applications, (d) fracture in foam rather than in bond in a foam/Cu-clad-Mo join during tension test, (e-g) showing brazed structures of C-C composites with Ti tubes and carbon foam for thermal management applications.

energy for production and improve qualities of products [8]. The examples include stalk tubes and heating element protection tubes used in aluminum casting, transfer rolls in the steel industry, and cylinders in rotary kilns. These components in many cases have a maximum size of more than 10 m. Such huge ceramic components are very difficult to produce as one body because huge production facilities are needed and the available forming techniques are limited. An alternative approach is to join several ceramic units together to form such a huge component [9]. When employing a joining technique for this purpose, however, it is essential to join the units by locally heating the joint region in order to minimize energy consumption and the cost required by the production process. In this section, we focus on two local heating techniques recently developed for joining silicon nitride ceramic units to make large components. One uses microwave radiation [10], and the other uses a special electric furnace [11].

3.1.1 Microwave Local Heating

Microwave heating is a unique technique in which the materials themselves are heated directly and volumetrically by electromagnetic radiation in the microwave spectrum, causing polarization of the molecules in the materials. In addition to rapid and uniform heating, this heating technique requires only a simple chamber consisting of stainless-steel plates, a magnetron, and an insulator. Microwave absorption for heating depends on a material's specifics, indicating that local heating can be achieved by using an appropriate combination of materials; an example is a combination of silicon nitride, an alumina fiberboard insulator, and a silicon carbide susceptor as shown in Figure 3. Two silicon nitride pipes (90 and 60 mm in length) with outer and inner diameters of 28 and 16 mm, respectively, which are commercially available thermocouple protection tubes used for aluminum casting (Hitachi Metals, Ltd., Japan), were used as a parent material for joining. As an insert material, glass was prepared by mixing raw powders, Si_3N_4 (SN-E10, Ube Industries Ltd., Japan), Y_2O_3 (Shin-Etsu Chemical Co. Ltd., Japan), Al_2O_3 (AL160SG4, Showa Denko K.K., Japan), and SiO_2 (Kojundo Chemical Lab. Co., Ltd., Japan) with a composition (in mass %) of 30.1 Si_3N_4, 43.4 Y_2O_3, 11.8 Al_2O_3, and 14.7 SiO_2, following the previous study by Xie et al. [12]. Alumina fiberboard with a cylindrical hole with a diameter of 30 mm, where the two silicon nitride pipes and the insert were put, was employed as the insulator. Silicon carbide granules were adopted as the susceptor and were placed surrounding the joint with a thickness of 5 mm and a length of 40 mm (i.e., 20 mm from the joint). Silicon carbide readily absorbs microwaves [13–15] while silicon nitride and alumina are poor microwave absorbers; this difference in microwave absorption ability can generate local heating around the susceptor.

Heating was conducted using a microwave heating furnace with a magnetron (frequency: 2.45 GHz; maximum output power: 6 kW; Takasago Industry Co., Ltd., Japan) in N_2 gas flow. Figure 4 shows a typical heating profile for the joining using microwave local heating [16], compared with that using a resistance

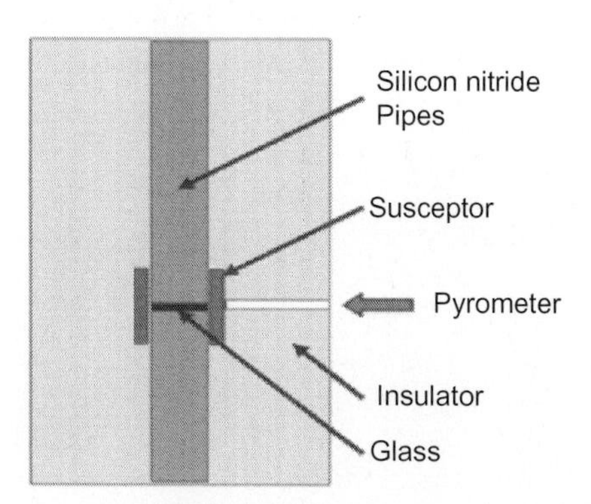

FIGURE 3 Schematic illustration of the arrangement of microwave local heating.

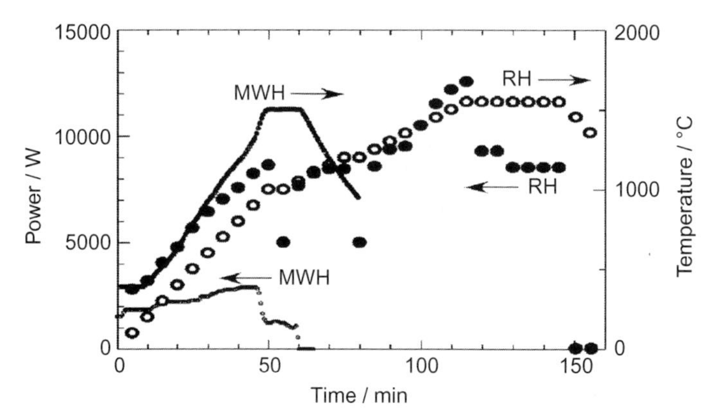

FIGURE 4 Typical heating profiles of microwave heating (MWH) and resistance heating (RH) for obtaining equivalent joining properties.

heating furnace (High-multi-10000, Fujidempa Kogyo Co., Ltd., Japan), which is required to obtain the mechanical properties equivalent to those of the microwave heating, as is stated later. As for the microwave local heating, in order to avoid thermal shock fractures due to rapid heating, the radiation power was relatively slowly increased so that the joint temperature increased to 1500 °C in 40 min. After soaking at 1500 °C for 10 min, the power was switched off so that the joined pipe was cooled down to room temperature. The maximum power required during heating was about 3000 W. Despite the very gradual increase of the microwave radiation power during heating up to 1500 °C, the temperature increase was substantially enhanced above 1300 °C. It has been known that the microwave absorption ability of liquid-phase sintered silicon nitrides increases sharply when the temperature exceeds the softening temperature of the grain-boundary glassy phase [15], leading to the rapid increase of temperature above 1300 °C. Figure 5 shows the two silicon nitride pipes successfully joined using the glass, with a cross-sectional view of the joint. The softened glass filled the gap between the pipes well. The specimens (3 x 4 x 40 mm) were cut from the joined pipes for the four-point flexural strength measurements with outer and

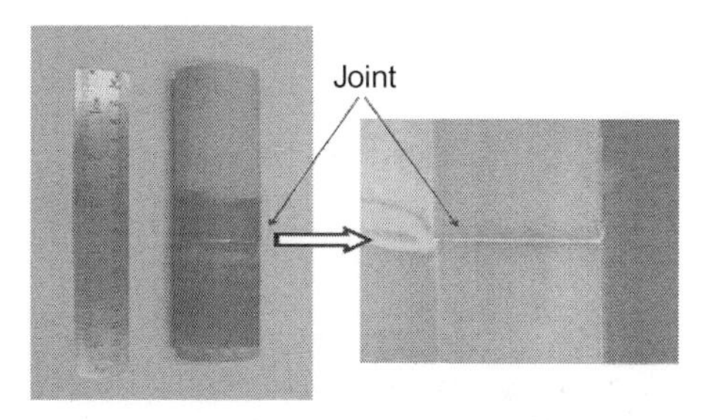

FIGURE 5 Joined silicon nitride pipe (left) and cross-sectional image of the joint (right).

inner spans of 30 and 10 mm, respectively, in accordance with JIS R1601. The average value and standard deviation were 446 MPa and 35 MPa, respectively. Most of silicon nitrides joined by similar glass compositions with no or low pressure (<0.01 MPa) exhibited strengths in the range of 400-600 MPa [17–20], and thus the obtained strength here is comparable to them.

On the other hand, to obtain the equivalent mechanical properties, the resistance heating required a heating time as long as 110 min along with a maximum power exceeding 10000 W (just before reaching 1500 °C) as shown in Figure 4. The electric power consumption needed for one heating cycle of joining is 2100 W and 33,000 W for the microwave and resistance heating, respectively. Thus, microwave local heating requires less than 10% of the electric power and less than half the time of resistance heating, demonstrating that it is a promising energy-saving technique for joining silicon nitride ceramics.

3.1.2 Local Heating Using an Electric Furnace

As mentioned in the previous section, microwave heating is a unique technique to achieve local, rapid, and energy efficient heating for joining. However, it has some difficulties to overcome for practical use, such as localization of the microwave radiation; the combination of bulk, insert, and susceptor materials; temperature control; selective heating; and vaporization of the liquid phase. An electric resistance heating system is the same heating system used in industrial furnaces. Though it is inferior to microwave heating in terms of of local heating and energy efficiency, local heating using an electric resistance furnace seems to be useful for many industries. In this section, the development of special equipment employed in a local heating furnace to join silicon nitride pipes is discussed [11,15]. Joined pipes and their properties are also described.

A schematic drawing of the equipment is shown in Figure 6. It shows a heating furnace with a graphite heating element of 300 mm in length at the center of a movable rail. Two chucks, each of which grips a silicon nitride pipe with a maximum outer diameter of ~60 mm, are located at the ends of the rail. One end of the pipe is gripped by a chuck, and the other end, the end of the pipe to be joined, is placed in the furnace. Therefore, only the joint sections of the pipes are inside the furnace, and the joint section is heated locally. The joint section is heated up to 1700 °C in a nitrogen gas flow atmosphere. Mechanical pressure

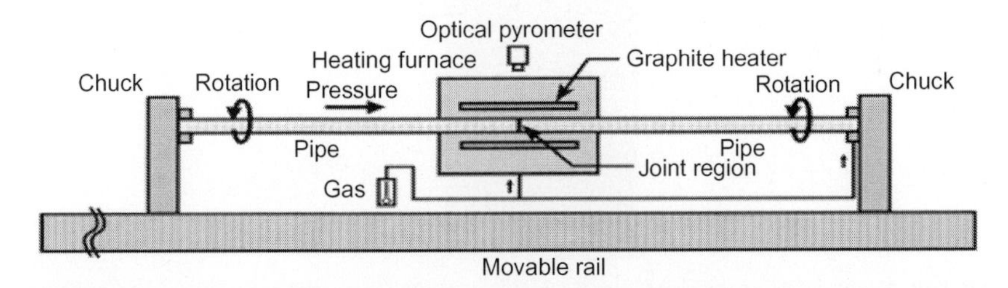

FIGURE 6 Schematic drawing of local heating equipment for joining ceramic pipes.

of a few MPa can be applied to the pipes for joining. Rotation of the pipes at ~5 rpm is necessary to avoid flowing down of the softened glass when glass is used as the insert material. Silicon nitride pipes can be joined by using this equipment. As the chucks grip the ends of the pipes, one long pipe can be made by joining the two shorter pipes.

An example of a joined pipe of 3 m in length is shown in Figure 7. This long pipe was made from three pipes of 1 m in length and 28 and 18 mm of outer and inner diameters, respectively, which were joined at two points.

Here, optimization of the joining conditions is briefly discussed. [11,15]. Silicon nitride pipes containing 5 mass% Y_2O_3 and 5 mass% Al_2O_3 as sintering additives were used. A slurry containing 30.1 mass% Si_3N_4, 43.4 mass% Y_2O_3, 11.8 mass% Al_2O_3, and 14.7 mass% SiO_2, which was previously reported by Xie et al. [12], was used as the insert. The slurry was brush coated on the surface to be joined. The temperatures, soaking time, and mechanical pressure needed for joining were investigated.

Mechanical pressures from 1 to 5 MPa were examined, and a higher pressure of 5 MPa was preferred to obtain a strong joint. A thicker glass-rich joint of ~10 μm remained in the specimen joined at a lower pressure. The thickness of the joint was reduced to ~5 μm at 5 MPa. Thus, not enough of the glass insert was squeezed out at a lower pressure. The existence of the glass-rich joint reduced the strength of the part.

FIGURE 7 Overview of joined silicon nitride pipe of 3 m in length. The long pipe was made from three pipes of 1 m in length. Joint regions were indicated by arrows.

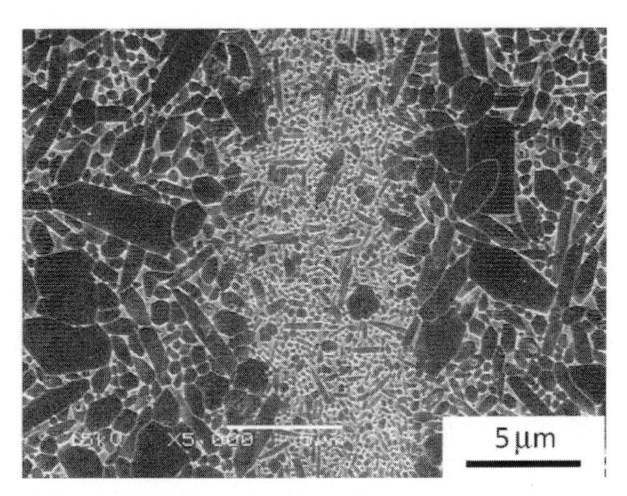

FIGURE 8 Microstructure around the joint. Joining condition was 1600°C, 1h and 5 MPa. Left and right regions with larger silicon nitride grains were the original pipes, and center region with smaller grains was the joint.

The temperature was changed from 1500 to 1700°C, and the soaking time was fixed at 1 h. Suitable temperatures to obtain strong joints were found to be 1600–1650°C. Typical microstructures around the joint are shown in Figure 8. The thickness of the joint was less than 5 μm. Diffusion and grain growth occurred; therefore, rod-like silicon nitride grains were formed in the joint. Some grains were elongated across the interface between original pipe and the joint. Void formation was limited. Strengths measured by flexural tests (size of bend specimen: 3 x 4 x 40 mm, bend condition: four-point flexural with outer and inner spans of 30 and 10 mm) were 677 MPa and 682 MPa joined at 1600 and 1650 °C, respectively. Lower temperatures of less than 1550°C could not make a strong joint because there was not enough diffusion and grain growth. The joint strength was 412 MPa at both 1500 and 1500 °C. A higher temperature of 1700 °C also reduced the strength. Diffusion and grain growth substantially occurred at this temperature; however, some voids were found in the joint as well as in the original pipe. The temperature was too high and caused the formation of voids by vaporization of the glass phase at the grain boundaries. The existence of a void reduced the strength to 529 MPa.

The effect of the soaking time was also examined. At a temperature of 1600 °C, voids were found to form after soaking for 4 h. The joint strength was slightly reduced from 677 MPa (1 h) to 644 MPa (4 h). Therefore, a longer soaking time is not recommended.

As mentioned in this section, local heating by an electric resistance furnace is a useful technique to join silicon nitride. This technique is expected to be applied to making large ceramic components from various oxide and non-oxide ceramics.

4 DIFFUSION BONDING

Silicon carbide (SiC)-based materials are among the most common ceramics that have been diffusion bonded. They have excellent high-temperature mechanical properties, oxidation and heat resistance, and thermo-chemical stability. These materials are leading candidates for various thermostructural applications at high temperatures in harsh environments in the aerospace and energy sectors. Common silicon carbide-based ceramics include CVD SiC, sintered SiC, hot-pressed SiC, and SiC fiber-bonded ceramic (SA-Tyrannohex™). Sintered, hot-pressed, and CVD silicon carbide ceramics have been diffusion bonded using refractory metal interlayers of titanium, nickel, molybdenum, tantalum, tungsten, niobium, zirconium, and the nickel (Ni)-based superalloy, Inconel 600. The diffusional transformation of a metal interlayer at high temperatures and under high mechanical pressures into carbides, silicides, or complex ternary and higher-order compounds produces strong joints. Carbides and some silicide compounds of refractory metals are thermodynamically more stable than SiC. As a result, diffusive conversion of a metal insert into carbides and silicides provides a pathway for bond formation.

Over the years, detailed structural and mechanical characterization of diffusion-bonded SiC ceramics has been undertaken [21–30]. Scanning electron microscopy (SEM), X-ray diffraction (XRD) analysis, energy dispersive spectroscopy (EDS), transmission electron microscopy (TEM), and a variety of mechanical testing methods have been used to investigate the diffusion-bonded ceramics with a view to optimizing the bonding conditions to maximize the joint strength and other properties. Thick interlayers lead to a significantly higher density of the cracks in the bond region; this requires optimization not only of the diffusion bonding conditions but also interlayer thickness. Figure 9 shows an example of a diffusion-bonded silicon carbide with a thin Ti interlayer that yielded sound bond quality for NASA's lean direct fuel injector assemblies. Work is currently in progress on the use of multiple interlayers of Ti, Cu, Mo-B, and other materials in diffusion-bonded

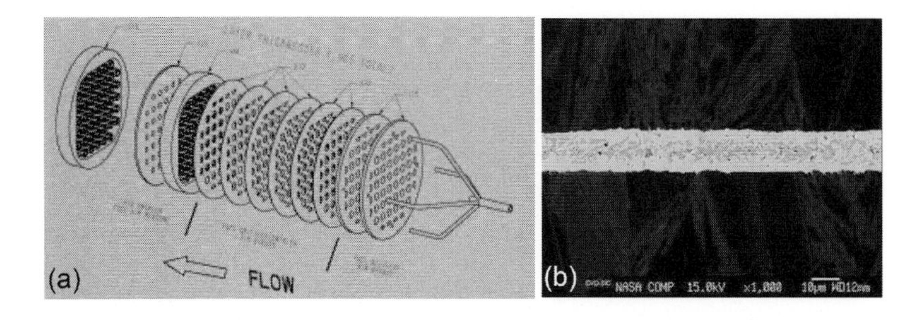

FIGURE 9 (a) A lean direct injector design and (b) a CVD SiC diffusion bonded joint from 10 μm PVD Ti coated SiC substrates.

joints to reduce the joining temperature, control the reaction phases and reaction layer thickness, and generally produce defect-free joints for advanced technology applications.

5 REACTION BONDING

In the late 1990s, a robust ceramic joining technology, named ARCJoinT, was developed at the NASA Glenn Research Center [31–34] to join SiC-based ceramics and composites. The basic approach involved applying a carbonaceous mixture to the joint surfaces followed by curing at a low temperature. Silicon or a Si alloy in tape, paste, or slurry form was then applied in the joint region and heated to 1250–1450°C to allow Si to react with the carbon to form SiC together with controlled amounts of Si and other phases in the material. Joints with good high-temperature strength were produced for a wide variety of applications. In chemical vapor infiltrated (CVI)-C/SiC composite joints made using the ARCJoinT technology, the shear strength exceeds that of the as-received C/SiC at elevated temperatures up to 1350°C. Figure 10 shows typical microstructure and joint strength data on reactively bonded silicon carbide ceramics. Microstructurally sound joints form when one C/SiC surface is machined and the other is in an as-received (not machined) state. Inferior quality joints containing voids and microcracks form when either both mating surfaces are machined or neither is machined. Similar results were obtained on C/C joints created using an interlayer of carbon fabric or felt impregnated with a

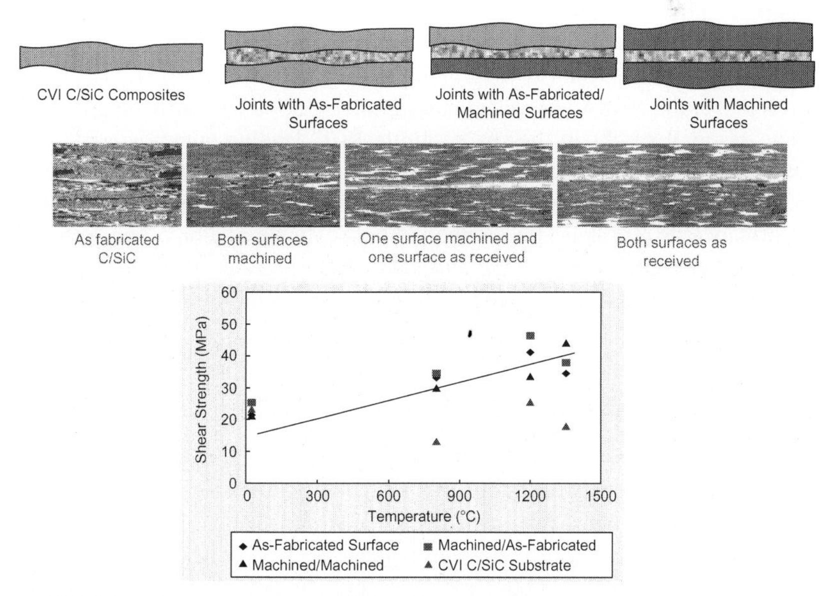

FIGURE 10 (a) Microstructures of bonded C/SiC composites and (b) Compression double notch shear strength of joined CVI SiC composites at different temperatures.

Joined C/SiC composites | Carbon-carbon composite valves for race car engines | Attachment for sensors

FIGURE 11 Examples of components joined using ARCJoinT.

carbonaceous bonding paste and liquid silicon. The highest shear strength was obtained in joints with a combination of a ground and an unground C/C test coupon mated with the carbon fabric (impregnated with the bonding paste), although the greatest improvements in shear strength were achieved when unground specimens were joined using the impregnated fabric as an interlayer. Some of the components fabricated using the reaction bonding technology are shown in Figure 11. Some of the joining technologies developed to manufacture large and complex components can be adapted to repair ceramic matrix composite (CMC) components damaged in service.

One of the authors has been actively involved in the development of a technology for advanced in-space repair of the reinforced C/C (RCC) composite thermal protection systems of the space shuttle. A new material, called Glenn Refractory Adhesive for Bonding and Exterior Repair (GRABER) has been developed at the NASA Glenn Research Center [35,36] for multi-use in-space repair of small cracks in the space shuttle RCC leading edge material. The new material has well-characterized and controllable properties such as viscosity, wetting behavior, working life, and so on, as well as excellent plasma performance as revealed in a number of lab-scale simulations. The material does not require post processing and converts to high-temperature ceramics during re-entry conditions. Initial performance evaluation tests under plasma conditions have been encouraging.

4 SUMMARY AND FUTURE DEVELOPMENTS

Manufacturing of complex and net-shape parts can be achieved via application of robust and reliable ceramic joining and integration technologies. These integration technologies are critically needed for the successful development and manufacturing of ceramic components in a wide variety of aerospace and ground-based applications. However, major efforts are needed to develop joint design methodologies, understand the size effects, and improve thermomechanical performance of integrated systems in service environments. In addition, life prediction models for integrated components must be developed for these components to be successfully implemented. Global efforts to standardize integrated ceramic testing and develop standard test methods are also required.

REFERENCES

[1] R. Asthana, M. Singh, Active metal brazing of advanced ceramic composites to metallic systems, in: D. Sekulic (Ed.), Advances in Brazing, Woodhead Publishing, Cambridge (UK), 2013, pp. 323–360.

[2] M. Singh, T. Ohji, R. Asthana, S. Mathur (Eds.), Ceramic Integration and Joining Technologies, John Wiley & Sons, Hoboken, New Jersey, 2011, p. 816.

[3] M. Singh, R. Asthana, Integration of C/C composite to metallic systems for thermal management applications, in: M. Singh, T. Ohji, R. Asthana, S. Mathur (Eds.), Ceramic Integration and Joining Technologies, John Wiley & Sons, Hoboken, New Jersey, 2011, pp. 163–191.

[4] M. Singh, R. Asthana, Advanced joining and integration technologies for ceramic-matrix composite systems, in: W. Krenkel (Ed.), Fiber-Reinforced Ceramic Composites, Wiley-VCH Publishers, Weinheim, Germany, 2008, pp. 303–325.

[5] J. Janczak-Rusch, Ceramic component integration by advanced brazing technologies, in: M. Singh, T. Ohji, R. Asthana, S. Mathur (Eds.), Ceramic Integration and Joining Technologies, Wiley, Boston, 2011.

[6] M.M. Schwartz, Joining of Composite Matrix Materials, ASM International, Materials Park, OH, 1994.

[7] M. Singh, R. Asthana, Brazing of advanced ceramic composites: Issues and challenges, Ceram. Trans. 198 (2007) 9–14.

[8] H. Kita, H. Hyuga, N. Kondo, T. Ohji, Exergy consumption through the life cycle of ceramic parts, Int. J. Appl. Ceram. Technol. 5 (4) (2008) 373–381.

[9] H. Kita, H. Hyuga, N. Kondo, Stereo fabric modeling technology in ceramics manufacture, J. Eur. Ceram. Soc. 28 (2008) 1079–1083.

[10] N. Kondo, H. Hyuga, H. Kita, K. Hirao, Joining of silicon nitride by microwave local heating, J. Ceram. Soc. Japan 118 (2010) 959–962.

[11] M. Hotta, N. Kondo, H. Kita, T. Ohji, Y. Izutsu, Joining of silicon nitride by local heating for fabrication of long ceramic pipes, Int. J. Appl. Ceram. Technol. 11 (1) (2014) 164–171.

[12] R.J. Xie, M. Mitomo, L.P. Huang, X.R. Fu, Joining of silicon nitride ceramics for high-temperature applications, J. Mater. Res. 15 (2000) 136–141.

[13] T.N. Tiegs, J.O. Kiggans, H.D. Kimrey, Microwave sintering of silicon nitride, Ceram. Eng. Sci. Proc. 12 (1991) 981–992.

[14] G. Xu, H. Zhuang, W. Li, F. Wu, Microwave sintering of α/β-Si_3N_4, J. Eur. Ceram. Soc. 17 (1997) 977–981.

[15] M.I. Jones, M.C. Valecillos, K. Hirao, Role of specimen insulation on densification and transformation during microwave sintering of silicon nitride, J. Ceram. Soc. Japan 109 (2001) 761–765.

[16] N. Kondo, M. Hotta, H. Hyuga, K. Hirao, and H. Kita, Energy Saving Joining Technique for Silicon Nitride by using Microwave Local Heating, Abstract book, 2005, Fall Symposium of the Ceramic Society of Japan, (2011). (in Japanese).

[17] S.M. Johnson, D.J. Rowcliffe, Mechanical properties of joined silicon nitride, J. Am. Ceram. Soc. 68 (1985) 468–472.

[18] M.L. Mecartney, R. Sinclair, R.E. Loehman, Silicon nitride joining, J. Am. Ceram. Soc. 68 (1985) 472–478.

[19] S.J. Glass, F.M. Mahoney, B. Quillan, J.P. Pollinger, R.E. Loehman, Refractory oxynitride joints in silicon nitride, Acta Mater. 46 (1998) 2393–2399.

[20] M. Hotta, N. Kondo, H. Kita, T. Ohji, Y. Izutsu, T. Arima, Effect of joining conditions on microstructure and flexural strength of long silicon nitride pipes fabricated by local heat-joining technique, J. Asian Ceram. Soc. 1 (2013) 308–313.

[21] M. Singh, M.C. Halbig, Bonding and Integration of Silicon Carbide Based Materials for Multifunctional Applications, Key Eng. Mater. 352 (2007) 201–206.

[22] M.C. Halbig, M. Singh, Development and characterization of the bonding and integration technologies needed for fabricating silicon carbide-based injector components, Ceram. Eng. Sci. Proc. 29 (9) (2009) 1–14.

[23] M.C. Halbig, M. Singh, Diffusion bonding of silicon carbide for the fabrication of complex shaped ceramic components, in: M. Singh, T. Ohji, R. Asthana, S. Mathur (Eds.), Ceramic Integration and Joining Technologies, John Wiley & Sons, New York, 2011, pp. 143–162.

[24] M.C. Halbig, M. Singh, H. Tsuda, Integration technologies for silicon carbide-based ceramics for micro-electro-mechanical systems-lean direct injector fuel injector applications, Int. J. Appl. Ceram. Technol. 9 (2012) 677–687.

[25] H. Tsuda, S. Mori, M.C. Halbig, M. Singh, TEM observation of the Ti interlayer between SiC substrates during diffusion bonding, Cer. Eng. Sci. Proc. 33 (8) (2012) 81–89.

[26] H. Tsuda, S. Mori, M.C. Halbig, M. Singh, Interfacial Characterization of Diffusion-Bonded Monolithic and Fiber-Bonded Silicon Carbide Ceramics, in: ICACC, 2013.

[27] M. Naka, J.C. Feng, J.C. Schuster, Phase reaction and diffusion path of the SiC/Ti system, Metall. Mater. Trans. 28A (1997) 1385–1390.

[28] B. Gottselig, E. Gyarmati, A. Naoumidis, H. Nickel, Joining of ceramics demonstrated by the example of SiC/Ti, J. Eur. Ceram. Soc. 6 (1990) 153–160.

[29] A.E. Martinelli, R.A.L. Drew, Microstructural development during diffusion bonding of α SiC to molybdenum, Mater. Sci. Eng. A 19 (1995) 239–247.

[30] B.V. Cockeram, The Diffusion Bonding of Silicon Carbide and Boron Carbide Using Refractory Metals, Report B-T-3255, USDOE Contract No. DE-ACI 1-98 PN38206, 1999, Bettis Atomic Power Laboratory, West Mifflin, PA, USA.

[31] M. Singh, A reaction forming method for joining of silicon carbide-based ceramics, Scr. Mater. 37 (8) (1997) 1151–1154.

[32] M. Singh, S.C. Farmer, J.D. Kiser, Joining of silicon carbide based ceramics by reaction forming approach, Cer. Eng. Sci. Proc. 18 (3) (1997) 161–166, The American Ceramic Soc., Westerville, OH.

[33] J. Martinez-Fernandez, A. Munoz, F.M. Varela-Feria, M. Singh, Interfacial and thermomechanical characterization of reaction formed joints in SiC-based materials, J. Eur. Ceram. Soc. 20 (2000) 2641–2648.

[34] M. Singh, E. Lara-Curzio, Design, fabrication and testing of ceramic joints for high-temperature SiC/SiC composites, Trans ASME 123 (2001) 288–292.

[35] On-orbit shuttle repair takes shape, www.aiaa.org/aerospace/images/articleimages/pdf/Iannottaaugust04.pdf.

[36] NASA One Step Closer to Shuttle Repair in Orbit, New Scientist, May 22, 2004, 24, www.newscientist.com.

Chapter 13

Three-Dimensional Sustainable Printing of Functional Ceramics

Soshu Kirihara
Joining and Welding Research Institute, Center of Excellence for Advanced Structural and Functional Materials Design, Osaka University, Osaka, Japan

1 INTRODUCTION

Three-dimensional printing to create microcomponents with geometrically designed patterns composed of functional ceramics using stereolithography and nanoparticle sintering techniques will be described in this chapter. Nanometer-sized ceramic particles with dielectric, biological, or electronic properties were dispersed into photosensitive liquid resins, and the mixed slurries were solidified by laser scanning or micropattering. These composite precursors were dewaxed and sintered carefully, and microlattice structures were successfully obtained. These components showed dendrite structures with periodic arrangements of microlattices to effectively control and modulate electromagnetic wave propagations and liquid materials fluid flows. Technological details of the ceramics' free forming and applications of the functional dendrite structures are also reviewed.

2 THREE-DIMENSIONAL PRINTING

2.1 Laser Scanning Stereolithography

Geometric patterns in three dimensions are modeled by a computer-aided design application. These graphic models are converted automatically into the stereolithography format, and sliced into a series of two-dimensional cross-sectional planes with a uniform layer thickness of 50 μm. The numerical data are transferred automatically into the stereolithography equipment, and raster patterns for laser scanning are created automatically. Figure 1 shows an illustration of the fabrication process. Photosensitive acrylic resin, including ceramic particles 200 nm in diameter at 40% volume fraction, is spread on a flat metal stage using a mechanical knife edge. The thickness is controlled automatically at the same value (50 μm) as in the model. An ultraviolet laser of 355 nm wavelength is used to scan the ceramics slurry in order to create cross-sectional

Green and Sustainable Manufacturing of Advanced Materials. http://dx.doi.org/10.1016/B978-0-12-411497-5.00013-8

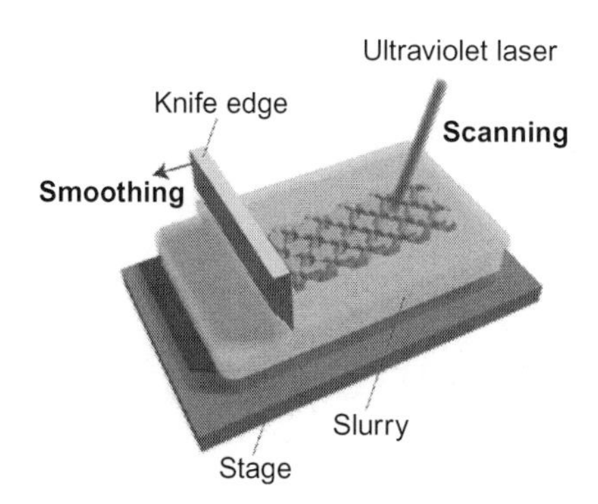

FIGURE 1 An illustration of a stereolithography system of computer-aided design and manufacturing processes.

planes with a 5 μm edge part accuracy. The laser beam is adjusted to 100 μm in spot size and 100 mW in power. After the formation of the solid pattern, the elevator stage is moved downward 50 μm in the layer thickness, and the next cross-section is stacked. Three-dimensional structures are fabricated by stacking all the two-dimensional layers. The part accuracies of green bodies can be measured and observed by using a digital optical microscope. The formed models are dewaxed at 600°C for 2 h with a heating rate of 1.0°C/min in the air, and full ceramics components are obtained after sintering. Microstructures of the sintered components can be observed by using a scanning electron microscope (SEM). The relative densities of these ceramic components can be measured by the Archimedian method.

2.2 Micropatterning Stereolithography

Three-dimensional micropatterns are designed using computer graphic software. The designed models are converted into the stereolithography file format and sliced into a series of two-dimensional cross-sectional data with a layer thickness of 10 μm. These data are then transferred into the microstereolithography equipment to create bit map images for automatic micropattering. Figure 2 shows an illustration of the microstereolithography system. Photosensitive acrylic resins, including ceramic nanoparticles of 200 nm average diameter at 40% volume content, are applied to a glass substrate from a dispenser nozzle using air pressure. This paste is spread uniformly by a mechanically controlled knife edge. The thickness of each layer is 10 μm. Two-dimensional solid patterns are obtained on the slurry surface by light-induced photo polymerization. High-resolution imaging is achieved by using a digital micromirror device. In this optical device, 1024 × 768 aluminum micromirrors of 14 μm edge length are assembled. Each mirror can be tilted independently by piezoelectric actuating. The ultraviolet ray of 405 nm is introduced into the digital micromirror device, and the cross-sectional image is reduced to 1/5 through an objective lens set

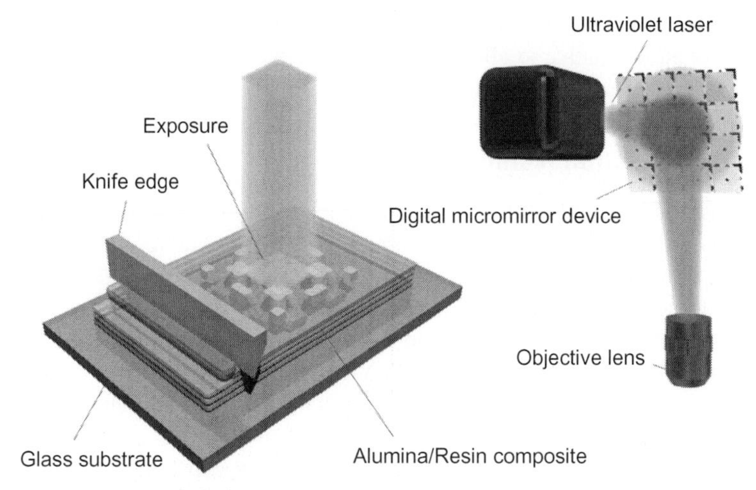

FIGURE 2 A microstereolithography system.

and concentrated into the exposed area, which is 1.3 × 1.7 mm in size. Through the computer-controlled layer stacking, the acrylic resin component with the dispersed ceramic particles is obtained. The composite precursor is dewaxed at 600°C for 2 hrs at the heating temperature and holding time in the air. The full ceramics microcomponents are obtained using the sintering heat treatment. The part accuracy and ceramics microstructures are of the sintered components can be observed by using a SEM.

3 ARTIFICIAL BONE

3.1 Bioceramics Formation

Tissue scaffolds are required to repair bone defects resulting from illnesses. To encourage osteoconductivity and tissue regeneration, prosthetics that mimic bone porosity and optimized flow behaviors are very important. Various techniques to fabricate artificial bone structures have been investigated, for example, polycaprolactone scaffolds with variable pore sizes and porosities ranging from 63% to 79% created using selective laser sintering [1], the modified hydroxyapatite ones created by a printing process [2], and the periodic microarrays created by using direct ink writing [3]. The scaffold structure requires suitable porosity and pore size to foster tissue regeneration in the human body [4].

Artificial bone must emulate natural bone with graded porosities from 50 to 90% volume fraction. However, the conventional artificial bones have porosities of nearly 75% [5,6]. In this section, the creation of novel artificial bones composed of hydroxyapatite ceramics with effective biocompability and high mechanical strength will be demonstrated [7], and the creation of graded porous structures composed of four coordinate lattices using laser scanning stereolithography will also be discussed. Evaluating osteogenesis requires long-term clinical experiments. As an alternative to clinical experiments, the biofluid flow behaviors in different types of scaffold structures that have the same porosity can be studied.

3.2 Coordination Number

The dendritic lattice structures in the biological scaffolds with four, six, eight, and 12 coordination numbers are designed as shown in Figure 3. The porosity of the scaffold models can be controlled in the range from 50 to 90% by adjusting the aspect ratios of the rod length to diameter, as shown in Figure 4. The porosity of all skeletal structures was 75%; this is the same porosity as a human bone. These scaffolds have perfect interconnected pores. Fluid circulation in the various dendrite scaffolds was visualized with the fluid dynamic solver. Flow velocity in the spatial grids in the scaffold models was calculated using the finite element method. The following values of these parameters were used

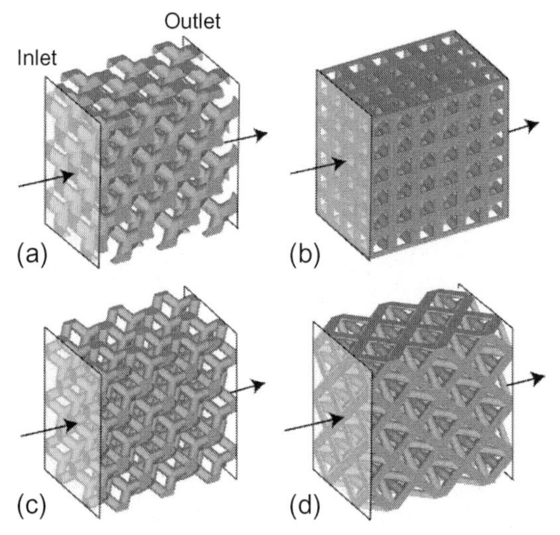

FIGURE 3 Dendrite scaffolds models. (a) Four coordinate. (b) Six coordinate. (c) Eight coordinate. (d) Twelve coordinate.

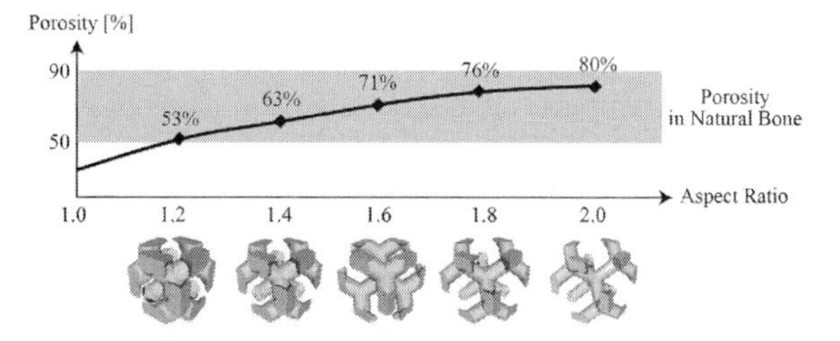

FIGURE 4 A chart relating the porosity to the aspect ratio of the rod.

in this simulation [8,9]. The fluid phase was represented as an incompressible Newtonian fluid with a viscosity of 1.45×10^{-3} Pa·s. The inlet velocity applied to the scaffolds was constant at 0.235 mm/s, and the pressure was zero at the outlet. No-slip surface conditions were assumed.

Figure 5 shows the streamline behavior in the dendrite scaffolds. Figure 5a shows a four-coordinate scaffold structure and indicates that inordinate flow at a low velocity was obtained. This also indicates the simulated biofluid flow to the whole structure, which is expected to provide active tissue regeneration. The fluid velocity in the six coordination number scaffolds is the highest, at above 1.0 mm/s, as shown in Figure 5b. There are no blockades from the inlet to the outlet, and the flow becomes linearly stable. The high fluid velocity area above 1.0 mm/s in the scaffold is subjected to shear stress, which can assume the difficulty of cell attachment on the scaffold's surface [10]. In the case of an eight coordination number scaffold, random fluid flow and high velocity are exhibited in some parts of the structure as shown in Figure 5c. Smooth fluid flows and propagations is seen in the four coordination number scaffold without obstacle lattices compared with the larger number of rod coordinates. In the 12 coordination number structure, moderate velocities and flow behaviors are exhibited, as shown in Figure 5d. However, there are no active flows in some areas because of the many rods in the architecture.

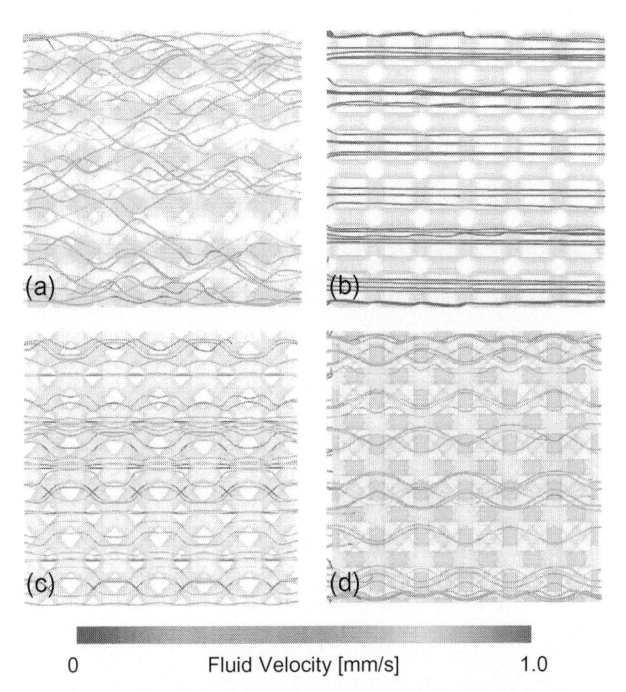

FIGURE 5 Fluid flow behaviors in the dendrite scaffolds seen using the finite element method. (a) Four coordinate. (b) Six coordinate. (c) Eight coordinate. (d) Twelve coordinate.

3.3 Biological Scaffold

The designed scaffold model with the graded porous structure is shown in Figure 6. The porosity is distributed gradually through modulations of the aspect ratio in the four coordination number lattices, as shown in Figure 6a. The acrylic scaffolds with hydroxyapatite particles are shown in Figure 6b. Micrometer-order ceramic lattices are successfully fabricated using laser scanning stereolithography. A photosensitive acrylic resin, including hydroxyapatite particles with a diameter of 10 μm at 45% volume, was used. The porosity of the scaffold form was approximately 75%. The part accuracies of the lattices were measured to have size differences of less than 50 μm. The formed precursor was dewaxed at 600°C for 2 hrs with a heating rate of 1.0°C/min and was sintered at 1250°C for 2 hrs with a rate of 5°C/min in the air [11,12]. The relative density of the sintered hydroxyapatite lattice was measured at 98% using the Archimedian method. The sintered scaffold model of the hydroxyapatite ceramic with the graded lattices is shown in Figure 6c. The linear shrinkage ratios for the horizontal and vertical axes were 23% and 25%, respectively. The smaller lattice structures could be obtained effectively through the controlled body shrinkages in the optimized sintering process. The microstructure of the sintered scaffold is observed using the SEM, as shown in Figure 7. Cracks and pores were not observed. The grain size was about 4 μm, and the relative density reached 98%. The formed scaffold is considered to have effective biocompability and high mechanical strength.

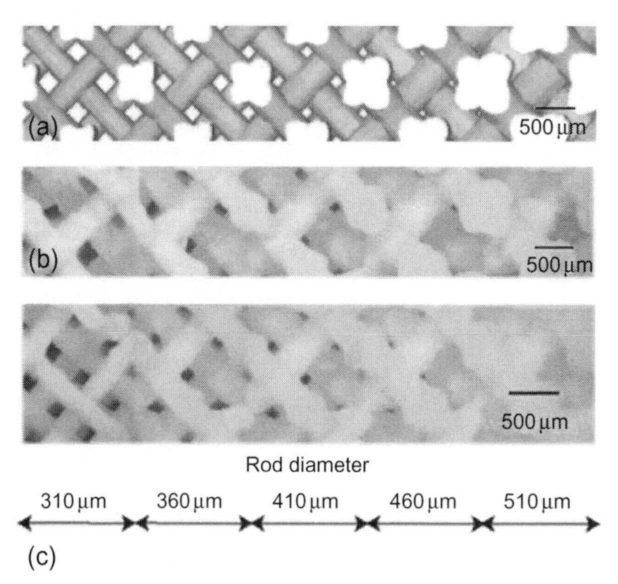

FIGURE 6 Four-coordinate lattices with graded porous structures. (a) The computer-designed graphic model, (b) the acrylic lattices, including hydroxyapatite particles fabricated by stereolithography, and (c) the sintered ceramic scaffold.

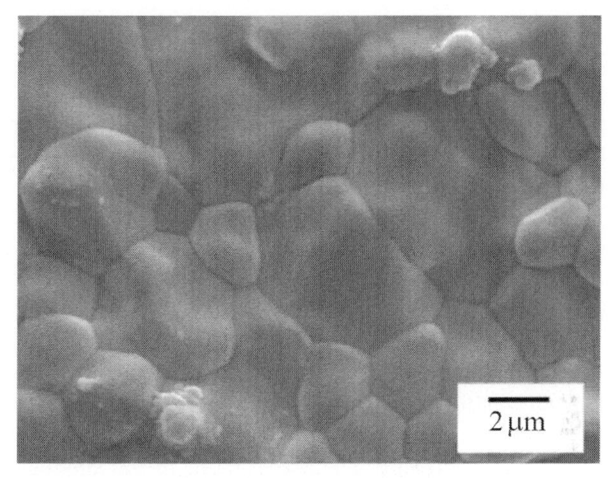

FIGURE 7 Microstructure of a sintered hydroxyapatite lattice obtained by dewaxing and sintering heat treatments.

4 SOLID OXIDE FUEL CELLS

4.1 Energy Generation

Solid oxide fuel cells (SOFCs) are promising next-generation energy conversion systems due to their higher power density and energy efficiency. Yttria-stabilized zirconia (YSZ) with added nickel (Ni) has many desirable properties for SOFC anodes, such as high electronic/ionic conductivity and chemical/mechanical stability at high operating temperatures. At the anode site, fuel gas diffusions and electrochemical reactions on the electrode surfaces, which are composed of a YSZ/Ni/gas triple-phase boundary (TPB), proceed simultaneously. In addition, activation and diffusion overpotentials should cause a decrease in the SOFC's operating voltage and energy efficiency. Therefore, anode microstructure design is essential to achieve better performance and miniaturization of SOFCs.

Porous YSZ-Ni anodes have been fabricated to realize large surface areas and high activation of the electrode reactions [13–15]. Relationships between the dispersion ratios of YSZ and Ni particles and SOFC output characteristics, such as energy densities and overpotentials, were investigated [16–18]. In addition, numerical analyses dealing with materials transportation in porous structures were conducted to evaluate the electrode structures [19,20]. However, it has been verified that random vacancy structures in the traditional porous anode materials obstruct fuel and produce H_2O gas diffusions and electrode reactions.

In this section, we discuss how the SOFC anode structure was optimized through computer simulations using finite element methods (FEMs), and solid electrodes with wide surface areas and smooth fluid permeability were successfully fabricated using microstereolithography. The dendritic structures constructed from micrometer-order ceramic rods propagating spatially are thought to be desirable electrode structures because they have large surface areas and cyclical vacancy structures. The aspect ratios of rod lengths to diameter were optimized to exhibit the maximum surface area, and then streamlines, velocities, and stress distributions in the dendritic structures were visualized to evaluate fluid permeability and mechanical strength. The dendritic structures with large surface areas and smooth gaseous diffusion properties are expected to activate electrode reactions and lower activation and diffusion overpotentials.

4.2 Porous Structures

The electrode structure was optimized through computer simulations. The dendritic lattices with a lattice constant of 100 μm constructed from micrometer-order rods with four, six, eight, and 12 coordination numbers were designed using computer graphic software. The aspect ratios of rod lengths to diameter were varied from 0.75 to 3.00 to investigate the relationships between aspect ratios and surface areas. These relationships, which are determined by rod diameters and lengths, are shown in Figure 8. Each dendritic lattice showed the maximum surface area when the aspect ratio was 0.90, 1.17, 2.18, and 2.34, respectively. The dendritic lattice with 12 coordination numbers and an aspect ratio of 2.18 was verified to exhibit the largest surface area, and the electrode texture was expected to increase the TPB points and lower the activation overpotential in the electrode.

Gaseous fluid permeability and stress distributions in the dendritic structures were simulated by the finite volume method and FEM calculations. An analysis model is shown in Figure 9. The numbers of arranged dendritic lattice unit cells were $5 \times 5 \times 1$. The static pressure of the inflow and outflow sides was 1.01 and 1.00 atm, respectively. In the mechanical analysis, the bottoms of the dendritic structures were fixed. Streamlines and velocities in the dendritic structures are

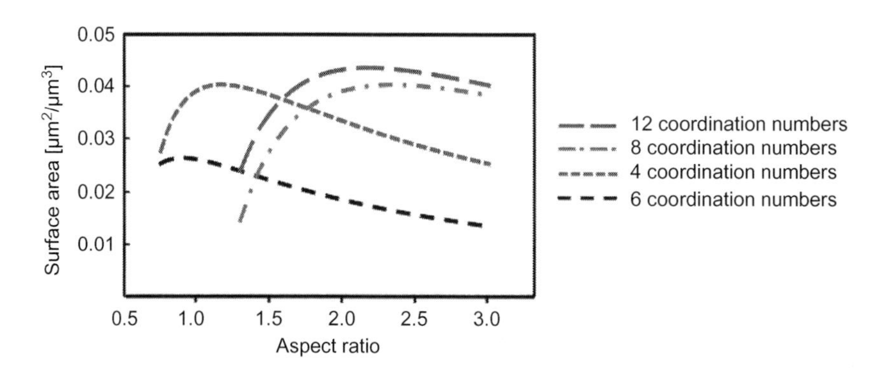

FIGURE 8 Calculated variations of surface areas in dendritic structures according to aspect ratios and coordination numbers.

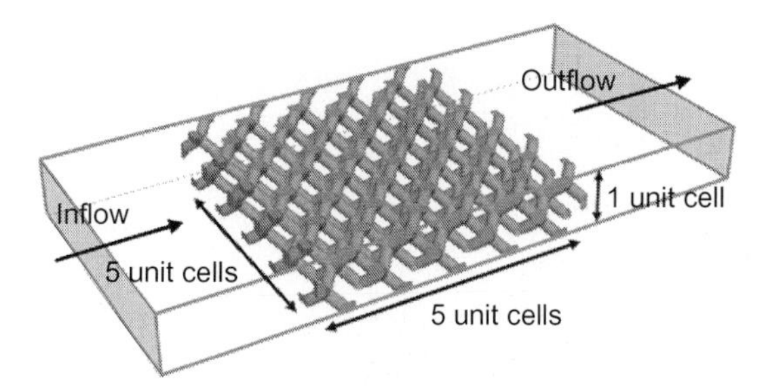

FIGURE 9 A dendritic structure model for fluid and mechanical property analysis.

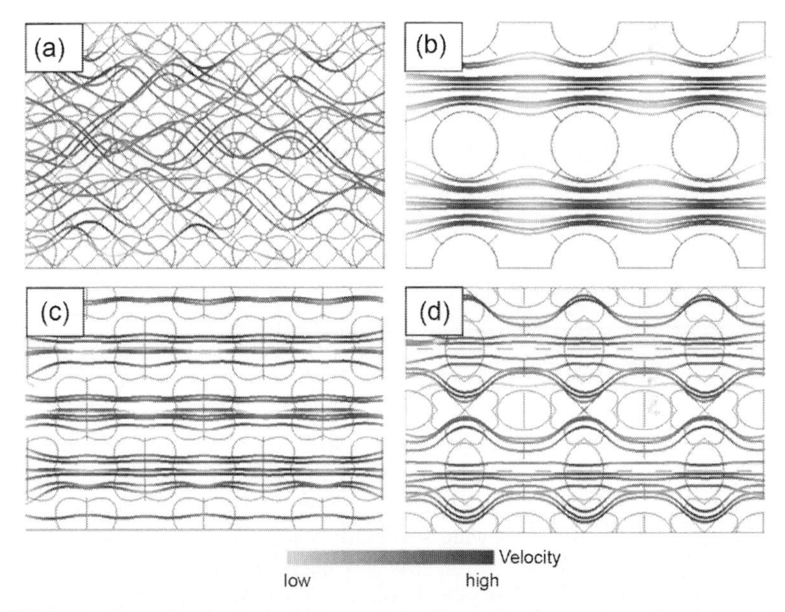

FIGURE 10 Streamlines in the dendritic structure with coordination numbers of (a) 4, (b) 6, (c) 8, and (d) 12 calculated using finite volume methods.

shown in Figure 10. Smooth streamlines according to cyclical vacancies were exhibited in the dendritic structure with six, eight, and 12 coordination numbers. The prompt fuel gas flows can realize reductions of diffusion overpotentials. However, there were rectilinear streamlines and large flow velocity differences in the dendritic structure with six and eight coordination numbers. These could be causes of inhomogeneous fuel gas diffusion, for example, back water or vortex. On the other hand, smooth and homogeneous fluid diffusion was indicated in the dendritic structure with 12 coordination numbers. In the SOFC anode, H_2O produced by the electrode reactions must be discharged from the electrodes promptly to avoid lowering the concentrations and partial pressures of the fuel

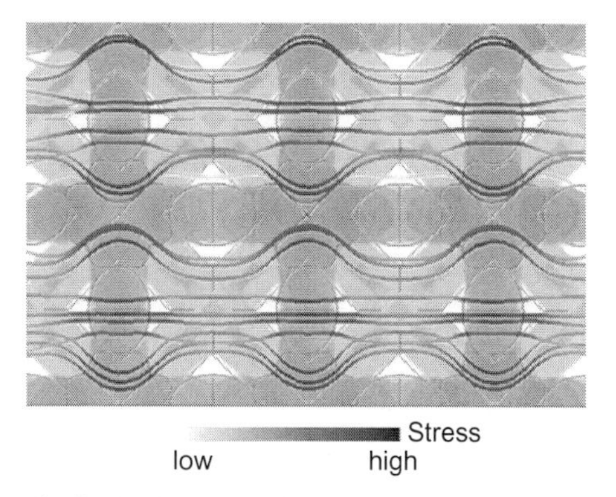

FIGURE 11 A stress distribution in the dendritic lattices with 12 coordination numbers calculated using a finite element method (FEM).

gases. A stress distribution applied by the fluid flow in the dendritic structure with 12 coordination numbers is shown in Figure 11. The stress concentration was not observed, and the dendritic structure can exhibit high mechanical strength through its isotropic lattice structures.

4.3 Dendritic Electrode

The optimized dendritic structure is formed using microstereolithography. A slurry paste of a photosensitive acrylic resin with nanometer-sized YSZ and Ni particles dispersed throughout was applied to a substrate from a dispenser nozzle using air pressure, and it was spread uniformly at a 5 μm layer thickness by a mechanical knife edge. A fabricated dendritic structure composed of YSZ and nickel oxide (NiO) is shown in Figure 12a. Micrometer-order ceramic lattices with 12 coordination numbers were successfully formed. These composite precursors were dewaxed at 600°C for 2 hrs and sintered at 1400°C for 2 hrs in air. The lattice constant of the sintered sample was 98.5 μm. Figure 12b shows the microstructure of the YSZ-NiO electrode surface observed by a SEM. The YSZ and NiO particles were connected successfully and distributed homogeneously. Microvoids or pores were not observed. X-ray diffraction (XRD) patterns of the

FIGURE 12 (a) A sintered dendritic structure with 12 coordination numbers composed of YSZ and nickel oxide and (b) a SEM image of the electrode surface.

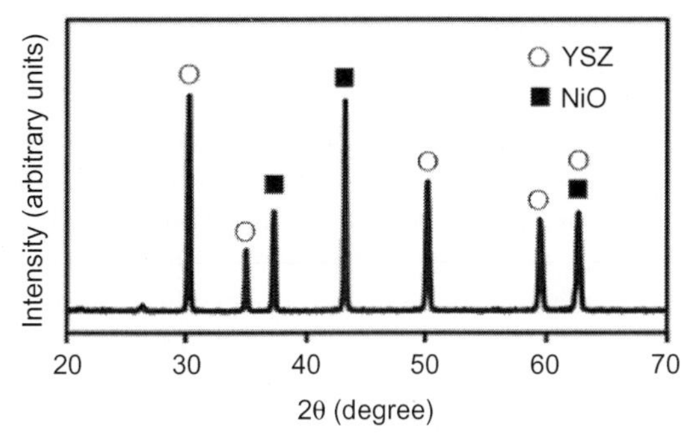

FIGURE 13 XRD patterns of a sintered YSZ-NiO composite.

sintered sample are shown in Figure 13c. The diffraction peaks of the YSZ and NiO are shown clearly. The ideal TPB could be created in the sintered dendritic lattices. The fabricated dendritic electrode with a large surface area and smooth fluid permeability is considered to activate electrode reactions effectively and lower activation and diffusion overpotentials.

5 PHOTONIC CRYSTALS

5.1 Band Gap Formation

Photonic crystals with periodic arrangements of dielectric lattice can form band gaps to reflect electromagnetic waves through Bragg diffraction [21,22]. In particular, diamond lattice structures can create perfect band gaps to totally reflect electromagnetic waves in all directions. By introducing an artificial defect of air cavities into the periodic arrangement, localized modes can be formed in the band gaps. These structural defects can localize and amplify electromagnetic waves at the specific frequencies and wavelengths corresponding to the sizes and dielectric constants of the resonation domains. The photonic crystal could be used in various microwave devices for resonators, filters, and directional antennas [23,24]. In this section, we discuss the design and fabrication processes of micrometer-order alumina lattices created using stereolithography as well as the electromagnetic properties in the terahertz frequency ranges [25].

Terahertz frequency waves have attracted considerable attention as novel analytical light sources. Because the electromagnetic wave frequencies from 0.1 to 10 THz can be synchronized with collective vibration modes of saccharide or protein molecules, terahertz wave spectroscopy is expected to be applied to various types of sensors for detecting harmful substances in human blood, skin cancer cells, and microbacteria in vegetables [26–30]. Moreover, the terahertz sensing technologies for aqueous phase environments to detect dissolved matters directly are extremely interesting. The alumina photonic crystals with diamond lattices can localize and strongly resonate terahertz waves into the cavities of structural defects, including various water solvents, in order to realize greater analytical precision.

5.2 Dielectric Lattices

Photonic crystals can realize the single-mode resonance without electromagnetic losses through the appropriate structural design. The three dimensional diamond lattices are regarded as the ideal photonic crystal structures. Figures 14a and b show a unit cell of a diamond photonic crystal and an electromagnetic band diagram calculated using a plane wave expansion method simulator [31]. The artificial crystal can totally reflect the terahertz wave with the corresponding wavelength to the lattice spacing through Bragg diffraction. The complete photonic band gap is formed to prohibit electromagnetic wave expansion in all crystal directions. The electromagnetic band properties of the diamond photonic crystals were calculated theoretically using the plane wave expansion method to determine the geometric parameters. Figure 15

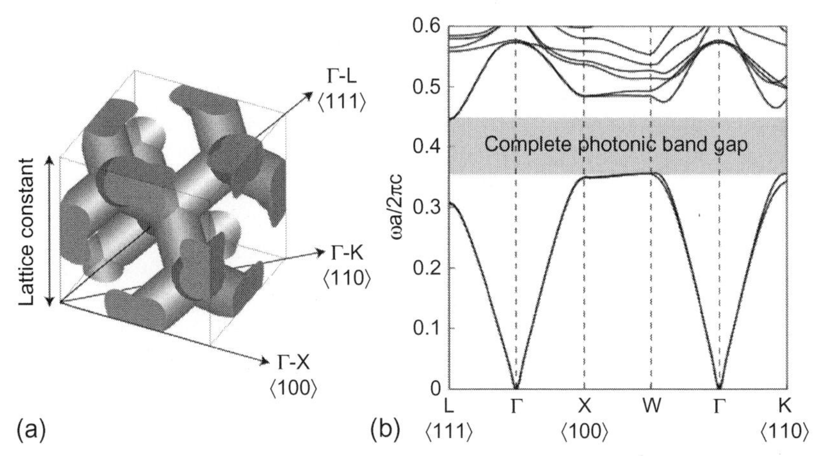

FIGURE 14 (a) A CAD image of a unit cell of diamond structure, and (b) a photonic band diagram of a diamond lattice structure calculated using the plane wave expansion (PWE) method.

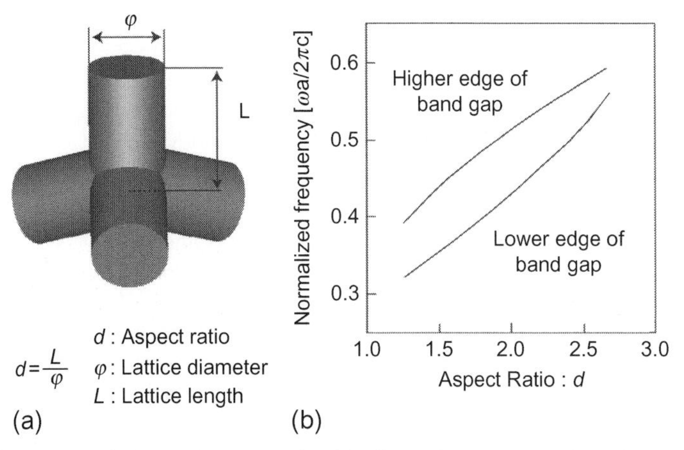

FIGURE 15 (a) Definition of the aspect ratio of the diamond structure, and (b) the band gap width as a function of the aspect ratio.

shows the variations of complete band gap widths as a function of the aspect ratio in the dielectric lattice. In the calculation, the alumina ceramics with a dielectric constant of 9.8 was assumed as the lattice material. The aspect ratio was optimized as 1.5 to create a wider band gap. Subsequently, the gap frequency can be shifted for the lower range, in inverse proportion to the lattice spacing, as shown in Figure 16. To open the band gaps in the terahertz frequencies, microscale structural periods need to be created. The lattice constant was designed as 375 μm, corresponding to the band gap frequencies from 0.3 to 0.6 THz.

The resonance efficiencies of the terahertz wave resonator with the diamond lattices were optimized by using a transmission line modeling simulator of a finite difference time-domain method. Figure 17 shows a computer model of the terahertz wave resonator. The microglass cell, including the water, was sandwiched as the plane defect between two alumina photonic crystals with the diamond lattice. The defect thickness of the resonator and the period numbers of the diamond units were selected as the principle parameters to control the resonance characteristics. The defect thickness enables researchers to tune the resonance frequencies in the band gap as shown in Figure 18. In this investigation, the plane defect was designed as the water cell with a thickness of 470 μm that was composed of two quartz plates of 160 μm and an aqueous cavity of 150 μm. The period numbers of the diamond lattices enables researchers to adjust the resonance qualities as shown

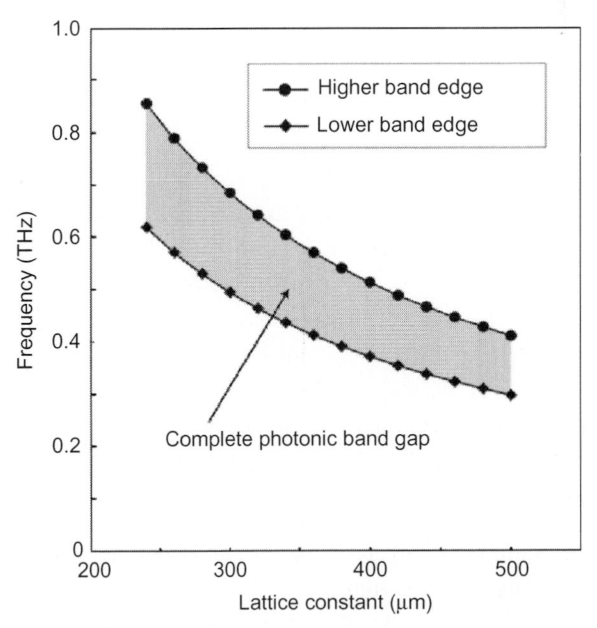

FIGURE 16 Variations of electromagnetic bands as functions of lattice constants in diamond structures.

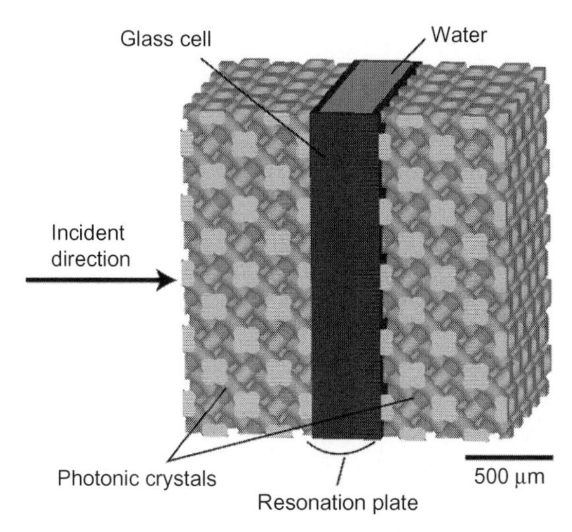

FIGURE 17 A computer model of a terahertz wave microresonator.

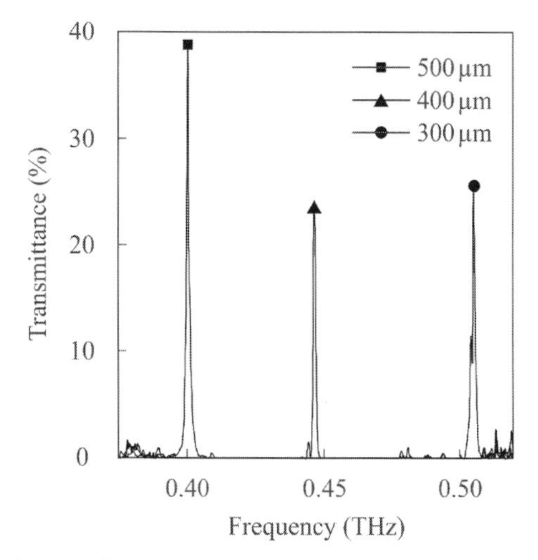

FIGURE 18 Calculated localized mode formations through electromagnetic wave resonations in microcells with various thicknesses.

in Figure 19. The resonance qualities can be enhanced by increasing the period numbers; however, the localized mode of the transmission peak becomes lower through the perfect confinement of the electromagnetic wave in the defect domain. The diamond lattices composed of two units in the period numbers were optimized and designed to detect the sharp localized mode peak in the transmission spectrum.

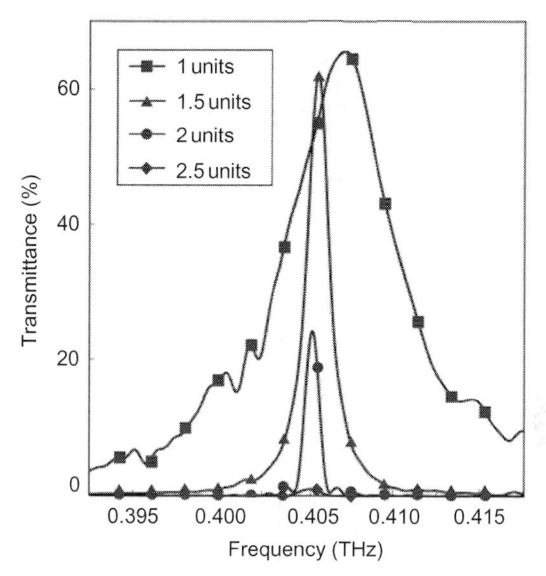

FIGURE 19 Calculated peak profiles of multiple resonation modes in water cells of plane defects put between photonic crystals with various numbers of layers.

5.3 Terahertz Wave Resonator

Figure 20a shows a photonic crystal composed of an acrylic lattice, including alumina particles, formed by microstereolithography. The tolerance between the design and the formed object was successfully reduced to within 5 μm. The lattice constant of the diamond structure and the aspect ratio of the dielectric lattice are 500 μm and 1.5, respectively. The whole size of the crystal component is 5×5×1 mm, consisting of 10×10×2 unit cells. The composite precursor was dewaxed at 600°C for 2 hrs and sintered at 1500°C for 2 hrs in the air. Figure 20b shows the alumina lattice with the diamond structure fabricated by ceramic powder sintering. No cracks or deformations were observed in the resulting components. The average linear shrinkage was 25%. The lattice constant of the sintered sample was 375 μm. Figure 20c shows the microstructures of the sintered sample observed by a SEM. The relative density of the sample reached 97.5%.

In order to obtain a plane defect between the two diamond structures, a microglass cell was also fabricated using microstereolithography. Figure 21a shows the components of the resonance cells. The quartz plates with thicknesses of 160 μm were inserted into the photosensitive acrylic resins in the stacking and exposing process. Finally, the microresonator cell was placed between the diamond photonic crystals, and the terahertz wave resonator was integrated successfully using acrylic resin flames, as shown in Figure 21b. These flames were glued together using the photosensitive liquid resin and ultraviolet exposure

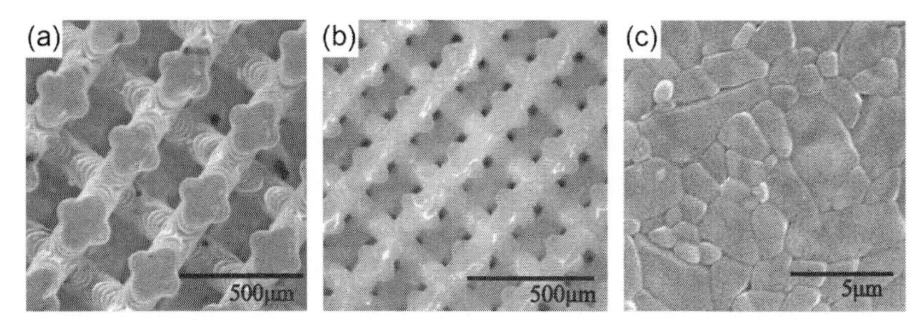

FIGURE 20 (a) A diamond-type acrylic lattice with dispersed alumina particles formed by microstereolithography, (b) a sintered alumina photonic crystal, and (c) a SEM micrograph of the sintered alumina lattice.

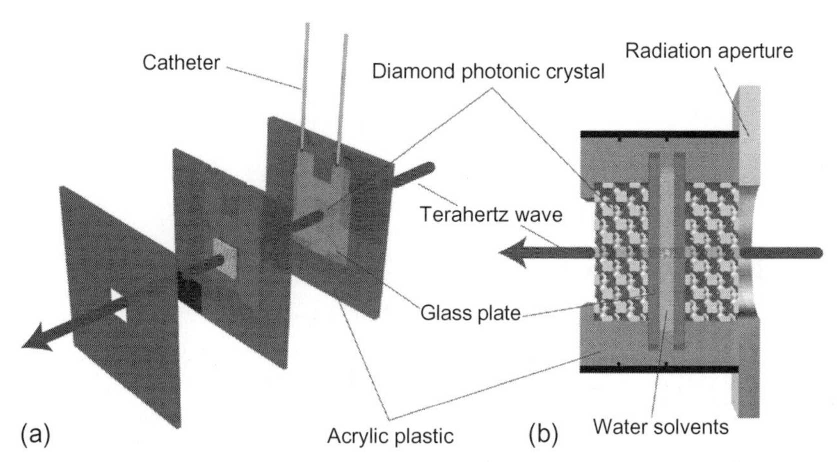

FIGURE 21 (a) An illustration of the parts of the resonance cell, and (b) the fabricated THz wave resonator. The resonator was successfully created by integrating the diamond structures into the cells to realize multiple resonations of the THz wave in an aqueous phase.

solidification. Water solutions were infused through catheters connected to the top of the resonance cell. The integrated terahertz wave resonator is shown in Figure 22. The two diamond lattice components were attached to the quartz glass plates, and these two plates were arranged in parallel with an interval of 150 μm. The tolerance for the transmission direction of the electromagnetic wave was reduced to within 5 μm. The cell capacity was 0.02 ml.

Figure 23a shows the measured transmission spectra for the resonators. The transmission properties of the incident terahertz waves were analyzed using terahertz time-domain spectroscopy. The distributions of the electric field intensities in the resonator were simulated and visualized at the localized frequency using transmission line modeling. Distilled water or ethanol was infused into the microcells. In the case of distilled water, two localized modes of transmission peaks were observed at frequencies of 0.410 and 0.491 THz in the

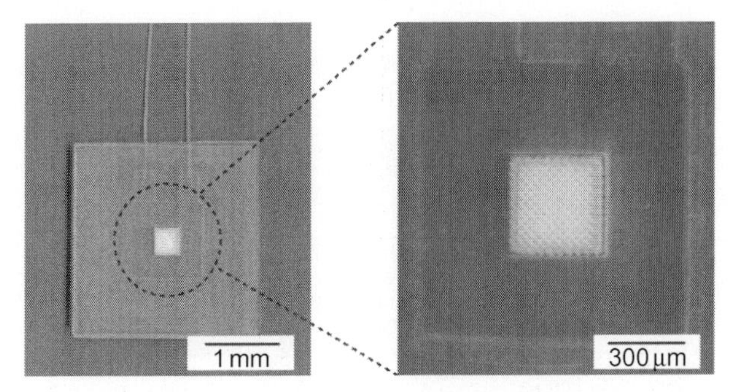

FIGURE 22 A terahertz wave resonator with a microliquid cell sandwiched between diamond photonic crystals. Distilled water and ethanol were infused through catheters implanted in the top of the cells.

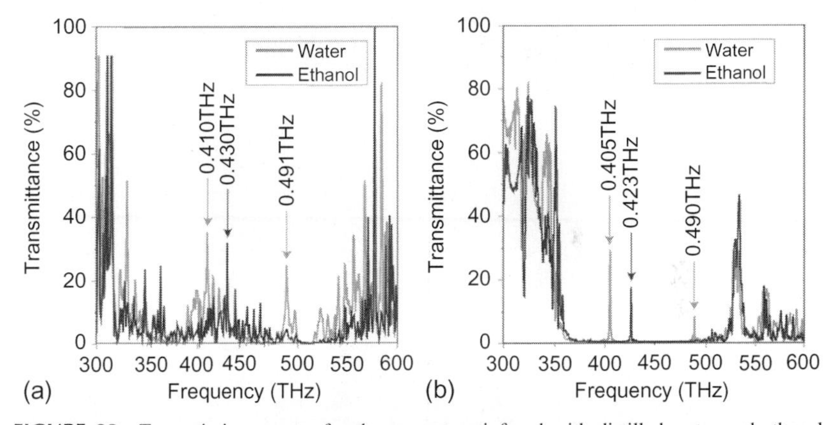

FIGURE 23 Transmission spectra for the resonators infused with distilled water and ethanol. The spectra (a and b) are measured and the properties calculated using terahertz wave time-domain spectroscopy and a transmission line modeling method, respectively.

photonic band gap. In the case of ethanol, an amplification peak was observed at 0.430 THz. The measured band gap ranges and the localized mode frequencies have good agreement with the simulated results obtained by transmission line modeling, as shown in Figure 23b. In the transmission spectrum through the photonic crystal resonator using the water, the localized modes of the higher and lower peak frequencies are defined as modes A and B, respectively. And, the localized mode peak in the transmission spectrum through the ethanol is defined as mode C.

The cross-sectional profiles of the electric field intensity corresponding to the localized modes A, B, and C were simulated and visualized theoretically, as shown in Figures 24a–c, respectively. The terahertz wave was propagated from the left to the right side. The white and black areas show that the electric

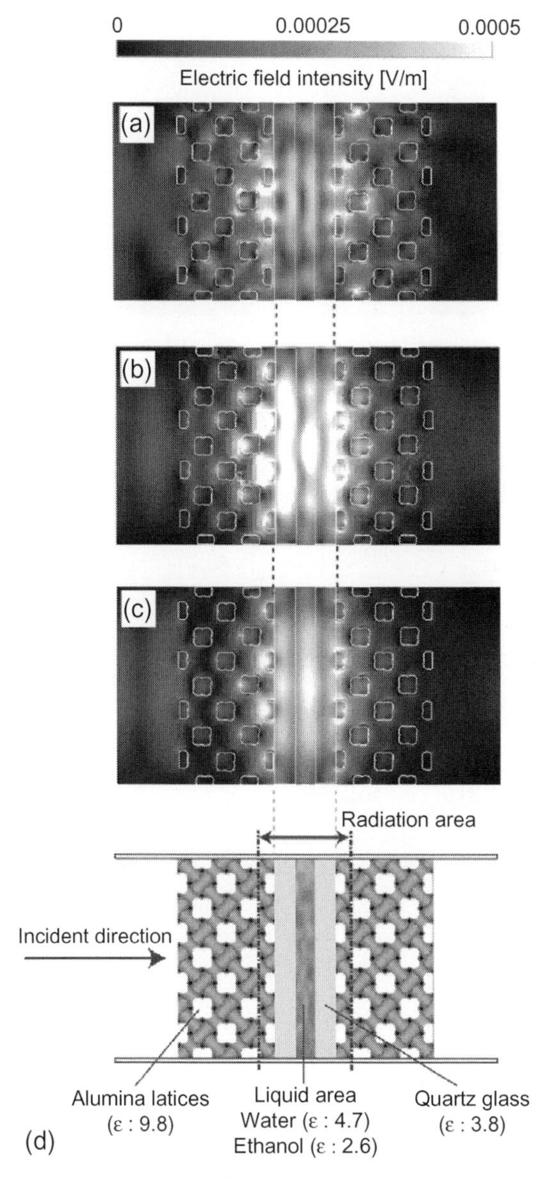

FIGURE 24 Electric field distributions of localized modes simulated using the transmission line modeling (TLM) method of a finite difference time domain (FDTD). The localized modes A, B, and C, are shown in (a), (b), and (c), respectively. Dielectric constants and composite dimensions of the photonic crystal resonator are shown in (d).

field intensity is high and low, respectively. The incident terahertz waves were resonated and localized strongly through the multiple reflections in the liquid cell between two diffraction lattices with the diamond structures. For the localized mode A, shown in Figure 24a, the standing wave with five nodes was observed in the vicinity of the glass cell. The localized mode B (with a lower frequency and a longer wavelength) is composed of the standing wave with four nodes between the diffraction lattices, as shown in Figure 24b. For the localized mode C (ethanol), shown in Figure 24c, the standing wave with four nodes has a resonance profile between the diffraction lattices that is similar to the localized mode B (water). Figure 24d shows the dielectric constants and composite dimensions of the photonic crystal resonator. The dielectric constants of water and ethanol are 4.7 and 2.6 in the terahertz wave frequency range [32,33]. Using terahertz wave time-domain spectroscopy, the dielectric constants of alumina and quartz glass were 9.8 and 3.8, respectively. Compared with the resonance areas of the localized modes B and C, the effective dielectric constants decrease from 3.02 to 2.59, and the optical lengths decrease from 1.13 to 1.05 mm, respectively. Therefore, the resonance peak frequencies of the localized modes B and C (with the similar standing wave profiles) are clearly shifted from 0.410 to 0.430 THz by replacing the water with the ethanol in the resonator. From these results, we can consider the fabricated phonic crystal resonator to be a promising novel analytical device to determine the dissolved components in an aqueous solution using terahertz spectroscopy.

REFERENCES

[1] J.M. Williams, A. Adewunmi, R.M. Schek, C.L. Flanagan, P.H. Krebsbach, S.E. Feinberg, S.J. Hollister, S. Das, Bone tissue engineering using polycaprolactone scaffolds fabricated via selective laser sintering, Biomaterials 26 (2005) 4817–4827.

[2] H. Seitz, W. Rieder, S. Irsen, B. Leukers, C. Tille, Three-dimensional printing of porous ceramic scaffolds for bone tissue engineering, J. Biomed. Mater. Res. B Appl. Biomater. 74 (2) (2005) 782–788.

[3] J.L. Simon, S. Michna, J.A. Lewis, E.D. Rekow, V.P. Thompson, J.E. Smay, A. Yampolsky, J.R. Parsons, J.L. Ricci, In vivo bone response to 3D periodic hydroxyapatite scaffolds assembled by direct ink writing, J. Biomed. Mater. Res. A 83A (3) (2007) 747–758.

[4] V. Karageorgiou, D. Kaplan, Porosity of 3D biomaterial scaffolds and osteogenesis, Biomaterials 26 (27) (2005) 5474–5491.

[5] S. Cai, J. Xi, Comput. Aided Des. 40 (2008) 1040–1050.

[6] S. Itoh, S. Nakamura, M. Nakamura, K. Shinomiya, K. Yamashita, Biomaterials 27 (2006) 5572–5579.

[7] I. Sopyan, M. Mel, S. Ramesh, K.A. Khalid, Sci. Technol. Adv. Mater. 8 (2007) 116–123.

[8] A.L. Olivares, E. Marsal, J.A. Planell, D. Lacroix, Finite element study of scaffold architecture design and culture conditions for tissue engineering, Biomaterials 30 (2009) 6142–6149.

[9] A.J.F. Stops, K.B. Heraty, M. Browne, F.J. O'Brien, P.E. McHugh, A prediction of cell differentiation and proliferation within a collagen–glycosaminoglycan scaffold subjected to mechanical strain and persuasive fluid flow, J. Biomech. 43 (2010) 618–626.

[10] C. Jungreuthmayer, M.J. Jaasma, A.A. Al-Munajjed, J. Zanghellini, D.J. Kelly, F.J. O'Brien, Deformation simulation of cells seeded on a collagen-GAG scaffold in a flow perfusion bioreactor using a sequential 3D CFD-elastostatics model, Med. Eng. Phys. 31 (2009) 420–427.
[11] M. Ishikawa, S. Kirihara, Y. Miyamoto, T. Sohmura, Advanced processing and manufacturing technologies for structural and multifunctional materials II, Ceram. Eng. Sci. Proc. 29 (2008) 131–138.
[12] B. Chen, T. Zhang, J. Zhang, Q. Lin, D. Jiang, Ceram. Int. 34 (2008) 359–364.
[13] Y. Li, Y. Xie, J. Gong, Y. Chen, Z. Zhang, Preparation of Ni/YSZ materials for SOFC anodes by buffer-solution method, Mater. Sci. Eng. B86 (119–122) (2001) 1.
[14] J.-H. Lee, H. Moon, H.-W. Lee, J. Kim, J.-D. Kim, K.-H. Yoon, Quantitative analysis of microstructure and its related electrical property of SOFC anode, Ni-YSZ cermet, Solid State Ionics 148 (2002) 15–26.
[15] K.-R. Lee, S.H. Choi, J. Kim, H.-W. Lee, J.-H. Lee, Viable image analyzing method to characterize the microstructure and the properties of the Ni/YSZ cermet anode of SOFC, J. Power Sources 140 (2005) 226–234.
[16] H. Koide, Y. Someya, T. Yoshida, T. Maruyama, Properties of Ni/YSZ cermet as anode for SOFC, Solid State Ionics 132 (2000) 253–260.
[17] T. Fukui, S. Ohara, M. Naito, K. Nogi, Performance and stability of SOFC anode fabricated from NiO-YSZ composite particles, J. Power Sources 110 (2002) 91–95.
[18] J.H. Yu, G.W. Park, S. Lee, S.K. Woo, Microstructural effects on the electrical and mechanical properties of Ni-YSZ cermet for SOFC anode, J. Power Sources 163 (2007) 926–932.
[19] Y. Suzue, N. Shikazono, N. Kasagi, Numercial simulation of mass transfer and electrochemical reaction for microscopic structure design of SOFC porous electrode, JSME Annu. Meet. 7 (2006) 181–182.
[20] M. Koyama, K. Ogiya, T. Hattori, H. Fukunaga, A. Suzuki, R. Sahnoun, H. Tsuboi, N. Hatakeyama, A. Endou, H. Takaba, M. Kubo, C.A. Del Carpio, A. Miyamoto, Development of three-dimensional porous structure simulator POCO2 for simulations of Irregular porous materials, J. Coput. Chem. Jpn. 7 (2008) 55–62.
[21] E. Yablonovitch, Inhibited spontaneous emission in solid-state physics and electronics, Phys. Rev. Lett. 58 (1987) 2059–2062.
[22] S. John, Strong localization of photons in certain disordered dielectric superlattices, Phys. Rev. Lett. 58 (1987) 2486–2489.
[23] S. Noda, Three-dimensional photonic crystals operating at optical wavelength region, Physica B 279 (2000) 142–149.
[24] S. Noda, Full three-dimensional photonic bandgap crystals at near-infrared wavelengths, Science 289 (2000) 604–606.
[25] W. Chen, S. Kirihara, Y. Miyamoto, Three-dimensional microphotonic crystals of ZrO2 toughened Al2O3 for terahertz wave applications, Appl. Phys. Lett. 91 (2007) 153507.
[26] M. Yamaguchi, F. Miyamaru, K. Yamamoto, M. Tani, M. Hangyo, Terahertz absorption spectra of L-, D-, and DL-alanine and their application to determination of enantiometric composition, Appl. Phys. Lett. 86 (2005) 053903.
[27] B.M. Fischer, M. Walther, P.U. Jepsen, Far-infrared vibrational modes of DNA components studied by terahertz time-domain spectroscopy, Phys. Med. Biol. 47 (2002) 3807–3814.
[28] Y. Oyama, L. Zhen, T. Tanabe, M. Kagaya, Sub-terahertz imaging of defects in building blocks, NDT&E Int. 42 (2009) 28–33.
[29] V.P. Wallace, A.J. Fitzgerald, S. Shankar, N. Flanagan, et al., Terahertz pulsed imaging of basal cell carcinoma ex vivo and in vivo, Br. J. Dermatol. 151 (2004) 424–432.

[30] M. Hineno, H. Yoshinaga, Far-infrared spectra of mono-, di- and tri-saccharides in 50-16 cm-1 at liquid helium temperature, Spectrochim. Acta 30A (1974) 411–446.

[31] K.M. Ho, C.T. Chen, C.M. Soukoulis, Existence of a photonic gap in periodic dielectric structures, Phys. Rev. Lett. 65 (1990) 3152–3155.

[32] H. Yada, M. Nagai, K. Tanaka, The intermolecular stretching vibration mode in water isotopes investigated with broadband terahertz time-domain spectroscopy, Chem. Phys. Lett. 473 (2009) 279–283.

[33] Y. Yomogida, Y. Sato, R. Nozaki, T. Mishina, J. Nakahara, Dielectric study of normal alcohols with THz time-domain spectroscopy, J. Mol. Liq. 154 (2010) 31–35.

Chapter 14

Future Development of Lead-Free Piezoelectrics by Domain Wall Engineering

S. Wada
Materials Science and Technology, Interdisciplinary Graduate School of Medical and Engineering, University of Yamanashi, Yamanashi, Japan

1 INTRODUCTION

Domain configurations in the ferroelectric materials can determine ferroelectric and related properties such as piezoelectricity, pyroelectricity, and dielectricity. Thus, one of the most interesting investigations for ferroelectric-related applications is the control of the desirable domain configuration, a technique called *domain engineering technique*. Domain engineering is the most important technique to obtain the enhanced ferroelectric-related properties for conventional ferroelectric materials. To date, the following domain engineering techniques have been proposed and established. The inhibited domain wall motion by the acceptor doping into $Pb(Ti,Zr)O_3$ (PZT) ceramics, namely, "hard" PZT, is one of the domain engineering techniques, and the hard PZT ceramics are used for the piezoelectric transformer application. [1] The enhanced domain wall motion by the donor doping into PZT ceramics, namely, "soft" PZT, is also one of the domain engineering techniques, and the soft PZT ceramics are used for the actuator application. [1] The induction of a periodic domain-inverted structure into lithium tantalate ($LiTaO_3$) and lithium niobate ($LiNbO_3$) single crystals is a typical domain engineering technique, and this device is used for the surface acoustic wave application [2,3] and the nonlinear optical application [4,5]. Recently, for the T-bit memory application, the writing and reading techniques of nanodomain on a $LiTaO_3$ single crystal plate were reported by Cho *et al.*, [6,7] and for the 3-D photonic crystal application, the writing and etching techniques of nanodomain of a $LiNbO_3$ single crystal plate were reported by Kitamura *et al.* [8] These techniques are very important domain engineering techniques, and by using these techniques, enhanced ferroelectric-related properties and new properties can be expected.

Green and Sustainable Manufacturing of Advanced Materials. http://dx.doi.org/10.1016/B978-0-12-411497-5.00014-X

Among these domain engineering techniques, the one using the crystallographic anisotropy of the ferroelectric single crystals is known as *engineered domain configuration*. [9–11] This engineered domain configuration can induce enhanced piezoelectric property in the ferroelectric single crystals. However, there are still many unknown factors such as the mechanism of the enhanced property. It is particularly important to investigate the most suitable engineered domain configuration for the piezoelectric applications. In this study, various engineered domain configurations were induced in $BaTiO_3$ single crystals, and their piezoelectric properties were investigated as a function of (1) the crystal structure, (2) the crystallographic orientation, and (3) the domain size (domain wall density). Moreover, for the tetragonal $BaTiO_3$ crystals with the engineered domain configurations, it was recently found that the piezoelectric properties significantly improved with decreasing domain sizes (domain wall density). [12,13] These results suggested that the domain walls in the engineered domain configuration could contribute significantly to the piezoelectric properties. In this chapter, we try to point out that a significant contribution of domain wall region to piezoelectric, elastic, and dielectric properties, and engineered domain configurations can be very helpful to fix the domain wall region in piezoelectric crystals, in other words, to prepare a composite between (a) a distorted domain wall region and (b) a normal tetragonal domain region. Thus, an increase in the volume fraction of the distorted domain wall region can result in crystals with ultrahigh piezoelectric properties.

2 HISTORY OF ENGINEERED DOMAIN CONFIGURATION

Engineered domain configuration was found in $Pb(Zn_{1/3}Nb_{2/3})O_3$-$PbTiO_3$ (PZN-PT) single crystals for the first time. In $[001]_c$ oriented rhombohedral PZN-PT single crystals, ultrahigh piezoelectric activities were reported by Park *et al.* [9,14,15] and Kuwata *et al.* [16,17] with a strain over 1.7%, a piezoelectric constant d_{33} over 2500 pC/N, an electromechanical coupling factor k_{33} over 93% and a hysteresis-free strain versus electric-field behavior. The $(1-x)$ PZN-xPT single crystals with $x<0.08$ have rhombohedral $3m$ symmetry at room temperature, and their polar directions are $[111]_c$. [16,17] However, the unipolar electric-field (E-field) drive along the $[111]_c$ direction showed a large hysteretic strain versus E-field behavior, a low d_{33} of 83 pC/N, and a low k_{33} of 38%. To explain the strong anisotropy in piezoelectric properties, *in situ* domain observation was carried out using $[111]_c$ and $[001]_c$ oriented pure PZN and 0.92PZN-0.08PT single crystals. [10,11] The result showed that when the E-field was applied along the $[001]_c$ direction, a very stable domain structure appeared, and domain wall motion was undetectable under DC-bias, resulting in the hysteresis-minimized strain versus electric-field behavior.

The $[001]_c$ poled $3m$ crystals have four equivalent domains with four polar vectors along the $[111]_c$, $[-111]_c$, $[1-11]_c$, and $[-1-11]_c$ directions because their polar directions are along $<111>_c$. Therefore, the components of each polar vector along the $[001]_c$ direction are completely equal with each other, so that each domain wall cannot move under the E-field drive along the $[001]_c$ direction owing to the equivalent domain energy changes. [9–11,14,15] This suggests the possibility of controlling domain configuration in single crystals using the crystallographic orientation, and the appearance of a new technology in the domain engineering field. Thus, this special domain structure in ferroelectric single crystals using the crystallographic orientation was called the engineered domain configuration. [9–11]

3 EFFECT OF ENGINEERED DOMAIN CONFIGURATION ON PIEZOELECTRIC PROPERTY

The engineered domain configuration is expected to possess the following five features in terms of piezoelectric performance: (1) hysteresis-free strain versus E-field behavior owing to the inhibition of domain wall motion, (2) higher piezoelectric constant along a nonpolar direction owing to the easy tilt of a polar vector by the E-field, (3) change of macroscopic symmetry in crystals with engineered domain configuration, [10,11] (4) complex contribution of other high shear piezoelectric constants such as d_{15}, and (5) contribution of non-180° domain wall region to piezoelectric property. Features (1) to (4) can be useful for both single-domain and multidomain crystals while feature (5) is expected only for multidomain crystals. The engineered domain configuration is defined as the domain structure composed of some equivalent polar vectors along the E-field drive direction. Therefore, the engineered domain configuration is basically a multidomain configuration, and many kinds of engineered domain configurations are possible depending on (a) the crystal structure and (b) the crystallographic orientation. [18]

For the rhombohedral $3m$ ferroelectric crystal with the polar directions of $<111>_c$, when the E-field is applied along the $[001]_c$ and $[110]_c$ directions, the two types of the engineered domain configurations formed are shown in Figure 1. For the orthorhombic $mm2$ ferroelectric crystal with the polar directions of $<110>_c$, when the E-field is applied along the $[001]_c$ and $[111]_c$ directions, two kinds of the engineered domain configurations are formed, as shown in Figure 2. Moreover, for the tetragonal $4mm$ ferroelectric crystal with the polar directions of $<001>_c$, when the E-field is applied along $[111]_c$ and $[110]_c$ directions, two kinds of the engineered domain configurations are formed, as shown in Figure 3. In general, most of the practical ferroelectric materials belong to the rhombohedral $3m$, orthorhombic $mm2$ or tetragonal $4mm$ symmetry. Therefore, in this section, only these three crystal structures will be considered as the conventional engineered domain configurations. On the other hand, in each of the three crystal structures, three different crystallographic directions of $[001]_c$, $[110]_c$, and $[111]_c$ can be considered to induce the aforementioned engineered domain configurations. However, it is expected that the $[110]_c$ poled

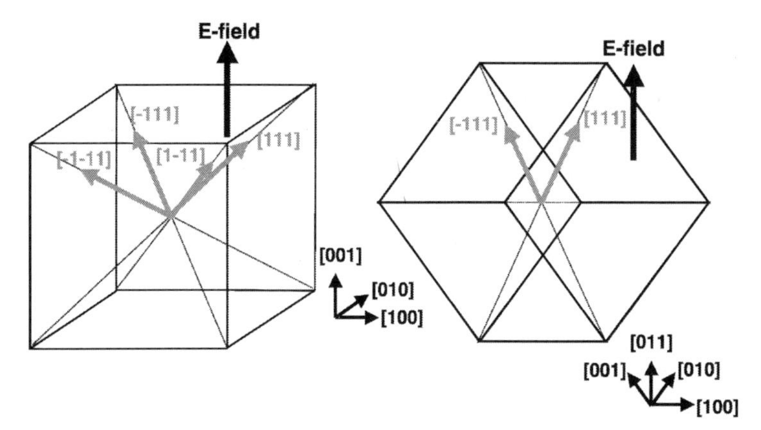

FIGURE 1 Schematic model of the engineered domain configurations for the [001]- and [011]-poled rhombohedral 3*m* ferroelectric crystals.

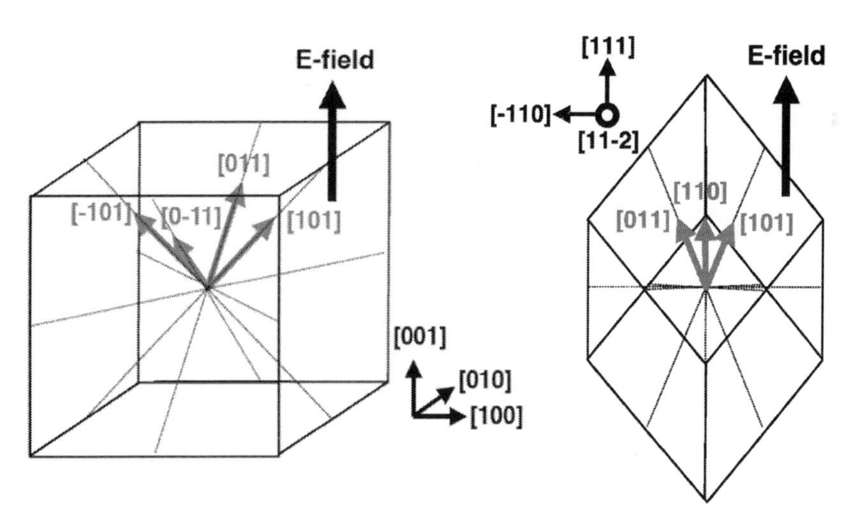

FIGURE 2 Schematic model of the engineered domain configurations for the [001]- and [111]-poled orthorhombic *mm*2 ferroelectric crystals.

3*m* and 4*mm* ferroelectric crystals may exhibit a much smaller influence on the ferroelectric-related properties than the $[001]_c$ poled 3*m* and *mm*2 ferroelectric crystals and the $[111]_c$ poled *mm*2 and 4*mm* ferroelectric crystals. This is because the engineered domain configuration in the $[110]_c$ poled 3*m* and 4*mm* ferroelectric crystals is composed of just two kinds of polar vectors. Thus, in this section, we consider only the two kinds of the crystallographic directions of $[001]_c$ and $[111]_c$.

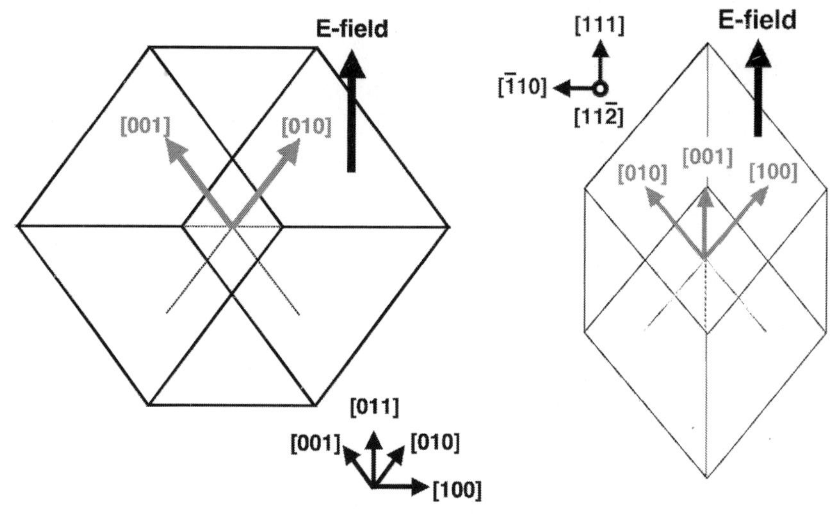

FIGURE 3 Schematic model of the engineered domain configurations for the [011]- and [111]-poled tetragonal 4*mm* ferroelectric crystals.

4 CRYSTAL STRUCTURE AND CRYSTALLOGRAPHIC ORIENTATION DEPENDENCE OF $BaTiO_3$ CRYSTALS WITH VARIOUS ENGINEERED DOMAIN CONFIGURATIONS

If the concept of engineered domain configuration for PZN-PT single crystals can be applied to other ferroelectric crystals, the enhanced piezoelectric properties are expected for lead-free ferroelectrics. $BaTiO_3$ single crystal is one of the typical lead-free ferroelectrics and has the tetragonal 4*mm* symmetry at room temperature. [19] The piezoelectric properties were investigated using the $[111]_c$ oriented tetragonal $BaTiO_3$ single crystals, as shown in Figure 4. [20–22] Note that the piezoelectric constant d_{33} was directly measured from the strain versus E-field curves. Two kinds of engineered domain configurations were formed: (a) the $[111]_c$ poled tetragonal engineered domain configuration and (b) the $[111]_c$ poled orthorhombic engineered domain configuration (Figure 4). This is because for the tetragonal $BaTiO_3$ crystals, the higher E-field drive over 10 kV/cm along the $[111]_c$ direction led to the E-field induced phase transition from the tetragonal 4*mm* to orthorhombic *mm2* phase. The original d_{33} below 10 kV/cm was 203 pC/N whereas the d_{33} in the induced phase above 10 kV/cm reached 295 pC/N. The d_{33} value of the $[001]_c$ poled tetragonal $BaTiO_3$ single crystals was 90 pC/N. [19] Therefore, the d_{33} of the $[111]_c$ poled tetragonal $BaTiO_3$ was almost twice as large as, and the d_{33} of the $[111]_c$ poled orthorhombic $BaTiO_3$ was almost three times as large as, that of the $[001]_c$ poled tetragonal $BaTiO_3$. [20,22]

We must consider the difference between the d_{33} values of 203 and 295 pC/N. As shown in Figure 4, both the $[111]_c$ poled tetragonal and orthorhombic

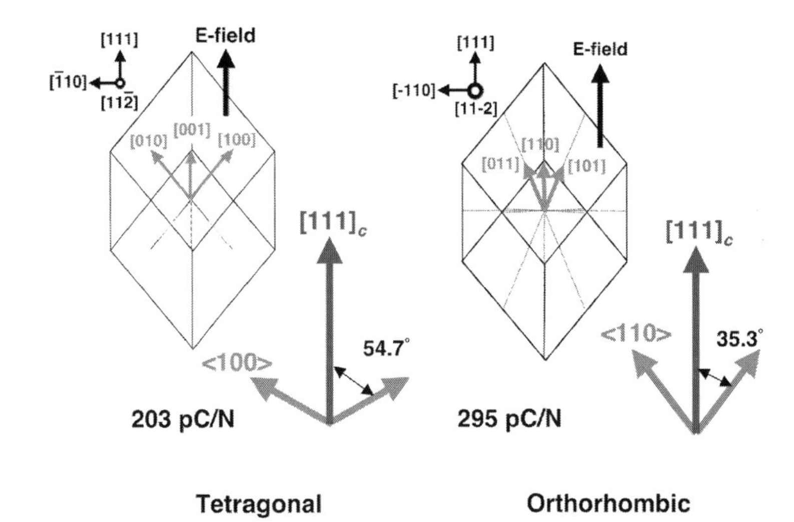

FIGURE 4 Two engineered domain configurations for the $[111]_c$ poled (a) tetragonal and (b) orthorhombic $BaTiO_3$ single crystals.

engineered domains were composed of three kinds of domains, but the angle θ between the polar direction and the E-field direction was quite different, that is, $\theta = 54.7°$ for the d_{33} of 203 pC/N, and $\theta = 35.3°$ for the d_{33} of 295 pC/N. It is very important to clarify whether this angle θ is effective for the piezoelectric performance or not. Thus, the piezoelectric properties of $BaTiO_3$ single crystals were investigated as a function of the crystallographic orientations of $[001]_c$ and $[111]_c$, and the crystal structures of tetragonal, orthorhombic, and rhombohedral phases, as shown in Figure 5. [20,22] The temperature was changed from −100 to 200°C. It should be noted that the piezoelectric properties were measured by two kinds of methods: (a) measurement under a high E-field over 100 V/cm (estimation of piezoelectric constants from the strain versus E-field curves), and (b) resonance-antiresonance measurement under a low ac E-field below 10 V/cm (estimation of piezoelectric properties from the impedance versus frequency curves).

4.1 Piezoelectric Properties Measured Under High E-Field

4.1.1 $[001]_c$ Oriented $BaTiO_3$ Single Crystals

Under high E-fields, the strain versus E-field curves of the $[001]_c$ oriented $BaTiO_3$ single crystals were measured from −100 to 200°C. [18,21] On the basis of the slope of the strain behaviors over 20 kV/cm, the piezoelectric constant d_{33} was calculated directly, as shown in Figure 6. It should be noted from Figure 6 that for the $[001]_c$ oriented orthorhombic $BaTiO_3$ single crystals, the maximum d_{33} of 500 pC/N was obtained at 0°C. This is a very large value that is almost comparable to that of PZT ceramics. The d_{33} of the $[001]_c$ poled orthorhombic

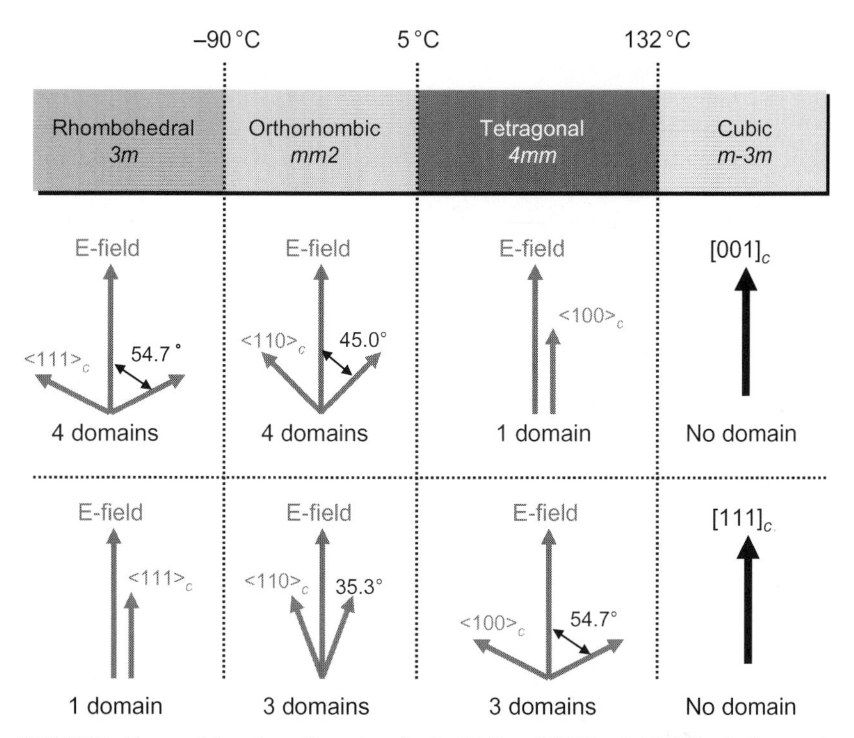

FIGURE 5 Expected domain configurations for the $[111]_c$ and $[001]_c$ poled $BaTiO_3$ single crystals with three kinds of the crystal structures.

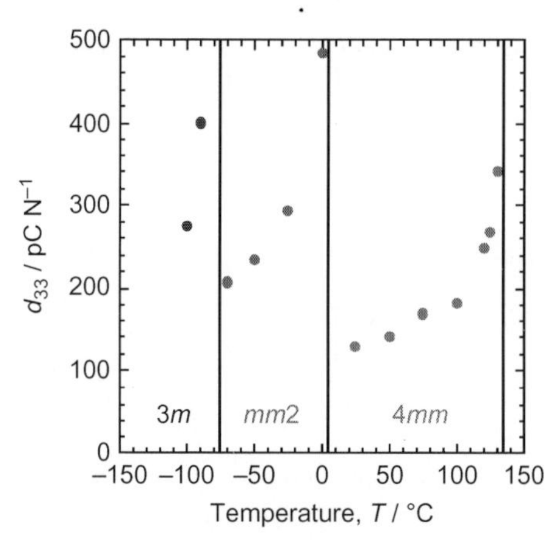

FIGURE 6 Relationship between the d_{33} and temperature for the $[001]_c$ oriented $BaTiO_3$ single crystals from −100°C to 150°C.

$BaTiO_3$ single crystals was much higher than that of the $[001]_c$ poled rhombohedral $BaTiO_3$ single crystals. When the E-field was applied along $[001]_c$ direction, two kinds of engineered domain configurations could be formed: (a) the $[001]_c$ poled orthorhombic $BaTiO_3$ crystal with the four equivalent domains and a θ of 45.0°, and (b) the $[001]_c$ poled rhombohedral $BaTiO_3$ crystal with the four equivalent domains and a θ of 54.7°.

It was expected that if the number of the equivalent domains constructing the engineered domain configuration is the same, the smaller angle θ can cause larger piezoelectric properties. The result in Figure 6 supported the aforementioned hypothesis. Therefore, when the E-field was applied along the $[001]_c$ direction of $BaTiO_3$ single crystals, the orthorhombic phase was very important for the higher piezoelectric performance.

4.1.2 $[111]_c$ Oriented $BaTiO_3$ Single Crystals

The strain versus E-field curves of the $[111]_c$ oriented $BaTiO_3$ single crystals were also measured from −100 to 200°C. [18,21] On the basis of the slope of the strain behaviors below 10 kV/cm, the piezoelectric constant d_{33} was calculated and shown in Figure 7. The strain versus E-field curves at 25°C underwent discontinuous changes owing to two kinds of the E-field induced phase transitions: (1) the first transition from tetragonal to orthorhombic phase around 4-6 kV/cm, and (2) the second transition from orthorhombic to rhombohedral phase around 30 kV/cm. [20–22] In Figure 7, it should be noted that for the $[111]_c$ oriented orthorhombic $BaTiO_3$ single crystals, the maximum d_{33} of 260 pC/N was obtained at −70°C. As previously mentioned, the d_{33} of the $[111]_c$ poled orthorhombic $BaTiO_3$ crystal over 10 kV/cm was 295 pC/N at 25°C (Figure 4). We believe that this discrepancy in d_{33} (35 pC/N) may be caused by the temperature difference. Moreover, in Figure 7, the d_{33} of the $[111]_c$ poled orthorhombic $BaTiO_3$

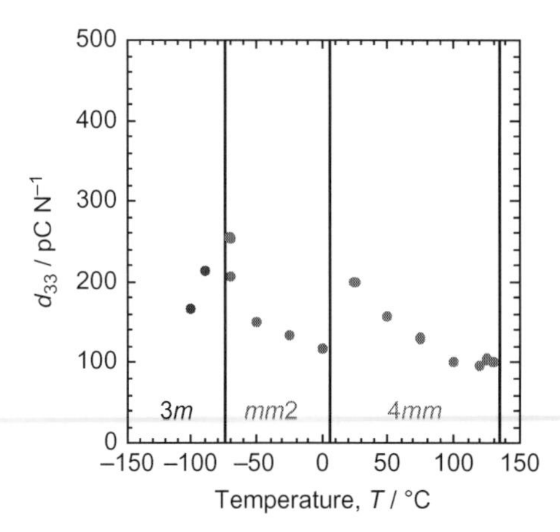

FIGURE 7 Relationship between the d_{33} and temperature for the $[111]_c$ oriented $BaTiO_3$ single crystals from −100°C to 150°C.

single crystals was much higher than that of the $[111]_c$ poled tetragonal $BaTiO_3$ single crystals. In Figure 5, when the E-field was applied along the $[111]_c$ direction, two kinds of engineered domain configurations could be obtained: (a) the $[111]_c$ poled orthorhombic $BaTiO_3$ crystal with the three equivalent domains and θ of 35.3°, and (b) the $[111]_c$ poled tetragonal $BaTiO_3$ crystal with the three equivalent domains and θ of 54.7°. Therefore, it was confirmed that when the number of the equivalent domains constructing the engineered configuration is the same, a smaller angle θ can cause larger piezoelectric properties. Therefore, when the E-field was applied along $[111]_c$ direction of $BaTiO_3$ single crystals, the orthorhombic phase was also very important for the higher piezoelectric performance.

4.1.3 Crystal Structure and Crystallographic Orientation for the Best Engineered Domain Configuration in $BaTiO_3$ Single Crystals

We have suggested that of the factors that can be responsible for the piezoelectric performance, the following two are the most important: (1) the number of the equivalent domains constructing the engineered domain configuration and (2) the angle θ between the polar direction and the E-field direction. [18] Based on Figure 5, we use the piezoelectric constants in Figures 6 and 7 to check these factors. As mentioned, when the number of the equivalent domains constructing the engineered domain configuration was the same, the smaller angle θ caused larger piezoelectric properties. Next, the role of the number of the equivalent domains constructing the engineered domain configuration can be clarified by comparing the $[001]_c$ poled rhombohedral $BaTiO_3$ crystal that had four equivalent domains and θ of 54.7° with the $[111]_c$ poled tetragonal crystal that had three equivalent domains and the same θ of 54.7°. The d_{33} of the $[001]_c$ poled rhombohedral $BaTiO_3$ crystal was about 400 pC/N, which is twice as high as that of the $[111]_c$ poled tetragonal $BaTiO_3$ crystal (about 200 pC/N). This suggests that the effect of the number of the equivalent domains on the piezoelectric properties is significant larger than that of θ. [18] This observation is confirmed by comparing the $[001]_c$ poled orthorhombic $BaTiO_3$ crystal that had four equivalent domains and a θ of 45.0° with the $[111]_c$ poled orthorhombic $BaTiO_3$ crystal had three equivalent domains and a θ of 35.3°: the d_{33} of the former was twice as high as that of the latter. [18]

Our earlier discussion indicates a new direction to obtain the best engineered domain configuration for the piezoelectric application. The first step is to identify the engineered domain configuration with the largest number of equivalent domains. For the normal perovskite-type ferroelectric single crystals, only the $[001]_c$ poled orthorhombic and rhombohedral crystals can satisfy this requirement. The second step is to find the engineered domain configurations with the smallest angle θ. For the normal perovskite-type ferroelectric single crystals, only the $[001]_c$ poled orthorhombic crystals can satisfy the second request. Therefore, the best engineered domain configuration for the piezoelectric application can be found in the $[001]_c$ poled orthorhombic single crystals. From this reasoning, potassium niobate ($KNbO_3$) crystals appear to be one of the promising candidates with ferroelectric orthorhombic symmetry [23].

4.2 Piezoelectric Properties Measured Under Low ac E-Field

It was shown in the previous section that the high piezoelectric properties can be realized in the $[001]_c$ poled orthorhombic and rhombohedral $BaTiO_3$ crystals. Using the IEEE resonance technique, other piezoelectric properties such as the electromechanical coupling factor and dielectric constant were measured for the $[001]_c$ oriented $BaTiO_3$ crystals. [18,21] Figures 8 and 9 show the d_{33} and k_{33} versus temperature curves of the $[001]_c$ oriented $BaTiO_3$ crystals, respectively, which were measured using a low ac E-field of 1 V/cm and a high dc E-field of 6 kV/cm. As expected, the $[001]_c$ poled orthorhombic $BaTiO_3$ crystals exhibited the maximum d_{33} of 415 pC/N and k_{33} of 85% at −5°C. These values were much higher than those of PZT ceramics. Figure 10 shows the dc E-field dependence of k_{33} at −5°C using a low ac E-field of 1 V/cm. Before poling treatment (start point), k_{33} was just 70%, and during the poling treatment at 6 kV/cm, k_{33} increased to 85%. After poling treatment (final point), k_{33} still kept a high value of 79%. This result suggested that poling treatment is very important for high piezoelectric properties. On the other hand, Figures 11 and 12 show the d_{31} and k_{31} versus temperature curves of the $[001]_c$ oriented $BaTiO_3$ crystals. As expected, the $[001]_c$ poled orthorhombic $BaTiO_3$ crystals exhibited the maximum d_{31} of −280 pC/N and k_{31} of 65% at 0°C.

These results indicated that for the $BaTiO_3$ single crystals, the combination of the orthorhombic *mm*2 phase and the $[001]_c$ crystallographic direction

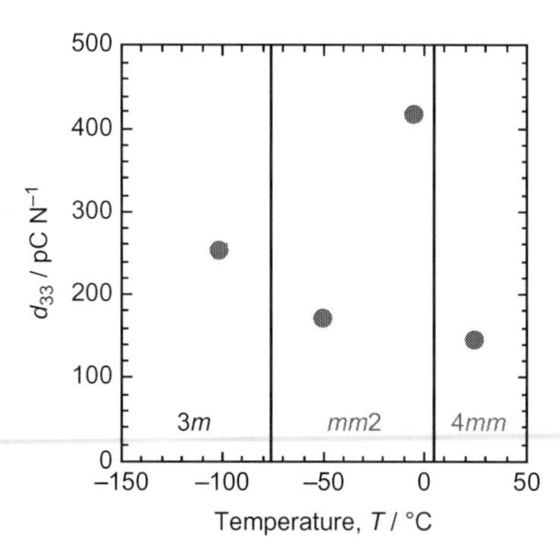

FIGURE 8 Relationship between the d_{33} and temperature for the $[001]_c$ oriented $BaTiO_3$ single crystals from −100°C to 25°C.

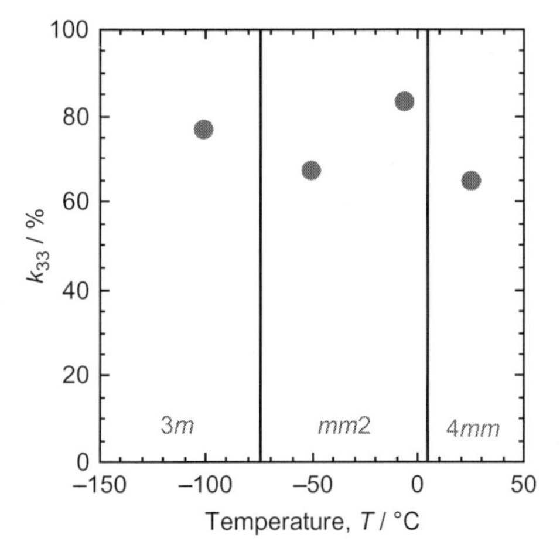

FIGURE 9 Relationship between k_{33} and temperature for the $[001]_c$ oriented $BaTiO_3$ single crystals from −100°C to 25°C.

exhibited the best piezoelectric properties. It is well known that the $[001]_c$ poled tetragonal $BaTiO_3$ crystals show a d_{33} of 90 pC/N and a k_{33} of 56%. [19] Therefore, the introduction of the best engineered domain configuration into the $BaTiO_3$ crystals resulted in a fivefold increase in d_{33} and a one-and-a-half-fold increase in k_{33}. Moreover, if the $[001]_c$ poled orthorhombic single crystals could be obtained at room temperature, we would expect much higher piezoelectric properties. Thus, as mentioned previously, $KNbO_3$ may exhibit a higher potential for piezoelectric applications because of its orthorhombic *mm2* phase at room temperature [23].

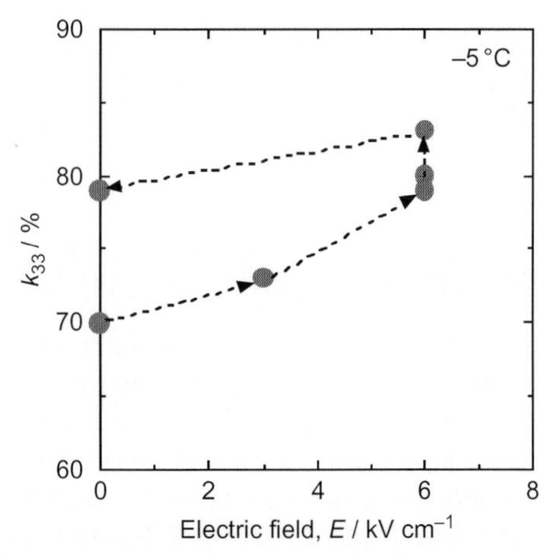

FIGURE 10 Relationship between the k_{33} and E-field for the $[001]_c$ oriented $BaTiO_3$ single crystals measured at −5°C.

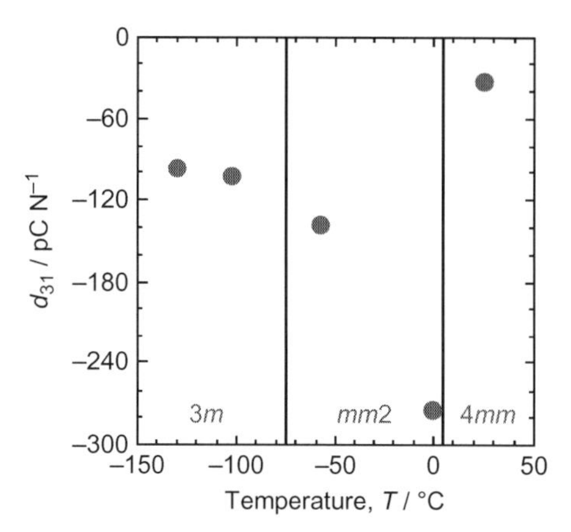

FIGURE 11 Relationship between the d_{31} and temperature for the $[001]_c$ oriented $BaTiO_3$ single crystals from −100°C to 25°C.

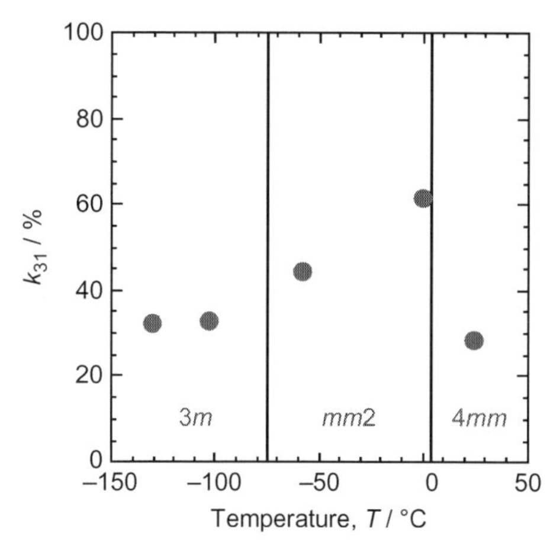

FIGURE 12 Relationship between the k_{31} and temperature for the $[001]_c$ oriented $BaTiO_3$ single crystals from −100°C to 25°C.

5 DOMAIN SIZE DEPENDENCE OF $BaTiO_3$ CRYSTALS WITH ENGINEERED DOMAIN CONFIGURATIONS

$BaTiO_3$ single crystals have a tetragonal *P4mm* phase at room temperature. To induce an engineered domain configuration in the tetragonal crystals, the E-field should be applied along the $[111]_c$ direction. The piezoelectric measurements showed that the d_{33} of the $[111]_c$ poled tetragonal $BaTiO_3$ crystal with the engineered domain configuration was almost 203 pC/N, [20] and this value was more than twice as large as the 90 pC/N of the $[001]_c$ poled $BaTiO_3$ single-domain crystal. [19] On the other hand, the d_{33} of the $[001]_c$ poled rhombohedral PZN-PT crystal with the engineered domain configuration was almost 30 times as large as the 83 pC/N of the $[111]_c$ poled PZN-PT single-domain crystal. [15] To explain this huge difference, we consider the domain size, that is, the domain wall density, for the engineered domain configuration. This is because the domain structure of the $[001]_c$ poled PZN-PT crystal was composed of the fiber-like domains with a domain length of 130 μm and a domain width of around 1 μm [10,11] whereas that of the $[111]_c$ poled $BaTiO_3$ crystal was made of very coarse domain with a wide domain width of 300 ~ 400 μm [20]. This result suggested that non-180° domain wall region might contribute to piezoelectric properties significantly owing to its structure gradient region. Thus, when the fine domain structure is induced in the $[111]_c$ poled tetragonal $BaTiO_3$ crystals with the engineered domain configuration, it is possible to obtain much enhanced piezoelectric property. Therefore, the piezoelectric properties of $BaTiO_3$ single crystals were investigated as a function of domain size, namely, domain wall density. Especially in the $[111]_c$ oriented tetragonal $BaTiO_3$ crystals with an engineered domain configuration, the domain size dependence on E-field and the temperature was investigated in detail.

5.1 Domain Size Dependence on E-Field and Temperature

To understand domain size dependence on E-field and temperature for the $[111]_c$ oriented $BaTiO_3$ crystals, the domain structures were observed at various temperatures from 0 to 200°C and various E-fields from 0 to 16 kV/cm. Prior to any domain observation, the $BaTiO_3$ crystals were heated up to 160°C, and then cooled down to the observation temperatures at a cooling rate of 0.4°C/min without E-field. At the constant temperature, the dc E-fields were applied along the $[111]_c$ direction very slowly.

Figure 13 shows the domain size dependence on the E-field and temperature for the $[111]_c$ oriented $BaTiO_3$ crystals with the engineered domain configuration. Figure 14 shows the typical domain structures observed in the corresponding regions A to G presented in Figure 13 (P and A indicate the crossed polarizer and analyzer). In Figure 13, the “A” and “B” regions were assigned to the orthorhombic *mm*2 phase. At 25°C, the dc E-field drive along the $[111]_c$ direction for the tetragonal $BaTiO_3$ crystals resulted in the E-field induced phase transition from 4*mm* to *mm*2, and this result was consistent with previous reports [18,20–22]. Figure 14a and b shows that these domain structures were composed of the fine domains. The fine domains in the A region were induced by a normal phase transition from the tetragonal phase to the orthorhombic phase without E-field. On the other hand, the fine domains in the B region resulted from the E-field induced phase transition from the tetragonal to orthorhombic phase. Thus, when the E-field was removed, the domain structure in the B region easily returned to the normal tetragonal domain configuration as shown in Figure 14c. Thus, the fine domain structure in the A and B regions cannot exist at room temperature without E-field. The C region was assigned to the tetragonal 4*mm* symmetry. By the poling treatment in this region, the domain structure was slightly changed, but the observed domain walls were all completely assigned to 90°

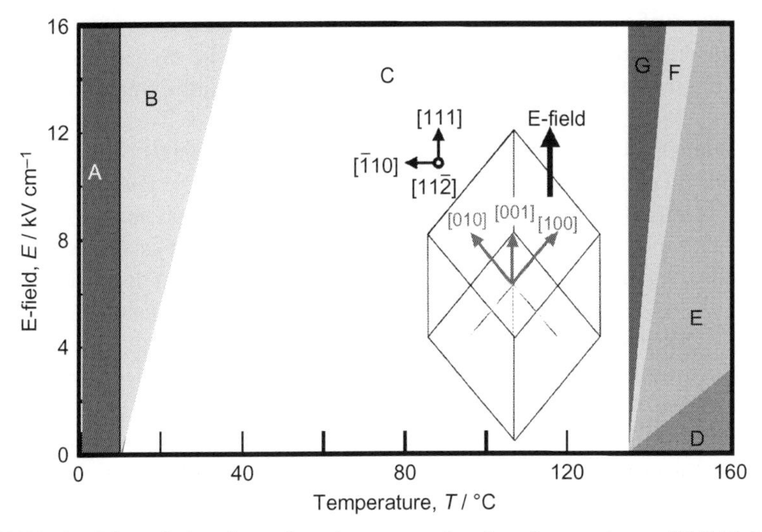

FIGURE 13 Schematic domain configuration map as a function of temperature and E-field along the $[111]_c$ direction for the $[111]_c$ oriented $BaTiO_3$ single crystals with the engineered domain configuration.

domain walls of the $\{110\}_c$ planes as shown in Figure 14c. [24] The D region was assigned to the optical isotropic state with the cubic *m*-3*m* symmetry as shown in Figure 14d. However, the E region was very unclear and abnormal. The same brightness in the E region under crossed-polarizers suggested that this domain state was not optical isotropic, but an anisotropic state.[25] When the E-field was applied along the $[111]_c$ direction for the cubic *m*-3*m* symmetry, it is expected that the cubic *m*-3*m* symmetry can be slightly distorted to become the rhombohedral or monoclinic symmetry. However, at present, the origin of this birefringence is unknown. In the F region, the coexistence of the bright state without the domain walls and the fine domain state was observed. In the G region, only fine domain structure was clearly observed, all the domain walls of which were assigned to 90° domain walls of the $\{110\}_c$ planes.

In the A, B, C, and D regions, these symmetries were assigned on the basis of some reports [18,20–22]. However, there is no information about the E, F, and G regions. Thus, these symmetries were measured using the *in situ* Raman scattering measurement, which was combined with the polarizing microscopic observation. As a result, the change from D to G by E-field was assigned to an E-field induced phase transition from the cubic to tetragonal phase [26].

5.2 Domain Size Dependence of the Piezoelectric Property Using 31 Resonators

On the basis of the result of the domain size dependence on the E-field and temperature, various kinds of domain sizes were induced in the $[111]_c$ oriented

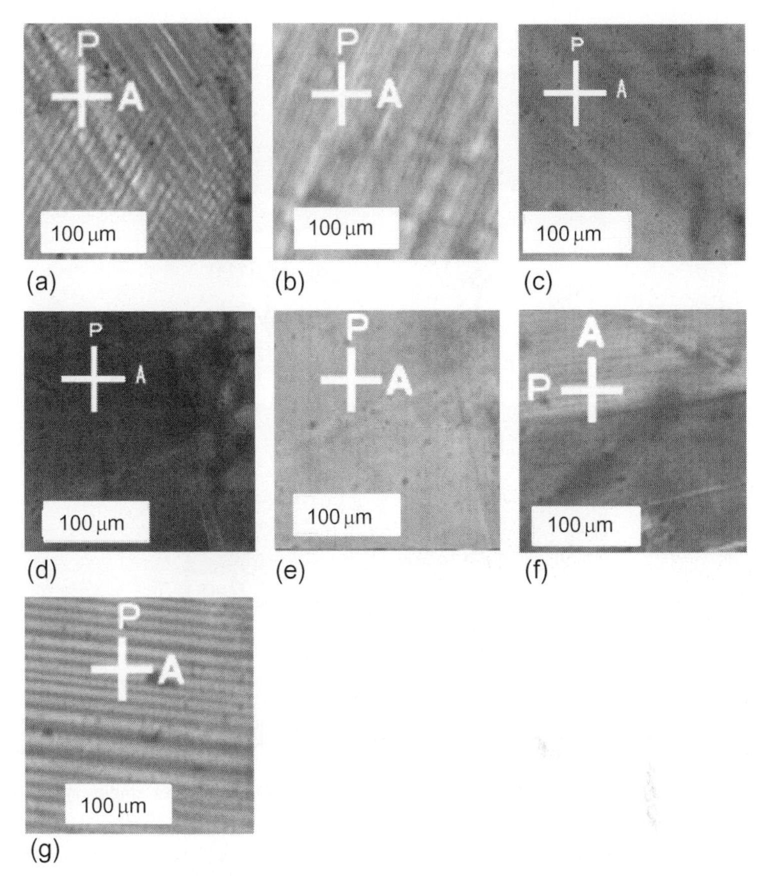

FIGURE 14 Various domain configurations in (a) the "A" region, (b) the "B" region, (c) the "C" region, (d) the "D" region, (e) the "E" region, (f) the "F" region and (g) the "G" region as shown in Figure 13. P and A on each graph indicate the crossed polarizer and analyzer.

$BaTiO_3$ single crystals with the engineered domain configuration. The engineered domain $BaTiO_3$ crystal with a large domain size was poled at just below the Curie temperature (T_c) whereas that with the finer domain size was poled at just above T_c. Figure 15 shows the $BaTiO_3$ single crystals with four kinds of domain sizes: (a) over 40 μm, (b) of 13.3 μm, (c) of 6.5 μm, and (d) of 5.5 μm. All of the domains observed in Figure 15 were assigned to 90° domain walls of $\{110\}_c$ planes. The domain configurations for all the resonators prepared in this study were composed of the same 90° domain walls [24], and the difference between these domain configurations was just domain size, specifically, domain wall density, as shown in Figure 16.

Using these 31 resonators with the different domain sizes, the piezoelectric properties were measured at 25°C by a conventional resonance-antiresonance method [27] using a weak ac E-field of 125 mV/mm. Figures 17 and 18 show

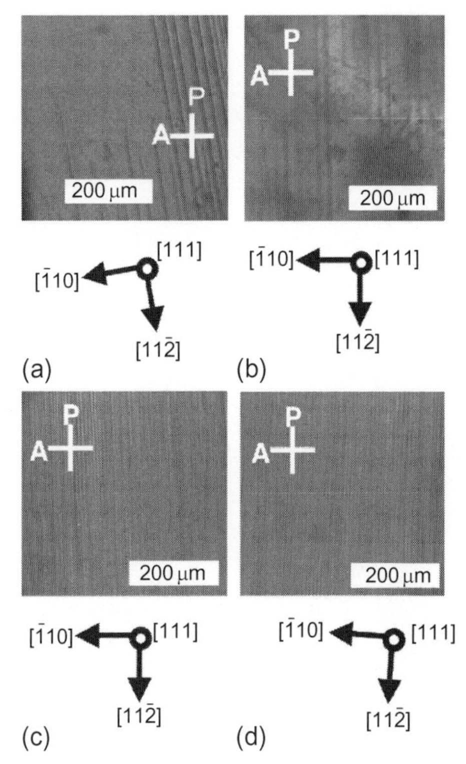

FIGURE 15 Engineered domain configurations of the $[111]_c$ oriented $BaTiO_3$ single crystals with the different average domain sizes of (a) greater than 40 μm, (b) 13.3 μm, (c) 6.5 μm and (d) 5.5 μm.

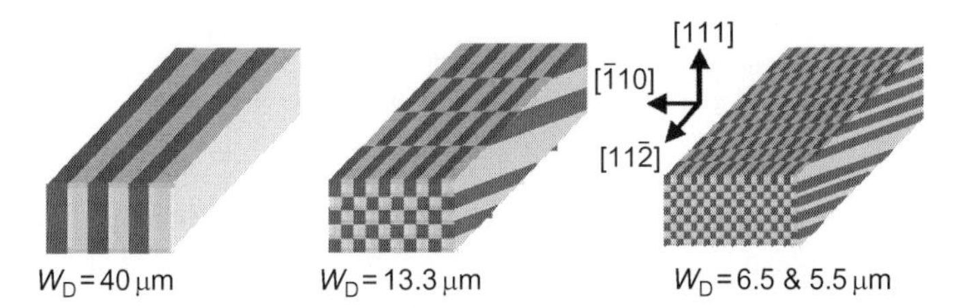

FIGURE 16 Schematic 31 resonators composed of the engineered domain configuration with different domain sizes.

the frequency dependence of the impedance and the phase for the 31 resonators with a domain size over 40 μm and that with a domain size of 6.5 μm, respectively. If the poling treatment is completely successful, the phase between resonance and antiresonance frequencies should be close to +90°. However, in these figures, the maximum phase angle was only +50° and +15°, respectively. These low phase angles suggested an insufficient poling treatment. Using these resonance and antiresonance frequencies, the piezoelectric-related constants

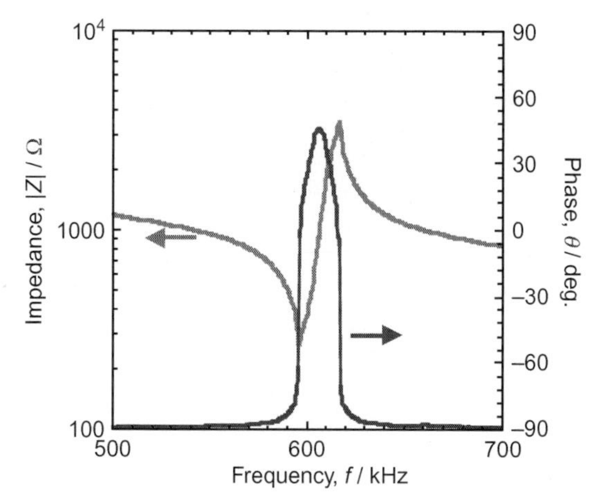

FIGURE 17 Frequency dependence of the impedance and phase for the $[111]_c$ oriented $BaTiO_3$ single crystals with an average domain size of greater than 40 μm, measured at 25°C.

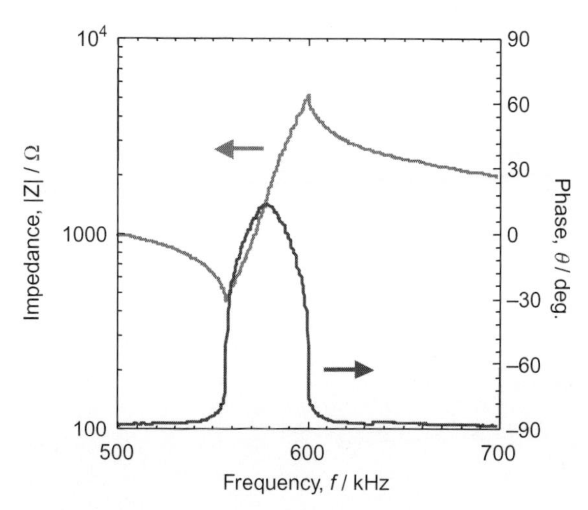

FIGURE 18 Frequency dependence of the impedance and phase for the $[111]_c$ oriented $BaTiO_3$ single crystals with an average domain size of 6.5 μm, measured at 25°C.

were estimated. Table 1 shows the domain size dependence of the properties using these 31 resonators. The d_{31} of the $[001]_c$ oriented $BaTiO_3$ single-domain crystal was reported as −33.4 pC/N whereas the effective d_{31} of the $[111]_c$ oriented $BaTiO_3$ single-domain crystal calculated using the d_{33} of 90 pC/N, d_{31} of −33.4 pC/N and d_{15} of 564 pC/N reported by Zgonik *et al.* [19] was −62 pC/N. On the other hand, the d_{31} of the $[111]_c$ poled $BaTiO_3$ single crystal with a domain size over 40 μm was estimated to be −97.8 pC/N while that of the $[111]_c$ poled $BaTiO_3$ single crystal with a domain size of 6.5 μm was found to be −180 pC/N.

TABLE 1 Domain Size Dependence of Piezoelectric Properties for the 31 $BaTiO_3$ Crystals Resonators

$BaTiO_3$ Single Crystals	ε_{33}^T	S_{11}^E (pm^2/N)	d_{31} (pC/N)	k_{31} (%)
[001][a] (single-domain)	129	7.4	−33.4	—
[111][b] (single-domain)	—	—	−62.0	—
[111], charged (domain size of 80 μm)	1299	10.9	−85.3	24.1
[111], charged (domain size of 50 μm)	2117	7.80	−98.2	25.7
[111], charged (domain size of 40 μm)	2185	7.37	−97.8	25.9
[111], charged + neutral (domain size of 20.0 μm)	2117	8.30	−102.7	26.0
[111], charged + neutral (domain size of 15.0 μm)	2186	8.20	−112.5	28.2
[111], charged + neutral (domain size of 13.3 μm)	2087	7.68	−134.7	35.7
[111], charged + neutral (domain size of 12.0 μm)	1921	8.20	−137.6	36.8
[111], charged + neutral (domain size of 10.0 μm)	2239	9.30	−140.5	32.8
[111], charged + neutral (domain size of 8.0 μm)	2238	9.10	−159.2	37.5
[111], charged + neutral (domain size of 7.0 μm)	2762	9.30	−176.2	36.9
[111], charged + neutral (domain size of 6.5 μm)	2441	8.80	−180.1	41.4
[111], charged + neutral (domain size of 5.5 μm)	2762	9.58	−230.0	47.5
"soft" PZT ceramics[c] $Pb_{0.988}(Ti_{0.48}Zr_{0.52})_{0.976}Nb_{0.024}O_3$	1,700	16.4	−171.0	34.4

[a] *Measured by Zgonik* et al. *[19].*
[b] *Calculated using the values measured by Zgonik* et al. *[19].*
[c] *Measured by Jaffe* et al. *[1].*

Especially, the 31 resonator with a domain size of 5.5 μm showed much higher d_{31} of −230 pC/N and k_{31} of 47.5% than those (d_{31} of −171 pC/N and k_{31} of 34.4%) reported for soft PZT ceramics [1]. Therefore, the $[111]_c$ oriented engineered domain $BaTiO_3$ crystal exhibit much higher piezoelectric properties than the $[001]_c$ oriented $BaTiO_3$ single-domain crystal.

To date, it was believed that the highest piezoelectric property must be obtained in single-domain crystals, and it is impossible for the material constants to be beyond the values of single-domain crystal. However, this study reveals that the 90° domain walls in the engineered domain configuration significantly enhance the piezoelectric effects, giving rise to much higher piezoelectric constants than those from single-domain crystals.

In general, under a high E-field drive, the domain walls can move very easily, and this domain wall motion made an intrinsic contribution to the piezoelectric properties very unclear. However, in the engineered domain configuration shown in Figure 3, the 90° domain walls cannot move with or without unipolar dc E-field drive. [20,21] This means that in the engineered domain configuration, the 90° domain walls can exist very stably with or without a unipolar E-field drive. Therefore, the contribution of the domain walls to the piezoelectric properties has been clarified for the first time using the engineered domain configuration. In other words, the engineered domain configuration is considered as domain-wall engineering among the domain engineering techniques. [28]

5.3 Domain Size Dependence of the Piezoelectric Property Using 33 Resonators

By the poling treatment at various electric fields and temperatures, [12,13] the 33 resonators of $BaTiO_3$ crystals with different domain sizes were prepared. The average domain sizes in the engineered domain configuration (Figure 3) were changed from 100 μm to 6 μm. The domain configurations for all the 33 resonators prepared in this study were composed of the same 90° domain walls, [24] and the difference between these domain configurations was just the domain size, as shown in Figure 19.

The piezoelectric properties were measured at 25°C using a weak ac E-field of 50 mV/mm. Table 2 shows the domain size dependence of piezoelectric-related properties using these 33 resonators. As a reference, the calculated d_{33} piezoelectric constant for the $[111]_c$ oriented $BaTiO_3$ single-domain crystal using the material constants reported by Zgonik *et al.* [19] was also listed. With decreasing domain size, all the piezoelectric-related properties increased. Surprisingly, the piezoelectric properties significantly depend on the domain size regardless of the 31 or 33 mode. Thus, we must consider the possible effect of non-180° domain walls on the piezoelectric property. Liu *et al.* proposed a model for the role of domain walls in the engineered domain configurations. [29] Here, we consider this useful model to explain the enhanced piezoelectric property with increasing non-180° domain wall density.

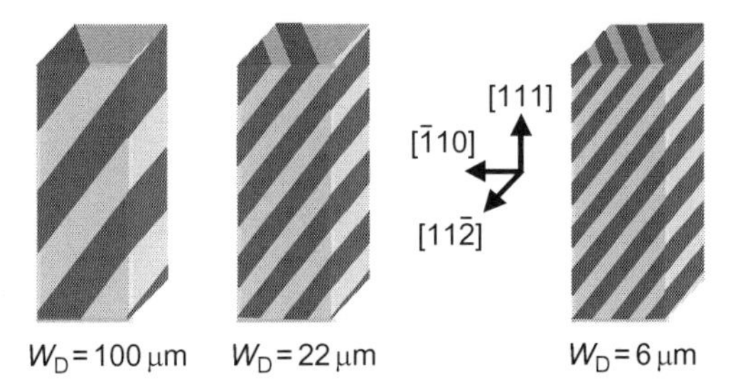

FIGURE 19 Schematic 33 resonators composed of the engineered domain configuration with different domain sizes.

TABLE 2 Domain Size Dependence of Piezoelectric Properties for the 33 $BaTiO_3$ Crystals Resonators

$BaTiO_3$ Single Crystals	ε_{33}^T	S_{33}^E (pm²/N)	d_{33} (pC/N)	k_{33} (%)
[001][a] (single-domain)	—	—	90	—
[111][b] (single-domain)	—	—	224	—
[111] neutral (domain size of 100 μm)	1984	10.6	235	54.4
[111], neutral (domain size of 60 μm)	1959	10.7	241	55.9
[111], neutral (domain size of 22 μm)	2008	8.8	256	64.7
[111], neutral (domain size of 15 μm)	2853	6.8	274	66.1
[111], neutral (domain size of 14 μm)	1962	10.8	289	66.7
[111], neutral (domain size of 6 μm)	2679	10.9	331	65.2

[a]*Measured by Zgonik* et al. *[19]*
[b]*The apparent d_{31} and d_{33} along [111] direction were calculated by using "Transformation of axes" with "d_{33} of 90 pC/N, d_{31} of −33.4 pC/N, and d_{15} of 564 pC/N" measured by Zgonik* et al. *[19], assuming that there were no domain wall contribution for piezoelectric properties.*

6 ROLE OF NON-180° DOMAIN WALL REGION ON PIEZOELECTRIC PROPERTIES

It is well known that a region near the 90° domain walls is gradually distorted to relax the strain between domains with different polar directions, as shown in Figure 20. [30–34] Moreover, for $BaTiO_3$ crystals, the crystal structure near the 90° domain walls gradually changed from normal tetragonal with *c/a* ratio of

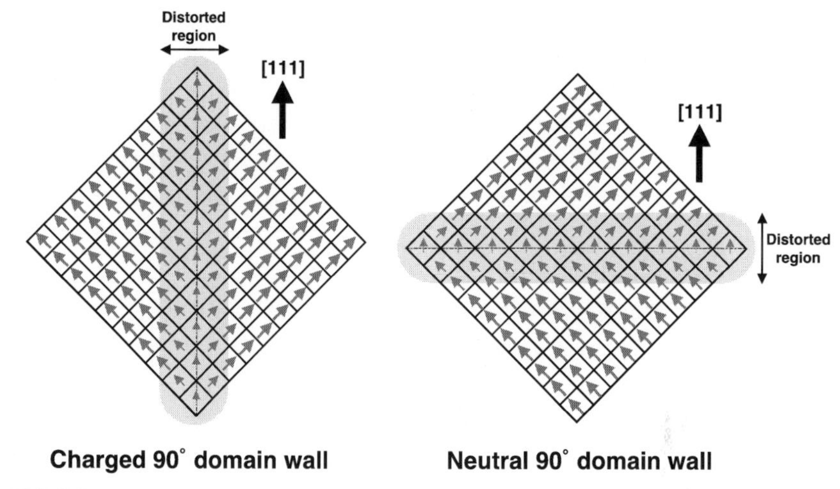

FIGURE 20 Schematic structure near the 90° domain walls for $BaTiO_3$ crystals.

1.011 to tetragonal with *c/a* ratio close to 1.0. Thus, it can be expected that the region near the 90° domain walls with pseudo-cubic structure exhibits the material constants of $BaTiO_3$ single crystals near T_c, as reported by Budimir *et al.* [34] In this study, the $BaTiO_3$ crystals with engineered domain configuration were regarded as a composite of (a) normal tetragonal region and (b) distorted 90° domain wall region. On the basis of this two-phase model, a volume fraction of the distorted domain wall region over the normal tetragonal region was estimated as follows. For a simplification of calculation, one-dimensional model was applied. [35] First, a thickness of the distorted 90° domain wall region (W_{DW}) was assumed using various sizes from 1 to 100 nm. [30–34] Next, using this thickness (W_{DW}) and the domain size (W_D), the volume fraction of the distorted 90° domain wall region (F) was estimated using the following equation,

$$F = \frac{W_{DW}}{W_D} \tag{1}$$

W_D can be measured from the experiment in this study whereas W_{DW} has unknown values. Thus, we must use valid W_{DW} values for the above calculation. To date, many researchers tried to estimate the domain wall thickness using Landau-Ginzburg-Devonshire (LGD) theories and transmission electron microscopic (TEM) observation, and their estimated values from 1 to 10 nm. [30–34] Recently, based on the developed measurement equipment and theories, some new methods were proposed to estimate the domain wall thickness. [35–37] Especially, the domain wall thickness was related to point defect, and the defect was responsible for the broadening of the domain wall thickness. [36] This means that it is very difficult to determine the 90° domain wall thickness for $BaTiO_3$. Thus, in this study, F values were calculated using the various domain wall thickness from 1 to 100 nm. Using the F values, the relationships between d_{31} and F and d_{33} and F were plotted in Figures 21 and 22, respectively, in which the slope of a line indicates the piezoelectric constant expected from the corresponding distorted 90° domain wall region. Using the estimated 90° domain wall thickness of 10 nm, d_{31} and d_{33} can be expressed using the following equations:

$$d_{31} = -82{,}676F - 69 \quad (2)$$

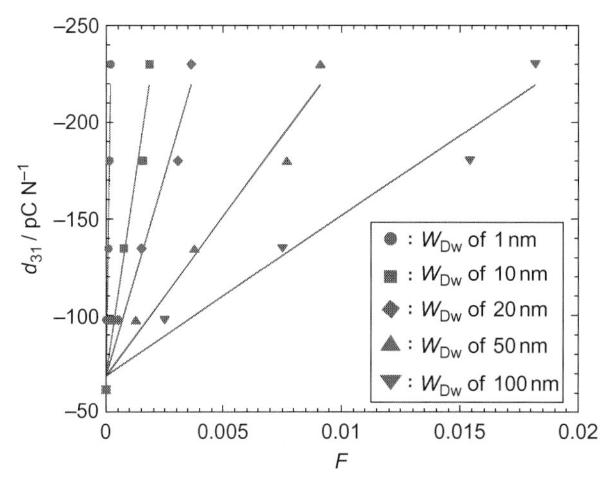

FIGURE 21 Relationship between d_{31} and F calculated using various W_{DW} values from 1 to 100 nm.

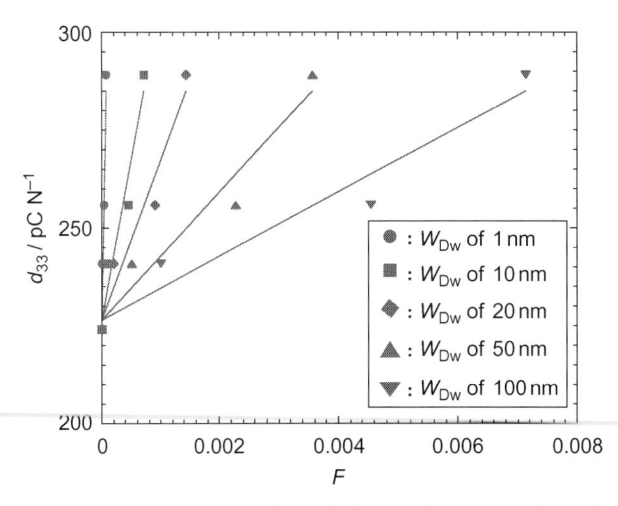

FIGURE 22 Relationship between d_{33} and F calculated using various W_{DW} values from 1 to 100 nm.

$$d_{33} = 81,744F + 227 \quad (3)$$

That is, d_{31} and d_{33} from the distorted 90° domain wall region are calculated to be 82,676 and 81,744 pC/N, respectively. Similarly, if the 90° domain wall thickness is 3 nm, d_{31} and d_{33} from the distorted 90° domain wall region are estimated to be 275,590 and 272,480 pC/N, respectively. If the 90° domain wall thickness is 1 nm, d_{31} of 826,760 and d_{33} of 817,440 can be expected from the distorted 90° domain wall region. The above values reveal that the piezoelectric constants resulting from the distorted 90° domain wall region are significantly high. Based on a recent theoretical calculation, the domain wall width is estimated to be around 3 nm [31], whereas the maximum domain wall width is found from recent experiments to be 10 nm. [38] Therefore, the d constants arising from domain wall region should higher than 80,000 pC/N.

This study shows that the distorted 90° domain walls can contribute significantly to the piezoelectric properties. When the domain wall density continues to increase, how do the piezoelectric constants change? The domain size dependences of d_{31} and d_{33} can also be expressed as follows:

$$d_{31} = \frac{-826,760}{W_D} - 69 \quad (4)$$

$$d_{33} = \frac{817,440}{W_D} + 227 \quad (5)$$

It should be noted that these Equations (4) and (5) are independent of W_{DW}. Thus, using the various W_D values, the relationships between d_{31} and W_D and between d_{33} and W_D were plotted in Figures 23 and 24, respectively. It can be seen

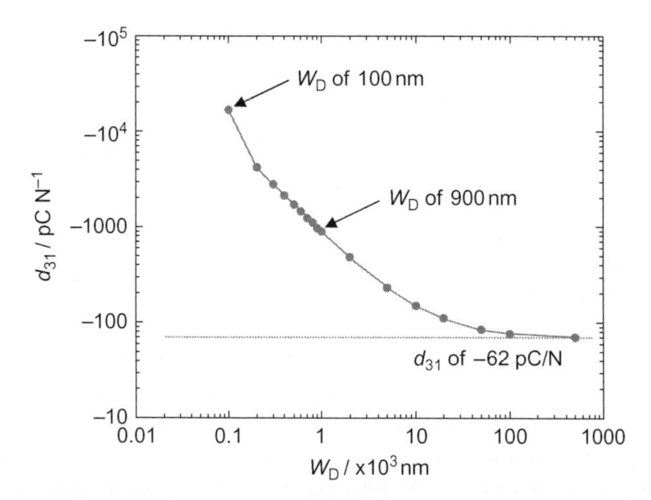

FIGURE 23 Relationship between d_{31} and W_D calculated using Equation (4).

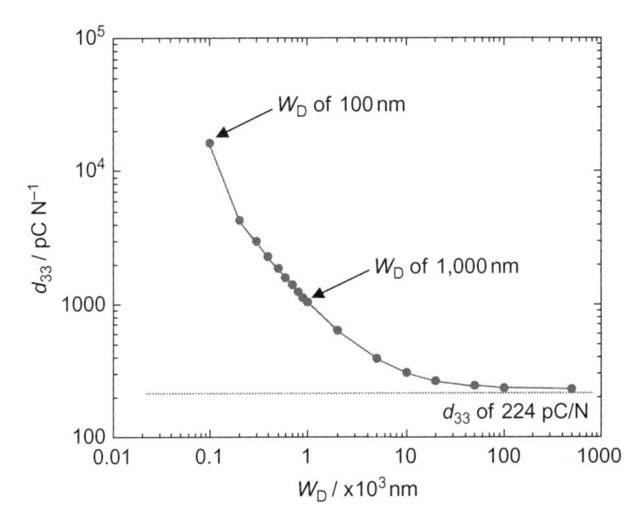

FIGURE 24 Relationship between d_{33} and W_D calculated using Equation (5).

that above the W_D of 20 μm, the piezoelectric coefficients were almost constant at the calculated single-domain values, whereas below the W_D of 10 μm, piezoelectric constants drastically increased with decreasing domain sizes. Moreover, when the domain size decreased down to 1 μm, both d_{31} and d_{33} reached around 1000 pC/N.

Park and Shrout reported that $[001]_c$ poled PZN single crystal exhibited the ultrahigh d_{33} of 1100 pC/N, [39] and the domain size in this PZN crystal was found to be around 1 μm. [10] They also reported that for $[111]_c$ poled PZN single-domain crystal had the d_{33} of just 83 pC/N [39], similar to that of $BaTiO_3$. Therefore, it can be expected that when the domain size of around 1 μm is induced in the $[111]_c$ poled $BaTiO_3$ crystals, high performance lead-free piezoelectrics with ultrahigh piezoelectric constants over 1000 pC/N can be created.

7 NEW CHALLENGE OF DOMAIN WALL ENGINEERING USING PATTERNING ELECTRODE

In the previous section, it was clarified that in $BaTiO_3$ crystals with engineered domain configuration, the piezoelectric constants significantly increased with increasing domain wall densities, that is, decreasing domain sizes. [12,13] In the poling treatments, the whole plane electrode was always used, and the minimum domain size was limited to greater than 5 μm. Therefore, it is important to establish a new poling method to induce much smaller domain sizes (below 5 μm) in the $BaTiO_3$ crystals. Recently, Urenski *et al.* reported that for the $KTiOPO_4$ single crystals, the periodic domain structure was successfully induced using patterned electrodes. [40] The objective in this section is to prepare the 31 resonators of $BaTiO_3$ crystals with a high piezoelectric constant (d_{31}) by introducing finer engineered domain configurations. For this purpose, a new poling method using a patterning electrode was proposed to induce the engineered domain configuration with smaller domain sizes below 5 μm, and the piezoelectric properties were investigated as a function of domain size.

The patterning electrode with gold strip line of 3 μm width per 6 μm spacing parallel to $[110]_c$ direction was prepared on the top surface using a photolithography technique while the whole electrode was prepared on the bottom surface. First, on the top surface of the resonator, a photoresist (Kayaku Microchem, SU-8 3000) layer with 2 μm in thickness was coated. Then, mask alignment, UV radiation and development were performed. After this process, gold electrodes were sputtered on both surfaces with an area of $1.2 \times 4.0\,mm^2$. The fine engineered domain configuration was induced by using the patterning electrodes at various electric fields (0-10 kV/cm) and temperatures (20-140°C). Figure 25 shows the desirable engineered domain configuration for the tetragonal $BaTiO_3$ single crystals. This domain configuration is composed of just two kinds of polarizations along the $[010]_c$ and $[100]_c$ directions, as shown in Figure 26. Moreover, a previous study revealed that in the engineered domain configuration, if the engineered domain configuration with average domain size of 3 μm was induced in $BaTiO_3$ crystals, the d_{31} is expected to be −337.7 pC/N using Equation (4). In general, it is known that d_{33} value can be expressed as $|2^*d_{31}|$. Thus, for a d_{31} of −337.7 pC/N, the d_{33} is expected to be 675.4 pC/N. Thus, a poling treatment was performed to induce the engineered domain configuration with a domain size of 3 μm.

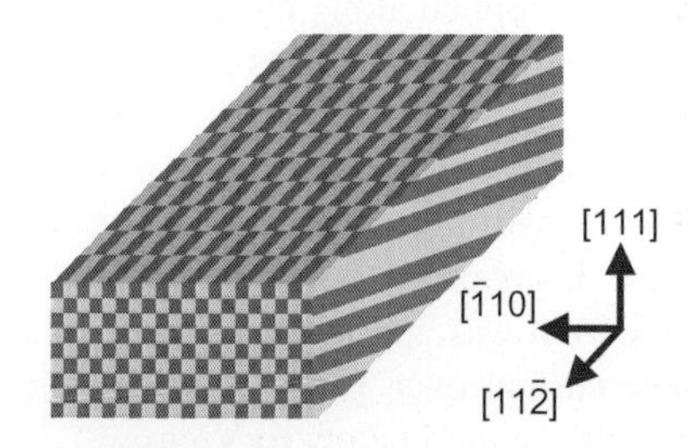

FIGURE 25 Schematic desirable finer domain configuration.

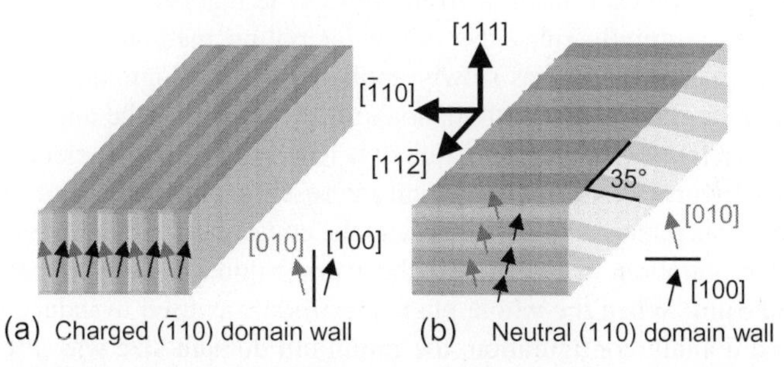

FIGURE 26 Domain configuration with (a) 90° charged and (b) 90° neutral domain walls.

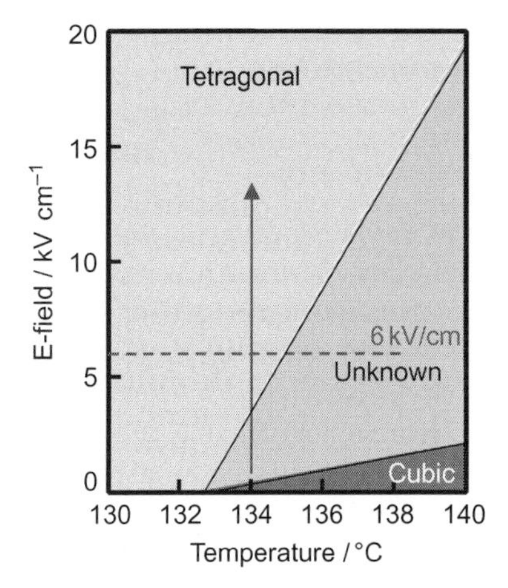

FIGURE 27 Phase transition diagram as functions of temperature and E-field for $[111]_c$ oriented $BaTiO_3$ crystals.

Figure 27 shows the temperature-electric field phase diagram used for the new poling method using patterning electrode. The phase transition temperature from tetragonal to cubic was 132.2°C for $BaTiO_3$ crystals. Thus, first, temperature increased up to 140.0°C without E-field, and the appearance of the optical isotropic state was confirmed. Then, the temperature decreased slowly down to 134.0°C, and an E-field was applied along the $[111]_c$ direction at 134.0°C. The patterning electrode prepared in this study was composed of (1) photoresist strip line of 3 μm width per 6 μm spacing and (2) gold electrode deposited on both the patterned photoresist and the crystal surface. Thus, it should be noted that in the patterning electrode, photoresist still remained between crystal surface and gold electrode. This is because this photoresist is stable up to 250°C. However, when an electric field over 6 kV/cm was applied at 134.0°C, photoresist was burned out with a large leakage current over 300 μA. Thus, in this study, the electric field applied at 134.0°C was limited below 6 kV/cm.

To induce the engineered domain configuration in $BaTiO_3$ crystals, the E-field induced phase transition from cubic to tetragonal phases above T_c of 132.2°C was required. Thus, the following poling method was applied. At 134.0°C, an electric field was slowly applied up to 2 kV/cm, and then rapidly increased up to 5.6 kV/cm. Without soaking at 5.6 kV/cm, the temperature decreased down to 50°C at a cooling rate of −10°C/min under the electric field of 5.6 kV/cm. Figure 28 shows the optical microscope photographs of patterning electrode with a pattern width of 3 μm and the consequently induced engineered domain configuration. In Figure 28b, the average domain size was estimated to be around 3 μm. When the whole plane electrode was used to induce the finer engineered domain configuration, the minimum domain size was always limited to be above 5 μm. Therefore, using the pattering electrode with a gold strip

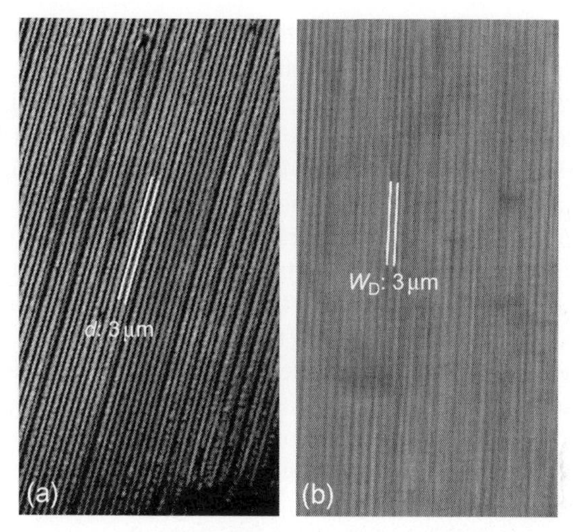

FIGURE 28 Photograph of (a) the patterning electrode with a gold electrode width of 3 μm and (b) the poled domain structure with an average domain size of 3 μm.

width of 3 μm, the engineered domain configuration with domain size of 3 μm was successfully induced in $BaTiO_3$ crystals. This reveals that in the poling treatment of $BaTiO_3$ crystals, the patterning electrode is a very useful technique.

Finally, for this 31 resonator with 3 μm domain width, the piezoelectric properties were measured using resonance-antiresonance technique. As mentioned before, a d_{31} of −337.7 pC/N was expected for the 31 resonator of the engineered domain configuration with domain width of 3 μm. [15] The measured d_{31} was −243.2 pC/N, and if regarding d_{33} as $|2*d_{31}|$, the d_{33} was estimated to be 486.4 pC/N. This measured value of d_{31} was only 70% of the expected value of −337.7 pC/N. Thus, the origin of this lower-than-expected value of d_{31} was investigated. Figure 29 shows the domain structures near the top and bottom electrodes of the 31 resonator. The domain size near the top patterning electrode (high voltage side) was estimated to be 3 μm while that near the bottom electrode (ground side) was 8-9 μm. The gradient domain sizes from 3 to 8-9 μm

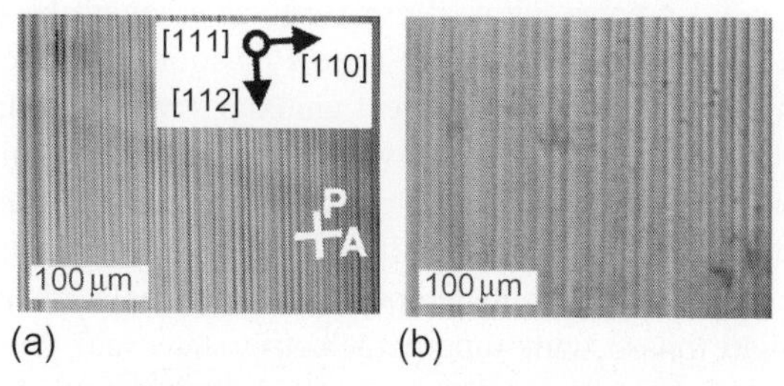

FIGURE 29 Domain structure of the 31 $BaTiO_3$ resonator near (a) the top surface (high voltage side) and (b) the bottom surface (ground side).

along thickness direction ($[111]_c$) were first observed. As mentioned previously, when the whole plane electrode was used for the poling treatment, the minimum domain size was limited to 5 μm, but there was no gradient domain size. [12,13] This difference might be originated from different E-fields. For the homogeneous domain size of 5.5 μm using the whole plane electrode, the applied E-field was 10.1 kV/cm. This E-field was almost twice as high as that of 5.6 kV/cm used in this study. Urenski *et al.* reported that for the introduction of the expected periodic domain structure similar to the patterning electrode, a much higher E-field than the coercive E-field (E_c) was required. [40] Therefore, if the high E-field over 10.1 kV/cm is applied to the 31 resonator using the patterning electrode with 3 μm width, it is expected that a homogeneous domain size of 3 μm can be induced in the resonator. At present, the patterning electrode without photoresist is designed for this purpose. In the near future, we should be able to induce much finer homogeneous domain sizes in the resonator to prepare a lead-free piezoelectrics with a higher d_{31} (over −500 pC/N). Moreover, applying uniaxial stress field can help reduce E-field using the above patterned electrodes in the poling method. This new approach will be discussed in the next section.

8 NEW CHALLENGE OF DOMAIN WALL ENGINEERING USING UNIAXIAL STRESS FIELD

It is well known that the ferroelectric phase transition is affected by temperature, E-field, and stress field. Thus, application of uniaxial stress field to poling treatment in addition to temperature and E-field may reduce the value of E-field below 10 kV/cm at above T_c. In this section, the phase transition behavior of the $[111]_c$ oriented $BaTiO_3$ crystals was investigated as functions of temperature, uniaxial stress, and E-fields. Moreover, on the basis of the results, a new poling method for the $BaTiO_3$ crystals will be proposed using control of temperature, uniaxial stress, and E-fields.

In the new poling system, temperature, E-field, and uniaxial stress field must be controlled independently. A new poling attachment was designed as shown in Figure 30. The temperature can be changed from −190 to 600°C while the E-field and uniaxial stress field are applied along the $[111]_c$ directions, independently. Because the uniaxial stress field was applied using a *z*-axis stage and a precise uniaxial pressure value cannot be measured, the moving length (μm) of the *z*-axis stage from a position at contact with sample without pressure was defined as apparent uniaxial stress field throughout this chapter. First, the phase transition behavior of the $BaTiO_3$ crystals by temperature only was investigated without E-field and uniaxial stress field by domain observation under crossed-polarizers. It was clearly observed that from temperature above T_c, with decreasing temperature, the paraelectric phase changed to an intermediate phase with superparaelectric state, and finally change to the ferroelectric phase with randomly oriented spontaneous polarization, as shown in Figure 31. The crystal structure of the intermediate phase with superparaelectric state is still unknown, but under crossed-polarizers, the crystal

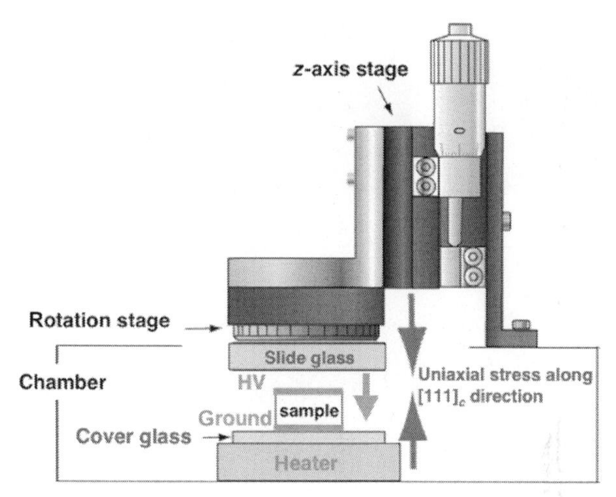

FIGURE 30 Schematic new poling attachment with controlled temperature, uniaxial stress-field and E-field.

remained always birefringent when being rotated, which suggested that this phase is not the paraelectric cubic phase. Next, above T_c, the phase transition behavior of the $BaTiO_3$ crystals induced only by E-field applied along the $[111]_c$ direction was investigated without uniaxial stress field by domain observation under crossed polarizers. With increasing E-field, the paraelectric phase changed to the intermediate phase with superparaelectric state, and finally to the ferroelectric tetragonal phase with three oriented polar directions along $[100]_c$, $[010]_c$, and $[001]_c$, as shown in Figure 32. It should be noted

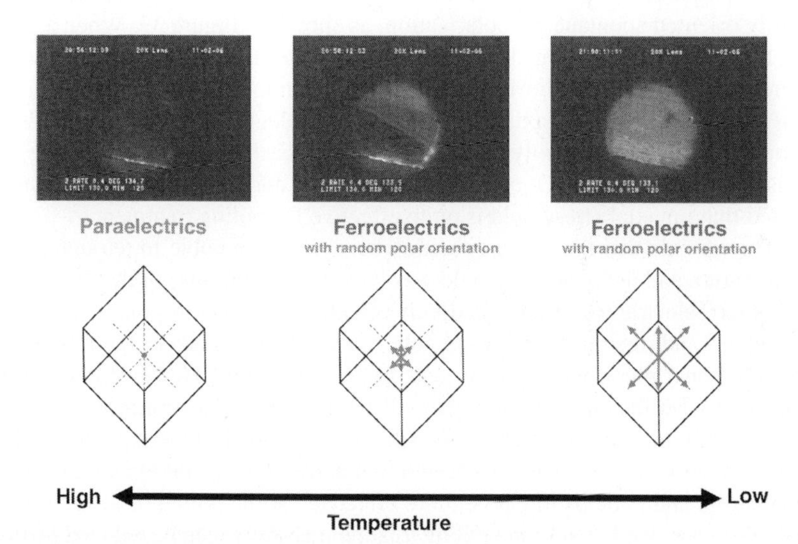

FIGURE 31 Phase transition behavior of the $BaTiO_3$ crystals by temperature without E-field and uniaxial stress-field.

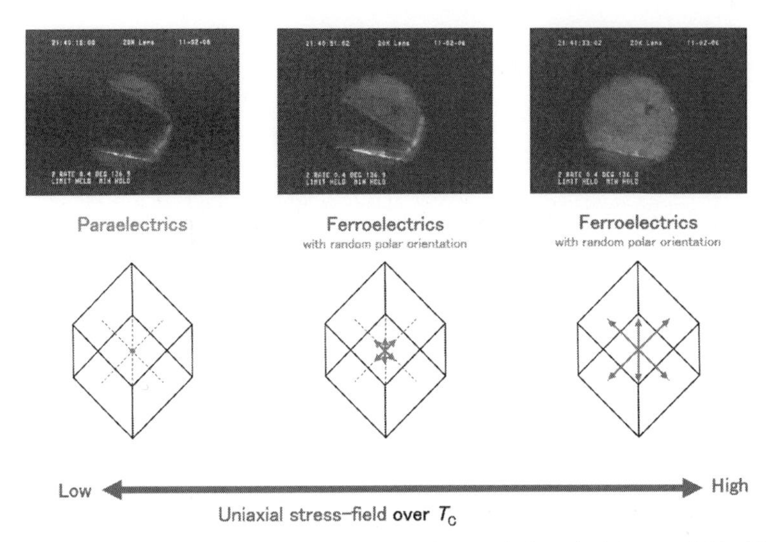

FIGURE 32 Phase transition behavior of the $BaTiO_3$ crystals by uniaxial stress-field without E-field at $(T_c + 1.5)$°C.

that at the temperature of $(T_c + 1.5)$°C, the electric field over 10 kV/cm is required to induce the tetragonal phase. Finally, above T_c, the phase transition behavior of the $BaTiO_3$ crystals induced only by uniaxial stress field applied along the $[111]_c$ direction was investigated without E-field. With increasing uniaxial stress field, the paraelectric phase changed to the intermediate phase with superparaelectric state, and finally to the ferroelectric phase with randomly oriented spontaneous polarization, as shown in Figure 33. When a poling treatment was performed using E-field, a higher E-field was required with increasing temperature above T_c. On the other hand, when a poling treatment was performed using uniaxial stress field, a smaller uniaxial stress field was enough for obtaining as fully poled state with increasing temperature above T_c. This is because the $BaTiO_3$ crystal is ferroelectric and ferroelastic, which suggested that uniaxial stress field is quite effective for poling treatment.

This reveals that the phase transition behavior from cubic to tetragonal by temperature and uniaxial stress field was quite similar because of the formation of the ferroelectric phase with randomly oriented spontaneous polarization. On the other hand, the phase transition behavior by E-field leads to the ferroelectric tetragonal phase with three oriented polar directions. However, the phase transition from the cubic to the intermediate phase with superparaelectric state induced by temperature, uniaxial stress field, and E-field shows completely the same behavior. These results suggested that above T_c, the combination of uniaxial stress and E-fields might be more effective for the poling of ferroelectric crystals. Thus, the E-field for a poling treatment above T_c can be reduced by the combination of uniaxial stress field and E-field drives.

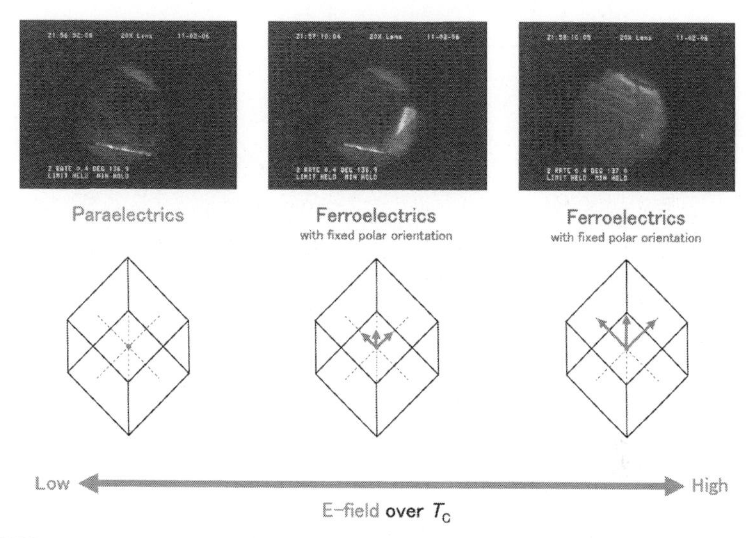

FIGURE 33 Phase transition behavior of the $BaTiO_3$ crystals by E-field without uniaxial stress-field at $(T_c + 1.5)$°C.

In this poling process, $(T_c + 1.5)$°C was chosen as the poling temperature. This is because at this temperature, it is impossible to pole the $BaTiO_3$ crystals using only E-field owing to electric breakdown. Thus, two kinds of poling treatments at $(T_c + 1.5)$°C were performed as follows: (a) a lower uniaxial stress field below 10 μm and then higher electric field above 10 kV/cm, and (b) a higher uniaxial stress field above 10 μm and then a lower electric field below 10 kV/cm. Poling treatment (a) was performed first; that is, an apparent uniaxial stress field of 9 μm was applied to induce the intermediate phase only, and after that, an electric field of 14 kV/cm was applied to induce the ferroelectric phase with the oriented polar direction. As a result, an almost fully poled state was achieved in the $[111]_c$ poled $BaTiO_3$ crystals, as shown in Figure 34. The piezoelectric properties were measured from Figure 34. Figure 35 shows the domain configuration of the $[111]_c$ poled $BaTiO_3$ crystals. An average domain size in Figure 35 was over 50 μm, and two kinds of 90° domain configuration were clearly observed. Poling treatment (b) was performed next, in which an apparent uniaxial stress field of 17 μm was applied to induce the coexistence of the intermediate and the ferroelectric tetragonal phases, and after that, an electric field of 9.5 kV/cm was applied to induce the ferroelectric phase. As a result, an almost fully poled state was achieved for the $[111]_c$ poled $BaTiO_3$ crystals as shown in Figure 36. The piezoelectric properties were measured From Figure 36. The domain configuration by the poling treatment (b) was completely the same as that shown in Figure 35. On the basis of the two impedance curves, the piezoelectric constants for the $[111]_c$ poled $BaTiO_3$ crystals were determined as shown in Table 3, when the d_{33} value was measured using a d_{33} meter. The d_{33} along the $[111]_c$ direction was calculated to be 224 pC/N for $BaTiO_3$ single-domain crystal. [41] In this

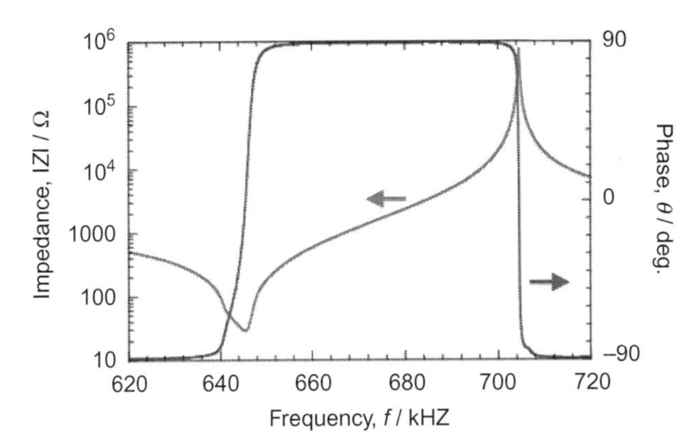

FIGURE 34 Frequency dependence of the impedance and the phase for the $[111]_c$ oriented $BaTiO_3$ single crystals poled using the poling treatment (a).

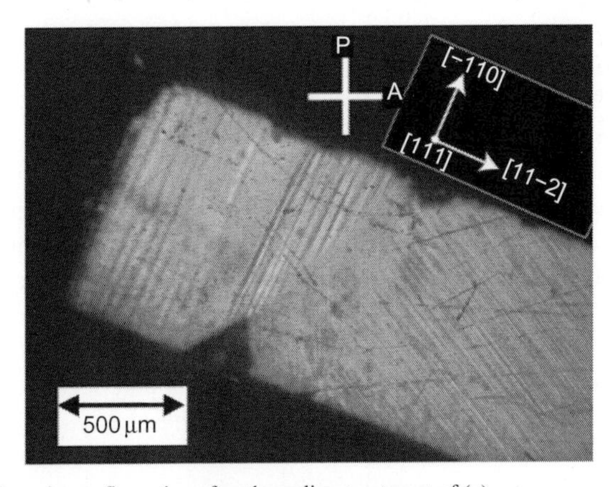

FIGURE 35 Domain configuration after the poling treatment of (a).

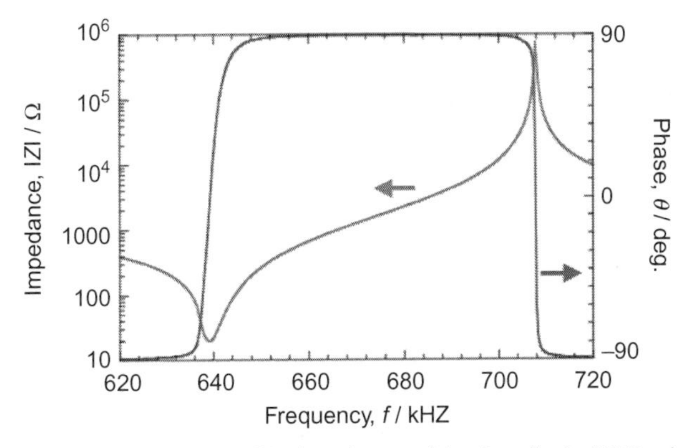

FIGURE 36 Frequency dependence of the impedance and the phase for the $[111]_c$ oriented $BaTiO_3$ single crystals poled using the poling treatment (b).

TABLE 3 Piezoelectric Properties for the $[111]_c$ Poled $BaTiO_3$ Crystals Using the Poling Treatment of (a) and (b).

Poling Treatment	e_{33}^T	S_{11}^E (pN/m²)	d_{31} (pC/N)	k_{31} (%)	d_{33}* (pC/N)
(a)	2114	8.7	−136	34	230
(b)	1983	8.6	−144	37	230

d_{33}*: measured by d_{33} meter.

study, the d_{33} was measured to be 230 pC/N, which is consistent with the previously calculated value. This revealed that the uniaxial stress field is very effective for the poling treatment above T_c to reduce the E-field below 10 kV/cm. This poling treatment can be universal for all of ferroelectric materials including PZT ceramics.

9 WHAT IS DOMAIN WALL ENGINEERING?

To enhance piezoelectric property, we must consider two contributions: intrinsic and extrinsic effects. The intrinsic effect is dependent on unit cell with symmetry and chemical composition. In the early sections, we discussed optimum crystal structure and crystallographic direction for the best engineered domain configurations. This discussion should be related to the intrinsic contribution to piezoelectric property. On the other hand, in the late sections, it was revealed that 90° domain wall regions with distorted and structure graduation had ultrahigh piezoelectric constant above 80,000 pC/N. This contribution is an extrinsic effect, and this technique should be called domain wall engineering. In this technique, we must prepare domain wall region fixed in the crystals, and engineered domain configuration is a technique to make the domain wall stable in the crystals. As a result, we can obtain a composite of domain wall regions and normal domain region.

Thus, to obtain the highest piezoelectric crystals, a combination of intrinsic and extrinsic effects is required. To maximize the intrinsic contribution, we should use the $[001]_c$ oriented orthorhombic crystals, and if this crystal has transitioned in phase from orthorhombic to tetragonal at around −10 to 0°C, we can expect more enhancement of piezoelectric properties. In addition, to maximize the extrinsic contribution, we should induce domain size of around 150 nm in the $[001]_c$ oriented orthorhombic crystals. Arlt *et al.* reported that in $BaTiO_3$ ceramics, the dielectric constant increased with decreasing grain size and at a grain size of 800 nm, dielectric constant reached a maximum value of 5000.[42] They also depicted the relationship between the grain size and domain size, and the domain size of $BaTiO_3$ ceramics with a grain size of 800 nm was 150 nm. This result suggested that for domain sizes above 150 nm, the dielectric and piezoelectric properties can increase with decreasing domain sizes, while for the domain size below 150 nm, both properties can decrease with decreasing domain sizes.

To induce the fine domain size around 150 nm, a combination of patterned electrode and uniaxial stress fields can be used in the new poling system above T_c. We consider that domain wall engineering includes a universal concept of ultrahigh domain wall contribution to piezoelectricity and any techniques to induce fine domain size of about 150 nm in the $[001]_c$ oriented orthorhombic crystals. It is expected that the development of the domain wall engineering will allow us to create new lead-free piezoelectrics with ultrahigh piezoelectric properties.

10 CONCLUSIONS AND FUTURE TRENDS

The various engineered domain configurations were induced in the $BaTiO_3$ single crystals, and their piezoelectric properties were investigated as a function of the crystal structure, crystallographic orientation, and domain size (domain wall density). As a result, the $BaTiO_3$ crystals with the orthorhombic *mm*2 phase showed the highest piezoelectric properties among three kinds of $BaTiO_3$ crystals of the tetragonal 4*mm*, orthorhombic *mm*2, and rhombohedral 3*m* phases. On the other hand, the $BaTiO_3$ crystals oriented along the $[001]_c$ direction always exhibited the larger piezoelectric properties than those oriented along the $[111]_c$ direction. In particular, the highest piezoelectric properties (d_{33} of 500 pC/N and k_{33} of 85%) were obtained for the $[001]_c$ poled orthorhombic $BaTiO_3$ crystals, and these values were much larger than those of the conventional PZT ceramics. Moreover, the domain size dependence of the piezoelectric properties was also discussed, and the result revealed that the piezoelectric properties were strongly dependent on the domain sizes (domain wall density); in other words, the piezoelectric properties significantly increased with decreasing domain size. The calculated d_{31} of the $[111]_c$ oriented tetragonal $BaTiO_3$ single-domain crystal was −62 pC/N, while the measured value of the $[111]_c$ poled tetragonal $BaTiO_3$ crystal with a domain size of 3 μm was −243.2 pC/N, that is, a fourfold increase. When the much finer domain size (below 1 μm) can be induced in the $[001]_c$ poled orthorhombic $BaTiO_3$ crystals, the significantly enhanced piezoelectric properties can be expected. However, the $BaTiO_3$ crystals have the tetragonal 4*mm* symmetry at room temperature and the desirable orthorhombic *mm*2 phase is stable below 5°C. Therefore, ferroelectric single crystals with an orthorhombic *mm*2 phase at room temperature are very desired, and the $[001]_c$ poled orthorhombic $KNbO_3$ single crystals with domain sizes below 1 μm will be one of the promising candidates for much enhanced piezoelectric properties.

ACKNOWLEDGMENTS

We would like to thank Mr. O. Nakao of Fujikura Ltd. for preparing the TSSG-grown $BaTiO_3$ single crystals with excellent chemical quality. We also would like to thank Dr. S.-E. "Eagle" Park, Dr. T. R. Shrout and Dr. L. E. Cross of MRL, Pennsylvania State University for their helpful suggestion and many discussions about the engineered domain configurations. Moreover, we would like to thank Dr. Y. Ishibashi of Aichi-shukutoku University, Dr. D. Damjanovic of EPFL, Dr. A. J. Bell of University of Leeds and Dr. L. E. Cross of Pennsylvania State University for their helpful discussions about the domain wall contribution to the piezoelectric properties. We would like to thank Dr. J. Erhart and Dr. J. Fousek of ICPR, Technical University of Liberec for their helpful discussions about the analysis of the domain configuration and calculation of the d_{31} surface. This study was partially supported by (1) a Grant-in-Aid for Scientific Research (11555164 and 16656201) from the Ministry of Education, Culture, Sports, Science, and Technology, Japan, (2) TEPCO Research Foundation, (3) the Japan Securities Scholarship Foundation, (4) Toray Science and Technology Grant, (5) the Kurata Memorial Hitachi Science and Technology Foundation, (6) the Electro-Mechanic Technology Advanced Foundation, (7) the Tokuyama Science Foundation and (8) the Yazaki Memorial Foundation for Science and Technology.

REFERENCES

[1] B. Jaffe, W.R. Cook Jr., H. Jaffe, Piezoelectric Ceramics, Academic Press, New York, 1971.

[2] K. Nakamura, H. Shimizu, Poling of ferroelectric crystals by using interdigital electrodes and its application to bulk-wave transformer, in: Proc. 1983 IEEE Ultrasonic Symp, 1983, pp. 527–530.

[3] K. Nakamura, H. Ando, H. Shimizu, Partial domain inversion in LiNbO3 plates and its applications to piezoelectric devices, in: Proc. 1986 IEEE Ultrasonic Symp, 1986, pp. 719–722.

[4] E.J. Lim, M.M. Fejer, R.L. Byer, W.J. Kozlovsky, Blue light generation by frequency doubling in periodically poled lithium niobate channel waveguide, Electron. Lett. 25 (1989) 731–732.

[5] J. Webjorn, F. Laurell, G. Arvidsson, Blue light generated by frequency doubling of laser diode light in a lithium niobate channel waveguide, IEEE Photon. Technol. Lett. 1 (1989) 316–318.

[6] Y. Hiranaga, K. Fujimoto, Y. Cho, Y. Wagatsuma, A. Onoe, K. Terabe, K. Kitamura, Ferroelectric data storage based on scanning nonlinear dielectric microscopy, Integr. Ferroelectr. 49 (2002) 203–209.

[7] Y. Cho, Y. Hiranaga, K. Fujimoto, Y. Wagatsuma, A. Ones, K. Terabe, K. Kitamura, Tbit/inch^2 ferroelectric data storage based on scanning nonlinear dielectric microscopy, Trans. Mater. Res. Soc. Jpn. 28 (2003) 109–112.

[8] K. Terabe, M. Nakamura, S. Takekawa, K. Kitamura, S. Higuchi, Y. Gotoh, Y. Cho, Nanoscale domain patterning in a stoichiometric $LiNbO_3$ crystal, Appl. Phys. Lett. 82 (2003) 433.

[9] S.-E. Park, M.L. Mulvihill, P.D. Lopath, M. Zipparo, T.R. Shrout, Crystal Growth and Ferroelectric Related Properties of $(1\text{-}x)Pb(A_{1/3}Nb_{2/3})O_3\text{-}xPbTiO_3$ ($A=Zn^{2+}$, Mg^{2+}), in: Proc. 10th IEEE Int. Symp. Applications of Ferroelectrics, Vol. 1, 1996, pp. 79–82.

[10] S. Wada, S.-E. Park, L.E. Cross, T.R. Shrout, Domain configuration and ferroelectric related properties of relaxor based single crystals, J. Korean Phys. Soc. 32 (1998) S1290–S1293.

[11] S. Wada, S.-E. Park, L.E. Cross, T.R. Shrout, Engineered domain configuration in rhombohedral PZN-PT single crystals and their ferroelectric related properties, Ferroelectrics 221 (1999) 147–155.
[12] S. Wada, T. Tsurumi, Enhanced piezoelectric property of barium titanate single crystals with engineered domain configurations, Br. Ceram. Trans. 103 (2004) 93–96.
[13] S. Wada, K. Yako, H. Kakemoto, T. Tsurumi, J. Erhart, Enhanced piezoelectric property of barium titanate single crystals with the different domain sizes, Key Eng. Mater. 269 (2004) 19–22.
[14] S.-E. Park, T.R. Shrout, Relaxor based ferroelectric single crystals for electro-mechanical actuators, Mater. Res. Innov. 1 (1997) 20–25.
[15] S.-E. Park, T.R. Shrout, Ultrahigh strain and piezoelectric behavior in relaxor based ferroelectric single crystals, J. Appl. Phys. 82 (1997) 1804–1811.
[16] J. Kuwata, K. Uchino, S. Nomura, Phase transition in the Pb(Zn1/3Nb2/3)O3-PbTiO3 system, Ferroelectrics 37 (1981) 579–582.
[17] J. Kuwata, K. Uchino, S. Nomura, Dielectric and piezoelectric properties of $0.91Pb(Zn_{1/3}Nb_{2/3})O_3$-$0.09PbTiO_3$ Single crystals, Jpn. J. Appl. Phys. 21 (1982) 1298–1302.
[18] S. Wada, H. Kakemoto, T. Tsurumi, S.-E. Park, L.E. Cross, T.R. Shrout, Enhanced ferroelectric related behaviors of ferroelectric single crystals using the domain engineering, Trans. Mater. Res. Soc. Jpn. 27 (2002) 281–286.
[19] M. Zgonik, P. Bernasconi, M. Duelli, R. Schlesser, P. Gunter, M.H. Garrett, D. Rytz, Y. Zhu, X. Wu, Dielectric, elastic, piezoelectric, electro-optic, and elasto-optic tensors of $batio_3$ crystals, Phys. Rev. B 50 (1994) 5941–5949.
[20] S. Wada, S. Suzuki, T. Noma, T. Suzuki, M. Osada, M. Kakihana, S.-E. Park, L.E. Cross, T.R. Shrout, Enhanced piezoelectric property of barium titanate single crystals with engineered domain configurations, Jpn. J. Appl. Phys. 38 (1999) 5505–5511.
[21] S.-E. Park, S. Wada, L.E. Cross, T.R. Shrout, Crystallographically engineered $BaTiO_3$ single crystals for high-performance piezoelectrics, J. Appl. Phys. 86 (1999) 2746–2750.
[22] S. Wada, T. Tsurumi, Domain configurations of ferroelectric single crystals and their piezoelectric properties, Trans. Mater. Res. Soc. Jpn. 26 (2001) 11–14.
[23] B.T. Matthias, New ferroelectric crystals, Phys. Rev. 75 (1949) 1771.
[24] J. Fousek, Permissible domain walls in ferroelectric species, Czech. J. Phys. B21 (1971) 955–968.
[25] E.E. Wahlstrom, Optical Crystallography, John Wiley and Sons, New York, 1979.
[26] S. Wada, K. Yako, T. Kiguchi, H. Kakemoto, T. Tsurumi, Enhanced piezoelectric properties of barium titanate single crystals with the different engineered domain sizes, J. Appl. Phys. 98 (2005) 014109.
[27] IEEE Standard on Piezoelectricity, American National Standard Institute, 1976.
[28] (a)J. Fousek, D.B. Litvin, L.E. Cross, Domain geometry engineering and domain average engineering of ferroics, J. Phys. Condens. Matter 13 (2001) L33–L38. .(b)J. Fousek, L.E. Cross, Open issues in application aspects of domains in ferroic materials, Ferroelectrics 293 (2003) 43–60.
[29] S.-F. Liu, S.-E. Park, L.E. Cross, T.R. Shrout, E-field dependence of piezoelectric properties of rhombohedral PZN-PT single crystal, Ferroelectrics 221 (1999) 169–174.
[30] D. Damjanovic, Rep. Prog. Phys. 61 (1998) 1267.
[31] Y. Ishibashi, E. Salje, A theory of ferroelectric 90 degree domain wall, J. Phys. Soc. Jpn. 71 (2002) 2800–2803.
[32] N. Setter, Piezoelectric materials in devices, ed. N. Setter (N. Setter, Lausanne, 2002) p.1.

[33] B. Meyer, D. Vanderbilt, Ab initio study of ferroelectric domain walls in $PbTiO_3$, Phys. Rev. B 65 (2002) 104111.

[34] M. Budimir, D. Damjanovic, N. Setter, Piezoelectric anisotropy-phase transition relations in perovskite single crystals, J. Appl. Phys. 94 (2003) 6753–6761.

[35] H. Chaib, F. Schlaphof, T. Otto, L.M. Eng, Electric and optical properties of the 90° ferroelectric domain wall in tetragonal barium titanate, J. Phys. Condens. Matter 15 (2003) 1–14.

[36] D. Shilo, G. Ravichandran, K. Bhattacharya, Investigation of twin-wall structure at the nanometer scale using atomic force microscopy, Nat. Mater. 3 (2004) 453–457.

[37] T. Tsuji, H. Ogiso, J. Akedo, S. Saito, K. Fukuda, K. Yamanaka, Evaluation of domain boundary of piezo/ferroelectric material by ultrasonic atomic force microscopy, Jpn. J. Appl. Phys. 43 (2004) 2907–2913.

[38] Y. Ishibashi, Private Communication (2004).

[39] S.-E. Park, T.R. Shrout, Characteristics of relaxor-based piezoelectric single crystals for ultrasonic transducers, IEEE Trans. Ultrason. Ferroelectr. Freq. Control 44 (1997) 1140.

[40] P. Urenski, M. Lesnykh, Y. Rosenwaks, G. Rosenman, M. Molotskii, Anisotropic domain structure of KTiOPO4 crystals, J. Appl. Phys. 90 (2003) 1950.

[41] S. Wada, K. Yako, K. Yokoo, H. Kakemoto, T. Tsurumi, Domain wall engineering in barium titanate single crystals for enhanced piezoelectric properties, Ferroelectrics 334 (2006) 17–27, in press.

[42] G. Arlt, D. Hennings, G. de With, Dielectric properties of fine-grained barium titanate ceramics, J. Appl. Phys. 58 (1985) 1619–1625.

Chapter 15

Nanostructuring of Metal Oxides in Aqueous Solutions

Yoshitake Masuda[1], Kazumi Kato[1], Tatsuki Ohji[1] and Kunihito Koumoto[2]
[1]National Institute of Advanced Industrial Science and Technology (AIST), Nagoya, Japan,
[2]Department of Applied Chemistry, Graduate School of Engineering, Nagoya University, Nagoya, Japan

1 INTRODUCTION

Metal oxides have been prepared with high-temperature annealing for a long time. A solid-state reaction was the basic technique of ceramics. Biomineralization was studied in late twentieth century. It showed that metal oxide nano-/microstructures were fabricated in animals or plants. They were formed in aqueous solutions at ordinary temperatures and pressures with common elements. Nature-guided materials processing was proposed. It was rapidly developed in the past ten years. In particular, aqueous solution processes were developed to crystallize metal oxides [1–5]. Nucleation and crystal growth were controlled by the surface condition of substrates. Nano-/micropatterns of metal oxides were fabricated on self-assembled monolayers (SAMs) [6–9]. Functional groups of SAMs effectively accelerated or suppressed nucleation and crystal growth of metal oxides. The technique also was applied to the patterning of colloidal crystals [10–14]. Essential ideas of "crystallization in aqueous solutions" and "control of nucleation and crystal growth" were proposed in these studies. Aqueous solution processes proceeded to the next stage. "Nanostructuring" or "morphology control" was proposed to control Nano-/microstructures of metal oxides; precise control of crystal growth was essential for them.

Here we report the nanostructuring or morphology control of metal oxide in aqueous solutions. They are expected as next-generation science and technology. Precise control of crystal growth can produce a variety of novel nano-/microstructures. Metal oxide devices can be formed on various substrates, including polymers and biomaterials. These devices perform well. In addition, aqueous solution processes have an important advantage in green chemistry. Metal oxide nano-/microstructures can be fabricated at ordinary temperatures and pressures with common elements. Low energy consumption and low environmental load contribute to a sustainable society. They will be the standard ceramic science in the next generation.

Green and Sustainable Manufacturing of Advanced Materials. http://dx.doi.org/10.1016/B978-0-12-411497-5.00015-1

2 NANOSTRUCTURING OF BARIUM TITANATE

2.1 Acicular Crystals [15]

Acicular barium titanate ($BaTiO_3$) particles were developed using solution systems [15]. The morphology of barium oxalate ($BaC_2O_4{\cdot}0.5H_2O$) was controlled to an acicular shape. Its phase transition to $BaTiO_3$ was achieved by introducing titanium (Ti) ions from the coprecipitated amorphous phase. Acicular $BaTiO_3$ particles have an aspect ratio as high as 18, and the particle size can be controlled by varying the growth period of $BaC_2O_4 \cdot 0.5H_2O$, which governs the size of $BaC_2O_4 \cdot 0.5H_2O$ particles. Acicular particles of crystalline $BaTiO_3$ can be used for ultrathin, multilayer ceramic capacitors (MLCCs).

MLCCs are indispensable electronic components for advanced electronic technology [16–26], but larger capacity and smaller size are needed for future electronic devices. To meet these needs, $BaTiO_3$ particles were downsized, but ferroelectric ceramics lose their ferroelectricity when their particle size is reduced, and they lose ferroelectricity entirely at a critical size [16–25]. This is known as the "size effect," and it impedes the progress of MLCCs, so a novel solution has been eagerly anticipated.

Here we propose an MLCC using acicular $BaTiO_3$ particles [27]. An ultrathin ferroelectric layer and high capacity can be realized by acicular particles with a high aspect ratio. The short side provides an ultrathin ferroelectric layer, and the large volume caused by the long side avoids the loss of ferroelectricity at the critical size. Anisotropic $BaTiO_3$ particles are thus a candidate for MLCCs. $BaTiO_3$ has, however, an isotropic cubic or tetragonal structure, and its morphology is extremely difficult to control because of its isotropic crystal faces. We focused on triclinic $BaC_2O_4 \cdot 0.5H_2O$, which has an anisotropic crystal structure; we controlled the morphology of these particles by precisely controlling crystal growth. We also achieved phase transition of $BaC_2O_4 \cdot 0.5H_2O$ to crystalline $BaTiO_3$ by introducing Ti ions from the coprecipitated amorphous phase. Having developed several key technologies, we were able to successfully produce anisotropic acicular $BaTiO_3$ particles.

Control of $BaTiO_3$ morphology to a rod shape was reported previously. In addition, metal oxalates (MC_2O_4) have been used for synthesis of rod-shaped oxides or hydroxides. Hayashi *et al.* [28] reported preparation of rod-shaped $BaTiO_3$ from rod-shaped TiO_2-nH_2O and $BaCO_3$ in molten chloride at a high temperature. Li *et al.* [29] reported preparation of nanoflakes and nanorods of $Ni(OH)_2$, $Co(OH)_2$, and Fe_3O_4 by hydrothermal conversion at 160°C for 12h from $MC_2O_4 \cdot 2H_2O$ in NaOH solutions. Sun *et al.* [30] prepared flower-like SnC_2O_4 submicrotubes in ethanol solutions containing $SnCl_2$ and oxalic acid. They were annealed at 500°C for 2h in an ambient atmosphere to obtain flower-like SnO_2 submicrotubes.

Oxalic acid (252mg) was dissolved into isopropyl alcohol (4mL) [15] (Figure 1). Butyl titanate monomer (0.122mL) was mixed with the oxalic acid solution, and the solution then was mixed with distilled water (100mL). The pH

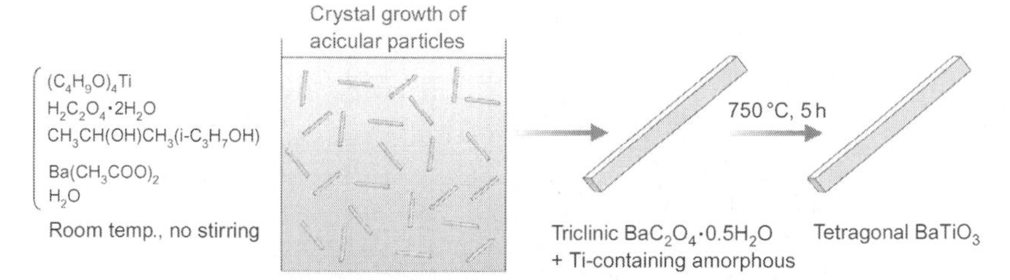

FIGURE 1 Conceptual process for fabricating acicular $BaTiO_3$ particles. Morphology control of $BaC_2O_4\cdot 0.5H_2O$ particles and phase transition to $BaTiO_3$. *(Reprinted with permission from Masuda et al. [15]. Copyright © American Chemical Society.)*

of the solution was increased to 7 by adding NaOH (1 mol/L) and distilled water; the volume of the solution was adjusted to 150 mL by these additions. The aqueous solution (50 mL) with barium acetate (39.3 mg) was mixed with the oxalic acid solution. The mixed solution containing barium acetate (0.77 mmol/L), butyl titanate monomer (2 mmol/L), and oxalic acid (10 mmol/L) was kept at room temperature for several hours with no stirring, and the solution gradually became cloudy. Stirring causes the collision of homogeneously nucleated particles and destruction of large grown particles and so was avoided in this process. The size of the precipitate was easily controlled from nanometer order to micrometer order by changing the growth period. Large particles were grown by immersion for several hours to evaluate in detail the morphology and crystallinity.

Oxalate ions ($C_2O_4^{2-}$) react with barium ions (Ba^{2+}) to form $BaC_2O_{41}\cdot 0.5H_2O$. $BaC_2O_4\cdot 0.5H_2O$ is dissolved in weak acetate acid provided by barium acetate ($(CH_3COO)_2Ba$); however, it can be deposited at pH 7, which is adjusted by adding NaOH. $BaC_2O_4\cdot 0.5H_2O$ thus was successfully precipitated from the solution.

Acicular particles were homogeneously nucleated and precipitated from the solution. They were, on average, 23 μm (ranging from 19 to 27 μm) in width and 167 μm (ranging from 144 to 189 μm) in length, giving a high aspect ratio of 7.2 (Figure 2). They had sharp edges and clear crystal faces, indicating high crystallinity. A gel-like solid also was coprecipitated from the solution as a second phase.

X-ray diffraction (XRD) patterns for the mixture of acicular particles and a gel-like solid showed sharp diffraction peaks of crystalline $BaC_2O_4\cdot 0.5H_2O$, with no additional phase. Acicular particles were crystalline $BaC_2O_4\cdot 0.5H_2O$, and the gel-like solid would be an amorphous phase.

Fortunately, $BaC_2O_4\cdot 0.5H_2O$ has a triclinic crystal structure, as shown by the model calculated from structure data [31] (Figure 2, first step in XRD), and thus anisotropic crystal growth was allowed to proceed to produce an acicular shape. Each crystal face has a different surface energy, surface nature (such as zeta potential), and surface groups. Anisotropic crystal growth is induced by minimizing the total surface energy in ideal crystal growth. In addition,

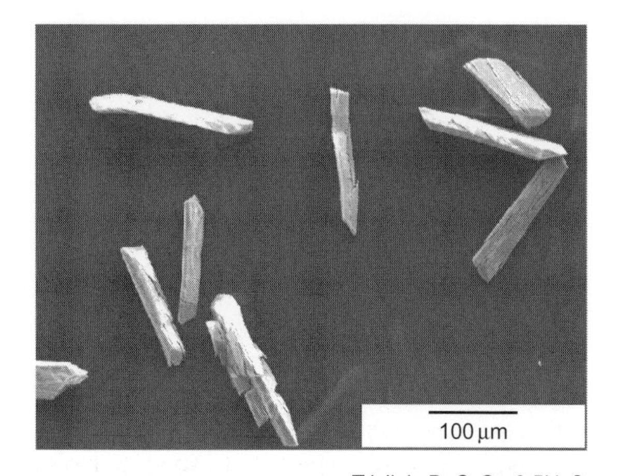

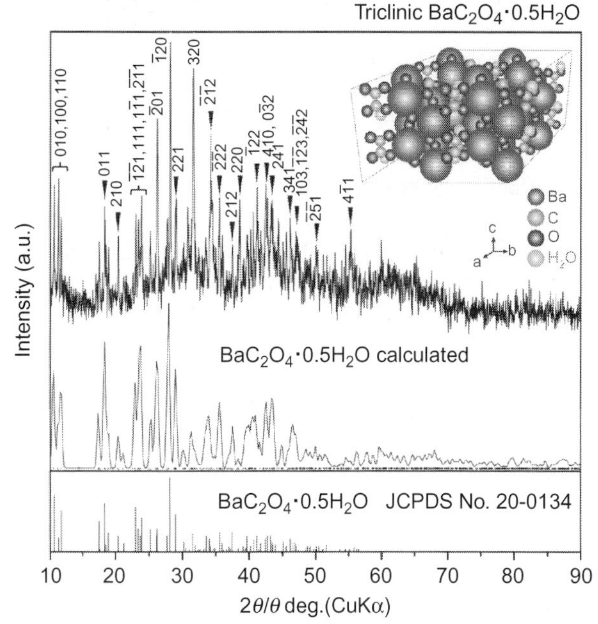

FIGURE 2 Scanning electron micrograph (top) and X-ray diffraction (XRD) patterns (bottom) of acicular $BaC_2O_4 \cdot 0.5H_2O$ particles precipitated from an aqueous solution at pH 7. XRD measurement data (first step), XRD pattern calculated from crystal structure data16 (second step), and XRD pattern of JCPDS no. 20-134 (third step) are shown for triclinic $BaC_2O_4 \cdot 0.5H_2O$. a.u., arbitrary units. *(Reprinted with permission from Masuda* et al. *[15]. Copyright © American Chemical Society.)*

site-selective adsorption of ions or molecules on specific crystal faces suppresses crystal growth perpendicular to the faces, thereby inducing anisotropic crystal growth. These factors cause anisotropic crystal growth of $BaC_2O_4 \cdot 0.5H_2O$ and hence allow us to control morphology and fabricate acicular $BaC_2O_4 \cdot 0.5H_2O$ particles. The positions of diffraction peaks corresponded with that of JCPDS no. 20-0134 (Figure 2, third step of XRD) and that calculated from crystal structure data [31] (Figure 2, second step of XRD). Several diffraction peaks, however, especially 320 and 201, were enhanced strongly compared to their relative intensity. The enhancement of diffraction intensity from specific crystal faces is related to anisotropic crystal growth; a large crystal size in a specific crystal orientation increases the XRD intensity for the crystal face perpendicular to the crystal orientation.

EDX elemental analysis indicated the chemical ratio of the precipitate, which included acicular particles and the gel-like solid, as ~1:1.5 (barium [Ba] to Ti). The chemical ratio indicated that the coprecipitated amorphous gel contained Ti ions. Additional Ba ions can be transformed into $BaCO_3$ by annealing and removed by hydrogen chloride (HCl) treatment in the next step. The ratio thus was controlled to slightly higher than a Ba-to-Ti ratio of 1 by adjusting the volume ratio of acicular particles and the gel-like solid. Consequently, acicular particles of crystalline $BaC_2O_4 \cdot 0.5H_2O$ with the gel-like solid containing Ti were successfully fabricated in an aqueous solution process.

By comparison, isotropic particles of barium titanyl oxalate ($BaTiO(C_2O_4)_2 \cdot 4H_2O$) were precipitated at pH 2. $TiOC_2O_4$ was formed by the following reaction, in which the reaction of oxalic acid ($H_2C_2O_4 \cdot 2H_2O$) with butyl titanate monomer ($(C_4H_9O)_4Ti$) and hydrolysis can take place simultaneously [32].

$$(C_4H_9O)_4 Ti + H_2C_2O_4{\cdot}2H_2O \rightarrow TiOC_2O_4 + 4C_4H_9OH + H_2O \qquad \text{(a)}$$

$TiO(C_2O_4)$ then was converted to oxalotitanic acid ($H_2TiO(C_2O_4)_2$) by the reaction

$$TiO(C_2O_4) + H_2C_2O_4 \cdot 2H_2O \rightarrow H_2TiO(C_2O_4)_2 + 2H_2O \qquad \text{(b)}$$

An alcoholic solution containing oxalotitanic acid ($H_2TiO(C_2O_4)_2$) formed by reaction (b) was subjected to the following cation exchange reaction by rapidly adding an aqueous solution of barium acetate at room temperature:

$$H_2TiO(C_2O_4)_2 + Ba(CH_3COO)_2 \rightarrow BaTiO(C_2O_4)\downarrow + 2CH_3COOH \qquad \text{(c)}$$

$BaTiO(C_2O_4)_2$ isotropic particles were formed by reaction (c).

On the other hand, neither $BaC_2O_4 \cdot 0.5H_2O$ nor $BaTiO(C_2O_4)_2$ was precipitated at a pH of 3 to 6. A gel-like solid was formed in the solution, and the XRD spectra showed no diffraction peaks. The amorphous gel that precipitated at a pH of 3 to 6 is the same as the amorphous gel coprecipitated at pH 7. These comparisons show that the crystal growth and morphology control of $BaC_2O_4 \cdot 0.5H_2O$ are sensitive to the solution conditions.

The precipitate was annealed at 750°C for 5 h in air. Acicular $BaC_2O_4 \cdot 0.5H_2O$ particles reacted with an amorphous gel containing Ti to introduce Ti ions to transform into crystalline $BaTiO_3$. XRD of the annealed precipitate showed crystalline $BaTiO_3$ and an additional barium carbonate phase ($BaCO_3$). Excess precipitation of $BaC_2O_4 \cdot 0.5H_2O$ generated a $BaCO_3$, as expected.

The annealed precipitate was further immersed in an HCl solution (1 mol/L) to dissolve $BaCO_3$. Acicular particles of crystalline $BaTiO_3$ were successfully fabricated, with no additional phase. Particles showed an acicular shape ($2.8 \times 10 \times 50\,\mu m$) and XRD of single-phase crystalline $BaTiO_3$ (Figure 3). The high aspect ratio of the particles (17.8 = 50/2.8) was provided by that of the $BaC_2O_4 \cdot 0.5H_2O$ particles. The particle size of acicular $BaTiO_3$ can be easily controlled by the growth period and the solution concentration for $BaC_2O_4 \cdot 0.5H_2O$ precipitation, which determines the size of $BaC_2O_4 \cdot 0.5H_2O$ particles.

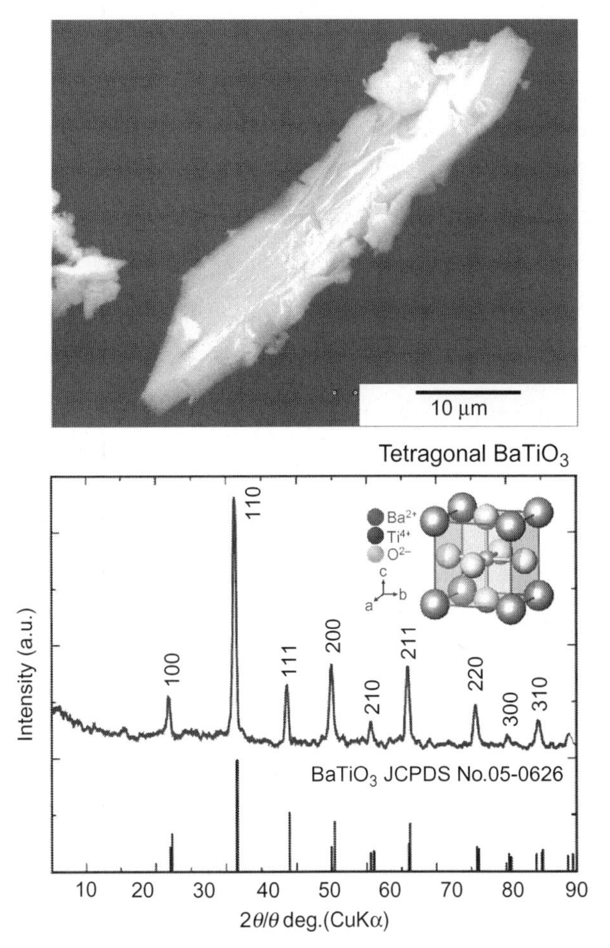

FIGURE 3 Scanning electron micrograph (top) and X-ray diffraction (XRD) pattern (bottom) of acicular $BaTiO_3$ particles after annealing at 750°C for 5 h and HCl treatment. XRD diffraction measurement data (first step) and XRD pattern of JCPDS no. 05-0626 (second step) are shown for tetragonal $BaTiO_3$. *(Reprinted with permission from Masuda* et al. *[15]. Copyright © American Chemical Society.)*

$BaTiO_3$ has a cubic crystal structure at high temperatures, above that of phase transition, and has a tetragonal crystal structure at room temperature. The cubic crystal structure is completely isotropic; the tetragonal crystal structure results from stretching a cubic lattice along one of the lattice vectors. It is difficult to control the anisotropic growth of both crystal structures; however, with our newly developed process we could successfully control the morphology and fabricate acicular particles. This was achieved by controlling the morphology of triclinic $BaC_2O_4 \cdot 0.5H_2O$ to an acicular shape and the phase transition to $BaTiO_3$ by introducing Ti ions from the coprecipitated amorphous phase. This novel concept can be applied to the control of a wide variety of morphology and crystal growth for advanced electronic devices comprising crystalline materials.

2.2 Platy Crystals, Polyhedron Crystals, and Multineedle Crystals [33]

Oxide material fabrication in solution processes has attracted much attention in recent years, both from a scientific point of view and with regard to future device applications [34–37]. Particles and films have been synthesized, and their microfabrication—such as morphology control [38–41], two-dimensional patterning [7,8,42], and three-dimensional manufacturing processes—is greatly required for next-generation devices. One of the advantages of the solution process is the use of organic-inorganic interface science; thus control of inorganic materials by organic molecules is receiving widespread scientific and industrial attention. Creatures in nature actually fabricate miniaturized, inorganic nano-/microstructures through the use of organic-inorganic interactions in aqueous solution systems [36].

Morphology control of oxide particles has many applications in future devices, especially electronic ceramic devices. MLCC [20,26] are one advanced ceramic device that requires innovative technologies for next-generation versions. $BaTiO_3$ particles were downsized to achieve miniaturization and high capacity; however, the ferroelectricity of ferroelectric ceramics is reduced when their particle size is decreased, and they lose ferroelectricity at a critical size [20]. Platy particles of $BaTiO_3$ or its precursor is a candidate technique for future ultrathin capacitor layers with ferroelectric particles larger than the critical size. $BaTiO_3$ has an isotropic cubic or tetragonal structure, and controlling its morphology is extremely difficulty because of the isotropic crystal faces. The $BaTiO_3$ precursor thus provides an advantage for morphology control. The precursor should have an anisotropic crystal structure and a 1:1 Ba-to-Ti chemical ratio with no additional metal ions to be transformed into the single phase of $BaTiO_3$.

Here we focus on the $BaTiO_3$ precursor, that is, $BaTiOF_4$ [43,44], which has an anisotropic orthorhombic structure and a Ba-to-Ti chemical ratio of 1:1, to control its morphology through the use of organic-inorganic interactions [33]. We first prepared multineedle $BaTiOF_4$ particles and then changed their morphology to a polyhedron shape by adjusting the solution's pH [45]. Organic molecules such as polyacrylamide (PAA) then were added to further control its morphology to a platy particle shape. The large, flat crystal face of polyhedron $BaTiOF_4$ was used for adsorption of PAA on the flat face to suppress crystal growth perpendicular to the flat face. Platy particles were successfully obtained by the effect of the addition of PAA from an aqueous solution [45]. $BaTiOF_4$ platy particles were transformed to a single phase of $BaTiO_3$ particles by annealing [45]. $BaTiOF_4$ platy particles, that is, the $BaTiO_3$ precursor, had a high potential for application as MLCCs. Control of particles' morphology by organic-inorganic interaction through the precise control of the growth of each crystal face contributes to the development of crystal growth science and the fabrications of next-generation devices.

PAA (powder, $[–CH_2CHCONH_2–]_n$; Cas no. 9003-05-8; Kishida Chemical Co., Ltd) is a cationic polymer with a degree of polymerization of ~125,000-140,000 (molecular weight, 900,000-1,000,000) and has a hardy viscosity in an aqueous solution. Polyvinyl alcohol (PVA; $[–CH_2CH(OH)–]_n$; degree of polymerization, 1400; Cas no. 9002-89-5; Kishida Chemical Co., Ltd) is anionic in an aqueous solution. Polyethylene glycol (PEG; $[–H(OCH_2CH_2)_nOH–]_n$; degree of polymerization, 22-24; molecular weight, 950-1050; Cas no. 25322-68-3; Kishida Chemical Co., Ltd) is nonionic in an aqueous solution. L-aspartic acid (L-Asp; $HOOCCH_2CH[NH_2]COOH$; isoelectric point, 2.77; Cas no. 56-84-8; Kishida Chemical Co., Ltd) is an acidic amino acid. Glycine (Gly; H_2NCH_2COOH; isoelectric point, 5.97; Cas no. 56-40-6; Kishida Chemical Co., Ltd) is a neutral amino acid. L-arginine (L-Arg; $NH_2C[:NH]NH[CH_2]_3CH[NH_2]COOH$; isoelectric point, 10.76; Cas Nn. 74-79-3; Kishida Chemical Co., Ltd) is a basic amino acid. Ammonium hexafluorotitanate ($[NH_4]_2TiF_6$, 99%; Mitsuwa Chemicals Co., Ltd), barium nitrate ($Ba[NO_3]_2$; Kishida Chemical Co., Ltd), boric acid (H_3BO_3; Kishida Chemical Co., Ltd), and nitric acid (HNO_3; Kishida Chemicals Co., Ltd.) were used as received.

$(NH_4)_2TiF_6$ (0.05 mol/L), $Ba(NO_3)_2$ (0.05 mol/L), and H_3BO_3 (0.15 mol/L) were dissolved in deionized water. The solution became clouded by the homogeneous nucleation of $BaTiOF_4$ particles. The solution had a pH of 2.8 and was kept at 70°C for 20 h. Deposition of $BaTiOF_4$ proceeded by the following mechanisms:

$$(NH_4)_2TiF_6 \rightarrow 2NH_4^+ + TiF_6^{2-} \tag{a}$$

$$Ba(NO_3)_2 \rightarrow Ba^{2+} + 2NO_3^- \tag{b}$$

$$TiF_6^{2-} + 2H_2O \rightleftarrows TiF_4(OH)_2^{2-} + 2H^+ + 2F^- \tag{c}$$

$$H_3BO_3 + 3H^+ + 4F^- \rightarrow BF_4^- + 3H_2O \tag{d}$$

$$Ba^{2+} + TiF_4(OH)_2^{2-} \rightarrow BaTiF_4(OH)_2 \rightarrow BaTiOF_4 + H_2O \tag{e}$$

The single phase of $BaTiOF_4$ was formed in reaction (e) from Ba^{2+} and $TiF_4(OH)_2^{2-}$. The synthesis of the $BaTiO_3$ precursor (i.e., $BaTiO_4$), which has a Ba-to-Ti chemical ratio of 1:1, is the excellent advantage of this system. It allows the single phase of $BaTiO_3$ with no additional phase to be fabricated by annealing.

XRD measurements showed that precipitates were a single phase of crystalline orthorhombic $BaTiOF_4$ [43,44] (Figure 4a). The XRD diffraction pattern corresponds to that of $BaTiOF_4$ JCPDS no. 28-0161. The intensity of JCPDS data was shown using a common logarithm to easily understand the $BaTiOF_4$ system. Three strong diffractions from precipitates were shown as a blue bar in JCPDS data. $BaTiOF_4$ particles were observed as having a multineedle shape with a 3- to 5 μm diameter upon scanning electron microscopy (SEM) evaluation (Figure 5a). A rapid crystal growth rate provides multineedle particles in this system. Ba, Ti, oxygen (O), and fluorine (F) were observed from particles, and the Ba-to-Ti chemical ratio was estimated at 1:1 by EDX analysis built into SEM. The estimation of chemical composition is consistent with XRD analysis.

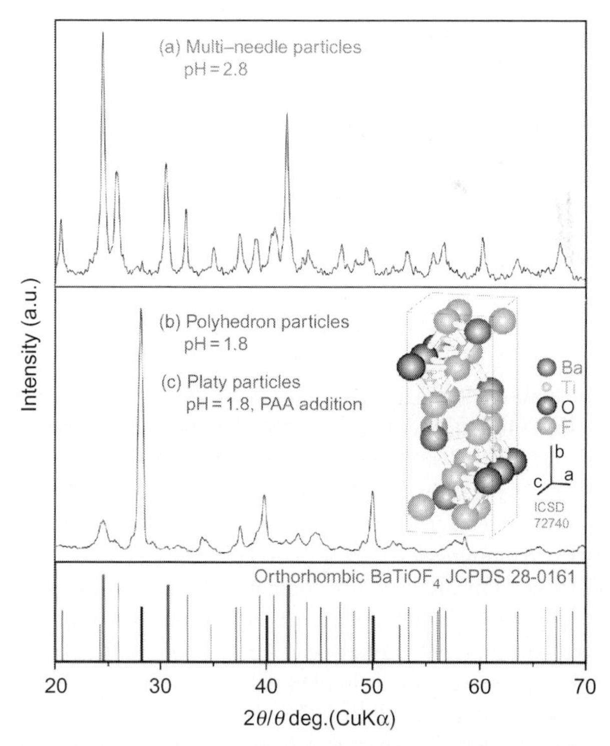

FIGURE 4 X-ray diffraction patterns of multineedle particles (a), polyhedron particles (b), and platy particles (c) of $BaTiOF_4$ and JCPDS no. 28-0161 using a common logarithm. a.u., arbitrary units; PAA, polyacrylamide. *(Reprinted with permission from Masuda* et al. *[33]. Copyright © Elsevier.)*

FIGURE 5 Scanning electron micrographs of multineedle particles (a), polyhedron particles (b), and platy particles (c1, c2) of $BaTiOF_4$. *(Reprinted with permission from Masuda* et al. *[33]. Copyright © Elsevier.)*

2.2.1 Synthesis of Polyhedron $BaTiOF_4$ Particles

We tried to synthesize $BaTiOF_4$ particles with clear, flat facets by using a slow crystal growth rate. $(NH_4)_2TiF_6$ (0.05 mol/L), $Ba(NO_3)_2$ (0.05 mol/L), and H_3BO_3 (0.15 mol/L) were dissolved in deionized water. The solution became clouded by the homogeneous nucleation of $BaTiOF_4$ particles. HNO_3 (1 mol/L) was added to the solution to completely dissolve the depositions. The solution became transparent, and the pH was changed from 2.8 to 1.8. The solution was kept at 70°C for 20 h. Deposition of $BaTiOF_4$ proceeded by the same mechanisms as those occurring for the multineedle particles. A high concentration of H^+ at a low pH (1.8) suppressed the formation of $TiF_4(OH)_2^{2-}$ in reaction (c) [4], and thus $BaTiO_4$ crystals grew slowly compared with growth at pH 2.8.

Precipitates were the single phase of crystalline orthorhombic $BaTiOF_4$ based on XRD measurement (Figure 4b). Three strong diffractions of precipitates were shown as the red bar in JCPDS data. Peak positions of XRD diffractions were the same as those of $BaTiOF_4$ JCPDS no. 28-0161; however, the relative intensity was different from that from JCPDS data. Particles had a high degree of orientation. This would be caused by the difference of preferential growth faces between rapid crystal growth at pH 2.8 and slow crystal growth at pH 1.8. SEM evaluation showed the $BaTiOF_4$ particles had a polyhedral shape with a 5- to 10-μm diameter(Figure 5b). Ba, Ti, O and F was observed from particles, and the chemical ratio was estimated to be 1 Ba:1 Ti based on EDX analysis, which is consistent with XRD analysis.

2.2.2 Synthesis of Platy $BaTiOF_4$ Particles

We tried to synthesize platy $BaTiOF_4$ particles by adding organic molecules. $(NH_4)_2TiF_6$ (0.05 mol/L), $Ba(NO_3)_2$ (0.05 mol/L), and H_3BO_3 (0.15 mol/L) were dissolved in deionized water. The solution became clouded by homogeneous nucleation of precursor particles. Nitric acid (1 mol/L) was added into the solution to completely dissolve the depositions. The solution became transparent and the pH changed from 2.8 to 1.8. Organic molecules such as PAA (0.1 g/L) was added to the solution and it was kept at 70°C for 5-20 h.

Peak positions and the relative intensity of the precipitate were the same as that of polyhedron particles based on XRD measurement (Figure 4c). The precipitate was the single phase of crystalline orthorhombic $BaTiOF_4$ with a high degree of orientation. Three strong diffractions of the precipitate were shown as a red bar in JCPDS data. A high degree of orientation could be caused by a difference in the preferential growth faces between rapid crystal growth at pH 2.8 and slow crystal growth at pH 1.8. $BaTiOF_4$ particles were observed to form a platy shape, $2.5 \times 2.5 \times 1$ to $12 \times 12 \times 4$ μm upon SEM evaluation (Figure 5c). Particle size can be controlled by the deposition period. Ba, Ti, O, and F were observed in platy particles, and the chemical ratio (Ba to Ti = 1:1) was consistent with XRD analysis.

PAA was adsorbed on a flat crystal face to suppress crystal growth perpendicular to the flat crystal face. Crystals thus grew parallel to a flat crystal face to form platy shape particles. Organic-inorganic interaction was effective for control of the morphology of particles by adsorption of organic molecules on particular crystal faces.

2.2.3 Influence of Organic Molecules on $BaTiOF_4$ Particles

Organic molecules such as PVA (0.1 g/L), PEG (0.1 g/L), L-Asp (0.05 mol/L), Gly (0.05 mol/L), or L-Arg (0.05 mol/L) was added to the solution containing $(NH_4)_2TiF_6$ (0.05 mol/L), $Ba(NO_3)_2$ (0.05 mol/L), H_3BO_3 (0.15 mol/L), and nitric acid (1 mol/L) instead of PAA for comparison. The solution was kept at 70°C for 5 h.

Platy particles could be obtained from the aqueous solution containing PAA, but not from the solution containing other organic molecules. Additions except for PAA revealed no influence on particle morphology, crystal structure, or chemical composition.

PVA and PEG are also water-soluble polymers, but they are anionic and nonionic, respectively. L-Asp, Gly, and L-Arg are acidic, neutral, and basic amino acids, respectively.

PAA has a positive zeta potential at pH 1.8; PVA, PEG, L-Asp, Gly, and L-Arg, however, have a negative zeta potential at pH 1.8. The adsorption of positive PAA on negative crystal faces probably suppresses crystal growth perpendicular to the flat crystal face to form platy particles. The sensitive combination of PAA and solution conditions, such as pH, allowed us to control crystal growth and particle morphology.

2.2.4 Phase Transition to $BaTiO_3$ by Annealing

Platy $BaTiOF_4$ particles were annealed at 600-1400°C for 3 h in air. $BaTiOF_4$ began a phase transition to $BaTiO_3$ at a temperature above 800°C, and diffraction peaks of $BaTiOF_4$ disappeared completely above 1100°C (Figure 6). XRD diffraction showed the annealed particles were a single phase of crystalline tetragonal $BaTiO_3$. High-temperature cubic $BaTiO_3$ phase transformed to low-temperature tetragonal $BaTiO_3$ phase during the cooling process before XRD measurement. $BaTiOF_4$ has a chemical ratio of 1:1 (Ba to Ti), and thus it transformed into the single phase of $BaTiO_3$, which also has the Ba-to-Ti chemical ratio 1:1. $BaTiOF_4$ has the advantages of chemical composition and particles shape for the fabrication of thin MLCC green sheets. Platy $BaTiOF_4$ particles aggregate and cannot keep their platy shape during annealing because of their phase transition. These evaluations indicate that $BaTiOF_4$ platy particles should be used for preparing thin green sheets of MLCC before high-temperature annealing. A minimum annealing temperature and duration should be used to suppress excess aggregation of platy particles for MLCC applications.

In summary, we prepared $BaTiO_4$ in an aqueous solution and successfully controlled the morphology of $BaTiOF_4$ particles using organic-inorganic interaction. Multineedle, polyhedron, and platy $BaTiOF_4$ particles were fabricated by changing the solution pH and adding PAA. Adsorption of positive PAA on negative crystal faces of $BaTiOF_4$ was used to control crystal growth. These

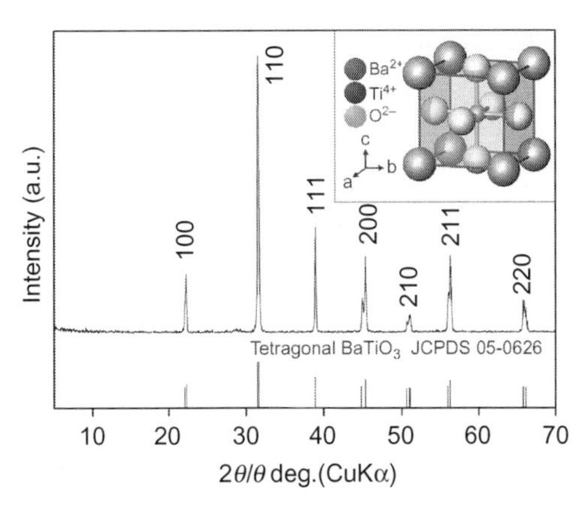

FIGURE 6 X-ray diffraction pattern of $BaTiO_3$ particles after annealing at 1100°C for 3 h. a.u., arbitrary units. Insert is a crystal structure of $BaTiO_3$. *(Reprinted with permission from Masuda* et al. *[33]. Copyright © Elsevier.)*

concepts and processes can be used to control the morphology of crystals in wide filed. $BaTiOF_4$ platy particles can be transformed into a single phase of $BaTiO_3$ and have high potential for use in an MLCC system because of their platy shape, chemical ratio, and phase transformation. Control of the morphology of crystalline materials has wide application and high potential for future materials and should be developed further based on basic scientific investigations.

3 NANOSTRUCTURING OF ZINC OXIDE

3.1 Hexagonal Cylinder Crystals, Long Ellipse Crystals, and Hexagonal Symmetry Radial Whiskers [46]

Ethylenediamine ($H_2N–CH_2CH_2–NH_2$; 15-45 mmol/L; Sigma-Aldrich) was added to a zinc acetate aqueous solution ($Zn[CH_3COO]_2$; 15 mmol/L; Kishida Chemical Co., Ltd) to promote the deposition of zinc oxide (ZnO) [46]. Zinc chelate ($Zn[H_2N–CH_2CH_2–NH_2]^{2+}$) was formed from zinc acetate and ethylenediamine in reaction (a). ZnO was crystallized from zinc chelate and hydroxide ion (OH^-) in reaction (c).

$$Zn(CH_3COO)_2 + 2H_2NCH_2CH_2NH_2 \rightleftarrows Zn(H_2NCH_2CH_2NH_2)_2^{2+} + 2CH_3COO^- \cdots \quad (a)$$

$$H_2O \rightleftarrows OH^- + H^+ \cdots \quad (b)$$

$$Zn(H_2NCH_2CH_2NH_2)_2^{2+} + 2OH^- \rightarrow ZnO + 2H_2NCH_2CH_2NH_2 + H_2O \cdots \quad (c)$$

The solution became turbid shortly after adding ethylenediamine. The molar ratio of ethylenediamine to zinc was 1:1 (a), 2:1 (b), and 3:1 (c). The pH of the solutions were 7.3, 8.0, and 8.7, respectively. Control of crystal growth rate and deposition of ZnO were attempted to change particle morphology. Silicon (Si) substrate (Newwing Co., Ltd) was immersed to evaluate deposited ZnO particles and particulate films. The solution was kept in a glass beaker at 60°C for 3 h using a water bath. The silicon substrate was cleaned before immersion, as described in the references [38,42]. The substrate was rinsed with distilled water after immersion.

ZnO particles with a hexagonal cylinder shape were homogeneously nucleated and deposited in the aqueous solution containing 15 mmol/L ethylenediamine (ethylenediamine-to-Zn ratio 1:1) (Figure 7a). XRD patterns showed the deposition to be well-crystallized ZnO (Figure 8a). The relative intensity of (10-10) and (0002) is similar to that of randomly deposited ZnO particles, indicating the random orientation of deposited ZnO hexagonal cylinders, which is consistent with SEM observations. Crystals showed hexagonal facets of about 100-200 nm in diameter and about 500 nm in length. ZnO has a hexagonal crystal structure, and thus hexagonal cylinders can be obtained by sufficiently slow

FIGURE 7 Scanning electron micrographs of ZnO hexagonal cylindrical particles (a), ZnO long ellipsoid particles (b), and ZnO hexagonal, symmetrical radial whiskers (c1). (c2) shows a magnified area of (c1). *(Reprinted with permission from Masuda* et al.*[46]. Copyright © American Scientific Publisher.)*

crystal growth. A slow crystal growth rate allows enough ions to diffuse to form a complete crystal structure.

ZnO with a hexagonal cylindrical shape showed strong photoluminescence intensity in the UV spectrum at about 370-400 nm and weak intensity in the visible light spectrum at about 530-550 nm by 350-nm excitation light, which appears in the UV spectrum (Figure 9a). ZnO crystals were reported to show UV luminescence (around 390 nm) attributed to band-edge luminescence and visible light luminescence caused by oxygen vacancy (450-600 nm) [47,48]. Oxygen vacancies are generated in ZnO during crystallization to show visible light luminescence.

The concentration of ethylenediamine was increased twice to an ethylenediamine-to-Zn ratio of 2:1. ZnO particles with a long ellipsoid shape were deposited homogeneously from the solution (Figure 7b). ZnO particles were about 100-200 nm in diameter and about 500 nm in length and were similar to those with a hexagonal cylindrical shape. XRD showed the deposition to be well crystallized ZnO (Figure 8b). Relative intensity of (10-10) diffraction is much stronger than that of (0002), indicating that mainly ZnO particles with a long ellipsoid shape were laid on the silicon substrate. This also was observed in SEM micrographs (Figure 7b). The deposition speed of ZnO with a long

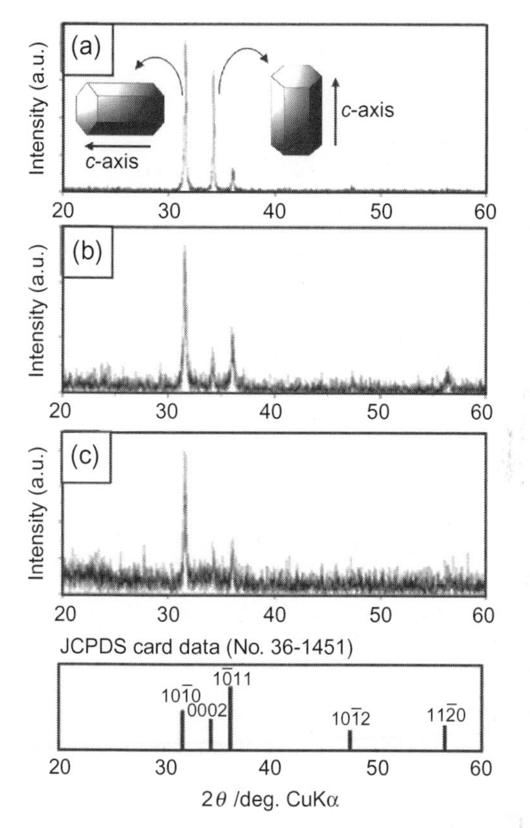

FIGURE 8 X-ray diffraction patterns of ZnO hexagonal cylindrical particles (a), ZnO long ellipsoid particles (b), and ZnO hexagonal symmetrical radial whiskers and ZnO (JCPDS no. 36-1451) (c). a.u., arbitrary units. *(Reprinted with permission from Masuda* et al. *[46]. Copyright © American Scientific Publisher.)*

ellipsoid shape was slightly faster than that of ZnO with a hexagonal cylindrical shape because of the high concentration of ethylenediamine. Ethylenediamine accelerates the crystallization of ZnO. In other words, the degree of supersaturation of the solution was increased by increasing the ethylenediamine concentration. As a result, ZnO grew slightly faster and formed not sharp hexagonal facets but rounded hexagonal cylinders, that is, a long ellipsoid shape. The photoluminescence spectrum of ZnO with a long ellipsoid shape (Figure 9b) was similar to that of ZnO with a hexagonal cylindrical shape (Figure 9a). The luminescence property was clearly stable and not influenced by the synthesis conditions at a range of ethylenediamine-to-Zn ratios of 1:1 to 2:1. High repeatability and stability of the photoluminescence property without being affected by the deposition conditions are major advantages of this system for large-scale production.

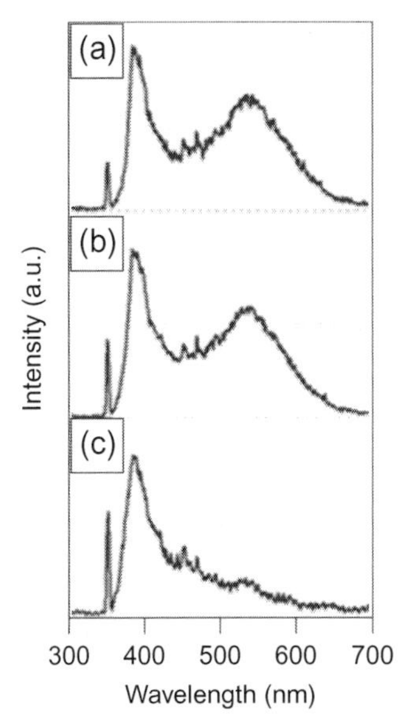

FIGURE 9 Photoluminescence spectra of ZnO hexagonal cylindrical particles (a), ZnO long ellipsoid particles (b), and ZnO hexagonal symmetrical radial whiskers (c). a.u., arbitrary units. *(Reprinted with permission from Masuda* et al. *[46]. Copyright © American Scientific Publisher.)*

Further control of morphology was attempted by increasing the degree of supersaturation. Ethylenediamine was added to the solution to obtain a ratio of 3:1 ethylenediamine to Zn in the deposition process. ZnO whiskers were successfully grown and deposited from the solution. The morphology was drastically changed by the precise investigation and control of solution conditions. The whiskers were about 10-100 nm in diameter and about 1000-2000 nm in length (Figure 7c1). A high aspect ratio was achieved by the fast growth rate of crystal end faces to elongate the whiskers. Details of morphology were further evaluated. They were dispersed in water and dropped on a silicon substrate. Whiskers were connected to form radial whiskers with a hexagonal symmetry (Figure 7c2). Six whiskers connected to form one particle. They had hexagonal symmetry. Tips of whiskers were finer than those at the center of the particles. XRD showed the whiskers to be well crystallized ZnO (Figure 8c). The relative intensity of (10-10) diffraction is much stronger than that of (0002), showing that mainly ZnO whiskers were laid on the substrate. This also was observed in SEM micrographs (Figure 7c). Deposition of ZnO whiskers with a high aspect ratio on the substrate would provide a ZnO network film with a large specific surface area. These whisker films can be applied to gas sensors [49,50] or solar cells [51] that require a large specific surface area. The whisker films also have high conductivity per unit volume [52] compared with conventional particulate films or mesoporous materials because the whiskers carry an electric current for a long distance without grain boundaries. Photoluminescence intensity in the visible light spectrum was quite different from that of ZnO with a hexagonal cylindrical shape or a long ellipsoid shape (Figure 8c). The large change in morphology indicates that the crystal growth mechanism differs greatly between ZnO particles with a hexagonal cylindrical or long ellipsoid shape and ZnO whiskers. Basically, ethylenediamine increased the crystal growth rate to generate oxygen vacancies. The oxygen vacancies increased photoluminescence intensity in the visible light spectrum. However, radial whiskers with a hexagonal symmetry prepared in an ethylenediamine/Zn solution (3:1) showed very weak photoluminescence intensity in the visible light spectrum compared with hexagonal cylinders or long ellipsoid particles. Excess ethylenediamine would decrease the photoluminescence intensity in the visible light spectrum. Ion concentration of the 3:1 ethylenediamine/Zn solution would drastically change at the initial stage. The 1:1 and 2:1 ethylenediamine/Zn solutions became turbid shortly after adding ethylenediamine and they gradually became transparent. The 3:1 ethylenediamine/Zn solution also became turbid shortly after adding ethylenediamine, but it became transparent rapidly compared with the 1:1 and 2:1 solutions. This indicates that ions were rapidly consumed to form particles in the 3:1 solution, and ion concentration would decrease drastically, with a color change. It can be assumed that a high concentration of ethylenediamine increased the crystal growth rate to form ZnO particles in the first stage. The particles generated in the first stage would not be whiskers but small ZnO particles. Ion concentration decreased rapidly to make the solution transparent by the formation of small particles because Zn ions were consumed in the formation of the particles. ZnO whiskers grow slowly on the small particles in dilute solutions at the second stage. Consequently, radial whiskers with a hexagonal symmetry were formed in the 3:1 ethylenediamine/Zn solution. This phenomenon was consistent with reported ZnO whiskers that had high crystallinity, high photoluminescence intensity in the UV spectrum, and low photoluminescence intensity in the visible spectrum. Novel properties such as unique morphology, large specific surface area, high conductivity per unit volume, low photoluminescence intensity in the visible spectrum, and high photoluminescence intensity in the UV spectrum may pave the way to a new age of ZnO devices. Furthermore, they can be fabricated on materials with little resistance to heat, such as polymers, paper, or organic materials for flexible devices.

3.2 Multineedle ZnO Crystals and Their Particulate Films [53]

ZnO has attracted much attention as a next-generation gas sensor for CO [54–56], NH_3 [57], NO_2 [58], H_2S [59], H_2 [55,60], ethanol [60,61], SF_6 [60], C_4H_{10} [60], and gasoline [60] and dye-sensitized solar cells [52,62–65]. Sensitivity directly depends on the specific surface area of the sensing material. ZnO particles, particulate films, and mesoporous material with a large specific surface area are thus strongly required.

ZnO has been crystallized to a hexagonal cylindrical shape for gas sensors or solar cells in many studies [52] using the hexagonal crystal structure of ZnO at a low degree of supersaturation. However, strategic morphology design and precise morphology control for a large specific surface area should be developed to improve the properties. ZnO particles should be controlled to have multiple needles or high surface asperity to increase the specific surface area by crystallization at a high degree of supersaturation.

Morphology control [38–41] and nano-/micromanufacturing [7,8,42,66] of oxide materials have recently been proposed in solution systems. Solution systems have the advantage of adjusting the degree of supersaturation and high uniformity in the system for particle morphology control. However, many factors affect the system compared with gas phase systems or solid-state reactions. Solution chemistry for oxide materials is therefore being developed, and many areas remain to be explored.

Control over the morphology of ZnO particles to a hexagonal cylindrical shape, an ellipsoid shape, and a multineedle shape was recently developed [38]. Photoluminescence was improved by changing the morphology and oxygen vacancy volume in this system. Control of ZnO morphology also has been proposed based on control of crystal growth [67–70]. Peng *et al.* [67] reported flower-like bunches synthesized on indium-doped tin oxide (SnO) glass substrates through a chemical bath deposition process. Lin *et al.* [68] fabricated nanowires on a ZnO-buffered Si substrate using a hydrothermal method. Zhang *et al.* [69] prepared flower-like, disc-like, and dumbbell-like ZnO microcrystals through a capping molecule-assisted hydrothermal process. Liu *et al.* [70] reported a hierarchical polygon prismatic Zn-ZnO core-shell structure grown on Si by combining liquid-solution colloidals together with the vapor-gas growth process. These studies showed highly controllable morphology of ZnO; however, morphology should be optimized to have a large specific surface area to apply to solar cells or gas sensors.

In the design of ZnO particle morphology was proposed for solar cells or gas sensors in which a large specific surface area, high electrical conductivity, and high mechanical strength are required. Multineedle ZnO particles with an ultrafine surface relief structure, as well as particulate films constructed from multineedle particles and thin sheets, were fabricated [53]. Morphology was controlled based on a new idea inspired by the morphology change in our former study [38]. A high degree of supersaturation of the solution was used for fast crystal growth, which induces the formation of multineedle particles, and low supersaturation was used for the formation of ZnO thin sheets.

Zinc nitrate hexahydrate ($Zn(NO_3)_2 \cdot 6H_2O$; $>99.0\%$; molecular weight, 297.49 Kanto Chemical Co., Inc.) and ethylenediamine ($H_2NCH_2CH_2NH_2$; $>99.0\%$; molecular weight, 60.10; Kanto Chemical Co., Inc.) were used as received. Glass (S-1225; Matsunami Glass Ind., Ltd) was used as a substrate. Zinc nitrate hexahydrate (15 mmol/L) was dissolved in distilled water at 60°C, and ethylenediamine (15 mmol/L) was added to the solution to induce the formation of ZnO. The glass substrate was immersed in the middle of the solution at an angle, and the solution was kept at 60°C using a water bath for 80 min with no stirring. The solution became clouded shortly after the addition of ethylenediamine. Ethylenediamine plays an essential role in the formation of crystalline ZnO. ZnO was homogeneously nucleated and grown to form a large amount of particles to make the solution cloudy. ZnO particles were gradually deposited and further grown on a substrate. Homogeneously nucleated particles precipitated gradually, and the solution became light white after 80 min. The degree of supersaturation of the solution was high at the initial stage of the reaction for the first 1 h and decreased as the color of the solution changed.

ZnO particulate films constructed from ZnO particles and thin sheets were fabricated by immersion for 48 h. The glass substrate was immersed in the middle of the solution at an angle, and the solution was kept at 60°C using a water bath for 6 h with no stirring. The solution then was left to cool for 42 h in the bath. The solution became cloudy shortly after the addition of ethylenediamine and clear after 6 h. The bottom of the solution was covered with white precipitate after 6 h. The degree of supersaturation of the solution was high at the initial stage of the reaction for the first 1 h and then decreased as the color of the solution changed.

Morphology of ZnO particles and particulate films was observed by a field emission SEM (JSM-6335FM; JEOL Ltd) after heating at 150°C for 30 min in a vacuum to dry the carbon paste (vacuum oven, VOS-201SD, EYELA; Tokyo Rikakikai Co., Ltd) and platinum coating for 3 nm (Quick Cool Coater, SC-701MCY; Sanyu Electronic Company). Crystal phases were evaluated by an X-ray diffractometer (RINT-2100V; Rigaku) with CuKα radiation (40 kV, 40 mA).

3.2.1 Design of ZnO Particle and Particulate Film Morphology

The morphology of ZnO particles was designed to increase the specific surface area, electrical conductivity, and mechanical strength of the base material of solar cells and sensors (Figure 10). Typical ZnO particles grown at a low degree of supersaturation are shown in Figure 10a [38]. The particles show edged hexagonal faces and elongate parallel to the *c*-axis. Because of the hexagonal crystal structure of ZnO, ZnO particles grow slowly to have a hexagonal cylindrical shape. The morphology of ZnO particles was controlled to have a multineedle shape in an aqueous solution [38] (Figure 10b). Multineedle particles have a large specific surface area compared with hexagonal cylindrical particles, but particulate films constructed from small particles have many grain boundaries, which reduce the electrical conductivity (Figure 10c). Particles should thus have a large grain size to decrease grain boundaries and increase electrical conductivity (Figure 10d). Furthermore, the specific surface area of ZnO particles should be increased to improve the sensing performance of sensors or generate efficient solar cells (Figure 10e). An ultrafine surface relief structure on ZnO particles

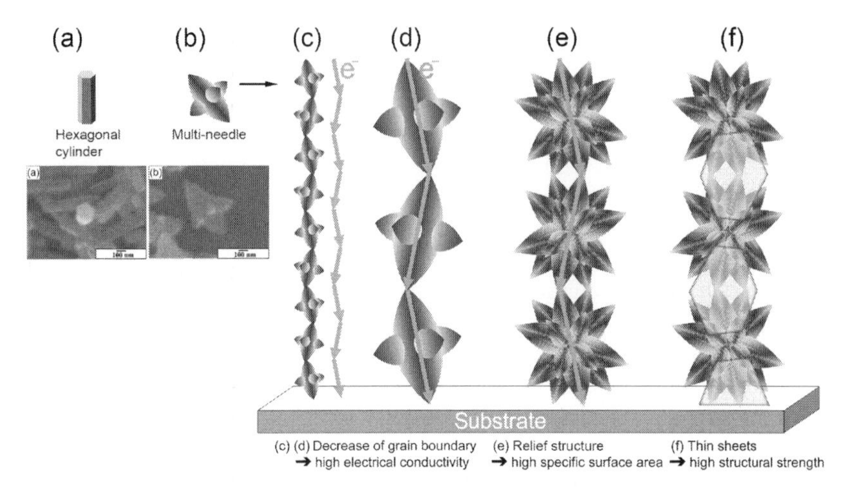

FIGURE 10 Design and morphology control of ZnO particles and particulate films for a large specific surface area, high electrical conductivity, and high mechanical strength. *(Reprinted with permission from Masuda and Kato [53]. Copyright © American Chemical Society.)*

is a candidate morphology for increasing the specific surface area. In addition, particles should be connected to each other and with a substrate by a combined member such as ZnO thin sheets (Figure 10f). The thin sheets increase the mechanical strength of particulate films and help increase the electrical conductivity and specific surface area. Control of ZnO particle morphology was attempted based on the strategic morphology design for application to sensors.

3.2.2 Control of ZnO Particle Morphology

After being immersed in the solution for 80 min, the substrate covered with ZnO particles was evaluated by SEM and XRD. ZnO particles had a multineedle shape in which many needles grew from the center of the particles (Figure 11). The particles have more needles compared with previously reported particles, which were constructed from two large needles and several small needles [38]. The size of particles was in the range of 1-5 μm, which is larger than the particles prepared previously [38] (Figure 10b). Needles were constructed from an assembly of narrow acicular crystals, and thus the side surfaces of the needles were covered with arrays of pleats. The tips of the needles had a rounded *V*-shape with many asperities. Edged hexagonal shapes were observed at the tips of needles, clearly showing high crystallinity and the direction of the *c*-axis. The *c*-axis is the long direction of multineedles and narrow acicular crystals. Elongation of the *c*-axis observed by SEM is consistent with a high diffraction intensity of (0002) (Figure 12). The (0002) diffraction intensity of multineedle ZnO particles was much stronger than $(10\bar{1}0)$ or $(10\bar{1}1)$ peaks, though (0002) diffraction is weaker than $(10\bar{1}0)$ or $(10\bar{1}1)$ diffractions in randomly oriented ZnO particles (JCPDS no. 36-1451). High diffraction intensity from (0002) planes that are perpendicular to the *c*-axis was caused by the crystalline ZnO

FIGURE 11 Scanning electron micrographs (a–d) of multineedle ZnO particles with an ultrafine surface relief structure. *(Reprinted with permission from Masuda and Kato [53]. Copyright © American Chemical Society.)*

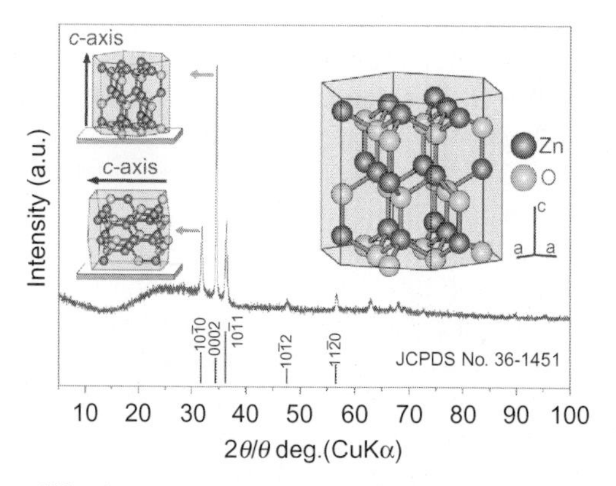

FIGURE 12 X-ray diffraction pattern of multineedle ZnO particles with an ultrafine surface relief structure. a.u., arbitrary units. Right insert is a crystal structure of ZnO. Left top insert is a crystal structure in which *c*-axis is perpendicular to a substrate. X-ray diffraction from the crystals contributes 0002 peak. Left bottom insert is a crystal structure in which *c*-axis is parallel to a substrate. X-ray diffraction from the crystals contributes 10–10 peak. *(Reprinted with permission from Masuda and Kato [53]. Copyright © American Chemical Society.)*

particles that grew to elongate the *c*-axis. The particles have more stacks of (0002) crystal planes compared with that of $(10\bar{1}0)$ planes, which are parallel to the *c*-axis, or $(10\bar{1}1)$ planes, and hence the intensity from (0002) planes was stronger than that from $(10\bar{1}0)$ or $(10\bar{1}1)$ planes.

ZnO grows into a hexagonal cylindrical shape at a low degree of supersaturation because of its hexagonal crystal structure; however, ZnO grows into a multineedle shape at a high degree of supersaturation, which induces fast crystal growth. ZnO thus grew into a multineedle shape in our solution in spite of its hexagonal crystal structure. The growth of ZnO was halted by a rapid decrease in supersaturation and removal of particles from the solution to obtain ZnO multineedle particles with an ultrafine surface relief structure. The morphology of the ZnO particles was controlled by the fast crystal growth as a result of the high degree of supersaturation and by the suppression of crystal growth caused by the rapid decrease of supersaturation and removal of particles from the solution.

3.2.3 Control of ZnO Particulate Film Morphology

ZnO particulate films showed a multineedle shape and were connected to each other by thin sheets (Figure 13a–c). The morphology of the particles was similar to that of the particles prepared by immersion for 80 min to have a large specific surface area. Thin sheets had a thickness of 10-50 nm and width of 1-10 μm and were closely connected to particles, with no clearance. The particulate films had continuous open pores ranging from several nanometers to 10 μm in diameter. The particulate films showed XRD patterns of ZnO crystal with no additional phase (Figure 14). Diffraction peaks were very sharp, showing high crystallinity of the particulate films. The high intensity of (0002) would be caused by the elongation of multineedle particles in the direction of the *c*-axis, which increases the stacks of (0002) crystal planes.

ZnO multineedle particles with an ultrafine surface relief structure were prepared at 60°C in the white solution during the initial 80 min (Figure 10e). Supersaturation was high at the initial stage of the reaction because of the high concentration of ions. ZnO particles then were precipitated, making the bottom of the solution white and the solution itself clear. Ions were consumed to form ZnO particles, and thus the ion concentration of the solution decreased rapidly. Thin sheets were formed at 25°C in the clear solution after the formation of multineedle particles (Figure 10f). Solution temperature and supersaturation influenced precipitates. Consequently, the particulate films constructed from multineedle particles and thin sheets were successfully fabricated by the two-step growth (Figure 10f).

For comparison, assemblies of thin sheets were prepared at air-liquid interfaces of the same solution we used in this study [53,71]. XRD patterns of the sheets were assigned to ZnO. The sheets had a *c*-axis orientation parallel to the sheets, that is, an in-plane *c*-axis orientation. Transmission electron microscopy (TEM) observations showed the sheets were dense polycrystals comprising nano-sized ZnO crystals. The electron diffraction pattern showed

FIGURE 13 Scanning electron micrograph (a–d) of ZnO particulate films constructed from ZnO multi-needle particles and thin sheets. *(Reprinted with permission from Masuda and Kato [53]. Copyright © American Chemical Society.)*

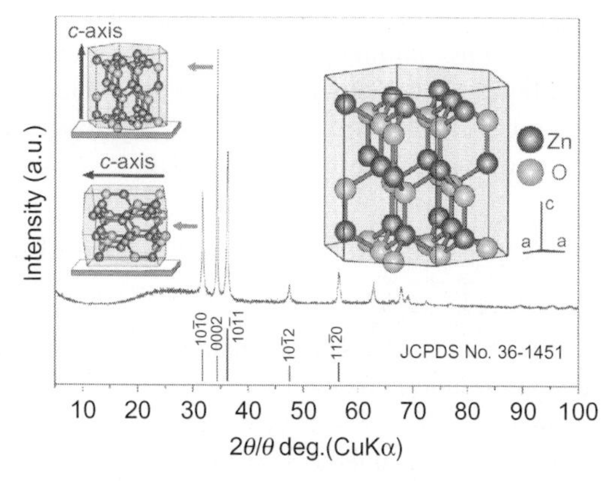

FIGURE 14 X-ray diffraction pattern of ZnO particulate films (a–d) constructed from ZnO multineedle particles and thin sheets. Right insert is a crystal structure of ZnO. Left top insert is a crystal structure in which *c*-axis is perpendicular to a substrate. X-ray diffraction from the crystals contributes 0002 peak. Left bottom insert is a crystal structure in which *c*-axis is parallel to a substrate. X-ray diffraction from the crystals contributes 10–10 peak. *(Reprinted with permission from Masuda and Kato [53]. Copyright © American Chemical Society.)*

a strong isotropic diffraction ring from (0002) planes. It suggested in-plane *c*-axis orientation of ZnO crystals, which was consistent with XRD evaluations. The mechanical strength and electrical property would be affected by crystal orientation and microstructures. The sheets would have a stronger mechanical strength compared with porous sheets because of their dense structure. They would have different electrical properties from randomly oriented sheets as a result of the in-plane *c*-axis orientation, because ZnO has anisotropic electrical properties caused by the anisotropic hexagonal crystal structure. The sheets prepared at air-liquid interfaces [71] would be similar to those prepared in the solutions (Figure 13) because both were prepared from the same solution and showed XRD patterns assigned to ZnO [71] (Figure 14). The sheets prepared in this study (Figures 13 and 14) would have mechanical and electrical properties similar to the sheets prepared at air-liquid interfaces [71].

Thin sheets were transformed to particles and porous particulate films by annealing at 500°C for 1 h in air. The sheets did not maintain their thin sheet shape because of the high slimness and/or phase transformation. Thin sheets would be inorganic films containing Zn ions, such as crystalline ZnO, amorphous ZnO, or zinc hydroxide, and were transformed to porous ZnO particulate films during annealing. Further investigation of the thin sheets would contribute to more precise control of ZnO structure morphology and further improvement of specific surface area. In addition, precise evaluation of mechanical strength and electrical properties should be performed to clarify the potential of ZnO particulate films for sensors or solar cells and to produce guidelines for improving their properties.

Multineedle ZnO particles with an ultrafine surface relief structure were successfully fabricated by the precise control of crystal growth in an aqueous solution. The morphology of ZnO was further controlled for ZnO particulate films constructed from ZnO multineedle particles and thin sheets. The thin sheets connected particles to each other and with a substrate. The morphology design and control described here will facilitate the progress of crystal science for developing future advanced materials and devices.

3.3 High *c*-Axis-Oriented, Standalone, Self-Assembled Films [71]

Crystalline ZnO [38,52,67–70,72–79] has recently been synthesized to utilize the high potential of the solution process for future devices and to realize green chemistry for a sustainable society. For instance, ZnO nanowire arrays have been synthesized using seed layers in aqueous solutions for dye-sensitized solar cells [52,72]. Full sun efficiency of 1.5% was demonstrated in this study. O'Brien *et al.* [73] prepared specular ZnO films consisting of clumps of elongated triangular crystals, small ZnO spherical clumps consisting of particles of ca. 100 nm [73], ZnO films consisting of random rod-shaped particles up to 1000 nm in length [73], and ZnO films consisting of very thin, random rod-shaped particles of ca. 1000 nm [73], and Saeed and O'Brien [79] prepared specular films consisting of flowers with well-formed triangular features in aqueous solutions.

ZnO films have usually been prepared on substrates [74,77,80–96], however; in particular, crystalline ZnO films with high *c*-axis orientation require expensive substrates such as single crystals or highly functional substrates. A simple and low-cost process for self-supporting crystalline ZnO films is expected to be used for a wide range of applications such as windows of optical devices or low-value-added products. Self-supporting crystalline ZnO films can also be applied by being pasted on a desired substrate such as polymer films with low heat resistance, glasses, metals, or paper.

This section describes highly *c*-axis-oriented, standalone ZnO self-assembled films fabricated using an air-liquid interface [71]. ZnO was crystallized from an aqueous solution without heat treatment or a catalyst. The ZnO film was fabricated at the air-liquid interface without using ammonia vapor.

Zinc nitrate hexahydrate ($Zn[NO_3]_2 \cdot 6H_2O$; > 99.0%; molecular weight, 297.49; Kanto Chemical Co., Inc.) and ethylenediamine ($H_2NCH_2CH_2NH_2$; > 99.0%; molecular weight, 60.10; Kanto Chemical Co., Inc.) were used as received. Zinc nitrate hexahydrate (15 mmol/L) was dissolved in distilled water at 60°C and ethylenediamine (15 mmol/L) was added to the solution to induce the formation of ZnO. The solution was kept at 60°C in a water bath for 6 h with no stirring. The solution then was left in the bath to cool for 42 h. Polyethylene terephthalate (PET) film, glass (S-1225; Matsunami Glass Ind., Ltd), and an Si wafer (p-type Si [97]; NK Platz Co., Ltd.) were used as substrates.

Morphology of the ZnO film was observed by a field emission SEM (JSM-6335FM; JEOL Ltd) and a transmission electron microscope (H-9000UHR, 300 kV; Hitachi). The crystal phase was evaluated by an X-ray diffractometer (RINT-2100V; Rigaku) with CuKα radiation (40 kV, 40 mA). An Si wafer was used as the substrate for XRD evaluation. The crystal structure model and diffraction pattern of ZnO were calculated from Inorganic Crystal Structure Database (ICSD) data no. 26170 (FIZ, Karlsruhe, Germany, and National Institute of Standards and Technology, Gaithersburg, MD) using FindIt and ATOMS (Hulinks Inc.).

The solution became cloudy shortly after the addition of ethylenediamine by homogeneous nucleation and growth of ZnO particles. ZnO particles were gradually deposited to cover the bottom of the vessel, and the solution became light white after 1 h and clear after 6 h. The supersaturation of the solution was high at the initial stage of the reaction for the first 1 h and decreased as the color of the solution changed.

White films were formed at the air-liquid interface and they grew to large films. The films had sufficiently high strength to be obtained as standalone films. In addition, a film was scooped to paste onto a desired substrate such as a PET film, Si wafer, glass plate, or paper, and the pasted ZnO film then was dried to bond it to the substrate. Both sides of the film can be pasted onto a substrate, and the film physically adhered to the substrate. The film maintained its adhesion during immersion in lightly ultrasonicated water; however, it can be easily peeled off again by strong ultrasonication. The film can be handled easily from one substrate to another. It also can be strongly attached to a substrate by annealing or the addition of chemical regents such as a silane coupling agent to form chemical bonds between the film and the substrate.

The film grew to a thickness of about 5 μm after 48 h, that is, at 60°C for 6 h then left to cool for 42 h. The air side of the stand-alone film had a smooth surface over a wide area because of the flat air-liquid interface (Figure 15a1), whereas the liquid side of the film had a rough surface (Figure 15b1). The films consisting of ZnO nanosheets were clearly observed from the liquid side (Figure 15b2) and the fractured edge-on profile of the film (Figure 15c1 and c2). The nanosheets had a thickness of 5-10 nm and were 1-5 μm diameter size. They mainly grew forward to the bottom of the solution, that is, perpendicular to the air-liquid interface, such that the sheets stood perpendicular to the air-liquid interface. Thus, the liquid side of the film had many ultrafine spaces surrounded by nanosheet and had a large specific surface area. The air side of the film, on the other hand, had a flat surface that followed the flat shape of the air-liquid interface. The air-liquid interface was thus effectively used to form the flat surface of the film. This flatness contributed to the strong adhesion to substrates when pasting the film. The air side of the surface prepared for 48 h had holes 100-500 nm in diameter (Figure 15a2) and were hexagonal, rounded hexagonal, or round in shape. The air side of the surface prepared for 6 h, by contrast, had no holes on the surface. The air side of the surface was well crystallized to form a dense surface, and ZnO crystals could partially grow to a hexagonal shape because of the hexagonal crystal structure. Well-crystallized ZnO hexagons then were etched to form holes on the surface by decreasing the pH. The growth face of the film is the liquid side. ZnO nanosheets grow to form a large ZnO film because of the Zn ions supplied from the aqueous solution. Further investigation of the formation mechanism would contribute to the development of crystallography in the solution system and the creation of novel ZnO fine structures.

The film showed a very strong (0002) XRD peak of hexagonal ZnO at $2\theta=34.04°$ and a weak (0004) diffraction peak at $2\theta=72.16°$, with no other ZnO diffractions (Figure 16). The (0002) planes and (0004) planes were perpendicular to the *c*-axis, and the diffraction peak from only (0002) and (0004) planes indicates a high *c*-axis orientation of the ZnO film. The inset in Figure 16 shows that the crystal structure of hexagonal ZnO stands on a substrate to make the *c*-axis perpendicular to the substrate. Crystallite size parallel to the (0002) planes was estimated from the half-maximum full width of the (0002) peak to 43 nm. This is similar to the threshold limit value of our XRD equipment, and thus the crystallite size parallel to the (0002) planes is estimated to be ≥43 nm. Diffraction peaks from an Si substrate were observed at $2\theta=68.9°$ and $2\theta=32.43°$. Weak diffractions at $2\theta=12.5°$, 24.0°, 27.6°, 30.5°, 32.4°, and 57.6° were assigned to coprecipitated zinc carbonate hydroxide ($Zn_5[CO_3]_2[OH]_6$; JCPDS no. 19-1458).

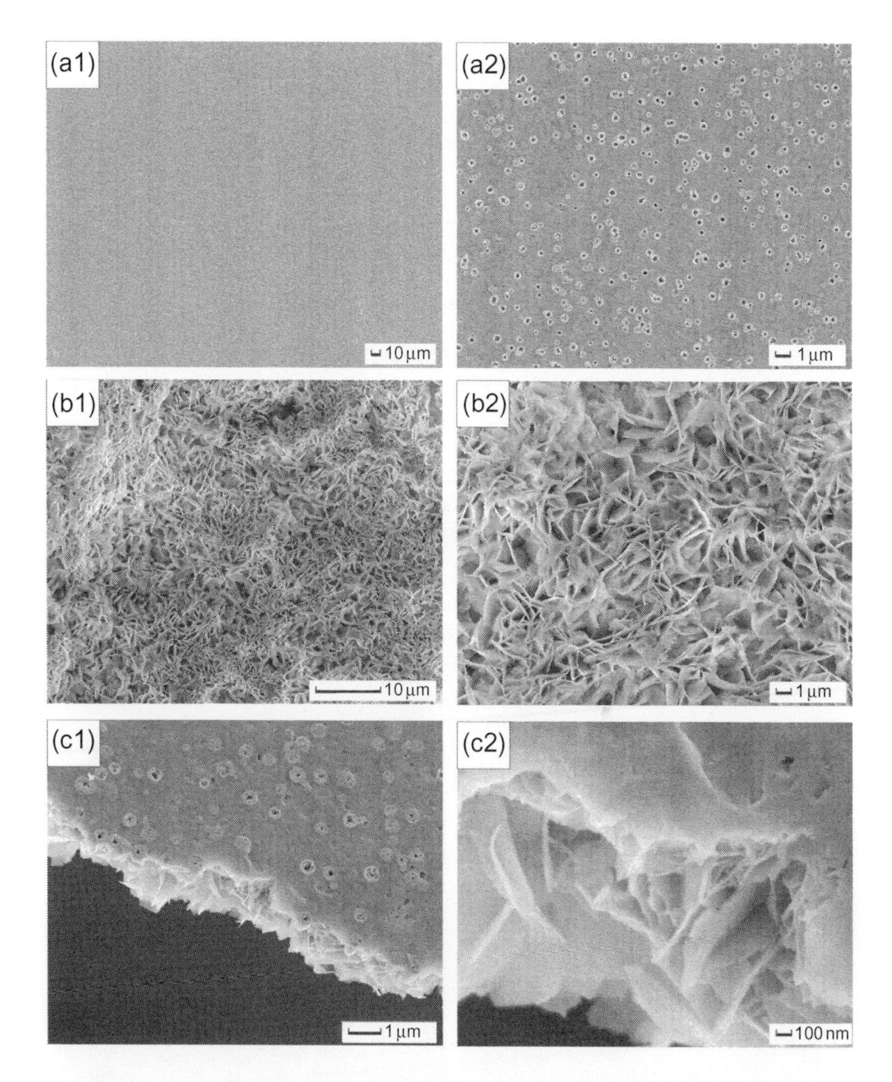

FIGURE 15 Scanning electron micrographs of highly *c*-axis-oriented, standalone, self-assembled ZnO film. (a1) The air-side surface of the ZnO film. (a2) A magnified area of (a1). (b1) The liquid-side surface of the ZnO film. (b2) A magnified area of (b2). (c1) Fracture cross section of the ZnO film from the air side. (c2) A magnified area of (c1). *(Reprinted with permission from Masuda and Kato [71]. Copyright © American Chemical Society.)*

Standalone ZnO film was further evaluated by TEM and electron diffraction. The film was crushed to sheets and dispersed in an acetone. The sheets at the air-liquid interface were skimmed by a cupper grid with a carbon supporting film. The sheets had a uniform thickness (Figure 17a) and were dense polycrystalline films constructed of ZnO nanoparticles (Figure 17b). A lattice image clearly showed the high crystallinity of the particles. The film was a

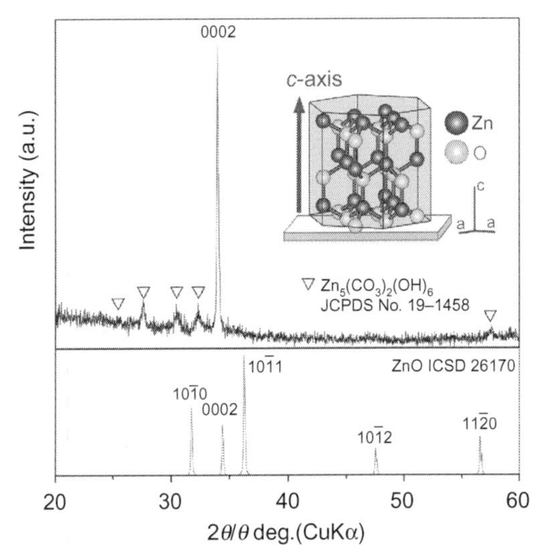

FIGURE 16 X-ray diffraction pattern of highly c-axis-oriented, standalone, self-assembled ZnO film. a.u., arbitrary units. Insert is a crystal structure of ZnO in which c-axis is perpendicular to a substrate. X-ray diffraction from the crystals contributes 0002 peak. *(Reprinted with permission from Masuda and Kato [71]. Copyright © American Chemical Society.)*

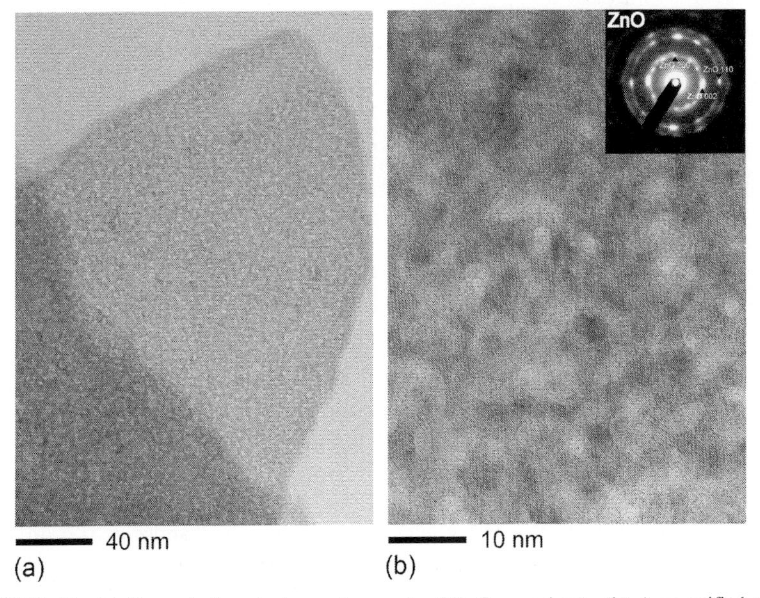

FIGURE 17 (a) Transmission electron micrograph of ZnO nanosheets. (b) A magnified area of (a). The inset shows the electron diffraction pattern of ZnO. *(Reprinted with permission from Masuda and Kato [71]. Copyright © American Chemical Society.)*

single phase of ZnO by electron diffraction pattern. These observations were consistent with XRD and SEM evaluations.

The film pasted on an Si wafer was annealed at 500°C for 1 h in air to evaluate the details of the films. The ZnO film maintained its structure during the annealing (Figure 18). The air side of the film showed a smooth surface (Figure 18a1) and the liquid side showed a relief structure with a large specific surface area (Figures 15b2 and 18b1). The air side showed the film consisted of densely packed small ZnO nanosheets, and the size of sheets increased toward the liquid side of the surface (Figure 18a2). ZnO sheets grew from the air side to the liquid side, that is, the sheets nucleated at the liquid-air interface and grew down toward the bottom of the solution because of the supply of Zn ions from the solution. Annealed film showed XRD of ZnO and the Si substrate with no additional phases. As-deposited ZnO nanosheets were crystalline ZnO because the sheets maintained their fine structure during annealing without any phase transition. High *c*-axis orientation also was maintained during annealing, showing a very strong (0002) diffraction peak.

The solution was kept at 25°C for 1 month to evaluate the details of the deposition mechanism. The film prepared at the air-liquid interface for 1 month was not hexagonal ZnO. The film showed strong XRD patterns of single-phase zinc carbonate hydroxide. ZnO dissolves as pH decreases. ZnO was crystallized

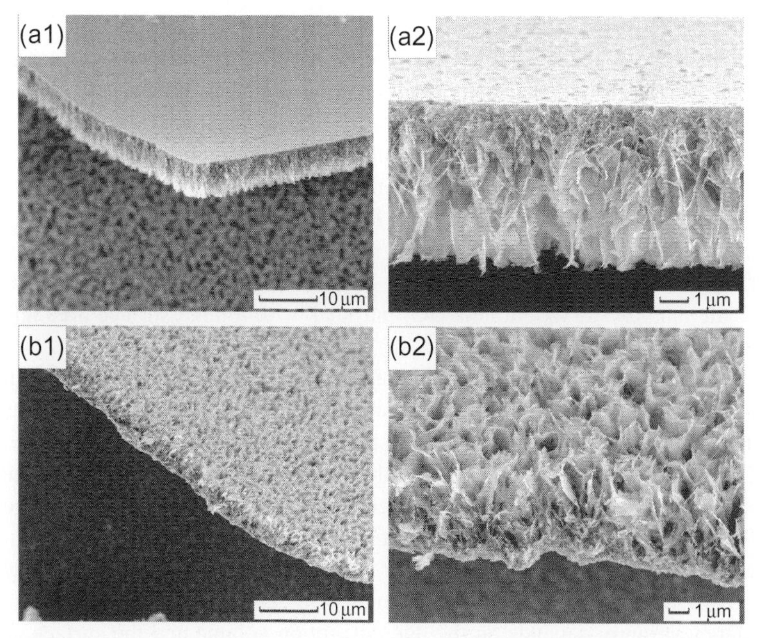

FIGURE 18 Scanning electron micrographs of highly *c*-axis-oriented, standalone, self-assembled ZnO film annealed at 500°C for 1 h in air. (a1) Fracture edge-on profile of the ZnO film from the air side. (a2) Cross-sectional profile of the ZnO film from the air side. (b1) Fracture edge-on profile of the ZnO film from the liquid side. (b2) Cross-sectional profile of the ZnO film from the liquid side. *(Reprinted with permission from Masuda and Kato [71]. Copyright © American Chemical Society.)*

at the initial reaction stage for the first 48 h then was gradually etched and dissolved by nitric acid; zinc carbonate hydroxide was crystallized using Zn ions, which were supplied by the dissolution of crystalline ZnO.

Highly *c*-axis-oriented, standalone, self-assembled ZnO film was fabricated using a simple solution process. The film consisted of ZnO nanosheets crystallized at the air-liquid interface. The nanosheets grew perpendicular to the film. The film had high *c*-axis orientation and showed a strong (0002) diffraction peak and a weak (0004) peak. The air side of the film had a flat surface, whereas the liquid side had a rough surface with many ultrafine spaces surrounded by ZnO nanosheets. The rough surface of the liquid side can be used for sensors or dye-sensitized solar cells. The ZnO film also was pasted on a desired substrate such as PET films, an Si substrate, or glass plates. The surface of a flexible polymer film with low heat resistance was modified with a highly *c*-axis-oriented crystalline ZnO film without heat treatment. This low-cost, low-temperature technique can be used for a wide range of applications including sensors, solar cells, electrical devices, and optical devices using the various properties of highly *c*-axis-oriented crystalline ZnO.

4 NANOSTRUCTURING OF TiO_2

4.1 Nanocrystal Assemblies [98]

Anatase TiO_2 particles, 100-200 nm in diameter, were developed in an aqueous solution at 50°C [98]. The particles were assemblies of nano TiO_2 crystals covered with nano-relief surface structures. The crystals grew anisotropically along the *c*-axis to form acicular crystals. The particles showed *c*-axis orientation because of high-intensity XRD from the (004) planes. The particles had a BET specific surface area of 270 m^2/g. BJH and DFT/Monte Carlo analysis of adsorption isotherm indicated the existence of pores ~3 nm and ~1 nm in diameter. Crystallization and self-assembly of nano TiO_2 were effectively used to fabricate nanocrystal assembled TiO_2 particles with a large surface area and a nano-relief surface structure.

Nanoporous TiO_2 architecture with micropores (<2 nm), mesopores (2-50 nm), and/or macropores (>50 nm) is of considerable interest for both scientific and technical applications. The latter include cosmetics, catalysts [99], photocatalysts [97,100–102], gas sensors [103,104], lithium batteries [105–107], biomolecular sensors [108], and dye-sensitized solar cells [109,110]. Crystalline anatase generally exhibits higher properties than rutile when used in photocatalysts, biomolecular sensors, and dye-sensitized solar cells. Electrons are obtained from dyes adsorbed on TiO_2 electrodes in sensors and solar cells. Photoelectric conversion efficiency strongly depends on the dye adsorption volume and the surface area of TiO_2. A large surface area is required to achieve high efficiency and sensitivity of the devices. In addition, the surface of TiO_2 should be covered with nano-/micro-relief structures to adsorb large amounts of dye, molecules, and DNA for biomolecular sensors and dye-sensitized solar cells.

TiO_2 nanoparticles have been prepared by flame synthesis [111,112], ultrasonic irradiation [113,114], chemical vapor synthesis [115], sol-gel methods [100,116–120], a sonochemical method [121], and liquid phase deposition of amorphous TiO_2 [122–125]. Use of a high temperature in the treatment processes, however, causes aggregation of nanoparticles and a decreased surface area. Formation of nano-relief structures on the surface of the particles is difficult to achieve in these processes.

Highly porous materials have been prepared via template-based methods, including soft templates (surfactants, chelating agents, block polymers, etc.) [126–129] and hard templates (porous anionic alumina, porous silica, polystyrene spheres, carbon nanotubes, etc.) [130,131]. However, the nanostructures of these materials usually change because of amorphous-phase crystallization to anatase TiO_2 during annealing. This decreases the surface area and damages the surface nanostructures.

This section described porous anatase TiO_2 particles developed in an aqueous solution. Nano-TiO_2 was crystallized in the solution to assemble into particles 100-200 nm in diameter. The surface of the particles was covered with nano-relief structures. The particles showed *c*-axis orientation because of the anisotropic crystal growth of TiO_2 along the *c*-axis. BET surface area of the particles was estimated to be 270 m^2/g [132]. BJH and DFT/Monte Carlo analysis of the adsorption isotherm indicated the existence of pores ~2.8 and ~3.6 nm in diameter, respectively. The existence of micropores ~1 nm also was indicated. Crystallization and self-assembly of acicular TiO_2 were effectively used to fabricate nanocrystal assembled TiO_2 particles with a large surface area and a nano-relief surface structure.

$(NH_4)_2TiF_6$ (FW, 197.95; purity, 96.0%; Morita Chemical Industries Co., Ltd) and H_3BO_3 (FW, 61.83; purity, 99.5%; Kishida Chemical Co., Ltd) were used as received. $(NH_4)_2TiF_6$ (12.372 g) and H_3BO_3 (11.1852 g) were separately dissolved in deionized water (600 mL) at 50°C. An H_3BO_3 solution was added to the $(NH_4)_2TiF_6$ solution at concentration of 0.15 and 0.05 mol/L, respectively.

The solution was kept at 50°C for 30 min using a water bath with no stirring, after which it was centrifuged at 4000 rpm for 10 min (Model 8920; Kubota Corp.). Preparation for centrifugation, centrifugation at 4000 rpm, deceleration from 4000 to 0 rpm, and preparation for the removal of supernatant solution took 4, 10, 10, and 8 min, respectively. The particles contacted the solution, the temperature of which was gradually lowered for 32 min after being maintained 50°C for 30 min. Precipitates were dried at 60°C for 12 h after removal of the supernatant solution.

The crystal phase of the particles was evaluated by XRD (RINT-2100V; Rigaku) with CuKα radiation (40 kV, 30 mA). Diffraction patterns were evaluated using JCPDS and ICSD data (FIZ and National Institute of Standards and Technology) and FindIt. The morphology of TiO_2 was observed by TEM (JEM2010, 200 kV; JEOL). The zeta potential and particle size distribution were measured by a electrophoretic light-scattering spectrophotometer (ELS-Z2;

Otsuka Electronics Co., Ltd) with an automatic pH titrator. Samples of 0.01 g were dispersed in distilled water (100 g) and ultrasonicated for 30 min before measurement. The pH of colloidal solutions was controlled by the addition of HCl (0.1 mol/L) or NaOH (0.1 mol/L). The zeta potential and particle size distribution were evaluated at 25°C and integrated 5 and 70 times, respectively.

Nitrogen adsorption-desorption isotherms were obtained using Autosorb-1 (Quantachrome Instruments), and samples of 0.137 g were outgassed at 110°C under 10^{-2} mm Hg for 6 h before measurement. Specific surface area was calculated by the BET method using adsorption isotherms. Pore size distribution was calculated by the BJH method using adsorption isotherms because an artificial peak was observed from the BJH size distribution calculated from desorption branches. Pore size distribution was further calculated by the DFT/Monte Carlo method (N_2 at 77 K on silica [cylinder/sphere, pore, nonlocal DFT adsorption model; adsorbent: oxygen) using adsorption branches.

Total pore volume (*V*) and average pore diameter (4*V*/*A*) were estimated from pores smaller than 230 nm (diameter) at $P/P_0=0.99$ through the use of the BET surface area (*A*). Adsorption volume increased drastically at high relative pressure in isotherms, indicating a large data error at a high relative pressure. Therefore, total pore volume and average pore diameter estimated from pores smaller than 11 nm at $P/P_0=0.80$ would provide useful information. A relative pressure P/P_0 of 0.80 was selected for estimations because it was lower than the drastic adsorption increase and higher than adsorption hysteresis. The data can be compared with those from TiO_2 with a similar morphology with small errors. Total pore volume (cumulative pore volume) also was estimated using the BJH method based on pores smaller than 154 nm. This usually has a small error compared with that estimated from isotherm data including a high relative pressure such as $P/P_0=0.99$ because estimation using the BJH method is not effected by adsorption volume errors at a high relative pressure.

4.1.1 Liquid Phase Crystal Deposition of Anatase TiO_2

The solution became cloudy about 10 min after mixing the $(NH_4)_2TiF_6$ solution and the H_3BO_3solution. The particles were homogeneously nucleated in the solution, turning it white.

Deposition of anatase TiO_2 proceeds by the following mechanisms [4, 42, 133]:

$$TiF_6^{2-} + 2H_2O \rightleftarrows TiO_2 + 4H^+ + 6F^- \quad (a)$$

$$BO_3^{3-} + 4F^- + 6H^+ \rightarrow BF_{4^-} + 3H_2O \quad (b)$$

Equation (a) is described in detail by the following two equations:

$$TiF_6^{2-} \xrightarrow{nOH^-} TiF_{6-n}(OH)_n^{2-} + nF^- \xrightarrow{(6-n)OH^-} Ti(OH)_6^{2-} + 6F^- \quad (c)$$

$$Ti(OH)_6^{2-} \rightarrow TiO_2 + 2H_2O + 2OH^- \quad (d)$$

Fluorinated titanium complex ions gradually change into titanium hydroxide complex ions in an aqueous solution, as shown in Equation (c). The increase in the F^- concentration displaces Equations (a) and (c) to the left; however, the F^- produced can be scavenged by H_3BO_3 (BO_3^{3-}), as shown in Equation (b), to displace Equations (a) and (c) to the right. Anatase TiO_2 is formed from titanium hydroxide complex ions ($Ti(OH)_6^{2-}$) in Equation (d).

4.1.2 Crystal Phase of TiO_2 Particles

XRD peaks for the particles were observed at $2\theta=25.1$, 37.9, 47.6, 54.2, 62.4, 69.3, 75.1, 82.5, and 94.0° after evaluation of N_2 adsorption. They were assigned to the 101, 004, 200, 105+211, 204, 116+220, 215, 303+224+312, and 305+321 diffraction peaks of anatase TiO_2 (JCPSD no. 21-1272, ICSD no. 9852) (Figure 19).

The 004 diffraction intensity of randomly oriented particles is usually 0.2 times the 101 diffraction intensity, as shown in JCPDS data (no. 21-1272). However, the 004 diffraction intensity of the particles deposited in our process was 0.36 times the 101 diffraction intensity. In addition, the integral intensity of the 004 diffraction was 0.18 times the 101 diffraction intensity, indicating the *c*-axis orientation of the particles. Particles were not oriented on the glass holder for XRD measurement. Therefore, TiO_2 crystals would have an anisotropic shape in which the crystals were elongated along the *c*-axis. The crystals would have a large number of stacks of *c* planes such as (001) planes compared to stacks of (101) planes. The diffraction intensity from the (004) planes would be enhanced compared with that from the (101) planes.

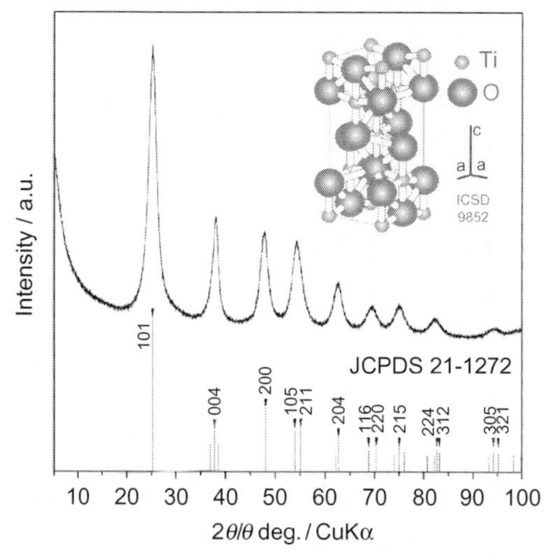

FIGURE 19 X-ray diffraction pattern of anatase TiO_2 particles. a.u., arbitrary units. *(Reprinted with permission from Masuda and Kato [98]. Copyright © American Chemical Society.)*

Crystallite size perpendicular to the (101) or (004) planes was estimated from the full-width half-maximum of the 101 or 004 peak to be 3.9 or 6.3 nm, respectively. Elongation of crystals in the *c*-axis direction also was suggested by the difference in crystallite size.

4.1.3 TEM Observation of TiO_2 Particles

The particles were shown to be assemblies of TiO_2 nanocrystals (Figure 20a). Particle diameter was estimated to be 100-200 nm. Relief structures had formed on the surfaces and open pores had formed inside because the particles were porous assemblies of nanocrystals.

Nanocrystals were shown to have acicular shapes (Figure 20b). They were about 5-10 nm in length. The longer direction of acicular TiO_2 is indicated by the black arrow. The inset FFT image shows the 101 and 004 diffractions of anatase TiO_2. Nanocrystals are assigned to the single phase of anatase TiO_2. It is notable that the diffraction from the (101) planes has a ring shape due to the random orientation, but diffraction from the (004) planes was observed only in the upper right and lower left regions of the FFT image. Anisotropic 004 diffractions indicated the direction of the *c*-axis, which was perpendicular to the (004) planes, as shown by the white arrow. It was roughly parallel to the longer direction of acicular TiO_2. These results suggest that acicular TiO_2 grew along the *c*-axis to enhance the diffraction intensity from the (004) planes. Crystal growth of anatase TiO_2 along the *c*-axis was previously observed in TiO_2 films [4]. Anisotropic crystal growth is one of the features of liquid-phase crystal deposition.

Acicular nanocrystals showed lattice images of anatase TiO_2 (Figure 20c). They were constructed of anatase TiO_2 crystals without amorphous or additional phases. Anatase crystals were not covered with amorphous or additional phases, even at the tips. Bare anatase crystal with a nano-sized structure is important to achieve high performance of catalysts and devices.

Crystallization of TiO_2 was effectively used to form assemblies of acicular nanocrystals in the process. Open pores and surface relief structures were successfully formed on the particles.

4.1.4 Zeta Potential and Particle Size Distribution

The dried particles were dispersed in water to evaluate the zeta potential and particle size distribution after evaluation of N_2 adsorption. The particles had a positive zeta potential of 30.2 mV at pH 3.1, which decreased to 5.0, −0.6, −11.3, and −36.3 mV at pH 5.0, 7.0, 9.0, and 11.1, respectively (Figure 21). The isoelectric point was estimated to be pH 6.7, slightly higher than that of anatase TiO_2 (pH 2.7-6.0) [133]. Zeta potential is very sensitive to particle surface conditions, ions adsorbed on particle surfaces, and the kind and concentration of ions in the solution. The variations in zeta potential were likely caused by the difference in the surface conditions of TiO_2 particles, affected by the interaction between particles and ions in the solution.

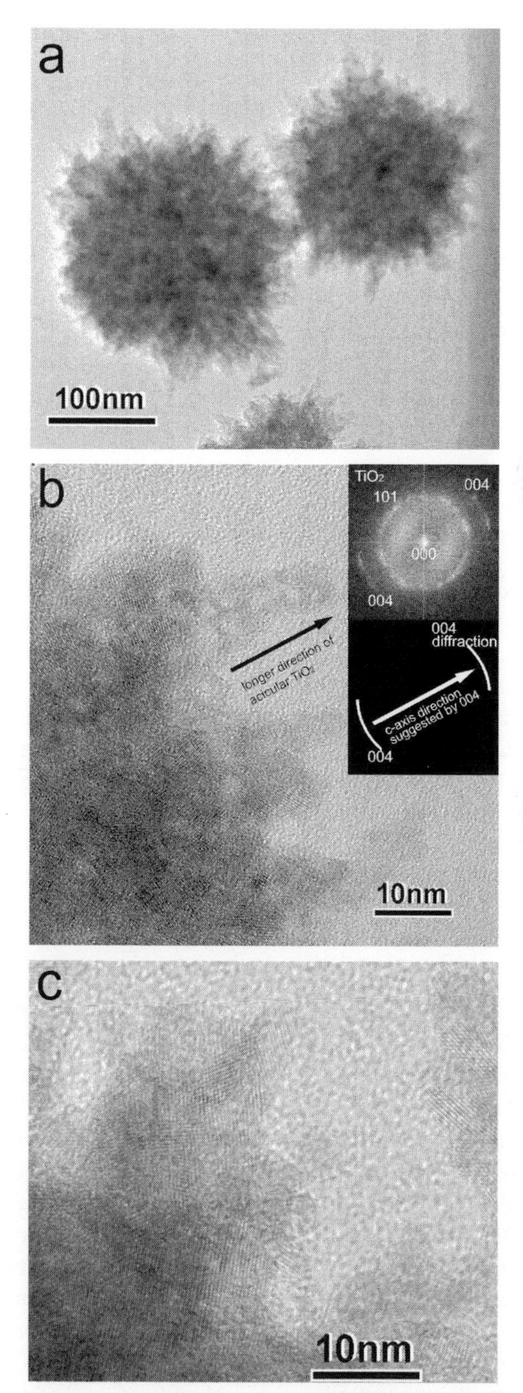

FIGURE 20 (a) Transmission electron micrograph of anatase TiO_2 particles. (b) A magnified area of (a) showing the morphology of acicular crystals. Insert in (b) shows the FFT image of anatase TiO_2. (c) A magnified area of (a) showing lattice images of anatase TiO_2. *(Reprinted with permission from Masuda and Kato [98]. Copyright © American Chemical Society.)*

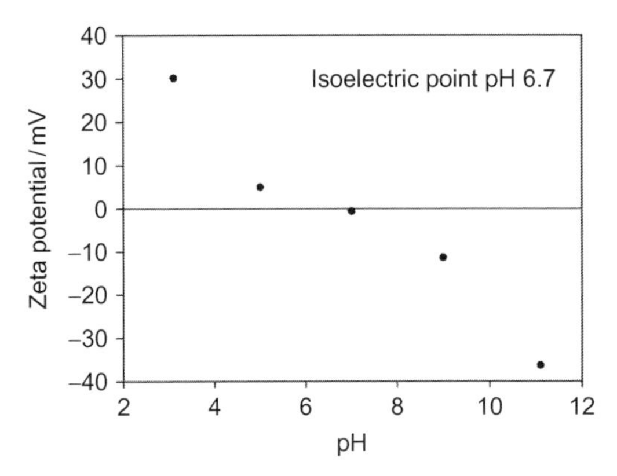

FIGURE 21 Zeta potential of anatase TiO_2 particles as a function of pH. *(Reprinted with permission from Masuda and Kato [98]. Copyright © American Chemical Society.)*

Mean particle size was estimated to be ~550 nm in diameter with a standard deviation of 220 nm at pH 3.1 (Figure 22a). This was larger than that observed by TEM. Slight aggregation occurred at pH 3 because the particles were dried completely before measurement. Particle size increased with pH and showed a maximum near the isoelectric point (550 nm at pH 3.1, 3150 nm at pH 5, 4300 nm at pH 7, 5500 nm at pH 9, or 2400 nm at pH 11.1) (Figure 22b). Strong aggregation resulted from the lack of repulsion force between particles near the isoelectric point.

The particles were generated in a solution at pH 3.8 in this study. It would be suitable to obtain repulsion force between particles for crystallization without strong aggregation.

4.1.5 N_2 Adsorption Characteristics of TiO_2 Particles

TiO_2 particles exhibited N_2 type IV adsorption-desorption isotherms (Figure 23a). The desorption isotherm differed from the adsorption isotherm in the relative pressure (P/P_0) range, from 0.4 to 0.7, showing mesopores in the particles. BET surface area of the particles was estimated to be 270 m^2/g (Figure 23b). This is higher than that of TiO_2 nanoparticles such as Aeroxide P25 (BET 50 m^2/g, 21 nm in diameter, 80% anatase + 20%rutile; Degussa), Aeroxide P90 (BET 90-100 m^2/g, 14 nm in diameter, 90% anatase + 10% rutile; Degussa), MT-01 (BET 60 m^2/g, 10 nm in diameter, rutile; Tayca Corp.), and Altair TiNano (BET 50 m^2/g, 30-50 nm in diameter; Altair Nanotechnologies Inc.) [134]. A large BET surface area cannot be obtained from particles with a smooth surface, even if the particle size is less than 100 nm. A large BET surface area would be realized by the unique morphology of TiO_2 particles constructed of nanocrystal assemblies.

Total pore volume and average pore diameter were estimated from pores smaller than 230 nm at P/P_0 = 0.99-0.431 cc/g and 6.4 nm, respectively. They were estimated to be 0.212 cc/g and 3.1 nm, respectively, from pores smaller

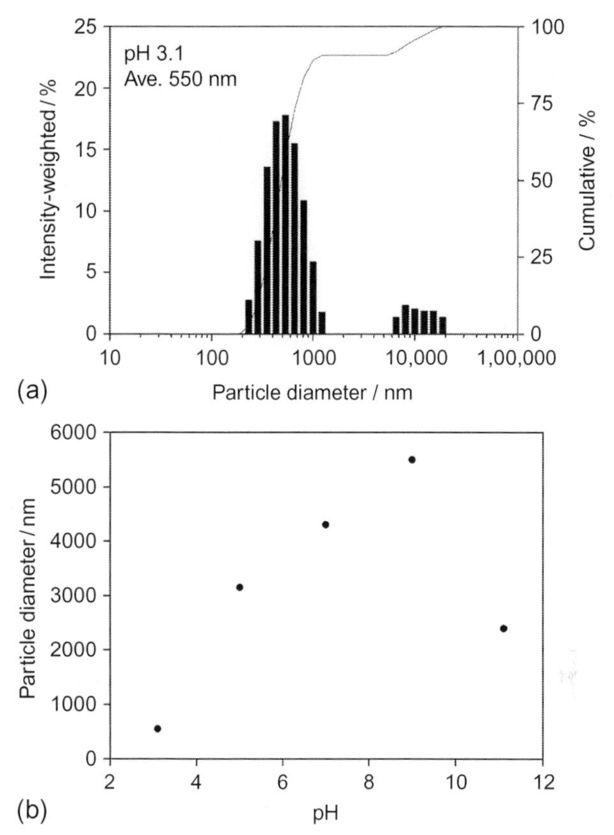

FIGURE 22 (a) Particle size distribution of anatase TiO_2 particles at pH 3.1. (b) Particle size of anatase TiO_2 particles as a function of pH. *(Reprinted with permission from Masuda and Kato [98]. Copyright © American Chemical Society.)*

than 11 nm at $P/P_0=0.80$. Total pore volume also was estimated to be 0.428 cc/g using the BJH method on pores smaller than 154 nm. Average pore diameter was estimated to be 6.3 nm using BET surface area.

Pore size distribution was calculated by the BJH method using adsorption isotherms (Figure 23c). It showed a pore size distribution curve with a peak at ~2.8 nm and pores larger than 10 nm. TiO_2 particles would have mesopores of ~2.8 nm surrounded by nanocrystals. Pores larger than 10 nm are considered to be interparticle spaces. The pore size distribution also suggested the existence of micropores smaller than 1 nm.

Pore size distribution was further calculated using the DFT/Monte-Carlo method. The model was in fair agreement with adsorption isotherms (Figure 23d). Pore size distribution showed a peak at ~3.6 nm that indicated the existence of mesopores of ~3.6 nm (Figure 23e). The pore size calculated by the DFT/Monte-Carlo method was slightly larger than that calculated by the BJH

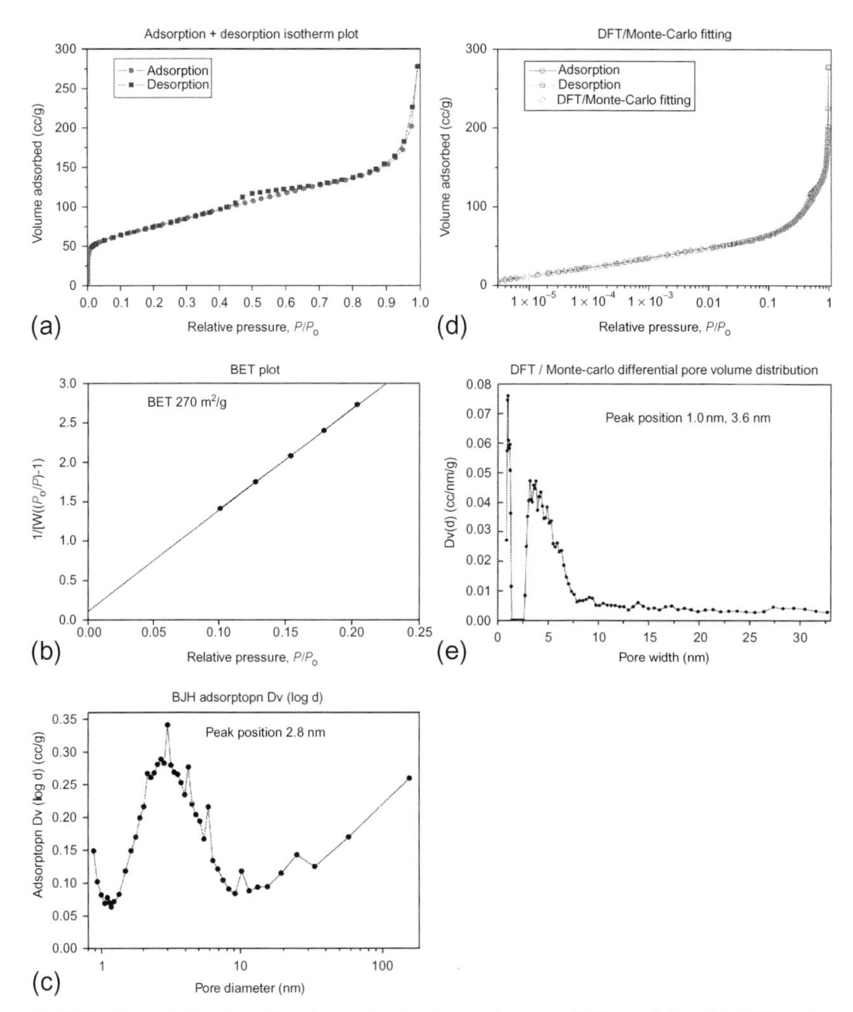

FIGURE 23 (a) N_2 adsorption-desorption isotherm of anatase TiO_2 particles. (b) BET surface area of anatase TiO_2 particles. (c) Pore size distribution calculated from the N_2 adsorption data of anatase TiO_2 particles using the BJH equation. (d) N_2 adsorption-desorption isotherm and the DFT/Monte-Carlo fitting curve of anatase TiO_2 particles. (e) Pore size distribution calculated from the N_2 adsorption data of anatase TiO_2 particles using the DFT/Monte-Carlo equation. *(Reprinted with permission from Masuda and Kato [98]. Copyright © American Chemical Society.)*

method because the latter is considered to produce an underestimation [135–137]. The pore size distribution also suggested the existence of micropores of ~1 nm, probably resulting from microspaces surrounded by nanocrystals and the uneven surface structure of nanocrystals.

The particles had a large surface area as well as micropores of ~1 nm, mesopores of ~2.8-3.6 nm, and pores larger than 10 nm, based on N_2 adsorption characteristics. Assembly of acicular nanocrystals resulted in unique features and large surface areas.

For comparison, TiO_2 particles were generated in solutions at 90°C for 1 h using an oil bath with no stirring. The solutions became cloudy after the addition of an H_3BO_3 solution into an $(NH_4)_2TiF_6$ solution. High temperature accelerated growth of TiO_2 crystals. HCl (0.6 mL) was added to the solutions (200 mL) to decrease the speed of TiO_2 crystallization. One hour after mixing the solutions the pH was 2.4. BET surface area of the particles was estimated at 18 m^2/g. This is much smaller than that of the particles prepared at 50°C and slightly smaller than those prepared at 90°C for 8 min in our previous work (44 m^2/g) [138]. Formation of TiO_2 was accelerated at a high temperature and it decreased the surface area. The particles grew in the solutions, decreasing surface area as a function of time. Crystallization of TiO_2 was strongly affected by growth conditions such as solution temperature and growth time.

Anatase TiO_2 particles, 100-200 nm in diameter, were successfully fabricated in an aqueous solution. They were assemblies of 5- to 10-nm nanocrystals that grew anisotropically along the *c*-axis to form acicular shapes. The particles thus had nano-relief surface structures constructed of acicular crystals. They showed *c*-axis orientation because of the high-intensity XRD from the (004) crystal planes. The particles had a large BET surface area of 270 m^2/g. Total pore volume and average pore diameter were estimated from pores smaller than 230 nm to be $P/P_0 = 0.99$-0.43 cc/g and 6.4 nm, respectively. They also were estimated from pores smaller than 11 nm to be $P/P_0 = 0.80$-0.21 cc/g and 3.1 nm, respectively. BJH and DFT/Monte-Carlo analysis of adsorption isotherms indicated the existence of ~2.8- and ~3.6-nm pores, respectively. In addition, the analyses suggested the existence of ~1-nm micropores. Crystallization and self-assembly of nano TiO_2 were effectively used to fabricate nanocrystal assembled TiO_2 particles with a large surface area and a nano-relief surface structure.

4.2 Multineedle TiO_2 Particles [139]

Flower-like, multineedle anatase TiO_2 particles were developed in aqueous solutions[139]. They were pure anatase TiO_2 crystals containing no cores, organic binders, or solvents. Furthermore, microstructured Si wafers were covered uniformly with the TiO_2 particles in the solutions. Their unique crystal growth and physicochemical profiles were precisely evaluated and discussed.

Si wafers were modified to have microstructures on the surfaces through cutting. They were cut using a precise diamond cutter under running water. Width and height of salient lines were 200 μm and 150 μm, respectively. They were formed at 500-μm intervals. They were blown with air to remove dust and were exposed to vacuum-UV (VUV) light (low-pressure mercury lamp PL16-110, air flow, 100 V, 200 W [SEN Lights Co.]; 14 mW/cm^2 for 184.9 nm at a distance 10 mm from the lamp, 18 mW/cm^2 for 253.7 nm at a distance 10 mm from the lamp) for 10 min in air. Bare Si surfaces were covered with a small amount of surface contamination. The VUV irradiation modified them to clean surfaces that showed super hydrophilic surfaces with a water contact angle ~0-1°.

$(NH_4)_2TiF_6$ (206.20 mg) and H_3BO_3 (186.42 mg) were separately dissolved in deionized hot water (100 mL) at 50°C. H_3BO_3solution was added to the $(NH_4)_2TiF_6$ solution at concentrations of 15 and 5 mmol/L, respectively. The Si wafers with patterned surfaces were immersed in the middle of the solutions with the bottom up at an angle. They were tilted at 15° to stand upright. The solutions were kept at 50°C for 19 h or 7 days using a drying oven with no stirring. The substrates were washed with running water and dried by blowing air. The solutions were centrifuged at 4000 rpm for 10 min. Precipitated particles were dried at 60°C for 12 h after removal of supernatant. The particles were dispersed in distilled water. They were centrifuged and dried again for purification.

4.2.1 Control of Multineedle TiO_2 Particle Morphology

Flower-like, multineedle TiO_2 particles were successfully formed in aqueous solutions (Figure 24a and b). Needle-shaped crystals grew from the center of the particles. Especially, the needles grew parallel to the direction of TEM observation, from the center of the particles (circled in red). TEM clearly showed that needles radiated in all directions to form a flower-like morphology. Each particle had about 6-10 tapered needles, with a width and length of about 200 and 100 nm, respectively. The aspect ratio was about 2 (200 nm/100 nm). The width of the needles became smaller in the direction of growth to form tips. The particles had no core or pore at the center of their bodies. Nucleation and crystal growth were well controlled to create a flower-like, multineedle morphology.

The electron diffraction pattern showed that the particles were a single phase of anatase TiO_2 crystals (Figure 24c). Interplanar spacing of (004), (200), and (204) planes were estimated at 0.243, 0.201, and 0.151 nm. Diffractions from (004), (200), and (204) planes were clear single spots (Figure 24c). Needle-shaped crystals was shown to be single crystals of anatase TiO_2. The long direction of the needle-shaped crystals was perpendicular to (004) crystal faces (Figure 24c), indicating that anatase TiO_2 crystals grew along the *c*-axis to form a needle-shaped morphology. They were thus surrounded by *a*-faces of anatase TiO_2 crystals.

Surfaces of the needle-like crystals were observed carefully (Figure 24c and d). There were no amorphous layers or second-phase layers on the surfaces. The particles had pure and bare anatase TiO_2 surfaces.

XRD analysis showed that the particles were a single phase of anatase TiO_2. XRD peaks were observed at $2\theta=25.12°$, 36.8°, 37.7°, 47.7°, 53.7°, 54.7°, 62.4°, 68.5°, 69.8°, 74.7°, 82.2°, 93.7°, and 94.4°. Pure anatase TiO_2 particles were obtained from the aqueous solutions in this study. Anatase phases have been prepared using high-temperature annealing in many reports. This causes the deformation of nano-/microstructures, aggregation of the particles, and a decrease of surface area. However, crystallization of anatase TiO_2 was realized at 50°C in this study to avoid degradation of the properties.

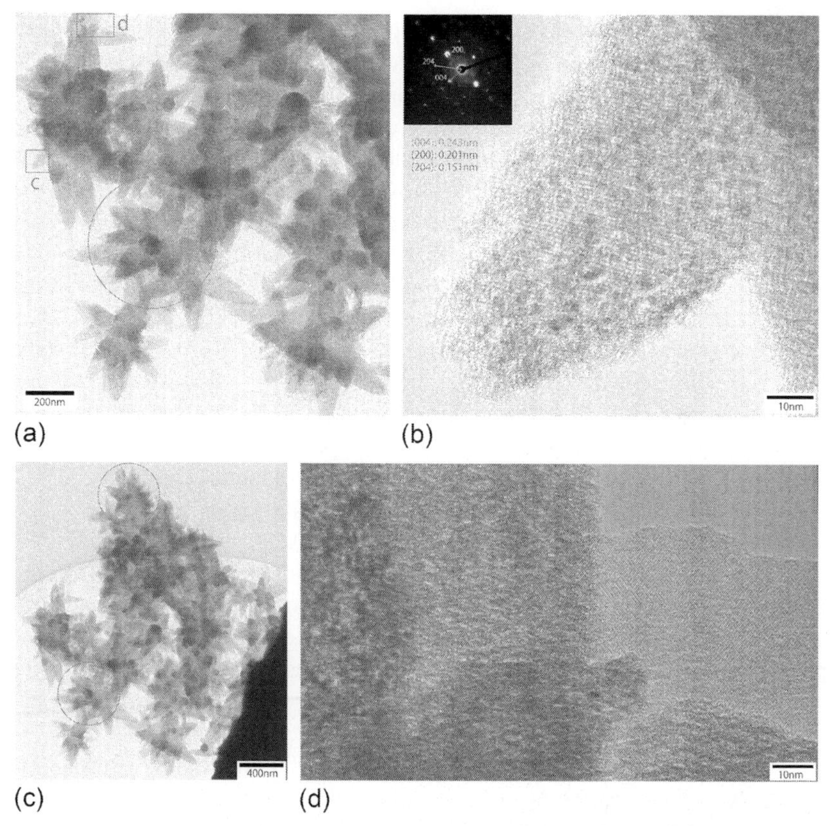

FIGURE 24 Transmission electron micrographs of multineedle TiO_2 particles. (a and b) Low-magnification images of the particles. (c) A high-resolution image of the needles that grew from the center of the particle, showing the tapered shape of the needle. (d) A high-resolution image of the surface of the needles, showing the bare TiO_2 surface. (c) and (d) are magnified areas of (a) and (b), respectively. Inset in (c) is the electron diffraction pattern indicating the crystal phase and interplanar spacing. Red circles indicate images typical of multineedle TiO_2 particles. *(Reprinted with permission from Masuda* et al. *[139]. Copyright © American Chemical Society.)*

Crystallite size perpendicular to the (101), (004), and (200) planes was estimated from the full-width half-maximum of the 101, 004, or 200 peak: 18.2, 29.8, and 19.2 nm, respectively. Crystallite size perpendicular to (004) was larger than that of others. Difference in crystallite size indicated anisotropic crystal growth along the *c*-axis.

For comparison, nanocrystal assembled TiO_2 particles were prepared in aqueous solutions at 50°C for 30 min in previous report [98]. Crystallite size perpendicular to the (101), (004), and (200) planes was estimated to be 3.9, 6.3, and 4.9 nm, respectively. The crystalline degree of the multineedle TiO_2 nanostructures was much improved compared with that of the previous nanocrystal [98]. The key to achieving a high crystalline degree even at a low temperature (i.e., 50°C) was slow and long-term growth. It was achieved by the low supersaturation of the solutions. Concentrations of $(NH_4)_2TiF_6$ and H_3BO_3 were one-tenth those used for previous nanocrystals [98]. Synthesis parameters of multineedle TiO_2 nanostructures were developed based on previous reports, as follows. In one case acicular nanocrystals were homogeneously formed immediately after mixing two solutions[98]. They aggregated into nanocrystal assembled particles. The particles were removed from the solutions 30 min after mixing to prevent further crystal growth. The crystallite size thus was small, and it contributed to the large specific surface area. In other studies, TiO_2 films were prepared in the same solutions at 50°C for several hours [140–142]. The films consisted of two layers. The underlayer comprised small nanocrystals. They were formed at an early stage of the immersion period in solutions with a high ion concentration. Ions were consumed gradually to form the crystals. The upper layer of acicular crystal assembly then formed. They grew in solutions with low ion concentrations. This indicated that low ion concentration and low supersaturation allowed the formation of acicular TiO_2 crystals. In addition, the acicular crystals were much larger than the initially deposited nanocrystals. We tried to form multineedle particles consisting of large acicular crystals on the basis of these results. Crystal growth in the solution with a low ion concentration and low supersaturation was used for anisotropic crystal growth. They were grown to large crystals over a long period, such as 7 days. The slow growth rate caused the formation of euhedral crystals that were affected by the crystal structure of tetragonal anatase.

The particles showed Raman peaks at 157, 412, 506, and 628 cm^{-1}. It was typical of the anatase TiO_2 phase. They were assigned to Eg ($\nu6$) mode (157 cm^{-1}), B1g mode (412 cm^{-1}), doublet of the A1g and B1g modes (506 cm^{-1}), and Eg ($\nu1$) mode (628 cm^{-1}), respectively. Additional peaks indicating rutile TiO_2 or other phases were not observed. It was notable that the Eg ($\nu6$) mode shifted with respect to those of bulk crystals or sintered powders. For the single crystal, Ohsaka *et al.* [143] determined the following allowed bands: 144 cm^{-1} (Eg), 197 cm^{-1} (Eg), 399 cm^{-1} (B1g), 513 cm^{-1} (A1g), 519 cm^{-1} (B1g), and 639 cm^{-1} (Eg). TiO_2 particles (P25; Degussa) showed peaks at 143.3 cm^{-1} (Eg), 196 cm^{-1} (Eg), 396 cm^{-1} (B1g), 516 cm^{-1} (A1g+B1g), and 638 cm^{-1} (Eg). Raman peak of Eg ($\nu6$) mode (144 cm^{-1}) has been reported to shift as a result of several factors, such as the effect of crystalline size (the quantum size confinement effect) [144,145], temperature [144], or pressure[146]. In addition, nitrogen-doped anatase TiO_2 particles, 8.40-9.80 nm in diameter, were reported to have an Eg ($\nu6$) mode at 151 cm^{-1} [145]. The multineedle particles showed a large peak shift of the Eg ($\nu6$) mode (from 144 to 157 cm^{-1}) because of the fluorine-doping effect and size effect. The large shift of Eg ($\nu6$) mode was one characteristic of the multineedle TiO_2 particles.

The particles showed absorption spectra in the infrared light region. The absorption bands related to TiO_2 were observed at 907 and 773 cm^{-1}. They were assigned to stretching vibrations of Ti=O and –Ti–O–Ti–, respectively.

They were consistent with TEM, XRD, and Raman analyses. The bands observed in the range of 1400-1750 cm^{-1} were attributed to bending vibrations of O–H [147]. Absorption bands in the frequency range of 600-450 cm^{-1} was reported to be attributed to stretching vibrations of Ti–F bonds in the TiO_2 lattice [148,149]. The bands were observed at 513, 532, 540, 558, or 568 cm^{-1} from $x TiOF_2 \cdot y BaF_2 \cdot z MnF_2$ glasses [148]. They indicated that F ions were partially replaced to O ions in the lattice [148]. Actually, absorption bands were observed at 455, 488, 505, 520, 552, and 567 cm^{-1} in the spectra of the multineedle TiO_2 particles. In addition, an absorption band at 1080 cm^{-1} was reported to be assigned to surface-fluorinated Ti–F species [150]. An absorption band was observed ~1058 cm^{-1} from the multineedle TiO_2 particles. These analyses suggested that F ions were partially doped into TiO_2 crystals to replace O ions.

The optical properties of the particles were evaluated with UV-Vis spectroscopy. Transparency gradually decreased as wavelength decreased. The bulk bandgap structures are known as direct transition for anatase titania and indirect transition for rutile titania. The bandgap of the particles was estimated at 3.20 eV (388 nm) by assuming a direct transition. It was similar to that of anatase TiO_2 nanostructures such as anatase TiO_2 nanorods (3.2 eV [388 nm]) [151], anatase TiO_2 nanowalls (3.2 eV [388 nm]) [151], anatase TiO_2 nanotubes (3.2 eV [388 nm]) [152], a single-layered TiO_2 nanosheet with a thickness <1 nm (3.15 eV [394 nm]) [153], stacked TiO_2 nanosheets (3.15 eV [394 nm]) [153], and anatase TiO_2 films (3.2 eV [388 nm]) [154]. It was higher than that of rutile TiO_2 films (2.9 eV [428 nm]) [154], rutile TiO_2 nanorods (3.0 eV [414 nm]) [151], rutile TiO_2 single crystals (3.0 eV [414 nm]; SHINKOSHA Co., Ltd). It was lower than that of amorphous TiO_2 films (3.5 eV [355 nm]) [154].

BET surface area of the particles was estimated to be 178 m^2/g from an adsorption branch in the range of P/P_0 = 0.1-0.29. Average pore diameter was estimated to be 11.1 nm using BET surface area. Surface area and pore diameter were estimated to be 166 m^2/g and 11.9 nm and 151 m^2/g and 17.6 nm from P/P_0 = 0.05-0.1 or P/P_0 = 0.02-0.07, respectively. BET surface area was larger than that of TiO_2 nanoparticles such as Aeroxide P25 (BET 50 m^2/g, 21 nm in diameter, 80% anatase + 20% rutile; Degussa), Aeroxide P90 (BET 90-100 m^2/g, 14 nm in diameter, 90% anatase + 10% rutile; Degussa), MT-01 (BET 60 m^2/g, 10 nm in diameter, rutile; Tayca Corp.), and Altair TiNano (BET 50 m^2/g, 30-50 nm in diameter; Altair Nanotechnologies Inc.) [134]. Pore size distribution calculated using the BJH method indicated that spaces of 2-3 nm existed in the particles. DFT/Monte-Carlo analysis showed several types of mesopores in the range of 3-10 nm (3.5, 4.9, 6.3, and 10 nm) and micropores of ~0.8 nm. Total pore volume and average pore diameter were estimated from pores smaller than 241 nm at P/P_0 = 0.99 to be 0.493 cc/g and 13 nm, respectively. They were estimated to be 0.158 cc/g and 4.2 nm, respectively, for pores smaller than 11 nm at P/P_0 = 0.81. Total pore volume also was estimated using the BJH method: For pores smaller than 156 nm, total pore volume was 0.505 cc/g using a desorption isotherm; for pores smaller than 160 nm this value was 0.628 cc/g using an adsorption isotherm.

4.2.2 Surface Coating of Microstructured Substrates with Multineedle TiO_2 Particles

Microstructured Si wafers were immersed in aqueous solutions at 50°C for 7 days. The substrates were successfully covered with multineedle TiO_2 particles (Figure 25a). Both salient regions and concave regions were uniformly modified with the particles (Figure 25b1 and c1). The original microstructure of the Si wafers was well maintained because the TiO_2 surface coatings were uniform thin layers about 200-600 nm thick (Figure 25b2 and c2). The coating layers consisted of flower-like multineedle TiO_2 particles of about 200-400 nm in diameter (Figure 25b3 and c3, red circles). The particles had several needles that grew from the center of the particles. The needles had a tapered shape along the direction of growth. The particle in the red circle of Figure 25b3 clearly indicates that a tetragonal crystal phase of anatase TiO_2 caused fourfold symmetry crystal growth of four needles parallel to the substrate. The *a*-axis and *c*-axis of the particle were parallel and perpendicular to the substrate, respectively. This shape was the basic morphology of the particles in this system. Other needles also grew from the center to form multineedle shapes, as shown in the red circles in Figure 25c3. Salient regions had a slightly trapezoidal geometry (Figure 25a and b1). TiO_2 particles were deposited on their walls. The area between the salient and concave regions was thus white in SEM images.

For comparison, microstructured Si wafers were immersed in aqueous solutions at 50°C for 19 h. The multineedle TiO_2 particles formed on both the salient regions and concave regions (Figure 26a). The microstructures of the substrates were maintained because the surface coatings were uniform thin layers (Figure 26b1 and c1). However, coverage of the surfaces was different from those immersed for 7 days. The TiO_2 particles covered approximately half of the salient regions and one-third of the concave regions (Figure 26b2 and c2). Coverage of the salient regions was higher than that of concave regions because concave regions are set back far from the salient region surfaces. The particles and ions were not well supplied to concave regions. The particles had several needles growing from the center (Figure 26b3 and c3). Their morphology was similar to that of particles immersed for 7 days; by contrast, however, their size was slightly smaller (~100-300 nm). These observations indicated the growth mechanism as follows. The particles were homogeneously nucleated from ions in the solutions. They then adhered on the substrates to form surface coatings. The particles grew further, increasing the number and size of the needles.

Surfaces of the TiO_2 coatings were analyzed with XPS. Ti, O, carbon, Si, and F were observed from the surface before (Figure 27, 1a) and after (Figure 27, 1b) Ar^+ sputtering. The Ti $2p_{3/2}$ spectrum was observed at 458.6 eV (Figure 27, 2a). The binding energy was higher than that of Ti metal (454.0 eV), TiC (454.6 eV), TiO (455.0 eV), TiN (455.7 eV), and Ti_2O_3 (456.7 eV) and similar to that of TiO_2 (458.4-458.7 eV) [6,155,156]. This suggested that the Ti atoms in the particles were positively charged relative to those of Ti metal by the formation of direct bonds with O. The Ti 2p spectrum changed its shape by the sputtering

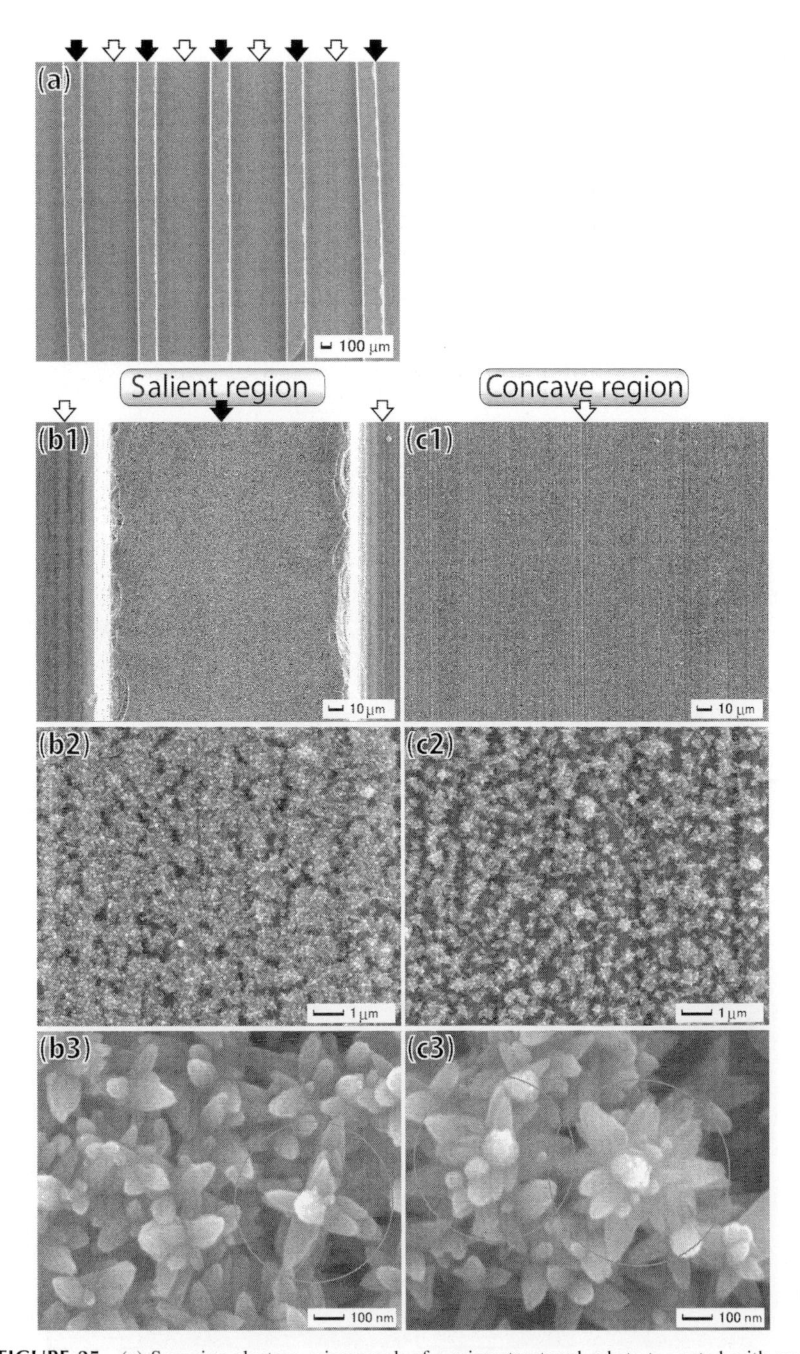

FIGURE 25 (a) Scanning electron micrograph of a microstructured substrate coated with multineedle TiO_2 particles deposited for 7 days. (b1) Salient region coated with the multineedle TiO_2 particles. (b2) Magnified area of (b1) showing uniform surface coating. (b3) Magnified area of (b2) showing the morphology of the multineedle TiO_2 particles. (c1) Concave region coated with the multineedle TiO_2 particles. (c2) Magnified area of (c1) showing uniform surface coating. (c3) Magnified area of (c2) showing the morphology of the multineedle TiO_2 particles. Black arrows indicate salient regions. White arrows indicate concave regions. Red circles indicate images typical of the multineedle TiO_2 particles. *(Reprinted with permission from Masuda* et al. *[139]. Copyright © American Chemical Society.)*

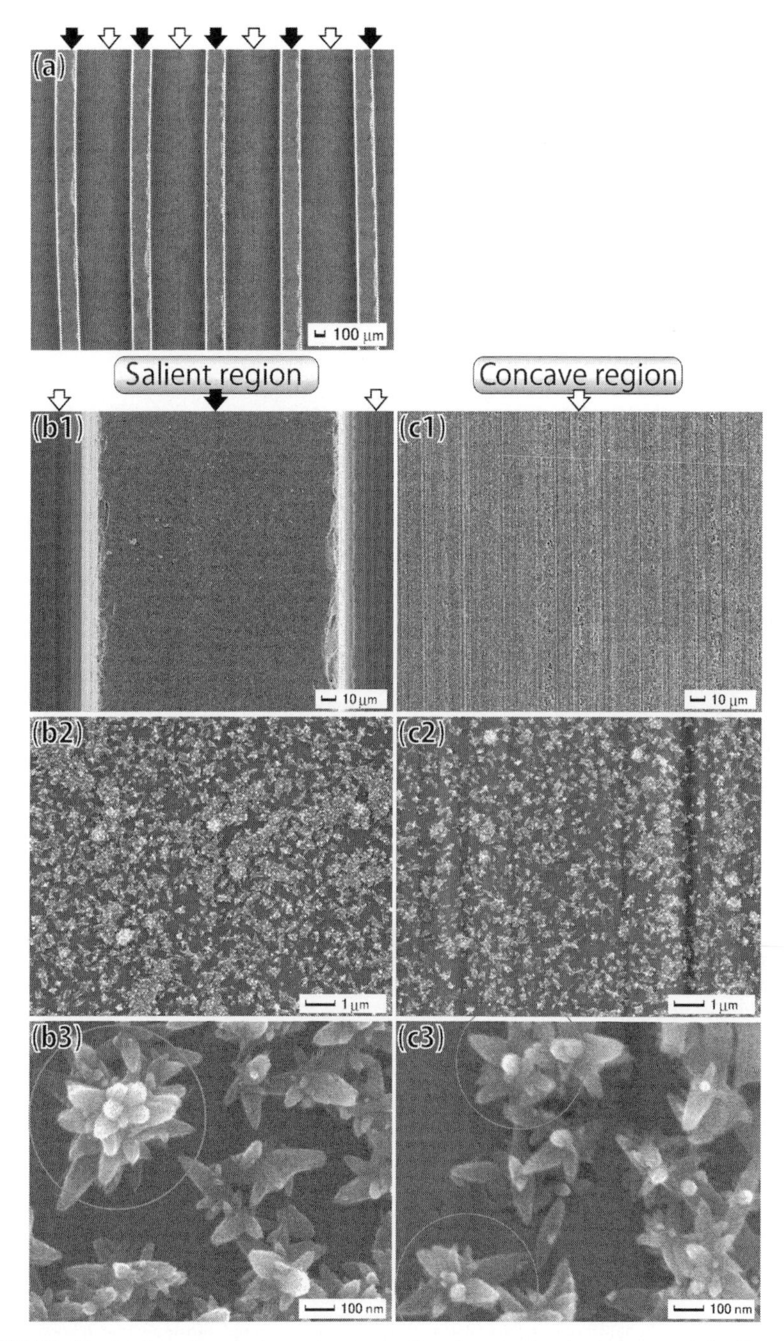

FIGURE 26 (a) Scanning electron micrograph of a microstructured substrate coated with multineedle TiO_2 particles deposited for 19 h. (b1) Salient region coated with the multineedle TiO_2 particles. (b2) Magnified area of (b1) showing uniform surface coating. (b3) Magnified area of (b2) showing the morphology of the multineedle TiO_2 particles. (c1) A concave region coated with the multineedle TiO_2 particles. (c2) Magnified area of (c1) showing uniform surface coating. (c3) Magnified area of (c2) showing the morphology of the multineedle TiO_2 particles. Black arrows indicate salient regions. White arrows indicate concave regions. Red circles indicate images typical of multineedle TiO_2 particles. *(Reprinted with permission from Masuda* et al. *[139]. Copyright © American Chemical Society.)*

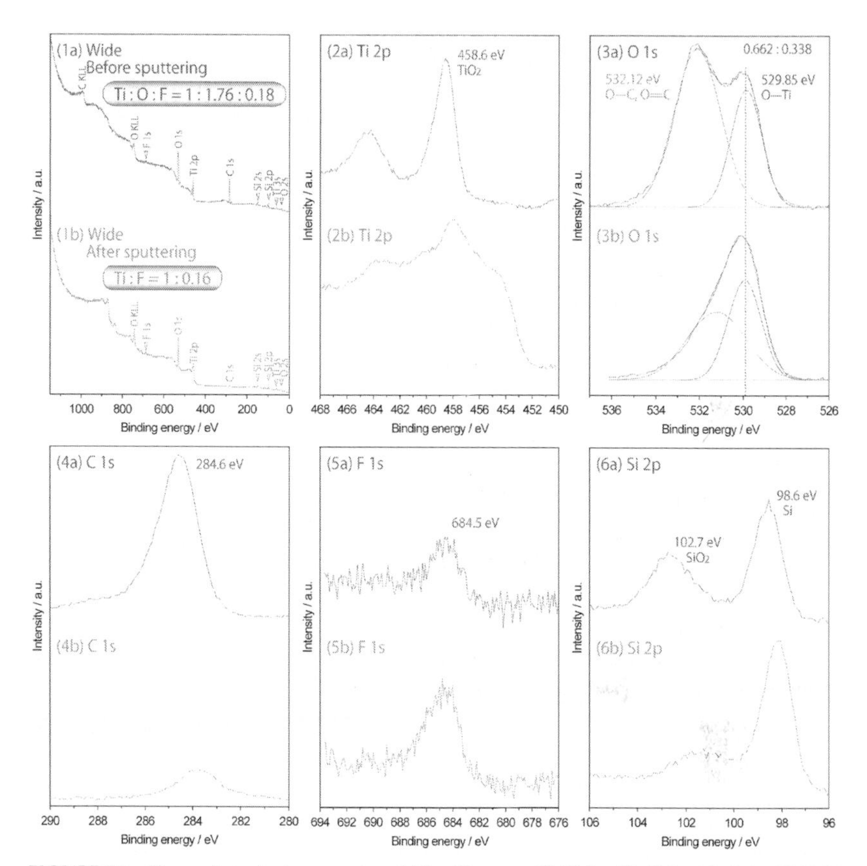

FIGURE 27 X-ray photoelectron spectra of (1) wide scan, (2) Ti 2p, (3) O 1s, (4) C 1s, (5) F 1s, and (6) Si 2p for a microstructured substrate coated with multineedle TiO_2 particles deposited for 7 days before (a) and after (b) Ar^+ sputtering. (3a) O1s spectra decomposed to two peaks after removing the background. a.u., arbitrary units. *(Reprinted with permission from Masuda* et al. *[139].. Copyright © American Chemical Society.)*

caused by the decrease of cation valence (Figure 27, 2b). The phenomenon was often observed in sputtering of TiO_2 [6]. The spectrum after sputtering was not suitable for estimating the chemical ratio because the ratio was slightly changed by decrease of the Ti cation valence. The chemical ratio of Ti to O thus was estimated from the spectra before sputtering.

The O 1s spectrum was decomposed into two Gaussian curves after removal of the background (Figure 27, 3a). The integral intensity ratio the O 1s peak at 532.12 eV to the O 1s peak at 529.85 eV was 0.662:0.338. The binding energy of the O 1s peak at 529.85 eV was similar to that of TiO_2 (529.9 eV [156], 530.1 eV [6,155]), showing that O was negatively charged compared with neutral O molecules (531.0 eV) through the formation of direct bonds with Ti. A component with high binding energy at 532.12 eV was assigned to O atoms combined with carbon atoms as C–O or C=O (532.8 eV). It was decreased by the sputtering for 10 s (Figure 27, 3b). It was included in surface contaminations. A drastic decrease of the C 1s spectrum by sputtering supported this ascription (Figure 27, 4a and b). The chemical ratio of Ti to O was estimated as 1:1.76 using Ti 2p at 458.6 eV (Figure 27, 2a) and O 1s at 529.85 eV (Figure 27, 3a). It was slightly smaller than that expected from TiO_2. Oxygen vacancy and doping of F ions decreased O volume. F was, in fact, observed from the surfaces at 684.5 eV (Figure 27, 5a). The Ti-to-O-to-F chemical ratio was estimated at 1:1.76:0.18. The sum of O and F was 1.94, which was similar to the value of 2 expected from TiO_2. The F spectrum intensity was not decreased by sputtering: Ti-to-F ratio = 1:0.16 (Figure 27, 5b). These indicated that F would not be included in surface contaminations but in the TiO_2 particles. F has been reported to improve propertics of TiO_2. Fluorination of TiO_2 increased surface acidity because of the strong electronegativity of F [157]. F doping into TiO_2 increased the surface OH radicals that were suitable for photocatalytic reactions and improved resistance to photocorrosion [158–163]. These properties expand the application areas of TiO_2.

Si $2p_{3/2}$ was observed at 98.6 eV and 102.7 eV (Figure 27, 6a). It was detected from Si regions not covered by TiO_2. The binding energies were similar to those in Si wafers (99.6 eV) and SiO_2 (103.4 eV). The latter component was decreased by sputtering because the native oxide layer on the surface of amorphous SiO_2 was removed (Figure 27, 6b).

Multineedle TiO_2 nanostructures with a large surface area, F doping, and novel physicochemical characteristics were successfully fabricated. Unique aqueous synthesis created their distinct morphologies and properties. The needle crystals grew along the *c*-axis from the center of the particles. They were surrounded by *a*-faces of anatase TiO_2 crystals. The width and length of the needles were about 200 and 100 nm, respectively. The aspect ratio was about 2 (200 nm/100 nm). Diffraction patterns showed that the particles were a single phase of anatase TiO_2. Interparticle spaces and micro-/mesopores of 1-10 nm allowed us to realize a large surface area of 178 m^2/g. Large Raman peak shift of Eg ($\nu 6$) mode suggested F doping and size effect. Bandgap was estimated at 3.20 eV (388 nm) with UV-Vis. These were characteristics of the multineedle TiO_2 particles. Furthermore, self-assembly surface coating of microstructured substrates was successfully achieved. The coating layers consisted of multineedle TiO_2 particles ~200-400 in diameter. XPS analyses indicated chemical bonds between Ti and O. The Ti-to-O-to-F chemical ratio was estimated to be 1:1.76:0.18, suggesting F doping in TiO_2. The multineedle particles and the surface coating of particles with nano-/microstructures may contribute to future metal oxide devices of solar cells, photo catalysts, and highly sensitive sensors.

4.3 Acicular Nanocrystal Coating and Its Patterning [140]

Morphology of TiO_2 crystals were controlled to create anisotropic acicular shapes [140]. They were formed on substrates in aqueous solutions. Furthermore, a micropattern of an acicular nanocrystal coating was successfully fabricated. Nucleation and crystal growth of TiO_2 were accelerated on the superhydrophilic SnO_2: F surface but were suppressed on the hydrophobic initial SnO_2:F surface. Consequently, liquid phase patterning of anatase TiO_2 was achieved on the substrate. TiO_2 crystals were directly deposited on the SnO_2:F surface without any insulating layers, which decrease the electrical conductivity between TiO_2 and the SnO_2:F substrate. The micropattern of anatase TiO_2 on the SnO_2:F substrates could be applied to electrodes of dye-sensitized solar cells or molecular sensors. This process can be used to form a flexible micropattern of anatase TiO_2 electrodes on conductive polymer films with low heat resistance. It contributes to the microfabrication of TiO_2 electrodes for dye-sensitized solar cells or molecular sensors.

4.3.1 Aqueous Synthesis of Acicular Nanocrystal Coating and Its Patterning

The initial SnO_2:F substrate showed a water contact angle of 96°. The UV-irradiated surface was, however, wetted completely (contact angle, 0-1°). The contact angle decreased with irradiation time (96°, 70°, 54°, 35°, 14°, 5°, and 0° for 0, 0.5, 1, 2, 3, 4, and 5 min, respectively). This suggests that a small number of adsorbed molecules on the SnO_2:F substrate were removed completely by UV irradiation. The surface of the SnO_2:F substrate would be covered by hydrophilic OH groups after irradiation. Consequently, the SnO_2:F substrate was modified to have a patterned surface with hydrophobic regions and superhydrophilic regions.

The SnO_2:F substrate with a patterned surface with hydrophobic regions and superhydrophilic regions was covered by an Si rubber sponge sheet (Silicosheet, SR-SG-S 5mmt RA grade; Shin-etsu Finetech Co., Ltd) to suppress deposition of TiO_2 at the initial stage. The substrate was immersed perpendicularly in the middle of the solution containing $(NH_4)_2TiF_6$ (0.05 mol/L) and H_3BO_3 (0.15 mol/L) (Figure 28). The solution was kept at 50°C with no stirring. The Si rubber sponge sheet was removed from the SnO_2:F substrate after 25 h, then the substrate was maintained at 50°C for another 2 h. The substrate was covered by the sheet instead of being immersed for 25 h to avoid agitation of the solution.

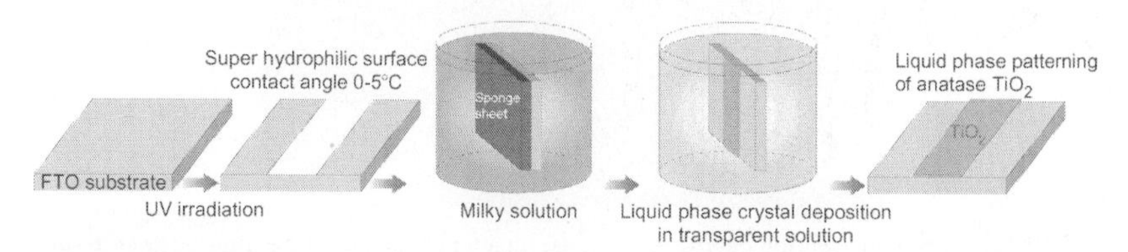

FIGURE 28 Conceptual process for liquid-phase patterning of anatase TiO_2 films using a superhydrophilic surface. UV, ultraviolet. *(Reprinted with permission from Masuda and Kato [140]. Copyright © American Chemical Society.)*

After being immersed in the solution, the substrate was rinsed with distilled water and dried in air. The initial FTO surface appeared blue-green under white light because of the light diffracted from the FTO layer. On the other hand, TiO_2 films deposited on the superhydrophilic surface appeared yellow-green. The color change could be caused by deposition of a transparent TiO_2 film that influenced the wavelength of the diffracted light.

The micropattern of TiO_2 was shown by SEM evaluation to be successfully fabricated (Figure 29). TiO_2 deposited on superhydrophilic regions showed black contrast, whereas the initial FTO regions without deposition showed white contrast, as seen in Figure 29. The average line width in Figure 29 is 55 μm. Line edge roughness [6], as measured by the standard deviation of the line width, is ~2.8 μm. This represents a ~5% variation (i.e., 2.8/55) in the nominal line width, similar to the usual 5% variation afforded by current electronics design rules. The minimum line width of the pattern depends on the resolution of the photomask and wavelength of the irradiated light (184.9 nm). It would be improved to ~1 μm by using a high-resolution photomask.

The FTO layer was a particulate film with a rough surface (Figure 30b1 and b2). Edged particles 100-500 nm in diameter were observed on the surface. The micropattern of the TiO_2 thin film was covered by an assembly of nanocrystals 10-30 nm in diameter (Figure 30a1 and a2). The nanocrystals would be anatase TiO_2, which grew anisotropically. The TiO_2 film also had a large structural relief 100-500 nm in diameter. Because the thin TiO_2 film was deposited on the edged particulate surface of the FTO layer, the surface of TiO_2 had a large structural relief.

The morphology of the TiO_2 layer and the FTO layer was further observed by fracture cross-section profiles (Figure 31). The polycrystalline FTO layer

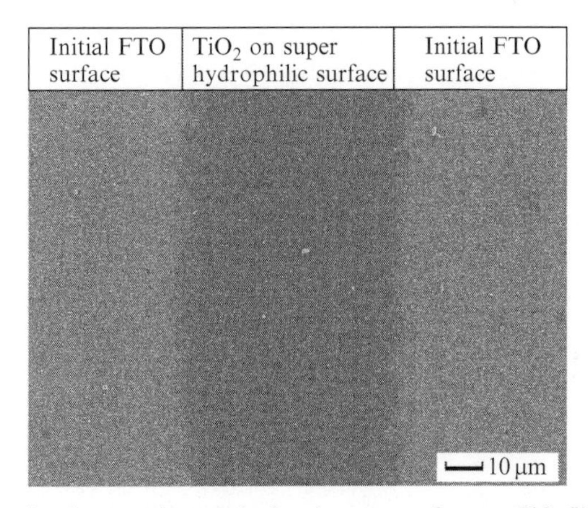

FIGURE 29 Scanning electron micrograph of a micropattern of anatase TiO_2 films on SnO_2:F substrates. *(Reprinted with permission from Masuda and Kato [140]. Copyright © American Chemical Society.)*

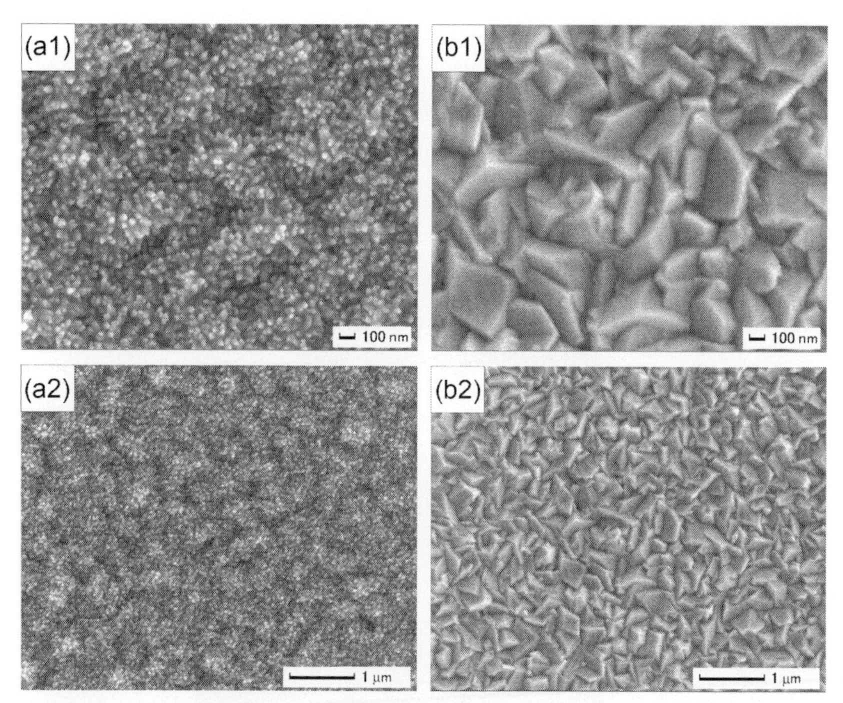

FIGURE 30 Scanning electron micrographs (SEMs) of a micropattern of anatase TiO_2 films and SnO_2:F substrates. All SEMs are a magnified area of the image in Figure 6. (a1) Surface of anatase TiO_2 films deposited on a superhydrophilic region. TiO_2 formed on the superhydrophilic region, which was cleaned by ultraviolet irradiation before immersion. (a2) Magnified area of (a1) showing the surface morphology of the anatase TiO_2 film. (b1) Surface of the SnO_2:F substrate without TiO_2 deposition. TiO_2 did not form on the unclean region. (b2) Magnified area of (b1) showing the surface morphology of the SnO_2:F substrate. *(Reprinted with permission from Masuda and Kato [140]. Copyright © American Chemical Society.)*

prepared on a flat glass substrate had a thickness of ~900 nm and a high roughness of 100-200 nm on the surface (Figure 31a). TiO_2 nanocrystals were deposited on the superhydrophilic FTO surface (Figure 31a), whereas no deposition was observed on the initial FTO surface. The superhydrophilic FTO surface was covered with an array of TiO_2 nanocrystals (Figure 31b and c), which had a long shape ~150 nm in length and ~20 nm in diameter. These observations were consistent with TEM and XRD evaluations. TiO_2 nanocrystals grew along the *c*-axis and thus enhanced the 004 XRD peak and the 004 electron diffraction peak. They formed a long shape with a high aspect ratio of 7.5 (150-nm length/20-nm diameter), as shown in the SEM fracture cross-section profile (Figure 31b and c) and TEM micrograph. The orientation of TiO_2 nanocrystals with their long axis perpendicular to the FTO layer (Figure 31b and c) would also enhance the 004 diffraction peak.

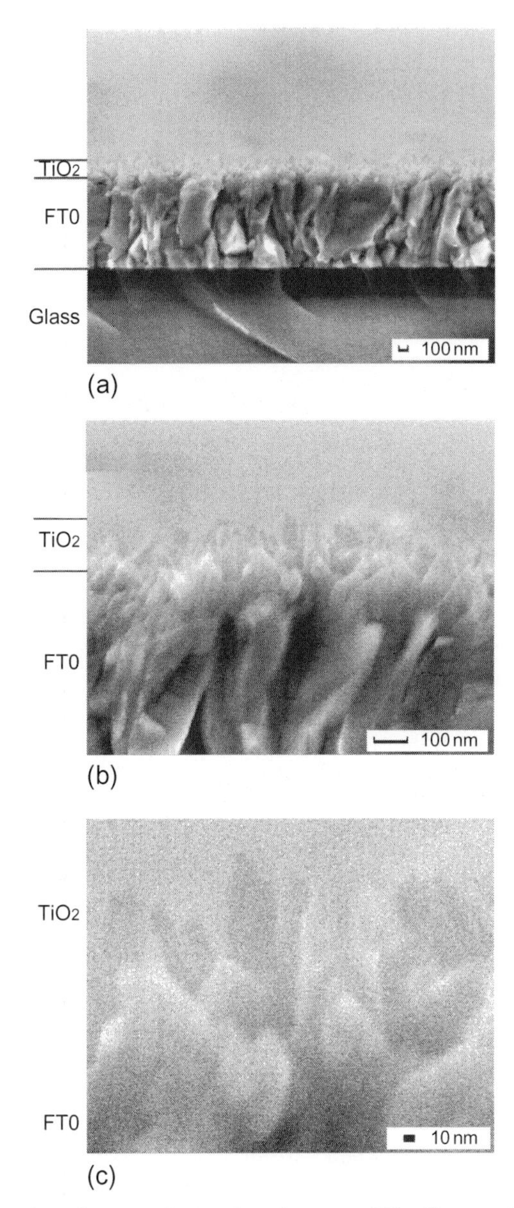

FIGURE 31 Scanning electron micrographs of anatase TiO_2 films on SnO_2:F substrates. (a) Fracture cross section of TiO_2 films. (b and c) Magnified areas of (a) showing the morphology of TiO_2 nanocrystals. *(Reprinted with permission from Masuda and Kato [140]. Copyright © American Chemical Society.)*

5 NANOSTRUCTURING OF SnO IN AQUEOUS SOLUTIONS

5.1 Nanosheet Assembled Crystals [33]

Tin oxide (SnO) particles were synthesized in aqueous solutions [33]. They consisted of nanosheets of SnO crystals. The sheets were about 50-100 nm diameter size and 5-10 nm thick. XRD analysis revealed that the particles were tin dioxide (SnO_2) and SnO crystals. The particles had a BET surface area of 85 m^2/g estimated using N_2 adsorption characteristics. BJH analysis indicated that mesopores 3.9 nm in size contributed to the increased surface area.

SnO_2, with a rutile-type crystalline structure, is an n-type wide bandgap (3.5 eV) semiconductor. It is an oxide of great interest for use in gas sensors [1–3], optical devices [164], lithium batteries [165–168], white pigments for conducting coatings, transparent conducting coatings for furnaces and electrodes [9], surge arrestors (varistors) [169,170], catalysts [171,172], opto-conducting coatings for solar cells [173], and dye-sensitized molecular sensors. Transparency, semiconductivity, and surfaced properties of SnO are suitable for these applications.

SnO_2 nanoparticles have been synthesized by several methods such as precipitation [168,174], hydrothermal synthesis [175,176], sol-gel [177,178], hydrolytic [179], carbothermal reduction [180], and polymeric precursor [181] methods. A variety of SnO_2 nanostructures, including nanowires, nanobelts [182], nanotubes [167,183,184], nanorods [185,186], spirals, nanorings [187], zigzag nanobelts [188], grains [189], flakes [189], plates [189], meshes [190], and columnar thin films [166], were synthesized.

Tin chlorides ($SnCl_2$ or $SnCl_4$) were commonly used in many reports. However, chlorine ions were difficult to remove from the systems and seriously altered superficial and electrical properties. For instance, chlorine ions degraded the sensitivity of gas sensors [[191]] and the aggregation of particles [192] and increased sintering temperatures [193]. Chloride problems can be avoided through the use of organic tin compounds such as alkoxides. These reagents are costly, however, which makes industrial syntheses implementation hardly attainable. In addition, sol-gel processes using alkoxides formed amorphous phases. High-temperature annealing, which increases cost, energy consumption, and CO_2 emissions, was necessary to obtain SnO crystals.

Aqueous syntheses of metal oxide crystals, including SnO, have recently been developed [15,38,194]. Control of metal oxide crystal morphology was realized by anisotropic crystal growth. They were induced by crystal growth control or organic additives. Thermodynamically stable crystal faces, for instance, depend on crystal growth conditions. Organic molecules adsorbed on typical crystal faces suppress crystal growth perpendicular to the faces. Morphology control provided an improvement in the properties of metal oxide devices.

In one study, SnO particles were prepared in aqueous solutions [33]. They were an assembly of nanosheets. Morphology, crystal phases, and N_2 adsorption characteristics were evaluated. Specific surface area and size distribution of pores were analyzed using isotherms.

5.1.1 Aqueous Synthesis of Nanosheet Assembled SnO Particles

SnF_2 (no. 202-05485, FW: 156.71, purity 90.0%; Wako Pure Chemical Industries, Ltd) was used as received. Distilled water in polypropylene vessels (200 mL) were capped with polymer films and kept at 90°C. SnF_2 (870.6 mg) was dissolved in the distilled water at 90°C to 5 mmol/L. The solutions were kept at 90°C for 30 min using a drying oven (DKN402; Yamato Scientific Co., Ltd), with no stirring. The solutions became cloudy shortly after the addition of SnF_2. The bottoms of the vessels were covered with white precipitates. The solutions were centrifuged at 4000 rpm for 10 min (Model 8920; Kubota Corp.). Precipitated particles were dried at 60°C for 12 h after removing the supernatant.

The solutions became cloudy shortly after the addition of SnF_2 because of the homogeneous nucleation and growth of SnO particles. SnO_2 and SnO are formed in aqueous solutions as follows [195,196]:

$$SnF_2 \xrightarrow{OH^-} SnF(OH) + F^- \xrightarrow{OH^-} Sn(OH)_2 + F^- \tag{a}$$

$$Sn(OH)_2 \rightarrow SnO + H_2O \tag{b}$$

$$Sn(OH)_2 \rightarrow SnO_2 + H_2 \tag{c}$$

$$SnO + H_2O \rightarrow SnO_2 + H_2 \tag{d}$$

The particles were gradually deposited to cover the bottom of the vessels. The supersaturation of the solutions was high at the initial stage and decreased as the color of the solutions changed.

Precipitated particles showed broad XRD peaks at $2\theta = 27$, 34, 38, 52, and 57.5. They were assigned to SnO_2 (JDPDS no. 41-1445). The peaks had a large width because of the small crystallite size of SnO_2.

5.1.2 Morphology of Nanosheet Assembled SnO Particles

Nanosheet assembled SnO particles were successfully fabricated. Crystallization of SnO nanosheets in aqueous solutions allowed a unique morphology of SnO to be obtained. Particles 300-800 nm in diameter were observed with field emission SEM (Figure 32). They were not dense particles but assemblies of SnO nanosheets. The sheets were about 50-100 nm diamenter size and 5-10 nm thick. The aspect ratio was estimated to be about 10. They were randomly aggregated to form the particles. The particles thus had continuous open pores inside them. They contributed to the increased surface area of the particles.

The nanosheets were generated in the solutions homogeneously. They aggregated to form particles, which made solutions cloudy. They then were precipitated to cover the bottoms of the vessels. These were consistent with the color change observed of solutions.

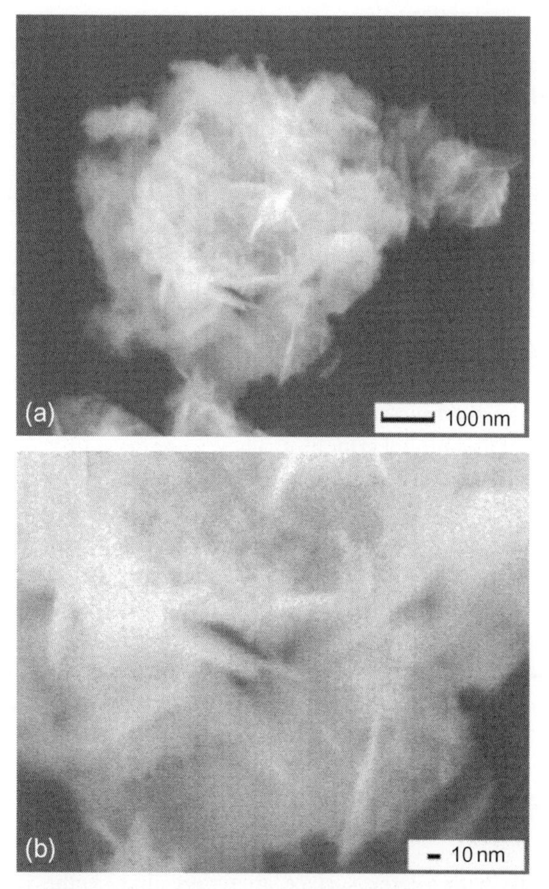

FIGURE 32 (a) Scanning electron micrograph of tin oxide particles. (b) A magnified area of (a). *(Reprinted with permission from Masuda and Kato [33]. Copyright © Elsevier.)*

5.1.3 N_2 Adsorption Characteristics of NanoSheet Assembled SnO Particles

The particles exhibited N_2 type IV adsorption-desorption isotherms. The desorption isotherm differed from adsorption isotherm in the relative pressure (P/P_0) range (from 0.45 to 0.97), showing mesopores in the particles. The BET surface area of the particles was estimated to be 85 m^2/g. This is higher than that of SnO_2 nanoparticles such as various types of SnO_2 (BET surface area of 47.2 m^2/g, 18.3 nm in diameter, no. 549657-25G [Aldrich]; BET surface area of 25.9 m^2/g, 34 nm in diameter [Yamanaka & Co., Ltd]; and BET surface area of 23 m^2/g, 26 nm in diameter, no. 37314-13 [NanoTek, C. I. Kasei Co., Ltd]) and In_2O_3-SnO_2 (BET surface area of 3-6 m^2/g, 100-300 nm in diameter; Sumitomo Chemical Co., Ltd). The particles were assemblies of nanosheets. Unique morphologies of the sheets contributed to the increased surface area of the particles.

Total pore volume and average pore diameter were estimated to be 0.343 cc/g and 16.1 nm, respectively for pores smaller than 259 nm at $P/P_0 = 0.9925$. They were estimated to be 0.088 cc/g and 4.1 nm, respectively, for pores smaller than 10.6 nm at $P/P_0 = 0.7994$. Using the BJH method, total pore volume also was estimated to be 0.354 cc/g for pores smaller than 174 nm.

Pore size distribution was calculated by the BJH method using desorption braches. It showed a pore size distribution curve with a strong peak at ~3.9 nm and a broad peak from 25 to 175 nm. The particles had a large number of mesopores ~3.9 and 25-175 nm in diameter. The pores were spaces surrounded by nanosheets and interparticle spaces. Micropores smaller than 2 nm were not suggested by BJH pore size distribution.

Nanosheet assembled SnO particles were fabricated by aqueous solution synthesis. Anisotropic crystal growth of SnO was effectively utilized to form a nanosheet assembled structure. They were crystallized at an ordinary temperature without annealing, allowing aggregation of the particles and a decrease in the surface area to be avoided. The particles were 300-800 nm in diameter and were SnO_2 and SnO crystals. The sheets were about 50-100 nm diameter size and 5-10 nm thick. The particles had a BET surface area of 85 m^2/g estimated based on N_2 adsorption characteristics. BJH analysis indicated that mesopores 3.9 nm in size contributed to the increased surface area. The particles were prepared by an environmentally friendly process. The system had the advantages of low cost, low energy consumption, and low CO_2 emission.

5.2 Morphology Control and Enhancement of Surface Area [197]

SnO nanocrystals with a large surface area were first synthesized in aqueous solutions at 50°C. BET surface area of 194 m^2/g was successfully reached [197]. It was much higher than that of SnO_2 with BET surface areas of 47.2 m^2/g (Aldrich), 25.9 m^2/g (Yamanaka & Co., Ltd), and 23 m^2/g (C. I. Kasei Co., Ltd) and In_2O_3-SnO_2 (BET surface area of 3-6 m^2/g; Sumitomo Chemical Co., Ltd). N_2 adsorption characteristics revealed that they had pores of 1-3 nm, which contributed to the large surface area. TEM, electron diffraction (ED), and XRD indicated the morphology, crystal structure, and chemical composition of nanocrystals. A novel process allowed sintering and deformation of the crystals to be avoided and hence achieved a large surface area and unique morphology.

SnO nanosheets were formed in solutions. A BET surface area of 194 m^2/g was successfully reached. The origin of the large surface area was discussed with pore size distribution and morphology observations.

SnF_2 (870.6 mg) was dissolved in distilled water (200 mL) at 50°C to 5 mmol/L. The solutions were kept at 50°C for 20 min and then at 28°C for 3 days without stirring. The nanocrystals precipitated to cover the bottom of the vessels. For comparison, the solutions were centrifuged at 4000 rpm for 10 min after being maintained at 50°C for 20 min. Precipitated particles were dried at 60°C for 12 h after removal of the supernatant.

The nanosheets synthesized at 50°C for 20 min and at 28°C for 3 days were a mixture of an SnO_2 main phase and an SnO additional phase (Figure 33a). XRD peaks at $2\theta=26.5$, 33, 51.4, 62, 64.5, 80, and 89 were assigned to 110, 101, 211, and so on of SnO_2 (JDPDS no. 41-1445). 101, 110, And 002 diffraction peaks of SnO (JDPDS no. 06-0395) were overlapped with peaks of SnO_2. For comparison, the diffraction pattern of the nanosheets synthesized at 50°C for 20 min is shown in Figure 33b. Half-maximum full width of the peaks was smaller than that shown in Figure 33a. Less SnO content resulted in sharp peaks of SnO_2 (Figure 33b). These observations indicated that SnO was mainly formed at 28°C rather than 50°C.

The nanosheets were well dispersed in ethanol. The nanosheets in supernatant solutions were skimmed with copper grids for TEM observations. They were 20-50 nm in diameter with a uniform thickness (Figure 34a). Similar structures were

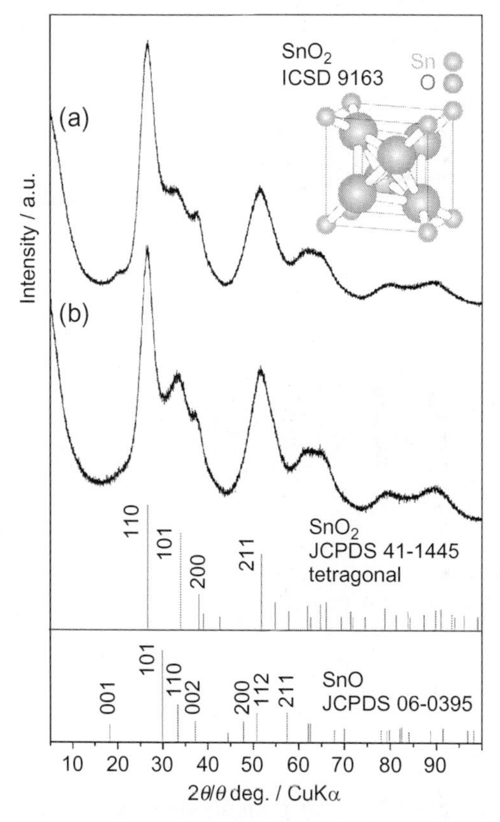

FIGURE 33 X-ray diffraction patterns of tin oxide nanosheets fabricated at 50°C for 20 min and at 28°C for 3 days (a) and tin oxide nanosheets fabricated at 50°C for 20 min (b). a.u., arbitrary units. *(Reprinted with permission from Masuda* et al. *[197]. Copyright © Wiley.)*

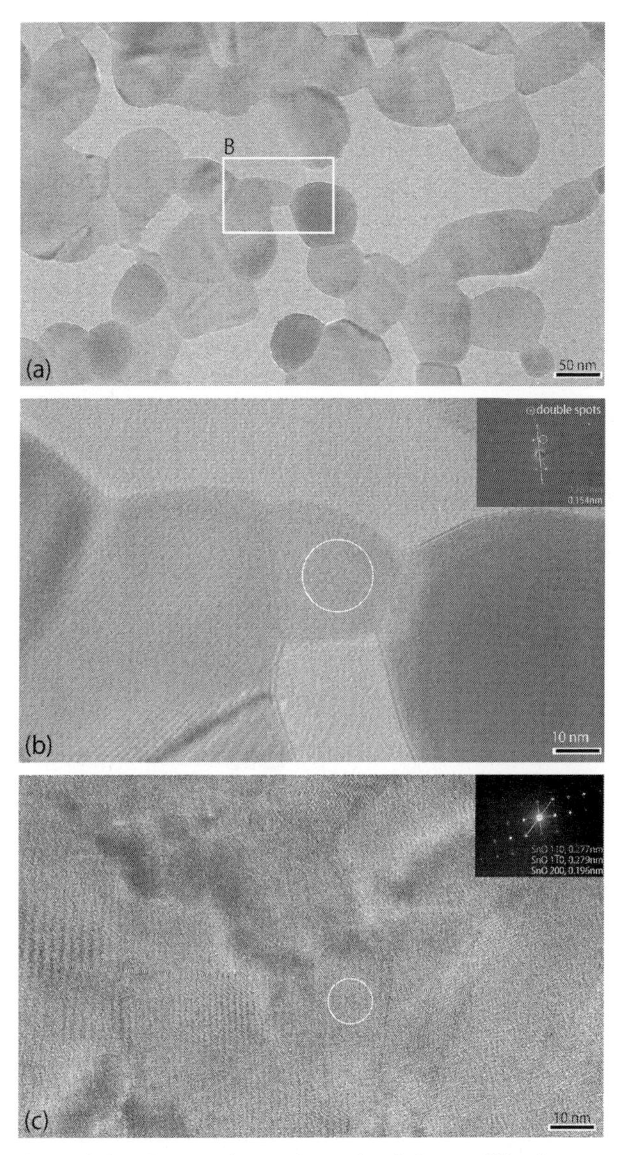

FIGURE 34 Transmission electron micrographs (a–c) and electron diffraction patterns (insets in [b] and [c]) of tin oxide nanosheets. (b) A high-magnification image of the square area in (a). (c) A high-magnification image of the circled area in (b). *(Reprinted with permission from Masuda* et al. *[197]. Copyright © Wiley.)*

observed in many areas. They curled up during long-term observation. It indicated that they had a sheet structure. Electron beams damaged them, transforming the structures. Some of them tightly connected with each other (Figure 34b). They had clear interfaces without pores or small grains. The nanosheet showed electron diffractions (Figure 34b, inset). Lattice spacing calculated from spots indicated with a red line and a yellow line were 0.283 and 0.154 nm, respectively. They can be assigned to SnO_2 and/or SnO. The former can be assigned to the 101 crystal plane of SnO_2 (0.264 nm) or the 110 crystal plane of SnO (0.269 nm). The latter can be assigned to the 310 crystal plane of SnO_2 (0.149 nm), the 221 crystal plane of SnO_2 (0.148 nm), the 202 crystal plane of SnO (0.149 nm), or the 103 crystal plane of SnO (0.148 nm). In addition, diffraction spots related to lattice spacing of 0.283 nm were observed in the upper part. They are marked with two white circles and a red line in the Figure 34b inset. The double spots indicate that the area shown in a white circle in Figure 34b comprised two crystals, that is, two stacked nanosheets. A high-magnification image also was obtained from another observation area (Figure 34c). The structure was thinner than that in Figure 34a and b. Clear image low-magnification images were not obtained because of the low contrast. However, there was a clear high-magnification image showing a lattice fringe (Figure 34c). The electron diffraction patterns showed lattice spacing of 0.277 nm (red line), 0.279 nm (green line), and 0.196 nm (yellow line). They were assigned to the 110, 110, and 200 crystal planes of SnO, respectively. Lattice spaces and their angles were well matched to those of SnO. Chemical composition was estimated from several points of a tightly-packed area that included nanosheets and spherical crystals. Chemical composition varied among each position in the range of 1:1.7-2.7 (Sn to O), which was more similar to that of SnO_2 than SnO. These observations indicated that the crystals were a mixture of SnO_2 and SnO.

BET surface area of the nanosheets was estimated from the N_2 adsorption isotherm (Figure 35a). The nanosheets had a large surface area of 194 m^2/g (Figure 35b). It was much higher than that of various SnO_2 nanoparticles (BET surface area 47.2 m^2/g, 18.3 nm in diameter, no. 549657-25G [Aldrich]; BET surface area 25.9 m^2/g, 34 nm in diameter [Yamanaka & Co., Ltd]; and BET surface area 23 m^2/g, 26 nm in diameter, no. 37314-13 [NanoTek, C. I. Kasei Co., Ltd]) and In_2O_3-SnO_2 (BET surface area 3-6 m^2/g, 100-300 nm in diameter; Sumitomo Chemical Co., Ltd). In addition, it was more than double that previously reported (85 m^2/g) [33]. Pore size distribution was analyzed by the BJH method using an adsorption isotherm (Figure 35c), indicating that the nanosheets included 1- to 2-nm pores. Micropore analysis was performed with DFT/Monte-Carlo fitting, which was completely consistent with isotherms (Figure 35d). Pores of 1-3 nm were in the nanosheets (Figure 35e). The micro- and mesopores contributed to the large surface area (194 m^2/g) in this system. For comparison, BET surface area of the nanosheets synthesized at 50°C for 20 min was estimated to be 146 m^2/g. It was also higher than that of nanoparticles described in previous studies.

This section describes nanosheets of SnOs fabricated in aqueous solutions at an ordinary temperature. BET surface area successfully reached 194 m^2/g, which is much higher than that of nanoparticles described in previous studies. The two-dimensional sheet structure was an ideal structure for a large surface area per unit weight. Nano-sized thickness directly contributed to the large surface area. Crystalline nanosheets were prepared without high-temperature

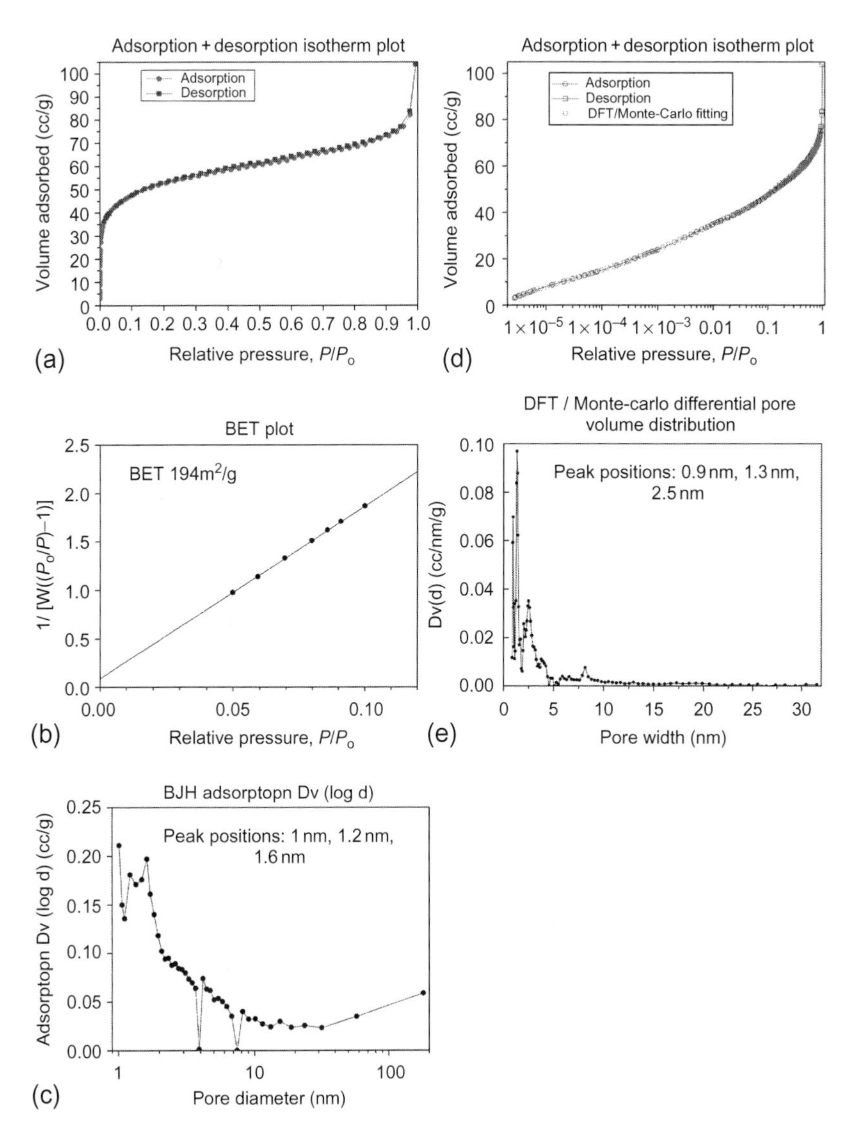

FIGURE 35 (a) N_2 adsorption-desorption isotherm of tin oxide nanosheets. (b) BET surface area of tin oxide nanosheets. (c) Pore size distribution calculated from the N_2 adsorption data of tin oxide nanosheets using the BJH equation. (d) N_2 adsorption-desorption isotherm and DFT/Monte-Carlo fitting curve of tin oxide nanosheets. (e) Pore size distribution calculated from the N_2 adsorption data of tin oxide nanosheets using the DFT/Monte-Carlo equation. *(Reprinted with permission from Masuda* et al. *[197]. Copyright © Wiley.)*

annealing, which degrades the surface area and nanostructures. A large surface area and the unique nanostructure of the sheets can be applied in gas sensors, dye-sensitized solar cells, and molecular sensors.

5.3 SnO Coatings on Polytetrafluoroethylene Films [198]

Tin polytetrafluoroethylene (PTFE) films were successfully coated with SnO in aqueous solutions [198]. SnO was crystallized in the solution and formed nanocrystal coatings on the polymer films. The coatings comprised SnO_2 and SnO crystals. They were assemblies of SnO nanosheets ~10-50 nm diameter size and about 5-nm thick. The nanocrystal films can be exfoliated from the PTFE substrates. SnO nanocrystal films had a rough liquid surface and a dense substrate surface. The transparency of PTFE films coated with SnO was the same as that of bare PTFE films in the range of 400-800 nm. SnO decreased the transparency about 25% at 320 nm. PTFE films coated with SnO nanocrystals can be pasted on desired substrates.

Organic-inorganic hybrid materials such as metal oxide electronics on polymer flexible films have received considerable attention in recent years for application in light-weight flexible sensors, displays, and dye-sensitized solar cells, among others. Polymer films offer the advantages of flexibility, light weight, low cost, and impact resistance.

SnO is an important semiconductor with a wide bandgap (3.6 eV) at room temperature. It has been widely used in gas sensors [184,199], optical devices [164], and lithium batteries [165–167]. A novel type of biosensor was proposed to detect environmental toxins such as bisphenol-A or dioxin [108,200]. SnO is a candidate material for the sensor because of its suitable bandgap, surface characteristics, and high transparency.

Metal oxides including SnO have been synthesized with high-temperature processes (several hundred degrees) for many years. The aqueous synthesis of metal oxides has recently attracted much attention as a next-generation science and technology [15,71,140,194]. Aqueous systems are environmentally friendly and have the advantages of low energy consumption, low cost, and a process free of organic solvents.

In recent years metal oxide films and their microstructures have been fabricated on organic surfaces such as PET films [9,201] and SAMs [4,38,202,203]. However, metal oxide formation on PTFE was difficult compared with PET films. PTFE films are widely used in electronic applications because of their low chemical reactivity, low coefficient of friction, high melting point (327°C), high resistance to corrosion, a high dielectric strength over many different frequencies, a low dissipation factor, and a high surface resistivity. They were selected as substrates in this study [198].

SnO nanocrystals were prepared on PTFE films in aqueous solutions [198,200]. They were crystallized in a solution containing tin ions to form films consisting of nanosheets. The process realized SnO film formation without high-temperature annealing and with a unique morphology of SnO crystals.

SnF_2 (870.6 mg) was dissolved in distilled water (200 mL) at 90°C to 5 mmol/L. PTFE films (thickness, 50 μm; ASF-110; Chukoh) with a silicone adhesive (thickness, 30 μm) were pasted on quartz substrates (25 × 50 × 1 mm). They were immersed in the middle of the solution, with the bottom up at an angle or with the top up at an angle. They were tilted 15° to stand upright. The solutions were kept at 90°C using a drying oven (DKN402; Yamato Scientific Co., Ltd) for 2 h with no stirring. The solutions became slightly cloudy after 2 h. The as-deposited nanocrystals on substrates were rinsed under running water and dried with a strong air spray. In addition, the solutions kept at 90°C for 2 h were centrifuged at 4000 rpm for 10 min (Model 8920; Kubota Corp.). Precipitated particles were dried at 20°C for 12 h after the removal of supernatant.

Bare PTFE films pasted on quartz substrates had cracks about 50 to 200 nm in length (Figure 36). The direction of longer cracks was perpendicular to the extensional direction. They formed during the adhesive processes. The quartz substrates and the PTFE films were transparent or slightly white, respectively. PTFE films can be pasted on desired substrates such as quartz, metals, and polymers with silicone adhesive.

The surface of PTFE films was completely covered with assemblies of nanosheets (Figure 37a). Uniform formation of SnO coatings is one of the advantages of the solution processes. Large sheets also were observed from the surfaces (Figure 37b). They were about 200 to 300 nm diameter size and about 10 nm thick (Figure 37b). Some of them stood perpendicular at an angle to the PTFE films. They had an angular outline, which was connected by straight lines. They were created from the anisotropic crystal growth of SnO, reflected in their crystal structure. Large sheets connected in a cross shape also were observed at the lower right (Figure 37b). Nanosheets were estimated to be roughly 10-50 nm diameter size and about 5 nm thick (Figure 37c). They connected to each other to form continuous films on the PTFE surfaces. The SnO films were exfoliated

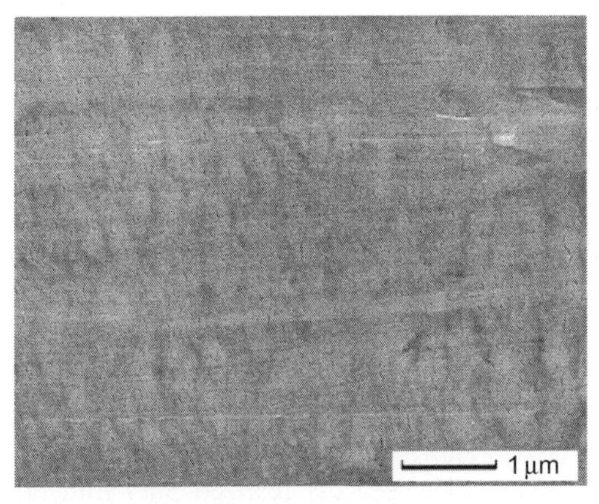

FIGURE 36 Scanning electron micrograph of a bare polytetrafluoroethylene film. *(Reprinted with permission from Masuda and Kato [198]. Copyright © Wiley.)*

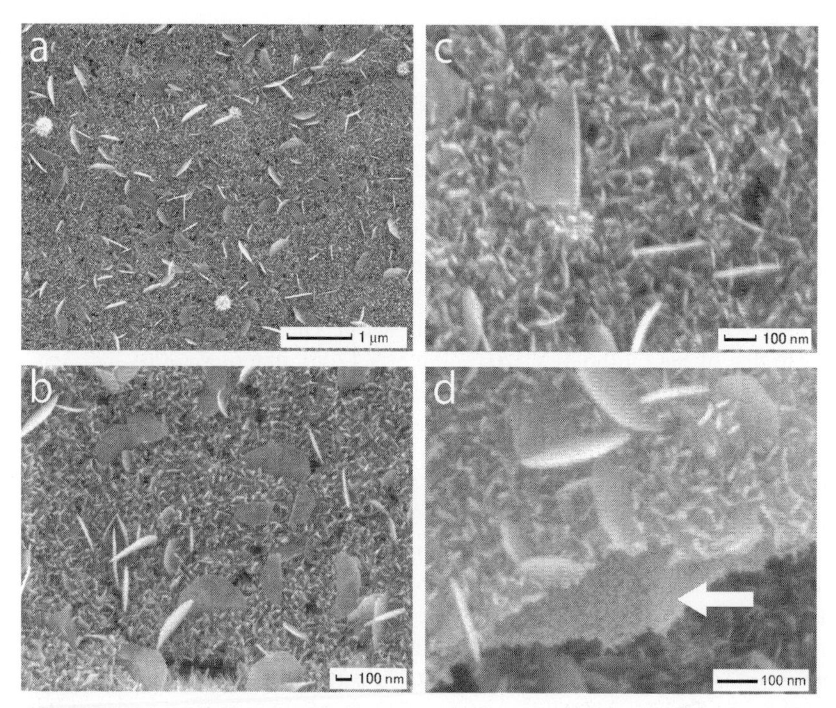

FIGURE 37 (a) Scanning electron micrograph of tin oxide nanocrystals on a polytetrafluoroethylene film. (b and c) Magnified areas of (a) showing the morphology of the nanocrystals. (d) Exfoliated tin oxide nanocrystal films showing the liquid-side and substrate-side surfaces. The substrate-side surface is indicated by a white arrow. *(Reprinted with permission from Masuda and Kato [198]. Copyright © Wiley.)*

from PTFE films by scratching using a metal spatula. Exfoliated SnO films were placed on a substrate for SEM observation. Three exfoliated SnO films were partially overlapping (Figure 37d). The top film and the bottom film showed their liquid surfaces. They were similar to that shown in Figure 37a–c. The middle film showed a substrate surface, which was in contact with the PTFE film during immersion. The substrate surface of the middle film is indicated by a white arrow. It had a dense surface, which consisted of nanocrystals ~5-10 nm diameter size. These observations indicated that these dense films formed on the PTFE films during the first stage of SnO film formation. SnO then grew to sheet shapes to form assemblies of nanosheets. In addition, large sheet crystals grew on the dense SnO films.

The XRD pattern obtained from a PTFE film coated with SnO was similar to that from a bare PTFE film. Diffraction peaks assigned to SnO were not observed because the film was thin. Precipitated particles were evaluated after drying. They showed broad XRD peaks at $2\theta=27$, 34, 38, and 52. They were assigned to 110, 101, 200, or 211 diffraction peaks from SnO_2 (JDPDS no. 41-1445). The peaks were wide because of the small crystallite size of SnO_2.

Peaks also were observed at $2\theta=29.5$, 32.0, 37.4, 48.0, 50.4, and 57.4. They were assigned to 101, 110, 002, 200, 112, and 211 diffraction peaks from SnO, respectively (JDPDS no. 06-0395). XRD analysis indicated that the particles obtained from the solutions consisted of SnO_2 crystals and SnO crystals.

Carbon, F, and O were detected from bare PTFE films (Figure 38b). Their chemical ratio (C to F to O) was estimated at 1:2.05:0.02. It was consistent with chemical composition of PTFE (C-to-F ratio of 1:2). A small amount of O was detected from surface contamination. PTFE films coated with SnO showed spectra of Sn, O, F, and carbon (Figure 38a). A spectral peak corresponding to Sn $3d_{5/2}$ was observed at 487.2 eV (Figure 38c1). The binding energy was similar to that of SnO_2 (486.3 eV [204], 486.5 eV [205], 486.6 eV [206], 487.3 eV) and higher than that of Sn metal (484.8, 484.85, 484.87, 484.9, and 485.0 eV), which suggested Sn atoms in surface coatings were positively charged by forming direct bonds with O. Binding energy of O 1s centered at about 531.2 eV corresponds to that of SnO_2 (Figure 38c2). The chemical ratio of the coatings (Sn to O to F to carbon) was estimated at 1:1.88:7.25:3.82; the carbon-to-F ratio was 1:1.90. The surface coatings consisted mainly of SnO_2 (Sn-to-O ratio of 1:2). They formed on PTFE films (carbon-to-F ratio of 1:2). Differences and similarities between XPS and XRD analyses suggest deposition mechanism of tin oxides. The crystal phases and chemical compositions of the surface coatings were different from those of the precipitated particles. SnO_2 crystallized on the films to form surface coatings, whereas SnO_2 and SnO homogeneously crystallized to form particles in the solutions. Pure SnO_2 coatings thus were successfully formed on PTFE films.

Quartz substrates had high transparency in the range of 200-850 nm (Figure 39, black line). PTFE films pasted on quartz substrates were visually observed to be slightly white. Their transparency was lower than that of quartz (Figure 39, red line). In particular, transparency decreased as wavelength decreased below 350 nm. PTFE films coated with SnO showed the same transparency as bare PTFE films, in the range from 400 to 850 nm (Figure 39, blue line). The decreased transparency about 25% at 320 nm. SnO particles precipitated from the solution were evaluated for comparison. They had an absorption peak centered at 320 nm. This absorption was caused by SnO. These analyses indicated that SnO was deposited on PTFE films immersed in the solutions and their transparency decreased at 320 nm.

PTFE films were successfully coated with SnO nanocrystals. SnO was crystallized in an aqueous solution to form nanosheet assembled films. They were about 10 to 50 nm diameter size and about 5 nm thick. Large sheets ~200 to 300 nm diameter size and about 10 nm thick also crystallized on the surfaces. XRD analysis indicated that SnO was a mixture of SnO_2 and SnO. PTFE films coated with SnO_2 were transparent in the range from 400 to 850 nm. SnO_2 on the films had absorption centered at 320 nm. SnO-coated PTFE films can be pasted onto desired substrates. Hybrid SnO-PTFE composites may be useful for increasing the potential application of SnO film in flexible electronics.

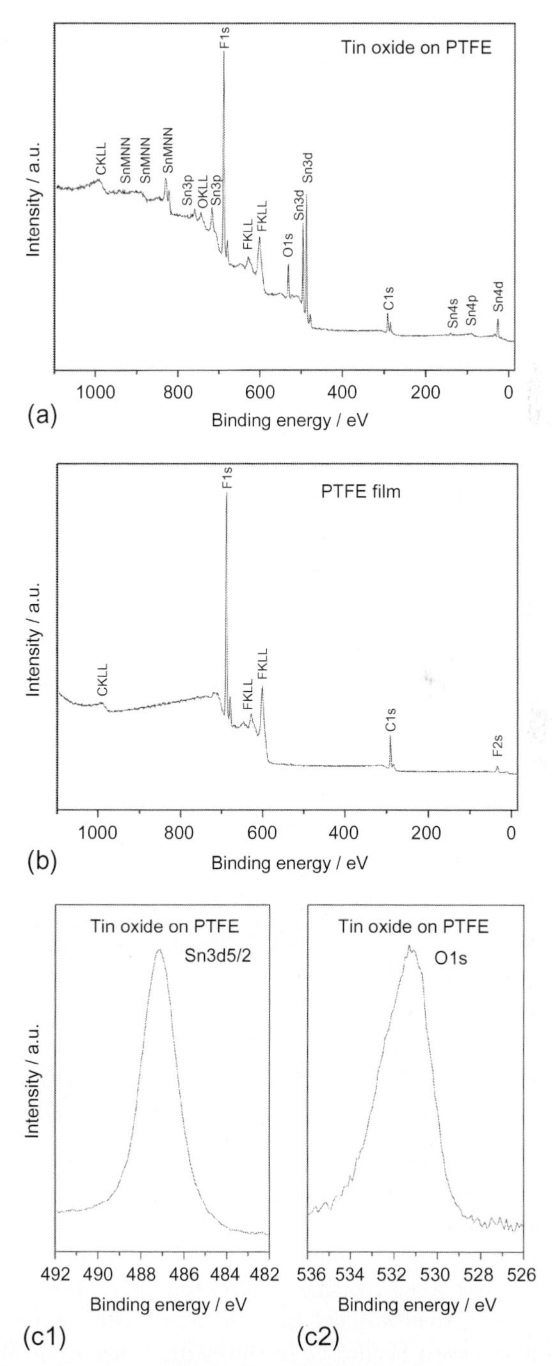

FIGURE 38 X-ray photoelectron spectra of tin oxide nanocrystals on a polytetrafluoroethylene (PTFE) film (a) and a bare PTFE film (b). The Sn $3d_{5/2}$ (c1) and O 1s (c2) spectra of tin oxide nanocrystals on a PTFE film. *(Reprinted with permission from Masuda and Kato [198]. Copyright © Wiley.)*

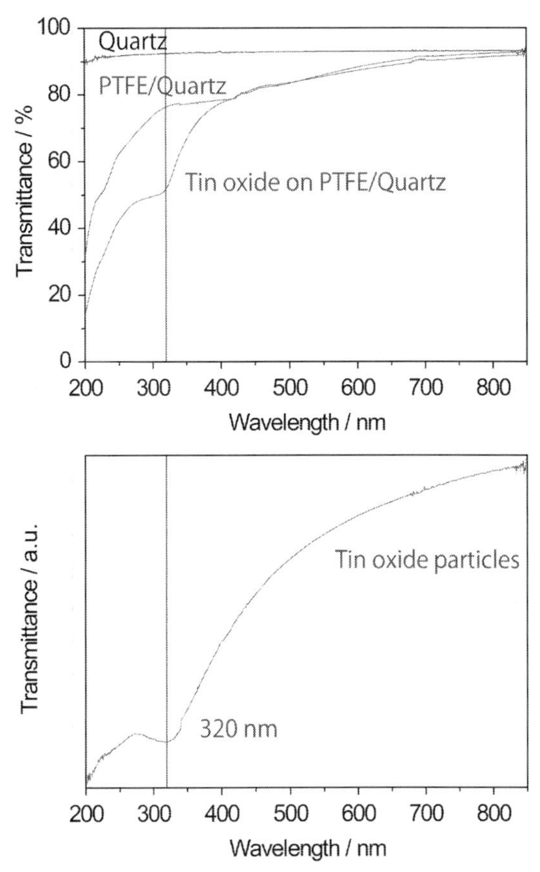

FIGURE 39 Ultraviolet/visible spectra of tin oxide particles on a quartz substrate (black line) (a), a polytetrafluoroethylene (PTFE) film pasted on a quartz substrate (red line) (b), and tin oxide nanocrystals on a PTFE film pasted on a quartz substrate (blue line) (c). *(Reprinted with permission from Masuda and Kato [198]. Copyright © Wiley.)*

5.4 Nanosheet Assembled Films for Superhydrophobic/ Superhydrophilic Surfaces and Cancer Sensors [207]

Superhydrophobic surfaces have recently attracted significant attention because of their water repellency, oil repellency, antifouling property, antifogging property, self-cleaning property, and their potential applications in various fields.

Hydrophobic surfaces are found in nature. Leaves of *Nelumbo nucifera* (i.e., sacred lotus) have a self-cleaning property because of their superhydrophobic surfaces [208,209]. Water striders walk on water using water-repellent legs [210,211]. Namib desert beetles collect drinking water from fog-laden wind using the hydrophilic/superhydrophobic patterned surface of their wings [212]. A combination of nano- and microscale roughness and hydrophobic materials create superhydropholic surfaces. It was reported that the maximum water contact angle of a flat surface was <120°. This indicated that control of nano- and microstructure is important for superhydrophobic surfaces.

Fluorocarbon layers are widely used for hydrophobic coatings. However, they are less resistant to high temperatures. Chemically and thermally stable coatings are required for long-term durability, high-temperature use, and voltage resistance in various applications. Metal oxides are candidate materials for stable surface coatings. They have high resistance to strong acids, strong alkalis, high temperature (hundreds of degrees Celsius), wear and abrasion, and high electric power. In addition, the process for fluorocarbon coatings is costly and difficult to coat on large substrates, complex shapes, and nanomaterials including particles and tubes. A simple and low-cost process is required to be used widely.

Transparent conductive oxide electrodes have been applied in many electronic devices. Surface properties and surface modification of the electrodes are key factors in the development of sophisticated devices such as molecular sensors, gas sensors, dye-sensitized solar cells, and electronic devices. Superhydrophobic or superhydrophobic coatings of transparent conductive fluorine-doped tin oxide (FTO) electrodes is required. The superhydrophobic surface of FTO can suppress surface contamination in high-precision devices, water molecular adhesion in gas sensors, and needless molecular adhesion in molecular sensors. The hydrophobic surface is effective for chemical molecular modification using hydrophobic interactions. In addition, electrodes with a superhydrophobic surface are required for surface modification with organic functional groups, biomolecules, or SAMs in molecular sensors and biosensors.

A unique simple process for superhydrophobic surface coatings has been reported [207]. These coatings were formed on transparent conductive FTO substrates. In addition, superhydrophobic surfaces were modified to superhydrophilic surfaces by light irradiation in air. Water contact angle, transparency, reflectance, microstructure, crystal phase, chemical ratio, and chemical state of the surface coatings were investigated. Furthermore, they were modified with a dye-labeled monoclonal antibody for cancer sensing. Photoluminescence and photocurrent from the biomolecule/semiconductor electrodes were evaluated.

5.4.1 Synthesis of SnO Nanosheet Coatings on FTO Substrates

Glass substrates coated with transparent conductive FTO films (FTO, SnO_2:F [Asahi Glass Co., Ltd]; 9.3-9.7 Ω/□, 26×50×1.1 mm) were blown by air to remove dust and were exposed to VUV (air flow, 100 V, 200 W; 14 mW/cm^2 for 184.9 nm and 18 mW/cm^2 for 253.7 nm at a 10-mm distance from lamp; low-pressure mercury lamp PL16-110 [SEN Lights Co.]) for 10 min. The initial FTO substrates showed a water contact angle of 96°. The VUV-irradiated surfaces, however, were wetted completely (contact angle, 0-1°). Various functional groups such as octadecyl or phenyl groups of SAMs formed on films are reported to be modified to OH groups by VUV irradiation using low-pressure mercury lamps [213,214]. These suggest that a small amount of adsorbed molecules on the FTO substrates was completely removed by VUV irradiation. Since the original FTO surfaces were covered by hydrophilic OH groups, the surfaces are modified to become superhydrophilic after irradiation. SnO can be crystallized on FTO substrates without organic molecules or contamination at the interface between the SnO and the FTO surfaces.

SnF_2 (FW, 156.71; 90.0% purity; no. 202-05485; Wako Pure Chemical Industries, Ltd) was used as received. Distilled water (200 mL) was heated at 90°C in a capped Teflon vessel using a drying oven (DKN402; Yamato Scientific Co., Ltd.). SnF_2 (870.6 mg) was dissolved in the distilled water at 94°C to 25 mmol/L. The FTO substrates were immersed in the middle of the solution with the bottom up, tilted at 15° to stand upright. The oven containing the vessels was heated to 200°C for 15 min. The solutions reached 200°C 35 min after immersion of the FTO substrates. They were kept at 200°C for 20 min. The vessels were withdrawn from the oven and naturally cooled for 100 min. The FTO substrates were removed from the solution and cleaned with distilled water and strong air blow. The substrates were kept at 200°C for 20 min in the above procedure, but they were in contact with the solutions for 155 min in total.

5.4.2 Surface Modification of SnO Nanosheets with Prostate-Specific Antigen

SnO nanosheets were formed on FTO substrates. They were modified with biomolecules as follows. Prostate-specific antigen,—a monoclonal antibody of NB013 (antihuman α-fetoprotein [AFP] monoclonal antibody; Human-AFPMoAb, no. NB013, IgG1 [Nippon Biotest Laboratory, Tokyo, Japan)—was labeled with Cy5 dye. It was dissolved in phosphate buffer solution (10 mmol/L, pH 7.4).

Monoclonal antibody of NB013 was used in this sensing system because it selectively reacted with human AFP. AFP was not detected in the blood serum of a normal adult. However, hepatocellular cancer (hepatocellular carcinoma) and germ cell teratoblastoma, as well as hepatic regeneration, viral hepatitis, and hepatocirrhosis (cirrhosis), caused an increase in the AFP concentration. Reaction of monoclonal antibody NB013 with AFP can be used for sensing of hepatocellular cancer. The sensing system can detect human AFP in adult blood serum for early detection of hepatocellular cancer.

Cy5 was selected because it emitted photoluminescence under excitation of usual red light. Combination of SnO semiconductor and Cy5 dye was suitable to generate a photocurrent because of the oxidation-reduction potential and bandgap engineering.

The dye-labeled antibody (Cy5-NB013) is a candidate material for our newly developed sensor. Cy5 is expected to adsorb on the substrates in the presence of human AFP through an antigen-antibody reaction. A red laser of 650 nm

(97 mW) was selected because it is suitable for the generation of electrons in the combination of Cy5 dye and an SnO_2 semiconductor. An electric signal was created by laser light irradiation on Cy5 at the surface of the electrodes. Generation of a photocurrent under irradiation was an essence of the sensing mechanism. The electrodes covered with the antibody that contained no dye were evaluated to measure noise current under irradiation for comparison.

Polyvinyl chloride (PVC) tape $(CH_2–CHCl)_n$, 26×22 mm, 100-μm thickness) was perforated with 9 holes (3 holes×3 rows) 25 mm in diameter using a flatbed cutting plotter (CG-60ST; Mimaki Engineering Co., Ltd). The middle areas of the FTO surfaces were covered with PVC tape after the deposition of SnO nanosheets.

Dye-labeled monoclonal antibody (Cy5-NB013) was dissolved in sterile water to 10 or 1 μg/mL. Monoclonal antibody (NB013) without dye was dissolved in sterile water to be 10 μg/mL to evaluate noise current for comparison.

Dye-labeled monoclonal antibody (Cy5-NB013, 10 μg/mL) solution (5 μL) was dropped with pipettes onto the three holes in the upper row on the PVC tape. Dye-labeled monoclonal antibody (Cy5-NB013, 1 μg/mL) solution (5 μL) or a solution (5 μL) of monoclonal antibody without dye (NB013, 0 μg/mL) was dropped onto the three holes in the middle and bottom rows, respectively. The substrates were kept at 37°C for 10 min in air. They were shaken in distilled water for 10 min using a slide washer (SW-4; Juji Field Inc., Tokyo, Japan).

The FTO substrates with SnO nanosheets were kept in a plastic case for a month after the deposition of SnO. Static water contact angle, advancing water contact angle, or receding water contact angle were evaluated with a contact angle meter (Easy Drop, DSA20SS, Kruss, Germany). They were exposed to VUV light for 10 min. Contact angles of the substrates were evaluated again after the irradiation.

Surfaces that have a static water contact angle >90° or <5° are known as superhydrophobic or superhydrophilic surfaces, respectively. Contact angle is governed by an energy equilibrium and is given by the Young equation (Equation (a)) (Figure 40). Surface chemistry and surface topology influence the contact angle and contact angle hysteresis. Hysteresis is defined as the difference between the advancing and receding water contact angles. Contact angle measurements give information about surface properties.

$$\gamma_{\mathrm{ls}} = \gamma_{\mathrm{cs}} + \gamma_{\mathrm{lc}} \cos\theta$$

where γ is the interfacial surface energy, θ is the contact angle, l is the liquid, s the substrate, and c the crystal.

$$V \equiv \frac{\pi h}{6}\left(3r^3 + h^2\right)$$

where V is the volume of a droplet, 1 mm^3 (1 μL), r is the radius of a droplet (5 mm), and h is the height of a droplet (in millimeters).

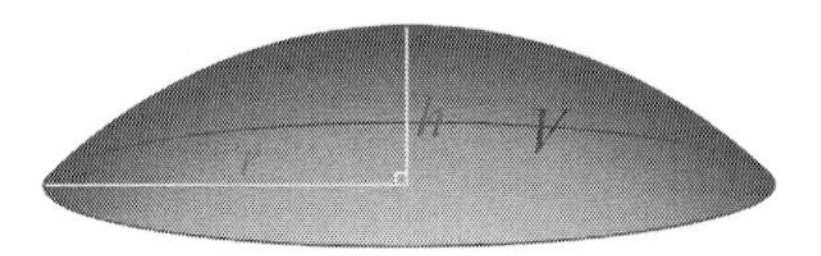

FIGURE 40 Model of a droplet for estimation of height. *(Reprinted with permission from Masuda* et al. *[207]. Copyright © American Chemical Society.)*

5.4.3 Morphology, Crystal Structure, and Chemical Composition of SnO Nanosheet Coatings on FTO Substrates

SnO nanosheets were formed on FTO substrates in an aqueous solution (Figure 41). FTO substrates had surface relief structures of about 500-1000 nm. They were polycrystals of SnO_2. Their surfaces were covered uniformly with SnO nanosheets (Figure 42a). They consisted of two kinds of crystals (Figure 42b). Larger nanosheets had a plane size of ~100-500 nm and a thickness of ~10-50 nm (Figure 42b). Smaller crystals had a plane size of ~10-100 nm and a thickness of ~1-10 nm. Their aspect ratios were roughly estimated at 10. Anisotropic crystal growth allowed us to synthesize SnO nanosheets.

Cross-sectional images of SnO coatings were observed with TEM. The FTO layer (800-900 nm thick) was on a glass substrate (Figure 43e). SnO nanosheets were deposited on the FTO layer (Figure 43a). A large nanosheet (~300-nm plane size and ~20 nm thick) was observed on the surface. The aspect ratio was roughly estimated to ~15 (300 nm/20 nm). It stood perpendicular to the substrate, with a tilted angle of about 10°. An electron diffraction pattern was obtained from the nanosheet in area 1 of Figure 43a. It was assigned to the single phase of an SnO_2 crystal without an additional phase. The pattern was not diffraction rings but electron diffraction spots. This indicated that the nanosheet was a single crystal. The crystal lattice distance parallel to the SnO_2 110 crystal planes was estimated at 0.346 nm. Both crystal lattice distances that were parallel to the SnO_2 101 or 011 planes were estimated at 0.293 nm. They were same lattice distance because SnO_2 had a tetragonal crystal shape with an *a*-axis, *a*-axis, and *c*-axis. The chemical composition (Sn to O) of the nanosheet was estimated to 1:1.8 using the EDS analysis system. It was similar to the chemical composition of SnO_2 rather than SnO.

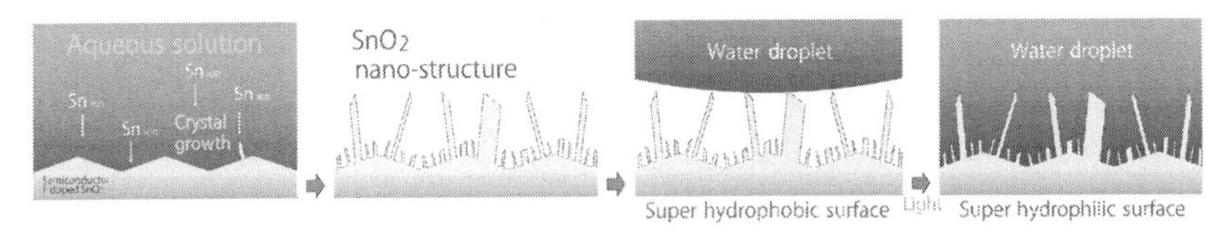

FIGURE 41 Concept of a tin oxide nanosheet coating on a FTO substrate with a hydrophobic or hydrophilic surface. *(Reprinted with permission from Masuda* et al. *[207]. Copyright © American Chemical Society.)*

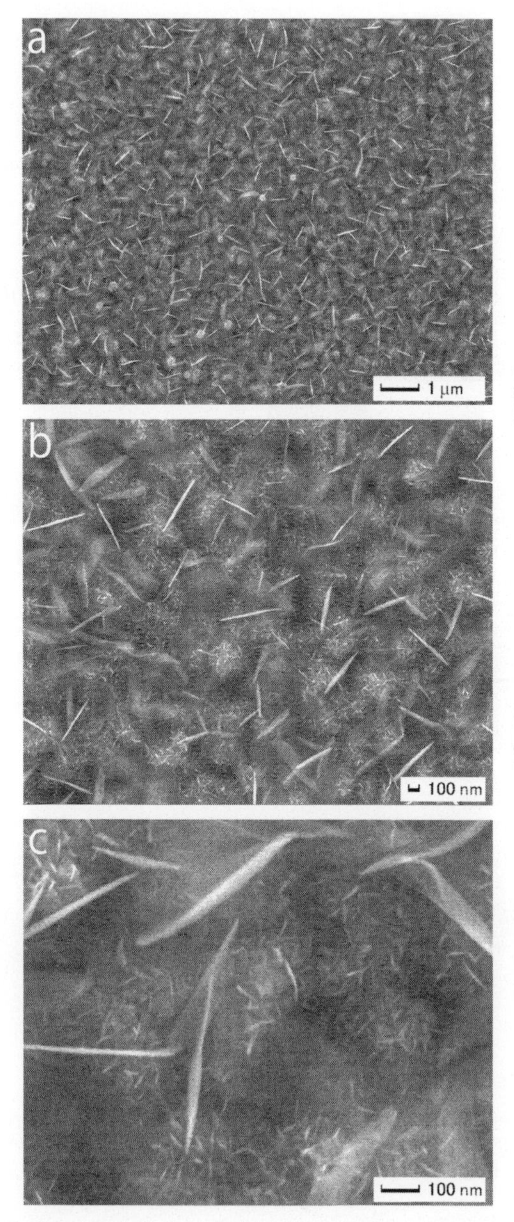

FIGURE 42 (a) Surface morphology of a tin oxide nanosheet coating on a FTO substrate. (b and c) Magnified areas of (a). *(Reprinted with permission from Masuda* et al. *[207]. Copyright © American Chemical Society.)*

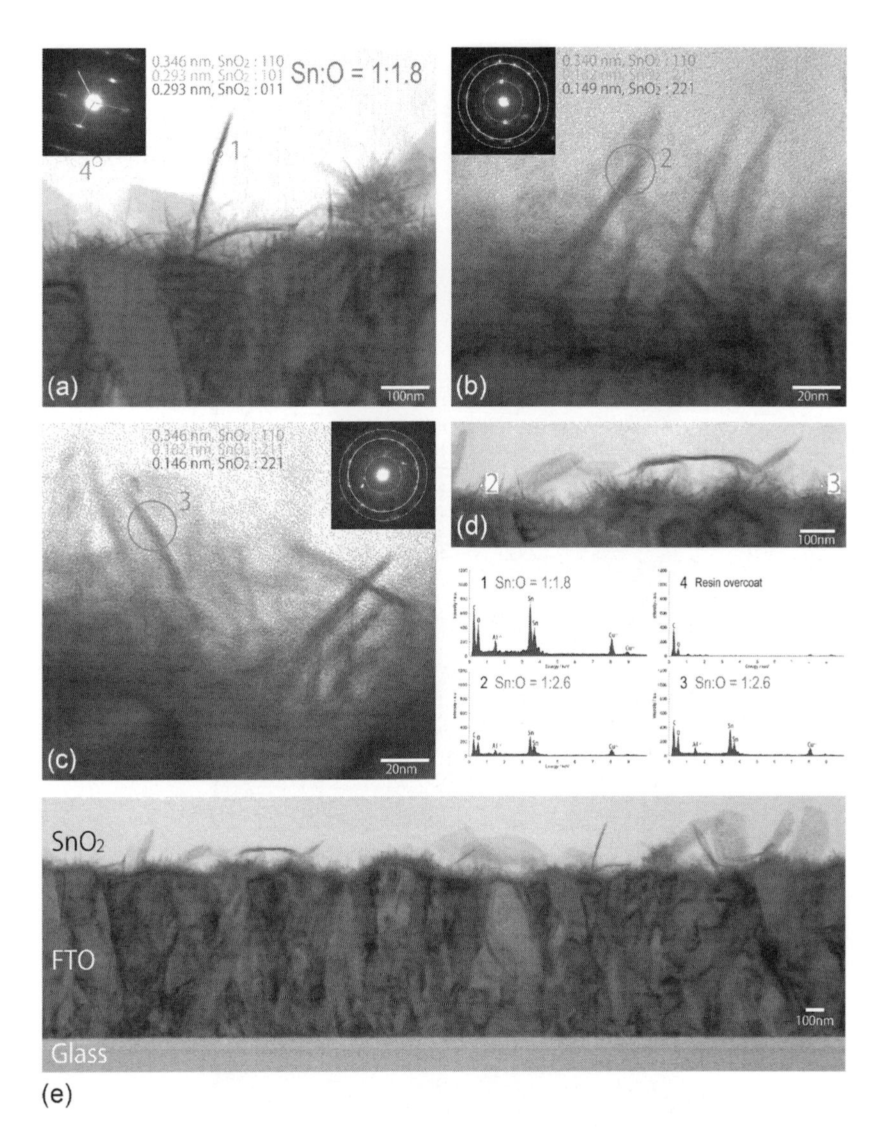

FIGURE 43 Cross-sectional images (a–e), electron diffractions (insets in [a]–[c]), and the chemical ratio (d, bottom) of a tin oxide nanosheet coating on an FTO substrate. (a–d) Magnified areas of the image in (e). *(Reprinted with permission from Masuda* et al. *[207]. Copyright © American Chemical Society.)*

Nanosheets in area 2 or 3 (Figure 43b and c) showed electron diffraction patterns. They were assigned to a single phase of SnO_2. Some nanosheets were included in the evaluation areas. Therefore, electron diffraction patterns had a few diffraction spots for each lattice distance. Two diffraction spots were, for instance, assigned to 110 of SnO_2 in the diffraction pattern from area 2. It suggested that two single crystals of SnO_2 were included in area 2. The crystal lattice distance of SnO_2 110 was estimated to 0.340 nm. Those of SnO_2 211 and 221 were 0.182 and 0.149 nm, respectively. The electron diffraction pattern from area 3 also was assigned to a single phase of SnO_2, with some spots for each lattice distance. The crystal lattice distance of SnO_2 110, 211, and 221 were estimated at 0.346, 0.182, and 0.146 nm, respectively. The chemical composition (Sn to O) in areas 2 and 3 was estimated at 1:2.6. It was larger than that of SnO_2. Oxygen and carbon were detected from a resin overcoat that was used to fix the substrate for the thinning process and TEM observation (Figure 43a, area 4). Resin was included in area 2 or 3. It increased the amount of oxygen in the EDS analysis.

5.4.4 XPS Analysis of SnO Nanosheet Coating on FTO Substrate

Sn, O, F, and carbon were detected on the surface of an SnO nanosheet coating using XPS analysis (Figure 44). Sn $3d_{5/2}$, Sn$3d_{3/2}$, O 1s, F 1s, and C 1s spectra were evaluated in detail. Chemical composition (Sn to O) was estimated at 1:1.85. It was more similar to the chemical composition of SnO_2 than SnO. It was consistent with EDS analysis and TEM observation. Sn $3d_{5/2}$ and Sn$3d_{3/2}$ spectra were observed at 486.9 and 495.3 eV, respectively. The binding energy of Sn $3d_{5/2}$ was similar to that of Sn^{4+} in SnO_2 (486.6 eV [216]) and higher than that of the Sn^{2+} site in SnO (485.9 eV [215]) or Sn metal (484.8 eV [216], 484.85 eV [217], 484.87 eV [218], 484.9 eV [219], and 485.0 eV [220]). This suggests that Sn atoms are positively charged by forming direct bonds with oxygen.

The binding energy of O 1s was 530.8 eV. It was similar to that in SnO_2 (530.5 eV [215]), in which O ions connected to Sn^{4+} ions, and higher than that

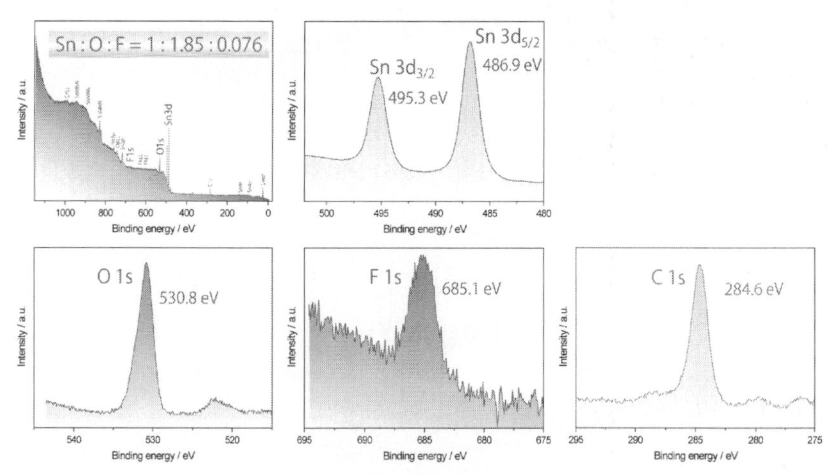

FIGURE 44 X-ray photoelectron spectra of tin/oxygen/fluorine together (a), tin (b), oxygen (c), fluorine (d), and carbon (d) for a tin oxide nanosheet coating on an FTO substrate. *(Reprinted with permission from Masuda* et al. *[207]. Copyright © American Chemical Society.)*

in SnO (529.8 eV [215]), in which O ions connected to Sn^{2+} ions. Sn $3d_{5/2}$ and O 1s spectra indicated that nanosheets were SnO_2 rather than SnO or Sn metal.

The fluorine 1s spectrum was observed at 685.1 eV. The binding energy was similar to that of F atoms, which were doped in SnO (684.4 eV [221]). This suggests that F was doped in the nanosheets. The chemical composition (Sn to F) was estimated at 1:0.076. Fluorine doping of about 0.03-0.1 is effective in improving the conductivity of SnO [222]. SnO nanostructures with high conductivity are strongly required for sensors and dye-sensitized solar cells.

Carbon was detected on the surface because of surface contamination. All spectra were corrected using the standard binding energy of C–C bonds (C 1s, 284.6 eV) in surface contaminations.

5.4.5 XRDD of SnO Nanosheet Coatings on FTO Substrate

An SnO nanosheet coating on an FTO substrate showed XRD peaks at $2\theta = 26.5°$, 33.8°, 37.8°, 38.9°, 42.6°, 51.6°, 54.6°, 61.7°, 64.6°, 65.8°, 71.1°, 78.4°, 80.8°, 83.4°, 89.5°, 90.6°, 92.9°, and 107.9° (Figure 45). They were assigned to diffraction from 110, 101, 200, 111, 210, 211, 220, 310, 112, 301, 202, 321, 400, 222, 312, 411, 420, and 213 crystal planes, respectively. XRD data of JCPDS

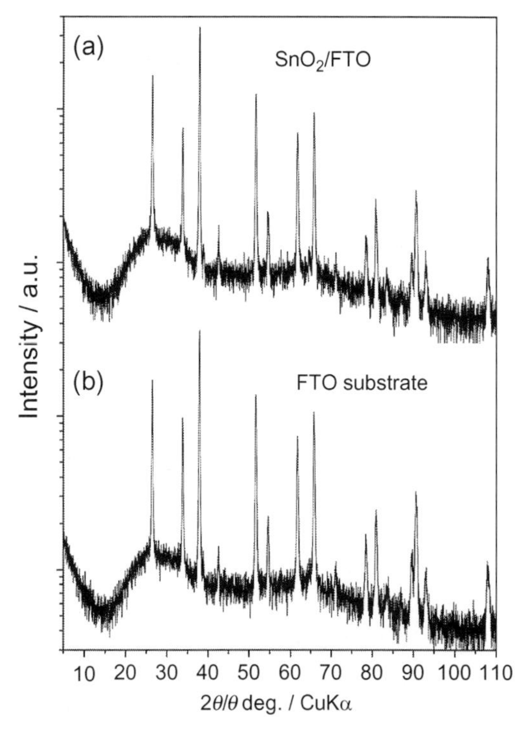

FIGURE 45 X-ray diffraction patterns of a tin oxide nanosheet coating on an FTO substrate (a) and on a bare FTO substrate (b). a.u., arbitrary units. *(Reprinted with permission from Masuda* et al. *[207]. Copyright © American Chemical Society.)*

no. 41-1445 (SnO_2) was used for identification. They also were observed from a bare FTO substrate and were assigned to SnO_2. A small peak from the bare FTO substrate was observed at 57.7°, but it was not observed from the SnO nanosheet coating on the FTO substrate. The peak at 57.7° was assigned to diffraction from the 200 crystal plane of SnO_2. The crystal structure of SnO_2 is tetragonal, which has an *a*-axis, *a*-axis, and *c*-axis. A decrease in the diffraction intensity from an *a*-plane such as the 200 crystal plane indicated a relative increase in diffraction intensity from the *c*-plane. This was not consistent with anisotropic crystal growth of the nanosheets along the *c*-axis.

5.4.6 Transparency of an SnO Nanosheet Coating on an FTO Substrate

The bare FTO substrate showed transmittance of 72.0%, 78.8%, 78.2%, 77.4%, and 76.4% at 400, 500, 600, 700, and 800 nm, respectively (Figure 46). Transmittance increased with SnO nanosheet coating to 72.3%, 81.6%, 82.8%, 83.4%, and 82.8% at 400, 500, 600, 700, and 800 nm, respectively. The increase of transmittance was explained in terms of reflectance. Reflectance of the bare FTO substrate was 14.3%, 11.4%, 12.6%, 12.0%, and 11.0 % at 400, 500, 600, 700, and 800 nm, respectively. It decreased to 10.6%, 9.3%, 9.3%, 7.7%, and 6.6% at 400, 500, 600, 700, and 800 nm, respectively, with SnO nanosheet coating. This indicated that the SnO nanosheet coating had an antireflection effect. A decrease in reflectance caused an increase in transmittance. The SnO nanosheet coating can be used for antireflection coating.

5.4.7 Superhydrophobicity of an SnO Nanosheet Coating on an FTO Substrate

An SnO nanosheet coating on an FTO substrate showed superhydrophobicity. A static water contact angle (125°) was reached. The advancing contact

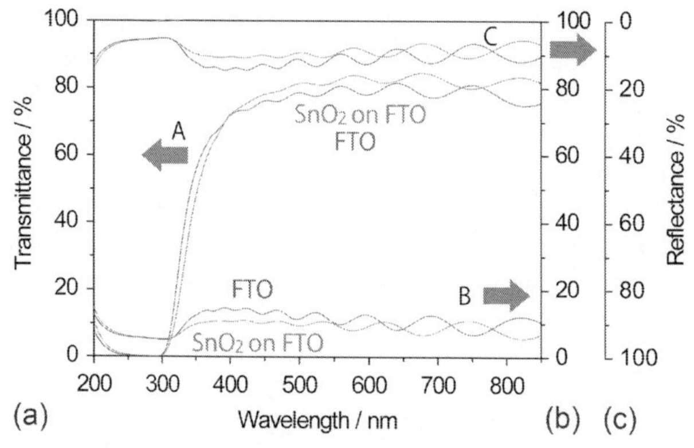

FIGURE 46 Optical transmittance(a) and reflectance (b and c) of a tin oxide nanosheet coating on a FTO substrate (red) or a bare FTO substrate (blue). *(Reprinted with permission from Masuda* et al. *[207]. Copyright © American Chemical Society.)*

angle and receding contact angle were 140° or 120°, respectively. The coatings had micro-sized and nano-sized structures. It is well known that micro and nanostructures enhance the hydrophobicity of a hydrophobic surface or the hydrophilicity of a hydrophilic surface. The superhydrophobicity of SnO nanosheet coatings was realized by combining the nature of the SnO surface and micro-/nanostructures. SnO is quite resistant to high temperatures up to several hundred degrees, oxidation, acids, bases, organic solvents, salt water, and sunlight because it is an oxidized metal. An SnO nanosheet coating was formed on FTO substrates at 90° in aqueous solutions. The coatings can be formed on various materials such metals, polymers, paper, and biomaterials. In addition, a uniform surface coating is one of the advantages of the simple immersion process. The coatings can be formed on not only flat substrates but also nano-sized materials, particles, fibers, meshes, tubes, structures with a complex shape, for control of pollution from medical equipment, automobile windshields, and the body of a ship.

5.4.8 Superhydrophilicity of an SnO Nanosheet Coating on an FTO Substrate

The coating was irradiated with VUV light for 10 min in air. Light energy and ozone, which was generated from oxygen gas by the irradiation, broke chemical bonds and decomposed adhesive organic molecules on the surface of the SnO nanosheets. The surface became superhydrophilic through irradiation. A static water contact angle was below the detection limit of 1°. A water droplet appeared above the surface before contact (Figure 47c1). It rapidly spread on the surface (Figure 47c2) and was not observed after the moment of contact (Figure 47c3).

Figure 47 is a composite image consisting of two captured images. The two images are shown in Figure 47c2 because of quickly changing shape of the droplet. The interval between the two images in Figure 47c2 was 0.00635 s; that is, the droplet spread on the surface in less than 0.00635 s. This indicates the rapid spread of the droplet and the high hydrophilicity of the surface. The change in the size of the droplet was estimated. A 2-μL droplet contacted the surface. Water (1 μL) remained at the tip of a needle. Water (1 μL = 1 mm^3) spread on the surface. The diameter of the droplet was about 10 mm. The thickness (i.e., height) of the droplet was estimated at 25 μm using a model of a spherical crown (spherical sector, spherical segment of one base) (Figure 40) and Equation (a). This indicates that the droplet spread widely to form a thin layer of water.

$$V \equiv \frac{\pi h}{6}\left(3r^3 + h^2\right) \tag{a}$$

where V is the volume of the 1-μL droplet over 1 mm^3, r is the radius of the droplet (5 mm), and h is the height of the droplet (in millimeters).

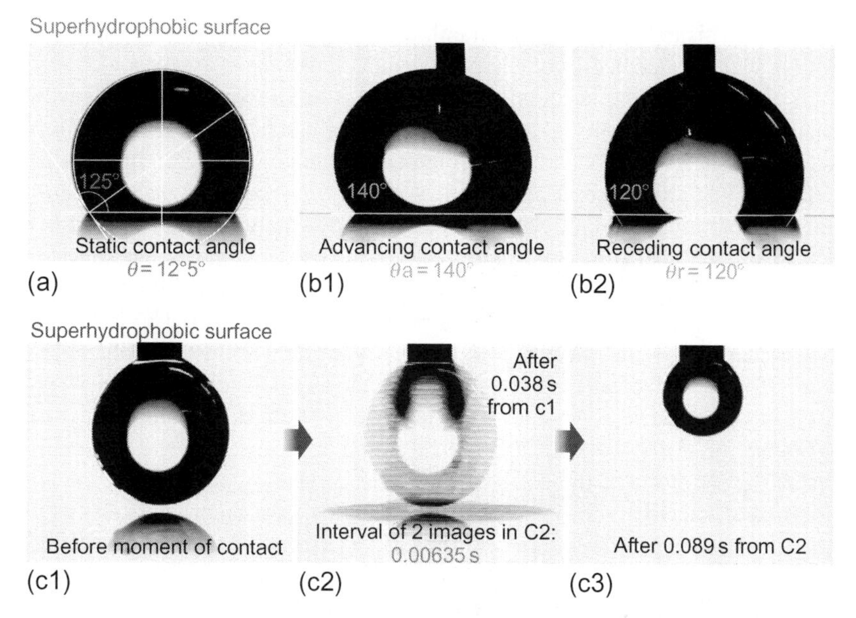

FIGURE 47 Water contact angle of a tin oxide nanosheet coating on an FTO substrate: static contact angle (a), advancing contact angle (b1), and receding contact angle of a hydrophobic surface (b2). (c1–c3) High-speed movement of the rapid spread of a water droplet on a hydrophilic surface after light irradiation. *(Reprinted with permission from Masuda* et al. *[207]. Copyright © American Chemical Society.)*

5.4.9 Photoluminescence and Photocurrent from SnO Nanosheets With dye-Labeled Prostate-Specific Antigen

SnO nanosheets were formed on FTO substrates. They were modified with the dye-labeled monoclonal antibody mentioned earlier. Photoluminescence intensity was evaluated with a Typhoon Trio scanner (GE Healthcare UK Ltd, United Kingdom) using excitation light of 632.8 nm (helium/neon laser). Detection sensitivity was set to PMT 520. Photoelectrochemical experiments were performed using sandwich-type cells. They consisted of the substrates (working electrodes), platinum counter electrodes, and electrolyte solutions containing the redox pair (I^-/I^{3-}) to fill the cells. A red laser (97 mW, with a wavelength of 650 nm) served as a light source. The photocurrent was determined by measuring the short-circuit photocurrents at excitation wavelengths of the red laser. All the measurements were made at room temperature. The light was irradiated from the glass side (back) of the substrates.

The SnO nanosheets with dye-labeled prostate specific antigen (10 μg/mL) showed strong photoluminescence: 2600 times the luminescence from the SnO nanosheets with unlabeled prostate-specific antigen and 4 times the luminescence from the SnO nanosheets with dye-labeled prostate-specific antigen

(1 μg/mL). This indicates that photoluminescence increased upon surface modification with dye-labeled prostate-specific antigen.

A photocurrent was successfully generated from the SnO nanosheets with dye-labeled prostate-specific antigen (monoclonal antibody of NB013, 10 μg/mL Cy5-NB013). It reached 603.7 nA under irradiation with red light (97 mW, 650 nm). This indicates that photon energy was converted to an electrical current by combining the semiconductor SnO_2 and dye on the surface of the SnO nanosheets. Photoelectric conversion effect was used for molecular sensing. In addition, the prostate-specific antigen of monoclonal antibody of NB013 was applicable to the photocurrent conversion sensing system. The influence of dye-labeled prostate-specific antigen concentration was further evaluated in detail. The SnO nanosheets with dye-labeled prostate-specific antigen (1 μg/mL) or with unlabeled prostate-specific antigen showed a photocurrents of 303.0 and 327.3 nA, respectively. They were mainly caused by the noise current in the system. On the basis of these measurements, the signal-to-noise ratio was estimated at ~2 (603.7 nA/327.3 nA). The SnO nanosheets with dye-labeled prostate-specific antigen can be used as electrodes in prostate cancer sensors.

We developed nanocrystalline superhydrophobic and superhydrophilic surface coatings. TCO substrates of F-doped SnO were coated with SnO nanosheets by simple immersion into an aqueous solution. Nanostructures were crystallized at an ordinary temperature and atmospheric pressure to modify the surface of the TOC, which is a key material for modern electronic devices. The surface coatings were F-doped tin dioxide nanosheets with a chemical ratio (Sn to O to F) of 1:1.85:0.076. The transparency of the substrates was improved by the surface coating because of the suppression of diffuse reflection on the surface. The advancing, receding, or static water contact angles of the surface coatings were measured to be 140°, 120°, and 125°, respectively. Water contact angle hysteresis occurred at 20°. Furthermore, superhydrophobic surfaces were modified to superhydrophilic surfaces with a static water contact angle <1° by simple irradiation from a low-pressure mercury lamp in air. This easy, low-cost procedure can be applied to superhydrophobic or superhydrophilic surface coatings of various materials including metals, glass, polymers, paper, and biomaterials.

Furthermore, nanosheet-coated TCO substrates were modified with a dye-labeled monoclonal antibody. Monoclonal antibody reacted with human AFP in the blood serum of a patient with hepatocellular cancer. Photoluminescence and photocurrent were detected from the electrodes under excitation with red light. A high photocurrent (603.7 nA) was obtained under irradiation of red light (650 nm) from the combination of an SnO semiconductor and dye. The reaction of the monoclonal antibody with AFP can be used as a detection mechanism. The biomolecule/semiconductor electrode will open the door for dye-sensitized molecule sensors for cancer sensing.

6 SUMMARY

Nanostructuring of metal oxides was developed in aqueous solutions. Various unique nano-/microcrystals were fabricated in the solutions. They showed large surface areas, highly hydrophobic/hydrophilic surfaces, an antireflection effect, and action as a cancer sensor. Precise control of nucleation and crystal growth are essential in this innovative proposal. Aqueous solution processes will open the door to a new generation of ceramics.

REFERENCES

[1] Y. Masuda, K. Koumoto, Advances in nanointegration methodologies: patterning, positioning, and self-assembly, in: M. Singh, T. Ohji, R. Asthana, S. Mathur (Eds.), Ceramic Integration and Joining Technologies: From Macro to Nanoscale, John Wiley & Sons, Inc., Hoboken, NJ, 2011.

[2] Y. Masuda, Y.F. Gao, P.X. Zhu, N. Shirahata, N. Saito, K. Koumoto, Site-selective deposition of ceramic thin films using self-assembled monolayers, J. Ceram. Soc. Jpn. 112 (2004) 1495–1505.

[3] Y. Masuda, Liquid phase patterning of ceramics, J. Ceram. Soc. Jpn. 115 (2007) 101–109.

[4] Y. Masuda, T. Sugiyama, W.S. Seo, K. Koumoto, Deposition mechanism of anatase TiO_2 on self-assembled monolayers from an aqueous solution, Chem. Mater. 15 (2003) 2469–2476.

[5] Y. Masuda, http://staff.aist.go.jp/masuda-y/index.html.

[6] Y. Masuda, T. Sugiyama, H. Lin, W.S. Seo, K. Koumoto, Selective deposition and micropatterning of titanium dioxide thin film on self-assembled monolayers, Thin Solid Films 382 (2001) 153–157.

[7] Y. Masuda, Y. Jinbo, T. Yonezawa, K. Koumoto, Templated site-selective deposition of titanium dioxide on self-assembled monolayers, Chem. Mater. 14 (2002) 1236–1241.

[8] Y. Masuda, S. Ieda, K. Koumoto, Site-selective deposition of anatase TiO2 in an aqueous solution using a seed layer, Langmuir 19 (2003) 4415–4419.

[9] J.H. Xiang, Y. Masuda, K. Koumoto, Fabrication of super-site-selective TiO2 micropattern on a flexible polymer substrate using a barrier-effect self-assembly process, Adv. Mater. 16 (2004) 1461–1464.

[10] Y. Masuda, M. Itoh, T. Yonezawa, K. Koumoto, Low-dimensional arrangement of SiO2 particles, Langmuir 18 (2002) 4155–4159.

[11] Y. Masuda, K. Tomimoto, K. Koumoto, Two-dimensional self-assembly of spherical particles using a liquid mold and its drying process, Langmuir 19 (2003) 5179–5183.

[12] Y. Masuda, T. Itoh, M. Itoh, K. Koumoto, Self-assembly patterning of colloidal crystals constructed from opal structure or NaCl structure, Langmuir 20 (2004) 5588–5592.

[13] Y. Masuda, T. Itoh, K. Koumoto, Self-assembly patterning of silica colloidal crystals, Langmuir 21 (2005) 4478–4481.

[14] Y. Masuda, T. Itoh, K. Koumoto, Self-assembly and micropatterning of spherical particle assemblies, Adv. Mater. 17 (2005) 841–845.

[15] Y. Masuda, T. Yamada, K. Koumoto, Synthesis of acicular BaTiO3 particles using acicular barium oxalates, Cryst. Growth Des. 8 (2008) 169–171.

[16] K. Tanaka, K. Suzuki, D.S. Fu, K. Nishizawa, T. Miki, K. Kato, Grain size effect on dielectric and piezoelectric properties of alkoxy-derived BaTiO3-based thin films, Jpn. J. Appl. Phys. 43 (2004) 6525–6529.

[17] S. Aoyagi, Y. Kuroiwa, A. Sawada, H. Kawaji, T. Atake, Size effect on crystal structure and chemical bonding nature in BaTiO3 nanopowder, J. Therm. Anal. Calorim. 81 (2005) 627–630.
[18] W.L. Luan, L. Gao, J.K. Guo, Size effect on dielectric properties of fine-grained BaTiO3 ceramics, Ceram. Int. 25 (1999) 727–729.
[19] X.H. Wang, R.Z. Chen, Z.L. Gui, L.T. Li, The grain size effect on dielectric properties of BaTiO3 based ceramics, Mater. Sci. Eng. B 99 (2003) 199–202.
[20] Y. Sakabe, Y. Yamashita, H. Yamamoto, Dielectric properties of nano-crystalline BaTiO3 synthesized by micro-emulsion method, J. Eur. Ceram. Soc. 25 (2005) 2739–2742.
[21] A.V. Polotai, A.V. Ragulya, C.A. Randall, Preparation and size effect in pure nanocrystalline barium titanate ceramics, Ferroelectrics 288 (2003) 93–102.
[22] M. Yashima, T. Hoshina, D. Ishimura, S. Kobayashi, W. Nakamura, T. Tsurumi, S. Wada, Size effect on the crystal structure of barium titanate nanoparticles, J. Appl. Phys. 98 (2005) 014313–1–8.
[23] M.H. Frey, Z. Xu, P. Han, D.A. Payne, The role of interfaces on an apparent grain size effect on the dielectric properties for ferroelectric barium titanate ceramics, Ferroelectrics 206 (1998) 337–353.
[24] M.H. Frey, D.A. Payne, Grain-size effect on structure and phase transformations for barium titanate, Phys. Rev. B 54 (1996) 3158–3168.
[25] S. Wada, T. Suzuki, T. Noma, Role of lattice defects in the size effect of barium titanate fine particles—A new model, J. Ceram. Soc. Jpn 104 (1996) 383–392.
[26] Y. Masuda, T. Koumura, T. Okawa, K. Koumoto, Micropatterning of Ni particles on a BaTiO3 green sheet using a self-assembled monolayer, J. Colloid Interface Sci. 263 (2003) 190–195.
[27] Y. Masuda, K. Koumoto, R. Ueyama, Acicular BaTiO3 crystals, their precursor, manufacturing process of them and green sheets. Japanese Patent Application Number: P 2007–113579, April 23, (2007).
[28] Y. Hayashi, T. Kimura, T. Yamaguchi, Preparation of rod-shaped BaTiO3 powder, J. Mater. Sci. 21 (1986) 757–762.
[29] X.L. Li, J.F. Liu, Y.D. Li, Low-temperature conversion synthesis of M(OH)2 (M=Ni, Co, Fe) nanoflakes and nanorods, Mater. Chem. Phys. 80 (2003) 222–227.
[30] H. Sun, S.Z. Kang, J. Mu, Synthesis of flowerlike SnO2 quasi-square submicrotubes from tin (II) oxalate precursor, Mater. Lett. 61 (2007) 4121–4123.
[31] J.C. Mutin, Y. Dusausoy, J. Protas, Structural description of endothermic decompositions in the form solid-1- solid-2+gas.1. Crystal-structure of barium oxalate, $2Bac_2O_4{\cdot}H_2O$, J. Solid State Chem. 36 (1981) 356–364.
[32] H.S. Potdar, S.B. Deshpande, S.K. Date, Alternative route for synthesis of barium titanyl oxalate: molecular prescursor for microcrystalline barium titanate powders, J. Am. Ceram. Soc. 79 (1996) 2795–2797.
[33] Y. Masuda, K. Kato, Aqueous synthesis of nano-sheet assembled tin oxide particles and their N2 adsorption characteristics, J. Cryst. Growth 311 (2009) 593–596.
[34] M. Yoshimura, Importance of soft solution processing for advanced inorganic materials, J. Mater. Res. 13 (1998) 796–802.
[35] M. Yoshimura, J. Livage, Soft processing for advanced inorganic materials, MRS Bull. 25 (2000) 12–13.
[36] S. Mann, Biomimetic Materials Chemistry, VCH Publishers, Weinheim, 1996.
[37] T.P. Niesen, M.R. DeGuire, Deposition of ceramic thin films at low temperatures from aqueous solutions, J. Electroceram. 6 (2001) 169–207.

[38] Y. Masuda, N. Kinoshita, F. Sato, K. Koumoto, Site-selective deposition and morphology control of UV- and visible-light-emitting ZnO crystals, Cryst. Growth Des. 6 (2006) 75–78.

[39] T. Kasuga, M. Hiramatsu, A. Hoson, T. Sekino, K. Niihara, Titania nanotubes prepared by chemical processing, Adv. Mater. 11 (1999) 1307.

[40] Y. Oaki, H. Imai, Amplification of chirality from molecules into morphology of crystals through molecular recognition, J. Am. Chem. Soc. 126 (2004) 9271–9275.

[41] Y.N. Xia, P.D. Yang, Y.G. Sun, Y.Y. Wu, B. Mayers, B. Gates, Y.D. Yin, F. Kim, Y.Q. Yan, One-dimensional nanostructures: synthesis, characterization, and applications, Adv. Mater. 15 (2003) 353–389.

[42] Y. Masuda, N. Saito, R. Hoffmann, M.R. De Guire, K. Koumoto, Nano/micro-patterning of anatase TiO2 thin film from an aqueous solution by site-selective elimination method, Sci. Technol. Adv. Mater. 4 (2003) 461–467.

[43] M.P. Crosnier, J.L. Fourquet, Synthesis and crystal-structure of $BaTiOF_4$, Eur J. Solid State Inorg. Chem. 29 (1992) 199–206.

[44] M. Wiegel, G. Blasse, M.P. Crosnierlopez, J.L. Fourquet, Luminescence properties of the barium fluorotitanate compounds Ba2TiOF6 and BaTiOF4, Eur J. Solid State Inorg. Chem. 30 (1993) 895–900.

[45] Y. Masuda, K. Koumoto, R. Ueyama, Platy BaTiO3 crystals, their precursor, manufacturing process of them and green sheets. Japanese Patent Application Number: P 2007–113580, April 23, (2007).

[46] Y. Masuda, N. Kinoshita, K. Koumoto, Hexagonal symmetry radial whiskers of ZnO crystallized in aqueous solution, J. Nanosci. Nanotechnol. 9 (2009) 522–526.

[47] X.L. Wu, G.G. Siu, C.L. Fu, H.C. Ong, Photoluminescence and cathodoluminescence studies of stoichiometric and oxygen-deficient ZnO films, Appl. Phys. Lett. 78 (2001) 2285–2287.

[48] J.S. Kang, H.S. Kang, S.S. Pang, E.S. Shim, S.Y. Lee, Investigation on the origin of green luminescence from laser-ablated ZnO thin film, Thin Solid Films 443 (2003) 5–8.

[49] N. Golego, S.A. Studenikin, M.J. Cocivera, Sensor photoresponse of thin-film oxides of zinc and titanium to oxygen gas, J. Electrochem. Soc. 147 (2000) 1592–1594.

[50] G. Sberveglieri, Recent developments in semiconducting thin-film gas sensors, Sens. Actuators, B 23 (1995) 103–109.

[51] T. Pauporte, D. Lincot, Electrodeposition of semiconductors for optoelectronic devices: results on zinc oxide, Electrochem. Acta 45 (2000) 3345–3353.

[52] M. Law, L.E. Greene, J.C. Johnson, R. Saykally, P.D. Yang, Nanowire dye-sensitized solar cells, Nat. Mater. 4 (2005) 455–459.

[53] Y. Masuda, K. Kato, Morphology control of zinc oxide particles at low temperature, Cryst. Growth Des. 8 (2008) 2633–2637.

[54] H. Gong, J.Q. Hu, J.H. Wang, C.H. Ong, F.R. Zhu, Nano-crystalline Cu-doped ZnO thin film gas sensor for CO, Sens. Actuators, B 115 (2006) 247–251.

[55] W.J. Moon, J.H. Yu, G.M. Choi, The CO and H-2 gas selectivity of CuO-doped SnO2-ZnO composite gas sensor, Sens. Actuators, B 87 (2002) 464–470.

[56] J.F. Chang, H.H. Kuo, I.C. Leu, M.H. Hon, The effects of thickness and operation temperature on ZnO: Al thin film CO gas sensor, Sens. Actuators, B 84 (2002) 258–264.

[57] H.X. Tang, M. Yan, H. Zhang, S.H. Li, X.F. Ma, M. Wang, D.R. Yang, A selective NH3 gas sensor based on Fe2O3-ZnO nanocomposites at room temperature, Sens. Actuators, B 114 (2006) 910–915.

[58] S.T. Shishiyanu, T.S. Shishiyanu, O.I. Lupan, Sensing characteristics of tin-doped ZnO thin films as NO2 gas sensor, Sens. Actuators, B 107 (2005) 379–386.

[59] M.S. Wagh, L.A. Patil, T. Seth, D.P. Amalnerkar, Surface cupricated SnO2-ZnO thick films as a H2S gas sensor, Mater. Chem. Phys. 84 (2004) 228–233.
[60] J.Q. Xu, Q.Y. Pan, Y.A. Shun, Z.Z. Tian, Grain size control and gas sensing properties of ZnO gas sensor, Sens. Actuators, B 66 (2000) 277–279.
[61] D.F. Paraguay, M. Miki-Yoshida, J. Morales, J. Solis, L.W. Estrada, Influence of Al, In, Cu, Fe and Sn dopants on the response of thin film ZnO gas sensor to ethanol vapour, Thin Solid Films 373 (2000) 137–140.
[62] J.B. Baxter, E.S. Aydil, Nanowire-based dye-sensitized solar cells, Appl. Phys. Lett. 86 (2005) 53114.
[63] R. Katoh, A. Furube, K. Hara, S. Murata, H. Sugihara, H. Arakawa, M. Tachiya, Efficiencies of electron injection from excited sensitizer dyes to nanocrystalline ZnO films as studied by near-IR optical absorption of injected electrons, J. Phys. Chem. B 106 (2002) 12957–12964.
[64] S. Karuppuchamy, K. Nonomura, T. Yoshida, T. Sugiura, H. Minoura, Cathodic electrodeposition of oxide semiconductor thin films and their application to dye-sensitized solar cells, Solid State Ionics 151 (2002) 19–27.
[65] K. Keis, C. Bauer, G. Boschloo, A. Hagfeldt, K. Westermark, H. Rensmo, H. Siegbahn, Nanostructured ZnO electrodes for dye-sensitized solar cell applications, J. Photochem. Photobiol., A 148 (2002) 57–64.
[66] T. Nakanishi, Y. Masuda, K. Koumoto, Site-selective deposition of magnetite particulate thin films on patterned self-assembled monolayers, Chem. Mater. 16 (2004) 3484–3488.
[67] W.Q. Peng, S.C. Qu, G.W. Cong, Z.G. Wang, Synthesis and structures of morphology-controlled ZnO nano- and microcrystals, Cryst. Growth Des. 6 (2006) 1518–1522.
[68] Y.R. Lin, S.S. Yang, S.Y. Tsai, H.C. Hsu, S.T. Wu, I.C. Chen, Visible photoluminescence of ultrathin ZnO nanowire at room temperature, Cryst. Growth Des. 6 (2006) 1951–1955.
[69] H. Zhang, D.R. Yang, D.S. Li, X.Y. Ma, S.Z. Li, D.L. Que, Controllable growth of ZnO microcrystals by a capping-molecule-assisted hydrothermal process, Cryst. Growth Des. 5 (2005) 547–550.
[70] K.H. Liu, C.C. Lin, S.Y. Chen, Growth and physical characterization of polygon prismatic hollow Zn-ZnO crystals, Cryst. Growth Des. 5 (2005) 483–487.
[71] Y. Masuda, K. Kato, High c-axis oriented stand-alone ZnO self-assembled film, Cryst. Growth Des. 8 (2008) 275–279.
[72] L.E. Greene, M. Law, D.H. Tan, M. Montano, J. Goldberger, G. Somorjai, P.D. Yang, General route to vertical ZnO nanowire arrays using textured ZnO seeds, Nano Lett. 5 (2005) 1231–1236.
[73] P. O'Brien, T. Saeed, J. Knowles, Speciation and the nature of ZnO thin film from chemical bath deposition, J. Mater. Chem. 6 (1996) 1135–1139.
[74] S. Yamabi, H. Imai, Growth conditions for wurtzite zinc oxide films in aqueous solutions, J. Mater. Chem. 12 (2002) 3773–3778.
[75] L. Vayssieres, K. Keis, S.E. Lindquist, A. Hagfeldt, Purpose-built anisotropic metal oxide material: 3D highly oriented microrod array of ZnO, J. Phys. Chem. B 105 (2001) 3350–3352.
[76] L. Vayssieres, Growth of arrayed nanorods and nanowires of ZnO from aqueous solutions, Adv. Mater. 15 (2003) 464–466.
[77] N. Saito, H. Haneda, T. Sekiguchi, N. Ohashi, I. Sakaguchi, K. Koumoto, Low-temperature fabrication of light-emitting zinc oxide micropatterns using self-assembled monolayers, Adv. Mater. 14 (2002) 418–421.
[78] N. Saito, H. Haneda, W.S. Seo, K. Koumoto, Selective deposition of ZnF(OH) on self-assembled monolayers in Zn-NH4F aqueous solutions for micropatterning of zinc oxide, Langmuir 17 (2001) 1461–1469.

[79] T. Saeed, P. O'Brien, Deposition and characterisation of ZnO thin films grown by chemical bath deposition, Thin Solid Films 271 (1995) 35–38.
[80] K. Kakiuchi, E. Hosono, T. Kimura, H. Imai, S. Fujihara, Fabrication of mesoporous ZnO nanosheets from precursor templates grown in aqueous solutions, J. Sol-Gel Sci. Technol. 39 (2006) 63–72.
[81] B. Schwenzer, J.R. Gomm, D.E. Morse, Substrate-induced growth of nanostructured zinc oxide films at room temperature using concepts of biomimetic catalysis, Langmuir 22 (2006) 9829–9831.
[82] H.D. Yu, Z.P. Zhang, M.Y. Han, X.T. Hao, F.R. Zhu, A general low-temperature route for large-scale fabrication of highly oriented ZnO nanorod/nanotube arrays, J. Am. Chem. Soc. 127 (2005) 2378–2379.
[83] K.S. Choi, H.C. Lichtenegger, G.D. Stucky, E.W. McFarland, Electrochemical synthesis of nanostructured ZnO films utilizing self-assembly of surfactant molecules at solid-liquid interfaces, J. Am. Chem. Soc. 124 (2002) 12402–12403.
[84] M. Yin, Y. Gu, I.L. Kuskovsky, T. Andelman, Y. Zhu, G.F. Neumark, S. O'Brien, Zinc oxide quantum rods, J. Am. Chem. Soc. 126 (2004) 6206–6207.
[85] X.J. Feng, L. Feng, M.H. Jin, J. Zhai, L. Jiang, D.B. Zhu, Reversible super-hydrophobicity to super-hydrophilicity transition of aligned ZnO nanorod films, J. Am. Chem. Soc. 126 (2004) 62–63.
[86] T. Yoshida, M. Tochimoto, D. Schlettwein, D. Wohrle, T. Sugiura, H. Minoura, Self-assembly of zinc oxide thin films modified with tetrasulfonated metallophthalocyanines by one-step electrodeposition, Chem. Mater. 11 (1999) 2657–2667.
[87] J.Y. Lee, D.H. Yin, S. Horiuchi, Site and morphology controlled ZnO deposition on pd catalyst prepared from Pd/PMMA thin film using UV lithography, Chem. Mater. 17 (2005) 5498–5503.
[88] K. Kopalko, M. Godlewski, J.Z. Domagala, E. Lusakowska, R. Minikayev, W. Paszkowicz, A. Szczerbakow, Monocrystalline ZnO films on GaN/Al2O3 by atomic layer epitaxy in gas flow, Chem. Mater. 16 (2004) 1447–1450.
[89] R. Turgeman, O. Gershevitz, M. Deutsch, B.M. Ocko, A. Gedanken, C.N. Sukenik, Crystallization of highly oriented ZnO microrods on carboxylic acid-terminated SAMs, Chem. Mater. 17 (2005) 5048–5056.
[90] R. Turgeman, O. Gershevitz, O. Palchik, M. Deutsch, B.M. Ocko, A. Gedanken, C.N. Sukenik, Oriented growth of ZnO crystals on self-assembled monolayers of functionalized alkyl silanes, Cryst. Growth Des. 4 (2004) 169–175.
[91] E. Mirica, G. Kowach, P. Evans, H. Du, Morphological evolution of ZnO thin films deposited by reactive sputtering, Cryst. Growth Des. 4 (2004) 147–156.
[92] Y.R. Lin, Y.K. Tseng, S.S. Yang, S.T. Wu, C.L. Hsu, S.J. Chang, Buffer-facilitated epitaxial growth of ZnO nanowire, Cryst. Growth Des. 5 (2005) 579–583.
[93] Y.F. Gao, M. Nagai, Morphology evolution of ZnO thin films from aqueous solutions and their application to solar cells, Langmuir 22 (2006) 3936–3940.
[94] X.D. Wu, L.J. Zheng, D. Wu, Fabrication of superhydrophobic surfaces from microstructured ZnO-based surfaces via a wet-chemical route, Langmuir 21 (2005) 2665–2667.
[95] R.B. Peterson, C.L. Fields, B.A. Gregg, Epitaxial chemical deposition of ZnO nanocolumns from NaOH, Langmuir 20 (2004) 5114–5118.
[96] T.Y. Liu, H.C. Liao, C.C. Lin, S.H. Hu, S.Y. Chen, Biofunctional ZnO nanorod arrays grown on flexible substrates, Langmuir 22 (2006) 5804–5809.
[97] R. Wang, K. Hashimoto, A. Fujishima, Light-induced amphiphilic surfaces, Nature 388 (1997) 431–432.

[98] Y. Masuda, K. Kato, Nanocrystal assembled TiO2 particles prepared from aqueous solution, Cryst. Growth Des. 8 (2008) 3213–3218.
[99] T. Carlson, G.L. Giffin, Photooxidation of methanol using V2O5/TiO2 and MoO3/TiO2 surface oxide monolayer catalysts, J. Phys. Chem. 90 (1986) 5896–5900.
[100] Z.B. Zhang, C.C. Wang, R. Zakaria, J.Y. Ying, Role of particle size in nanocrystalline TiO2-based photocatalysts, J. Phys. Chem. B 102 (1998) 10871–10878.
[101] W.Y. Choi, A. Termin, M.R. Hoffmann, The role of metal-ion dopants in quantum-sized TiO2—correlation between photoreactivity and charge-carrier recombination dynamics, J. Phys. Chem. 98 (1994) 13669–13679.
[102] Y.M. Sung, J.K. Lee, W.S. Chae, Controlled crystallization of nanoporous and core/shell structure titania photocatalyst particles, Cryst. Growth Des. 6 (2006) 805–808.
[103] N. Kumazawa, M.R. Islam, M. Takeuchi, Photoresponse of a titanium dioxide chemical sensor, J. Electroanal. Chem. 472 (1999) 137–141.
[104] M. Ferroni, M.C. Carotta, V. Guidi, G. Martinelli, F. Ronconi, M. Sacerdoti, E. Traversa, Preparation and characterization of nanosized titania sensing film, Sens. Actuators, B 77 (2001) 163–166.
[105] M. Wagemaker, A.P.M. Kentgens, F.M. Mulder, Equilibrium lithium transport between nanocrystalline phases in intercalated TiO2 anatase, Nature 418 (2002) 397–399.
[106] A.S. Arico, P. Bruce, B. Scrosati, J.M. Tarascon, W. Van Schalkwijk, Nanostructured materials for advanced energy conversion and storage devices, Nat. Mater. 4 (2005) 366–377.
[107] Y.G. Guo, Y.S. Hu, J. Maier, Synthesis of hierarchically mesoporous anatase spheres and their application in lithium batteries, Chem. Commun. 26 (2006) 2783–2785.
[108] H. Tokudome, Y. Yamada, S. Sonezaki, H. Ishikawa, M. Bekki, K. Kanehira, M. Miyauchi, Photoelectrochemical deoxyribonucleic acid sensing on a nanostructured TiO2 electrode, Appl. Phys. Lett. 87 (2005) 213901–213903.
[109] M.K. Nazeeruddin, F. De Angelis, S. Fantacci, A. Selloni, G. Viscardi, P. Liska, S. Ito, B. Takeru, M.G. Gratzel, Combined experimental and DFT-TDDFT computational study of photoelectrochemical cell ruthenium sensitizers, J. Am. Chem. Soc. 127 (2005) 16835–16847.
[110] P. Wang, S.M. Zakeeruddin, J.E. Moser, R. Humphry-Baker, P. Comte, V. Aranyos, A. Hagfeldt, M.K. Nazeeruddin, M. Gratzel, Stable new sensitizer with improved light harvesting for nanocrystalline dye-sensitized solar cells, Adv. Mater. 16 (2004) 1806–1811.
[111] P.W. Morrison, R. Raghavan, A.J. Timpone, C.P. Artelt, S.E. Pratsinis, In situ Fourier transform infrared characterization of the effect of electrical fields on the flame synthesis of TiO2 particles, Chem. Mater. 9 (1997) 2702–2708.
[112] G.X. Yang, H.R. Zhuang, P. Biswas, Characterization and sinterability of nanophase titania particles processed in flame reactors, Nanostruct. Mater. 7 (1996) 675–689.
[113] J.C. Yu, J.G. Yu, W.K. Ho, L.Z. Zhang, Preparation of highly photocatalytic active nano-sized TiO2 particles via ultrasonic irradiation, Chem. Commun. 19 (2001) 1942–1943.
[114] W.P. Huang, X.H. Tang, Y.Q. Wang, Y. Koltypin, A. Gedanken, Selective synthesis of anatase and rutile via ultrasound irradiation, Chem. Commun. 15 (2000) 1415–1416.
[115] S. Seifried, M. Winterer, H. Hahn, Nanocrystalline titania films and particles by chemical vapor synthesis, Chem. Vap. Depos. 6 (2000) 239–244.
[116] E. Scolan, C. Sanchez, Synthesis and characterization of surface-protected nanocrystalline titania particles, Chem. Mater. 10 (1998) 3217–3223.
[117] C.C. Wang, J.Y. Ying, Sol-gel synthesis and hydrothermal processing of anatase and rutile titania nanocrystals, Chem. Mater. 11 (1999) 3113–3120.
[118] S.D. Burnside, V. Shklover, C. Barbe, P. Comte, F. Arendse, K. Brooks, M. Gratzel, Self-organization of TiO2 nanoparticles in thin films, Chem. Mater. 10 (1998) 2419–2425.

[119] H.Z. Zhang, M. Finnegan, J.F. Banfield, Preparing single-phase nanocrystalline anatase from amorphous titania with particle sizes tailored by temperature, Nano Lett. 1 (2001) 81–85.

[120] Y.M. Sung, J.K. Lee, Controlled morphology and crystalline phase of poly(ethylene oxide)-TiO2 nanohybrids, Cryst. Growth Des. 4 (2004) 737–742.

[121] N. Perkas, V. Pol, S. Pol, A. Gedanken, Gold-induced crystallization of SiO2 and TiO2 powders, Cryst. Growth Des. 6 (2006) 293–296.

[122] S. Deki, Y. Aoi, O. Hiroi, A. Kajinami, Titanium(IV) oxide thin films prepared from aqueous solution, Chem. Lett. 6 (1996) 433–434.

[123] S. Deki, Y. Aoi, H. Yanagimoto, K. Ishii, K. Akamatsu, M. Mizuhata, A. Kajinami, Preparation and characterization of Au-dispersed TiO2 thin films by a liquid-phase deposition method, J. Mater. Chem. 6 (1996) 1879–1882.

[124] S. Deki, Y. Aoi, Y. Asaoka, A. Kajinami, M. Mizuhata, Monitoring the growth of titanium oxide thin films by the liquid-phase deposition method with a quartz crystal microbalance, J. Mater. Chem. 7 (1997) 733–736.

[125] H. Kishimoto, K. Takahama, N. Hashimoto, Y. Aoi, S. Deki, Photocatalytic activity of titanium oxide prepared by liquid phase deposition (LPD), J. Mater. Chem. 8 (1998) 2019–2024.

[126] S.Y. Choi, M. Mamak, N. Coombs, N. Chopra, G.A. Ozin, Thermally stable two-dimensional hexagonal mesoporous nanocrystalline anatase, meso-nc-TiO2: Bulk and crack-free thin film morphologies, Adv. Funct. Mater. 14 (2004) 335–344.

[127] D.M. Antonelli, J.Y. Ying, Synthesis of hexagonally packed mesoporous TiO2 by a modified sol-gel method, Angew. Chem. Int. Ed. 34 (1995) 2014–2017.

[128] H. Shibata, T. Ogura, T. Mukai, T. Ohkubo, H. Sakai, M. Abe, Direct synthesis of mesoporous titania particles having a crystalline wall, J. Am. Chem. Soc. 127 (2005) 16396–16397.

[129] H. Shibata, H. Mihara, T. Mlikai, T. Ogura, H. Kohno, T. Ohkubo, H. Sakait, M. Abe, Preparation and formation mechanism of mesoporous titania particles having crystalline wall, Chem. Mater. 18 (2006) 2256–2260.

[130] B.Z. Tian, X.Y. Liu, H.F. Yang, S.H. Xie, C.Z. Yu, B. Tu, D.Y. Zhao, General synthesis of ordered crystallized metal oxide nanoarrays replicated by microwave-digested mesoporous silica, Adv. Mater. 15 (2003) 1370.

[131] R. Ryoo, S.H. Joo, M. Kruk, M. Jaroniec, Ordered mesoporous carbons, Adv. Mater. 13 (2001) 677–681.

[132] Y. Masuda, K. Kato, Nano crystal assembled TiO2 and method of manufacturing same. Japanese Patent Application Number: P 2007-240236, Sep 14, (2007).

[133] D.N. Furlong, G.D. Parfitt, Electrokinetics of titanium dioxide, J. Colloid Interface Sci. 65 (1978) 548–554.

[134] R.K. Wahi, Y.P. Liu, J.C. Falkner, V.L. Colvin, Solvothermal synthesis and characterization of anatase TiO2 nanocrystals with ultrahigh surface area, J. Colloid Interface Sci. 302 (2006) 530–536.

[135] M. Kruk, M. Jaroniec, Gas adsorption characterization of ordered organic-inorganic nanocomposite materials, Chem. Mater. 13 (2001) 3169–3183.

[136] P.I. Ravikovitch, S.C. Odomhnaill, A.V. Neimark, F. Schuth, K.K. Unger, Capillary hysteresis in nanopores: theoretical and experimental studies of nitrogen adsorption on MCM-41, Langmuir 11 (1995) 4765–4772.

[137] C. Lastoskie, K.E. Gubbins, N. Quirke, Pore-size distribution analysis of microporous carbons—a density-functional theory approach, J. Phys. Chem. 97 (1993) 4786–4796.

[138] K. Katagiri, K. Ohno, Y. Masuda, K. Koumoto, Growth behavior of TiO2 particles via the liquid phase deposition process, J. Ceram. Soc. Japan 115 (2007) 831–834.

[139] Y. Masuda, T. Ohji, K. Kato, Multineedle TiO2 nanostructures, self-assembled surface coatings, and their novel properties, Cryst. Growth Des. 10 (2010) 913–922.

[140] Y. Masuda, K. Kato, Liquid-phase patterning and microstructure of anatase TiO2 films on SnO2:F substrates using superhydrophilic surface, Chem. Mater. 20 (2008) 1057–1063.

[141] Y. Masuda, K. Kato, Anatase TiO2 films crystallized on SnO2:F substrates in an aqueous solution, Thin Solid Films 516 (2008) 2547–2552.

[142] Y. Masuda, K. Kato, Acicular crystal-assembled TiO2 thin films and their deposition mechanism, J. Cryst. Growth 311 (2009) 512–517.

[143] T. Ohsaka, F. Izumi, Y. Fujiki, Raman spectrum of anatase, TiO2, J. Raman Spectrosc. 7 (1978) 321–324.

[144] K. Gao, Strong anharmonicity and phonon confinement on the lowest-frequency Raman mode of nanocrystalline anatase TiO2, Phys. Status Solidi B 244 (2007) 2597–2604.

[145] Y. Cong, J. Zhang, F. Chen, M. Anpo, Synthesis and characterization of nitrogen-doped TiO2 nanophotocatalyst with high visible light activity, J. Phys. Chem. C 111 (2007) 6976–6982.

[146] T. Ohsaka, S. Yamaoka, O. Shimomura, Effect of hydrostatic pressure on the ramanspectrum of anatase (TiO2), Solid State Commun. 30 (1979) 345–347.

[147] J.G. Yu, H.G. Yu, B. Cheng, X.J. Zhao, J.C. Yu, W.K. Ho, The effect of calcination temperature on the surface microstructure and photocatalytic activity of TiO2 thin films prepared by liquid phase deposition, J. Phys. Chem. B 107 (2003) 13871–13879.

[148] L.N. Ignat'eva, S.A. Polishchuk, T.F. Antokhina, V.T. Buznik, IR spectroscopic study of the structure of glasses based on titanium oxyfluoride, Glas. Phys. Chem. 30 (2004) 139–141.

[149] K. Nakamoto, Infrared Spectra of Inorganic and Coordination Compounds, Wiley, New York, 1970.

[150] S.C. Padmanabhan, S.C. Pillai, J. Colreavy, S. Balakrishnan, D.E. McCormack, T.S. Perova, Y. Gun'ko, S.J. Hinder, J.M. Kelly, A simple sol-gel processing for the development of high-temperature stable photoactive anatase titania, Chem. Mater. 19 (2007) 4474–4481.

[151] J.J. Wu, C.C. Yu, Aligned TiO2 nanorods and nanowalls, J. Phys. Chem. B 108 (2004) 3377–3379.

[152] S. Perathoner, R. Passalacqua, G. Centi, D.S. Su, G. Weinberg, Photoactive titania nanostructured thin films: Synthesis and characteristics of ordered helical nanocoil array, Catal. Today 122 (2007) 3–13.

[153] H. Sato, K. Ono, T. Sasaki, A. Yamagishi, First-principles study of two-dimensional titanium dioxides, J. Phys. Chem. B 107 (2003) 9824–9828.

[154] V.M. Naik, D. Haddad, R. Naik, J. Benci, G.W. Auner, Paper presented at the Symposium on Solid-State Chemistry of Inorganic Materials IV held at the 2002 MRS Fall Meeting, Boston, Ma, Dec 02-06 2002.

[155] D. Huang, Z.-D. Xiao, J.-H. Gu, N.-P. Huang, C.-W. Yuan, TiO2 thin films formation on industrial glass through self-assembly processing, Thin Solid Films 305 (1997) 110–115.

[156] F. Zhang, Y. Mao, Z. Zheng, Y. Chen, X. Liu, S. Jin, Surface characterization of titanium oxide films synthesized by ion beam enhanced deposition, Thin Solid Films 310 (1997) 29–33.

[157] D. Li, H. Haneda, S. Hishita, N. Ohashi, N.K. Labhsetwar, Fluorine-doped TiO2 powders prepared by spray pyrolysis and their improved photocatalytic activity for decomposition of gas-phase acetaldehyde, J. Fluor. Chem. 126 (2005) 69–77.

[158] C.M. Wang, T.E. Mallouk, Photoelectrochemistry and interfacial energetics of titanium-dioxide photoelectrodes in fluoride-containing solutions, J. Phys. Chem. 94 (1990) 423–428.

[159] C. Minero, G. Mariella, V. Maurino, E. Pelizzetti, Photocatalytic transformation of organic compounds in the presence of inorganic anions. 1. Hydroxyl-mediated and direct electron-transfer reactions of phenol on a titanium dioxide-fluoride system, Langmuir 16 (2000) 2632–2641.

[160] C. Minero, G. Mariella, V. Maurino, D. Vione, E. Pelizzetti, Photocatalytic transformation of organic compounds in the presence of inorganic ions. 2. Competitive reactions of phenol and alcohols an a titanium dioxide-fluoride system, Langmuir 16 (2000) 8964–8972.

[161] M.S. Vohra, S. Kim, W. Choi, Paper presented at the International Symposium on Photochemistry at Interfaces, Sapporo, Japan, Aug 09-11 2002.

[162] H.B. Fu, L.W. Zhang, S.C. Zhang, Y.F. Zhu, J.C. Zhao, Electron spin resonance spin-trapping detection of radical intermediates in N-doped TiO2-assisted photodegradation of 4-chlorophenol, J. Phys. Chem. B 110 (2006) 3061–3065.

[163] H. Irie, Y. Watanabe, K. Hashimoto, Nitrogen-concentration dependence on photocatalytic activity of TiO2-xNx powders, J. Phys. Chem. B 107 (2003) 5483–5486.

[164] D.S. Ginley, C. Bright, Transparent conducting oxides, MRS Bull. 25 (2000) 15–21.

[165] Y. Idota, T. Kubota, A. Matsufuji, Y. Maekawa, T. Miyasaka, Tin-based amorphous oxide: a high-capacity lithium-ion-storage material, Science 276 (1997) 1395–1397.

[166] Y.L. Zhang, Y. Liu, M.L. Liu, Nanostructured columnar tin oxide thin film electrode for lithium ion batteries, Chem. Mater. 18 (2006) 4643–4646.

[167] Y. Wang, J.Y. Lee, H.C. Zeng, Polycrystalline SnO2 nanotubes prepared via infiltration casting of nanocrystallites and their electrochemical application, Chem. Mater. 17 (2005) 3899–3903.

[168] A.C. Bose, D. Kalpana, P. Thangadurai, S. Ramasamy, Synthesis and characterization of nanocrystalline SnO2 and fabrication of lithium cell using nano-SnO2, J. Power Sources 107 (2002) 138–141.

[169] S.A. Pianaro, P.R. Bueno, E. Longo, J.A. Varela, A new SnO2-based varistor system, J. Mater. Sci. Lett. 14 (1995) 692–694.

[170] P.R. Bueno, M.R. de Cassia-Santos, E.R. Leite, E. Longo, J. Bisquert, G. Garcia-Belmonte, F. Fabregat-Santiago, Nature of the Schottky-type barrier of highly dense SnO2 systems displaying nonohmic behavior, J. Appl. Phys. 88 (2000) 6545–6548.

[171] T. Tagawa, S. Kataoka, T. Hattori, Y. Murakami, Supported SnO2 catalysts for the oxidative dehydrogenation of ethylbenzene, Appl. Catal. 4 (1982) 1–4.

[172] P.W. Park, H.H. Kung, D.W. Kim, M.C. Kung, Characterization of SnO2/Al2O3 lean NOx catalysts, J. Catal. 184 (1999) 440–454.

[173] K.L. Chopra, S. Major, D.K. Pandya, Transparent conductors—a status reviEW, Thin Solid Films 102 (1983) 1–46.

[174] N. Sergent, P. Gelin, L. Perier-Camby, H. Praliaud, G. Thomas, Preparation and characterisation of high surface area stannic oxides: structural, textural and semiconducting properties, Sens. Actuators, B 84 (2002) 176–188.

[175] N.S. Baik, G. Sakai, N. Miura, N. Yamazoe, Preparation of stabilized nanosized tin oxide particles by hydrothermal treatment, J. Am. Ceram. Soc. 83 (2000) 2983–2987.

[176] M. Ristic, M. Ivanda, S. Popovic, S. Music, Dependence of nanocrystalline SnO2 particle size on synthesis route, J. Non-Cryst. Solids 303 (2002) 270–280.

[177] L. Broussous, C.V. Santilli, S.H. Pulcinelli, A.F. Craievich, SAXS study of formation and growth of tin oxide nanoparticles in the presence of complexing ligands, J. Phys. Chem. B 106 (2002) 2855–2860.

[178] J.R. Zhang, L. Gao, Synthesis and characterization of nanocrystalline tin oxide by sol-gel method, J. Solid State Chem. 177 (2004) 1425–1430.

[179] Z.X. Deng, C. Wang, Y.D. Li, New hydrolytic process for producing zirconium dioxide, tin dioxide, and titanium dioxide nanoparticles, J. Am. Ceram. Soc. 85 (2002) 2837–2839.
[180] E.R. Leite, J.W. Gomes, M.M. Oliveira, E.J.H. Lee, E. Longo, J.A. Varela, C.A. Paskocimas, T.M. Boschi, F. Lanciotti, P.S. Pizani, P.C. Soares, Synthesis of SnO2 nanoribbons by a carbothermal reduction process, J. Nanosci. Nanotechnol. 2 (2002) 125–128.
[181] E.R. Leite, A.P. Maciel, I.T. Weber, P.N. Lisboa, E. Longo, C.O. Paiva-Santos, A.V.C. Andrade, C.A. Pakoscimas, Y. Maniette, W.H. Schreiner, Development of metal oxide nanoparticles with high stability against particle growth using a metastable solid solution, Adv. Mater. 14 (2002) 905–908.
[182] Z.R. Dai, J.L. Gole, J.D. Stout, Z.L. Wang, Tin oxide nanowires, nanoribbons, and nanotubes, J. Phys. Chem. B 106 (2002) 1274–1279.
[183] Y. Liu, M.L. Liu, Growth of aligned square-shaped SnO2 tube arrays, Adv. Funct. Mater. 15 (2005) 57–62.
[184] J. Huang, N. Matsunaga, K. Shimanoe, N. Yamazoe, T. Kunitake, Nanotubular SnO2 templated by cellulose fibers: synthesis and gas sensing, Chem. Mater. 17 (2005) 3513–3518.
[185] Y.K. Liu, C.L. Zheng, W.Z. Wang, C.R. Yin, G.H. Wang, Synthesis and characterization of rutile SnO2 nanorods, Adv. Mater. 13 (2001) 1883–1887.
[186] J.Q. Sun, J.S. Wang, X.C. Wu, G.S. Zhang, J.Y. Wei, S.Q. Zhang, H. Li, D.R. Chen, Novel method for high-yield synthesis of rutile SnO2 nanorods by oriented aggregation, Cryst. Growth Des. 6 (2006) 1584–1587.
[187] R.S. Yang, Z.L. Wang, Springs, rings, and spirals of rutile-structured tin oxide nanobelts, J. Am. Chem. Soc. 128 (2006) 1466–1467.
[188] J.H. Duan, S.G. Yang, H.W. Liu, J.F. Gong, H.B. Huang, X.N. Zhao, R. Zhang, Y.W. Du, Single crystal SnO2 Zigzag nanobelts, J. Am. Chem. Soc. 127 (2005) 6180–6181.
[189] H. Ohgi, T. Maeda, E. Hosono, S. Fujihara, H. Imai, Evolution of nanoscale SnO2 grains, flakes, and plates into versatile particles and films through crystal growth in aqueous solutions, Cryst. Growth Des. 5 (2005) 1079–1083.
[190] H. Uchiyama, H. Imai, Tin oxide meshes consisting of nanoribbons prepared through an intermediate phase in an aqueous solution, Cryst. Growth Des. 7 (2007) 841–843.
[191] D.E. Niesz, R.B. Bennett, M.J. Snyder, Strength characterization of powder aggregates, Am. Ceram. Soc. Bull. 51 (1972) 677–680.
[192] A. Roosen, H. Hausener, Techniques for agglomeration control during wetchemical powder synthesis, Adv. Ceram. Mater. 3 (1988) 131.
[193] O. Vasylkiv, Y. Sakka, Synthesis and colloidal processing of zirconia nanopowder, J. Am. Ceram. Soc. 84 (2001) 2489–2494.
[194] T.P. Niesen, M.R. DeGuire, Review: deposition of ceramic thin films at low temperatures from aqueous solutions, J. Electroceram. 6 (2001) 169–207.
[195] C.F. Baes, R.E. Mesiner, The Hydrolysis of Cations, John Wiley & Sons, Inc., Wiley-Interscience, New York, 1976.
[196] C. Ararat Ibarguena, A. Mosqueraa, R. Parrab, M.S. Castrob, J.E. Rodríguez-Páeza, Synthesis of SnO2 nanoparticles through the controlled precipitation route, Mater. Chem. Phys. 101 (2007) 433–440.
[197] Y. Masuda, T. Ohji, K. Kato, Highly enhanced surface area of tin oxide nanocrystals, J. Am. Ceram. Soc. 93 (2010) 2140–2143.
[198] Y. Masuda, K. Kato, Tin oxide coating on polytetrafluoroethylene films in aqueous solutions, Polym. Adv. Technol. 21 (2010) 211–215.
[199] S. Shukla, S. Seal, L. Ludwig, C. Parish, Nanocrystalline indium oxide-doped tin oxide thin film as low temperature hydrogen sensor, Sens. Actuators, B 97 (2004) 256–265.

[200] Y. Masuda, K. Kato, S.,Sonezaki, M. Ajimi, M. Bekki, Surface-treated material, method of manufacturing same, electrode for sensor and sensor. Japanese Patent Application Number: P 2008-227389, (2008).
[201] J.H. Xiang, P.X. Zhu, Y. Masuda, K. Koumoto, Fabrication of self-assembled monolayers (SAMs) and inorganic micropattern on flexible polymer substrate, Langmuir 20 (2004) 3278–3283.
[202] Y. Masuda, M. Yamagishi, K. Koumoto, Site-selective deposition and micropatterning of visible-light-emitting europium-doped yttrium oxide thin film on self-assembled monolayers, Chem. Mater. 19 (2007) 1002–1008.
[203] Y. Masuda, S. Wakamatsu, K. Koumoto, Site-selective deposition and micropatterning of tantalum oxide thin films using a monolayer, J. Eur. Ceram. Soc. 24 (2004) 301–307.
[204] J.M. Themlin, M. Chtaib, L. Henrard, P. Lambin, J. Darville, J.M. Gilles, Characterization of tin oxides by X-ray-photoemission spectroscopy, Phys. Rev. B 46 (1992) 2460–2466.
[205] L. Yan, J.S. Pan, C.K. Ong, XPS studies of room temperature magnetic Co-doped SnO2 deposited on Si, Mater. Sci. Eng. B 128 (2006) 34–36.
[206] C.D. Wagner, Quantification of AES and XPS, in: D. Briggs, M.P. Seah (Eds.), second ed., in: Practical Surface Analysis, vol. 1, John Wiley, New York, 1990.
[207] Y. Masuda, T. Ohji, K. Kato, Tin oxide nanosheet assembly for hydrophobic/hydrophilic coating and cancer sensing, ACS Appl. Mater. Interfaces 4 (2012) 1666–1674.
[208] W. Barthlott, C. Neinhuis, Purity of the sacred lotus, or escape from contamination in biological surfaces, Planta 202 (1997) 1–8.
[209] Y.T. Cheng, D.E. Rodak, C.A. Wong, C.A. Hayden, Effects of micro- and nano-structures on the self-cleaning behaviour of lotus leaves, Nanotechnology 17 (2006) 1359–1362.
[210] X.F. Gao, L. Jiang, Water-repellent legs of water striders, Nature 432 (2004) 36.
[211] F. Shi, J. Niu, J.L. Liu, F. Liu, Z.Q. Wang, X.Q. Feng, X. Zhang, Towards understanding why a superhydrophobic coating is needed by water striders, Adv. Mater. 19 (2007) 2257–2261.
[212] L. Zhai, M.C. Berg, F.C. Cebeci, Y. Kim, J.M. Milwid, M.F. Rubner, R.E. Cohen, Patterned superhydrophobic surfaces: toward a synthetic mimic of the Namib desert beetle, Nano Lett. 6 (2006) 1213–1217.
[213] W.J. Dressick, J.M. Calvert, Patterning of self-assembled films using lithographic exposure tools, Jpn. J. Appl. Phys. 32 (1993) 5829–5839.
[214] W.J. Dressick, C.S. Dulcey, J.H. Georger, J.M. Calvert, Photopatterning and selective electroless metallization of surface-attached ligands, Chem. Mater. 5 (1993) 148–150.
[215] M. Kwoka, L. Ottaviano, M. Passacantando, S. Santucci, G. Czempik, J. Szuber, XPS study of the surface chemistry of L-CVD SnO2 thin films after oxidation, Thin Solid Films 490 (2005) 36–42.
[216] A.C. Parry-Jones, P. Weightman, P.T. Andrews, The M4,5N4,5 N4,5 Auger spectra of Ag, Cd. In and Sn, J. Phys. C Sol. State Phys. 12 (1979) 1587–1600.
[217] C.D. Wagner, W.M. Riggs, L.E. Davis, J.F. Moulder, G.E. Muilenberg, Handbook of X-ray photoelectron spectroscopy, Perkin-Elmer Corp., Physical Electronics Div, Eden Prairie, Minnesota, 1979.
[218] M. Pessa, A. Vuoristo, M. Vulli, S. Aksela, J. Väyrynen, T. Rantala, H. Aksela, Solid-state effects in M4,5N4,5N4,5 Auger spectra of elements from 49In to 52Te, Phys. Rev. B: Condens. Matter Mater. Phys. 20 (1979) 3115–3123.
[219] A.W.C. Lin, N.R. Armstrong, T. Kuwana, X-ray photoelectron auger electron spectroscopic studies of tin and indium metal foils and oxides, Anal. Chem. 49 (1977) 1228–1235.

[220] C.D. Wagner, Chemical shifts of Auger lines, and the Auger parameter, Faraday Discuss. 60 (1975) 291–318.
[221] A.I. Martinez, L. Huerta, J. de Leon, D. Acosta, O. Malik, M. Aguilar, Physicochemical characteristics of fluorine doped tin oxide films, J. Phys. D. Appl. Phys. 39 (2006) 5091–5096.
[222] C.H. Han, S.D. Han, I. Singh, T. Toupance, Micro-bead of nano-crystalline F-doped SnO2 as a sensitive hydrogen gas sensor, Sens. Actuators, B 109 (2005) 264–269.

Chapter 16

Green Manufacturing of Photocatalytic Materials for Environmental Applications

Toshihiro Ishikawa
Tokyo University of Science, Yamaguchi, Japan

1 INTRODUCTION

Since Fujishima and Honda [1] published a work regarding photo-induced water splitting at titanium dioxide (TiO_2) electrodes, TiO_2 has been the most investigated photocatalytic material. Because electrons and holes photo-generated on the surface of TiO_2 photocatalysts show strong reduction and oxidation activity, respectively, they can drive a variety of reactions. The photo-generated electron and hole pairs without recombination can migrate to the TiO_2 surface to participate in redox reactions with adsorbed species, with the possible formation of superoxide radical anions and hydroxyl radicals, respectively. These reactive oxygen species play an important role in the degradation of organic pollutants in water. The mechanism of photo-induced reaction on the surface of a TiO_2 photocatalyst is schematically shown in Figure 1. In this reaction most of the organic chemicals are susceptible to oxidation by TiO_2, and they can form intermediate radicals that may subsequently trigger a series of radical reactions in this process. Because of the excellent oxidation ability of hydroxyl radicals, the resulting intermediates can finally be decomposed into carbon dioxide (CO_2) and water (H_2O).

Using this property, many environmental applications have been considered and practiced:

(i). Degradation of volatile organic compounds contained in air
(ii). Water purification and disinfection (water and wastewater treatment)
(iii). Solar hydrogen production from water (photo-induced water splitting)
(iv). Photocatalytic reduction of CO_2 with H_2O

Of these, water and wastewater treatment are the most attractive technology for human health and the global environment [2–14]. Of course, wastewater treatment contains many steps, including preliminary treatment (screening grit removal), primary treatment (sedimentation), secondary treatment (biological

Green and Sustainable Manufacturing of Advanced Materials. http://dx.doi.org/10.1016/B978-0-12-411497-5.00016-3

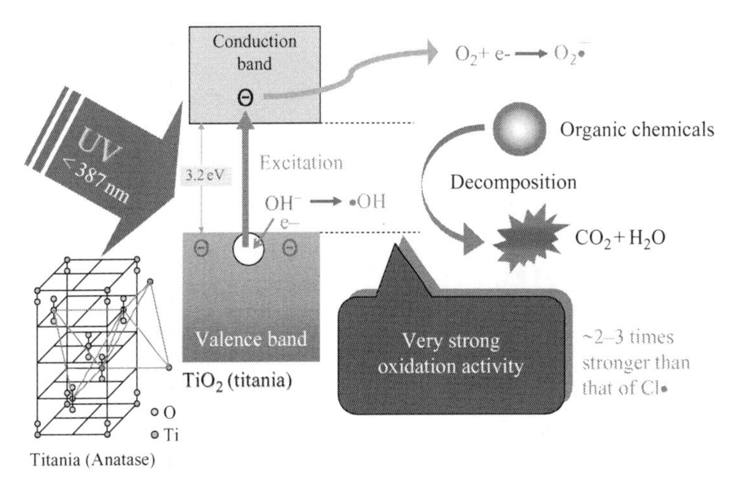

FIGURE 1 The mechanism of a photo-induced reaction on the surface of a titanium dioxide (TiO_2) photocatalyst. UV, ultraviolet.

treatment), and tertiary treatment (disinfection). Of these treatments, photocatalytic water treatments are addressed as part of the tertiary treatment process. As mentioned above, hydroxyl radicals, which can be formed on the surface of photocatalysts, have a strong oxidation ability. Because of the existence of the hydroxyl radicals, most organic chemicals can finally be decomposed into CO_2 and H_2O. Some conventional water treatment methods involve the addition of chemicals, which may lead to secondary waste pollution problems. In addition, some chemicals are risky for human health. Compared with conventional water treatment methods, the photocatalytic water treatment method is a promising technology for the water purification process. To date, various semiconductor photocatalysts such as TiO_2, ZnO, ZnS, Fe_2O_3, CdS, CeO_2, WO_3, SnO_2, and other complex oxide materials have been examined for use in the water purification process. An ideal photocatalyst should be chemically and biologically inert, photoactive, photostable, inexpensive, nontoxic, and excited with visible and/or ultraviolet (UV) light. Among the semiconductor photocatalysts listed above, a large number of studies has concentrated on TiO_2 because it is highly photostable, nontoxic, inexpensive, photoreactive, and chemically and biologically inert.

This chapter introduces mainly TiO_2-based photocatalytic materials that can be used for environmental applications (especially photocatalytic water treatment) and their important morphologies and simple production processes, along with actual applications.

2 TiO_2-BASED PHOTOCATALYTIC MATERIALS WITH VARIOUS MORPHOLOGIES

As mentioned above, TiO_2 photocatalysis has been recognized as one of the most promising technologies for water purification using the high reactivity of hydroxyl radicals, which form on the surface of TiO_2 crystals by irradiation of UV light. To obtain the highest efficiency, however, not only the material but also its crystalline structure and morphology have a considerable influence on the photocatalytic performance of TiO_2 crystals. From this viewpoint, various types of TiO_2-based photocatalytic materials with different morphologies have been developed (Table 1).

Of course, although the aforementioned TiO_2-based photocatalytic materials with various morphologies can be exploited to tackle many challenges involving pollutant degradation, photo-organic synthesis, and water splitting, the most notable characteristic for environmental applications is the decomposition ability of inorganic and organic compounds by photo-induced hydroxyl radicals on the surface of TiO_2 crystals. In this case, to achieve the highest reaction efficiency, the larger surface area and crystalline structure are very important. The aforementioned nanostructures have been actively developed based on these characteristics. That is to say, various types of one-dimensional (1D) morphology

TABLE 1 Various Types of TiO_2 Materials with Different Morphologies

Morphologies of TiO_2-based Photocatalytic Materials Other than Simple Powders	Synthesis Methods	References
TiO_2 nanorods	Sol-gel template method	[15]
	Hydrothermal method	[16]
	Chemical vapor disposition	[17]
TiO_2 nanotubes	Electrochemical disposition	[18]
	Template method	[19,20]
	Hydrothermal method	[21]
TiO_2 nanowires/nanofibers	Hydrothermal method	[22]
	Electrospinning	[23]
TiO_2 nanobelts	Solvothermal method	[24]
	Chemical vapor disposition	[25]
	Hydrothermal method	[26]
TiO_2 fibers	Electrospinning	[27]
TiO_2/SiO_2 fibers	In situ production process making the best use of the bleed-out phenomenon using a silicon-based polymer precursor	[28]

Some information of this table was quoted from Paramasivam et al. [29].

(nanorods, nanotubes, nanowires, nanofibers, and so on) and some other fibrous morphologies have been created. In the following sections some actual results are introduced.

3 TiO_2-BASED PHOTOCATALYSTS WITH ONE-DIMENSIONAL MORPHOLOGY

To date, much research on 1D TiO_2-based photocatalysts (1D-TiO_2) has been performed, and several reviews of this research have been published [29–31]. As can be seen in Table 1, different types of 1D-TiO_2 materials have been synthesized by various methods, and their morphologies depend on the preparation conditions. Of these, TiO_2 nanorods as one 1D-TiO_2 material synthesized using the sol-gel template method, chemical vapor deposition, or the hydrothermal method [30]. In general, TiO_2 nanorods have a relatively small number of grain boundaries and can act as single crystals, which can exhibit higher photocatalytic activity than nanoparticles. Also, because of their slender structure, TiO_2 nanorods show relatively less recombination of electrons and holes compared with nanoparticles. Consequently, this enhances their photocatalytic activity. One example of synthesized TiO_2 nanorod arrays are shown in Figure 2. These nanorod arrays have diameter of 160-250 nm and a smooth surface, along with a dense core structure.

On the other hand, by other production process using electrochemical deposition (the anodization method), very ordered TiO_2 nanotubes can be synthesized [31]. The typical production process is as follows: titanium (Ti) foil is used as an anode and the same size platinum foil as a cathode. The distance between two electrodes is fixed at 2 cm. A mixture containing 50 mL ethylene glycol, 0-8.00 mol L^{-1} acetic acid, and 0.05-0.50 mol L^{-1} NH_4F is used as an electrolyte. Then, under 30 V voltage (anodization), the TiO_2 nanotubes propagate on the Ti substrate. The TiO_2 nanotubes obtained are shown in Figure 3.

FIGURE 2 Field emission scanning electron microscopy image of synthesized TiO_2 nano-rod arrays. *(Reprinted from Feng et al. [30].)*

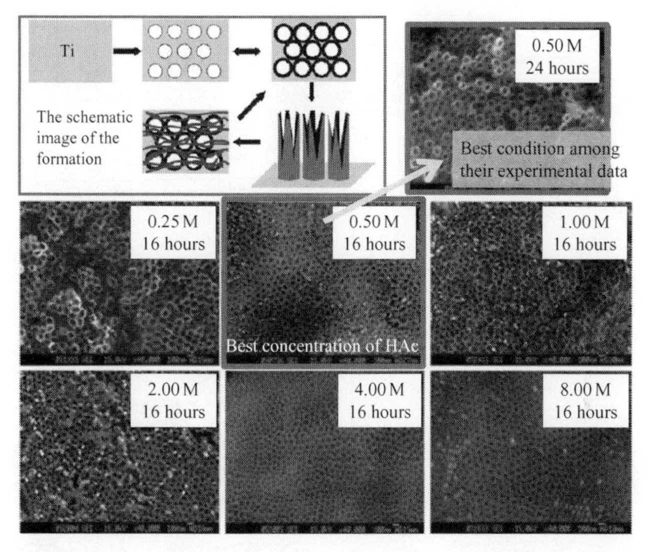

FIGURE 3 The SEM images of TiO_2 nanotubes synthesized by electrochemical deposition. Each panel shows differences in the morphology of the TiO2 nanotubes with various concentrations of acetic acid (HAc). The nanotubes were prepared at 30 V for 15 hours in ethylene glycol solution containing 0.20 mol L^{-1} NH_4F with the aforementioned HAc concentrations. *(Images reprinted from Nie et al. [31].)*

As can be seen in Figure 3, this type of TiO_2 nanotube shows a honeycomb structure with a hole size that can be controlled. This architecture results in highly efficient electron transport and suppression of the recombination of photo-generated electrons and holes assisted by an external electric field.

4 FIBROUS PHOTOCATALYST (TiO_2/SiO_2 PHOTOCATALYTIC FIBER)

As mentioned in the Introduction, anatase–TiO_2 is well known as a semiconductor catalyst; it exhibits better photocatalytic activity by UV irradiation with wavelength shorter than 387 nm (3.2 eV). The decomposition of harmful substances using the photocatalytic activity of anatase–TiO_2 has attracted a great deal of attention. This effect is attributed to the strong oxidant (hydroxyl radical) that is generated on the surface of the TiO_2 crystals by the UV irradiation. At present, most research has been performed using a powdery material or a coated material on the substrate. Of these, powdery photocatalysts have some difficulties in practical use [32]. For example, they have to be filtered from treated water. Coated photocatalysts on the substrate cannot provide sufficient area for contact with harmful substances [32]. In addition, the coated layer is easily peeled from the substrate during use. To avoid these problems, other types of research concerning fibrous photocatalysts have been conducted [33]. Up to the present, however, a combination of excellent photocatalytic activity and high fiber strength has not

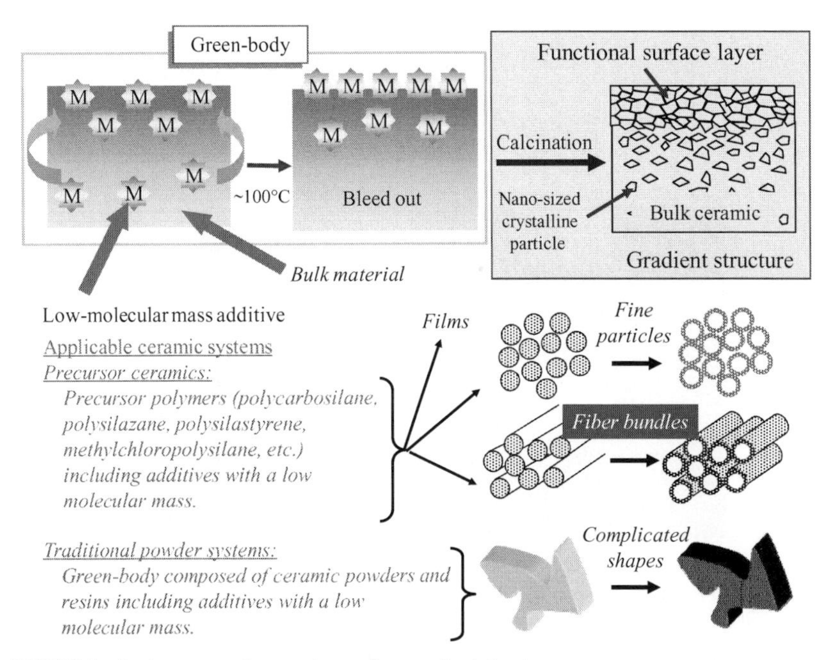

FIGURE 4 In situ process for creating surface gradient structures.

been achieved using the sol-gel method or a simple polymer blend [34]. To avoid these problems, a unique in situ process by which a gradient-like surface functional layer was easily formed was developed [28] (Figure 4).

Making the best use of this process, a strong (2.5 GPa) and continuous photocatalytic fiber comprising an SiO_2-based fiber with a small diameter (~5-7 μm) covered with a gradient-like surface TiO_2 layer has been developed. This fiber is a TiO_2-dispersed, SiO_2-based fiber with a sintered anatase–TiO_2 layer on the surface. The surface gradient layer comprising nanoscale TiO_2 crystals (8 nm) was strongly sintered and exhibited excellent photocatalytic activity, which can lead to the efficient decomposition of harmful substances and any bacterium contained in air and/or water by UV irradiation. In this section the abovementioned photocatalytic fiber produced by this unique in situ process and its actual applications are described.

4.1 Synthesis of the TiO_2/SiO_2 Photocatalytic Fiber

Polytitanocarbosilane containing an excess amount of titanium alkoxide was synthesized by the mild reaction of polycarbosilane $(–SiH(CH_3)–CH_2–)_n$ (20 kg) with titanium(IV) tetra-n-butoxide (20 kg) at 220 °C in a nitrogen atmosphere. The precursor polymer obtained was continuously melt-spun at 150 °C using melt-spinning equipment. The spun fiber, which contained an excess amount of unreacted titanium alkoxide, was pretreated with heat at 100 °C and subsequently fired up to 1200 °C in air to obtain a continuous, transparent fiber (5- to 7–micron diameter). In the initial stage of the heat treatment, effective bleeding of the excess amount of unreacted titanium(IV) tetra-n-butoxide from inside the spun fiber occurred to form the surface gradient layer containing a large amount of titanium(IV) tetra-n-butoxide. During the next firing process, the heat-treated precursor fiber was converted into a TiO_2-dispersed, SiO_2-based fiber covered with a gradient-like TiO_2 layer (photocatalytic fiber). The fundamental concept of the unique production process for this TiO_2/SiO_2 photocatalytic fiber is shown in Figure 5.

Figure 6 shows the surface appearance and the cross section of the obtained TiO_2/SiO_2 photocatalytic fiber. As shown in this figure, the surface of the fiber is densely covered with nanoscale anatase–TiO_2 particles (8 nm), which are strongly sintered with each other directly or through an amorphous SiO_2 phase. The thickness of the surface TiO_2 layer is ~100-200 nm. The tensile strength of

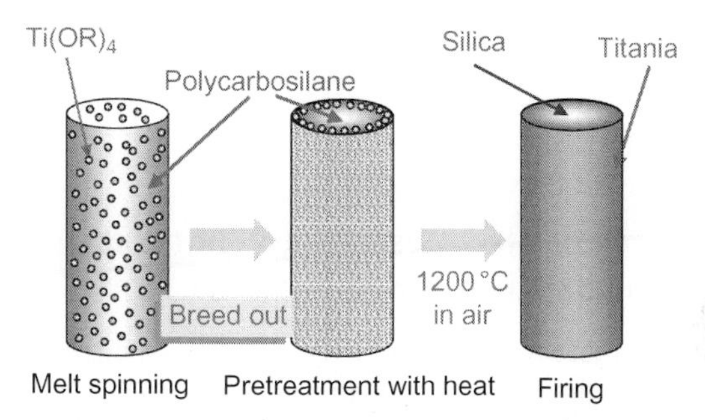

FIGURE 5 Fundamental concept of the TiO_2/SiO_2 photocatalytic fiber.

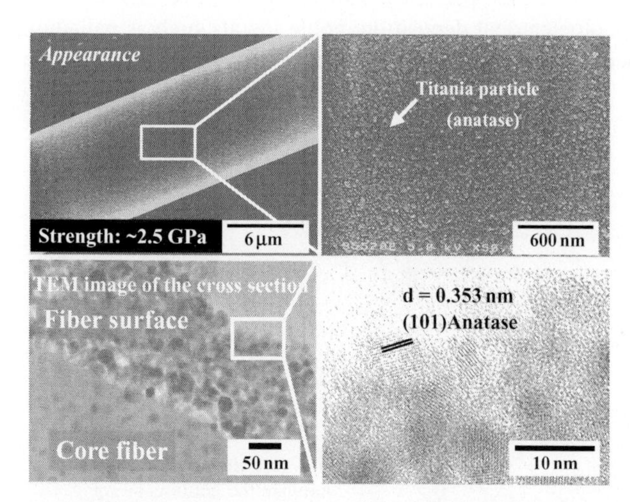

FIGURE 6 The surface and cross-section of the TiO_2/SiO_2 photocatalytic fiber. TEM, transmission electron microscopy.

this fiber as measured by the single filament method was, on average, 2.5 GPa using an Orientec UTM-20 with a gauge length of 25 mm and a cross-head speed of 2 mm/min. This mechanical strength is markedly superior to that of existing photo-catalytic TiO_2 fibers (<1 GPa) that were produced using a sol-gel method [33] or polytitanosiloxanes [34]. The high strength of the aforementioned TiO_2/SiO_2 photocatalytic fiber is closely related to the dense structure without pores, which is created by its higher firing temperature compared with other TiO_2 fibers.

The photocatalytic activity of the TiO_2/SiO_2 photocatalytic fiber is caused by the anatase–TiO_2 existing on the surface of each fiber. It is well known that anatase–TiO_2 converts into rutile at temperatures ranging from 700 °C to 1000 °C [33]. In particular, pure nanocrystalline anatase–TiO_2 easily converts into rutile at lower temperatures (~500 °C) [35]. In the aforementioned case (firing temperature of 1200 °C), it is thought that the surrounding SiO_2 phase stabilized the anatase–TiO_2 phase. At the interface between TiO_2 and SiO_2, atoms constructing TiO_2 are substituted into the tetrahedral SiO_2 lattice, forming tetrahedral Ti sites [36]. The interaction between the tetrahedral Si species and the tetrahedral Ti sites in the anatase–TiO_2 is thought to prevent the transformation into rutile (Figure 7).

4.2 Palladium-Deposited Mesoporous Photocatalytic Fiber with High Photocatalytic Activity

By removing interfacial SiO_2 phases between TiO_2 crystalline arrays existing in the surface region of the aforementioned TiO_2/SiO_2 photocatalytic fiber using hydrofluoric acid solution, one can generate 2- to 20-nm mesopores between coherently bonded TiO_2 particles (8 nm) in the surface TiO_2 layer.

Using photoelectrical deposition, noble metals such as palladium (Pd) can be deposited into interstices in the surface TiO_2 layer. The deposition of Pd

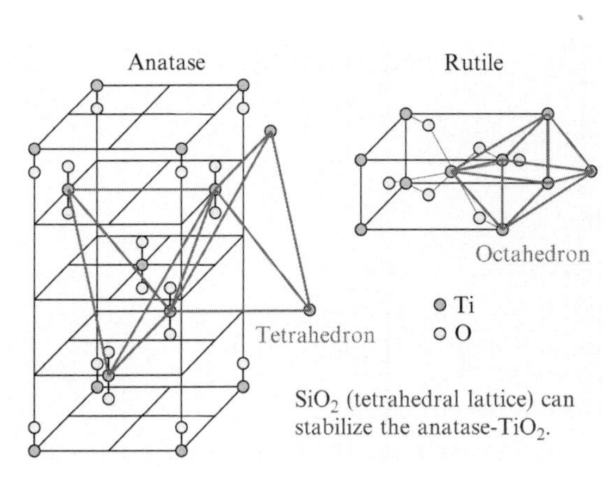

FIGURE 7 Crystalline structure of titania.

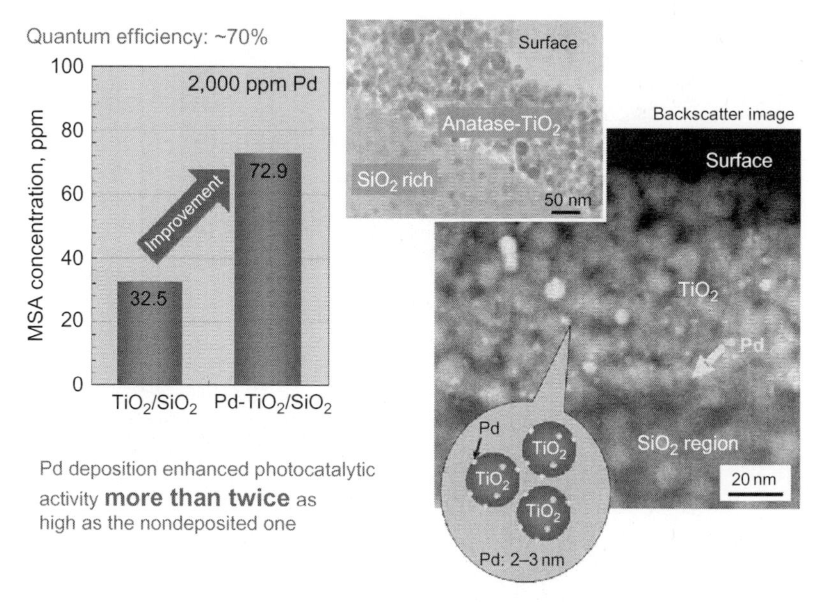

FIGURE 8 Significant improvement in photocatalytic activity by palladium (Pd) deposition.

selectively formed in the interstices of the TiO_2 layer between photoactive nanocrystal arrays readily proceeds [37]. The remarkable improvement in photocatalytic activity by the aforementioned Pd deposition is shown in Figure 8.

As can be seen in Figure 8, the photocatalytic activity of the Pd-deposited mesoporous photocatalytic fiber is about twice as high as that of the fiber without any deposits. The quantum efficiency is about ~70%. In this case the surface TiO_2 crystals play an important role as an oxidation point for the organic materials contained in the water, whereas the Pd inside plays an important role as a reduction point that transfers the electron to water. Figure 9 shows an image of the charge separation on the surface region of the aforementioned Pd-deposited photocatalytic fiber.

Moreover, the higher photocatalytic activity of the Pd-deposited photocatalytic fiber remained after a long period of use. This result implies that the Pd metals do not peel off from the surface region. Therefore, Pd-deposited mesoporous photocatalytic fibers must be useful and effective for actual water purification.

4.3 Environmental Application of TiO_2/SiO_2 Photocatalytic Fibers

Fundamental technologies concerning many types of water purification systems have been developed using the aforementioned TiO_2/SiO_2 photocatalytic fiber and Pd-deposited mesoporous photocatalytic fiber (Figure 10). Some basic information about these applications follows.

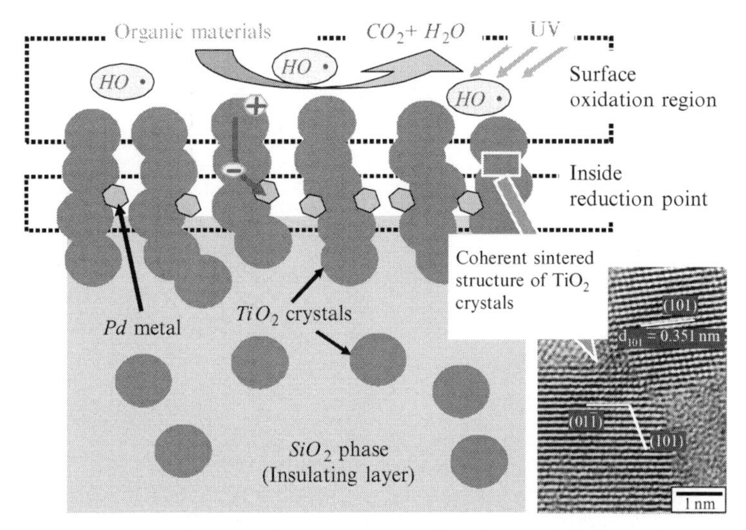

FIGURE 9 The charge separation on the surface region of the palladium (Pd)-deposited photocatalytic fiber. UV, ultraviolet light.

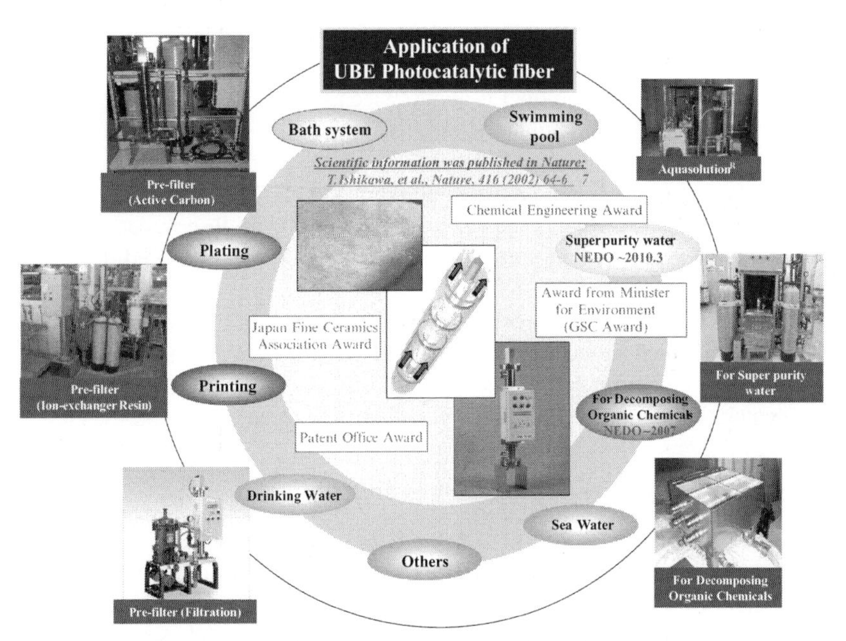

FIGURE 10 Many types of applications using the photocatalytic fibers.

As the first example, water in collective bathtubs and swimming pools was purified. This test used a simple purifier with a module comprising cone-shaped felt material (made of the TiO_2/SiO_2 photocatalytic fiber) and a UV lamp. The average intensity of UV light on the TiO_2/SiO_2 photocatalytic fibers was 10 mW/cm^2. The muddiness of the pool water was remarkably improved by several passes through the purifier. Organic filth and chloramines also decreased after passage through the purifier. In addition, many bacteria (common bacterium, *Legionella pneumophila*, and coliform) that existed in the initial bath water were effectively decomposed into CO_2 and H_2O. The experimental data on the sterilization of *L. pneumophila* is shown in Figure 11.

As shown in Figure 11, the *L. pneumophila* was effectively decomposed by a single passage through the water purifier using the aforementioned TiO_2/SiO_2 photocatalytic fiber. In this case the retention time for the single passage through the purifier was only 5 sec. Although this time may seem too short, it is sufficiently long compared with the lifetime of hydroxyl radicals (10^{-6} seconds). Each oxidation reaction ought to proceed within the aforementioned short lifetime of the hydroxyl radicals. Accordingly, the decomposition reaction can be accomplished like this, as long as the number of photons is sufficient during the passage.

The result of the decomposition of colon bacillus by the TiO_2/SiO_2 photocatalytic fiber irradiated by UV light is shown in Figure 12, along with the comparative data obtained after only UV irradiation. As can be seen in this figure, use of the TiO_2/SiO_2 photocatalytic fiber led to effective decomposition of colon bacillus accompanied by the generation of CO_2. On the other hand,

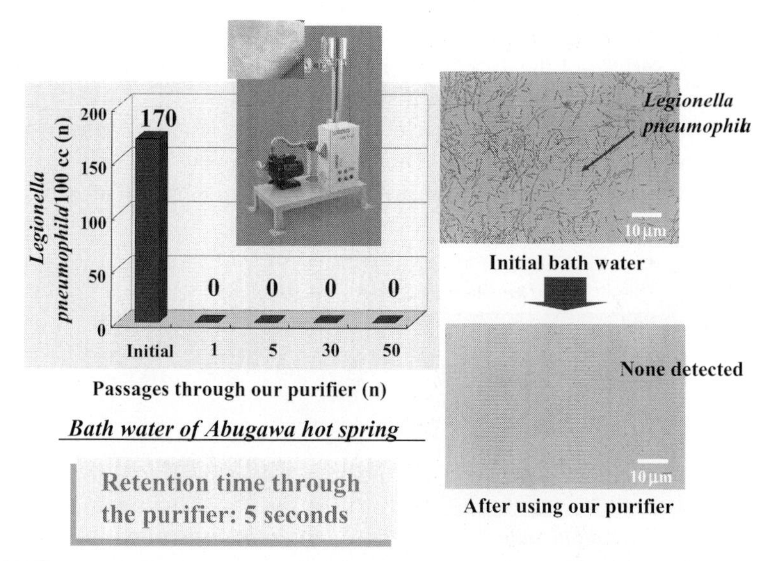

FIGURE 11 Sterilization of *Legionella pneumophila* existing in circulating bath water.

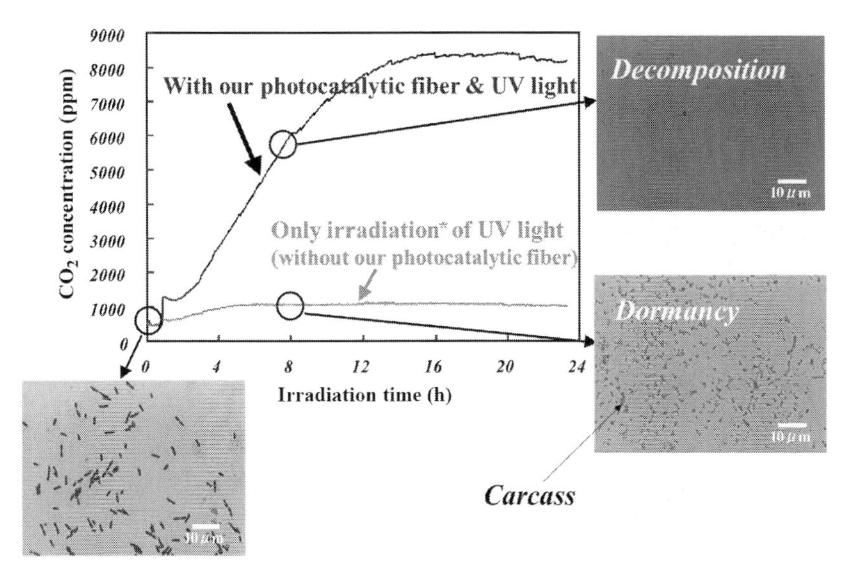

FIGURE 12 Decomposition of colon bacillus accompanied by the release of carbon dioxide using the TiO_2/SiO_2 photocatalytic fiber, along with the comparative result using only ultraviolet (UV) light. *Irradiation at 254 nm (12 mW/cm^2).

UV irradiation alone resulted in many dead colon bacillus cells, with no apparent decomposition.

Damaged colon bacillus is shown in Figure 13. Although almost all of the colon bacillus was perfectly decomposed into CO_2 and H_2O, some residual bacteria also were remarkably damaged. This type of decomposition reaction proceeds on the surface of the TiO_2/SiO_2 photocatalytic fiber when colon bacillus contacts each fiber irradiated by UV light.

Fundamentally, this can be applied to purification systems containing organic chemicals, including dioxin. Dioxin is one of many persistent organic pollutants. It has been recognized that the perfect decomposition of persistent

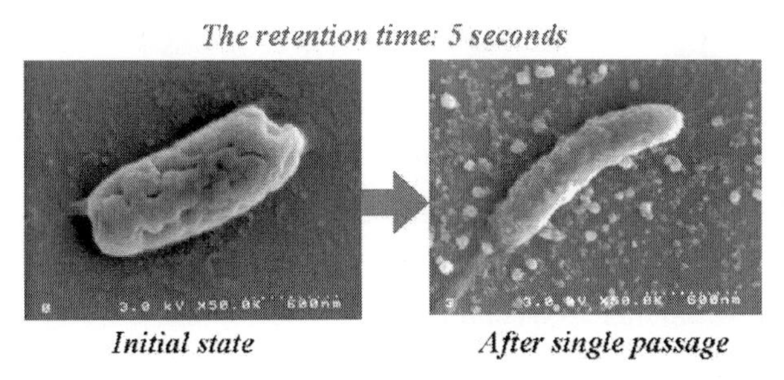

FIGURE 13 Change in the morphology of colon bacillus after passage through the water purifier using the TiO_2/SiO_2 photocatalytic fiber.

organic pollutants is very difficult. However, the use of the aforementioned photocatalytic water purifier partly enables the oxidation of dioxin into CO_2 and H_2O. This result is caused by the oxidation activity of hydroxyl radicals generated on the surface of the photocatalytic fiber irradiated by UV light. Furthermore, this water purification system using TiO_2/SiO_2 photocatalytic fibers can decompose the hard outer shell structure of *Bacillus subtilis*, which is covered with a hard shell out of the body. Figure 14 shows evidence that the hard shell of *B. subtilis* was actually damaged by contacting the TiO_2/SiO_2 photocatalytic fiber irradiated by UV light.

It is well known that botulinus and anthrax are a variety of *B. subtilis*. This type of bacteria is hardly diminished by boiling H_2O or chlorine. Accordingly, the water purifier using TiO_2/SiO_2 photocatalytic fibers is very effective for avoiding this type of hazard. Regarding drinking water, some people are endangered if they use a common water purifier containing an active carbon filter. Many bacteria can propagate in an active carbon filter. We detected the number of heterotrophic bacteria in H_2O put out by an active carbon filter. According to experimental results, a tremendous number of heterotrophic bacteria were found in H_2O from the active carbon filter. This is not good for human health. Even such H_2O as this could be effectively purified by the use of the aforementioned photocatalytic fiber.

This undesirable phenomenon also was confirmed in the field of medical water. Active carbon filters are actually used in this field (water for dialysis). In this case a tremendous number of bacteria were detected in the H_2O from the active carbon filter. Using a TiO_2/SiO_2 photocatalytic fiber behind the active carbon filter, however, the bacteria were effectively decomposed, as expected.

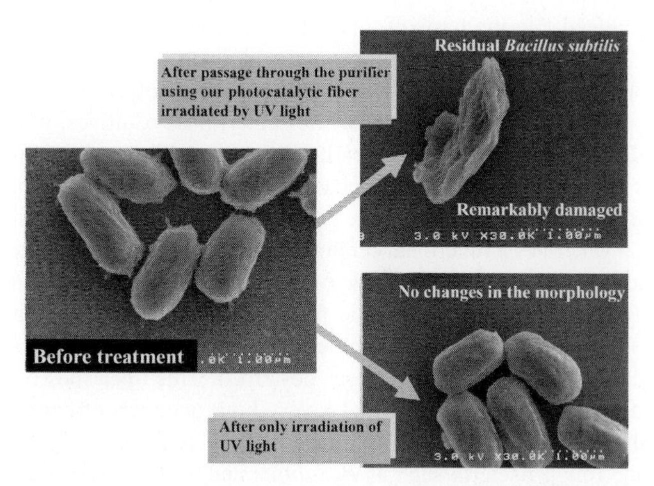

FIGURE 14 Damaged *Bacillus subtilis* after a single passage through the water purifier using TiO_2/SiO_2 photocatalytic fibers, along with the comparative result (no change) using only ultraviolet (UV) irradiation.

As this information demonstrates, we need to reconsider the use of active carbon filters, which often are believed to be a safe purification system. In fact, active carbon filters should be recognized as a kind of breeding zone for bacteria, although these filters plays an important role for adsorption of various impurities. To avoid such an undesirable situation, a combination with the photocatalytic technologies described here is a good solution.

5 SUMMARY

Since 1972, photo-induced reaction using photocatalysts has been considered a promising technology for environmental applications, especially as water purification technology. In this case the most important thing is the higher reactivity under irradiation of effective light. From this viewpoint, many photocatalysts have been developed to date. To obtain a larger surface area and an effective crystalline structure, various types of TiO_2-based photocatalytic materials with different morphologies (nanorods, nanotubes, nanofibers, TiO_2/SiO_2 photocatalytic fibers, and so on) have been developed. In this chapter basic information on the photocatalytic reaction and the actual materials were introduced. TiO_2 nanorods, one type of 1D-TiO_2 material, have a relatively small number of grain boundaries and can act as single crystals, which can exhibit higher photocatalytic activity than nanoparticles. In addition, ordered TiO_2 nanotubes can be synthesized by electrochemical deposition, and the hole size can be controlled. TiO_2 nanotubes with a honeycomb structure show the higher efficiency of electron transport and suppression of the recombination of photo-generated electrons and holes by an external electric field.

Furthermore, the TiO_2/SiO_2 photocatalytic fiber and its actual applications were introduced. On the basis of a precursor method using a polycarbosilane, a new in situ process for preparing functional ceramic fibers with a gradient surface layer was developed. This process treated a polycarbosilane containing an excess amount of selected low-molecular-mass additives, which can be converted into functional ceramics by heat treatment. Thermal treatment of the precursor fiber leads to controlled phase separation ("bleed out") of the low-molecular-mass additives from inside to outside the fiber. After that, subsequent calcination generates a functional ceramic fiber with a gradient surface structure during the production of bulk ceramic components. As the embodied functional ceramic fiber, a strong photocatalytic fiber comprising a surface gradient-like TiO_2-crystalline layer and an SiO_2-based core structure was developed. A Pd-deposited photocatalytic fiber also was described. This fiber showed excellent photocatalytic activity, that is, it had a very high quantum efficiency (~70%) and the ability to decompose organic chemicals and any kind of bacteria by UV irradiation. Tough bacteria such as *B. subtilis* and anthrax, which are covered with a hard shell and hardly decomposed by chlorine or boiled H_2O, can also be effectively decomposed by TiO_2/SiO_2 photocatalytic fibers and UV irradiation. In the field of bacterial sterilization, these photocatalytic technologies are very profitable.

REFERENCES

[1] A. Fujishima, K. Honda, Electrochemical photolysis of water at a semiconductor electrode, Nature 238 (1972) 5551.

[2] T. Bora, J. Dutta, Applications of Nanotechnology in Wastewater Treatment, J Nanosci Nanotechnol 14 (2014) 613–626.

[3] T. Mano, S. Nishimoto, Y. Kameshita, M. Miyake, Water treatment efficacy of various metal oxide semiconductors for photocatalytic ozonation under UV and visible light irradiation, Chem Eng J 264 (2015) 211–229.

[4] M. Muruganandham, R.P.S. Suri, Sh. Jafari, M. Sillanpaa, Gang-Juna Lee, J.J. Wu, M. Swaminathan, Recent developments in Homogeneous Advanced Oxidation Processes for Water and Wastewater Treatment, Int. J. Photoenergy 2014 (2014) 21, Article ID 821674.

[5] D. Kanakaraju, B.D. Glass, M. Oelgemoller, Titanium Dioxide Photocatalysis for Pharmaceutical Wastewater Treatment, Environ Chem Lett 12 (2014) 27–47.

[6] David G. Rickerby, Nanostructured Titanium Dioxide for Photocatalytic Water Treatment, in: B.I. Kharison, O.V. Kharissova, H.V. Rasika Dias (Eds.), first Ed., In: Nanomaterials for Environmental Protection, John Wiley & Sons, Inc, 2014, pp. 169–182.

[7] Ranjana Das, Application Photocatalysis for Treatment of Industrial Waste Water-A Short Review, Open Access Library J. 1 (e713) (2014), http://dx.doi.org/10.4236/oalib.1100713.

[8] K. Zoschke, H. Bornick, E. Worch, Vacuum-UV Radiation at 185 nm in Water Treatment, Water Res 52 (2014) 131–145.

[9] S. Helali, M.I. Polo-Lopez, P. Fernandez-Ibanez, B. Ohtani, F. Amano, S. Maloto, C. Guillard, Solar Photocatalysis: A green Technology for E. Coli Contaminated Water Disinfection. Effect of Concentration and Diffetent Types of Suspended Catalyst, J Photochem Photobiol A Chem 276 (2014) 31–40.

[10] X. Wu, X. Xu, C. Guo, H. Zeng, Metal Oxide Heterostructures for Water Purification, J Nanomater 2014 (2014), http://dx.doi.org/10.1155/2014/603096, Article ID 603096 2 pages.

[11] A. Ayati, A. Ahmadpour, F.F. Bamoharram, B. Tanhaei, M. Manttari, M. Sillanpa, A Review on Catalytic Applications of Au/TiO2 Nanoparticles in the Removal of Water Pollutant, Chemosphere 107 (2014) 163–174.

[12] M. Mehrjouei, S. Muller, D. Moller, Energy Consumption of Three Different Advanced Oxidation Methods for Water Treatment: A Cost-effectiveness Study, J Clean Prod 65 (2014) 178–183.

[13] M. Mehrjouei, S. Muller, D. Moller, A Review on Photocatalytic Ozonation Used for the Treatment of Water and Wastewater, Chem Eng J 263 (2015) 209–219.

[14] N. Wang, X. Zhang, Y. Wang, W. Yu, H.L.W. Chan, Microfluidic Reactors for Photocatalytic Water Purification, Lab Chip 14 (2014) 1074–1082.

[15] A.S. Attar, M.S. Ghamsari, F. Hajiesmaeilbaigi, S. Mirdamadi, K. Katagiri, K. Koumoto, Sol-gel template synthesis and characterization of aligned anatase-TiO_2 nanorod arrays with different diameter, Mater Chem Phys 113 (2-3) (2009) 856–860.

[16] X.J. Feng, J. Zhai, L. Jiang, The fabrication and switchable superhydrophobicity of TiO_2 nanorod films, Angewante Chemie International Edition 44 (32) (2005) 5115–5118.

[17] S. Feng, J. Yang, M. Liu, CdS quantum dots sensitized TiO_2 nanorod-array-film photoelectrode on FTO substrate by electrochemical atomic layer epitaxy method, Electrochemi. Acta 8 (2012) 321–326.

[18] P. Hoyer, Formation of a titanium dioxide nanotube array, Langmuir 12 (6) (1996) 1411–1413.

[19] D.V. Bavykin, J.M. Friedrich, F.C. Walsh, Protonated titanates and TiO_2 nanostructured materials: synthesis, properties, and applications, Adv Mater 18 (21) (2006) 2807–2824.

[20] T. Peng, A. Hasegawa, J. Qiu, K. Hirao, Fabrication of titania tubules with high surface area and well-developed mesostructural walls by surfactant-mediated templating method, Chem Mater 15 (10) (2003) 2011–2016.

[21] N. Liu, X. Chen, J. Zhang, J.W. Schwank, A review on TiO2-based nanotubes synthesized via hydrothermal method: formation mechanism, structure modification, and photocatalytic applications, Catal Today 225 (2014) 34–51.

[22] A. Fujishima, T.N. Rao, D.A. Tryk, Titanium dioxide photocatalysis, J Photochem Photobiol C 1 (1) (2000) 1–21.

[23] J. Lee, Y. Lee, H. Song, D. Jang, Y. Choa, Synthesis and characterization of TiO_2 nanowires with controlled porosity and microstructure using electrospinning method, Curr Appl Phys 11 (1) (2011) S210–S214.

[24] A. Hu, X. Zhang, K.D. Oakes, P. Peng, Y.N. Zhou, M.R. Servos, Hydrothermal growth of free standing TiO_2 nanowire membranes for photocatalytic degradation of pharmaceuticals, J Hazard Mater 189 (1-2) (2011) 278–285.

[25] N.T.Q. Hoa, Z. Lee, S.H. Kang, V. Radmilovic, E.T. Kim, Synthesis and ferromagnetism of Co-doped TiO2 nanobelts by metallorganic chemical vapor deposition, Appl Phys Lett 92 (12) (2008).

[26] Z.L. He, W.X. Que, J. Chen, X.T. Yin, Y.C. He, J.B. Ren, Photocatalytic degradation of methyl orange over nitrogen-fluorine codoped TiO2 nanobelts prepared by solvothermal synthesis, ACS Appl. Mater. Interface 4 (2012) 6815–6825.

[27] I.M. Szilagyi, D. Nagy, Review on one-dimensional nanostructures prepared by electrospinning and atomic layer deposition, J Phys Conf Ser 559 (2014) 1–13.

[28] T. Ishikawa, H. Yamaoka, Y. Harada, T. Fujii, T. Nagasawa, A general process for in situ formation of functional surface layers on ceramics, Nature 416 (2002) 64–67.

[29] I. Paramasivam, H. Jha, N. Liu, P. Schmuki, A review of photocatalysis using self-organized TiO2 nanotubes and other ordered oxide nanostructures, Small 8 (20) (2012) 3073–3103.

[30] T. Feng, G.S. Feng, L. Yan, J.H. Pan, One-dimensional nanostructured TiO2 for photocatalytic degradation of organic pollutants in wastewater, International Journal of Photoenergy 2014 (2014), Article ID 563879, 14 pages.

[31] X. Nie, J. Chen, G. Li, H. Shi, H. Zhao, Po-Keung Wong, T. An, Synthesis and characterization of TiO_2 nanotube photoanode and its application in photoelectrocatalytic degradation of model environmental pharmaceuticals, J. Chem. Technol. Biotechnol. 88 (2012) 1488–1497, http://dx.doi.org/10.1002/jctb.3992, Wileyonlinelibrary.com.

[32] T. Ishikawa, Ceramic Fiber with Decomposition Ability of Dioxine, Miraizairyo 3 (2) (2003) 26–33.

[33] H. Koike, Y. Oki, Y. Takeuchi, Preparing Titania fibers and their Photo-catalytic Activity, Mater RessocSymProc 549 (1999) 141–146.

[34] T. Gunji, I. Sopyan, Y. Abe, Synthesis of Polytitanosiloxanes and their Transformation to SiO2-TiO2 Ceramic Fibers, J Polym Sci, Part A Polym Chem 32 (1991) 3133–3139.

[35] P.I. Gouma, P.K. Dutta, M.J. Mills, Structural Stability of Titania Thin Films, Nano Struct. Mater. 11 (8) (1999) 1231–1237.

[36] C. Anderson, A.J. Bard, Improved Photocatalytic Activity and Characterization of Mixed TiO2/SiO2 and TiO2/Al2O3, J Phys ChemB 101 (1997) 2611–2616.

[37] T. Matsunaga, H. Yamaoka, S. Ohtani, Y. Harada, T. Fujii, T. Ishikawa, Appl. Catal. A: Genera 1351 (2008) 231–238.

Chapter 17

Solution Processing of Low-Dimensional Nanostructured Titanium Dioxide: Titania Nanotubes

T. Sekino
ISIR, Osaka University, Osaka, Japan

1 INTRODUCTION

Titanium dioxide (TiO_2, titania) is a well-known wide-gap semiconductor oxide that is inexpensive, chemically stable, and harmless. It has various polymorphisms, however; three types of crystalline structures—anatase (tetragonal), rutile (tetragonal), and brookite (orthorhombic)—are the most common and are used practically in many fields. Among them, anatase, which is a low-temperature phase in TiO_2, shows the most unique photochemical properties. Although it has no optical absorption in the visible light spectrum, it is responsive to ultraviolet (UV) light; when UV is irradiated to TiO_2, an electron and hole pair is generated within the crystal and induces various chemical reactions at the surface. Therefore, the most promising characteristic of TiO_2 is its photochemical properties, and thus it has been widely studied as a photocatalyst [1–4] for water and/or air purification, hydrogen/oxygen generation, as an antibacterial coating, and so on. In addition, extensive research has explored its use in a variety of applications, such as gas sensors [5,6], nonlinear optics [7], and photoelectrodes in low-cost, dye-sensitized solar cells (DSSCs) [8–10], among others, showing it is promising candidate as an environmentally friendly and energy-creating base material.

TiO_2-derived materials with low-dimensional structures and architectures, including fibrous, tubular, rod, wire, and sheet in the nano- to macro-scale, are expected to exhibit various improved photochemical properties because of their unique structural characteristics compared with colloidal or particulate titania. Thus much effort has been concentrated on fundamental investigations of their synthesis, formation mechanisms, physical and chemical properties, and modification of their physicochemical functions. Also, extensive research and

Green and Sustainable Manufacturing of Advanced Materials. http://dx.doi.org/10.1016/B978-0-12-411497-5.00017-5

development has been done to utilize titania for practical applications related to environmental cleaning systems, such as water purifiers and self-cleaning paintings. For this purpose, morphological/structural control is one of the key technologies.

Among the various nanostructured compounds, nanotube materials have attracted much attention because of their unique one-dimensional (1D) and hollow structures, of which the most well known is the carbon nanotube (CNT). Until now, various nonoxide nanotubes, such as carbon [11], boron nitride [12], and molybdenum disulfide [13] nanotubes, and oxide nanotubes such as vanadium oxide [14–16], aluminum oxide [16], silicon dioxide [16,17], TiO_2 [18–23], and natural minerals such as imogolite [24,25], have been reported. Among these nanotubular materials, the TiO_2 (titania) nanotube (TNT) is considered to be the most interesting and promising advanced material because of the synergy of titania's unique photochemical properties and 1D nanostructure.

In this chapter the formation of TNTs via low-temperature chemical processing; their nanostructural characteristics and fundamental physical, optical, and chemical properties; as well as their multiple functionalities are reviewed. Furthermore, novel methodologies of tuning the structures and various functions of TNTs from the viewpoints of environmental cleaning and energy creation applications are reviewed in relation to their processing and unique 1D structures.

2 FABRICATION OF 1D OXIDE NANOTUBES

2.1 Synthesis Route of Oxide Nanotubes

The artificial synthesis route of oxide nanotubes, including TNTs, is classified into two methods: the template (replica) and template-less (self-structuralization or self-organization) routes. The latter route is divided further into two routes: solution chemical and electrochemical methods. The synthesis methods of these categories are summarized in Table 1. Many efforts have been paid to fabricate tubular materials by using low-dimensional, nanometer-scale, organic and inorganic templates such as wire, fiber, whiskers, and tubular and hole arrays, such as anodized aluminum oxide films coupled with oxides coated on them via precipitation or a sol-gel method and removal of the templates afterward [14,15,18,22,23]. Preparing size-controlled nanotubes with homogeneous distribution of diameter and length using this processing route is easy because of the use of size-controlled templates. The as-synthesized nanotubular materials obtained are usually amorphous. They can be crystallized by heat treatment; however, the most of them are tubular polycrystalline consisting of nanocrystalline oxides.

On the other hand, the template-less route can realize self-structuralization or self-organization of nanotubular matter during chemical or electrochemical processes. Synthetic imogolite [25], sol-gel-derived silicon dioxide nanotubes [17],

TABLE 1 Comparison of Nanotube Fabrication Processes and Their Features

Synthesis Method	Process Characteristics	Nanotube Features
Template (replica) method	Scalable nanotubes morphology Complicated multiple-step process	Ordered or random alignment (depending on the template) Amorphous (NPs)
Electrochemical method (anodic oxidation)	Ordered alignment Complicated process conditions	Ordered array thin films Amorphous (NPs)
Solution chemical method (this work)	Easy to synthesize Low-temperature Template-less processing	Random alignment (powder form) Crystallites

chemically prepared TNTs [19,20], and nanotube/nanohole arrays such as aluminum oxide [26,27] and TiO_2 [21,28] prepared by electrochemical anodic oxidation of metal films are the typical systems fabricated by this self-organizing process. In this method no special nano-sized templates are necessary, and some of the materials obtained crystallize in the as-synthesized form. In this case the self-organized nanotubular formation route is considered to be suitable when one needs to use the fundamental physical and chemical properties of oxides because these are more sensitive and exploited in crystallized form [29].

2.2 Formation of TNTs

Among many nanostructured oxides, TNTs are one promising material. As mentioned before, TiO_2 is inexpensive, chemically stable, and harmless but is well known as a wide-gap semiconductor. Therefore, the most promising characteristic of TiO_2 lies in its photochemical properties, such as high photocatalytic activity under UV light; hence for the past 60 years it has been widely studied by many researchers as a photocatalyst [2,3,8,30], an electrode of DSSCs [8], a gas sensor [5], among others.

Various methods such as anodization of metal substrates [21,28] and replica [18,22] and template methods [23] have been investigated to prepare tubular TiO_2. However, a self-organized TNT was successfully prepared by Kasuga *et al.* [19,20] through of a template-less and simple chemical processing of a low-temperature solution (Table 1). The typical TNT has an open-ended structure with 8- to 10-nm and 5- to 7-nm outer and inner diameters, respectively. Based on this, the so-called Kasuga method, structural analysis, process optimization, and properties evaluation have been extensively explored by many investigations [31–34].

Because of the mutual and synergistic combination of various factors of a nanotubular TiO_2 semiconductor—such as crystal structure, chemical bonding, physical/chemical properties of the matter, low-dimensional nanostructures/nano-space/nano-surface, and self-organization/ordering of the structure—not only fundamental interests in the formation mechanism as well as the unique nanotubular structures but also function enhancements and novel functionalization hence are expected from TNTs [29].

2.3 Chemical Synthesis Using a Low-Temperature Solution

The typical route of synthesis of TNT using a high-concentration alkaline solution [19,20] is shown in Figure 1, together with the morphological changes of the products. Not only TiO_2 powders, including anatase, rutile, or their mixtures, but also titanium alkoxide, titanium metal [35], and natural minerals [36] can be used as a source material. The raw material is refluxed in a sodium hydroxide (NaOH) aqueous solution of 5 to 10 mol/L in concentration around 110°C, which corresponds to the boiling temperature of the high-concentration alkaline solution, for 24 h or longer. The product is washed repeatedly by distilled water (DW) to remove sodium ions. Then, 0.1 mol/L hydrogen chloride (HCl) aqueous solution is added to neutralize the solution, and again it is washed with DW until the solution's electrical

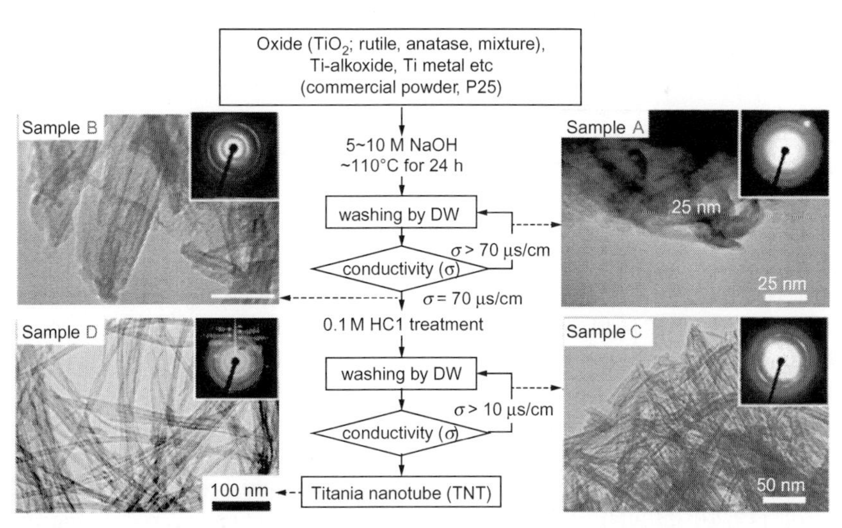

FIGURE 1 Typical synthesis procedure of titania nanotubes and transmission electron microscopy images of the products in each chemical synthesis step: sample A, after refluxing in alkaline (10 M NaOH, 110°C, 24 h); sample B, after 0.1 M HCl treatment; sample C, during the second was with distilled water (DW); and sample D, the final product.

conductivity is <10 μS/cm. The product then is separated by a filtering, centrifugation, or freeze-drying technique and finally dried.

Hydrothermal synthesis using an autoclave, which provides a closed reaction environment and higher pressure during the chemical treatment, can also be used to attempt to synthesize TNTs [34]. This route provides thicker TNTs because it uses a synthesis temperature higher than 110°C because of the increased pressure effect in the reaction environment. A size-controlled TNT, especially a thicker (larger) one, could also be obtained by process control using larger crystalline TiO_2 particles [37,38] and also using a mixed solution of water and ethanol (H_2O/C_2H_5OH) for synthesis [39]. Detailed morphology and properties are described in the section 3.4.

X-ray diffraction (XRD) analysis of products during the synthesis processing revealed that the product consisted of an amorphous and crystalline phase of sodium titanate (Na_2TiO_3) after the alkaline treatment. After the water and HCl treatment, Na_2TiO_3 disappeared completely and another crystalline phase with low crystallinity was observed. In this stage a nanometer-sized, sheet-like morphology was obtained (Figure 1B), which was considered to be the TiO_2 nanosheet. Further washing in water provided a fibrous product (Figure 1C and D) with a length of several hundred nanometers to several micrometers. A higher-magnification transmission electron microscopy (TEM) photograph (Figure 2a) clearly revealed that the outer and inner diameters of the final product are around 8-10 and 5-7 nm, respectively, and it had an open-ended structure. The size of the TNTs obtained does not depend on the kind of raw materials used; however, for some process parameters, such as the crystalline size of raw TiO_2, the addition of a small amount of ethanol provided a larger TNT diameter and smaller aspect ratio, respectively, as will be mentioned in the section 3.4. The surface area of a typical TNT is ~300-350 m^2/g, which is in good agreement with the calculated theoretical value of 345 m^2/g by assuming the tubular structure, observed size, and density of the TiO_2 crystal.

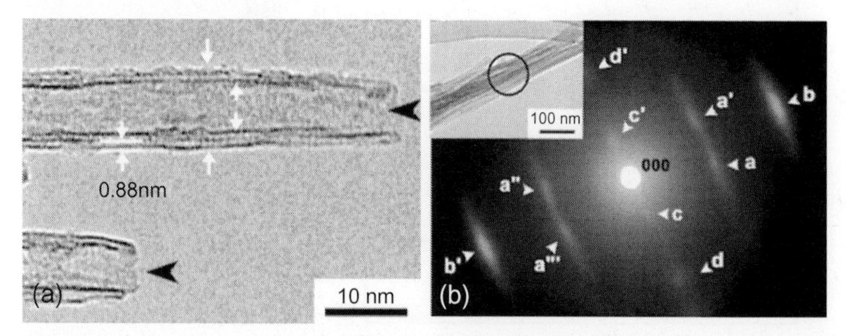

FIGURE 2 High-magnification transmission electron microscopy image of a synthesized titania nanotube (TNT) (a) and a selected area of the electron diffraction pattern (b) of the TNT bundle (inset).

2.4 Nanostructures and Their Formation Mechanism

TiO_2 has a rigid and isotropic crystal structure in which a lattice consisting of a TiO_6 octahedron spreads out three-dimensionally so that its crystal shape is usually equiaxial. However, the solution chemical synthesis described above gives an anisotropic and open-ended nanotubular form. To identify the structural characteristics and to understand the formation mechanism of TNTs in relation to the synthesis process, structural analyses have been explored using X-ray and neutron diffraction and high-resolution electron microscopy coupled with the electron diffraction technique [19,20,29,34–36,40–44].

The electron diffraction pattern of a selected area of a TNT bundle (Figure 2b) exhibited diffraction spots with belt-like spreading, which is typical for a fibrous compound. The interplanar spacing (*d*-spacing) of obtained spots corresponded to (101), (200), and (100) of the typical anatase crystal of TiO_2 [29], implying that the TNT basically has a crystal structure similar to that of the anatase-type of TiO_2. The longitudinal direction of the nanotube corresponds to the *a*-axis ([100] direction), whereas the cross section is parallel to the *b*-plane ((010) plane) of the anatase crystal.

On the other hand, the diffraction spot corresponding to the *d*-spacing of 0.87 nm, which also was found in the XRD pattern, at 2θ of ~9° (Figure 3). This is in good agreement with the spacing of 0.88 nm at the TNT wall seen in Figure 2a. This large interplanar distance is a typical characteristic of TNTs and is closely related to the formation of the tubular structure, as will be described in the later part of this section.

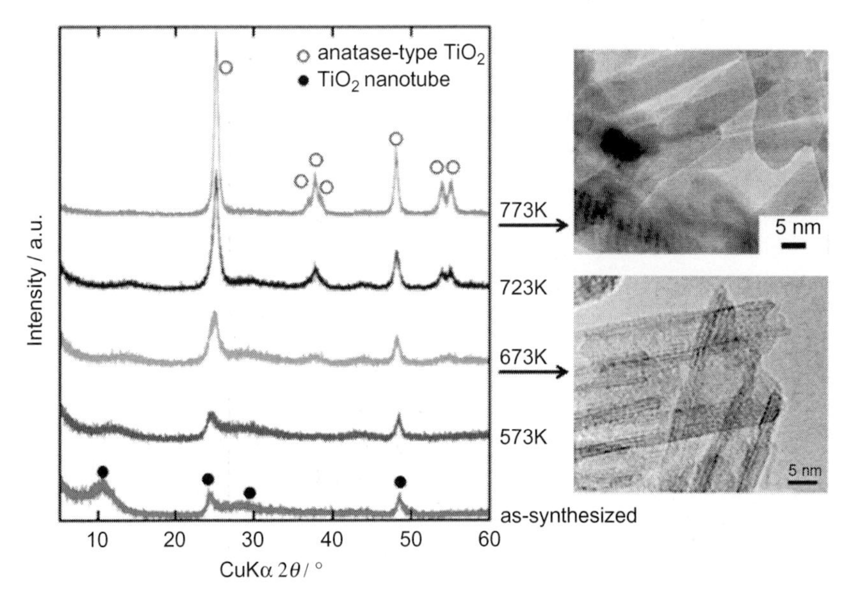

FIGURE 3 High-temperature X-ray diffraction patterns of synthesized TiO_2 nanotubes from room to 773 K (500°C) and corresponding structure change.

Thermogravimetry coupled with mass spectroscopic analysis for the as-synthesized TNT exhibited weight loss continuing up to approximately 350° C, where the major specie detected during heating was water (H_2O). High-temperature XRD (Figure 3) demonstrated that the typical diffraction peak intensity found at 2θ at ~9° decreased with heating up to around 400° C, whereas the anatase structure for TiO_2 became the major crystalline phase. The specific surface area for a pure TNT decreases with an increase in annealing temperature (Table 2 and Figure 3). A large surface area was maintained up to around 400° C, whereas a sudden decrease occurred above that temperature. It reached a value of approximately 100 m²/g at an annealing temperature higher than 450° C. TEM investigation of the annealed TNT revealed that the nanotubular structure was maintained up to around 450° C [29]. These facts imply that the as-synthesized TNT contains a hydroxyl group (–OH) and/or structure water (H_2O) together with the TiO_6 octahedral network, and hence it has a titanate-like structure. Heat treatment (annealing) of the as-synthesized TNT up to around 400°C released H_2O, and then the nanotube became a stoichiometric TNT with an anatase structure as its base crystal structure.

Extensive detailed structural analyses have been carried out. Using high-resolution TEM, Chen *et al.* [40] reported that the TNT was a form of titanate ($H_2Ti_3O_7$) and proposed a structural model. Ma *et al.* [41] showed that it was lepidocrocite, which is one of the defects containing titanate with the formula of $H_xTi_{2-x/4}V\ddot{o}_{x/4}O_4$, where *Vö* is the oxygen vacancy. Further, various chemical formulas were proposed: $Na_2Ti_2O_4(OH)_2$, or its protonated formula $H_2Ti_2O_4(OH)_2$ [42], and $H_2Ti_4O_9$ [35]. These compounds, however, can be basically described as $(TiO_2)_n \cdot (H_2O)_m$ and have a layered structure; thus it can reasonably be rationalized that H_2O is released by the heat treatment of as-synthesized TNTs, as mentioned previously [29].

Considering these facts, the formation of an alkaline titanate such as Na_2TiO_3 or its amorphous matter (Figure 1b) is an essentially important as a intermediate compound for the formation of nanosheets and subsequent nanotubes. First, alkali titanate is formed during the solution chemical treatment, then the alkali element is ion exchanged via water washing, and protonated titanate is formed as a nanosheet. In a later step of the chemical treatment, the nanosheet might convert to form a tubular structure (Figure 4) by a scrolling process.

TABLE 2 Variation of BET Surface Area[a] on the Annealing Temperature for the Titania Nanotubes

Temperature (°C)	Room Temperature	200	400	450	500	550
Surface area (m²/g)	322	308	228	123	101	95.0

[a]*The surface area is measured by the BET method.*

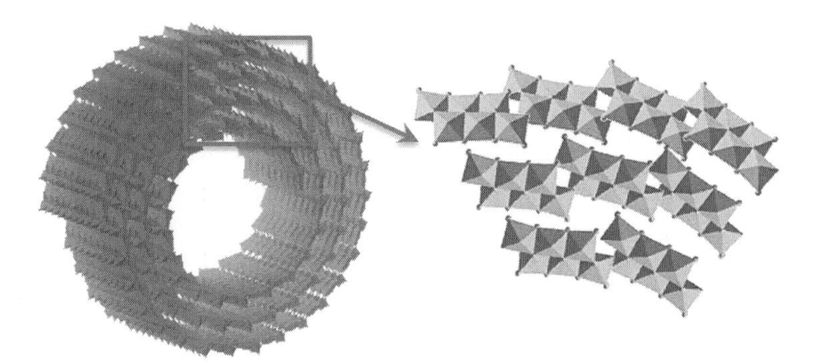

FIGURE 4 Schematic model of a titania nanotube structure and closed-up lattices based on the zigzag chain of TiO_6 octahedrons by assuming the chemical composition $H_2Ti_3O_7$ (illustrated based on the references 40 and 41).

Many discussions of the actual structure models as well as formation mechanisms for the TNTs are continuing. Nevertheless, it should be noted that the crystal structure based on the TiO_6 polyhedron framework and layered compound-like nature coupled with 1D tubular morphology of TNTs (see Figure 4) is quite unique and different from those of the CNTs and template-assisted nanotubular oxides.

3 PHOTOCATALYTIC AND PHYSICAL-PHOTOCHEMICAL FUNCTIONS

3.1 Fundamental Properties of Nanotubular Titania

Similar to particulate TiO_2 powder, nanotubular titania is a white powder. The optical bandgap energy calculated from the UV-visible light (UV-vis) absorption spectra is approximately 3.41-3.45 eV for chemically synthesized TNTs [29,45]; this value is slightly larger than that of common anatase (3.2 eV) and rutile (3.0 eV) TiO_2 crystals. This blue shift is attributed to the quantum size effect of the TiO_2 semiconductor [46] in TNTs as a results of the very thin nanotube wall thickness, ~1-2 nm (Figure 2). Recent materials design strategy for TiO_2 as a photocatalyst is generally focusing on adding visible light responsivity to TiO_2 [47]. Thus, the current enlarged bandgap seems to be disadvantageous; nevertheless, TNTs exhibit excellent photochemical properties, as will be described later.

To clarify the photochemical characteristics of TNTs, charge separation and recombination dynamics and kinetics through light irradiation on TNTs and the resultant photocatalytic one-electron oxidation reaction of an organic molecule were analyzed by Tachikawa *et al.* [48] using time-resolved diffuse reflectance spectroscopy. They revealed that the TNT generated radical cations with remarkably long lives—approximately 5 times or longer than those of the TiO_2 nanoparticles (NPs). Further, they observed that the electron generated by the steady-state irradiation of UV light could also exist longer on the TNT surface, but this phenomenon was not confirmed for the particulate TiO_2 nanopowder. These are considered because of the unique 1D nanostructure of TNTs and are the reason for their good photocatalytic properties; TNTs have such a thin wall, but a long axial length, that generated carriers can effectively move to the surface, and hence charge recombination is inhibited, clearly implying its morphological advantage in charge recombination dynamics. This characteristic might be also advantageous for the use of TNTs as photoelectrodes in DSSCs. In fact, the longer lifetime and resultant longer diffusion length of electrons in TNT photoelectrodes was confirmed when they were used for DSSCs [49].

In situ electron spin resonance (ESR) of TNTs and TiO_2 NPs during UV light irradiation at a cryogenic temperature (20 K) has recently been analyzed. For all cases, the generation of photo-induced superoxide radicals ($O_2\bullet^-$) was confirmed. As shown in Figure 5, however, the amount of radicals per UV irradiated area was the highest for the heat-treated TNT, second highest for the as-synthesized TNT, and the smallest for the NPs [50]. In addition, the ESR signal corresponding to the $O_2\bullet^-$ was clearly observed at room temperature for TNTs, whereas it could not be detected for TiO_2 NPs [51], probably because of the shorter lifetime of the radicals generated on the NPs than the detection limit of ESR. This fact is in good agreement with results showing the long lifetime of photogenerated radicals observed by laser-flush radiolysis [48], as mentioned above.

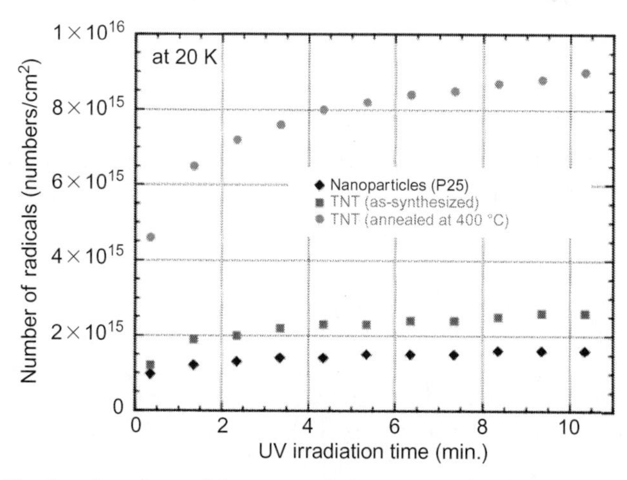

FIGURE 5 The time dependence of the amount of photo-generated radicals per irradiation area for TiO_2 nanoparticles (P-25; Degussa AG, Germany) and as-synthesized and heat-treated titania nanotubes (TNTs) [50]. The numbers of radicals were calculated from the electron spin resonance (ESR) signals measured using ESR equipment (ESP350E; Bruker Co., Germany) at 20 K under the following conditions: central magnetic field, 3400 G; magnetic field sweep width, 400 G; microwave frequency and power, 9.48 GHz, 0.1 mW (for P-25)/0.025 mW (for TNTs); modulation frequency, 100 kHz.

All these results also clearly exhibit that nanotubular titania has excellent photochemical properties owing to the synergy of its unique 1D structure and the semiconductive properties of TiO_2 crystals.

3.2 Photocatalytic Water-Splitting Performance

As mentioned, TiO_2 is a well-known water splitting photocatalyst used to produce hydrogen and oxygen. Figure 6 shows variations of H_2 generation by UV irradiation for TNTs and commercial TiO_2 NPs (P-25 and ST01; Ishihara Sangyo Kaishia, Ltd., Japan) in a mixed solution of water/methanol (the methanol was used as a hall scavenger) [29]. As-synthesized TNTs show lower photocatalytic activity than the commercial TiO_2 powders; however, annealed (400°C) TNTs can generate approximately 2 to 3 times more H_2 than NPs when compared with the amount of H_2 per unit mass of TiO_2 photocatalyst. The enhanced hydrogen generation by the heat-treated (annealed) TNTs is caused by the improved crystallinity (see Figure 3), in keeping with its nanotubular structure and larger surface area (228 m^2/g; see Table 2), over that of TiO_2 NPs (surface area ~50 m^2/g).

It should also be noted that this H_2 generation test was performed without loading any co-catalyst such as noble metals (Pt, Pd, etc.) that usually act as electron collectors and reduction sites of H^+ as well, showing TNT itself has potential as a water-splitting photocatalyst. As expected, TNTs loaded with noble metal (annealed) exhibited significant water-splitting H_2 generation under UV light, as shown in Figure 6, showing approximately a 10 times higher amount of H_2 generation for Pd-loaded TNTs than that of nonloaded annealed TNTs. These results imply that TNTs are a promising candidate as a next-generation high-performance hydrogen production photocatalyst.

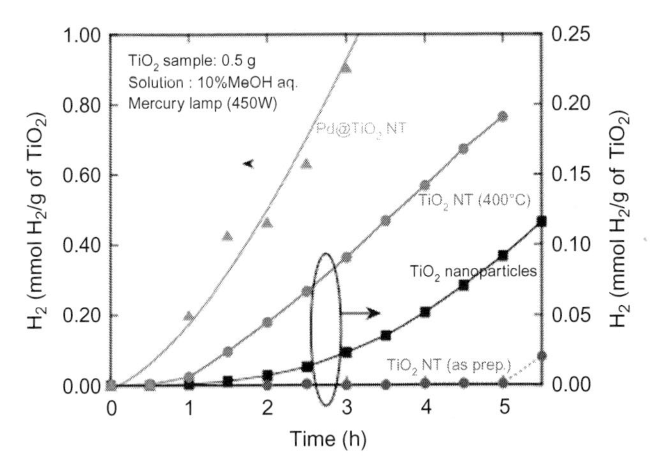

FIGURE 6 Hydrogen generation by water splitting under ultraviolet irradiation for various TiO_2 powders (P-25 and ST01; commercial TiO_2 nanopowders, as-prepared titania nanotubes [TNTs], and TNTs annealed at 400°C) (right axis) and for TNTs loaded with palladium (Pd) nanoparticles (left axis).

3.3 Novel Environmental Purification Multifunctions

To evaluate environmental purification performance, a test to remove organic molecules in a water system was carried out [52]. Figure 7 represents the variation of methylene blue (MB) concentration (MB bleaching test) of TiO_2 dispersed in water under dark and UV light irradiation conditions. In the case of commercial TiO_2 NPs, MB concentration is quickly decreased under UV irradiation, whereas it is not changed under the dark condition, indicating the excellent photocatalytic performance of TiO_2 NPs. A decrease in MB concentration is confirmed, even under the dark condition, for the TNTs, however, and this is accelerated further under UV irradiation. This fact indicates that TNTs have a molecule adsorption characteristic. When TNTs are annealed, MB adsorption under the dark condition is reduced, but the photo-degradation is much enhanced compared with that of as-synthesized TNTs.

In general, TiO_2 powder has low molecule adsorption capability compared with typical adsorbent materials such as zeolite, activated carbon, and clay minerals. Thus, the development of composite materials of TiO_2 photocatalysts and some other adsorbents such as mesoporous silica [53] has been investigated. By contrast, TNTs have both an excellent photocatalytic property and a high molecule adsorption capability as a single-phase material, and thus TNTs can be considered as real multifunctional nanomaterials. Further, TNTs act as a good water-splitting photocatalyst under UV light. The synergy of these properties in one material is attributed to its unique layered compound-like crystal and nanostructure as well as the photochemical characteristics schematically described in Figure 8. It is thus expected that TNTs are an excellent candidate for advanced high-performance environmental purification systems.

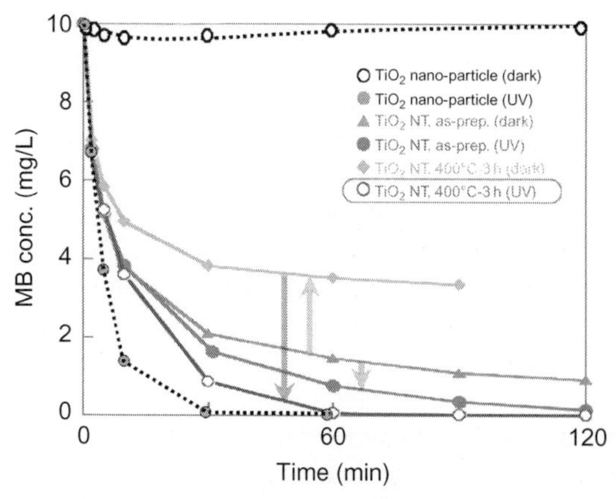

FIGURE 7 Variation of methylene blue concentration (MB conc.) under dark and ultraviolet (UV) light irradiation conditions for TiO_2 nanoparticles and as-synthesized and annealed titania nanotubes (TNTs) dispersed in a water system. as prep., as prepared.

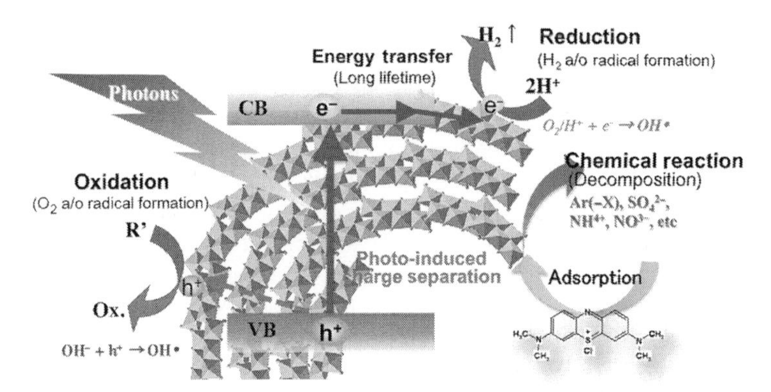

FIGURE 8 Schematic of the adsorption/photochemical multifunctionality of the titania nanotube.

3.4 Size Control of TNTs and Their Application to Solar Energy Conversion

Low-dimensional titania was demonstrated to be useful as the photoelectrode (photoanode) of DSSCs because there would be fewer particle contacts (grain boundary) among the oxide photoelectrode film, which might reduce impedance and then enhance charge transport properties in DSSCs. As expected, the conversion efficiency of a DSSC with a TNT electrode was enhanced compared with that of a common nanoparticulate TiO_2 photoelectrode [49], implying the use of 1D oxide might be suitable for developing high-performance DSSCs.

However, the typical diameter of chemically synthesized TNTs is approximately 10 nm in diameter, and the effect of titania photoelectrode size on the performance of DSSCs has not been clarified yet. As described briefly earlier, size-controlled TNTs with a slightly larger diameter [37,38] and aspect ratio [39] have recently been successfully prepared. In this section the influence of the particle size of raw TiO_2 powders on the morphological characteristics of TNTs prepared through the low-temperature chemical synthesis route are described. The relationship between material processing and structural characteristics, as well as the photovoltaic solar energy conversion performance of these size-controlled TNTs as an electrode of DSSCs, also is discussed.

First, anatase TiO_2 powder was annealed at 500, 1000, and 1300°C for 4 h to obtain different crystallite sizes of TiO_2 particles. The TNTs were synthesized by refluxing them at 110°C in a 10 mol/L NaOH aqueous solution. XRD and TEM (Figure 9) investigation revealed that the annealing of raw TiO_2 powder yielded different crystallite sizes of TiO_2, from 54.3 to 117 nm, and TNTs with various sizes and morphologies have been synthesized successfully (see also Table 3). The synthesized TNTs showed a strong dependence on the crystallite size of the raw materials, despite their crystal phase. The size of TNTs made from 117-nm TiO_2 was around 23.6 nm in diameter, whereas that of from 54.3-nm TiO_2 was 9.8 nm [37]. It is thus concluded that the desired morphology and size of TNTs can be achieved by controlling the starting material.

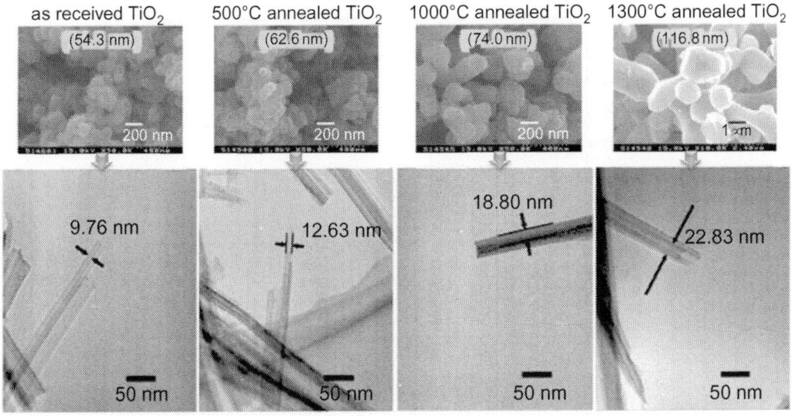

FIGURE 9 Transmission electron microscopy images of different sized titania nanotubes made from raw TiO_2 powders with different crystallite sizes: as-received (a) and annealed at 500°C (b), 1000°C (c), and 1300°C (d).

TABLE 3 Crystallite size of Raw Materials, Typical Diameter of Synthesized Titania Nanotubes (TNTs), Specific Surface Area, Cell Efficiency in a DSSC, and Amount of Adsorbed Dye for the Size-Controlled TNTs

Samples	Crystallite Size of Raw TiO_2 (nm)	Diameter of TNT (TiO_2) (nm)	Surface Area (m^2/g)	Cell Efficiency (%)	Amount of Adsorbed Dye (mol/cm^2)
TNT	54.3	9.8	282	5.19	1.46×10^{-7}
TNT-500	62.6	12.9	225	5.83	1.11×10^{-7}
TNT-1000	74.0	19.1	115	4.63	0.862×10^{-7}
TNT-1300	117	23.6	105	3.97	0.723×10^{-7}
TNT/P-25[a]	—	9.8/54.3	—	5.84	2.02×10^{-7}
TNT1300/P-25[a]	—	23.6/54.3	—	5.80	1.88×10^{-7}
P-25 (NPs)	(54.3)	54.3	52.0	5.42	0.765×10^{-7}

The sample ID corresponds to the annealing temperature (°C) of raw TiO_2 powder.
[a]*Double-layered photoelectrode film consisting of a TNT (upper layer) and TiO_2 NPs (P-25, lower layer).*

These TNTs were used as the photoelectrode in a DSSC using N719 dye [38]. The J-V properties were measured under AM1.5 conditions, and the result is shown in Figure 10 and summarized in Table 3. The electrodes made from modified TNTs showed a strong dependence on their specific surface area and the resultant amount of dye adsorption. The conversion efficiency of the cell

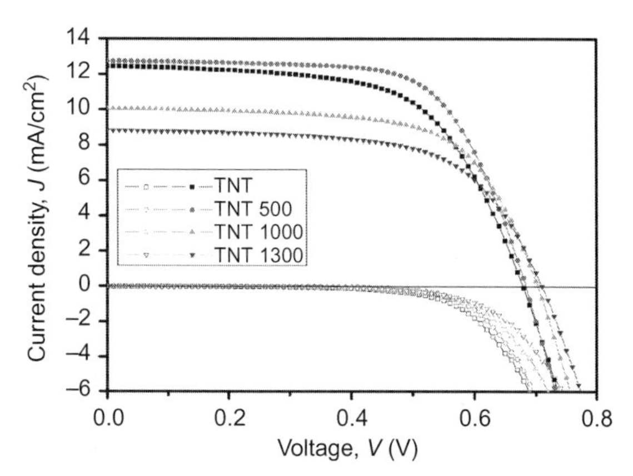

FIGURE 10 Photovoltaic J-V curve for a dye-sensitized solar cell using a size-controlled titania nanotube (TNT) photoelectrode measured under the AM1.5 condition.

made from TNTs with a 12.9-nm diameter (TNT-500) was enhanced by 12% compared with that of the smallest TNT (9.8-nm diameter). These results suggest that the PV performance improved by the suppression of photo-generated charge recombination in spite of a 25.3% reduction in the specific surface area. In addition, larger TNTs could be used as a scattering layer on the top of the TiO_2 nanoparticulate working electrode, where higher efficiency (>5.8%) was achieved for the electrode with multilayered TNT/TiO_2 NPs (Table 3) [38]. This controlled TiO_2 photoelectrode architecture exhibited enhanced conversion efficiency without $TiCl_4$ treatment for the advanced DSSC application.

4 STRUCTURE TUNING OF TNTs FOR FURTHER PROPERTY ENHANCEMENTS

To enhance properties and/or to add new functions, materials tuning is useful methodology. For instance, as mentioned before, loading of NPs to TNTs that provide small-dimensional nanocomposites is one promising way to enhance photocatalytic water-splitting performance. Doping some elements to the other material also is often adopted to control materials' structures and the resultant properties. Based on this strategy, structural tuning of TNTs by ion doping and composition with the other materials has been attempted. Physical, thermal, and photochemical properties, including environmentally friendly functions, are reviewed in relation to their synthesis processes and structural characteristics.

4.1 Lattice-Level Tuning of TNTs by Ion Doping

Various metal ions have been chosen as dopant elements and doped to TNT via the solution chemical synthesis route [29,45,54,55]. Transition metal ions such as Zn^{2+}, Cr^{3+}, Mn^{2+}, Mn^{3+}, Co^{2+}, Sn^{4+}, Nb^{5+}, and V^{5+} were added as oxide, nitrate, sulfate, or citrate to the chemical reaction vessel together with the TiO_2 raw material. For all cases, nanotubular morphology was confirmed, and large surface area of the doped TNTs were almost as same as those of pure TNTs. Electrical conductivity of the doped TNT films, however, which were prepared by squeegeeing sample slurry on glass substrate and sintering around 450°C, was around one to two orders of magnitude higher (e.g., 1.0×10^{-4} S/cm for 0.08 mol% chromium-doped TNT) than those of TiO_2 NPs (2.6×10^{-6} S/cm) and pure TNT (3.0×10^{-6} S/cm) [45].

The other advantage of ion doping TNTs was found in the improvement in thermal stability, that is, surface area and tubular morphology degradation of TNTs by heat treatment. The BET surface area of nondoped TNTs slightly decreased up to around 400°C, whereas it suddenly dropped off above this temperature when the morphology changed from a nanotube to a nanorod and/or nanoparticulate structures (see Figure 3 and Table 2), showing the critical temperature is around 400°C [29]. However, the critical temperature was improved approximately 50°C (Mn^{3+}, Co^{2+}, Nb^{5+}, V^{5+}) to 100° C (Cr^{3+}) for the doped TNTs [29]. Thus ion doping with transition metals is considered a suitable modification method for enhancing both the electrical conductivity and thermal stability of TNTs, and these facts are advantageous when using TNTs in various devices, such as sensors, electrodes, and catalysts in a film or bulky forms.

4.2 Multifunctionality in TNTs Doped with Rare Earth Elements

Contrary to transition metal doping, rare earth (RE) elements have various unique functions such as photoluminescent (PL) properties caused by the 4f electron orbits. Thus RE-doped TNTs are expected to exhibit multifunctionality, including optical function. One example was confirmed in samarium (Sm)-doped TNTs [54–56]. Anatase TiO_2 powder and $Sm(NO_3)_3\cdot 5H_2O$ (99.9%), where the Sm concentration in the TNT varied from 0 to 5 mol%, were subjected to the chemical synthesis method by refluxing in a 10 mol/L NaOH aqueous solution at 110°C, as mentioned before.

Nanotubular products were obtained for all the Sm concentrations, as seen in Figure 11; the length of the nondoped and Sm-doped TNTs was approximately 200-800 nm, whereas the diameter slightly increased to 16.4 nm when 1 mol% Sm

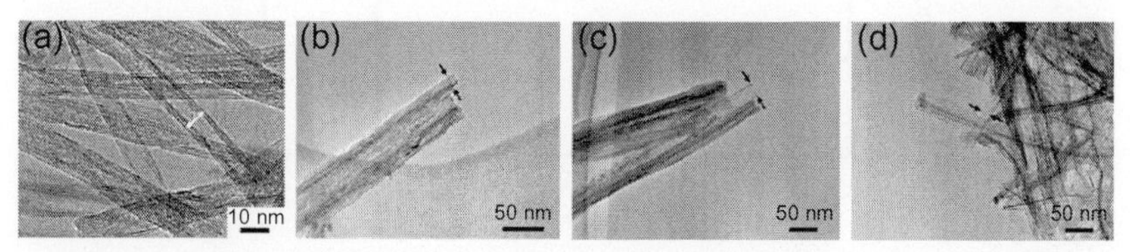

FIGURE 11 Transmission electron microscopy images of samarium (Sm)-doped titania nanotubes (TNTs) showing tubular structures: nondoped TNTs (a), 1 mol% Sm-doped after calcination at 400°C for 3 h (b), 3 mol% Sm-doped TNTs (c), and 5 mol% Sm-doped TNTs (d).

was added; the diameter reached ~21-23 nm for the TNT doped with 5 mol% Sm. The addition of Sm might enhance the growth of titanate nanosheets, which is the precursor of TNTs, during the alkaline solution process [54].

Molecular adsorption and photocatalytic properties were evaluated by the MB removal test in the dark and with UV irradiation; the result is shown in Figure 12. Detailed adsorption characteristics were investigated in an adsorption isotherm experiment by assuming a Langmuir adsorption model, and the results are summarized in Table 4, together with some physical properties.

The adsorption of MB on common TiO_2 powder in the dark is remarkably low, as mentioned before. The decrease in MB concentration for TNTs is greater for Sm-doped TNTs than for nondoped TNTs, however, showing the excellent adsorption of the Sm-doped TNTs. Saturated adsorption capacity was greatly increased with an increase in Sm content (Table 4), and the value was often compatible or higher than that of conventional adsorbents such as zeorite. This implies that the doping of ions to TNTs is a suitable way to develop high-performance environmental catalysts.

The molecular adsorption properties of TNTs is improved not only by an increase in surface area but also by the introduction of surface charge to the TNTs caused by doping with a trivalent cation (Sm^{3+}). Such an increased adsorption capability also was confirmed for doping with a trivalent transition metal (Cr^{3+}). The addition of M^{3+} to TNT (TiO_2) lattices also contributed to the formation of oxygen vacancies ($V\ddot{o}$) as $Ti^{4+}_{1-x}M^{3+}_{x}O_{2-x/2}V\ddot{o}_{x/2}$, resulting in an enhanced electrostatic effect between the adsorbent and ionic molecules such as MB. Thus the molecular adsorption capability enhancement of doped TNTs is attributed to the both electrostatic and physicochemical characteristics enhanced by trivalent ions such as Sm^{3+} cations [54].

On the other hand, Sm-doped TNTs exhibited slightly less photocatalytic MB degradation activity compared with the nondoped TNTs. Cation doping

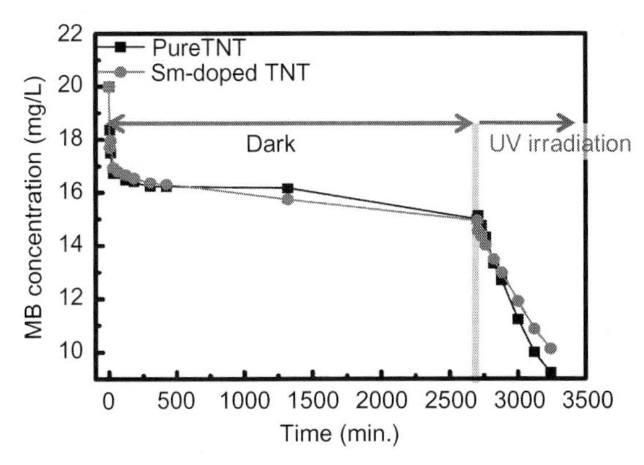

FIGURE 12 Methylene blue (MB) concentration degradation of nondoped and 5 mol% samarium (Sm)-doped titania nanotubes (TNTs) in the dark and under ultraviolet (UV) irradiation (Hg/Xe lamp with a center wavelength of 365 nm).

TABLE 4 Physical Properties and Methylene Blue (MB) Molecular Adsorption Capacity for TiO_2 Powder and Sm-doped Titania Nanotubes (TNTs)

Samples	Observed Sm Concentration (mol%)	BET Surface Area (m^2/g)	Saturated Adsorption Capacity (μg/mg)	Pore Size (nm)	Pore Volume (cc/g)
Anatase TiO_2	—	9.40	—	—	—
Pure TNT	—	279	88.4	3.06	1.24
Sm-doped TNT					
1 mol%	0.50	269	95.4	3.42	0.69
3 mol%	1.05	276	201	3.43	0.87
5 mol%	2.24	356	236	12.5	1.65
Zeorite (ref.)			~78-100		
Montmorillonite (ref.)			~340		

of a TiO_2 lattice often exhibited significant degradation of photocatalytic activity caused by the formation of impurities among the bandgap, as well as oxygen vacancies, which would act as recombination sites for photo-induced electron-hole pairs in the crystal. Nevertheless, as can be seen from Figure 12, the decrease in the photocatalytic activity of TNTs by Sm doping under UV irradiation is small, showing the advantage of 1D nanostructured oxide semiconductors such as long-lived photo-excited charges on the TNT's surface, as mentioned in the previous section. From these results, Sm-doped TNTs are considered to be environmentally friendly nanomaterials with excellent molecule adsorption capability coupled with photocatalytic property [54].

In relation to TNT modification by ion doping, we recently succeeded in developing visible light responsive TNT photocatalyst in which function originates from the formation of impurity level among the bandgap of TNTs and in part from unique energy transfer and subsequent redox reaction on the TNT's surface in relation to the high molecular adsorption property. Such a feature is rarely confirmed for common TiO_2 crystals; thus TNTs are considered to be advantageous for creating high-performance photocatalysts.

PL spectra of the Sm-doped TNTs were measured to understand photo-induced energy transfer characteristics. The PL emission spectrum shown in Figure 13 exhibits strong red fluorescence with 350-nm excitation, which could be observed clearly by the naked eye. This fact shows that the clear recombination of hole and electron takes place in this Sm-doped TNT.

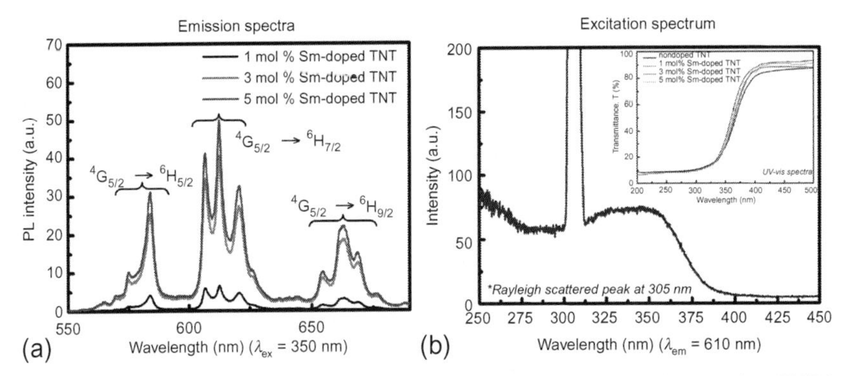

FIGURE 13 Photoluminescence (PL) spectra of samarium (Sm)-doped titania nanotubes (TNTs) with various emission spectra (excited at 350 nm) (a) and excitation spectra (monitored at 610 nm; a Rayleigh scattered peak is present at 305 nm) (b). The inset in (b) shows the ultraviolet-visible diffused reflectance spectra of Sm-doped TNTs. a.u., arbitrary units.

Three major peaks at around 580, 610, and 660 nm correspond to the emissions among f-f transitions, from $^4G_{5/2}$ to $^6H_{5/2}$, $^6H_{7/2}$, and $^6H_{9/2}$, respectively, of the Sm^{3+} in the $4f^5$ configuration. On the other hand, the excitation spectrum (Figure 13b) shows a broad curve without obvious peaks, which is quite similar to the UV-vis optical absorption spectrum (inset). In general, when the RE ion is directly excited, both the emission and excitation spectra should be symmetrical and have sharp peaks. However, this is not the case in this material. The following considerations are reasonable: Bandgap excitation in a TNT lattice occurs first by the incident UV light irradiation, and then the energy transfers from the TNT to the Sm^{3+} ions, which excites the ions; finally, an f-f transition in the ions takes place to emit the blight fluorescence. Such a unique energy transfer characteristic is caused by the synergy of the 1D semiconductor host matrix and the doped functional (RE) ions [55].

In conclusion, the TNTs functionalized by RE elements obtained by solution chemical synthesis exhibit excellent multifunctionality by combining physical, chemical, and optical functions. This class of multifunctional nanomaterials is expected to open a new window toward environmental, energy, and device applications in the near future [56].

4.3 Development of 0-Dimensional/1-Dimensional Nanocomposites Based on NP Loading and Their Multiple Functions

Composing nanometer-scale TNTs, which are unique 1D nanomaterials, with other low-dimensional materials such as 0-dimensional (0D) NPs is regarded as a promising methodology for tuning the structures and functions of TNTs, including multifunctionalization. According to this hypothesis, loading various metals and/or compounds inside the nanotubes and/or onto the surfaces, and

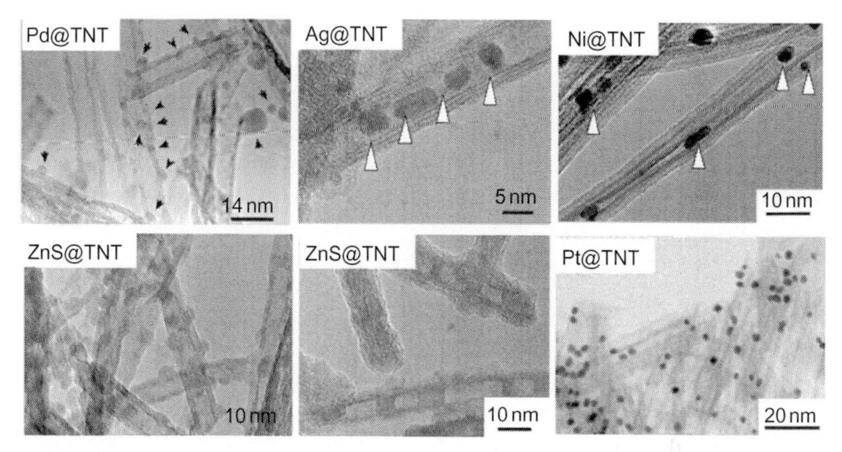

FIGURE 14 Various one-dimensional metals and compounds loaded titania nanotubes (TNT) nanocomposites: a palladium (Pd)-loaded TNT prepared by the sonochemical method, silver (Ag) nanoparticles formed inside of a TNT by chemical reduction, nickel (Ni) nanoparticles inside of a TNT prepared by chemical vapor deposition, a zinc sulfide (ZnS)-loaded TNT prepared by solution chemical and vapor-phase reaction, and platinum (Pt) loaded by the photo-reduction route.

hybridization of a TNT with another low-dimensional nanomaterial, have been investigated [52,57,58]. Figure 14 shows TEM images of TNT nanocomposites loaded with metals and sulfide compounds that were prepared using various physicochemical processes. Various NPs could successfully be loaded onto TNTs, exhibiting the formation of 0D/1D nanocomposites. 0D NPs could successfully be introduced not only on the surface but also inside of hollow TNTs by optimizing *ex situ* fabrication processes such as ion adsorption and chemical/thermal/photo reduction, vapor phase deposition, and sonochemistry.

Wang *et al.* [59] recently synthesized Pt NP-loaded TiO_2 nanowires (NWs) and reported that the Pt loading greatly decreased the adsorption amount of MB to TiO_2 NWs (from 58% to 7% for pure NWs and 2 at% Pt-loaded TiO_2 NWs, respectively) through the occupation of the adsorption site by precipitated NPs. In this case the saturated amount of MB adsorption was calculated based on adsorption isotherm experiments using MB in the dark. The observed absorption amount was slightly lower for the TNTs loaded with a small amount of Pt (66.4 μg MB/mg of catalyst for 0.045 mass% Pt) than that of the pure TNT (96.0 μg/mg), implying that NP loading degraded the molecular adsorption capability of the TNTs. Nevertheless, sufficiently more adsorption(~69-96% for the pristine TNT) was still maintained even after loading with the Pt NPs. Thus, the fundamental characteristic of TNTs owing to its large surface area and unique layered compound-like structure is also essential for the present Pt-loaded TNTs, showing their high potential as an adsorbent nanomaterial. Under UV irradiation conditions, Pt-loaded TNTs exhibited a level of MB degradation similar to that of the pure TNT; that is, they had sufficient photocatalytic properties.

As already mentioned, noble metal NP-loaded TNTs exhibited excellent hydrogen generation under the UV light as a result of the co-catalysis effect of the loaded noble metals. Considering all these issues, 1D/0D nanohybridization using functional oxide nanotubes and metal NPs is a promising method to develop multifunctional nanomaterials that would be applicable not only to high-performance environmental cleaning systems but also advanced energy creation systems.

5 CONCLUSION AND PROSPECTS

Optimization of solution chemical processing could realize the development of unique multifunctional semiconductor oxide nanotubes. In this chapter the chemical processing, nanostructure, and physical and chemical properties of TNTs were reviewed. Tuning the structure and function of TNTs is an advantageous way to develop next-generation high-performance nanomaterials. Even now, a large number and variety of fundamental studies, application-oriented research, and developments for this low-dimensional nanotube are being carried out worldwide; not only enhancement of various properties of TiO_2 but also multifunctionalization as a result of the harmonization of material properties and unique low-dimensional nanostructures are expected. In addition the TNTs, extensive challenges for developing various nanotubes also are continuing.

As for the application of TNTs, they have been used as the oxide electrode of DSSCs, and better cell efficiency and structure-related characteristics of the charge transport phenomenon have been reported [38,49,60]. Also reported recently were TNTs exhibiting proton intercalation/de-intercalation and resultant electrochromism, size-selective adsorption of molecules [61], anion doping to develop visible light-responsive TNTs [62], and biocompatible/biomedical functions [50,63–66]. All these facts imply that oxide nanotubes, including TNTs, are a promising multifunctional nanostructured material because of the structure-property correlations, and hence various applications in many fields are expected, as illustrated in Figure 15. Based on these facts, one of the future

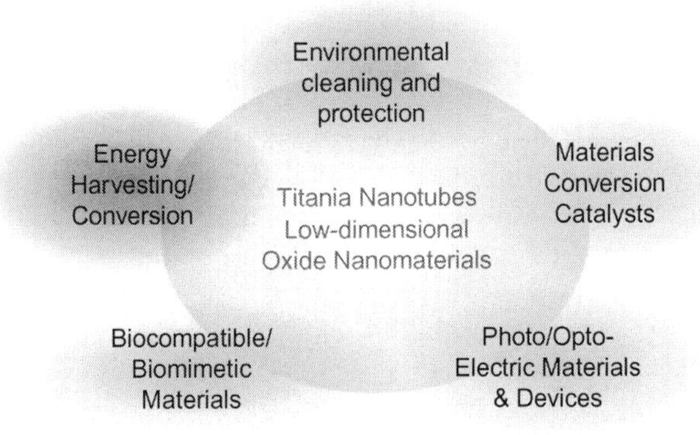

FIGURE 15 Multiple functions and applications for low-dimensional semiconductor oxides.

research directions relating to TNTs might lie toward their application in environmentally friendly and/or energy harvesting/creating systems, which would become more important in the near future.

REFERENCES

[1] S. Kato, F. Masuo, Kogyo Kagaku Zasshi 67 (8) (1964) 1136–1140 (in Japanese).
[2] A. Fujishima, K. Honda, Nature 238 (1972) 37–38.
[3] A.L. Linsebigler, G. Lu, J.T. Yates, Chem. Rev. 95 (3) (1995) 735–758.
[4] D. Beydoun, R. Amal, G. Low, S. McEvoy, J. Nanopart. Res. 1 (4) (1999) 439–445.
[5] S. Hasegawa, Y. Sasaki, S. Matsuhara, Sens. Actuators, B 13-14 (1993) 509–510.
[6] A.M. Ruiz, G. Sakai, A. Cornet, K. Shimanoe, J.R. Morante, N. Yamazoe, Sens. Actuators, B 103 (2004) 312–317.
[7] A.H. Yuwono, J. Xue, J. Wang, H.I. Elim, W. Ji, Y. Li, T.J. White, J. Mater. Chem. 13 (2003) 1475–1479.
[8] B. O'Regan, M. Grätzel, Nature 353 (24) (1991) 737–740.
[9] B. O'Regan, J. Moser, M. Anderson, M. Grätzel, J. Phys. Chem. 94 (24) (1990) 8720–8726.
[10] N.N. Bwana, J. Nanopart. Res. 11 (8) (2009) 1917–1923.
[11] S. Iijima, Nature 354 (1991) 56–58.
[12] N.G. Chopra, R.J. Luyken, K. Cherrey, V.H. Crespi, M.L. Cohen, S.G. Louie, A. Zettl, Science 269 (1995) 966–968.
[13] Y. Feldman, E. Wasserman, D.J. Srolovitz, R. Tenne, Science 267 (1995) 222–225.
[14] M.E. Spahr, P. Bitterli, R. Nesper, M. Müler, F. Krumeich, H.U. Nissen, Angew. Chem. Int. Ed. 37 (1998) 1263–1265.
[15] P.M. Ajayan, O. Stephan, Ph. Redlich, C. Colliex, Nature 375 (1995) 564–567.
[16] B.C. Satishkumar, A. Govindaraj, E.M. Vogl, L. Basumallick, C.N.R. Rao, J. Mater. Res. 12 (3) (1997) 604–606.
[17] H. Nakamura, Y. Matsui, J. Am. Chem. Soc. 117 (9) (1995) 2651–2652.
[18] P. Hoyer, Langmuir 12 (1996) 1411–1413.
[19] T. Kasuga, M. Hiramatsu, A. Hoson, T. Sekino, K. Niihara, Langmuir 14 (1998) 3160–3163.
[20] T. Kasuga, M. Hiramatsu, A. Hoson, T. Sekino, K. Niihara, Adv. Mater. 11 (1999) 1307–1311.
[21] D. Gong, C.A. Grimes, O.K. Varghese, W. Hu, R.S. Singh, Z. Chen, E.C. Dickey, J. Mater. Res. 16 (2001) 3331–3334.
[22] H. Masuda, K. Nishio, N. Baba, Jpn. J. Appl. Phys. 31 (1992) L1775–L1777.
[23] H. Imai, Y. Takei, K. Shimizu, M. Matsuda, H. Hirashima, J. Mater. Chem. 9 (1999) 2971–2972.
[24] N. Yoshinaga, S. Aomine, Soil Sci. Plant Nutr. 8 (3) (1962) 114–121.
[25] L.A. Bursill, J.L. Peng, L.N. Bourgeois, Philos. Mag. A 80 (1) (2000) 105–117.
[26] F. Keller, M.S. Hunter, D.L. Robinson, J. Electrochem. Soc. 100 (9) (1953) 411–419.
[27] S. Yamanaka, T. Hamaguchi, H. Muta, K. Kurosaki, M. Uno, J. Alloys Compd. 373 (1-2) (2004) 312–315.
[28] V. Zwilling, E. Darque-Ceretti, A. Boutry-Forveille, D. David, M.Y. Perrin, M. Aucouturier, Surf. Interface Anal. 27 (7) (1999) 629–637.
[29] T. Sekino, Bull. Ceram. Soc. Jpn 41 (4) (2006) 261–271 (in Japanese).
[30] S. Kato, F. Masuo, Kogyo Kagaku Zasshi 67 (1959) 1136–1140 (in Japanese).
[31] G.H. Du, Q. Chen, R.C. Che, Z.Y. Yuan, L.M. Peng, Appl. Phys. Lett. 79 (2001) 3702–3704.
[32] Q.H. Zhang, L.A. Gao, J. Sun, S. Zheng, Chem. Lett. 31 (2002) 226–227.
[33] Y. Suzuki, S. Yoshikawa, J. Mater. Res. 19 (2004) 982–985.
[34] Z.-Y. Yuan, B.-L. Su, Colloids Surf. A 241 (1-3) (2004) 173–183.

[35] A. Nakahira, W. Kato, M. Tamai, T. Isshiki, K. Nishio, H. Aritani, J. Mater. Sci. 39 (2004) 4239–4245.
[36] Y. Suzuki, S. Pavasupree, S. Yoshikawa, R. Kawahata, J. Mater. Res. 20 (2005) 1063–1070.
[37] J.Y. Kim, T. Sekino, D.J. Park, S.-I. Tanaka, J. Nanopart. Res. 13 (2011) 2319–2327.
[38] J.Y. Kim, T. Sekino, S.-I. Tanaka, J. Mater. Sci. 46 (2011) 1749–1757.
[39] T. Sekino, D.J. Park, J.Y. Kim, S.-I. Tanaka, Mater. Int. 25 (3) (2012) 17–24 (in Japanese).
[40] Q. Chen, W.Z. Zhou, G.H. Du, L.M. Peng, Adv. Mater. 14 (2002) 1208–1211.
[41] (a) R. Ma, Y. Bando, T. Sasaki, Chem. Phys. Lett. 380 (2003) 577–582. (b) R. Ma, K. Fukuda, T. Sasaki, M. Osada, Y. Bando, J. Phys. Chem. B 109 (2005) 6210–6214.
[42] J. Yang, Z. Jin, X. Wang, W. Li, J. Zhang, S. Zhang, X. Guo, Z. Zhang, Dalton Trans. 20 (2003) 3898–3901.
[43] B. Poudel, W.Z. Wang, C. Dames, J.Y. Huang, S. Kunwar, D.Z. Wang, D. Banerjee, G. Chen, Z.F. Ren, Nanotechnology 16 (2005) 1935–1940.
[44] D.V. Bavykin, J.M. Friedrich, F.C. Walsh, Adv. Mater. 18 (2006) 1–9.
[45] T. Sekino, T. Okamoto, T. Kasuga, T. Kusunose, T. Nakayama, K. Niihara, Key Eng. Mater. 317–318 (2006) 251–255.
[46] Y. Wang, N. Herron, J. Phys. Chem. 95 (1991) 525–532.
[47] R. Asahi, T. Morikawa, T. Ohwaki, K. Aoki, Y. Taga, Science 293 (2001) 269–271.
[48] T. Tachikawa, S. Tojo, M. Fujitsuka, T. Sekino, T. Majima, J. Phys. Chem. B 110 (2006) 14055–14059.
[49] Y. Ohsaki, N. Masaki, T. Kitamura, Y. Wada, T. Okamoto, T. Sekino, K. Niihara, S. Yanagida, Phys. Chem. Chem. Phys. 7 (2005) 4157–4163.
[50] O. Komatsu, H. Nishida, T. Sekino, K. Yamamoto, Nano Biomed. 6 (2) (2014) 63–72.
[51] O. Komatsu, H. Nishida, T. Sekino, in preparation (2015).
[52] T. Sekino, in: T. Kijima (Ed.), Inorganic and Metallic Nanotubular Materials- Recent Technologies and Applications, Springer-Verlag, Berlin, 2010, pp. 17–32.
[53] K. Inumaru, T. Kasahara, M. Yasui, S. Yamanaka, Chem. Commun. 16 (2005) 2131–2133.
[54] D.J. Park, T. Sekino, S. Tsukuda, S.-I. Tanaka, Res. Chem. Intermed. 39 (2013) 1581–1591.
[55] D.J. Park, T. Sekino, S. Tsukuda, A. Hayashi, T. Kusunose, S.-I. Tanaka, J. Solid State Chem. 184 (2011) 2695–2700.
[56] T. Sekino, S.-I. Tanaka, Mater. Jpn 53 (11) (2014) 546–549 (in Japanese).
[57] S.-I. Tanaka, T. Sekino, S. Tsukuda, Mater. Int. 24 (4) (2011) 86–91 (in Japanese).
[58] D.J. Park, T. Sekino, S. Tsukuda, S.-I. Tanaka, J. Ceram. Soc. Jpn 120 (2012) 307–310.
[59] C. Wang, L. Yin, L. Zhang, N. Liu, N. Lun, Y. Qi, Appl. Mater. Interfaces 2 (11) (2010) 3373–3377.
[60] J.Y. Kim, T. Sekino, S.-I. Tanaka, Int. J. Appl. Ceram. Technol. 8 (2011) 1353–1362.
[61] H. Tokudome, A. Shimai, Y. Mitsuya, Y. Tsuru, M. Miyauchi, Mater. Int. 18 (1) (2005) 31–35 (in Japanese).
[62] H. Tokudome, M. Miyauchi, Chem. Lett. 33 (2004) 1108–1109.
[63] T. Kasuga, Mater. Int. 18 (1) (2005) 26–30 (in Japanese).
[64] K. Sasaki, K. Asanuma, K. Johkura, T. Kasuga, Y. Okouchi, N. Ogiwara, S. Kubota, R. Teng, L. Cui, X. Zhao, Ann. Anat. 188 (2006) 137–142.
[65] H. Nishida, H. Egusa, T. Sekino, Y. Taguchi, S. Komasa, K. Kusumoto, M. Tanaka, K. Yamamoto, J. Jpn Assoc. Oral Rehabil. 24 (1) (2011) 52–57.
[66] T. Fujino, Y. Taguchi, S. Komasa, T. Sekino, M. Tanaka, J. Hard Tissue Biol. 23 (1) (2014) 63–69.

Chapter 18

Environmentally Friendly Processing of Transparent Optical Ceramics

Yan Yang[1], Yin Liu[1], Shunzo Shimai[2] and Yiquan Wu[1]
[1]*Kazuo Inamori School of Engineering, New York State College of Ceramics at Alfred University, Alfred, New York, USA,* [2]*Tokyo University of Agriculture and Technology, Tokyo, Japan*

1 INTRODUCTION TO GELCASTING

The processing methods of ceramics are generally divided into two categories: dry processing and wet processing. Dry processing usually starts with ceramic powders and uses high temperature and pressure for consolidation, whereas wet processing is based on colloidal effects, which can result in better control of particle interactions and particle packing. Unlike some dry processing routes requiring sophisticated facilities with the ability to control the atmosphere and enable high temperature and pressure, most wet-processing methods can be operated at room temperature. Much more attention has been focused on wet processing in recent years, with the development of advanced ceramics, which require the materials to possess high stability, a complex shape, and a low manufacturing cost. Conventional dry-processing routes, such as dry pressing and isostatic pressing, could introduce density gradients, pores, delamination, and other defects that may degrade the product's stability. In addition, some other issues may arise during dry processing, such as high cost, high energy consumption, and difficulties in mold release and secondary machining. In contrast to the dry process, the wet process is flexible in ceramics selection and shaping, with more cost-effective processing methods. Also, owing to the nature of colloidal forming in wet processing, the ceramic particles can be well dispersed in the wet stage and later solidified into green bodies with high uniformity and reduced defects [1–5]. All of these merits can meet the requirements for manufacturing advanced ceramics.

There have been many well-established wet-processing methods, such as tape casting, slip casting, injection mold coagulation casting, and gelcasting. Of these manufacturing techniques, gelcasting has been regarded as a high-potential and near-net-shape (NNS) method for making high-quality, complex

Green and Sustainable Manufacturing of Advanced Materials. http://dx.doi.org/10.1016/B978-0-12-411497-5.00018-7

ceramic parts. First developed in the 1900s by Omatete and Janney [6–9] from Oak Ridge National Laboratory, the gelcasting technique involves mixing ceramic phase (particles) and polymer components in an aqueous slurry to form a gel precursor. The suspended ceramic powders can be cross-linked by monomers in a freeze condition. To be more specific, a three-dimensional (3D) macromolecular gelling network is formed during the polymerization of monomers in support of ceramic particles. This network can provide sufficient strength to support its own weight and results in small shape distortion. A detailed gelcasting flowchart is provided in Figure 1. The ceramic powder is dispersed in aqueous or nonaqueous solvent, along with a monomer, a cross-linker, and other organic additives to initiate the polymerization. The ceramic-polymer slurry is well stirred in a vacuum to eliminate the residual bubbles and increase the fluidity of the suspension. Then the degassed slurry, with high solid-loading and low viscosity, is poured into the mold for casting. In some cases the casting is followed by a heating process to get the casted green body. After drying out the solvent, the green body usually has a microstructure to similar that in gel suspensions. The demolded green body is fired in a furnace to eliminate the remaining organic binders and then finally evolved into ceramics after sintering.

Compared with other wet-processing techniques, gelcasting owns some superior features because of its high solid-loading and low-viscosity nature. For instance, gelcast green bodies contain relatively small amounts of organic binders compared with those from injection molding, therefore resulting in much easier polymer removal [10]. In addition, flaw formation during binder thermolysis can be reduced using gelcasting [5]. In contrast to slip casting (which usually has lower solid-loading and uses a porous mold for drying), green bodies prepared by gelcasting possess higher uniformity and higher strength for green body machining. The advantages of the gelcasting technique can be summarized as the following aspects:

- Generic technique suitable for a wide range of ceramic powders, as well as metal or alloy powders

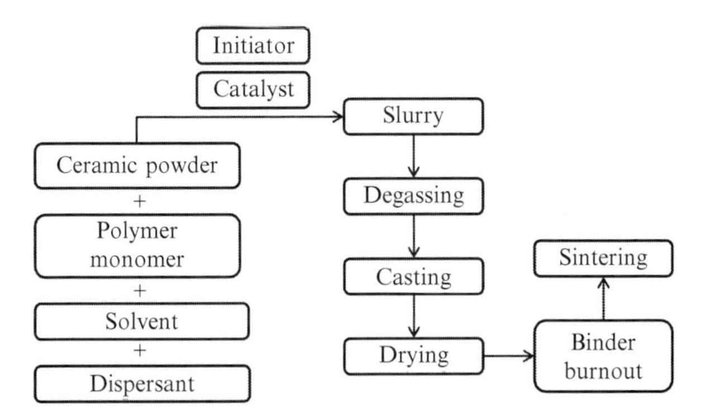

FIGURE 1 Schematic flowchart of the typical gelcasting process.

- High green body strength for machining
- Homogenous microstructure with no density difference
- Low shrinkage after sintering into ceramics
- Little organic content compared with other wet-processing techniques
- Complex shape of final product
- Few equipment requirements and low manufacturing cost
- Applicable for mass production

In recent years the environmental and health aspects of gelcasting have received special attention. Even though gelcasting has some superior merits over other techniques, the organic components from the nonaqueous solvent or monomers may need special precautions concerning flammability and toxicity [5]. The acrylamide gelling system is one of the most commonly used gelcasting systems because of its excellent gelation properties and the advantages provided by the aqueous slurry. This gelling agent is neurotoxic, however, and cannot be applied in industry. Other potential hazards (flammability, corrosiveness, and oxidation) may arise from some of the gelling chemicals, which can be reactive and irritating. Therefore, much of the current research on developing new gelcasting systems has focused on other gel systems that have similar properties to the acrylamide-based system, yet exhibit low or no toxicity [11,12].

2 GELCASTING PRINCIPLES

2.1 Solid Loading of Slurry

An overview of gelcasting slurry is given in this subchapter, including the generic ceramic powders, and the common rheological features for suspensions. In general, the solid-loading of the slurry turns into the green density of the casting part; therefore, it is crucial to use as much solid-loading as possible in the slurry. Thus, maximizing the solid-loading while maintaining slurry fluidity is one of the main tasks in gelcasting research. There is a maximum packing density (φ_m) associated with infinite viscosity. This volume fraction is a function of particle morphology and particle distribution. It also determines the relative viscosity (η_s), which can be described by a semi-empirical Kreiger-Dougherty model [13]:

$$\eta_s = \left(1 - \frac{\varphi}{\varphi_m}\right)^{-[\eta]\varphi_m} \tag{1}$$

where $[\eta]$ is the intrinsic viscosity. The product term of ($[\eta]\ \varphi_m$) depends on the shape of the particles, and the value is ~2 for spherical particles according to a modified Quemada model [14]. With the particle morphology deviating from sphericity, the maximum packing density (φ_m) usually decreases while the intrinsic viscosity increases; both exponentially affect the viscosity.

From another perspective, to obtain a homogenous microstructure and eliminate warping or cracking after sintering, narrowly dispersed and submicron ceramic powders are preferred in the suspensions. Using fine powders can effectively reduce the powder precipitation in the slurry and lower the sintering temperature. Since forming intricate parts requires the slurry to adequately fill molds with a complex geometry, the viscosity of the slurry should be kept below a certain level, which is usually controlled by the dispersant component. Dispersant molecules absorbed on the surface of the ceramic powders would increase the zeta potential value (over a pH range of ~8-10), thus improving the powder dispersion in the slurry. Typically, there is an optimal amount of dispersant for enhancing fluidity and plasticity. The apparent viscosity decreases with an increase in dispersant level at a relatively lower concentration. When the dispersant level exceeds a certain level, the viscosity increases as a result of the increased level of cation ionic strength and the compressed double layer of the surface charge of ceramic particles. In addition, most studies demonstrate that dispersion can be also affected by organic monomers. The organic monomers either degraded the suspension dispersion or enhanced it slightly [15–17]. For a standard acrylamide monomer, the dispersant is selected from a group containing acrylic or methacrylic acid salts [8].

Even though gelcasting stems from slip casting and polymer chemistry, solid-loading in gelcasting can be over 50 vol%, compared with 30-35 vol% used in slip casting [10]. To achieve castability and pourability in such high solid-loading, dispersant and ball mixing are applied. Various ceramics prepared via gelcasting process are summarized in Table 1.

Solid-loading is one of the macroscopic factors affecting the suspension rheology. The rheological properties of gelcasting premix suspension are fundamentally attributed to fluid mechanics between particle interactions. Different rheological responses are illustrated in Figure 2. Newtonian is the simplest flow mode; the apparent viscosity does not change with increased strain rate. The shear thickening or "dilatant" mode shows an increased viscosity with an increased strain rate. If the viscosity decreases as the strain rate increases, this behavior is called shear thinning or "pseudoplastic." Shear thinning is a common feature for gelcast suspensions because of the perturbation of the suspension by shearing [19,20]. However, the shear-thickening phenomenon may occur under high shear-strain in a colloidally stable system if the solid-loading is high. The causes of this phenomenon are the presence of clusters [21–23] and the particle order-to-disorder transition in which the particle arrangement in the shear-thinning region becomes less ordered above a critical shear rate [21,24]. The resultant high viscosity of the dilatant slurry may cause defects in the green body after casting.

Shear thinning is a common behavior for gelcasting slurries and occurs with an increasing shear rate (as shown in Figure 3). Low- and high-shear Newtonian plateaus are separated by the shear-thinning region. At relatively low shear rate, Brownian motion dominates the behavior of concentrated suspensions, in which the particles are randomly distributed and close to an equilibrium state. With an increased shear rate, viscous force becomes more dominant and shear thinning

TABLE 1 Various Ceramics Prepared by Gelcasting

Materials	Solid loading level	Dispersant	Product	Author
Al_2O_3	50 vol%	1 wt% sodium alginate	Dense and complex parts	Yu *et al.*
Al_2O_3	75 wt%	Polyacrylic acid ammonium	Translucent	Mao *et al.*
	76-78 wt%		Transparent alumina disk	Krell *et al.*
Hydroxyapatite	40 wt%	1 wt% citric acid	Tissue engineering scaffolds	Kim *et al.*
Hydroxyapatite	-	-	Transparent disc	Varma *et al.*
GYGAG(Ce)	25-45 vol%	Ammonium acrylate	Transparent ceramic scintillators	Seeley *et al.*
Al_2O_3–ZrO_2	55 vol%	1 wt% PAA	Zirconia-toughened alumina (ZTA)	Liu *et al.*
YSZ	50 vol%	-	Porous ceramics	Gu *et al.*
AlN	55.4 vol%	PEI	Dense and complex parts	Xue *et al.*
ZrO_2	50 vol%	0.3 wt% Dolapix PC75	Dense cast spheres	Adolfsson
SiC	51.3 vol%	0.15 wt% TMAH	Dense and complex parts	Dong *et al.*
Si_3N_4	45 wt%	TMAH	Macroporous	Huang *et al.* and Dai *et al.*
SiAlON	45 wt%	Dolapix A88	Dense ceramics	Ganesh *et al.*
$BaTiO_3$	40 vol%	Duramax D3005	Green body	Santacruz *et al.*
Cordierite	67.1 wt%	4.1 wt% Lupasol-HF	Porous ceramics	Park *et al.*
$LaCoO_3$	-	-	Nanopowders for cathod	Cheng *et al.*

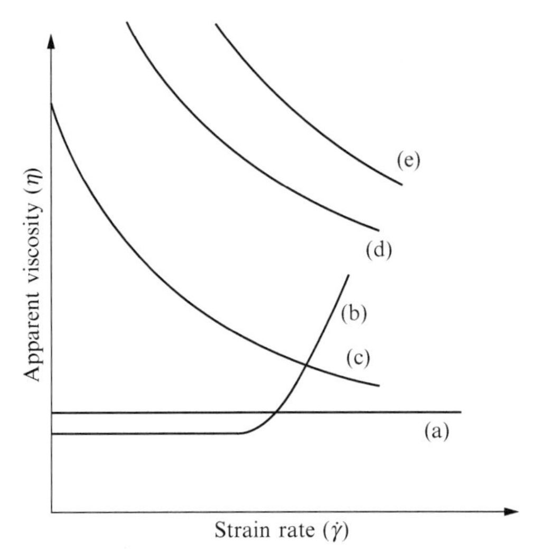

FIGURE 2 Rheological behavior categories under steady shear conditions, plotted as the viscosity versus the strain rate: Newtonian (a); shear thickening (b); shear thinning (c); Bingham plastic (d); and nonlinear plastic (e) [18].

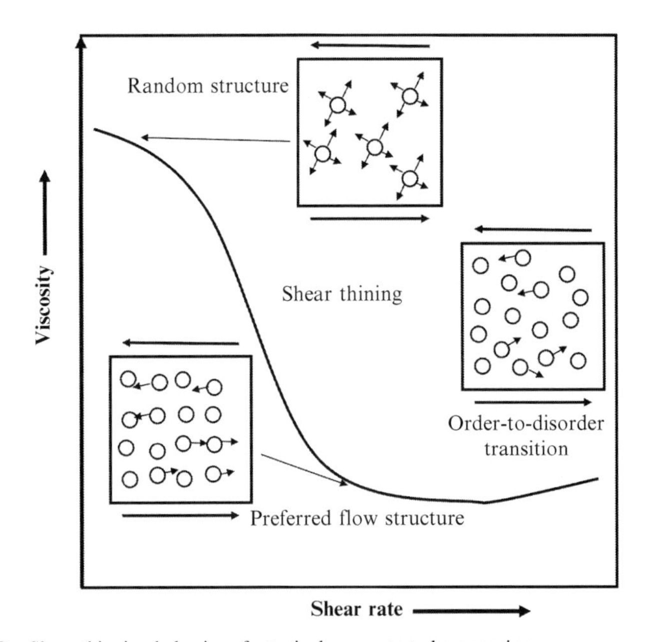

FIGURE 3 Shear-thinning behavior of a typical concentrated suspension.

is initiated. Under shear-stress and the original repulsive interparticle forces, the colloidally stable particles form a layered structure. In gelcast suspensions the polymer chains are aligned with the shear direction, and the viscosity reduces. The shear-thinning region follows a powder law relation according to some theoretical models such as Cross, Carreau, and Carreau-Yasuda [25–28]. For anisotropic particles, the degree of shear thinning would be higher. Shear thinning is also dependent on time and shows a thixotropy behavior in which the viscosity rises again when the shear ceases, and the polymer chains recover to a random state. At a high shear rate, hydrodynamic interaction is more significant, and an ordered particle structure is formed with lower viscosity. Shear-thickening phenomenon at higher shear rates is observed when the solid-loading is high, such as in SiC [29] and Al_2O_3 [30]. Shear thickening occurs at a critical shear rate when the viscous force overrules the repulsive interparticle forces [31,32]. The ordered structure breaks up, and the viscosity begins to increase as a result of the energy released from particle jamming.

Since gelcasting is operated in a concentrated ceramic suspension, interparticle forces as a result of the van der Waals (vdW) force and the repulsive barrier become very significant to rheological response. The repulsive barrier refers to a characteristic distance (δ) determined by an electric double layer (electrostatic force) and an adlayer (steric force). The effect of interparticle force on the rheological suspension property results in a new term of effective volume fraction (φ_{eff}), which can be defined by the following equations, depending on the particle shape:

$$\text{For spherical colloids: } \varphi_{eff} = \varphi\left(1+\frac{\delta}{R}\right)^3 \tag{2}$$

$$\text{For nonspherical colloids: } \varphi_{eff} = \varphi\left(1+\frac{\rho\delta A_s}{R}\right)^3 \tag{3}$$

where (φ) is the volume fraction of solid-loading, R is the radius of the particle, ρ is the powder density, and A_s is the specific area of the powder. These relations show that the absorbed layer can enhance the effective volume fraction. In other words, the addition of the thickness of the repulsive barrier releases the close packing and flocculation in the hard-sphere system. When the barrier thickness is smaller (e.g., by changing the dispersant level), the effective volume fraction will be slightly smaller, provided the particle size is the same. In addition, the packing densities are determined by the relative value between φ_{eff} and φ. A larger φ/R ratio would result in a lower packing density, whereas a lower ratio could cause aggregation.

2.2 Colloidal Stability

Gelcasting allows NNS ceramic parts to be fabricated from a premixed solution with colloidal stability and appropriate rheological behavior. Homogenous dispersion of ceramic powders is mainly determined by interparticle forces, including vdW, electrostatic (double-layer), and steric (polymeric) forces [19,33]. The vdW attraction is a ubiquitous interaction between two materials and the intervening media. The free-energy formula involving the Hamaker constant (A), radius (R), and surface separation (D) is given in following equation [34]:

$$V_{\mathrm{vdW}}(D) = -\frac{AR}{12D} \tag{4}$$

The Hamaker constant is determined by the dielectric properties of interacting macroscopic bodies, as well as the intervening medium, and is usually in the range of 10^{-19} to 10^{-20} J. The vdW force between two different bodies in a medium can be either attractive or repulsive. If the particle surfaces are identical, the vdW force is always attractive. When the surfaces are not identical and the Hamaker constant of the medium falls between those of the two surfaces, however, the force becomes repulsive. There are several reported material combinations that show dissimilar surfaces with a repulsive vdW force [35,36].

To achieve desirable suspension stability, the attractive vdW force should be alleviated by interparticle repulsion from either the electrostatic double-layer or polymer-induced steric stabilization. For example, using the DLVO theory, colloidal stability can be achieved by balancing the vdW attractive force resulting from the interparticle dipole and the electrical double-layer repulsion caused by the particle surface charge. When the ceramic powders are immersed in water (or another polar solvent used to prepare the suspension), the particle surfaces become charged, mainly by ionization or dissociation of surface groups. In addition, the surface charge can also be altered by adjusting the pH of the suspension. Counter ions are attracted to the primary charged particles and, in turn, repel other ions with the same charge, thus forming a layer-structured ion cloud around the primary absorbed ions. The double-layer structure is illustrated in Figure 4. The slipping plane defines the size of the double-layer cloud that can move as a whole unit, with the corresponding electrical potential called zeta potential. "Stern plane" refers to the first layer containing counter-ions with intermediate potential between the particle surface and the diffusion layer.

The double-layer thickness is determined by the valence and concentration of counter-ions. In a colloidal system with negatively charged particles, the increase of cations (higher ionic strength) compresses the double layer, which increases the viscosity of the suspension, allowing particles to flocculate.

In some cases it is difficult to reach colloidal stability by adjusting only the pH of the suspension. A polymeric dispersant is commonly used to increase repulsion and mitigate coagulation. The polymer-induced steric force results from the approach of two particles coated with organic molecules within a distance less than twice the polymeric layer thickness [19]. The repulsive steric force can be quantified using the De Gennes theory [37], which involves the thickness and affinity of the polymer layer, as well as the interaction range between

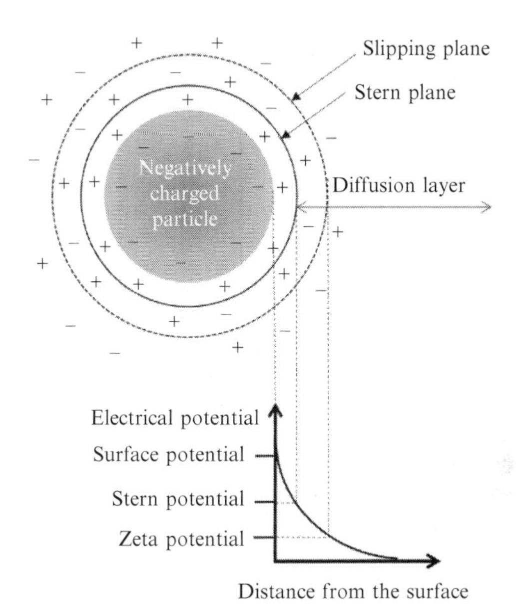

FIGURE 4 Electrostatic double layer around a negatively charged particle.

two particles. Another type of organic dispersant polyelectrolyte, known as super plasticizers, can introduce both electrostatic repulsion and steric repulsion [38]. Guldberg-Pedersen and Bergström [39] used atomic force microscopy to identify the relative importance of the electrostatic and steric contributions at various adsorption and solution conditions. By changing the pH and ionic strength, the dimension of the absorbed polyelectrolyte can be modified, which can further affect the degree of electrostatic and steric forces.

2.3 Gelation Process

The gelation process mainly involves polymerization of monomers to form a 3D hydrogel, as well as other reactions including cross-linking and catalysis. Gelation can proceed in both nonaqueous and aqueous solvents. Nonaqueous solvent usually refers to some organic solvents, such as dibutyl phthalate, dibasic ester, petroleum solvents, pyrrolidones, and long-chain ethanol. Various monomers, as well as other organic binders, should dissolve in nonaqueous solvents and exhibit a low viscosity and low vapor pressure at a cross-linking temperature. The drying process for the water-free gelling system is also more complex than the aqueous system because all of the organics must be burned out. The firing process is long and may need multiple steps to remove all of the binders and solvents. In addition, more polymer exhaust from nonaqueous solvents may cause environmental problems. Aqueous gelcasting, on the other hand, is often preferred and used in both laboratories and industry. The viscosity of the aqueous slurry is low enough for casting. Also, using water as the solvent does not make the gelcasting process differ much from other wet-forming techniques, and similar colloidal effects and drying process could be applied in gelcasting.

The monomer is one of the most distinguishable features of gelcasting and has two types: monofunctional and difunctional. Monofunctional monomers have a single bond and form a linear chain after polymerization. Difunctional monomers have at least two double bonds and form a cross-linked polymer network. Therefore, difunctional monomers are also called cross-linkers. The polymerization process can be achieved by two systems, according to the different functionalities of the monomer. The first system is acrylate-based. A commonly used monomer combination includes hydroxyethylmethacrylate (HEMA), methyl acrylate (MA), and N-vinylpyrrolidone (NVP) as nonfunctional monomers and ethyleneglycol dimethacrylate (EDMA) and diethylene glycol diacrylate (DEGDA) as difunctional monomers (cross-linkers). Since the two types of monomers are not miscible simultaneously in water, a co-solvent such as NMP is usually needed in the acrylate system. Therefore, the solvent in acrylate gelation is a mixture of water and an organic solvent instead of pure water, possibly resulting in the acrylate route having some limitations compared with the nonaqueous suspension. The second system is called the acrylamide system, in which the monomer is acrylamide and the cross-linker is *N,N′*-methylene bisacrylamide. This method is usually preferred because only water is used as the solvent. As seen in the flowchart in Figure 1, initiators and the catalyst are usually added into the premix. The main function of initiators is to provide free radicals in a radical polymerization system. The catalyst serves as an auxiliary additive to expedite the decomposition of the initiator; therefore, the free radicals can be generated more efficiently or at a lower temperature. There is a period of idle time before the beginning of polymerization. The viscosity stays constant until gelation takes place, and the organic network is formed.

The major problem with the acrylamide gelling system, as mentioned previously, is the toxic nature of acrylamide, which cannot meet the needs for environmentally friendly processing of advanced ceramics. Therefore, many research efforts have focused on developing low-toxicity or nontoxic monomers. Table 2 provides a brief list for potential low-toxicity or nontoxic monomers summarized from other research [3,6,40,41].

2.4 Degassing

Degassing is a crucial step after mixing (sometimes degassing is also required after casting) to eliminate residual pores in the slurry. These pores can be introduced during either mixing or the chemical reaction, or they can form as a result of entrapped air during casting. Another detrimental effect of entrapped air on gelcasting is its inhibition of the free radical polymerization reaction. Figure 5 shows the significance of degassing in reducing the flaw size of gelcast ceramics. The degassing time needs to be precise; it cannot be either too short or too

TABLE 2 Low and Nontoxic Monomers Used in Gelcasting

	Abbreviation	Gelation Function
Monofunctional monomer		
Methacrylic acid	MA	Acrylate
Methacrylamide	MAM	Acrylamide
N-hydroxymethylacrylamide	HMAM	Acrylamide
Ammonium acrylate	AA	Acrylate
Methoxy-poly(ethylene glycol) monomethacrylate	MPEGMA	Acrylate
N-vinyl pyrrolidone	NVP	Vinyl
Glycerol monoacrylate	GMA	Acrylate
2-Hydroxyethyl acrylate	HEA	Acrylate
Difunctional monomer		
N,*N*-methylenebisacrylamide	MBAM	Acrylamide
poly(ethylene glycol) dimethacrylate	PEGDMA	Acrylate
Di(ethylene glycol) diacrylate	DEGDA	Acrylate

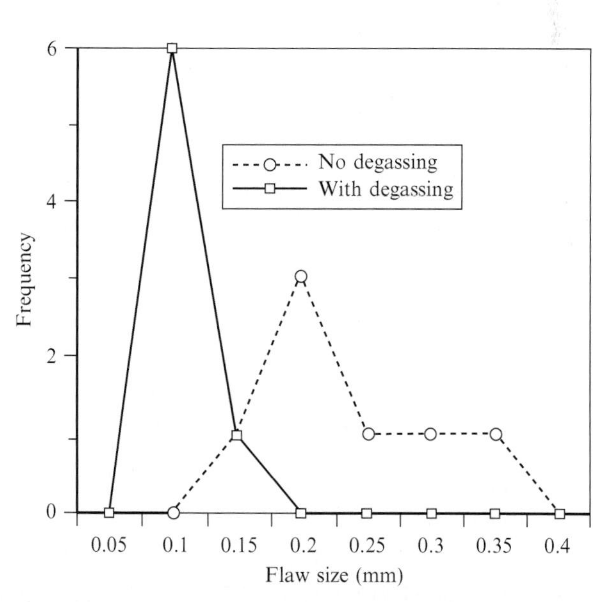

FIGURE 5 Distribution of flaw size in gelcast plates with and without degassing [8].

long. If the degassing time is too short, it will not be adequate to remove the pores, whereas too long a degassing time may cause unwanted evaporation of liquid and a change in slurry rheology.

2.5 Drying Process

The drying process is crucial to minimize the shape distortion of green bodies. Since high solid-loading and low binder content (usually less than ~5 wt% in the dried green body [9]) are used in gelcasting, less aqueous solvent is extracted during drying, and the gelcast green body is tough and machinable. Issues in terms of drying shrinkage are less crucial than those in slip casting or injection molding. Drying in gelcast green bodies is isotropic, with linear shrinkage of ~1-4%. Still, much research has been conducted to examine various parameters affecting the drying process and green body microstructure. Various experimental data demonstrate that there is almost no constant-rate period in the drying of gelcasting [7–9,15–17]. These researchers concluded that drying temperature and moisture are the dominant variables. The physical mechanism of gelcasting drying, on the other hand, is more complicated. A three-stage drying process for gel was proposed by Scherer [42] and Ghosal *et al.* [43] As shown in Figure 6, the first stage of the constant-rate period is relatively short compared with the sol-gel or slip casting parts. Free moisture is transported by capillary force to the superficial surface and evaporates into the air at a constant rate. The particle rearrangement is imitated, and the body shrinks at the same rate as the moisture evaporation rate. When the particles, together with the polymer gelling networks, are closely packed, the rearrangement stops, and the moisture transport to the surface is restrained. Then the shrinkage of gel begins around ceramic particles, yielding the shrinkage of the gelling network and the formation of interconnected pores between ceramic particles. Because the capillary force transporting moisture to the pores is smaller than that between the pores, the free moisture now migrates into the pores between the particles, which denotes the start of the second stage, called the falling rate period. This is a rate-controlled

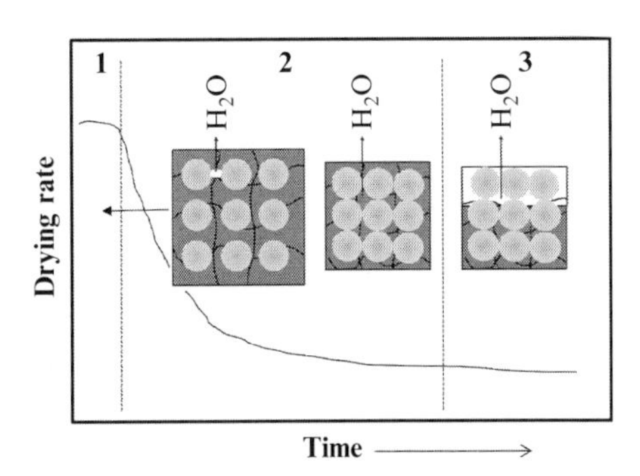

FIGURE 6 Schematic drying profile for a gelcast green body. stage 1: constant-rate period; stage 2: rate-controlled period; stage 3: diffusion-controlled period.

process dominated by the diffusion of moisture from the pores to the surface. When the partial pressure of the moisture in the pores reaches saturation, the moisture has to be removed through porous paths from the inner part to the superficial surface, under the driven force of decreasing concentration gradients. After the gel around the ceramic particles dries and the polymer network collapses, the transport of water by capillary force along the pores ceases, and the residual water has to diffuse through the dried polymer interfaces, which occurs much more slowly. This is why the drying rate in stage 3 is much lower than in the first two stages.

After drying, the gelcast green body undergoes a similar firing process as green bodies prepared by other routes. The binder burn out is carried out in air, and the final step of pressureless sintering is determined by the material and the purpose.

2.6 Inner Stress in the Gelcasting Process

Inner stress is a common issue for wet-processing of green bodies and occurs during the casting or drying process [44]. The root cause of inner stress can traced back to the casting process in the gelling suspension. This inner stress can deteriorate in the successive drying process in which green body shrinkage occurs and expedite the strain growth. Inner stress is more detrimental in large-scale or complex parts, in which microcracks could develop, more easily causing failure. A concentration gradient and free radical gradient in the initial stage of monomer polymerization could be induced or enlarged by the metallic impurities existing in raw ceramic powders [18]. The mechanism is interpreted as various metallic ions coordinating with monomers and going through a reduction reaction, then decomposing the initiator, such as ammonium per sulfate, via an oxidation reaction. The alternation of the initiator could result in a free-radical gradient, thus causing uneven contraction during solidification. Some other factors may cause free-radical gradients, such as inadequate mixing of initiators or catalysts for the cross-linking process in the gelling system.

Temperature gradient is another factor causing inner stress when heating is applied. The gelcasting green body undergoes polymer burn out and sintering. Sometimes these two firing procedures may be carried out at the same time by setting a multistep sintering profile, according to the purpose or need of the ceramic products. At this stage, the inner stress is introduced in a manner similar to conventional ceramic sintering. Uneven contraction resulting from nonsymmetrical shrinkage and warping during firing are the two major causes of inner stress. In some gelcasting cases heating is applied to the mold at an earlier stage, in which the wet green body needs to be heated in the drying process to remove the water or achieve a gelation reaction. This temperature is much lower than the firing temperature. There always exists a temperature gradient between the mold surface and the slurry. The asynchronous solidification associated with heating can result in asynchronous gelation between the outer surface and the inner part of the slurry. As a consequence, the microstructure is not uniform and inner stress is introduced. Because of the presence of a temperature gradient during drying, the interface between solid ceramic powders and the liquid solvent may migrate from the surface to the interior of the green body during the three steps of the drying process [43]. Inner stress is thus induced, constraining the shrinkage of the interior part during drying.

Even though gelcasting is featured with high solid-loading as an attempt to reduce shrinkage, the solid-loading level has a limit for desirable viscosity. The common sense method for solving the drying issue is to prolong the drying time and increase the drying humidity, but this method is not suitable for manufacturing purposes. Therefore, several methods to release the inner stress caused by temperature gradient in the drying process have been proposed. Huang *et al.* [45] developed pressure-assisted gelcasting by combining conventional injection molding and gelcasting of an aqueous ceramic suspension. The solidification rate can be controlled by hydrostatic pressure. A plasticizer or moderator (hydroxyethyl acrylate) added into the monomer solution also decreased the stiffness, thus reducing the magnitude of inner stress. Liquid desiccant is another alternative to overcome the limitations of humidity and atmosphere control during green body drying [44,46,47]. The chemical potential difference between the polymeric chains in the liquid desiccant and the solvent in the polymeric network in the green body plays the main role in eliminating nonuniform drying while expediting water retraction. Gelcasting alumina and silicon green bodies immersed in a liquid desiccant exhibited shortened de-binding time, free of cracks and controlled shrinkage.

3 TRANSPARENT CERAMICS PREPARED BY GELCASTING

Gelcasting was originally proposed to produce high-density ceramic parts, which is favorable for manufacturing transparent ceramics. Polycrystalline transparent ceramics have been widely studied because of their high potential in replacing glass and single-crystalline materials. Transparent ceramics possess superior mechanical properties, high resistance to wear, excellent chemical stability, and a low manufacturing cost compared with single-crystalline ceramics. Ceramics usually contain many scattering sources, however, such as grain boundary phases, residual pores, and secondary phases, which cause significant scattering losses. Therefore, improving the microstructure is one of the key challenges to improve transparency. Fabricating transparent ceramics through dry process routes usually requires high pressure, high temperature, and part dimensions restricted to small sizes and simple geometries. In addition, the requirements for raw ceramic powders, and their sintering conditions, are very harsh. In contrast to dry processing, wet-processing routes such as slip casting, injection molding, and extruding, have been applied to fabricate translucent or transparent ceramics [48–52]. Today, ceramics with complex shapes in a larger scale are required, and the above techniques cannot fulfill these needs. The gelcasting technique, on the other hand, can meet the current requirements and is promising as a means of fabricating transparent ceramics. Several research groups have successfully prepared transparent ceramics via gelcasting. The gelling system and fabrication routes of several cases are discussed in this section.

Alumina is one of the most commonly studied transparent ceramics, and the gelcasting forming method associated with other sintering routes has been tested to successfully obtain transparent or translucent alumina ceramics. The residual porosity of polycrystalline ceramics has to be lower than 0.05% to make the ceramics transparent, which means a relative density >99.5% should be achieved. Krell *et al.* [50] used a commercial corundum powder with a premix aqueous solvent (containing acrylamide and methylene-bis-acrylamide) to obtain a stable slurry with 70-78 wt% solid-loading. The slurries were milled under decreased pressure for the degassing process. After the gelcast bodies dried in ambient air for 2 days, the green body was fired at 800 °C to remove the organic binders, and it finally was sintered in a hot isostatic press at 1150-1400 °C in argon. A density >99.9% was obtained with ~60% real in-line transmittance.

Another type of transparent ceramic is yttrium aluminum garnet (YAG) or a related doped or undoped rare earth garnet, including yttrium scandium aluminum garnet and gadolinium gallium garnet. This class of ceramics have been widely studied because of their potential use in optic lasers and armor [53,54]. Various researchers used flame-spray pyrolysis to prepare $Gd_{1.49}Y_{1.49}Ce_{0.02}Ga_{2.2}Al_{2.8}O_{12}$ GYGAG(Ce) nanoparticles with sizes between 20 and 100 nm [55–57]. Then the fine nanoparticles were dispersed in deionized water with an adjusted pH of about 10.5. In the alkaline environment the zeta potential for an oxide suspension can be far from zero, which results in a colloidally stable suspension. The solid-loading was between 25 and 45 vol%. The slurry contained only polyethylene glycol (not necessary) and ammonium polymethacrylate (Darvan C-N).The unique aspect of this research is that there was no polymerization, and the cross-linking process was fulfilled by the gelling agent. If the particles are very fine (within the nano-scale), they may gel without the use of monomers or binders, but require the help of heating the mold. The gelcast green body was fired first at 1050 °C to remove the organics, followed by high vacuum ($<2 \times 10^{-6}$ torr) sintering to a relative density of 97%. The ceramic finally underwent hot isostatically pressing under 200 MPa in argon to become fully densified. Figure 7 shows the sample at different stages.

For all gelling systems developed to date, a suitable dispersant is needed for high solid-loading suspension [9,30,58–60]. In addition, 3D network gel components, including epoxy or hardeners, are expensive, volatile, toxic, and flammable, with poisonous exhaust gas burnt out into the environment [9,61,62]. Furthermore, nonaqueous dispersants, such as epoxy and hardeners, require an organic solvent, which complicates the operation process and exposes more harms to the environment, as well as increases the cost of the operation-protection instrument. For gelcasting, some gelling agents react through free radical polymerization under protection from the atmosphere because of oxygen inhibition, which is very complex. [9] With further studies of

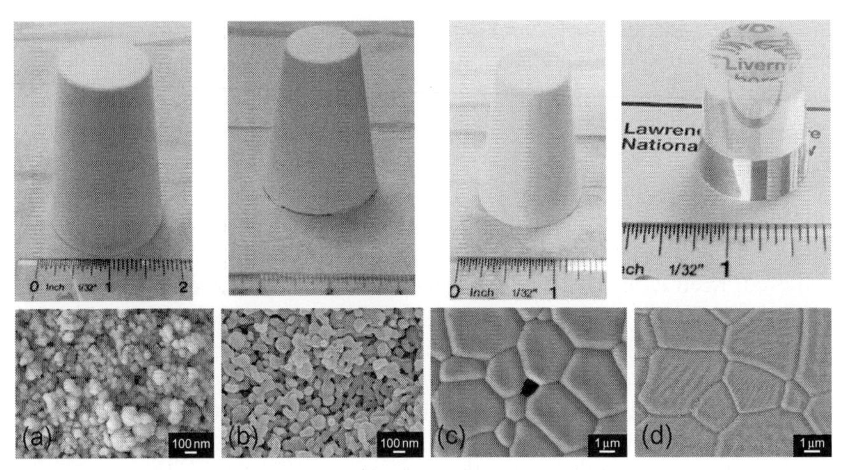

FIGURE 7 Sample digital images (top) and SEM (bottom) of gelcast (a), calcined (b), vacuum-sintered (c), and (d) ceramics via hot isostatic press (HIP).

gelcasting, the low-cost gelling agent polyvinyl alcohol (PVA) was adopted [30]. PVA is a high-viscosity aqueous solution, influencing the rheological properties and limiting the solid-loading of a slurry with a catalyst (acidic region with a pH of 1.5-2) and a high temperature (60-80 °C) [63]. As a result, control of the gelling and drying processes is difficult when using PVA. Room-temperature gelcasting could be finished only by using many kinds of additives (e.g., dispersant, monomer, cross-linking agent, catalyst, or initiator). The development of monomers with low toxicity without an external cross-linker, such as glycerol mono-acrylate [58] and 3-*O*-acrylic-D-glucose [59], also was applied for gelcasting of alumina. However, the addition of dispersant and initiators are still needed. In addition, many scientists have studied nontoxic natural polymers (agarose [60], alginate [64], protein [65], and starch [66]) for gelcasting techniques and succeeded preparing ceramics (ZrO_2, Al_2O_3), which need heating treatment at ~60 °C.

A new room-temperature spontaneous gelcasting system was recently developed at the Shanghai Institute of Ceramics, Chinese Academy of Science, using a multifunctional commercial copolymer of isobutylene and maleic anhydride (ISOBAM) as a spontaneous gelling agent [67]. This study is discussed in the next section.

4 INTRODUCTION TO A NEW WATER-SOLUBLE, SPONTANEOUS GELLING AGENT: ISOBAM

ISOBAM is the trade name of an alternative copolymer of isobutylene and maleic anhydride developed by Kuraray, Osaka, Japan. ISOBAM is a water-soluble, long-chain copolymer; four different grades of ISOBAM available (Figures 8–11). Various useful reactants are obtained by the

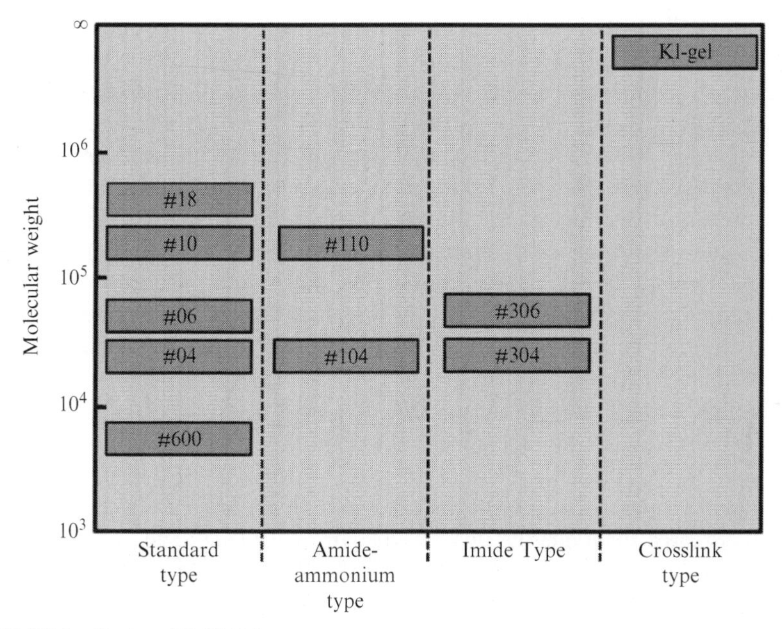

FIGURE 8 Grades of ISOBAM.

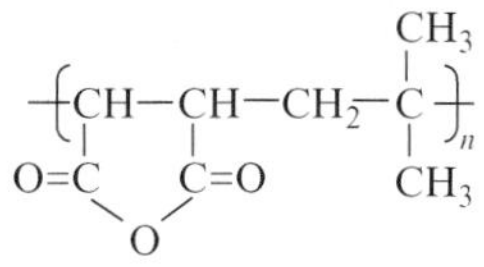

$$\left[\underset{\substack{|\\ O=C \diagdown}}{CH} - \underset{\substack{|\\ \diagup C=O}}{CH} - CH_2 - \underset{\substack{|\\ CH_3}}{\overset{\substack{CH_3\\|}}{C}} \right]_n \quad \text{(ring O bridging the two C=O)}$$

FIGURE 9 Molecular structure of standard ISOBAM.

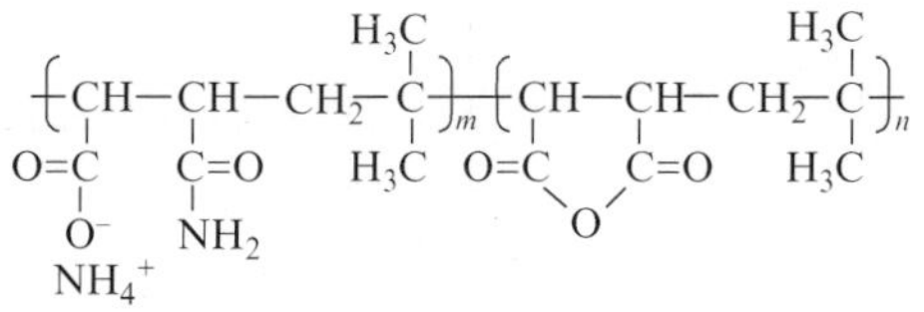

$$\left[\underset{\substack{|\\ O=C\\ |\\ O^-\ NH_4^+}}{CH} - \underset{\substack{|\\ C=O\\ |\\ NH_2}}{CH} - CH_2 - \underset{\substack{|\\ H_3C}}{\overset{\substack{H_3C\\|}}{C}} \right]_m \left[\underset{\substack{|\\ O=C \diagdown}}{CH} - \underset{\substack{|\\ \diagup C=O}}{CH} - CH_2 - \underset{\substack{|\\ H_3C}}{\overset{\substack{H_3C\\|}}{C}} \right]_n \quad \text{(ring O)}$$

FIGURE 10 Molecular structure of ISOBAM104.

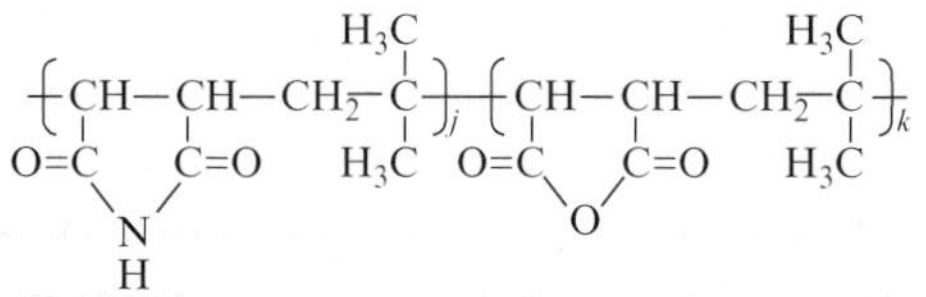

$$\left[\underset{\substack{|\\ O=C \diagdown}}{CH} - \underset{\substack{|\\ \diagup C=O}}{CH} - CH_2 - \underset{\substack{|\\ H_3C}}{\overset{\substack{H_3C\\|}}{C}} \right]_j \left[\underset{\substack{|\\ O=C \diagdown}}{CH} - \underset{\substack{|\\ \diagup C=O}}{CH} - CH_2 - \underset{\substack{|\\ H_3C}}{\overset{\substack{H_3C\\|}}{C}} \right]_k \quad \text{(first ring N–H, second ring O)}$$

FIGURE 11 Molecular structure of imide ISOBAM.

reaction of ISOBAM with alcohol, amine, and epoxy compounds, with the application of resins and plastics.

Shunzo Shimai introduced water-soluble, nontoxic, organic ISOBAM to Dr. Shiwei Wang's group (Shanghai Institute of Ceramics) as the gelling agent for epoxy. During the experimental process, the amazing results showed that ISOBAM104 could spontaneously gel by itself at room temperature in air, without any additions. This finding brings gelcasting techniques into an era of room-temperature spontaneous gelling. ISOBAM104 could also work as a dispersant for ceramic aqueous suspensions. The multifunctional properties of ISOBAM104 are attracting increasing attention for application in wet-forming techniques, working as both a dispersant and a spontaneous gelling agent.

ISOBAM104 (Figure 10) is an ammonium-modified product, based on standard ISOBAM amide-ammonium salt (Figure 9), with the appearance of white powder. The molecular weight of ISOBAM104 is ~55,000-65,000 g/mol, with a density of 1.3 g/cm^3. The molecular structure is shown in Figure 10 (m:n = 1:1). There are three functional groups: $-COO-NH_4^+$, $-CONH_2$, and $-CO-O-CO-$. The pH of the aqueous solution is neutral. The hydrocarbon chain serves as a backbone, threading its way along the entire molecule.

ISOBAM is a multifunctional organic and could be applied as:

(1) Dispersant for ceramics wet-forming techniques.
(2) A spontaneous gelling agent for ceramics gelcasting techniques at room temperature in air.
(3) Binders for ceramics forming (dry pressing, tape-casting techniques, and excellent membrane construction).

ISOBAM also has the following merits:

(1) Nontoxic (no harm to the environment or operators, as well as an easy production process).
(2) Water soluble (application in an aqueous casting system without any organic solvent).
(3) Forms a gel at room temperature in air with the addition of only low ISOBAM104 (~0.3 wt% for ceramic powders) [67].
(4) Easy operation.
(5) Low production costs.
(6) Easy control of the drying process and the subsequent de-bindering process.
(7) High strength of the green body after drying.

Concerning the application purposes and merits listed above, ISOBAM is a very promising organic that could be used in ceramics wet-forming techniques, such as slip casting, gelcasting, tape casting, injection casting, and pressure casting. A schematic flowchart of the typical room-temperature spontaneous gelcasting process is shown in Figure 12.

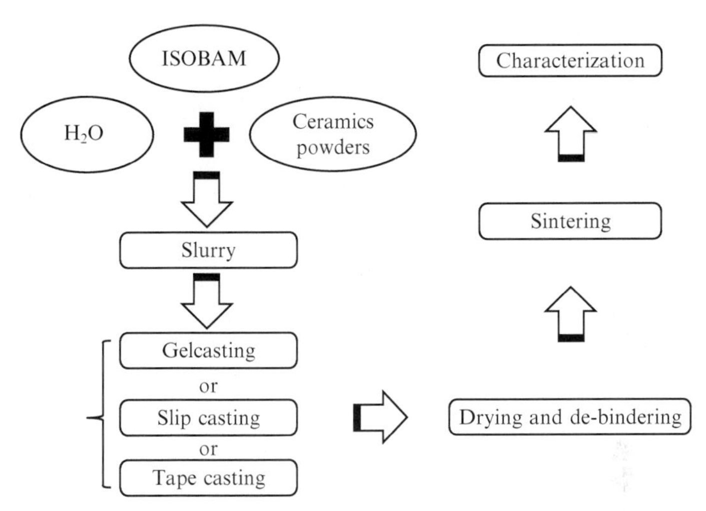

FIGURE 12 Flowchart of the ISOBAM gelcasting process.

5 ISOBAM ACTS AS DISPERSANT

5.1 Mechanisms of Dispersants

Solid ceramic particles, polymers, and a liquid medium comprise the whole system of a ceramic slurry. It is important to produce a well-dispersed suspension during the colloidal process, which is crucial for a high solid-loading slurry and homogeneous, high-density green bodies for ceramics with a complex shape without cracks during wet-forming techniques. For pure ceramic powders, primary particles can easily form agglomerations by vdW forces. Hard agglomerations can be broken down only by colloid milling. During the milling process, primary powders have to be stabilized to prevent particles from flocculating again. The stability of a slurry depends on the particle size, particle size distribution, the solubility of the solid in the dispersion medium for the specific ceramic powders, and dispersants (also known as deflocculants), which play an important role in dispersion by inhibiting the aggregation of the dispersed phase. Dispersants can be either organics (polymers [68–71], surfactants [19,72], or other small molecules [73,74]) or inorganics (silicate, phosphate) [19,70,75].

Dispersants should exhibit the following properties:

(1) Provide good wetting with the external surfaces of the ceramics powders.
(2) Decrease the viscosity of the system when wet milling.
(3) Stabiliz the dispersion.

So, the adsorption of the dispersant on the particle surface is required. The adsorption of the dispersant at the solid surface is associated with a decrease in free surface energy, which is charged by the particle-dispersant attraction and the attraction/repulsion between the adjacent dispersant molecules adsorbed on the surface. The former results in dense adlayers or multiple layers, whereas the latter leads to saturated monolayers [76]. The dispersant molecule adheres to the particle's surface by ionic, physical, or chemical forces [77], including the following:

(1) Ionic adsorption between the dispersant ions and the oppositely charged site on the particle's surface.
(2) Ion exchange between the dispersant ions and similarly charged ions in the particle's surface.
(3) Hydrogen bonds formed between the dispersant molecule and the particle.
(4) Physical adsorption by dispersion forces, dipole-dipole interactions, hydrophobic bonding, or charge transfer forces caused by donor-acceptor interactions [78].
(5) Chemisorption coming from the covalent chemical bond.

Nonionic dispersants are absorbed by mechanisms 3, 4, and/or 5 listed above. For ionic dispersants, all these mechanisms contribute to adsorption. The mechanisms of high-molecule polymeric dispersants are more complex than low-molecule dispersants [79]. The configuration of polymer adsorption on the particle surface was assumed by Jenckel and Rumbach [80], as shown in Figure 13. Some of the polymer directly attaches to the particle's surface ("train"), and some of the polymer has free ends extending into the liquid media, with the other end adsorbed on the particle's surface ("tail"). For other polymers, some segments extend into the medium, with some functional groups sticking to the particles' surfaces ("loop"). For naked particle surfaces, all polymers have the same possibility to adsorb on it. After that, other polymer molecules may attach to the adsorbed polymer because of the dynamic state of adsorption, or polymer molecules can detach, and their sites can be occupied by larger, adjacent molecules [77].

The mechanism of the dispersing role of ISOBAM could be determined from the three functional groups of its molecular structure: $-COO-NH_4^+$, $-CONH_2$, and $-CO-O-CO-$.

5.2 The Mechanism of ISOBAM to be Qualified as a Dispersant

ISOBAM is a long-chain organic with three functional groups (Figure 3), which are likely the source of its good dispersibility. As described by Davies and

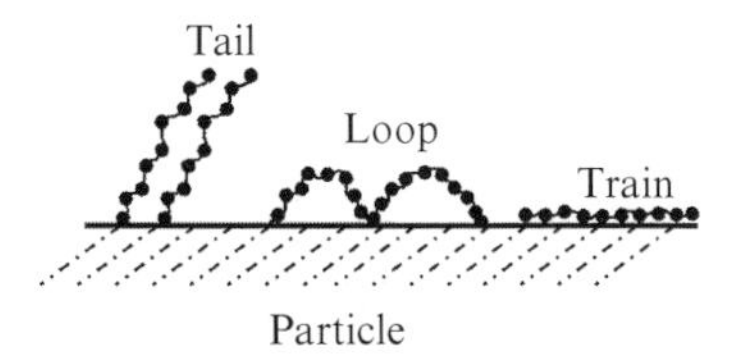

FIGURE 13 Configurations of polymer absorbed on the particle's surface.

Binner [81] the excellent dispersion role of ammonium polyacrylate (NH_4^+PAA) (Figures 14–16) can be attributed to the key functional group $-COO-NH_4^+$, which is stretched along the polymer length as a result of electrostatic repulsions (Figure 17a) and anchors on the powder's surface in a thin, flat train configuration. The formation of loops is suppressed (Figure 13), resulting in increased particle dispersion. Furthermore, the extension of the attached polymer may also introduce steric (Figure 17c) and electrosteric (Figure 17b) forces. According to the DLVO theory (Figure 4), after soluble ISOBAM is wetted with water, functional groups attach to the ceramic powder's surface, providing electrostatic and steric stabilization, leading to effective deflocculation of the ceramic powder [82].

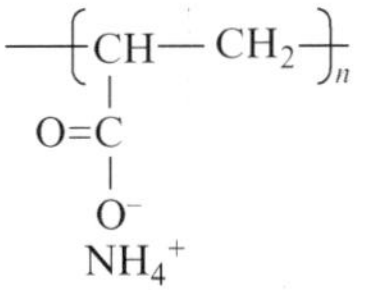

$$-\!\!\left[CH-CH_2\right]_n$$
O=C
O⁻
NH_4^+

FIGURE 14 Molecular of ammonium polyacrylate (NH_4^+PAA).

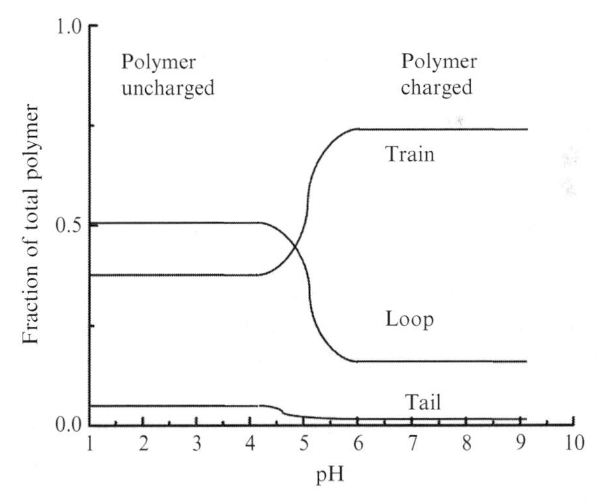

FIGURE 15 Train, loop, and tail configurations of a polyelectrolyte attached to ceramics powder surfaces in a suspension *(From J. Davies, et al., The role of ammonium polyacrylate in dispersing concentrated alumina suspensions,* J. Eur. Ceram. Soc., *Vol. 20, No. 10, P1539, 2000).*

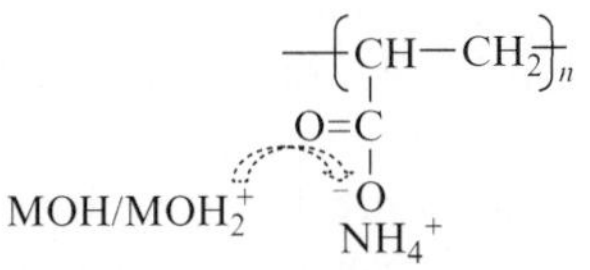

$$-\!\!\left[CH-CH_2\right]_n$$
O=C
MOH/MOH_2^+ O⁻
NH_4^+

FIGURE 16 Interaction of ammonium polyacrylate (NH_4^+PAA) with a ceramics particle surface.

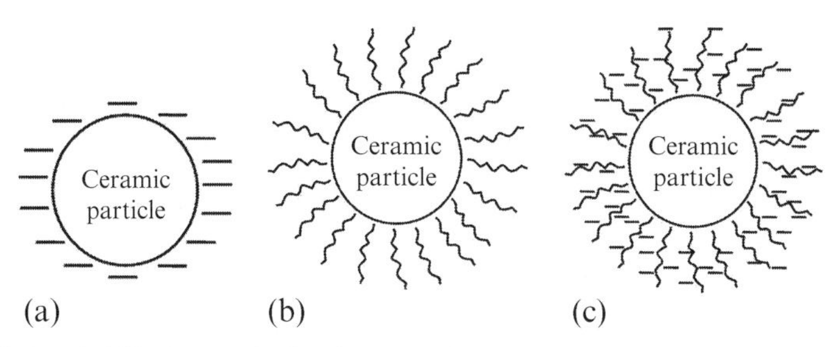

FIGURE 17 Stabilization mechanism in a ceramics suspension: electrostatic stabilization (a); electrosteric stabilization (b); steric stabilization (c).

```
 —CH—CH—                     —CH—CH—
  |    |                      |    |
O=C    C=O  +  H2O  <—>     O=C    C=O
  \   /                       |    |
    O                        HO    OH
```

FIGURE 18 H-bonding between an amide function group and oxide surface sites.

```
MOH/MOH2+  O-                                      O
           ||  |                                   ||   |
   H—O—C—CH—              MOH-  +H—O—C—CH—
               |                                          |
   H—O—C—CH—              MOH-  +H—O—C—CH—
           O-  |                                   ||   |
MOH/MOH2+                                          O
(a)                         (b)
```

FIGURE 19 Reaction of acid anhydride functional groups with water.

For dissolved ISOBAM, acid anhydride functional groups can react with water (Figure 18).

Two –COOH carboxyl acid groups are introduced into an ISOBAM polymer, making the molecular properties similar to those of polyacrylic acid (PAA). Some carboxylate groups coordinate with positively (Figure 19a) or negatively charged (Figure 19b) or covalent sites [83] on the solid surface, and some parts of the carboxylate groups extend into the solution to supply the molecular charge and a double layer with considerably greater latitude in expansion or contraction. This also enables adaptability to charge and structure sites on solid surfaces without inducing bond stress in the molecule. This improves the dispersive role of ISOBAM for ceramic powders. Meanwhile, –COOH groups are hydrolyzed, or at least partially dissociated, into carboxylate (COO^-) ions [52].

The nonionic, polymeric functional group $CO–NH_2$ also has been widely used for dispersing ceramic particles [84–89]. In the presence of nonionic polymers, the functional groups produce hydrogen bonding. The adsorption density, as well as the affinity of the nonionic polymers, is not affected by solution pH or electrolyte concentration, coming from the absent nonionic polymers electrostatic interactions [86]. As a result, the $–CO–NH_2$ brings in an excluded volume and mixing/osmotic components to compress the adsorbed polymer layer, causing an increase in polymer concentration and providing steric stabilization of the suspension [90].

In ceramic powder, when wetted with water, the metal species ("M") hydrolyzes on the particle's surface and forms hydroxyl-metal complexes as a result of the protonation or deprotonation process; this is expressed by:

$$MOH + H^{+} \leftrightarrow MOH_2^{+} \tag{5}$$

$$MOH \leftrightarrow MO^{-} + H^{+} \tag{6}$$

For the functional group CO–NH_2, Lee and Somasundaran [87] studied the adsorption of amide functional groups on a mineral oxide surface. Because of the greater electron affinity of oxygen, the carbonyl (C=O) is a stronger dipole than the C-N dipole, leading amide functional groups to act as hydrogen bonding acceptors. The electronegative C=O function of the amide acts as an H-bonding base, and the oxide surface hydroxyls act as an H-bonding acid. So, the neutral, undissociated MOH group (the MOH_2^+ group) can act as a proton donor (Figure 20a), even if H-bonding is not considered an electrostatic interaction. Compared with MOH, the MOH_2^+ group has a much easier time forming H-bonds with C=O. MOH^- bonds with weakly acidic –NH_2 for H-bonding (Figure 20b).

Based on the discussion above, the final functional groups are: –COO^+ and –CO–NH_2.

5.3 The Application of ISOBAM as a Dispersant

5.3.1 The Application of ISOBAM on Nonhydrated Ceramics Powder

Yang *et al.* [67] reported the dispersibility of ISBOAM104 by testing how the ζ-potential of Al_2O_3 slurry (Al_2O_3, AES-11; Sumitomo Chemical Co. Ltd.) (Figure 21), varies with pH values under different ISOBAM additions (powder/water = 0.15 g/L) (Figure 22), with the comparison with the well-known dispersant NH^{4+}PAA. It shows that Al_2O_3 without a dispersant has an isoelectric point (IEP) at pH 8.3. With the addition of 0.3 wt% ISOBAM104, the IEP moves to pH 5.8 and further decreases to pH 2.8 with the addition of 1.0 wt% ISOBAM104, which is very similar to the conventional dispersant NH^{4+}PAA (0.5 wt%).

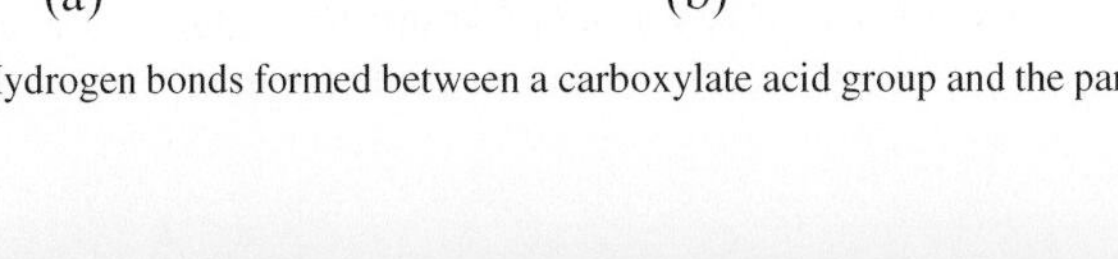

$$\text{(a)}\quad MOH/MOH_2^{+} \cdots {}^{-}O{=}C(NH_2){-}\!\!\left[CH{-}CH_2\right]_n \qquad \text{(b)}\quad MOH^{-} \cdots {}^{+}H_2N{-}C({=}O^{-}){-}\!\!\left[CH{-}CH_2\right]_n$$

FIGURE 20 Hydrogen bonds formed between a carboxylate acid group and the particle's surface.

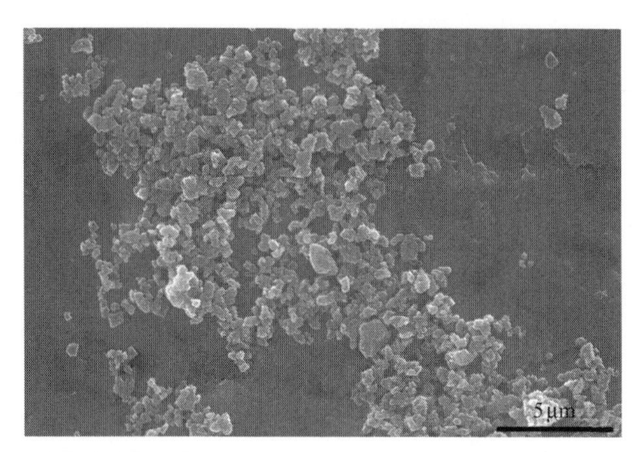

FIGURE 21 SEM of Al_2O_3 (AES-11) slurry.

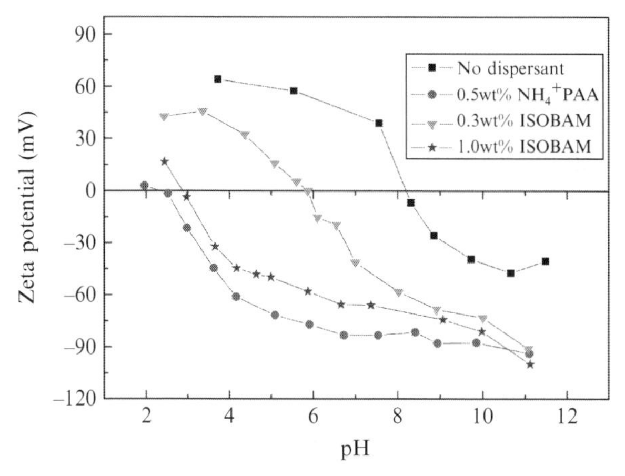

FIGURE 22 Effect of the dispersiveness of ISOBAM on ζ-potentials of an alumina slurry *(From Y. Yang, et al., Room-temperature gelcasting of alumina with a water-soluble copolymer,* J. Mater. Res., *Vol. 28, No. 11, P1512, 2013).*

The ζ-potential of alumina slurry is about 60 mV at pH 8.3 with the addition of 0.3 wt% ISOBAM104 and 75 mV for the addition of 1.0 wt%. This verifies the excellent dispersibility of ISOBAM in ceramics aqueous slurry. Shimai *et al.* [91] applied this ISOBAM to disperse finer Al_2O_3 (CR-10; Baikowski, Annecy, France) (Figure 23), with the addition of 0.5 wt%, with the ζ-potential −50 mV at pH 9.2 and IEP at pH 4.2, as shown in Figure 24.

5.3.2 The Application of ISOBAM on Waterproof Paint-Coated Ceramic Powders

ISOBAM is also used to disperse aluminum nitride (AlN) coated with polyurethane (waterproof paint from hydrating) (Grand F; Toduyama Soda Co. Ltd.), reported by Shu *et al.* [92], with the addition of 0.3 wt%, changing the IEP of

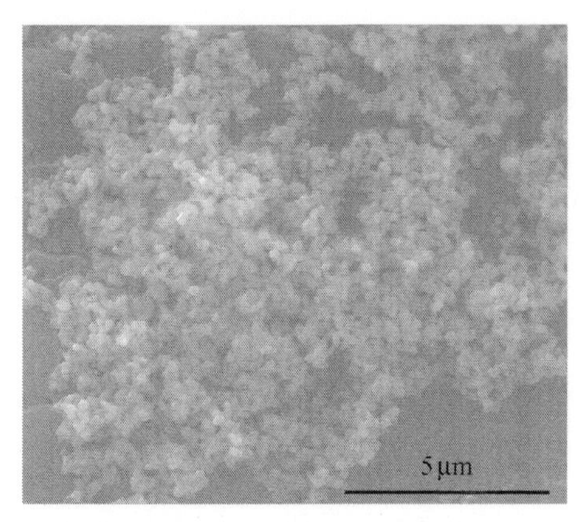

FIGURE 23 SEM of Al_2O_3 (CR-10).

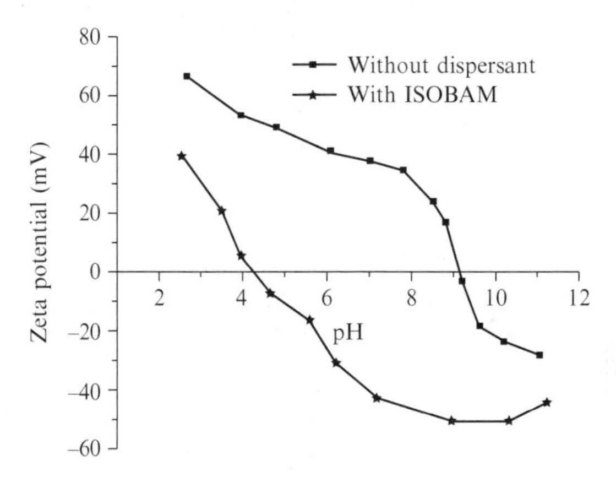

FIGURE 24 ζ-Potential of Al_2O_3 (CR-10) with ISOBAM (0.5 wt%) *(From S. Shunzo, et al., Spontaneous glecasting of translucent alumina ceramics,* Opt. Mater. Express, *Vol. 3, No. 8, P1000, 2013).*

polyurethane-coated AlN from pH 7.8 to 3.5 and increasing the ζ-potential from 0 to −40 mV at pH 7.8 (Figure 25).

6 ISOBAM ACTS AS SPONTANEOUS GELLING AGENT AT ROOM TEMPERATURE IN AIR

6.1 The Mechanism of the ISOBAM Spontaneous Gelling Process

After discussing the dispersive role of ISOBAM, it is now known that ceramic powder surfaces are covered with organic ISOBAM104 layers by steric/electrosteric/electrostatic effects. Some of the ISOBAM104 segments coordinate with the powder's surface, and other segments dissolve in aqueous medium (water).

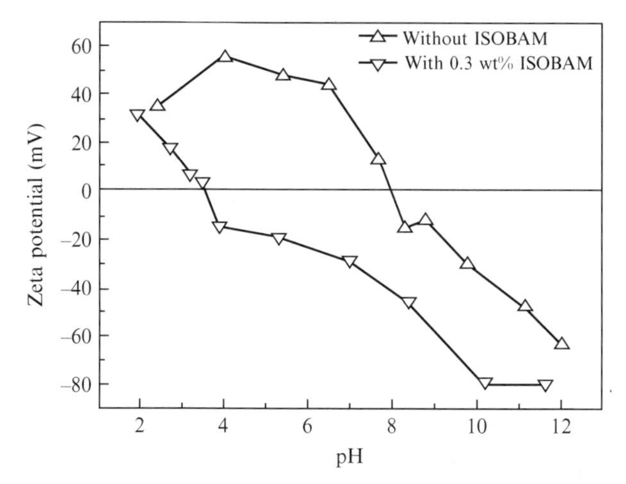

FIGURE 25 ζ-Potential of polyurethane coated AlN with ISOBAM (0.5 wt%) *(From X. Shu, Gelcasting of aluminum nitride using a water-soluble copolymer,* J. Inorg. Mater., *Vol. 29, No. 3, P327, 2014).*

However, the mechanism for the spontaneous gelling behavior of ISOBAM104 is still unclear. There are several proposed possibilities for the gelling mechanism.

6.1.1 Dissolution of ISOBAM in a Suspension Medium (Water)

ISOBAM is a water-soluble organic with the appearance of a solid and long C-chains aggregating and tangling together (Figure 26, left). When ISOBAM swells in water, the molecules stretch out freely, and the functional groups start to interact with water. Water also tends to invade the space between the ISOBAM molecules and builds a layer of hydration around the molecules, thus separating them (Figure 26, middle).The situation for ceramic powders added to the ISOBAM-water solution is shown on the right in Figure 26, which indicates the ISOBAM104 chains are absorbed onto the ceramics surface, with the other end freely swelling into H_2O or hand-in-hand/twinning with other ISOBAM chains.

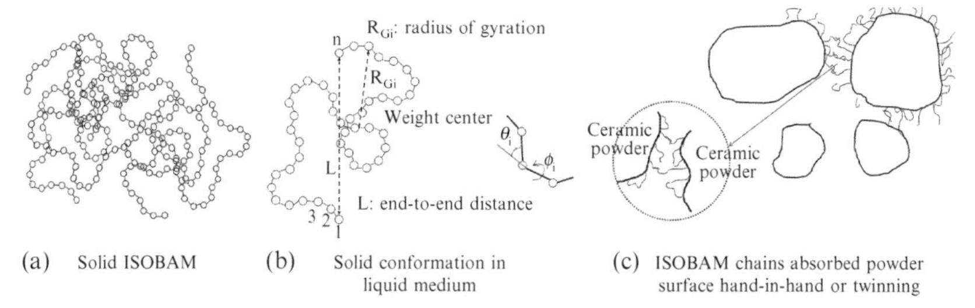

FIGURE 26 The behavior of ISOBAM at different periods: solid ISOBAM (a); ISOBAM dissolved in water (b); and ISOBAM dissolved in water and interacting with ceramics powders (c).

6.1.2 Possible Mechanism for ISOBAM104 Spontaneous Gelling Process

There are three possible coordinations between ISBOAM104 functional groups and water/ceramic powders during this process to prevent agglomeration during gel formation:

1. ISBOAM/ceramics (as discussed in dispersant parts)
2. ISOBAM/ISOBAM (Figure 27, ①⑤⑥)
3. ISOBAM/water (Figure 27, ②③④⑦)

The reaction and coordination of the ceramics powder surface/ISOBAM/water lead to the adsorption and hydrogen bonding between the ceramic powder/ISOBAM/water, which is beneficial for the formation of 3D networks in the ceramic powder suspension.

The molecular structure of ISOBAM104 is illustrated in Figure 10; the three main functional groups ($-COO^{-}-NH4^{+}$, $-CO-NH_2$, and $-CO-O-CO-$) may possess the chemical/physical forces required to combine the organic chains and particle surfaces. This may lead to hydrogen boding, hydrophobic interactions, and a decrease in free water because of the absorption of water molecules into the ISOBAM104 molecular gap. Further evidence and research are needed to fully understand the ISOBAM104 spontaneous gelation mechanism.

All of the reactions (Figure 27, ①-⑦), including hydration of acid anhydride and hydrogen bond formation, could finish at room temperature in air without the problems of oxygen inhibition [9], catalyst addition [63], or heating [60,63], These effects may impart to ISOBAM its spontaneous gelling ability in very simple surroundings (at room temperature in air), which simplifies the entire gelcasting process and decreases the equipment cost of production.

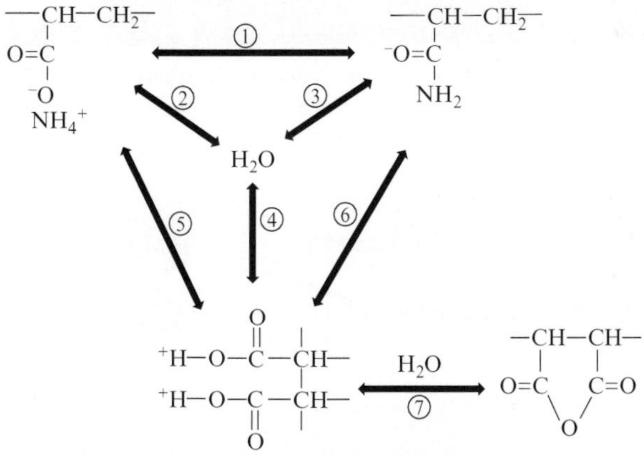

FIGURE 27 The coordination and reaction of ISOBAM functional groups with water.

6.2 Application of the Spontaneous Gelling Agent ISOBAM

6.2.1 Spontaneous Gelling of Traditional Al_2O_3 Ceramics Using ISOBAM

Yang *et al.* [67] succeeded in using the spontaneous gelling agent ISOBAM to make conventional ceramics. The wet gel could be bent and twisted without cracking (Figure 28).

The rheological property of slurry with ISOBAM working as the dispersant and gelling agent is examined by the viscosity and storage modulus as a function of the added ISOBAM and solid-loading. Figures 29 and 30 show separately the variation in the viscosities of slurries with varying ISOBAM104 content and solid-loading. They show that all the slurries of 50 vol% solid-loading and the addition of varying ISOBAM104 amounts show shear thinning under a higher shearing rate. The shear-thinning process benefits from the homogeneous distribution of powders and the elimination of agglomeration during the mixing process. The reason for the shear-thickening process may come from the assumption that microgelling may exist in the mixing process, bringing rheological changes to the slurry [67]. The same phenomena exist for slurries with 0.3 wt% ISOBAM, varying with different solid-loading (shown in Figure 30). The viscosities of all the slurries were <10 Pa s at a shear rate of $100 \cdot s^{-1}$, meaning the slurries are suitable for casting.

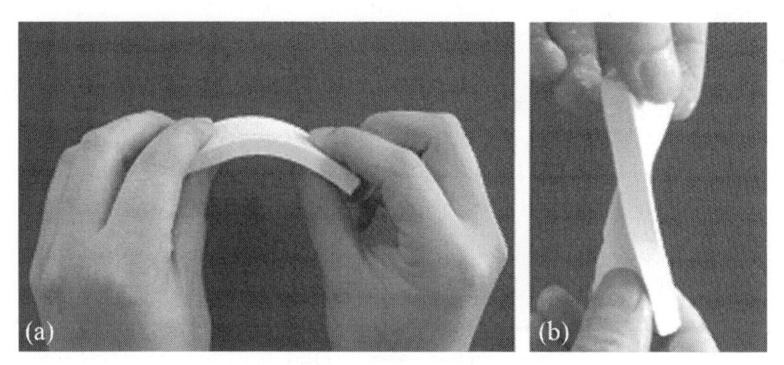

FIGURE 28 Photos of a wet Al_2O_3 gel (50 vol% solid-loading, 0.3 wt% ISOBAM) during bending (a) and twisting (b) *(From Y. Yang, et al., Room-temperature gelcasting of alumina with a water-soluble copolymer,* J. Mater. Res., *Vol. 28, No.11, P1512, 2013).*

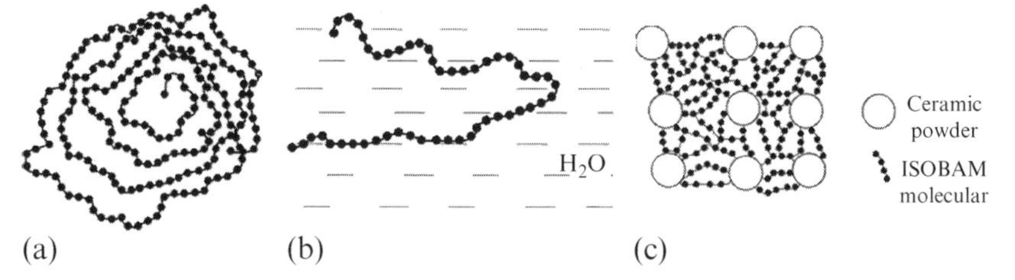

FIGURE 29 The appearance of ISOBAM at different periods (a) Solid ISOBAM; (b) ISOBAM dissolved into water; (3) ISOBAM dissolved into water and interacted with ceramic powders.

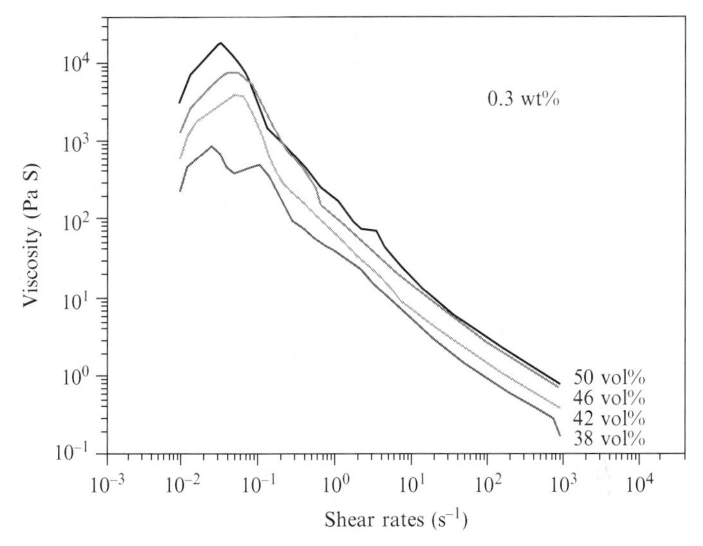

FIGURE 30 Viscosity versus solid loading for slurries (0.3 wt% ISOBAM) *(From Y. Yang, et al., Room-temperature gelcasting of alumina with a water-soluble copolymer,* J. Mater. Res., *Vol. 28 (11), P1512, 2013).*

The gelling process also was reported by comparing gelling times varying with different amounts of ISOBAM104 (Figure 31a), or different solid-loadings (Figure 31b) by testing the storage modulus as a function of time.

In Figure 31a, the storage modulus increased slowly with increasing time, revealing the formation and resultant growth of microgel with the characteristics of a fast increase in storage modulus. With increases in the ISOBAM104 concentration from 0.3 to 1.0 wt%, the gelation rate decreases, characterized by a slow increase of the storage modulus throughout the whole process. For example, for a 1.0 wt% addition, the storage of the slurries is lower than 500 Pa at the end of the test compared with that in 0.3 wt% slurry (>6000 Pa). The source of this phenomenon could potentially be an overconcentration of ISOBAM104 molecules, which results in the molecules not being able to fully extend, producing insufficient swelling of ISOBAM104 and, ultimately, causing difficulties in hand-in-hand and twinning of ISOBAM chains. For slurries with 0.3 wt% ISOBAM under different solid-loading values, the storage modulus increased slowly at the beginning and then increased rapidly until gels formed. In general, the observed trend is that higher solid-loading results in shorter gelling times. These results provide further evidence of the participation of alumina particles in gel formation. Higher solid-loading increases the amount of alumina particles and shortens the distance between gelling components (ISOBAM, Al_2O_3 particles, H_2O).

The microstructure of an ISOBAM gelcasted Al_2O_3 dry green body was given by Yang *et al.* [67] and is shown in Figure 32. It can be seen that after drying, the alumina particles are coated with a thin layer of ISOBAM.

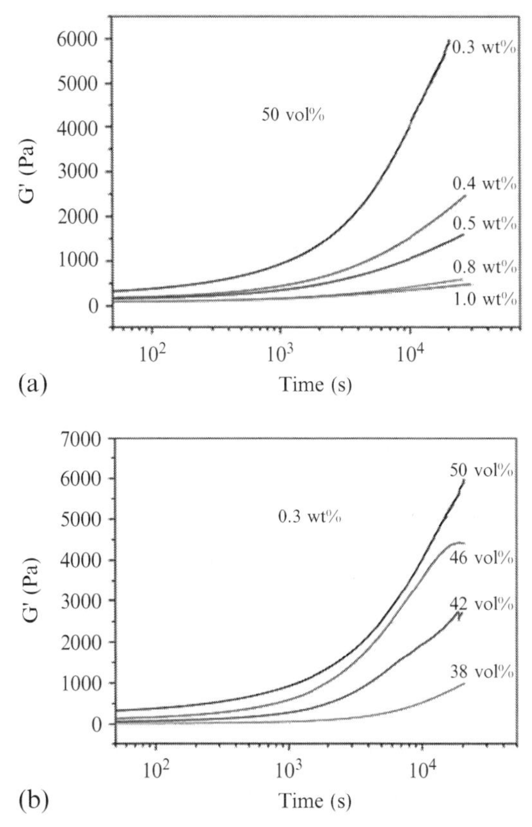

FIGURE 31 Storage modulus of slurries with different amounts of ISOBAM added (50 vol%) (a) or solid loading (0.3 wt% ISOBAM) (b) *(From Y. Yang, et al., Room-temperature gelcasting of alumina with a water-soluble copolymer,* J. Mater. Res., *Vol. 28, No.11, P1512, 2013).*

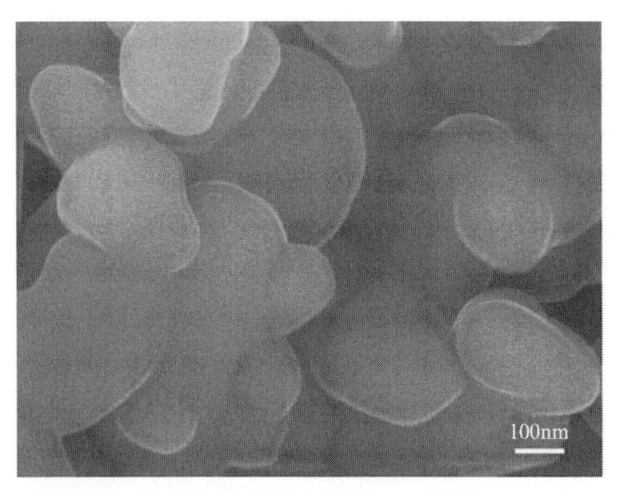

FIGURE 32 Cross section of a dried, gelcast Al_2O_3 green body (50 vol%, 0.3 wt% ISOBAM) *(From Y. Yang, et al., Room-temperature gelcasting of alumina with a water-soluble copolymer,* J. Mater. Res., *Vol. 28, No.11, P1512, 2013).*

6.2.2 Spontaneous Gelling of Porous Al_2O_3 Ceramic Using ISOBAM

Yang *et al.* [93] also succeeded in producing porous Al_2O_3 by using the spontaneous gelling agent ISOBAM104. The 3D networking structure is able to stabilize the foams, with a final porosity in the region of 22-89%. The gelling behavior of slurries (50 vol% solid-loading, 0.3 wt% ISOBAM104) also was observed by testing the storage modulus (G′) (Figure 33) with and without the addition of a surfactant. The results show that with more surfactant added, the gelling speed is increased, possibly coming from an interaction between ISOBAM104 and the surfactant (Surf-E) [93]. The viscosities of slurries (50 vol% solid-loading, 0.3 wt% ISOBAM) with the addition of varying amounts of surfactant (Surf-E) (Figure 34) are very similar to each other, exhibiting shear thinning at a high shear rate, which is beneficial for porous ceramics because mechanical foaming is improved at a high shear rate with a lower viscosity, and bubbles are stabilized at a low shear rate with a higher viscosity. This microstructure is shown in Figure 35.

6.2.3 Spontaneous Gelling of Translucent Al_2O_3 Ceramics Using ISOBAM104

Translucent Al_2O_3 was synthesized utilizing this spontaneous gelling system [91]. The same property tests (viscosity and storage modulus) were conducted for the Al_2O_3 slurry. Figure 36 shows the influence of the addition of ISOBAM104 on the viscosity of Al_2O_3 slurries (40 vol% solid-loading).

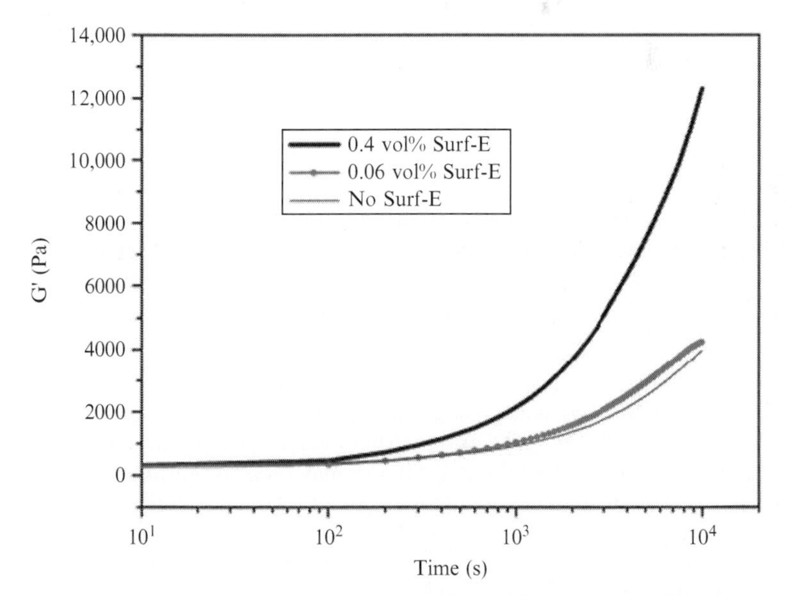

FIGURE 33 Storage modulus of slurries (50 vol%, 0.3 wt% ISOBAM) with different surfactant (Surf-E) concentrations *(From Y. Yang, et al., Fabrication of porous Al_2O_3 ceramics by rapid gelation and mechanical foaming,* J. Mater. Res., *Vol. 28, No. 15, P2012, 2013).*

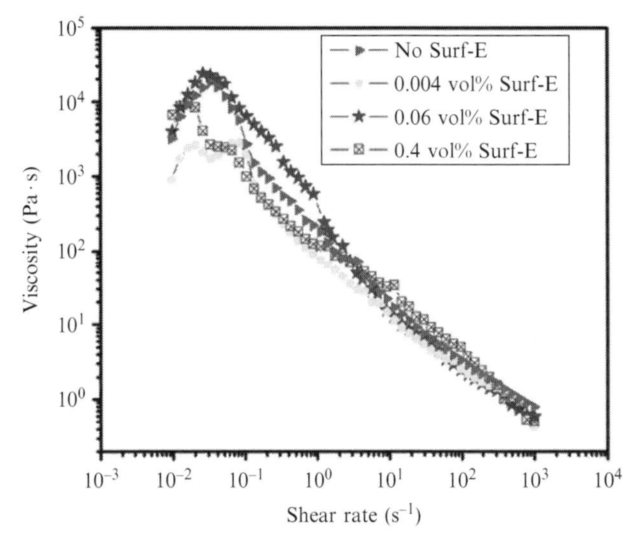

FIGURE 34 Viscosity of slurries (50 vol%, 0.3 wt% ISOBAM) with different surfactant (Surf-E) concentrations *(From Y. Yang, et al., Fabrication of porous Al_2O_3 ceramics by rapid gelation and mechanical foaming,* J. Mater. Res., *Vol. 28, No. 15, P2012, 2013).*

The viscosity increases slightly with increasing ISOBAM, from 0.3 to 2 wt%. Insufficient ISOBAM content in the slurry results in a higher viscosity, which is ascribed to the insufficient dispersal of Al_2O_3 powder in the slurry. With higher ISOBAM content (0.5-2.0 wt%), the viscosity does not significantly change. For ISOBAM104 content of 0.5 wt%, the viscosity increased with higher solid-loadings from 32 to 42 vol% (Figure 37). Regardless of the change in solid-loading or the ISOBAM added, all of the curves show similar trends, as illustrated by Figures 29, 30, and 34. Each of these figures exhibits shear thinning at a high shear rate. The viscosity of the slurries is lower than 2 Pa s at a shear rate of 100 s – 1, which is suitable for casting.

From the storage modulus data in Figure 38, it can be observed that the storage modulus increases slowly at the beginning, revealing the appearance of microgels, and then increases rapidly from 2600 s, indicating the formation of a 3D network. The final translucent Al_2O_3 product is shown in Figure 39, with its transmittance in the visible/near infrared spectra shown in Figure 40. These data suggest that the spontaneous gelling agent ISOBAM104 can be used for gelcasting of translucent or other transparent ceramics at room temperature in air. Yang *et al.* [94] applied ISOBAM104 to make translucent Al_2O_3 with complex shapes, which can be seen in Figures 41 and 42.

6.2.4 Application of a Spontaneous Gelling Agent in Water-Reactive Ceramics Powder

Shu *et al.* [92] studied gelcasting of AlN coated with waterproof paint using ISOBAM104 and succeeded in making complicated parts (Figure 36). AlN has

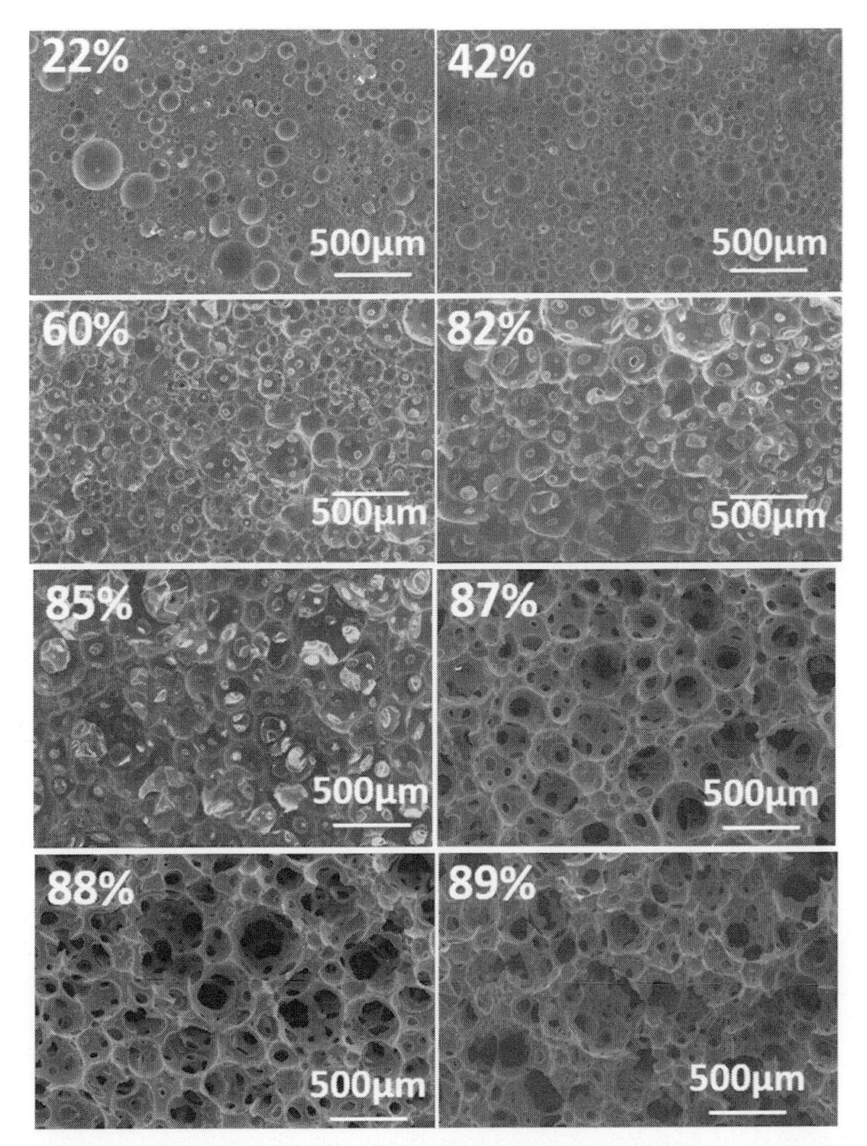

FIGURE 35 Microstructure of porous Al_2O_3 ceramics with different porosities made using the spontaneous gelling agent ISOBAM *(Modified from Y. Yang, et al., Fabrication of porous Al_2O_3 ceramics by rapid gelation and mechanical foaming,* J. Mater. Res., Vol. *28, No. 15, P2012, 2013).*

a high thermal conductivity and low dielectric property, making it a promising candidate as an electronic substrate material [95,96]. The application of AlN requires specific shapes, such as a sheets arrayed fin figuration [97] (Figure 43). The microstructure shows AlN produced by this technique is very dense and rarely contains bubbles (Figure 44).

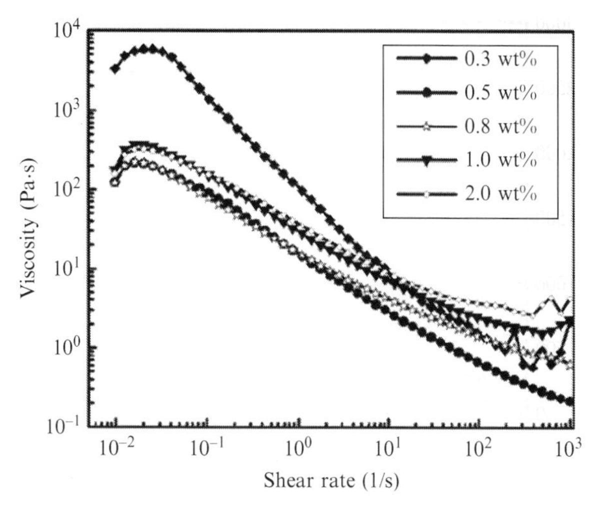

FIGURE 36 Viscosity of slurries (40 vol%) with different amounts of ISOBAM added *(From S. Shunzo, et al., Spontaneous glecasting of translucent alumina ceramics,* Opt. Mater. Express, *Vol. 3, No. 8, P1000, 2013).*

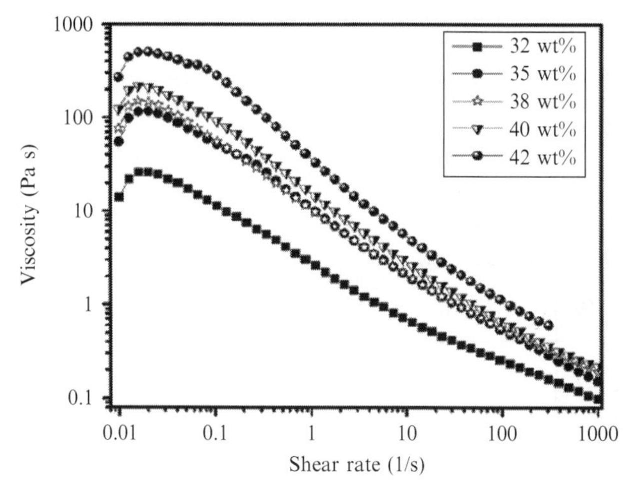

FIGURE 37 Viscosity of slurries (0.5 wt% ISOBAM) with different solid loading *(From S. Shunzo, et al., Spontaneous glecasting of translucent alumina ceramics,* Opt. Mater. Express, *Vol. 3, No. 8, P1000, 2013).*

Shu *et al.* [92] also studied the viscosities of slurries with different solid-loading values (Figure 45) and ISOBAM concentrations (Figure 46). All of the slurries exhibited shear-thinning behavior at a shear rate $< 100\,s-^{1}$, followed by a shear-thickening trend at the highest rates. The explanation provided for the shear-thickening phenomenon was that, as the shear rate increased, the hydrodynamic forces became increasingly dominant and thus affected the suspension structure [98]. This led to an increase in the number of collisions per unit time.

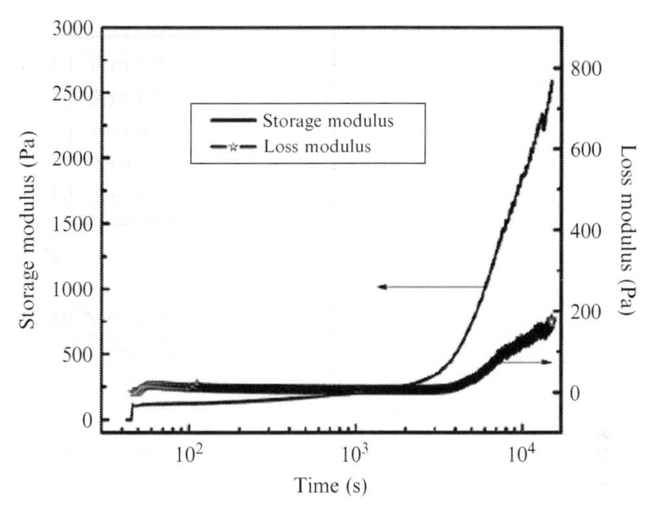

FIGURE 38 Storage and loss moduli of alumina slurry (42 vol%, 0.5 wt% ISOBAM) *(From S. Shunzo, et al., Spontaneous glecasting of translucent alumina ceramics,* Opt. Mater. Express, *Vol. 3, No. 8, P1000, 2013).*

FIGURE 39 Photo of translucent Al_2O_3 made using the spontaneous gelling agent ISOBAM *(From S. Shunzo, et al., Spontaneous glecasting of translucent alumina ceramics,* Opt. Mater. Express, *Vol. 3, No. 8, P1000, 2013).*

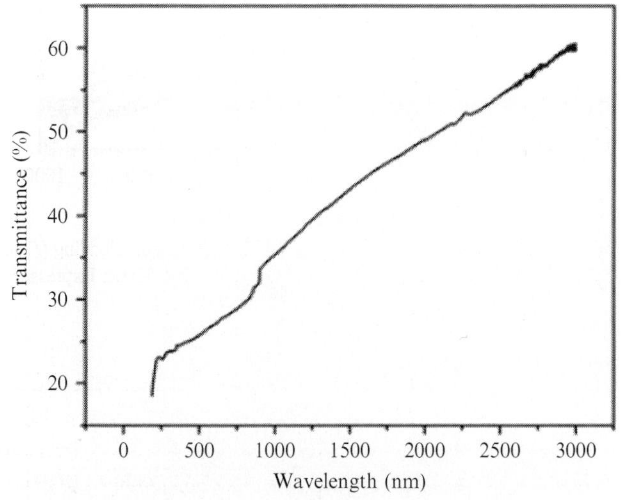

FIGURE 40 Transmittance of translucent Al_2O_3 made using the spontaneous gelling agent ISOBAM (thickness, 1 mm) *(From S. Shunzo, et al., Spontaneous glecasting of translucent alumina ceramics,* Opt. Mater. Express, *Vol. 3, No. 8, P1000, 2013).*

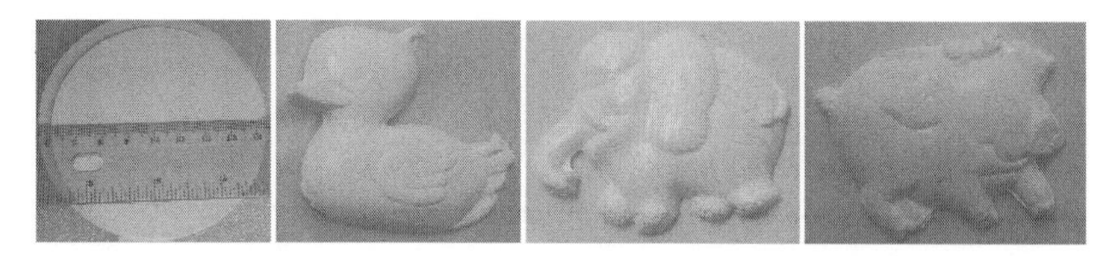

FIGURE 41 Photos of green bodies with complicated shapes gelcasted using ISOBAM.

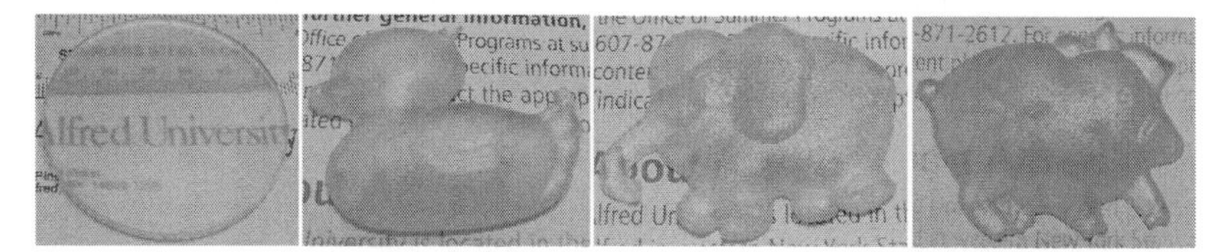

FIGURE 42 Photos of ISOBAM-gelcast and vacuum-sintered Al_2O_3 with complicated shapes.

FIGURE 43 Photo of AlN fabricated by spontaneous gelcasting (ISOBAM) *(From X. Shu, Gelcasting of aluminum nitride using a water-soluble copolymer,* J. Inorg. Mater., *Vol. 29, No. 3, P327, 2014).*

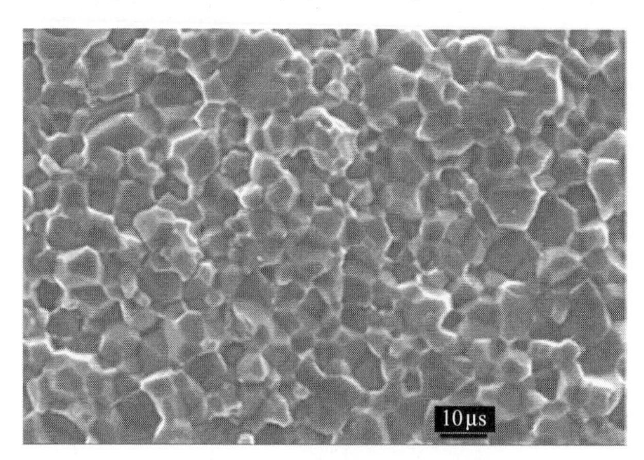

FIGURE 44 Cross section of the microstructure of sintered AlN fabricated by spontaneous gelcasting (ISOBAM) *(From X. Shu, Gelcasting of aluminum nitride using a water-soluble copolymer,* J. Inorg. Mater., *Vol. 29, No. 3, P327, 2014).*

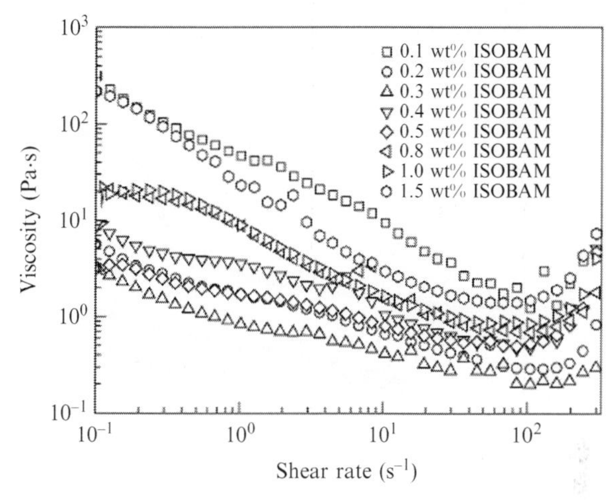

FIGURE 45 Viscosity of AlN slurries with various ISOBAM concentrations (52 vol% solid-loading) *(From X. Shu, Gelcasting of aluminum nitride using a water-soluble copolymer,* J. Inorg. Mater., *Vol. 29, No. 3, P327, 2014).*

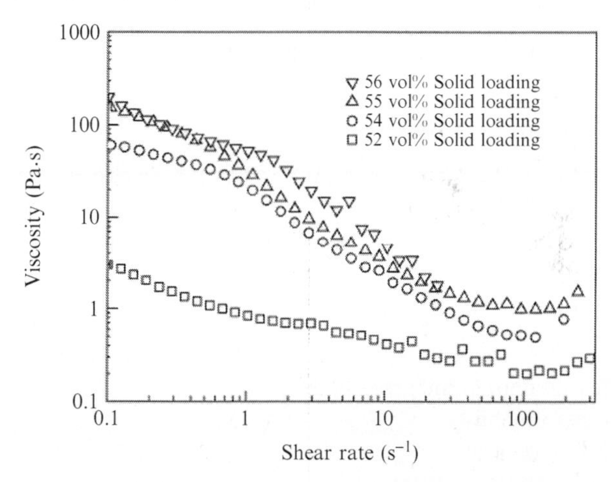

FIGURE 46 Viscosity of AlN slurries with different solid loading (0.3 wt% ISOBAM) *(From X. Shu, Gelcasting of aluminum nitride using a water-soluble copolymer,* J. Inorg. Mater., *Vol. 29, No. 3, P327, 2014).*

As a result, the particles slide over each other under higher shear stress as a result of increasing average separation distance, particularly when the particles are coarse and there is high solid-loading.

There is also an interesting result when testing the storage modulus: Y_2O_3, the minimally added sintering aid for AlN, has an obvious influence on the gelling process (Figure 47). Uncoated Y_2O_3 as the sintering aid shortens the gelling time for the whole process. The reason is that Y_2O_3 easily reacts with water to form $Y(OH)_3$, and the hydroxyl groups on the surface of Y_2O_3 powders can interact with functional groups of ISOBAM104, increasing the rate of 3D network forming.

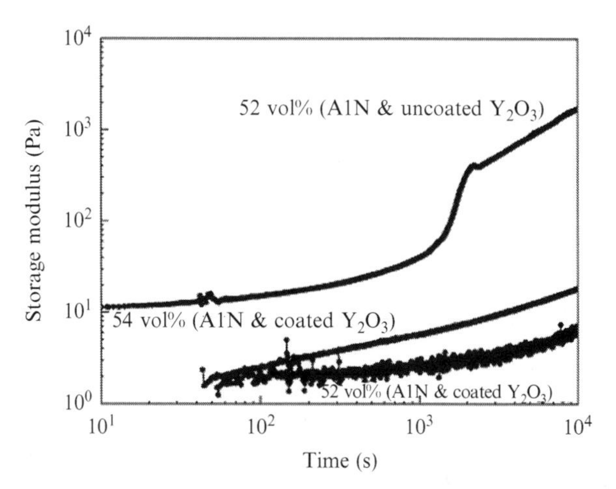

FIGURE 47 Change of storage modulus during gelation *(From X. Shu, Gelcasting of aluminum nitride using a water-soluble copolymer,* J. Inorg. Mater., *Vol. 29, No. 3, P327, 2014).*

7 APPLICATION OF ISOBAM AS BINDER FOR TAPE CASTING

7.1 Mechanism of a Binder

Organic binders are long-chain polymer molecules dissolved in aqueous solvents and used to process many commercial ceramics by providing strength to green bodies to be molded and kept in their desired shape before firing. After drying off the solvent, the binders are kept in the body to provide organic bridges between the ceramic particles. The ideal binder system for ceramic processing has the following properties:

(1) Provides strength to the green body
(2) Enhances formability
(3) Provides lubrication
(4) Minimizes contamination
(5) Is nontoxic and inexpensive

Atoms (mainly carbon) of long-chain binder molecules form covalent bonds with neighbors, producing a backbone. The functional groups attached to the backbone of these molecules play important roles during binder application. The characteristic features of binders include the number of active groups present, molecular chain length, molecular weight, and atomic structure. These characteristic features determine the resultant properties such as viscosity, cohesion, and bonding strength. The binder effect lies in the following aspects [99]:

1) *The coordination of functional groups on the solid surface.* The coordination effect is decided by the kinds and number of adsorbed active groups. The number of these active groups is roughly proportional to the chain length of

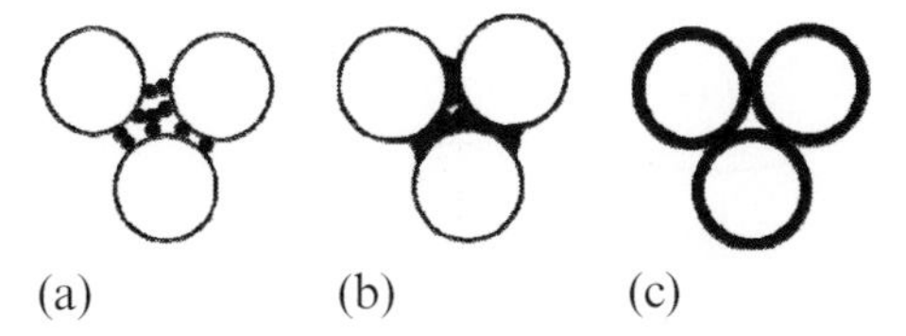

FIGURE 48 Wetting behavior of a binder on a solid surface: nonwetting (a); wetting (pendular ring state) (b); and coating liquid (c).

the molecules. As a result, a higher molecular weight is achieved, along with a stronger bonding ability [79,99].

2) *The wetting ability of the binder to the solid surface.* Onoda [100] introduces three kinds of binder situations between a binder and a solid surface: (1) a nonwetting state (Figure 48a); (2) a wetting (pendular ring) state (Figure 48b); (3) a coated state (Figure 48c). The pendular ring state produces the highest green strength.

The addition of a binder also has important influences on rheological properties, including viscosity, pseudo-plasticity, and gelation. Onoda [100] illustrated the relationships between the rheological variables and the forming properties:

(1) Viscosity is one of the primary considerations for binder application.
(2) The pseudo-plasticity of a binder helps prevent solid particles from settling in the suspension and also helps maintain sufficient fluidity during the process.
(3) The gelled structure maintains the homogeneous distribution of the binder and ceramics powder during the subsequent drying process.

The amount of binder added also influences the strength, drying, and debinder process. The addition of a binder increases from dry pressing (<3 wt%) to tape casting (~3-17 wt%) [101]. Enough binder must be loaded to guarantee the strength for the following production process. However, too much binder causes the particle separation distance to become too long and decreases the green body strength. In addition, green body defects such as deformation, sagging, or blistering, occur if there is too much binder added. Onoda [100] found that the best binder addition amount is between 8 and 15 vol% for particle volume.

7.2 Application of ISOBAM104 for Tape Casting of Traditional Al_2O_3, Translucent Al_2O_3, and Transparent YAG

7.2.1 Application of ISOBAM for Tape Casting of Traditional Al_2O_3

From various analyses it is known that ISOBAM104 can work as a dispersant. ISOBAM104 can also adsorb to ceramics powder surfaces and bind them together, working as a binder for ceramics processing. The dispersant properties

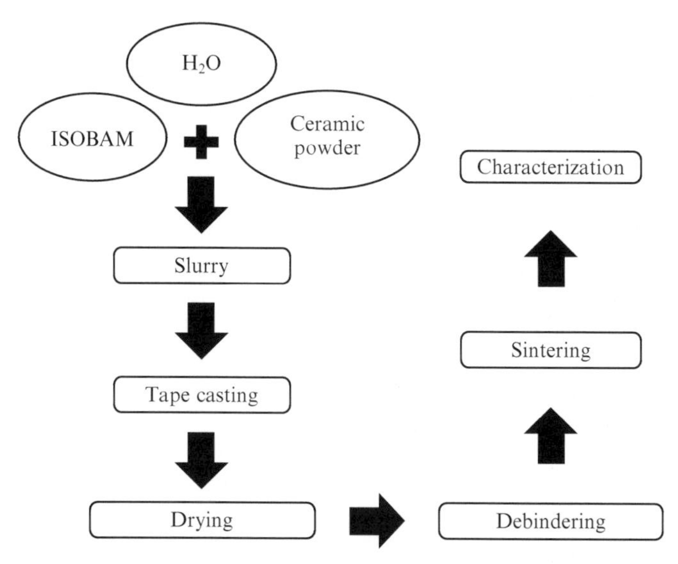

FIGURE 49 Tape casting procedure using ISOBAM as the binder.

and binder behavior applied in ceramic tape casting techniques simplify the whole production process (Figure 49).

Using ISOBAM104 as a binder for tape casting, Yan in Dr. Yiquan Wu's Group (Alfred University) carried out several studies [102][103] of binder addition and plasticizer choice and finally succeeded in making large ceramic tape with good plasticity (Figure 50).

7.2.2 Application of ISOBAM for Tape Casting of Translucent Al_2O_3

Yan Yang applied the same ISOBAM tape-casting techniques to fabricate translucent Al_2O_3 [102]. Photos and the microstructures are shown in Figures 51–54. Adopting these ISOBAM104 tape-casting techniques, the green ceramic tape, with a thickness of 0.71 mm and good plasticity, was synthesized (Figure 44). After vacuum sintering, translucent Al_2O_3 with a thickness of 0.66 mm was achieved (Figure 45). By examining the microstructure and cross section of the translucent Al_2O_3 surface (Figure 46), it can be observed that the ISOBAM

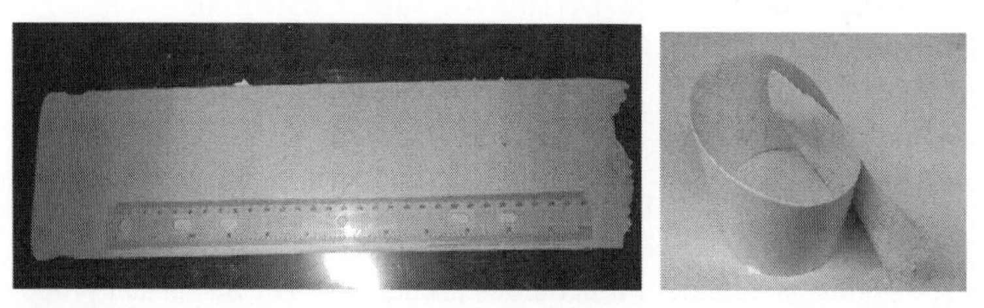

FIGURE 50 Photos of tape-casted Al_2O_3 green tape using ISOBAM.

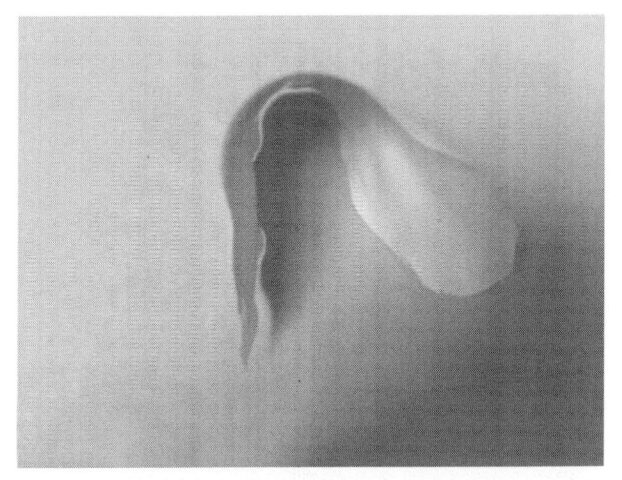

FIGURE 51 Photos of a tape-casted Al_2O_3 green body (thickness, 0.71 mm) using ISOBAM.

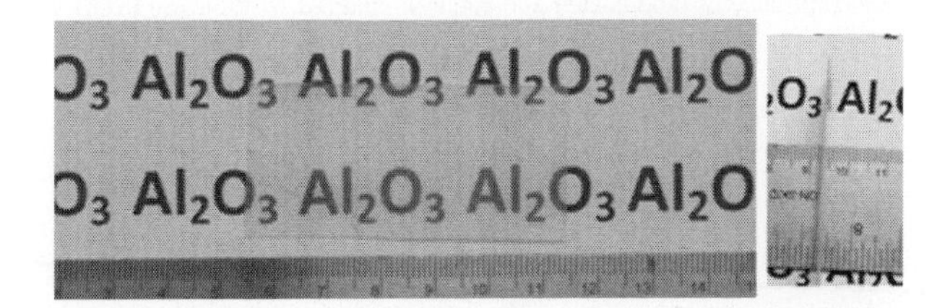

FIGURE 52 Photos of tape-casted and vacuum-sintered Al_2O_3 using ISOBAM.

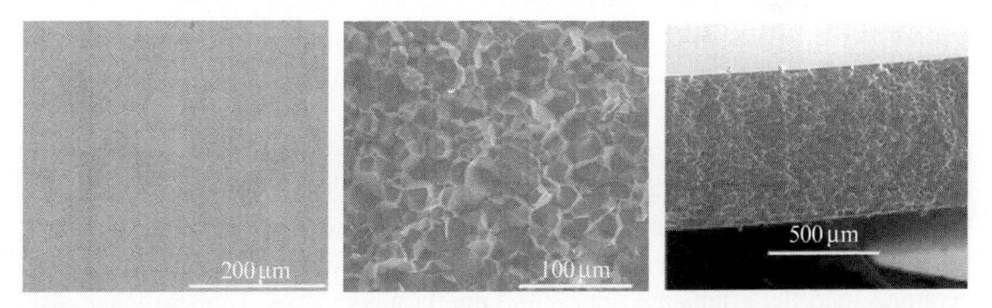

FIGURE 53 Microstructure of tape-casted and vacuum-sintered Al_2O_3 using ISOBAM: the surface (left) and the cross section (middle and right).

tape-casted Al_2O_3 is very dense, without any pores present on the surface or in the cross section. The grain size of ISOBAM104 tape-casted and vacuum-sintered Al_2O_3 is about 40-50 μm, with a transmittance higher than 70% at 5 μm; this decreases to zero when the wavelength approaches 7.5 μm.

7.2.3 Application of ISOBAM for Tape Casting of Traditional YAG

Yan Yang (Dr. Yiquan Wu's Research Group at Alfred University) [103] produced transparent YAG using ISOBAM tape-casting techniques. Extremely

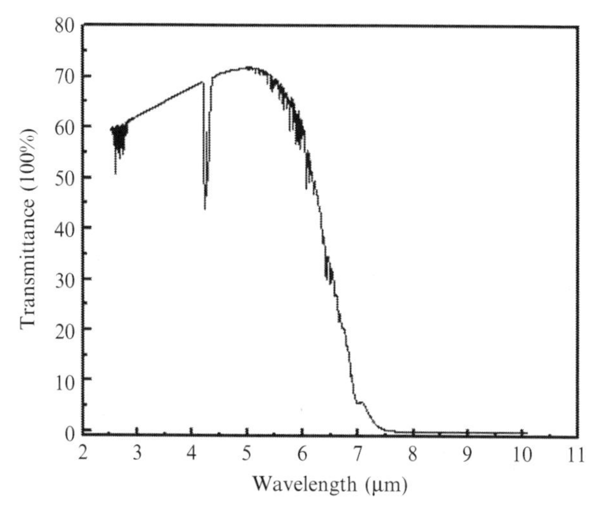

FIGURE 54 Transmittance of tape-casted and vacuum-sintered Al_2O_3 using ISOBAM.

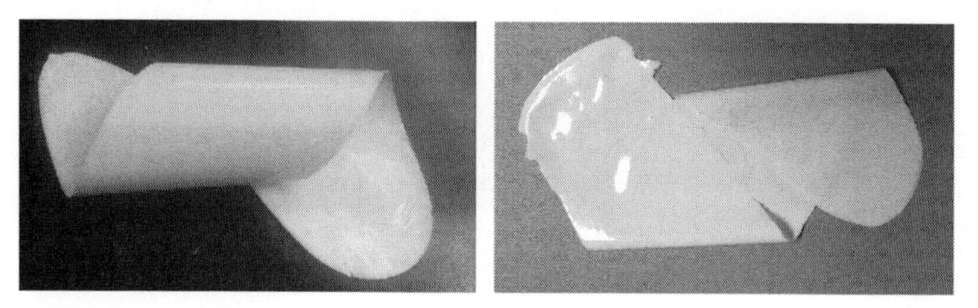

FIGURE 55 Photos of tape-casted YAG green body using ISOBAM at thicknesses of 0.71 mm (left) and 0.26 mm (right).

thin, plastic YAG tape (Figure 55) with a flat and shiny surface can be obtained. After vacuum sintering, transparent YAG (Figure 56) with the small grain size (2-6 μm) was made. As can be seen in the microstructure in Figure 57, there are no bubbles present on the surface or in the fractured bulk of transparent YAG ceramics, and the grains are very small and homogeneous. Transmittance also was examined (Figure 58). The results show that the transmittance is >90% at 2.45-5.9 μm and decreases to zero when the wavelength approaches 11 μm.

8 OTHER RESEARCH ON SPONTANEOUS GELLING SYSTEMS

Sun *et al.* [102] developed another short-molecule, environmentally friendly dispersant (ISOBAM600AF) for high-strength Al_2O_3. The high-strength Al_2O_3 ceramics were made by a high solid-loading (58 vol%) aqueous slurry with water-soluble copolymer ISOBAM104 combined with ISOABAM600AF, both acting as gelling agents and dispersants. ISOBAM600AF is a copolymer of

FIGURE 56 Photos of ape-casted and vacuum-sintered YAG using ISOBAM.

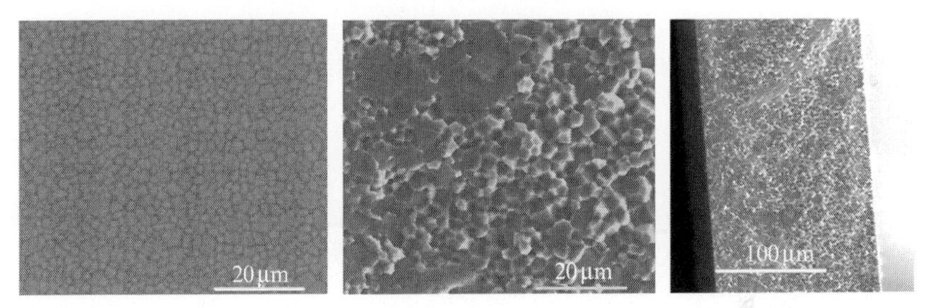

FIGURE 57 Microstructure of tape-casted and vacuum-sintered YAG using ISOBAM: the surface (left) and the cross section (middle and right).

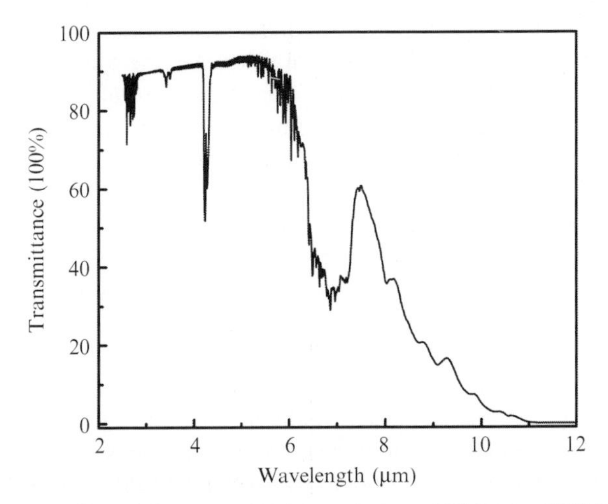

FIGURE 58 Transmittance of tape-casted and vacuum-sintered YAG using ISOBAM.

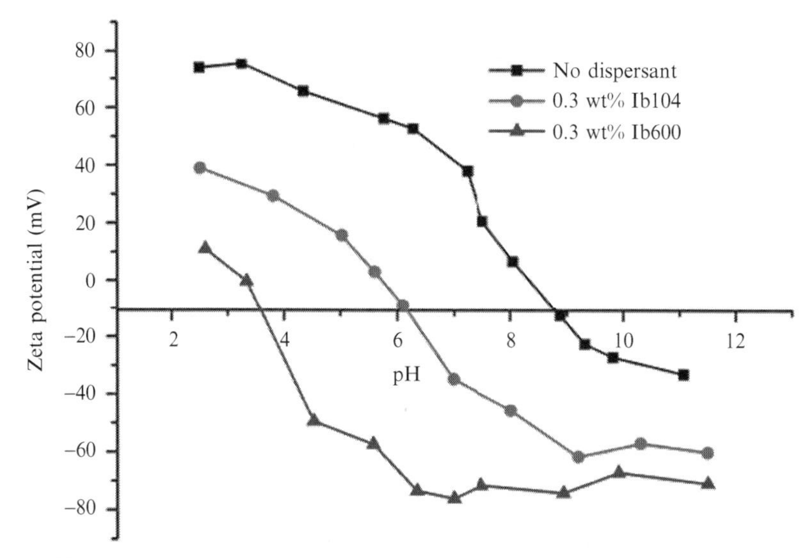

FIGURE 59 ζ-Potential of ISOBAM and ISOBAM600 in Al_2O_3 slurries *(From Y. Sun, et al., A method for gelcasting high strength alumina ceramics with low shrinkage,* J. Mater. Res., *Vol. 29, No. 2, P247, 2014).*

isobutylene and maleic anhydride, with a smaller molecular length (5000-6000) and better dispersibility than ISOBAM. The molecular structure is similar to that of ISOBAM104 (Figure 10).

The dispersibility of ISOBAM600AF is examined by ζ-potential varying with pH values, as shown in Figure 59. The IEP of pure Al_2O_3 with no dispersant was pH 8.7. With 0.3 wt% ISOBAM104, the IEP moved to pH 4.8, close to the results of Yang *et al.* [67] With the addition of 0.3 wt% ISOBAM600AF, the IEP moved further to pH 3.3, showing a more obvious influence on the alumina powder surfaces. The absolute ζ-potential value over the whole testing range is higher than that of ISOBAM104. The reason for these results may be explained by the shorter molecular chain of ISOBAM600AF, which as a smaller molecular weight, better dispersibility, and thinner covering layers on ceramics powder surfaces [105].

The viscosity test in Figure 60 shows that all the slurries have shear-thinning behavior with an increasing shear rate. ISOBAM600AF in the region of 0.15-0.25% with 56 vol% solid-loading produces the lowest viscosity, much lower than that of ISOBAM104 at 0.3 wt% and 50 vol% solid-loading. This result further exhibits the excellent dispersaive role of ISOBAM600AF in the aqueous slurry.

The gelling behavior of aqueous slurries using ISOBAM600AF/ISOBAM104 was characterized by the storage modulus in Figure 61. Just as explained by Sun *et al.* [106], the gelling time was compared for three slurries: (1) 50 vol%-0.3 wt% ISOBAM104; (2) 56 vol%-0.2 wt% ISOBAM600AF; and (3) 56 vol%-0.2 wt% ISOBAM600AF and 0.1 wt% ISOBAM104, with the following results:

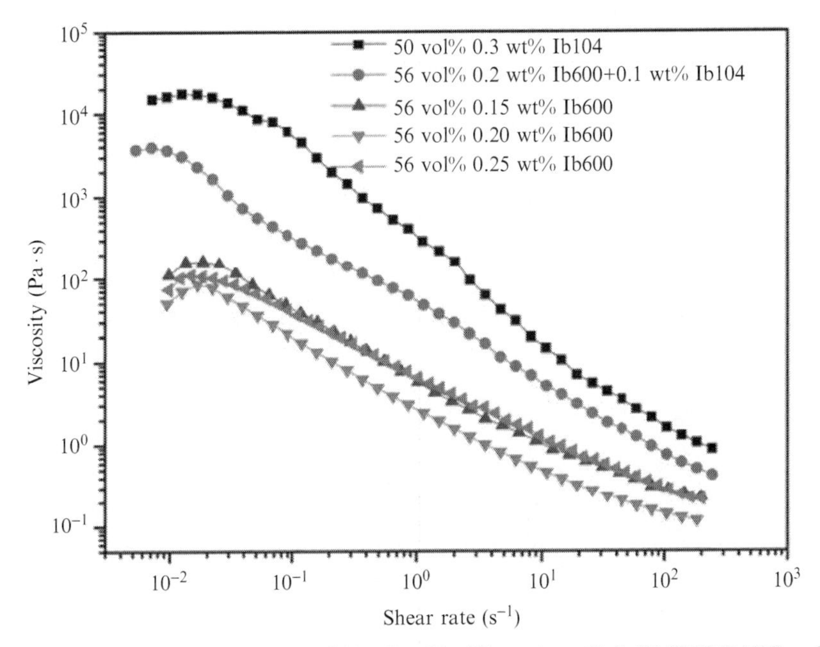

FIGURE 60 Viscosity of slurries (solid loading 56vol%) varying with Ib600(ISOBAM600) and Ib104 (ISOBAM104) *(From Y. Sun, et al., A method for gelcasting high strength alumina ceramics with low shrinkage,* J. Mater. Res., *Vol. 29, No. 2, P247, 2014).*

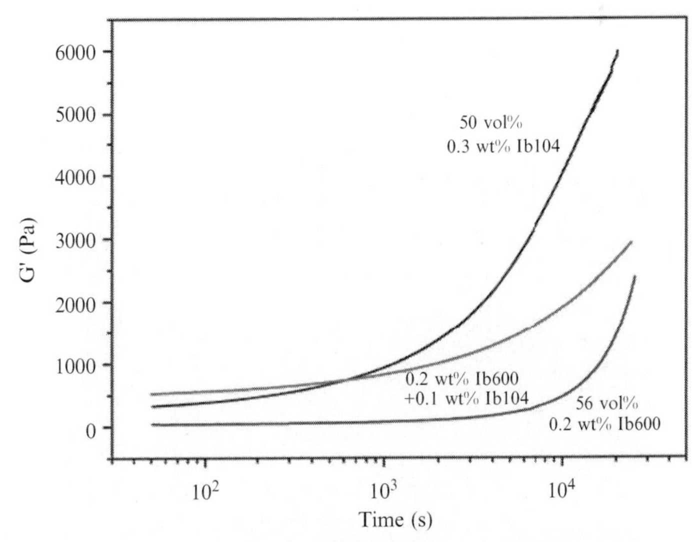

FIGURE 61 Storage modulus (G') of slurries varying with Ib600(ISOBAM600) and Ib104 (ISOBAM104) *(From Y. Sun, et al., A method for gelcasting high strength alumina ceramics with low shrinkage,* J. Mater. Res., *Vol. 29, No. 2, P247, 2014).*

$$50\,\text{vol\%} - 0.3\,\text{wt\%}\ \text{ISOBAM104} < 56\,\text{vol\%} - 0.2\,\text{wt\%}\ \text{ISOBAM600AF}$$
$$< 56\,\text{vol\%} - 0.2\,\text{wt\%}\ \text{ISOBAM600AF, and } 0.1\,\text{wt\%}\ \text{ISOBAM104}$$

These results indicate that the gelation ability of ISOBAM104 is better than that of ISOBAM600AF, resulting from the differences in molecular chain length.

A water-soluble, nontoxic, and environmentally friendly copolymer, ISOBAM104/ISOBAM600AF was used by Sun *et al.* [104] to produce transparent Y_2O_3. By adding ISOBAM600AF, slurry with 81 wt% solid-loading was made for preparing and casting. The final ceramics is shown in Figure 62, together with its microstructure (Figure 63). It can be seen that there are almost no pores on the fracture surface of the gelcasted and vacuum-sintered Y_2O_3. By testing the transmittance of fabricated Y_2O_3 (Figure 64), gelcasted and vacuum-sintered Y_2O_3 has very high transmittance (80.9%) at 1100 nm, close to the theoretical value of Y_2O_3.

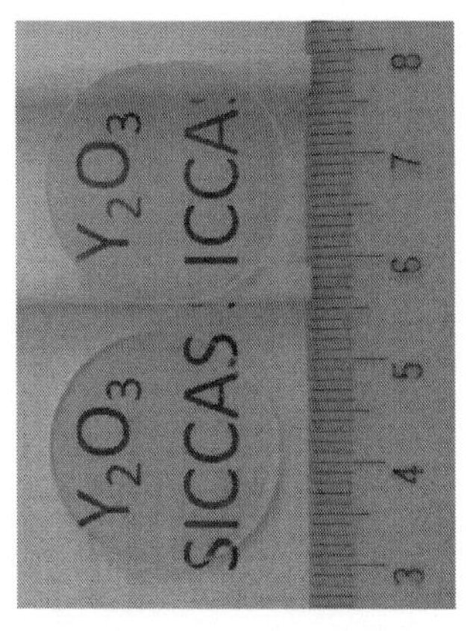

FIGURE 62 Photo of Y_2O_3 made by gelcasting and vacuum sintering using ISOBAM/ISOBAM600 *(From Y. Sun, et al., Fabrication of transparent Y_2O_3 ceramics via aqueous gelcasting,* Ceram. Int., *Vol. 40, No. 6, P8841, 2014).*

FIGURE 63 Microstructure of Y_2O_3 made by gelcasting and vacuum sintering using ISOBAM/ISOBAM600 *(From Y. Sun, et al., Fabrication of transparent Y_2O_3 ceramics via aqueous gelcasting,* Ceram. Int., *Vol. 40, No. 6,P8841, 2014).*

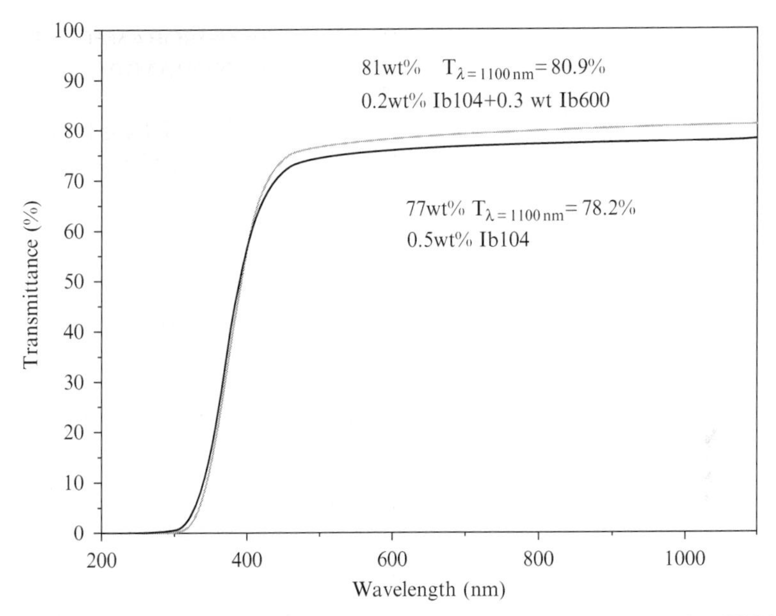

FIGURE 64 Transmittance of Y_2O_3 made by gelcasting and vacuum sintering using ISOBAM (Ib)/ISOBAM600 *(From Y. Sun, et al., Fabrication of transparent Y_2O_3 ceramics via aqueous gelcasting,* Ceram. Int., *Vol. 40, No. 6, P8841, 2014).*

REFERENCES

[1] P. Falkowski, P. Bednarek, A. Danelska, T. Mizerski, M. Szafran, Application of monosaccharides derivatives in colloidal processing of aluminum oxide, J. Eur. Ceram. Soc. 30 (14) (2010) 2805–2811.

[2] G.V. Franks, B.V. Velamakanni, F.F. Lange, Vibraforming and in situ flocculation of consolidated, coagulated, alumina slurries, J. Am. Ceram. Soc. 78 (5) (1995) 1324–1328.

[3] M. Kokabi, A.A. Babaluo, A. Barati, Gelation process in low-toxic gelcasting systems, J. Eur. Ceram. Soc. 26 (15) (2006) 3083–3090.

[4] K. Prabhakaran, R. Sooraj, A. Melkeri, N. Gokhale, S. Sharma, A new direct coagulation casting process for alumina slurries prepared using poly (acrylate) dispersant, Ceram. Int. 35 (3) (2009) 979–985.

[5] J. Yang, J. Yu, Y. Huang, Recent developments in gelcasting of ceramics, J. Eur. Ceram. Soc. 31 (14) (2011) 2569–2591.

[6] M.A. Janney, O.O. Omatete, C.A. Walls, S.D. Nunn, R.J. Ogle, G. Westmoreland, Development of low-toxicity gelcasting systems, J. Am. Ceram. Soc. 81 (3) (1998) 581–591.

[7] O.O. Omatete, M.A. Janney, S.D. Nunn, Gelcasting: from laboratory development toward industrial production, J. Eur. Ceram. Soc. 17 (2) (1997) 407–413.

[8] M.A. Janney, O.O. Omatete, Method for molding ceramic powders using a water-based gel casting, Google Patents, 1991.

[9] O.O. Omatete, M.A. Janney, R.A. Strehlow, Gelcasting: a new ceramic forming process, Am. Ceram. Soc. Bull. 70 (10) (1991) 1641–1649.

[10] R. Gilissen, J.P. Erauw, A. Smolders, E. Vanswijgenhoven, J. Luyten, Gelcasting, a near net shape technique, Mater. Des. 21 (4) (2000) 251–257.

[11] F. Ortega, F. Valenzuela, C. Scuracchio, V. Pandolfelli, Alternative gelling agents for the gelcasting of ceramic foams, J. Eur. Ceram. Soc. 23 (1) (2003) 75–80.
[12] C.H. Schilling, P. Tomasik, C. Li, M. Sikora, Protein plasticizers for aqueous suspensions of micrometric- and nanometric-alumina powder, Mater. Sci. Eng. A 336 (1) (2002) 219–224.
[13] I.M. Krieger, T.J. Dougherty, A mechanism for non-Newtonian flow in suspensions of rigid spheres, Trans. Soc. Rheol. 3 (1) (1959) 137–152.
[14] D. Quemada, J. Casas-Vasquez, G. Lebon, Stability of thermodynamic systems, Lecture Notes in Physics, Springer, Berlin, 1982, pp. 210-247.
[15] O.O. Omatete, A.C. Young, M.A. Janney, J. Adair, Investigation of Dilute Gelcasting Alumina Suspensions, Oak Ridge National Lab, TN, USA, 1990.
[16] A. Bleier, O. Omatete, C. Westmoreland, Rheology of zirconia-alumina gelcasting slurries, MRS Proceedings 271 (1992) 269.
[17] A. Bleier, O. Omatete, Rheology and microstructure of concentrated zirconia-alumina suspensions for gelcasting composites, MRS Proceedings 289 (1992) 109.
[18] L. Zhao, J.-L. Yang, L.-G. Ma, Y. Huang, Influence of minute metal ions on the idle time of acrylamide polymerization in gelcasting of ceramics, Mater. Lett. 56 (6) (2002) 990–994.
[19] J.A. Lewis, Colloidal processing of ceramics, J. Am. Ceram. Soc. 83 (10) (2000) 2341–2359.
[20] J. Ma, Z. Xie, H. Miao, Y. Huang, Y. Cheng, W. Yang, Gelcasting of alumina ceramics in the mixed acrylamide and polyacrylamide systems, J. Eur. Ceram. Soc. 23 (13) (2003) 2273–2279.
[21] L. Bergström, Shear thinning and shear thickening of concentrated ceramic suspensions, Colloids Surf. A Physicochem. Eng. Asp. 133 (1-2) (1998) 151–155.
[22] J. Reed, Principles of Ceramic Processing, second ed., John Wiley & Sons, New York, NY, 1995.
[23] J. Bender, N.J. Wagner, Reversible shear thickening in monodisperse and bidisperse colloidal dispersions, J. Rheol. 40 (1996) 899–916.
[24] R. Hoffman, Discontinuous and dilatant viscosity behavior in concentrated suspensions. I. Observation of a flow instability, Trans. Soc. Rheol. 16 (1) (1972) 155–173 (1957-1977).
[25] M. Cross, A. Kaye, Simple procedures for obtaining viscosity/shear rate data from a parallel disc viscometer, Polymer 28 (3) (1987) 435–440.
[26] M. Cross, Relation between viscoelasticity and shear-thinning behaviour in liquids, Rheol. Acta 18 (5) (1979) 609–614.
[27] A. Nojoomi, M.A. Faghihi-Sani, M. Khoshkalam, Shear-rate dependence modeling of gelcast slurries: effects of dispersant content and solid loading, Ceram. Int. 40 (1, Part A) (2014) 123–128.
[28] H. Watanabe, Viscoelasticity and dynamics of entangled polymers, Prog. Polym. Sci. 24 (9) (1999) 1253–1403.
[29] T. Zhang, Z. Zhang, J. Zhang, D. Jiang, Q. Lin, Preparation of SiC ceramics by aqueous gelcasting and pressureless sintering, Mater. Sci. Eng. A 443 (1-2) (2007) 257–261.
[30] S.L. Morissette, J.A. Lewis, Chemorheology of aqueous-based alumina-poly (vinyl alcohol) gelcasting suspensions, J. Am. Ceram. Soc. 82 (3) (1999) 521–528.
[31] W.H. Boersma, J. Laven, H.N. Stein, Shear thickening (dilatancy) in concentrated dispersions, AIChE J. 36 (3) (1990) 321–332.
[32] W.H. Boersma, J. Laven, H.N. Stein, Viscoelastic properties of concentrated shear-thickening dispersions, J. Colloid Interface Sci. 149 (1) (1992) 10–22.
[33] W.M. Sigmund, N.S. Bell, L. Bergström, Novel powder-processing methods for advanced ceramics, J. Am. Ceram. Soc. 83 (7) (2000) 1557–1574.
[34] L. Bergström, Colloidal processing of ceramics, in: Handbook of Applied Surface and Colloidal Chemistry, John Wiley & Sons Ltd, West Sussex, UK, 2001.

[35] J.N. Israelachvil, Intermolecular and Surface Forces, revised third ed., Academic Press, Amsterdam, 2011.

[36] L. Bergström, Hamaker constants of inorganic materials, Adv. Colloid Interface Sci. 70 (1997) 125–169.

[37] P. De Gennes, Polymers at an interface; a simplified view, Adv. Colloid Interface Sci. 27 (3) (1987) 189–209.

[38] J.A. Lewis, H. Matsuyama, G. Kirby, Polyelectrolyte effects on the rheological properties of concentrated cement suspensions, J. Am. Ceram. Soc. 83 (8) (2000) 1905–1913.

[39] H. Guldberg-Pedersen, L. Bergström, Stabilizing ceramic suspensions using anionic polyelectrolytes: adsorption kinetics and interparticle forces, Acta Mater. 48 (18-19) (2000) 4563–4570.

[40] C. Tallon, R. Moreno, M.I. Nieto, D. Jach, G. Rokicki, M. Szafran, Gelcasting performance of alumina aqueous suspensions with glycerol monoacrylate: a new low-toxicity acrylic monomer, J. Am. Ceram. Soc. 90 (5) (2007) 1386–1393.

[41] M. Potoczek, E. Zawadzak, Initiator effect on the gelcasting properties of alumina in a system involving low-toxic monomers, Ceram. Int. 30 (5) (2004) 793–799.

[42] G.W. Scherer, Theory of drying, J. Am. Ceram. Soc. 73 (1) (1990) 3–14.

[43] S. Ghosal, S. Ghosal, A. Emami Naeini, Y.P. Harn, B.S. Draskovich, J.P. Pollinger, A physical model for the drying of gelcast ceramics, J. Am. Ceram. Soc. 82 (3) (1999) 513–520.

[44] A. Barati, M. Kokabi, M.H.N. Famili, Drying of gelcast ceramic parts via the liquid desiccant method, J. Eur. Ceram. Soc. 23 (13) (2003) 2265–2272.

[45] Y. Huang, L. Ma, H. Le, J. Yang, Improving the homogeneity and reliability of ceramic parts with complex shapes by pressure-assisted gel-casting, Mater. Lett. 58 (30) (2004) 3893–3897.

[46] A. Barati, M. Kokabi, N. Famili, Modeling of liquid desiccant drying method for gelcast ceramic parts, Ceram. Int. 29 (2) (2003) 199–207.

[47] N. Kayaman, O. Okay, B.M. Baysal, Phase transition of polyacrylamide gels in PEG solutions, Polym. Gels Networks 5 (2) (1997) 167–184.

[48] E.R. Kupp, G.L. Messing, J.M. Anderson, V. Gopalan, J.Q. Dumm, C. Kraisinger, N. Ter-Gabrielyan, L.D. Merkle, M. Dubinskii, V.K. Simonaitis-Castillo, Co-casting and optical characteristics of transparent segmented composite Er: YAG laser ceramics, J. Mater. Res. 25 (03) (2010) 476–483.

[49] S.H. Lee, E.R. Kupp, A.J. Stevenson, J.M. Anderson, G.L. Messing, X. Li, E.C. Dickey, J.Q. Dumm, V.K. Simonaitis Castillo, G.J. Quarles, Hot isostatic pressing of transparent Nd: YAG ceramics, J. Am. Ceram. Soc. 92 (7) (2009) 1456–1463.

[50] A. Krell, P. Blank, M. Hongwei, T. Hutzler, Transparent sintered corundum with high hardness and strength, J. Am. Ceram. Soc. 86 (1) (2003) 12.

[51] L. Jin, G. Zhou, S. Shimai, J. Zhang, S. Wang, ZrO_2-doped Y_2O_3 transparent ceramics via slip casting and vacuum sintering, J. Eur. Ceram. Soc. 30 (10) (2010) 2139–2143.

[52] K.A. Appiagyei, G.L. Messing, J.Q. Dumm, Aqueous slip casting of transparent yttrium aluminum garnet (YAG) ceramics, Ceram. Int. 34 (5) (2008) 1309–1313.

[53] A. Ikesue, T. Kinoshita, K. Kamata, K. Yoshida, Fabrication and optical properties of high-performance polycrystalline Nd: YAG ceramics for solid-state lasers, J. Am. Ceram. Soc. 78 (4) (1995) 1033–1040.

[54] A. Ikesue, Y.L. Aung, Ceramic laser materials, Nat. Photonics 2 (12) (2008) 721–727.

[55] Z.M. Seeley, N.J. Cherepy, S.A. Payne, Homogeneity of Gd-based garnet transparent ceramic scintillators for gamma spectroscopy, J. Cryst. Growth 379 (2013) 79–83.

[56] N.J. Cherepy, J.D. Kuntz, Z.M. Seeley, S.E. Fisher, O.B. Drury, B.W. Sturm, T.A. Hurst, R.D. Sanner, J.J. Roberts, S.A. Payne, Transparent ceramic scintillators for gamma spectroscopy and radiography, Proc. SPIE 7805 (2010) 78050I–780505.

[57] J.P. Hollingsworth, J.D. Kuntz, Z.M. Seeley, T.F. Soules, Transparent ceramics and methods of preparation thereof, Google Patents, 2011.

[58] C. Tallon, D. Jach, R. Moreno, M.I. Nieto, G. Rokicki, M. Szafran, Gelcasting of alumina suspensions containing nanoparticles with glycerol monoacrylate, J. Eur. Ceram. Soc. 29 (5) (2009) 875–880.

[59] P. Bednarek, M. Szafran, Y. Sakka, T. Mizerski, Gelcasting of alumina with a new monomer synthesized from glucose, J. Eur. Ceram. Soc. 30 (8) (2010) 1795–1801.

[60] E. Adolfsson, Gelcasting of zirconia using agarose, J. Am. Ceram. Soc. 89 (6) (2006) 1897–1902.

[61] X. Mao, S. Shimai, M. Dong, S. Wang, Gelcasting and pressureless sintering of translucent alumina ceramics, J. Am. Ceram. Soc. 91 (5) (2008) 1700–1702.

[62] X. Mao, S. Shimai, S. Wang, Gelcasting of alumina foams consolidated by epoxy resin, J. Eur. Ceram. Soc. 28 (1) (2008) 217–222.

[63] F. Chabert, D.E. Dunstan, G.V. Franks, Cross-linked polyvinyl alcohol as a binder for gelcasting and green machining, J. Am. Ceram. Soc. 91 (10) (2008) 3138–3146.

[64] Y. Jia, Y. Kanno, Z.-P. Xie, Fabrication of alumina green body through gelcasting process using alginate, Mater. Lett. 57 (16) (2003) 2530–2534.

[65] O. Lyckfeldt, J. Brandt, S. Lesca, Protein forming—a novel shaping technique for ceramics, J. Eur. Ceram. Soc. 20 (14) (2000) 2551–2559.

[66] X. Mao, S. Wang, S. Shimai, Porous ceramics with tri-modal pores prepared by foaming and starch consolidation, Ceram. Int. 34 (1) (2008) 107–112.

[67] Y. Yang, S. Shimai, S. Wang, Room-temperature gelcasting of alumina with a water-soluble copolymer, J. Mater. Res. 28 (11) (2013) 1512–1516.

[68] R.G. Horn, Surface forces and their action in ceramic materials, J. Am. Ceram. Soc. 73 (5) (1990) 1117–1135.

[69] S. Liufu, H. Xiao, Y. Li, Adsorption of poly (acrylic acid) onto the surface of titanium dioxide and the colloidal stability of aqueous suspension, J. Colloid Interface Sci. 281 (1) (2005) 155–163.

[70] W.M. Carty, U. Senapati, Porcelain—raw materials, processing, phase evolution, and mechanical behavior, J. Am. Ceram. Soc. 81 (1) (1998) 3–20.

[71] D. Hotza, P. Greil, Review: aqueous tape casting of ceramic powders, Mater. Sci. Eng. A 202 (1) (1995) 206–217.

[72] P. Panya, O.-A. Arquero, G.V. Franks, E.J. Wanless, Dispersion stability of a ceramic glaze achieved through ionic surfactant adsorption, J. Colloid Interface Sci. 279 (1) (2004) 23–35.

[73] Y. KwongáLeong, Rheological evidence of adsorbate-mediated short-range steric forces in concentrated dispersions, J. Chem. Soc. Faraday Trans. 89 (14) (1993) 2473–2478.

[74] C.H. Schilling, M. Sikora, P. Tomasik, C. Li, V. Garcia, Rheology of alumina–nanoparticle suspensions: effects of lower saccharides and sugar alcohols, J. Eur. Ceram. Soc. 22 (6) (2002) 917–921.

[75] J.S. Reed, Principles of Ceramics Processing, Wiley, New York, 1995.

[76] F.R. Eirich, The conformational states of macromolecules adsorbed at solid-liquid interfaces, J. Colloid Interface Sci. 58 (2) (1977) 423–436.

[77] E. Kissa, Dispersions: Characterization, Testing, and Measurement, vol. 84, CRC Press, Boca Raton, USA, 1999.

[78] K. Holmberg, Mechanism of acid-catalyzed curing of alkyd-melamine resin systems, J. Oil Col. Chem. Assoc. 61 (9) (1978) 359–361.
[79] V.T. Crowl, Theoretical strength of dried green bodies with organic binders, J. Oil Col. Chem. Assoc. 50 (1967) 1023–1059.
[80] E. Jenckel, R. Rumbach, Z. Elecktrochem., 55 (1951) 612.
[81] J. Davies, J. Binner, The role of ammonium polyacrylate in dispersing concentrated alumina suspensions, J. Eur. Ceram. Soc. 20 (10) (2000) 1539–1553.
[82] S. Farrokhpay, G.E. Morris, D. Fornasiero, P. Self, Influence of polymer functional group architecture on titania pigment dispersion, Colloids Surf. A Physicochem. Eng. Asp. 253 (1) (2005) 183–191.
[83] R.F. Conley, R.F. Conley, Practical Dispersion: A Guide to Understanding and Formulating Slurries, VCH, New York, 1996.
[84] G.J. Fleer, Polymers at Interfaces, Springer, Berlin, 1993.
[85] V. Hackley, P. Somasundaran, J. Lewis, Polymers in Particulate Systems: Properties and Applications, Marcel Dekker, Inc., New York, 2001.
[86] S. Farrokhpay, G.E. Morris, D. Fornasiero, P. Self, Effects of chemical functional groups on the polymer adsorption behavior onto titania pigment particles, J. Colloid Interface Sci. 274 (1) (2004) 33–40.
[87] L.-T. Lee, P. Somasundaran, Adsorption of polyacrylamide on oxide minerals, Langmuir 5 (3) (1989) 854–860.
[88] P. Chong, G. Curthoys, Adsorption of partially hydrolyzed polyacrylamide on titanium dioxide, Int. J. Miner. Process. 5 (4) (1979) 335–347.
[89] E. Pefferkorn, Polyacrylamide at solid/liquid interfaces, J. Colloid Interface Sci. 216 (2) (1999) 197–220.
[90] D.H. Solomon, D.G. Hawthorne, Chemistry of Pigments and Fillers, Wiley, New York, 1983.
[91] S. Shimai, Y. Yang, S. Wang, H. Kamiya, Spontaneous gelcasting of translucent alumina ceramics, Opt. Mater. Express 3 (8) (2013) 1000–1006.
[92] X. Shu, J. Li, H.-L. Zhang, M.-J. Dong, S.-M. Shunzo, S.W. Wang, Gelcasting of aluminum nitride using a water-soluble copolymer, J. Inorg. Mater. 29 (3) (2014), 327–330.
[93] Y. Yang, S. Shimai, Y. Sun, M. Dong, H. Kamiya, S. Wang, Fabrication of porous Al2O3 ceramics by rapid gelation and mechanical foaming, J. Mater. Res. 28 (15) (2013) 2012–2016.
[94] Y. Yang, Y. Wu, New gelling systems to fabricate complex-shaped transparent ceramics, Proc. of SPIE, 8708 (2013) 8708D.
[95] L.M. Sheppard, Aluminum nitride—a versatile but challenging material, Am. Ceram. Soc. Bull. 69 (11) (1990) 1801–1812.
[96] J. Schulz-Harder, Advantages and new development of direct bonded copper substrates, Microelectron. Reliab. 43 (3) (2003) 359–365.
[97] Y. Kurokawa, K. Utsumi, H. Takamizawa, T. Kamata, S. Noguchi, AlN substrates with high thermal conductivity, IEEE Trans. Compon. Hybrids Manuf. Technol. 8 (2) (1985) 247–252.
[98] J.M. Ferreira, H.M. Diz, Effect of solids loading on slip-casting performance of silicon carbide slurries, J. Am. Ceram. Soc. 82 (8) (1999) 1993–2000.
[99] E.P. McNamara, J.E. Comeforo, Classification of natural organic binders*, J. Am. Ceram. Soc. 28 (1) (1945) 25–31.
[100] G.Y. Onoda, Jr., The rheology of organic binder solutions, in: G.Y. Onoda, L.L. Hench (Eds.), Ceramic Processing Before Firing, John Wiley & Sons, New York, 1978, pp. 235–251.
[101] J.A. Lewis, Binder removal from ceramics, Annu. Rev. Mater. Sci. 27 (1) (1997) 147–173.

[102] Y. Yang, Y. Wu, Tape-casted transparent alumina ceramic wafers, J. Mater. Res. 29 (19) (2014) 2312–2317.

[103] Y. Yang, Y. Wu, Environmentally benign processing of YAG transparnt wafers, Opt. Mater. (2015) In press.

[104] Y. Sun, S. Shimai, X. Peng, G. Zhou, H. Kamiya, S. Wang, Fabrication of transparent Y_2O_3 ceramics via aqueous gelcasting, Ceram. Int. 40 (6) (2014) 8841–8845.

[105] Y. Yar, F.Y. Acar, E. Yurtsever, M. Akinc, Reduction of viscosity of alumina nanopowder aqueous suspensions by the addition of polyalcohols and saccharides, J. Am. Ceram. Soc. 93 (9) (2010) 2630–2636.

[106] Y. Sun, S. Shimai, X. Peng, M. Dong, H. Kamiya, S. Wang, A method for gelcasting high-strength alumina ceramics with low shrinkage, J. Mater. Res. 29 (02) (2014) 247–251.

Chapter 19

A Perspective on Green Body Fabrication and Design for Sustainable Manufacturing

S. Gupta
University of North Dakota, Grand Forks, ND, USA

1 INTRODUCTION

Sustainability has become an integral component of research for the twenty-first century. The key aspects are (1) rapid urbanization and population growth, (2) the large amount of waste disposed to landfills yearly, (3) the global scarcity of natural resources such as fossil fuels, minerals, and water, (4) declining infrastructure, (5) the emergence of carbon dioxide (CO_2) emissions, and (6) climate change [1–6]. Moreover, sustainable research promises a societal evolution toward a more balanced world in which the natural environment and our cultural achievements are preserved for generations to come (main thought is from Ref. 1,2). The carrying capacity of natural systems sustainability is currently one of major challenges facing humanity [3–6].

In general, industrial manufacturing is responsible for around "35% of global electricity use, over 20% of CO_2 emissions, and over a quarter of primary resource extraction. Along with extractive industries and construction, manufacturing currently accounts for 23% of global employment. It also accounts for up to 17% of air pollution-related health damage. Estimates of gross air pollution damage range from 1% to 5% of the global gross domestic product. Moreover, if the life of all manufactured products were to be extended by 10%, then the volume of resources extracted could be cut by a similar amount" [7]. Currently, there are many efforts around the world to make manufacturing industry "greener." The U.S. Department of Commerce's Sustainable Manufacturing Initiative [8] defines "sustainable manufacturing" as: "the creation of manufactured products that use processes that minimize negative environmental impacts, conserve energy and natural resources, are safe for employees, communities, and consumers and are economically sound."

In particular, ceramic manufacturing is a highly energy-intensive process [9–11]. For example, energy accounts for 15% of the average direct manufacturing

Green and Sustainable Manufacturing of Advanced Materials. http://dx.doi.org/10.1016/B978-0-12-411497-5.00019-9

costs in ceramic tile manufacturing [11]. Ceramics are commonly produced by firing or sintering a porous particulate (green) body. During green body fabrication, organic substances such as binders, dispersants, plasticizers, or lubricating agents are often used [12]. These organic substances are driven from the green bodies as CO_2 and hydrocarbon gases before sintering. The green body also undergoes numerous endothermic and exothermic processes, including dehydration, decomposition, an phase formation (in the case of reactive systems) simultaneously [13]. Removal of organic substances from the green body raises environmental concerns [14] and requires prolonged heating, which can further lead to a decrease in manufacturing efficiency and an increase in energy consumption [9,12]. From an environmental and engineering perspective, reducing the amounts of organic additives during manufacturing should assist in minimizing such problems.

Clearly, the fabrication of green bodies is an important step in ceramic manufacturing. It is often the most critical step as well. The presintering process can potentially lead to the defect formation in ceramics, especially in porous ceramic products [13]. Current product losses from cracking and breakage of green tiles are "estimated at 3%, whereas losses in green tiles of fired products are estimated at about 2%. For example, in Europe this translates into a loss of €200 million, in addition to some 6.25×10^8 kg of solid wastes that mainly go to landfills" [15]. In addition, forming operations such as hard machining can account for as much as 80% of the overall manufacturing costs of a ceramic component due to low material removal rates [16,17]. For this reason, Jannsen et al [17] stated, "there are a number of cases where ceramic components have been designed and tested successfully but, despite their superior properties, have not been put into production because of their high price".

Currently, there are three innovative techniques that are very promising for increasing the efficiency of ceramic forming: (1) green ceramic machining (GCM) [18–20], (2) free-form manufacturing [17,21–26], and (2) near net shape (NNS) manufacturing [21–29]. Sua *et al.* [19] had further emphasized that GCM technique is very promising as green bodies can be machined into complicated structures by using computer numerical controlled (CNC) machining technology. The waste composed of green material can be reused. The mold-less manufacturing of ceramics using novel layer manufacturing and computer-aided design is generally referred to as solid freeform fabrication (SFF) or rapid prototyping [17,21,22]. Sigmund et al. [21] defined the process as development of ceramic components from gradual build up of thin sheets. The thin sheets were generated from simulation and cutting of three dimensional (3D) computer-aided design. The integral process of depositing thin layers of material, one after another, until the designed component is created is collectively referred to as additive manufacturing [29]. Using additive manufacturing, it is possible to fabricate NNS objects, which can significantly reduce the cost of machining.

Fundamentally, almost all the processes use polymeric binder systems to bond the ceramic particles to form green parts. Thus, fundamentally understanding the design of green bodies is critical for (1) faster and efficient manufacturing; (2) decreasing wastage of valuable raw materials resulting from product rejection; (3) GCM, SFF, and/or NNS manufacturing; and (4) minimizing the evolution of harmful gases such as CO_2 during organic evolution.

The aim of this chapter is to review (1) theoretical models, (2) different processing additives used during the fabrication of green bodies, (3) different methods for characterizing green bodies, and (4) experimental studies about the mechanical behavior of ceramic green bodies and (5) to explore novel ideas for designing green bodies to increase the efficiency of ceramics manufacturing processes.

2 THEORETICAL MODELS

2.1 Rumpf's Theory of Particle Adhesion

Rumpf [30] developed a model for calculating the strength of particulate bodies consisting of randomly packed spheres of uniform size:

$$\sigma = \left(\frac{9k\varphi}{32(1-\sigma)\pi R^2} \right) F, \tag{1}$$

where φ is the packing fraction or relative density of the particles, k is the average number of touching neighbors around each particle, R is the particle radius, and F is the interparticle bond strength.

Onada [31] calculated the theoretical dry strength of green ceramics using Rumpf's model. Uhland et al. [58] analyzed the model and stated "the force required to break the interparticle bonds is dependent on the concentration of binder at particle necks". By using the same approach, "the cross-sectional area (A) of the binder concentrated at the particle neck can be calculated by assuming that the binder forms pendular bridges because of capillarity" (Figure 1a):

$$A = \left(\frac{4\pi r^2 (1-\sigma)^{1/2}}{(3k)^{1/2}} \right) (v_B)^{1/2}, \tag{2}$$

where v_B is the volume fraction of the binder relative to the ceramic [31,58]; the author is using the approach developed in the latter reference. The force required to rupture the binder bridge is given by:

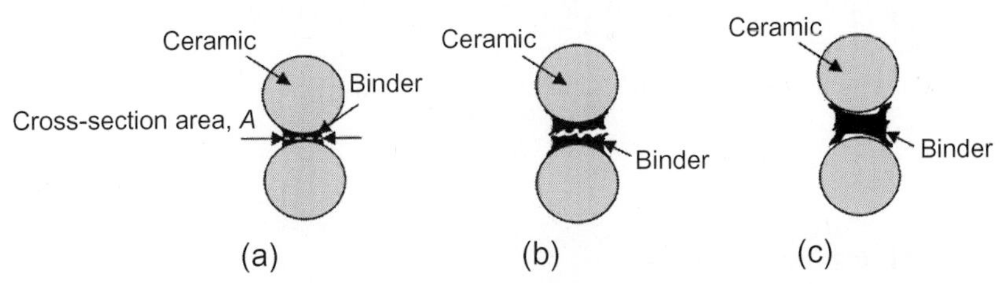

FIGURE 1 (a) Pendular state of the binder at the particle necks. (b) Fracture through the binder phase. (c) Fracture along the binder-ceramic interface [31,58].

$$F = \sigma_B A, \tag{3}$$

where σ_B is the strength of the binder. The theoretical strength of the green body can be obtained by combining Equations (1) through (3):

$$\sigma = \left(\frac{3\sqrt{3\pi}}{8} \times \frac{\varphi}{(1-\varphi)^{1/2}} \right) (v_B)^{1/2} \sigma_B \tag{4}$$

The green strength of ceramic green bodies is directly proportional to the strength of the binder. The strength of the binder can be represented by the polymer's cohesive strength, or the adhesive strength of the binder, depending on whether failure occurs through the binder (Figure 1b) or along binder-particle interface (Figure 1c). Thus, to design stronger green bodies, it is important to understand the binder (polymer) chemistry and the binder's interaction with the ceramic surface [31,58].

The model created by Rumpf [30] has been criticized because it did not take a fracture mechanics approach, but rather envisioned an entire plane of particles fracturing simultaneously. For this reason it cannot predict the marked effect of cracks and the wide variation of strength obtained from nominally identical samples [32–40]. Nevertheless, as discussed earlier, Rumpf's model gives a reasonable understanding of the origin of the strength of green bodies.

2.2 Fracture Mechanics

Kendall *et al.* [32,33] used linear fracture mechanics for a better understanding of the tensile strength of granular material. This model, like the Rumpf model, "accounts for the effect of particle size and packing; it also explains the influence of the flaws on green strength, describes the variability of strength, and interprets the effects of powder mixing, of environment, and of sintering" (Kendall Model is quoted from Ref. 32,33). Assuming a pure van der Waals attraction, the rupture energy (G_{Kendall}) depended roughly on the fourth power of packing density (φ):

$$G_{\text{Kendall}} = 56\varphi^4 \left(\frac{\Gamma^5}{E^2 D^2} \right)^{1/3}, \tag{5}$$

where E is the Young's modulus, relative density; D the average diameter of the particles; and Γ the particle-particle interfacial energy or the energy required to break the contact.

The toughness predicted by Kendall's model was [32–34]:

$$K_{\text{IC}} = 22\varphi^4 \frac{\Gamma}{R^{\frac{1}{2}}}, \tag{6}$$

where R is the particle radius.

Bortzmeyer *et al.* [35] further explored Kendall's model and observed that the dry pressed ceramic green samples (binderless zirconia powder) obeyed fairly well classical laws such as Weibull's statistics and the relationship between stress intensity factor and fracture energy. In other words, the fracture energy predicted with K_{IC} and E was very close to the experimental value. They also calculated the critical crack size of a zirconia body to be ~170 μm by assuming that the critical crack size could be modeled as a thorough-thickness surface crack. Subsequently, linear fracture mechanics was used successfully to understand the failure mechanism of ceramic green bodies with binder [39,40]. I will discuss some of these findings in the Section 4.

Cannon *et al.* [34] recently proposed a model to predict the toughness (K_{IC}) of green ceramics by considering the binding forces of menisci formed between particles near the tip of a critical flaw. The model considers capillary pressure, surface tension, and the viscous flow of binders, unlike Kendall's model [32,33], which considers particle surface energy forces binding the particles together. Cannon *et al.* reported that the predicted values of K_{IC} are lower than, but of the same magnitude as, measured values in binderless green bodies with two different specific areas. The authors also proposed that this model may be used to estimate the toughness of binder-containing green bodies. More fundamental studies are needed to fully explore this model.

2.3 Young's Modulus

Several models have been proposed to describe the elastic behavior of granular ceramic compacts. Using the model described by Kendall *et al.* [36,40], the Young's modulus of a granular compact can be expressed as:

$$E = 17.1\varphi^4\left(\frac{E_0^2\Gamma}{D}\right)^{1/3} \tag{7a}$$

Kendall *et al.* [36] also proposed that the particle coordination number (Z) is proportional to φ^2, and on the basis of adhesion theory, the diameter of particle contacts (d) can be expressed as:

$$d = \left(\frac{9\pi\Gamma D^2\left[1-\nu^2\right]}{2E_o}\right)^{1/3}, \tag{7b}$$

where ν is the Poisson's ratio and E_o is Young's modulus of original spheres.

Thus, from the abovementioned concept and Equation (7b), Equation (7a) can be written as (this approach is cited from Ref. 40):

$$\frac{E}{E_0} = \alpha Z\varphi^2\left(\frac{d}{D}\right), \tag{7c}$$

where α is a material constant [40].

Green *et al.* [37] suggested that their model of partially sintered materials can be applied to green bodies; the bulk modulus of the compact can be expressed by

$$\frac{B}{B_0} = \alpha Z\left(\frac{d}{D}\right). \tag{7d}$$

If the Poisson's ratio of the compact is insensitive to density changes, the relative bulk modulus (Equation (7d)) is similar to the relative Young's modulus (E/E_0). Arato *et al.* [38] also proposed a similar model:

$$\frac{E}{E_0} = \alpha Z\varphi\left(\frac{d}{D}\right)^2 \tag{7e}$$

Carneim and Green [40] integrated all the models and proposed that relative elastic modulus (E/E_0) can be expressed as:

$$\frac{E}{E_0} = \beta\varphi^z \tag{7f}$$

where z is the density component. Values of z are predicted to be 2, 3, or 4 by theoretical analysis.

By analyzing theoretical models, the Elastic modulus can be construed as dependent on the particle packing fraction or relative density. However, the experimental results showed much higher dependence on the green density compared with theoretical models. Carneim and Green [39] hypothesized that different factors—such as the coordination of particles involved in the transmission of force and the particle contact sizes, or removal of binder from between the particles in the consolidation process, or contact development between spray-dried granules—may explain the unexpected increase.

More important, by analyzing the different theoretical models presented in this section, we can appreciate the fact that ceramic green bodies have highly complex mechanical behavior that is dependent on various factors such as binder choice and content, compaction stress, forming technique, failure path, and moisture content, among others [30–40].

3 DIFFERENT PROCESSING ADDITIVES FOR FABRICATING CERAMIC GREEN BODIES

In this section I review a few types of processing additives for fabricating ceramic green bodies using dry and wet processing routes. It is important to critically understand the fundamental mechanisms of these additives for designing better green bodies.

3.1 Common Types of Binders

The most important role of the binder is to improve the strength of the green body during handling and machining before the sintering process. Table 1 summarizes common types of binders used during ceramic processing. (Please consult Reed [12] for detailed description of binders. Some of the critical binders

TABLE 1 Different Types of Binder Materials [Please check Ref. [12] for detailed list of references]

Binder Material	Examples
Colloidal particle type	
Organic	Microcrystalline cellulose
Inorganic	Kaolin, ball clay, bentonite
Molecular type (organic)	
Natural gums	Xanthum gum
Polysaccharides	Refined starch
Lignin extracts	Paper waster liquor
Refined alginate	Na, NH_4 alginate
Cellulose ethers	Methyl cellulose
	Hydroxyethyl cellulose
	Sodium carboxymethyl
	Cellulose
Polymerized alcohol	Polyvinyl alcohol
Polymerized butyral	Polyvinyl butyral
Acrylic resins	Polymethyl methacrylate
Glycols	Polyethylene glycol
Waxes	Paraffin, wax emulsions
	Microcrystalline wax
Molecular type (inorganic)	
Soluble silicates	Sodium silicate
Organic silicates	Ethyl silicate
Soluble phosphates	Alkali phosphates
Soluble aluminates	Sodium aluminate

are summarized in the next paragraph are cited from Ref. 12.) Colloidal clay binders are used in the processing of different types of traditional ceramics and advanced ceramics systems when alumina and silica are acceptable. Examples of clay binders are fine kaolin, ball clay, and bentonite (Table 1). Microcrystalline cellulose is an organic colloidal particle binder (Table 1).

Molecular binders may be natural or synthetic substances. In general, refined natural materials are more expensive than clay binders, but less expensive than refined and synthetic organic polymers [12]. Molecular binders are composed of low- to high-molecular-weight polymer molecules, which may absorb on the surfaces of particles or form a polymer-polymer bonded network (film) between particles. The former is referred to as adhesion binders, whereas the latter are film binders. Examples of adhesion binders are polyvinyl alcohol (PVA), cellulose binders, and polyethylene glycol (PEG). Examples of film binders are common waxes, for example, paraffin derived from petroleum, candelilla and carnauba waxes derived from plants, and beeswax of insect origin. Most polymer binders used in ceramics processing are nonionic or mildly anionic [12].

Inorganic molecular binders form a large amount of residue; thus they should be used only when the inorganic component is compatible with the particle composition. The detailed mechanisms of these binders are described in an excellent text by Reed [12].

3.2 Plasticizers and the Effect of Humidity and Glass Transition Temperature (T_g) of the Binder on Green Bodies

The role of a plasticizer is to reduce the glass transition temperature (T_g) of the polymer [12]. For example, water acts as a plasticizer for PVA (a water-soluble and hygroscopic binder) [12]. Plastisizers for thermoelastic polymers such as polyethylene and polystyrene are oils and waxes that can be melted at the molding temperature. However, a plasticized binder is of lower strength but is more deformable and resistant to failure upon impact [12,41–43].

Nies *et al.* [41] demonstrated, using a PEG-plasticized PVA binder that when the T_g of the binder is below the pressing pressure, then the densification is enhanced by the increased deformability of the binder system and the decrease in the granule strength. In another study, DiMilia *et al.* [42] also observed similar behavior. They showed that relative humidity altered the T_g of the organic binder phase (PVA) and consequently affected the compaction behavior of spray-dried samples. For example, it was necessary to press the low-humidity specimens at three times higher compaction pressures.

3.3 Lubricants, Dispersants, and Foaming and Antifoaming Agents

Reed [12] defined *lubricant* as an interfacial phase that reduces resistance to sliding in an effective manner. Reed [12] further classified lubricant as fluid-type ("provides a thick film of low viscosity between sliding particles, but these may migrate rapidly from an interface under a compressive stress"); boundary-type ("provides an adsorbed film of high lubricity that improves the surface smoothness and minimizes adhesion between surfaces"); and solid lubricants ("fine particles with a lamellar structure and smooth surfaces"). Boundary-type lubricants are most commonly used in dry pressing. For high-temperature applications, solid lubricants are sometimes mixed with molecular boundary lubricants [12].

Lubricants can be further classified as internal and external lubricants. Internal lubricants are added to a slurry before spray-drying. External lubricants are added to the "surface of spray-dried granules with the intention of reducing die-wall fraction during both compaction and ejection (definition of Internal and External Lubricant is quoted from Ref. 44, Ref. 45 also mentioned this). Balasubramanian *et al.* [44] showed that the addition of 2 wt% internal lubricant (ammonium stearate) increased the green body densities and had few defects at comparable compaction pressures. Similarly, Uppalapati and Green [45] also showed that the addition of 0.5 wt% external lubricant (zinc stearate) increased fracture strength and decreased strength variability.

Dispersants are added to ceramics suspensions to prevent agglomeration [12]. In general, polyelectrolytes are added to ceramic suspensions as dispersants [46–50]. A dispersant forms a diffuse electrical double layer around each particle. The interaction of these double layers provides a repulsive force between two approaching particles and, if sufficiently large, prevents particle agglomeration [47,48]. For example, ammonium or sodium polyacrylate is currently used for the dispersion of concentrated alumina suspensions [49,50].

A foaming agent reduces the surface tension of the foaming solution, "increases film elasticity, and prevents localized thinning". In general, a foaming agent contains a surfactant, which makes the particles hydrophobic and reduces surface tension, and a stabilizing component. For example, tall oil or sodium alkyl sulfate and polypropylene glycol ether are effective aqueous foaming agents. Antifoaming or defoaming aids are used to eliminate bubbles from a slurry. An effective antifoam is a surfactant with low surface tension. Commercial aqueous defoaming surfactants include fluorocarbons, dimethylsilicones, high-molecular-weight alcohols and glycols, and calcium and aluminum stearate [12].

4 METHODS OF CHARACTERIZING GREEN BODIES

4.1 Mechanical Behavior

Compressive strength measurement during uniaxial compression is the most common method of characterizing mechanical behavior [13,51]. Figure 2 shows the typical stress-versus-strain behavior of a ceramic green body composed of mullite formers, 8 vol% methylcellulose binder, and 1 vol% lubricant (Composition MO) [13]. Because of its relative simplicity, the split tensile test is the most common method for measuring the tensile behavior of ceramic green bodies.

Due to relative simplicity, split tensile test is the most common method for measuring the tensile behavior of ceramic green bodies. The tensile strength is calculated by [13, 52]:

$$\sigma = \frac{2F_{max}}{\pi Dt} \tag{8}$$

where F_{max} is the load applied at fracture, and D and t are the diameter and thickness of the sample, respectively.

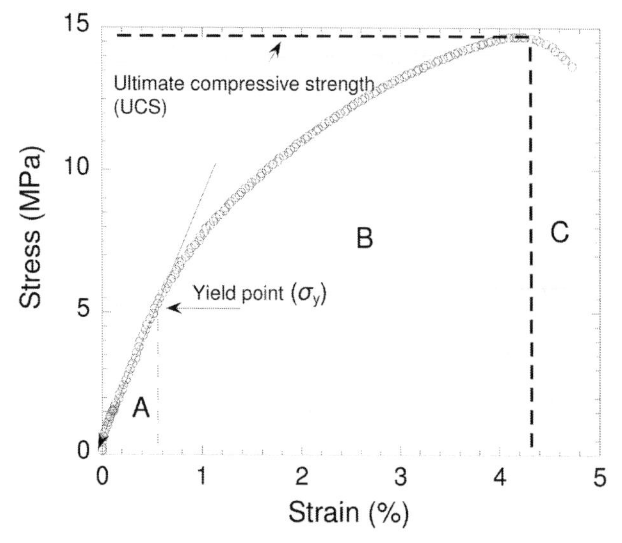

FIGURE 2 Typical stress-versus-strain profiles of MO green bodies during compression [13].

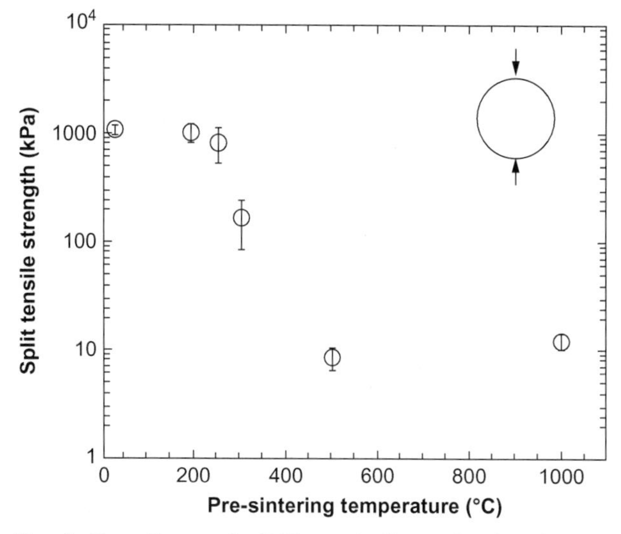

FIGURE 3 Plot of split tensile strength of MO green bodies as a function of temperature [13].

Figure 3 shows a plot of the split tensile strength of MO green bodies as a function of temperature.

The dispersion of strength data of green compact ($n=20$ samples), and then the defect size distribution, can be estimated by Weibull statistics [53–55]. The strength data are analyzed using the two-parameter empirical model relating the probability of failure P_f to the fracture strength σ_f:

$$P_f = 1 - \exp\left[-\left(\frac{\sigma_f}{\sigma_f}\right)^m\right]$$

with

$$P_f = 1 - \exp\left[-\left(\frac{\sigma_f}{\sigma_o}\right)^m\right]$$

where m is the Weibull modulus and σ_0 a normalizing factor. Tensile strengths are ranked in order and assigned a probability of failure according to the formula $P_f = i/(N+1)$, where i is the ith specimen. The Weibull modulus is the slope of a plot of $\ln[\ln(1/P_f)]$ vs $\ln[\sigma_f]$ [52].

The Young's modulus is measured using the pulse-echo method [54]. During the pulse-echo method, the longitudinal "long bar" mode was used to determine the Young's modulus. The "long bar" condition implies that the cross-sectional dimension is small compared to the wavelength λ. The phase velocity V_L, at which an ultrasonic compressional wave propagates, is given by [54]:

$$V_L = \left(\frac{E}{\rho}\right)^{1/2}, \tag{10}$$

where E is the Young's modulus and ρ is the density of the pressed sample.

Based on the measurement of the round-trip time t_i between two echoes corresponding to the propagation in the sample, the velocity is measured by

$$V_L = \left(\frac{2L}{t_i}\right), \tag{11}$$

where L is the length of the specimen.

This technique has two main advantages: (1) in highly porous media, the attenuation of the low-frequency ultrasonic waves is sufficiently low to allow propagation, and (2) only one measurement of ultrasonic velocity is needed from Equation (11) to determine the Young's modulus [54].

Fracture toughness can be measured using (1) notched diametral compression [40] or (1) a single-edge v notch beam [34]. During both methods, a notch of a measured size is introduced in the samples. For example, during notched diametral compression, if the notch is regarded as a through-thickness surface crack, the fracture toughness (K_{IC}) can be measured by [40]:

$$K_{IC} = \left(\frac{1.246(a)^{\frac{1}{2}} P_F}{Dt}\right), \tag{12}$$

where a is the crack length that was measured using optical microscopy of the fracture samples, and P_F corresponds to the maximum load at which the sample began to crack.

The critical crack radius c can be determined by [40]:

$$c = \left(\frac{K_{IC}}{\sigma_f}\right)\frac{\pi}{5.02}. \quad (13)$$

Zhang and Green [40] explored two different methods for making notches, namely, (1) a compaction plunger that possessed a prism-shaped extension that allowed the samples to be V-notched during compaction, and (2) a narrow saw used to notch the samples. These notching techniques gave fracture toughness values that differed, on average, by 13%. Method 2 is more promising because the notch created during method 1 can produce a stress concentration that can influence the local density during compaction [40].

4.2 Microstructure Analysis

Scanning electron microscopy is the most common method of studying the microstructure of green bodies [13,39,40,44,45,51,54,55]. Figure 4 shows a typical microstructure of an green body (a) and the green body after binder burnout (b) [13]. Microstructural features such as particle morphology, cracks, voids, and binder distribution can be visualized using this technique [13,39,40,44,45,51,54,55].

Table 2 summarizes defects in monolithic ceramics and their possible source in a green body [56]. Characterizing the defects in green bodies is critical for developing a high-quality final ceramic product. Advanced experimental techniques such as X-ray tomography [45], immersion microscopy [56], and infrared microscopy [57] can be used for detailed evaluation of the microstructure of green body samples. Because of space limitations here, readers can review the cited references for details about each type of characterization technique.

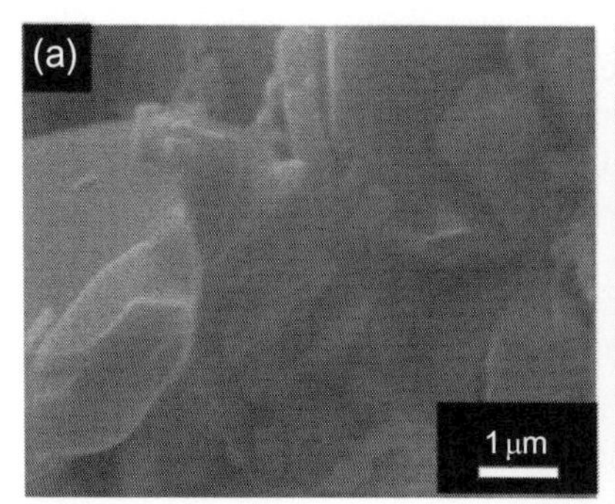

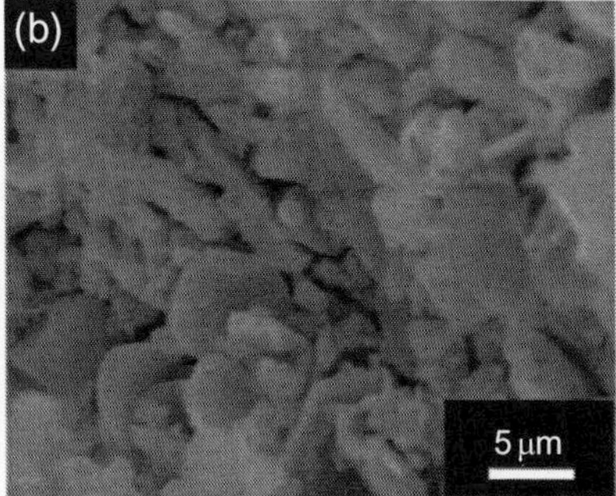

FIGURE 4 Field emission scanning electron micrographs of gold-coated fractured surfaces of green bodies composed of MO at room temperature (a) and 1000°C (b) [13].

TABLE 2 Defects in Monolithic Ceramics and Their Possible Source in a Green Body [56]

Defects	Possible Origin of the Defects in a Green Body
Large pores	Agglomerates, nonuniform powder packing
Large grains	Agglomerates
Internal strain	Orientation of grains, large particles, agglomerates, nonuniform particle packing
Deformation	Nonuniform powder packing of an especially large scale
Cracks	Extremely non-uniform powder packing of an especially large scale

5 MECHANICAL BEHAVIOR OF GREEN BODIES

Figure 2 shows typical stress versus strain behavior of the MO green bodies at room temperature during compression [13]. Initially, strain varied linearly with stress (stage A). Thereafter, the sample yielded at −5 MPa and there was a nonlinear region (stage B). After reaching an ultimate compressive strength (UCS) at −14.7 MPa, cracks were observed in the sample and the sample was crushed (stage C).

Similar stages in the deformation process also occur at higher temperatures, but the yield stress and UCS decrease with the increasing temperature, and the failure is initiated at lower strains because of the softening of the binder [13] (Figure 5a). Because of the viscous nature of the binder, it is difficult to pinpoint the exact point of initiation of a visible crack. After the binder is removed, however, these green bodies form a porous network of interlocked inorganic

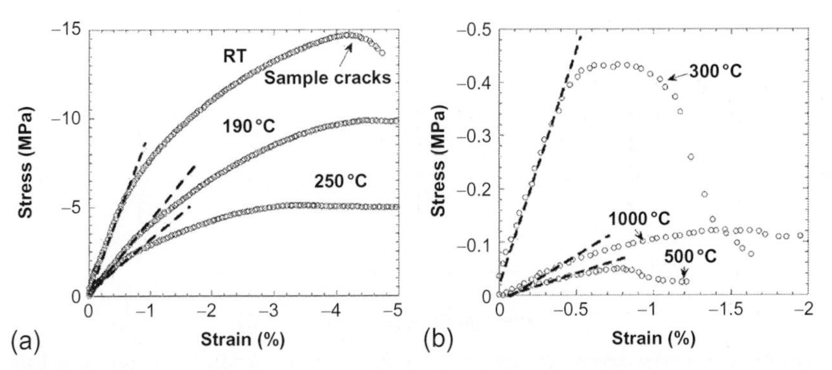

FIGURE 5 Stress-versus-strain plots of green bodies composed of mullite formers at room temperature (RT) and 190 and 250°C (a) and at 300, 500, and 1000°C (b) [13].

particles, and no major changes in the deformation mechanism with temperature are expected. Figure 5b compares the room-temperature strengths of specimens heated to 300, 50°, and 1000°C. As expected, Gupta *et al.* [13] observed decrease in strength at 300 and 500°C [13]. However, the strength started to increase again by 1000°C because of the initial stages of glass formation.

Gupta *et al.* [13] proposed that stage A (linear region) is the simultaneous effect of particle rearrangement and binder shear that causes particle interlocking. Stage B (nonlinear regime) is the result of further binder shear that causes de-cohesion or interparticle friction and is responsible for the decrease in slope. Finally, as the sample reaches the final stages of deformation, cracks initiate at the UCS. Stress-versus-strain plots during testing between room temperature and 250°C show deformation at both stages A and B (Figure 5a). During stage B, a gradual axial deformation of ~3% was observed during testing at all three temperatures before failure at the UCS. However, during testing at 300 and 500°C, stage B clearly was shortened, and <0.5% axial strain was observed during stage B (Figure 5b). This fact further supports that stage B is due to the binder shear between particles that causes binder de-cohesion, and at higher temperatures the sample collapses because of the absence of binders in the system. Figures 3 and 6 show the variation of the tensile strength and UCS as a function of temperature. The average compressive and tensile strengths of the sample at room temperature were −14.5 and 1.06 MPa, respectively. Both the compressive and tensile strengths decreased by two orders of magnitude with temperature (notice the log scale). Moreover, tensile strength is typically an order of magnitude lower than the UCS. Uppalapati and Green [51] also ob-

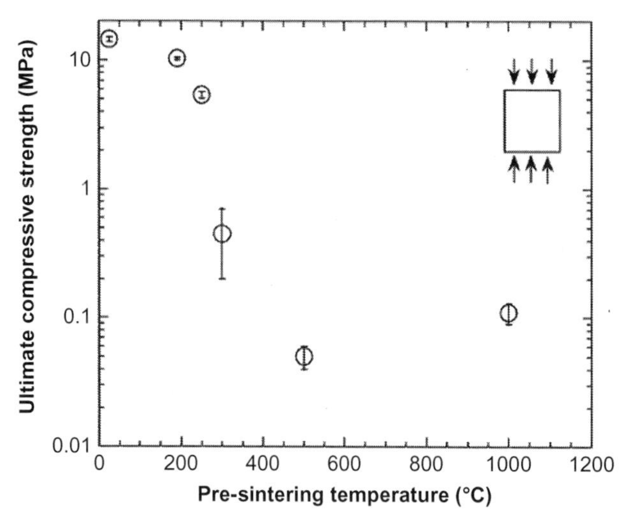

FIGURE 6 Ultimate compressive strength (UCS) of MO bodies at different temperatures. UCS was measured from stress-versus-strain plots in uniaxial compression. The inset shows the schematic of the test [13].

served similar stress-versus-strain behavior during mechanical testing, as well as viscoelastic properties such as stress relaxation, creep, and delayed elastic strain. These studies show that the mechanical behavior of green bodies is dependent on polymer-ceramic interaction, and the green bodies are vulnerable to low stresses once the binder is removed. Thus, it is of paramount importance to minimize the formation of defects in green bodies.

Baklouti *et al.* [55] studied the effect of two different binder systems (PEG and PVA; 3 wt% on a dry basis with respect to alumina) on the mechanical properties of dry pressed alumina compact. They observed that the binder and moisture content significantly influenced strength. Strength variability also decreased significantly with increasing compaction stress and were influenced by T_g of the binder. In a later work, the same group [54] observed that, during the spray-drying process, organic binders, especially water-soluble ones (e.g., PVA), tend to migrate with the solvent flow. The PVA forms a polymer-rich layer by segregating on the surface of granules. These authors also estimated the thickness of this layer to be only a small percentage of the granule radius. Zhang and Green [40] observed that the binder tends to segregate in intergranular regions. They also concluded that binder segregation is linked to the low toughness or strength of the regions between the granules. These interesting research studies guide us toward more fundamental research in which the use of a binder can be minimized by understanding the interaction between ceramic and polymer interphase.

6 NOVEL APPROACHES FOR DESIGNING GREEN BODIES

In this section, I explore different “out of the box” solutions for designing robust and environmentally friendly green bodies.

6.1 Low Binder Systems

Uhland *et al.* [58] demonstrated that 2.5 vol% of a cross-linkable binder, which is based on a soluble poly(acrylic acid) (PAA; molecular weight, 60,000) and glycerol, can increase the strength of silica (IEP = 2.3) and alumina (IEP = 10.4) green bodies by at least a factor of 8 and 25, respectively, compared with binder-free systems. The efficacy of the binder phase was proven to be dependent on the surface chemistry of the ceramic phase. The affinity of the PAA binder decreased as the ceramic's IEP decreased. Thus, higher relative green body strengths resulted from the adhesion of the binder to the surface of the ceramic particles.

Sato *et al.* [59] used photoreactive molecules on the surface of particles, which, upon subsequent radiation with ultraviolet light, formed bridges that bound the entire structure together (Figure 7). The chemical bonds formed during the process decreased the use of the binder to only ~0.5 wt%. Sato *et al.* [59] stated further that the Stronger bonds between the binder and the ceramic also prevented

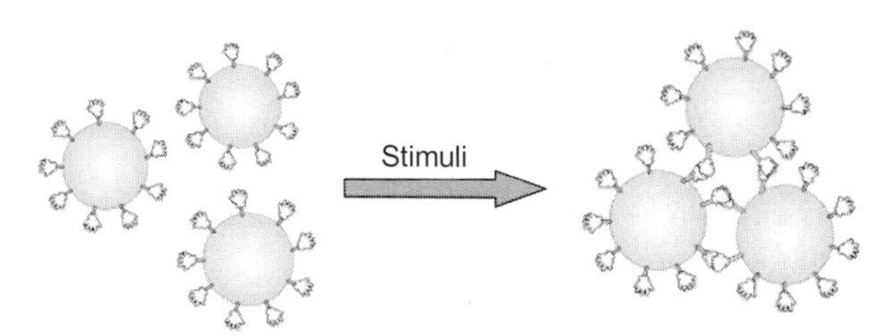

FIGURE 7 Schematic of the formation of mutual linkages between ceramics particles resulting from the application of external stimuli (photoreactivity or microwave radiation) on binders [59].

ceramic particles and the binder. Sato at al. [60] also performed a similar study using microwave radiation to activate binders to form mutual linkages between ceramic particles. These studies give an excellent fundamental basis for designing green bodies with low binder content by understanding the interactions between ceramics particles and polymer additives.

6.2 Multifunctional Additives

The use of different additives in a suspension can lead to competitive adsorption of these additives onto the surface of ceramic particles, which can further lead to a decrease in the efficiency of a binder and/or dispersants [61]. From a sustainable manufacturing perspective, better green bodies can be fabricated if an additive can perform multiple functions, hence increasing the efficiency of the manufacturing process.

Romdhane *et al.* [46] demonstrated that PVX binders containing both carboxylate (–COO$^-$) and hydroxyl groups (vinyl alcohol) can perform the multifunctional roles of dispersants and binders for dry-pressed green parts. These authors demonstrated that carboxylate groups are responsible for strong adsorption of copolymers onto the alumina surface and can promote sufficient electrostatic repulsive forces for >35% of the carboxylic group in the copolymer. The hydroxyl group, on the other hand, conferred higher mechanical strength compared with a green comprising only PVA. The addition of ~1.5 wt% PV35 (0.7 wt% adsorbed and 0.8 wt% nonadsorbed) can effectively perform the multiple functions of a dispersant and a binder simultaneously.

McNamara *et al.* [62] showed that silk can be used as a multifunctional biocohesive sacrificial binder. Initially, the silk derived from *Bombyx mori* was used to consolidate ceramic grains during green body formation; later it was used as a sacrificial polymer for imparting porosity during sintering. Similarly, Lyckfeldt *et al.* [63] used starch as a binder/consolidator and pore former.

6.3 Organic Binder-less Processing

Nagaoka *et al.* [64] used the reaction between hydraulic alumina and water to generate a 3D network of boehmite gel. Even without the addition of a binder, the green bodies showed a high compressive strength. Furthermore, water acted as a fugitive pore former. Shirai *et al.* [65] demonstrated that microwave irradiation applied to an alumina green body strongly enhanced the hydration reaction between water and particle surfaces, which results in the formation of a cementitious aluminum trihydroxide structure that tightly bound the particles together. Shirai *et al.* [65] further stated, "this process makes possible the manufacture of mechanically strong green bodies with excellent shape retention without the use of organic binders (Figure 8)".

During freeze casting, a ceramic slip is poured into a mold, and then it is frozen and the solvent is subjected to sublimative drying under vacuum. The frozen solvent temporarily acts as a binder to hold the parts together for demolding, which minimizes the additive concentration for enhanced solid purity and faster binder burnout cycles [66-68]; definition of freeze casting is quoted from Ref. 67 — it has appeared in different papers. Sofie and Dogan [67] fabricated highly dense alumina bodies with a uniform microstructure using the combined effects of slurries with high solid-loading (>57.5 vol%) and glycerol additions. These authors further postulated that the interaction between glycerol and the dispersant may result in the formation of a micelle structure to increase particle-particle separation (Figure 9). They also speculated that the excess glycerol in the slurry may interact with the binder and wetting agents, which can result in a

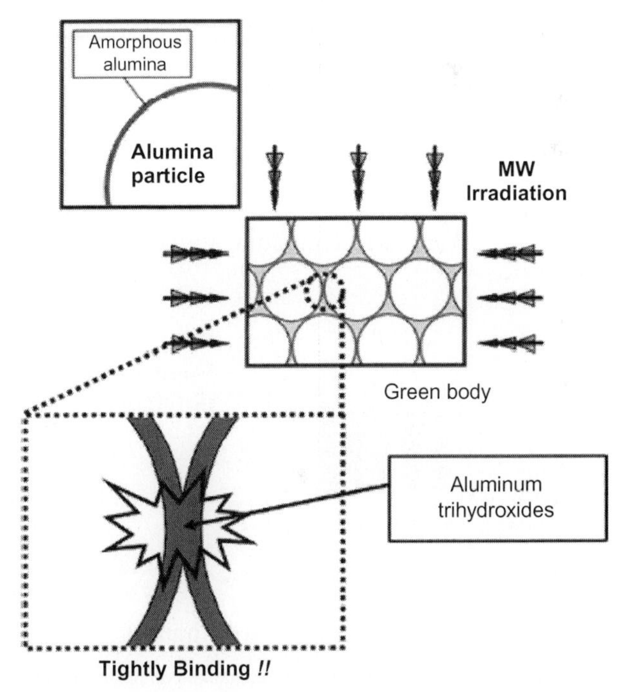

FIGURE 8 Schematic of enhanced hydration between connecting particles through microwave irradiation. MW, microwave. *(Adapted from Ref. [65], with permission from the publisher.)*

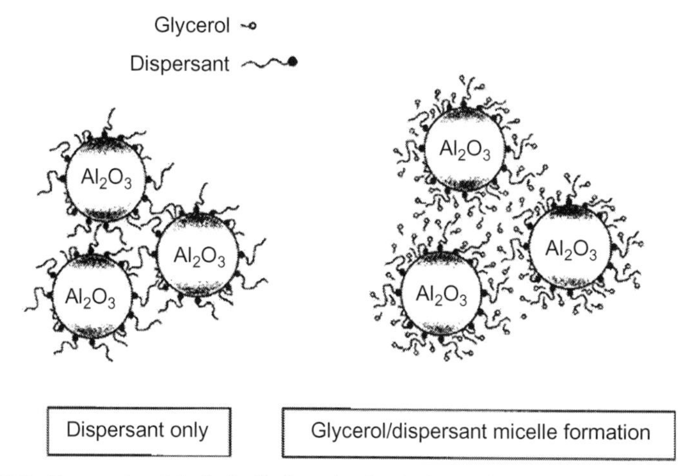

FIGURE 9 Proposed model of micelle formation by the interaction of a dispersant and glycerol. *(Adapted from Ref. [67], with permission from the publisher.)*

lubricating effect in highly loaded slurries. Green bodies containing ~4 wt% glycerol were sintered to highly dense materials at heating rates of 5°C/min. Because of the relatively small amount of organic additives, a rate-controlled organic burnout process was not necessary.

Fukasawa *et al.* [68] used a freeze-drying process for fabricating porous silicon nitride with macroscopically aligned channels. Initially, a water-based slurry of silicon nitride was frozen while unidirectionally controlling the direction of the ice's growth (Figure 10). Thereafter, pores were generated subse-

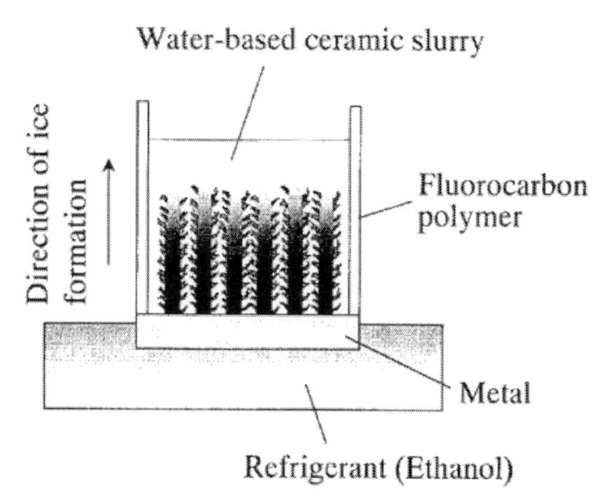

FIGURE 10 Schematic of the freezing process. The container comprises two parts: the bottom was made of metal with high thermal conductivity, and the side was made of a fluorocarbon polymer. *(Adapted from Ref. [68], with permission from the publisher.)*

limation of the columnar ice during freeze-drying. By sintering this green body, a porous silicon nitride with high porosity (>50%) was obtained. Moreover, the porosity was controllable by the slurry concentration. Freeze-form extrusion fabrication, an SFF process, uses a highly loaded aqueous ceramic paste (≥50 vol% solids loading) with a small quantity (~2 vol%) of "organic binder to fabricate a ceramic green part layer by layer with a computer-controlled 3D gantry machine at a temperature below the freezing point of the paste" [69].

Hydrolysis-assisted solidification (HAS) is a forming method that uses an inorganic reaction to consume the dispersing media [21,70]. During HAS, thermally induced decomposition of AlN in an aqueous medium results in the formation of aluminum hydroxides and NH_3 where slurry gels by absorbing water which further increases the viscosity of the system. Because of the precipitation of aluminum hydroxides, green bodies (chemical gels) with high strength are achieved. The high strength allows the production of delicate structures with minimal warpage. The main limiting factor is the hydrolysis reaction, which yields Al_2O_3 that is embedded in the sintered specimen; thus HAS can be used only in systems where the system can tolerate Al_2O_3. This technique can be used for both casting and injection molding (IM) [Sigmund et al. summarized these points in Ref. 21.The author has rephrased it. Please refer to Ref. 21 for details].

The plastic deformation of metal particles during compaction can be used to generate green bodies with no binder as additives. Gupta and Riyad [71] proposed a novel method of manufacturing macroporous ceramics by oxidation-induced sintering of porous compacts comprising Ti powders and fugitive pore formers (Figure 11). Using this novel manufacturing method, novel porous TiO_2 ceramics can be fabricated over a wide range of porosity at a relatively faster heating rate (4°C/min) and shorter sintering periods (≤4 h) at 1450°C. The porous ceramics were subsequently characterized by SEM, XRD, and compressive strength measurements. This process can be made greener by minimizing the use of organic pore formers.

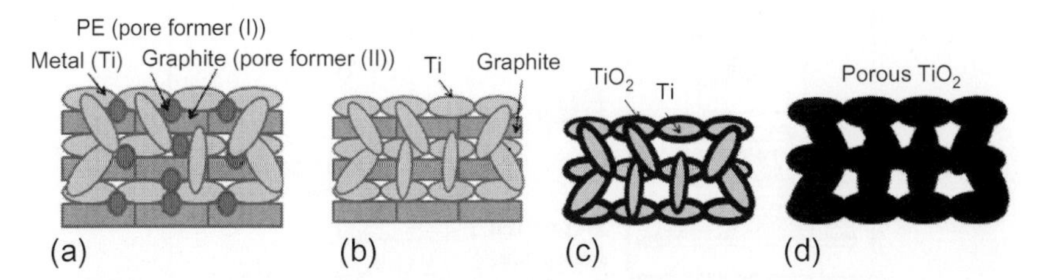

FIGURE 11 Schematics of the manufacturing of porous oxide ceramics using the oxidation-induced sintering method: packing of mixed powders of metal (Ti), polyethylene (PE; pore former I), and graphite (pore former II) into compacts (a); oxidation and removal of PE (pore former I) during heating (b); oxidation and removal of graphite (pore former II) during further heating and concomitant partial oxidation of Ti (c); and formation of porous TiO_2 ceramics at higher temperatures as a result of the complete oxidation of metal powders (d) [71].

6.4 IM and Related SFF Techniques

During IM, a premix of a relatively high concentration of inorganic powder with the polymer is molded [12]. The polymer provides the flow and strength for handling. This process has a few limitations, for example, long binder removal times (up to 7 days), thick section cracking, size limitations (<3 cm diameter), and green body defects such as knit lines, short shots, flashing, sink marks, and thermal strains (list of defects is from ref. 72). Low-pressure IM uses wax or an aqueous system [12,73]. During low-pressure IM, the debinding process is also critical and time-consuming [73].

Fused deposition of ceramics, a filament-based direct ink writing (DIW) SFF technique, is similar to IM because of the usage of suspended particle systems [21,74,75]. The mixed powder-binder feedstock is extruded into filaments, which then are deposited, and solidification occurs during cooling of the melt ([21, 75], Figure 12). The fused deposition green part is subjected to conventional binder removal and sintering processes to produce dense structural ceramic components. Green bodies can be fabricated rapidly using this method, but compared with powder injection-molded parts, the burnout process may take several days. (This is a generic description of the process. Please see Ref. 21 for details.)

Selective laser sintering is an SFF technique whereby powder layers are deposited by a roller or a scraper and a laser beam is used as a heating source to locally heat and sinter the deposited powder layer according to predetermined geometries. The sequence of powder deposition and laser scanning is repeated until the part is completed [21, 76, 77]; definition of SLS is from Ref. 77 (it has appeared in various texts)). Because of the high melting temperature, low or no

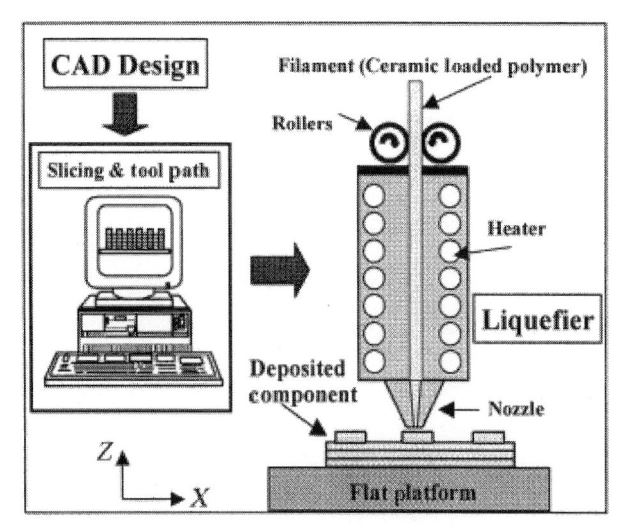

FIGURE 12 Schematic of the fused deposition process, including the computer-assisted design (CAD) file, slicing and tool-path generation, and deposition on a platform in a layer-by-layer fashion. *(Adapted from Ref. [75], with permission from the publisher.)*

plasticity and low thermal shock resistance of ceramics, selective laser sintering of ceramics is more challenging than for metals and polymers [77].

6.5 Colloidal Processing

Table 3 summarizes different types of colloidal processing methods. Lewis [22] classified them into different categories according to the consolidation mechanisms, namely, (1) fluid removal, (2) particle flow, and (3) gelation. Lewis *et al.* [21] had further pointed out that drying is the most vital process in colloidal processing as this step can lead to plethora of problems like cracking, segregation, dimensional mismatch, among others. There are excellent articles and reviews describing different types of colloidal processes in detail [12,21–24]. In this section I review a few promising colloidal forming methods from a green body fabrication perspective.

TABLE 3 Different Types of Colloidal-Forming Routes Classified by Consolidation Mechanism [21,22]

Forming Method	Consolidation Mechanism
Fluid removal	
Slip casting	Slip casting fluid flow into a porous mold; driven by capillary forces
Pressure filtration	Fluid flow through a porous filter; driven by an applied pressure
Tape casting, laminated object manufacturing[a]	Fluid removal by evaporation
3D printing,[a] inkjet printing,[a] robocasting[b]	Fluid removal by evaporation
Particle flow	
Centrifugal consolidation	Particle flow caused by an applied gravitational force
Electrophoretic deposition	Particle flow caused by an applied electric field
Gelation	
Aqueous injection molding	Physical organic gel forms in response to a temperature change
Gelcasting	Cross-linked organic network forms because of a chemical reaction
Direct coagulation casting	Colloidal gel forms because of flocculation
Robocasting,[b] stereolithography[a]	Colloidal gel formed because of flocculation or ultraviolet treatment

[a] *Solid free-form technique.*
[b] *Solid free-form technique and can be performed by fluid removal and gelation mechanisms.*

6.5.1 Consolidation by Fluid Removal

Reed [12] defined slip casting as a conventional casting of a slip or slurry in a porous gypsum mold where particles coagulates due to absorption of water by capillary action. Reed [12] further stated that during industrial casting - the control of the microstructure of the slip, gelation behavior, and the microstructure of the cast when using industrial minerals and unrefined water can make process cumbersome to control. Moreover, slurries that do not contain clay are difficult to cast by slip casting. To summarize, slip casting is a tedious process often associated elongated draining time which can further lead to inhomogeneity and concomitant defects. In pressure casting, slip pressure controls the casting time, and the mold acts as a shaped filtration support. This process results in both a larger differential pressure across the cast and a density gradient across the cast. The density gradient can cause different types of green body defects such as cracks upon drying, differential shrinkage, and shape distortion [12].

Tape casting is used for producing flexible green sheets of various ceramics composition. Slurries with a relative high binder and plastisizer content are formulated for producing a tape with satisfactory mechanical properties. Strict process controls are required to obtain a high yield of tape and laminated products [12]. Tape-cast layers also serve as feedstock for laminate-based SFF techniques, such as laminated object manufacturing (LOM) [78]. During LOM, parts are built through a layer-by-layer approach, but each layer is cut by a tool according to the cross section of a part. Layers then are bonded with each other using an additive. Thus, the LOM process can be considered a hybrid between additive and subtractive manufacturing.

Robocasting is a very promising computer-controlled, filament-based DIW SFF technique that deposits slurries into 3D component with well defined control of porosity by using low organic content (1 wt%) [22,23,79–86] (Figure 13). The ceramics paste used during robocasting should be carefully

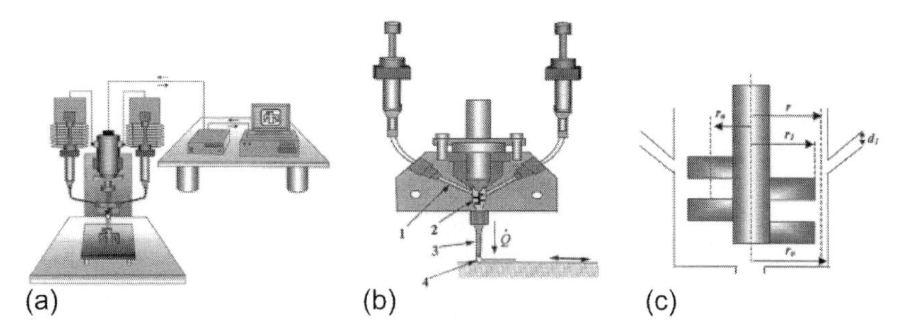

FIGURE 13 Schematic illustrations of the robocasting apparatus (a), a two-nozzle delivery system showing four shear zones in the mixing chamber: (1) pumping from the syringe, (2) mixing via the paddle mixer, (3) extrusion from the tip, and (4) deposition onto the moving *X-Y* table (b); and the paddle design, where $r = 1.651$ mm, $r_a = 0.89$ mm, $r_1 = 1.59$ mm, $r_0 = 1.72$ mm, and $d_1 = 1.52$ mm (c). *(Adapted from Ref. [81], with permission from the publisher.)*

engineered with proper viscosity and deposition kinetics. It is also critical to maintain enough viscosity of the paste so that it can maintain its current shape without yielding. Lewis [22] summarized robocasting as the "only SFF technique developed to date that uses colloidal systems of low organic content to directly write 3D bodies". Because of the low organic content and quick consolidation time, high furnace ramp rates (3-5°C/min for burnout and 10°C/min for firing) and short burnout times (2 h) can be achieved [83].

Droplet-based DIW SFF techniques, such as 3D printing, inkjet printing, and hot-melt inkjet printing, are currently in different stages of development [23,24]. These techniques rely on the pixel array approach for creating 3D ceramic components with a predefined shape and dimensions. Droplet size, the extent of droplet spreading upon deposition on the substrate (or underlying layers), and deformation of the droplet upon solidification determine the lateral and vertical resolution within the layer. The drying or solidification time required for the as-deposited droplets, as well as the rate of droplet delivery by the nozzle, controls the printing rate. (Definition of droplet based DIW is paraphrased from Ref. 23.)

6.5.2 Consolidation by Particle Flow

Green bodies can also be produced by the flow of particles in response to different types of applied force, for example, centrifugation [22,87–89] and electrophoretic deposition (EPD) [22,90]. Particles flow in response to gravitational forces during centrifugation. Huisman *et al.* [88] produced green bodies with very good particle packing (68% of theoretical density) and narrow pore size distributions. The largest flaws in the centrifuged parts are caused by agglomerates that are not eliminated in the slurry preparation sequence. In EPD, particles flow in response to an applied direct current electric field [22,90]. Besra and Liu [90] indicated that the "process needs further research and development to fully develop a quantitative understanding of the fundamental mechanisms of EPD and further optimize the working parameters for a broader use of EPD in materials processing".

6.5.3 Consolidation by Gelation

Aqueous IM is the IM of aqueous ceramic suspensions through the gelling of polysaccharides. Milan et al. [92] summarized that this technique strongly reduces the binder requirements and provides an environmentally friendly technology; the solvents and polymers are substituted by water and low concentrations (2-3 vol%) of noncontaminant additives [22,91,92]. During this process [91,92], a conventional aqueous suspension with a gelling additive (e.g., agarose) is prepared. Milan et al. [92] further summarized that because of the low viscosity of these suspensions, the colloidal chemistry can be better controlled, which results in the reduction of the size and number of defects and an improvement in reproducibility. Another significant advantage of aqueous IM process is that it is "compatible with existing commercial IM equipment and yields high-strength components that can be green machined" [22].

Gelcasting is a direct casting method [22,72,92–96]. It is based on the casting of a slurry containing powder, water, and water-soluble organic monomers [72,92–96]. The mixture is then polymerized to form green parts. Subsequent drying, burning out, and sintering complete the manufacturing route. The process is generic and can be used for fabricating NNS prototypes or small series using cheap molds for a wide range of ceramic and metallic powders [95, 96]. Gilissen et al. [95] further summarized that unlike slip casting, gelled parts are more homogeneous and have a much higher green strength. More important, Gilissen et al. [95] further summarized that gel-cast parts contain only a few percentages of organic components, making binder removal much less critical compared with IM [95]. However, the development of inner stress and surface exfoliation in ceramic green bodies during drying are two of the critical problems associated with gelcasting [96].

Stereolithography is one of the earliest SFF techniques [21,97,98]. During this process, a ceramic suspension comprising 40-55 vol% ceramics powder is photocured layer by layer to fabricate a chemically gelled 3D ceramic green body. The green body is subjected to subsequent binder removal and sintering. The inherent disadvantages of this process are that the ultraviolet-curable material (1) is expensive, odorous, and toxic and (2) must be shielded from light to avoid premature polymerization [21,98].

The direct coagulation casting process relies on destabilization of electrostatic stabilized ceramic suspensions by time-delayed in situ reactions. [Gaukler (99) defined this process] These reactions are catalyzed by enzyme or pH to produce H_3O^+ or OH^- ions or solubilized salt. During this process, the double-layer repulsive forces are minimized by either shifting pH toward the IEP or increasing the ionic strength of the system, thus resulting in destabilization [22,99–101]. This process is very attractive because of the use of minimal binder content (~1 vol%) and homogeneous packing densities. The direct coagulation process is severely restrictive due to, (a) slow coagulation rate, and (b) formation of green bodies with low strength and machinability. Their high salt content can also lead to problems during drying, and they can affect component performance [Lewis summarized these points in Ref. 22. Please refer to the text for further details.]. For example, wet coagulated bodies prepared from 50 vol% alumina slurry had a compressive strength of nearly 0.05 MPa [101].

6.6 Controlling the Thermomechanical Behavior of Green Bodies

As discussed earlier, the mechanical strength of ceramic green bodies after the binder burns out is low [13]. The ceramic body is susceptible to cracking and warpage during this stage. For this reason, binder burnout is a critical stage in the design of green bodies [13,102]. Organic additives should be tailored to engineer the mechanical strength of green bodies without causing microstructural defects in the ceramics pieces during the debinding process.

Gupta *et al.* [13] developed a method for characterizing the thermomechanical behavior of green bodies during the early stages of the sintering process [Summarized and paraphrased from Ref. 102]. By performing sinter forging, thermogravimetric analysis, and dilatometric measurements, the following critical temperatures were identified: (a) 190°C—softening of compact, (b) 200-600°C—binder burnout and gibbsite decomposition, and (c) >600°C—the body is essentially composed of particles held together by mechanical interlocking. The major strains during presintering were caused by binder removal and the quartz phase transformation. Gupta *et al.* also observed that the early stage of binder removal was sensitive to an applied stress. The average compressive and tensile strengths of the sample at room temperature were −14.5 and 1.06 MPa, respectively. Both the compressive and tensile strengths decrease significantly by two orders of magnitude with temperature. More important, during the presintering process, the compressive strengths were significantly higher than the tensile strengths, and both strengths decreased dramatically with increasing temperature. Baklouti *et al.* [102] also observed a huge decrease in the mechanical strength during binder burnout.

Slow heating rates during binder burnout can increase productivity by minimizing defect formation [12]. Another novel approach is to minimize the thermal strain during the presintering process [103]. Figure 14 shows a case study in which the addition of kaolinite decreased the shrinkage of designed wet-pressed green bodies (composition C) during heating [103] (Table 4). The four different stages during presintering are (1) drying and binder burnout up to 400°C, (2) dehydroxylation of kaolin between 400 and 600°C, (3) dimension change of samples during heating accompanied by no or negligible weight loss between 600 and 950°C, and (4) the onset of mullitization. The mechanical performance of the designed composition was also better than that of composition B (mullite formers) (Figure 15).

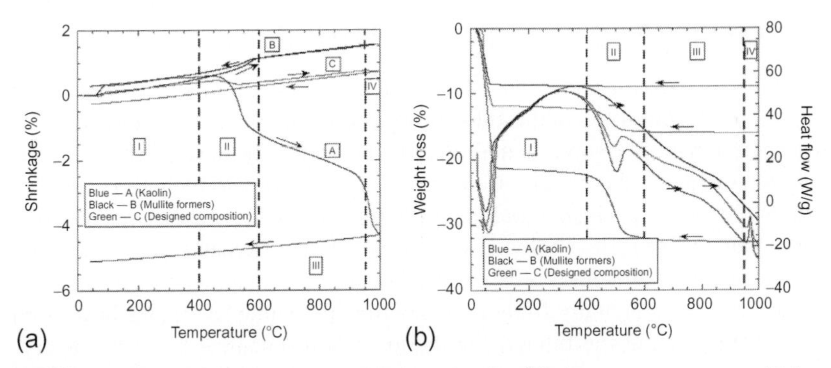

FIGURE 14 Plot of shrinkage versus temperature (a) and weight loss versus temperature (b) during heating at 10°C/min of three different types of green bodies [103].

TABLE 4 Summary of Compositions [103]

Starting Precursors	*A*	*B*	*C*
Aluminum oxide	0	62.9	48.4
Silicon dioxide	0	24.5	11.1
Kaolin	75.4	0	25.4
1 wt% methylcellulose solution	24.6	12.6	15.0

Values are wt%.

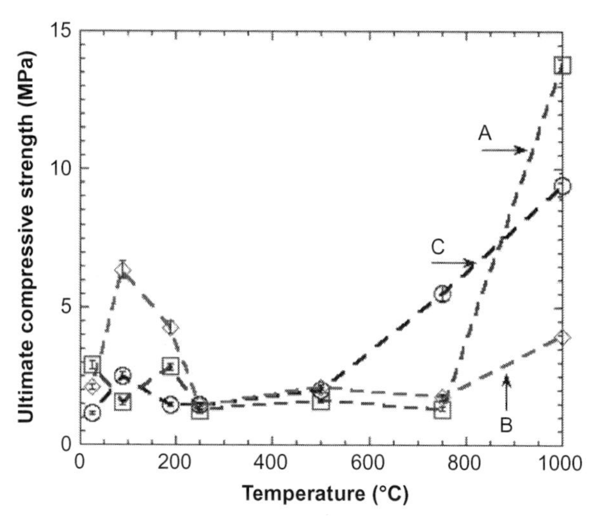

FIGURE 15 Plot of ultimate compressive strength (UCS) versus temperature of different green bodies [103].

7 CONCLUDING REMARKS AND FUTURE DIRECTIONS

Ceramics manufacturing is an energy-intensive process. For example, during the production of ceramic tiles, firing and drying have been identified as hot spots for most environmental effects considered because of the high energy requirements and emissions of acidic gases [104]. It is recommended that detailed life cycle analysis should be performed for a direct comparison of different manufacturing technologies discussed within this chapter.

Green body fabrication is especially important for making ceramics manufacturing environmentally friendly. Numerous fundamental studies have been done to understand the binder distribution and mechanical behavior of dry-pressed green bodies. Similar detailed studies are expected to be used for understanding the mechanical behavior of green bodies produced by novel processing methods. Some of the promising research areas for designing and fabricating green bodies are synthesizing multifunctional organic additives, understanding the mechanical and thermomechanical behavior of green bodies, low-binder and binderless processing, green machining, novel forming techniques such as SFF and/or NNS fabrication, and colloidal processing, among others. Thus, there is a huge potential that further interdisciplinary research will lead to significant faster developments in this technologically important research area.

As a final note, a "sustainable manufacturing paradigm" for designing green bodies for further research can be summarized as additive manufacturing of ceramic performs fabricated with optimized concentrations of environmentally friendly additives, which will allow for faster heating rates during drying and firing cycles with negligible rejection rates by using readily available (or potential for mass production) and highly energy-efficient equipment.

REFERENCES

[1] T. Dyllick, K. Hockerts, Beyond the business case for corporate sustainability, Bus. Strateg. Environ. 11 (2002) 130–141.

[2] S. Schrettle, A. Hinz, M. Scherrer-Rathje, T. Friedli, Turning sustainability into action: explaining firms' sustainability efforts and their impact on firm performance, Int. J. Prod. Econ. 147 (2014) 73–84.

[3] http://esa.un.org/unup/Analytical-Figures/Fig_overview.htm.

[4] http://www.oecd.org/dac/povertyreduction/36301078.pdf.

[5] http://www.fhwa.dot.gov/hfl/.

[6] http://climate.nasa.gov/scientific-consensus.

[7] http://www.unep.org/greeneconomy/Portals/88/documents/ger/GER_7_Manufacturing.pdf.

[8] http://www.trade.gov/competitiveness/sustainablemanufacturing/how_doc_defines_SM.asp.

[9] Reference document on best available techniques in the ceramic manufacturing industry, European Commission, August 2007 (http://eippcb.jrc.ec.europa.eu/reference/BREF/cer_bref_0807.pdf).

[10] J. Peng, Y. Zhao, L. Jiao, W. Zheng, L. Zeng, CO_2 emission calculation and reduction options in ceramic tile manufacture-the foshan case, Energy Procedia 16 (2012) 467–476.

[11] A. Mezquita, E. Monfort, V. Zaera, Ceramic tiles manufacturing and emission trading scheme: reduction of CO_2 emissions, European benchmarking, Bol. Soc. Esp. Ceram. Vidrio. 48 (2009) 211–222.

[12] J.E. Reed, Principles of Ceramics Processing, second ed., John Wiley and Sons Inc., New York, 1995.

[13] S. Gupta, D.J. Green, G.L. Messing, I.M. Peterson, Thermomechanical behavior of ceramic green bodies during presintering, J. Am. Ceram. Soc. 93 (2010) 2611–2616.

[14] http://www.epa.gov/ttnchie1/ap42/ch11/final/c11s07.pdf.

[15] J.L. Amoros, V. Cantavella, J.C. Jarque, C. Felıu, Green strength testing of pressed compacts: an analysis of the different methods, J. Eur. Ceram. Soc. 28 (2008) 701–710.

[16] F. Klocke, Modern approaches for the production of ceramic components, J. Eur. Ceram. Soc. 17 (1997) 457–465.

[17] R. Janssen, S. Scheppokat, N. Claussen, Tailor-made ceramic-based components—advantages by reactive processing and advanced shaping techniques, J. Eur. Ceram. Soc. 28 (2008) 1369–1379.

[18] R. Westerheide, K.A. Driisedau, T. Hollstein, T. Schwickert, H. Zipse, Advances in characterisation of machined green compacts, J. Eur. Ceram. Soc. 17 (1997) 467–472.

[19] B. Sua, S. Dhara, L. Wangb, Green ceramic machining: a top-down approach for the rapid fabrication of complex-shaped ceramics, J. Eur. Ceram. Soc. 28 (2008) 2109–2115.

[20] S. Dhara, Su. Bo, Green machining to net shape alumina ceramics prepared using different processing routes, Int. J. Appl. Ceram. Technol. 2 (2005) 262–270.

[21] W.M. Sigmund, N.S. Bell, L. Bergstrom, Novel powder-processing methods for advanced ceramics, J. Am. Ceram. Soc. 83 (2000) 1557–1574.

[22] J.A. Lewis, Colloidal processing of ceramics, J. Am. Ceram. Soc. 83 (2000) 2341–2359.

[23] J.A. Lewis, J.E. Smay, J. Stuecker, J. Cesarano III, Direct ink writing of three-dimensional ceramic structures, J. Am. Ceram. Soc. 89 (2006) 3599–3609.

[24] B.G. Compton, J.A. Lewis, 3D-printing of lightweight cellular composites. Adv. Mater. 26 (2014) 5930–5935, http://dx.doi.org/10.1002/adma.201401804.

[25] J. Rödel, A.B.N. Kounga, M. Weissenberger-Eibl, D. Koch, A. Bierwisch, W. Rossner, M.J. Hoffmann, R. Danzerf, G. Schneider, Development of a roadmap for advanced ceramics: 2010–2025, J. Eur. Ceram. Soc. 29 (2009) 1549–1560.

[26] X. Yin, N. Travitzky, P. Greil, Near-net-shape fabrication of Ti_3AlC_2-based composites, Int. J. Appl. Ceram. Technol. 4 (2007) 184–190.

[27] F. Niu, D. Wu, S. Zhou, G. Ma, Power prediction for laser engineered net shaping of Al_2O_3 ceramic parts, J. Eur. Ceram. Soc. 34 (2014) 3811–3817.

[28] S.A. Bernard, V.K. Balla, S. Bose, A. Bandyopadhyay, Direct laser processing of bulk lead zirconate titanate ceramics, Mater. Sci. Eng. B 172 (2010) 85–88.

[29] K.V. Wong, A. Hernandez, A review of additive manufacturing, ISRN Mech. Eng. 2012 (2012) 10, Article ID 208760.

[30] H. Rumpf, The strength of granules and agglomerates, in: International Symposium on Agglomeration. Interscience, London, U.K, 1962.

[31] G. Onoda, Theoretical strength of dried green bodies with organic binders, J. Am. Ceram. Soc. 59 (1976) 236–239.

[32] K. Kendall, N.M. Alford, J.D. Birchall, The strength of green bodies, Br. Ceram. Proc. 37 (1986) 255–265.

[33] K. Kendall, Agglomerate strength, Powder Metall. 31 (1988) 28–31.

[34] W.R. Cannon, P.A. Lessing, L.D. Zuck, Crack model for toughness of green parts with moisture or a fluid binder, J. Am. Ceram. Soc. 95 (2012) 2957–2964.

[35] D. Bortzmeyer, G. Langguth, G. Orange, Fracture mechanics of green products, J. Eur. Ceram. Soc. 11 (1993) 9–16.

[36] K. Kendall, N. McNAlford, J.D. Birchall, Elasticity of particle assemblies as a measure of the surface energy of solids, Proc. R. Soc. Lond. A412 (1987) 269–283.

[37] D.J. Green, R. Brezny, C. Nader, The elastic behavior of partially sintered materials, Mater. Res. Soc. Symp. Proc. 119 (1988) 43–48.

[38] P. Arato', E. Besenyei, A. Kele, F. Weber, Mechanical properties in the initial stage of sintering, J. Mater. Sci. 30 (1995) 1863–1871.

[39] T.J. Carneim, D.J. Green, Mechanical properties of dry-pressed alumina green bodies, J. Am. Ceram. Soc. 84 (2001) 1405–1410.

[40] Z. Zhang, D.J. Green, Fracture toughness of spray-dried powder compacts, J. Am. Ceram. Soc. 85 (2002) 1330–1332.

[41] C.W. Nies, G.L. Messing, Effect of glass-transition temperature of polyethylene glycol-plasticized polyvinyl alcohol on granule compaction, J. Am. Ceram. Soc. 67 (1984) 301–304.
[42] R.A. DiMilia, J.S. Reed, Dependence of compaction on the glass transition temperature and of the binder phase, Am. Ceram. Soc. Bull. 62 (1983) 484–488.
[43] R.A. DiMilia, J.S. Reed, Effect of humidity on the pressing characteristics of spray-dried alumina, forming of ceramics, in: J.A. Mangels (Ed.), in: Advances in Ceramics, vol. 9, American Ceramic Society, Columbus, OH, 1984, pp. 38–46.
[44] S. Balasubramanian, D.J. Shanefield, D.E. Niesz, Effect of internal lubricants on defects in compacts made from spray-dried powders, J. Am. Ceram. Soc. 85 (1) (2002) 134–138.
[45] M. Uppalapati, D.J. Green, Effect of external lubricant on mechanical properties of dry-pressed green bodies, J. Am. Ceram. Soc. 88 (2005) 1397–1402.
[46] M.R.B. Romdhane, T. Chartier, S. Baklouti, J. Bouaziz, C. Pagnoux, J.-F. Baumard, A new processing aid for dry-pressing: a copolymer acting as dispersant and binder, J. Eur. Ceram. Soc. 27 (2007) 2687–2695.
[47] E. Lee, K.T. Chou, J.P. Hsu, Sedimentation of a concentrated dispersion of composite colloidal particles, J. Colloid Interface Sci. 295 (2006) 279–290.
[48] Y.K. Leong, P.J. Scales, T.W. Healy, D.V. Boger, Interparticle forces arising from adsorbed polyelectrolytes in colloidal suspensions, Colloids Surf. A Physicochem. Eng. Asp. 98 (1995) 43–52.
[49] M.P. Albano, L.B. Garrido, A.B. Garcia, Ammonium polyacrylate adsorption on "aluminium hydroxides and oxyhydroxide" coated silicon nitride powders, Ceram. Int. 26 (2000) 551–559.
[50] J.P. Boisvert, J. Persello, J.C. Castaing, B. Cabanes, Dispersion mof alumina-coated TiO2 particles by adsorption of sodium polyacrylate, Colloids Surf. A Physicochem. Eng. Asp. 178 (2001) 187–198.
[51] M. Uppalapati, D.J. Green, Effect of relative humidity on the viscoelastic and mechanical properties of spray-dried powder compacts, J. Am. Ceram. Soc. 89 (4) (2006) 1212–1217.
[52] A.S.T.M. Standards Designation D 3967–95., Standard Testing Method for Splitting Tensile Strength of Intact Rock Core Specimens, American Society for Testing and Materials, Philadelphia, PA, 1995.
[53] K. Trustrum, A.S. Jayatilaka, Applicability of Weibull analysis for brittle materials, J. Mater. Sci. 18 (1983) 2765–2770.
[54] S. Baklouti, T. Chartier, C. Gault, J.F. Baumard, Young's modulus of dry-pressed ceramics: the effect of the binder, J. Eur. Ceram. Soc. 19 (1999) 1569–1574.
[55] S. Baklouti, T. Chartier, J.F. Baumard, Mechanical properties of dry-pressed ceramic green products: the effect of the binder, J. Am. Ceram. Soc. 80 (8) (1997) 1992–1996.
[56] K. Uematsu, Immersion microscopy for detailed characterization of defects in ceramic powders and green bodies, Powder Technol. 88 (1996) 291–298.
[57] K. Uematsu, N. Uchida, Z. Kato, S. Tanaka, T. Hotta, M. Naito, Infrared microscopy for examination of structure in spray-dried granules and compacts, J. Am. Ceram. Soc. 84 (2001) 254–256.
[58] S.A. Uhland, R.K. Holman, S. Morissette, M.J. Cima, E.M. Sachs, Strength of green ceramics with low binder content, J. Am. Ceram. Soc. 84 (2001) 2809–2818.
[59] Y. Sato, T. Hotta, K. Nagaoka, M. Asai Watari, S. Kawasaki, Mutual linkage of particles in a ceramic green body through photoreactive organic binders, J. Ceram. Soc. Jpn 113 (2005) 687–691.
[60] K. Sato, Y. Hotta, T. Nagaoka, K. Watari, Microwave-reactive organic binder for ceramics forming, in: Proceedings of International Symposium of EcoTopi Science 2007, ISETS07, 2007, pp. 736–738.

[61] P.C. Hidber, T.J. Graule, L.J. Gaukler, Competitive adsorption of citric acid and poly(vinyl alcohol) onto alumina an its influence on the binder migration during drying, J. Am. Ceram. Soc. 78 (1995) 1775–1780.

[62] S.L. McNamara, J. Rnjak-Kovacina, D.F. Schmidt, T.J. Lo, D.L. Kaplan, Silk as a biocohesive sacrificial binder in the fabrication of hydroxyapatite load bearing scaffolds, Biomaterials 35 (2014) 6941–6953.

[63] O. Lyckfeldt, J.M.F. Ferreira, Processing of porous ceramics by starch consolidation, J. Eur. Ceram. Soc. 18 (1998) 131–140.

[64] T. Nagaoka, T. Tsugoshi, Y. Hotta, K. Sato, K. Watari, Fabrication of porous alumina ceramics by new eco-friendly process, J. Ceram. Soc. Jpn 113 (2005) 87–91.

[65] T. Shirai, M. Yasuoka, K. Watari, Novel binder-free forming of Al_2O_3 ceramics by microwave-assisted hydration reaction, Mater. Sci. Eng. B 148 (2008) 221–225.

[66] S. Deville, Freeze-casting of porous ceramics: a review of current achievements and issues, Adv. Eng. Mater. 10 (2008) 155–169.

[67] S.W. Sofie, F. Dogan, Freeze casting of aqueous alumina slurries with glycerol, J. Am. Ceram. Soc. 84 (2001) 1459–1464.

[68] T. Fukasawa, M. Ando, T. Ohji, S. Kanzaki, Synthesis of porous ceramics with complex pore structure by freeze-dry processing, J. Am. Ceram. Soc. 84 (2001) 230–232.

[69] T. Huang, M.S. Mason, X. Zhao, G.E. Hilmas, M.C. Leu, Aqueous-based freeze-form extrusion fabrication of alumina components, Rapid Prototyp. J. 15 (2009) 88–95.

[70] T. Kosmai, S. Novak, M. Sajko, Hydrolysis-assisted solidification (HAS): a new setting concept for ceramic net-shaping, J. Eur. Ceram. Soc. 17 (1997) 427–432.

[71] S. Gupta, M.F. Riyad, Oxidation-induced sintering: an innovative method for manufacturing porous ceramics. Int. J. Appl. Ceram. Technol. 11 (2014) 1–7, http://dx.doi.org/10.1111/ijac.12282.

[72] O. Omatete, M.A. Janney, S.D. Nunn, Gelcasting: from laboratory development toward industrial production, J. Eur. Ceram. Soc. 17 (1997) 407–413.

[73] E. Medvedovski, M. Peltsman, Low pressure injection moulding mass production technology of complex shape advanced ceramic components, Adv. Appl. Ceram. 111 (2012) 333–344.

[74] S.C. Danforth, M. Agarwala, A. Bandyopadghyay, N. Langrana, V.R. Jamalabad, A. Safari, R. Weeren, Solid freeform fabrication methods. United States patents, 5, 738,817 (1998).

[75] M. Allahverdi, S.C. Danforth, M. Jafari, A. Safari, Processing of advanced electroceramic components by fused deposition technique, J. Eur. Ceram. Soc. 21 (2001) 1485–1490.

[76] J.C. Nelson, N.K. Vail, J.W. Barlow, J.J. Beaman, D.L. Bourell, Harris L. Marcus, Selective laser sintering of polymer-coated silicon carbide powders, Ind. Eng. Chem. Res. 34 (1995) 1641–1651.

[77] K. Shahzada, J. Deckersb, J.-P. Kruth, J. Vleugelsa, Additive manufacturing of alumina parts by indirect selective laser sintering and post processing, J. Mater. Process. Technol. 213 (2013) 1484–1494.

[78] L. Weisensel, N. Travitzky, H. Sieber, P. Greil, Laminated object manufacturing (LOM) of SiSiC composites. Adv. Eng. Mater. 6 (2004) 899–903, http://dx.doi.org/10.1002/adem.200400112.

[79] J. Cesarano, P.D. Calvert, Freeforming objects with low-binder slurry. United States Patent, 6,027,326 (2000).

[80] J. Cesarano, R. Segalman, P. Calvert, Robocasting provides moldless fabrication from slurry deposition, Ceram. Ind. 148 (1998) 94–102.

[81] S.L. Morissette, J.A. Lewis, J. Cesarano, D.B. Dimos, T. Baer, Solid freeform fabrication of aqueous alumina-poly(vinyl alcohol) gelcasting suspensions, J. Am. Ceram. Soc. 83 (2000) 2409–2416.
[82] B.A. Tuttle, J.E. Smay, J. Cesarano III, J.A. Voigt, T.W. Scofield, W.R. Olson, J.A. Lewis, Robocast $Pb(Zr_{0.95}Ti_{0.05})O_3$ ceramic monoliths and composites, J. Am. Ceram. Soc. 84 (2001) 872–874.
[83] J.N. Stuecker, J. Cesarano, D.A. Hirschfeld, Control of the viscous behavior of highly concentrated mullite suspensions for robocasting, J. Mater. Process. Technol. 142 (2003) 318–325.
[84] J.E. Smay, J. Cesarano III, J.A. Lewis, Colloidal inks for directed assembly of 3-D periodic structures, Langmuir 18 (2002) 5429–5437.
[85] J.E. Smay, G. Gratson, R.F. Shepard, J. Cesarano III, J.A. Lewis, Directed colloidal assembly of 3D periodic structures, Adv. Mater. 14 (2002) 1279–1283.
[86] J.G. Dellinger, J. Cesarano, R.D. Jamison, Robotic deposition of model hydroxyapatite scaffolds with multiple architectures and multiscale porosity for bone tissue engineering, J. Biomed. Mater. Res. A 82 (2007) 383–394.
[87] F.F. Lange, Forming a ceramic by flocculation and centrifugal casting. United States Patent, 4,624,808 (1986).
[88] W. Huisman, T. Graule, L.J. Gauckler, Alumina of high reliability by centrifugal casting, J. Eur. Ceram. Soc. 15 (1995) 811–821.
[89] L. Figiela, Z. Jaworska, P. Pedzich, P. Wyzga, P. Putyra, P. Klimczyk, Al_2O_3 and ZrO_2 powders formed by centrifugal compaction using the ultra HCP method, Ceram. Int. 39 (2013) 635–640.
[90] L. Besra, M. Liu, A review on fundamentals and applications of electrophoretic deposition (EPD), Prog. Mater. Sci. 52 (2007) 1–61.
[91] A.J. Fanelli, R.D. Silvers, W.S. Frei, J.V. Burlew, G.B. Mars, New aqueous injection molding process for ceramic powders, J. Am. Ceram. Soc. 72 (1989) 1833–1836.
[92] A.J. Millan, M.I. Nieto, R. Moreno, Aqueous injection moulding of silicon nitride, J. Eur. Ceram. Soc. 20 (2000) 2661–2666.
[93] O.O. Omatete, M.A. Janney, R.A. Strehlow, Gelcasting—a new ceramic forming process, Am. Ceram. Soc. Bull. 70 (1991) 1641–1649.
[94] A.C. Young, O.O. Omatete, M.A. Janney, P.A. Menchhofer, Gelcasting of alumina, J. Am. Ceram. Soc. 74 (1991) 612–618.
[95] R. Gilissen, J.P. Erauw, A. Smolders, E. Vanswijgenhoven, J. Luyten, Gelcasting, a near net shape technique, Mater. Des. 21 (2000) 251–257.
[96] J. Yang, J. Yu, Y. Huang, Recent developments in gelcasting of ceramics, J. Eur. Ceram. Soc. 31 (2011) 2569–2591.
[97] M.L. Griffith, J.W. Halloran, Freeform fabrication of ceramics via stereolithography, J. Am. Ceram. Soc. 79 (1996) 2601–2608.
[98] D.T. Pham, R.S. Gault, A comparison of rapid prototyping technologies, Int. J. Mach. Tool Manuf. 38 (1998) 1257–1287.
[99] L.J. Gaukler, Th. Graule, F. Baader, Ceramic forming using enzyme catalyzed reactions, Mater. Chem. Phys. 61 (1999) 78–102.
[100] K. Prabhakaran, A. Melkeri, N.M. Gokhale, T.K. Chongdar, S.C. Sharma, Direct coagulation casting of YSZ powder suspensions using MgO as coagulating agent, Ceram. Int. 35 (2009) 1487–1492.

[101] K. Prabhakaran, S. Raghunath, A. Melkeri, N.M. Gokhale, S.C. Sharma, Novel coagulation method for direct coagulation casting of aqueous alumina slurries prepared using a poly(acrylate) dispersant, J. Am. Ceram. Soc. 91 (2) (2008) 615–619.
[102] S. Baklouti, J. Bouaziz, T. Chartier, J.-F. Baumard, Binder burnout and evolution of the mechanical strength of dry-pressed ceramics containing poly(vinyl alcohol), J. Eur. Ceram. Soc. 21 (2001) 1087–1092.
[103] R. Johnson, T. Hammann, M. Sauka, H. Feilen, Intelligent design of ceramic green bodies for smart manufacturing. Senior Design Project, Advisor—Prof. S. Gupta (2012-13).
[104] V. Ibáñez-Forés, M.D. Bovea, A. Azapagic, Assessing the sustainability of best available techniques (BAT): methodology and application in the ceramic tiles industry, J. Clean. Prod. 51 (2013) 162–176.

Part IV

Sustainable Manufacturing—Polymeric and Composite Materials

Chapter 20

Adoption of an Environmentally Friendly Novel Microwave Process to Manufacture Carbon Fiber-Reinforced Plastics

Y. Hotta
National Institute of Advanced Industrial Science and Technology (AIST), Nagoya, Japan

1 MANUFACTURING METHODS OF CARBON FIBER-REINFORCED PLASTICS

In the transportation field, such as automobiles and aircrafts, efforts to decrease oil dependency have been accelerated. Technology to create light-weight structural materials used for automobiles or aircrafts is one of the most important in research and development. In particular, the replacement of metal with fiber-reinforced plastics is expected to be an effective means of weight reduction. In recent years, carbon fiber-reinforced plastic (CFRP) with light-weight and high mechanical properties has attracted attention, and the development of novel manufacturing technologies with the aim of improving productivity has become an important issue.

There are two types of CFRPs: carbon fiber-reinforced thermosetting (CFRTS) and carbon fiber-reinforced thermoplastics (CFRTPs). Figure 1 shows characteristics of thermosetting [1] and thermoplastic resins [2]. Most CFRPs, which are used for aircrafts, automobiles, and sports fields, are CFRTS that consist of carbon fiber (CF) and a thermosetting resin as an epoxy. In general, the manufacturing time for CFRTS is long because thermosetting resin is cured by a chemical reaction between the base resin and the curing agent (Figure 1a). Therefore, the development of novel manufacturing technology for promoting the curing of matrix resin is expected to reduce production costs and to improve productivity. On the other hand, thermoplastic resin is solid at ordinary temperatures and is melted by the movement of molecules with a linear chemical structure upon heating (Figure 1b). Therefore, forming a CFRTP consisting of CF and a thermoplastic resin is easier than forming CFRTS. Press, injection, and extrusion methods with superior productivity can be applied to form CFRTPs.

Green and Sustainable Manufacturing of Advanced Materials. http://dx.doi.org/10.1016/B978-0-12-411497-5.00020-5

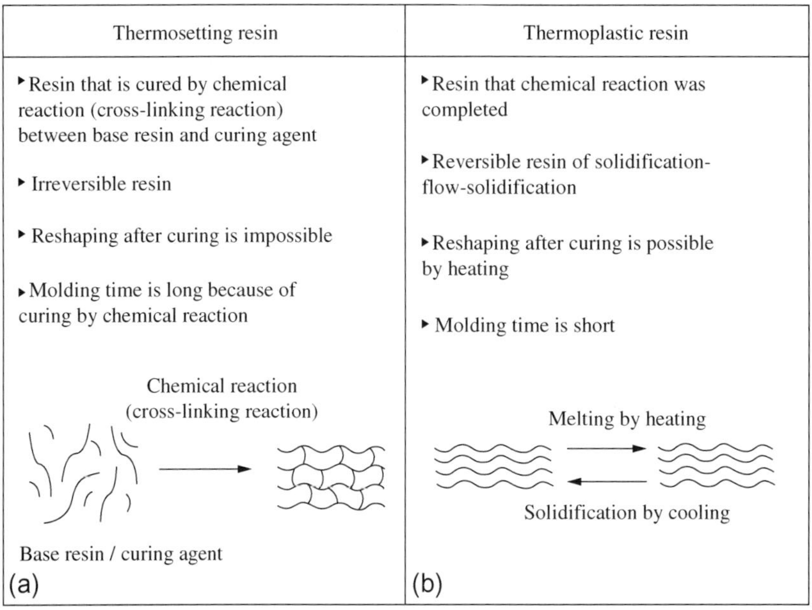

Thermosetting resin	Thermoplastic resin
▸ Resin that is cured by chemical reaction (cross-linking reaction) between base resin and curing agent ▸ Irreversible resin ▸ Reshaping after curing is impossible ▸ Molding time is long because of curing by chemical reaction Chemical reaction (cross-linking reaction) Base resin / curing agent (a)	▸ Resin that chemical reaction was completed ▸ Reversible resin of solidification-flow-solidification ▸ Reshaping after curing is possible by heating ▸ Molding time is short Melting by heating Solidification by cooling (b)

FIGURE 1 Characteristics of thermosetting (a) and thermoplastic (b) resins.

General CFRP manufacturing methods are shown in Figure 2. The various manufacturing methods are applied according to discontinuous or continuous CFs. When manufacturing CFRPs with high mechanical properties, CFRPs are prepared by compositing continuous CFs and resin, but the duration of the manufacturing time becomes a problem from the viewpoint of production costs. Because autoclave and oven heating systems are used in the curing process of a thermosetting resin, in particular for CFRP as a material for aircraft, , several hours are required for CFRP manufacturing with large energy consumption. Therefore, the development of novel CFRP manufacturing technologies without using the autoclave method has been explored recently.

2 RAPID RESIN-CURING OF CFRPs (A CF-REINFORCED EPOXY COMPOSITE) BY MICROWAVES

CFs have excellent strength and stiffness [3]. Therefore, CFs, which are a new breed of material, have become quite important in various industries as a kind of reinforcement for polymeric composites. Epoxy of a thermosetting resin is generally adopted as a matrix component of CFRP composite. Composite materials that consist of CFs and a thermosetting resin is called CFRTS. Epoxy resin is cured by chemical reactions between a base resin and a curing agent at a high temperature. CFRTS composites are usually cured thermally in autoclaves or closed cavity tools such as an electric oven. The curing process for epoxy resin must be done several times [4]. Moreover, energy consumption during the

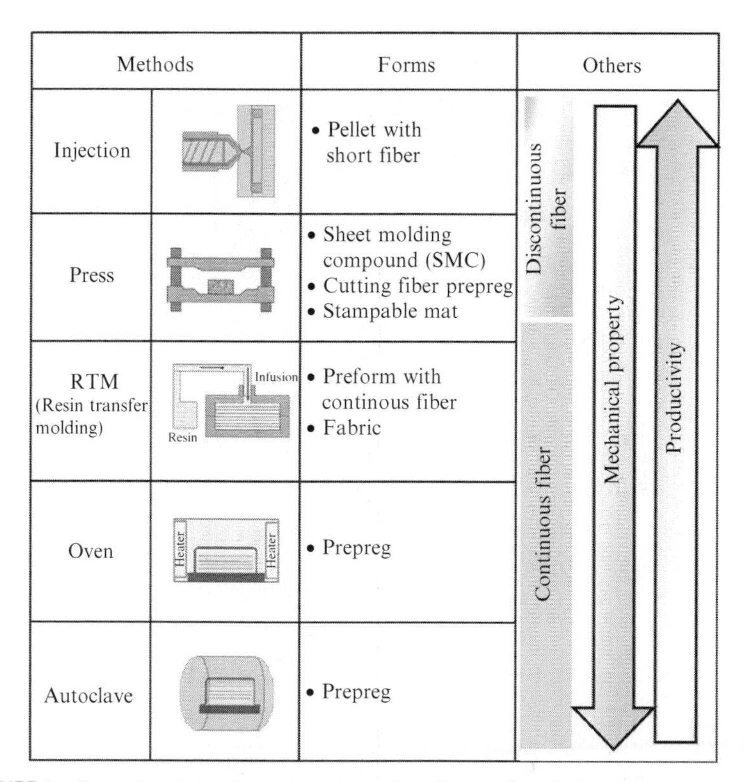

FIGURE 2 General methods of manufacturing carbon fiber-reinforced plastics.

curing process of an epoxy resin is enormous [5]. Therefore, the development of novel resin curing methods with rapid processing time and energy-saving properties has been expected in CFRTS fields [6,7].

Microwaves are being used in various technological and scientific fields to heat carbon-based composites [8,9]. Microwaves can penetrate materials and deliver energy directly to materials via molecular interaction with the electromagnetic field. Thus, it is more appropriate to consider heating by microwave irradiation as conversion of electromagnetic energy to thermal energy rather than heat transfer [8,10]. In general, the output power of microwaves is represented by Equation (1):

$$P = 1/2\sigma|E|^2 + \pi f \varepsilon_0 \varepsilon''|E|^2 \tag{1}$$

$$\varepsilon'' = \varepsilon' \tan\delta \tag{2}$$

where P is the output power, E is the electric field, f is the frequency, σ is the conductivity, δ is the dielectric loss angle, ε_0 is permittivity in a vacuum, ε' is the dielectric constant, and ε'' is dielectric loss of material. Assuming that all of output power is converted to heat generation, the heat generation is enhanced in case of high dielectric constant of materials. The dielectric constant

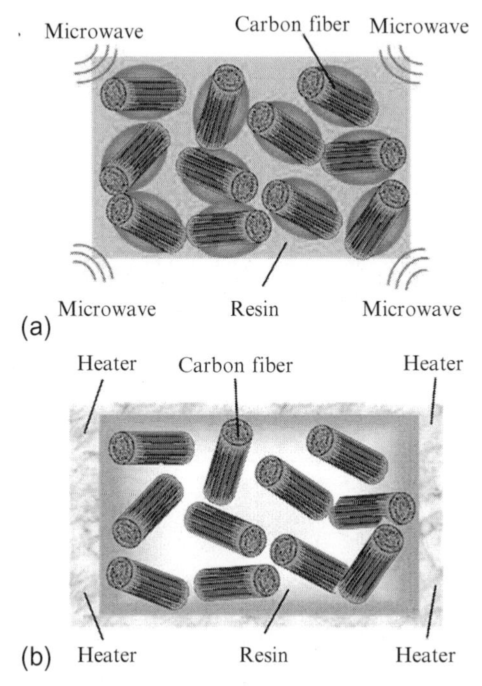

FIGURE 3 Schematic of the heating method of carbon fiber-reinforced plastics: internal heating by microwaves (a) and external heating by an autoclave or oven (b).

of a CF is high compared with that of an epoxy resin, indicating that CFs in CFRPs selectively adsorb microwaves. Thus, internal heating is caused rapidly by microwave irradiation (Figure 3a). The heating mechanism of CFRPs by microwave irradiation is significantly different from that using external heating such as an electric oven or autoclave (Figure 3b). Therefore, microwave processing has recently attracted attention as a rapid thermosetting resin-curing system of CFRTS composites. Attempts to cure a thermosetting resin in CFRTS composites using microwave processing have been reported [11–13], and selective heating gives unique properties to CFRTS composites.

There are two types of CFRP composites: those with continuous CFs and those with discontinuous CFs. Discontinuous CFRS composites are attractive because of their ease of use in fabrication and their economic efficiency [14]. In this section are described the effects of resin curing in discontinuous CF-reinforced epoxy composites using the microwave process, and the effectiveness of the microwave process in CFRTS manufacturing is introduced.

Discontinuous CF-reinforced epoxy composites, which were prepared using microwave irradiation, were reported to have similar mechanical properties after a shorter curing time as a composite cured by conventional heating using an electric oven [15]. Figure 4 shows the flexural modulus and flexural strength of CF-reinforced epoxy composites (5 vol% CF); 3-mm CFs were prepared

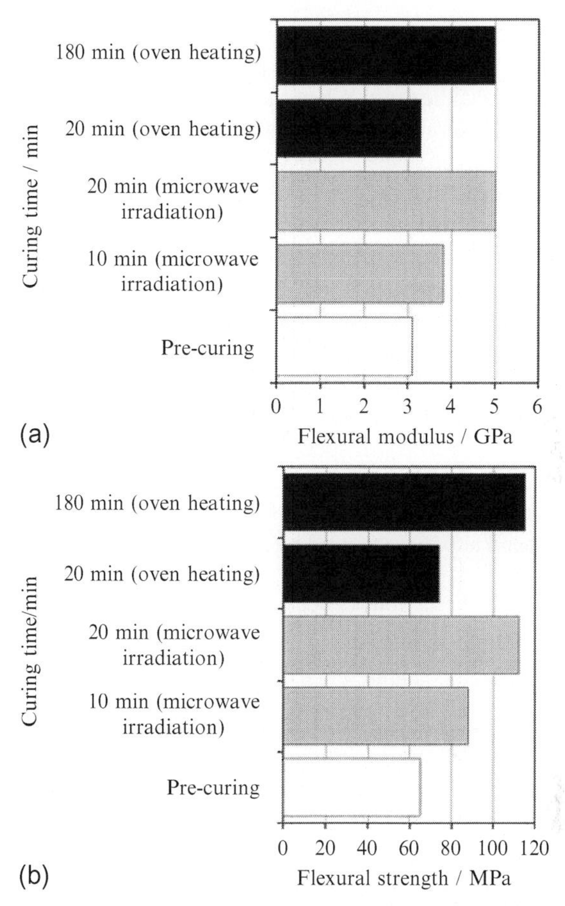

FIGURE 4 Flexural modulus (a) and flexural strength (b) dependence of curing time on microwave (MW)- and conventional heating-cured carbon fiber-reinforced thermoplastics (carbon fiber reinforced epoxy composites [5 vol% CFs] with 3-mm CFs) at 120°C. White, gray, and black bars show precured CF-reinforced thermosetting for 24 h at room temperature, microwave-irradiated CFRTS, and conventional heated CFRTS, respectively.

using microwave irradiation and conventional heating at 120°C, and Figure 4 shows the mechanical properties of a precured CF-reinforced epoxy composite that was cured for 1440 min (24 h) at room temperature. It can be seen that the flexural modulus of the precured CF-reinforced epoxy composite is 3.1 GPa. The flexural modulus of the CF-reinforced epoxy composite prepared by conventional heating for 20 min is 3.3 GPa, indicating that the curing epoxy does not progress at a curing time of 20 min. On the other hand, in the case of CF-reinforced epoxy composites irradiated by microwaves for 10 and 20 min, the flexural moduli are 3.8 and 5.0 GPa, respectively, indicating that the mechanical property is increased rapidly by microwave irradiation for 20 min. This suggests that the reaction between the epoxy and the curing agent in a CF-reinforced epoxy composite progresses under a short period of microwave irradiation compared with conventional heating. When the curing time with conventional heating is 180 min, the flexural modulus is 5.0 GPa. The value is similar to that of CF-reinforced epoxy composite cured with microwaves for 20 min at 120°C. Because flexural strength shows the same tendency as the flexural modulus, microwave irradiation can rapidly cure resins in a discontinuous CF-reinforced epoxy composite compared with conventional heating. The microwave process leads to the rapid resin-curing of CFRTS, with a short curing time and enhanced mechanical properties of discontinuous CF-reinforced plastic composites.

A diagnostic band from six-membered rings of an epoxy resin and an absorption feature ascribed to epoxide groups can be seen at 831 and 916 cm^{-1}, respectively, upon infrared spectra measurement [16,17]. While the former absorption band can be observed irrespective of the degree of reaction with the epoxy and curing agent, the latter diminishes as the reaction proceeds. The degree of the reaction can be estimated by absorbance ratios by dividing the integrated intensity of the band at 916 cm^{-1} band by that of the band at 831 cm^{-1}:

$$\left[\text{Degree of reaction between epoxy and curing agent}\right] = \begin{bmatrix}\text{Integrated intensity of} \\ 916\ \text{cm}^{-1}\ \text{band}\end{bmatrix} \Big/ \begin{bmatrix}\text{integrated intensity of} \\ 831\ \text{cm}^{-1}\ \text{band}\end{bmatrix}$$

Thus, the diminished absorbance ratio defined above can be used to monitor the progress of the curing reaction of the epoxy resin. The absorbance ratios of a CF-reinforced epoxy composite prepared using conventional heating for 180 min and microwave irradiation for 20 min are compared in Figure 5.

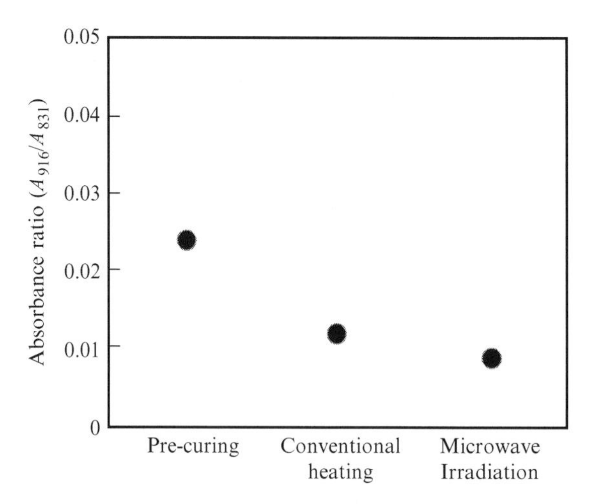

FIGURE 5 Calculated absorbance ratios for the infrared spectra of precured CFRTS and CFRTS composites prepared by conventional heating and microwave irradiation.

The curing reaction of epoxy in the CF-reinforced epoxy composite prepared by microwave irradiation is faster than the one prepared using conventional heating, indicating that it is effective to use microwave processing in CFRP manufacturing.

Figure 6 shows the fracture surfaces of CF-reinforced epoxy composites prepared by conventional oven heating and microwave irradiation. The surface of the CF is clear in the CF-reinforced epoxy composite that was cured by conventional heating, whereas the resin obviously remains on the surface of the CFs in the microwave-irradiated CF-reinforced epoxy composite. This is an indication that better adhesion between the CF surface and the epoxy resin occurs in the microwave-irradiated system than in the conventional heating system.

Duty output power dependence on curing temperature is shown in Figure 7 when discontinuous CF-reinforced epoxy composites (CFRTS), which were prepared from CFs of different lengths (130 μm or 3 mm), were irradiated with microwaves for 20 min at 80°C, 100°C, and 120°C. The duty output power is the quantity of microwaves needed to maintain the setting temperature. In the case of a CF-reinforced epoxy composite with short CFs (130 μm), the duty output power increases as the curing temperature increases. On the other hand, in the case of CF reinforced epoxy composites with 3-mm CF, the duty output power is almost constant be 6% at the 80°C, 100°C, and 120°C curing temperatures. Thus, this system suggests that the heating of the composite used only 6% of the

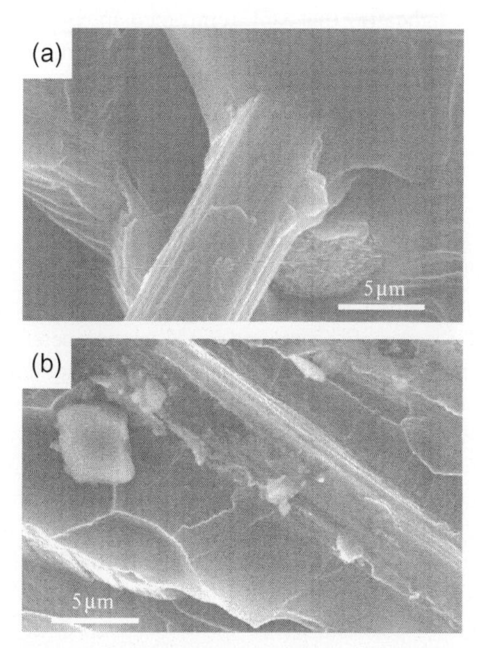

FIGURE 6 Fracture surfaces of carbon fiber-reinforced epoxy composites (CFRTS) prepared by conventional oven heating (a) and microwave irradiation (b).

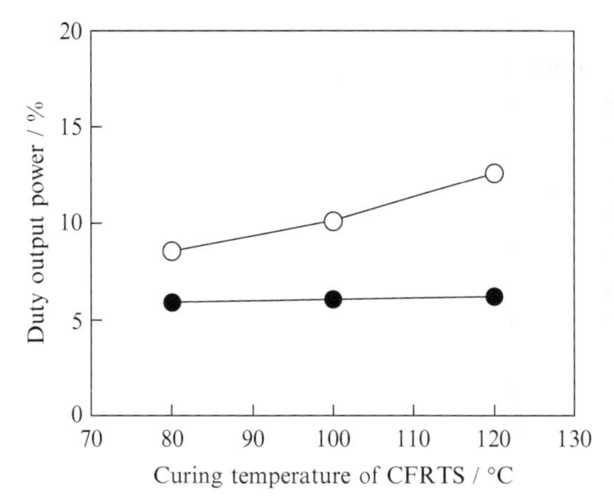

FIGURE 7 Duty output power of microwave generator dependence of CFRTS curing temperature. White and black points show carbon fiber-reinforced epoxy composites consisting of 130-μm (white circles) and 3-mm (black circles) carbon fibers, respectively.

microwave power. Furthermore, it can be seen that the duty output power in CF-reinforced epoxy composites with 3-mm CFs is lower than those with 130-μm CFs. At a 120°C curing temperature, the duty output power in CF-reinforced epoxy composites with 3-mm CFs is 50% lower than that of the composites with short CFs (130 μm), indicating that a CF-reinforced epoxy composite composed of long CFs is heated effectively by microwave irradiation with low energy. This means that energy consumption for resin curing in CFRTS by microwave irradiation is affected by the length of the CFs.

Furthermore, heating efficiency (*H*) can be estimated using Equation (3) [18,19]:

$$H = 1.66 \times 10^{-2} \times mC\Delta T / P \quad (3)$$

where P (kW) is output power, m (kg/min) is treatment amount per one minute, C (J/kg °C) is specific heat, and T (°C) is temperature. The relationship between heating efficiency by microwaves and curing temperature on CF-reinforced epoxy composites (CFRTS) with 130-μm and 3-mm CFs is shown in Figure 8. The heating efficiency (*H*) of CF-reinforced epoxy composites that consisted of 130-μm CFs is almost constant, whereas the heating efficiency (*H*) of microwave-irradiated CF-reinforced epoxy composites with 3-mm CFs is enhanced by increasing the curing temperature and is higher than a composite made of short fibers. This indicates that the length of the CFs in a CF-reinforced epoxy composite obviously contributes to heating efficiency during microwave processing.

The epoxy curing rate of a CF-reinforced epoxy composite during microwave processing is faster than one using conventional heating, and microwave

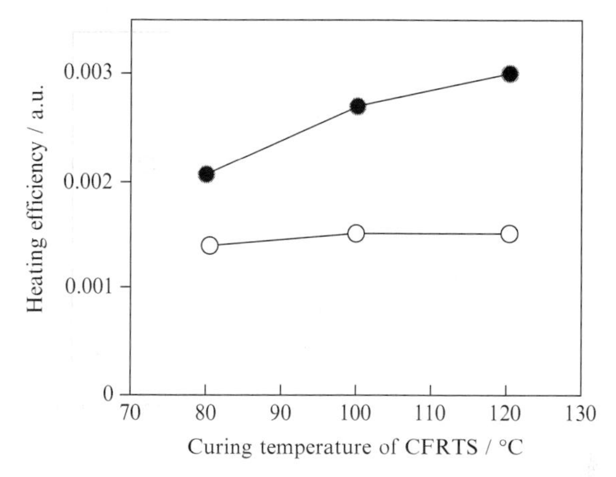

FIGURE 8 Relationship between heating efficiency by microwave irradiation and curing temperature on carbon fiber-reinforced epoxy composites with 5 vol% carbon fibers of 130 μm (white circles) and 3 mm (black circles). a.u., arbitrary units.

processing promotes resin curing in a shorter amount of time. Curing process kinetics, which can be estimated based on mechanical properties, is represented in Equation (4):

$$F^2 - F_0^2 = Kt + A \tag{4}$$

$$K = \left(8d\gamma V_{\mathrm{m}} / N_{\mathrm{A}} C\right)\exp\left(-\Delta G^* / RT\right) \tag{5}$$

where F is the flexural modulus at curing time t and F_0 is the initial flexural modulus, A is the constant value, d is the moving distance of the molecule, γ is the surface energy, V_{m} is the molar volume, N_{A} is Avogadro's number, C is Planck's constant, R is the gas constant, T is the curing temperature, and ΔG^* is the activation energy for the curing process. When $F \gg F_0$, the correlation between the flexural modulus and the curing time is represented as Equation (6):

$$F^2 = \left[\left(8d\gamma V_{\mathrm{m}} / N_{\mathrm{A}} C\right)\exp\left(-\Delta G^* / RT\right)\right]t + A \tag{6}$$

Figure 9 shows the square of the flexural modulus dependence on curing time for CF-reinforced epoxy composites with 3-mm CFs that were prepared using microwave and conventional curing at 120°C. By fitting these data into Equation (6), the activation energy for the curing process (ΔG^*) can be estimated. in general, a chemical reaction such as epoxy resin-curing contributes to the amount of activation energy (ΔG^*) as Gibbs free energy. From Figure 9, the activation energies for resin curing in a CF-reinforced epoxy composite undergoing microwave processing are 30% lower compared those undergoing conventional heating, indicating that resin-curing by microwave processing is

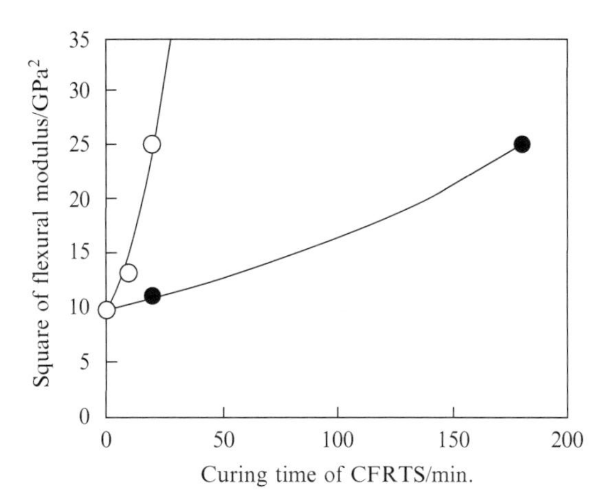

FIGURE 9 Square of the flexural modulus dependence on curing time for carbon fiber-reinforced epoxy composites with 3-mm CFs, which were prepared by microwave irradiation (white circles) and conventional curing (black circle) at 120°C. CFRTS, carbon fiber-reinforced thermosetting.

promoted compared with conventional heating. Therefore, microwave processing is suitable for rapid CFRTS manufacturing.

3 EFFECT OF MICROWAVE IRRADIATION ON CONTINUOUS CFRPs

In Section 2, the effectiveness of microwaves for rapid resin-curing of CFRTS was introduced from the viewpoint of resin reactive kinetics. CFRPs that are used for automobiles and aircrafts are required to have high mechanical properties, which are achieved by compositing continuous CFs and thermosetting resins. Furthermore, the reduction of CFRP manufacturing cost is important. From these perspectives, vacuum-assisted resin transfer molding (VaRTM) has recently attached attention as a low-cost manufacturing method without an autoclave [20–22]. A schematic of the VaRTM process is shown in Figure 10. In the VaRTM process, dry fabrics of continuous CFs are impregnated with a thermosetting resin using a vacuum bag and a vacuum pump. The impregnated

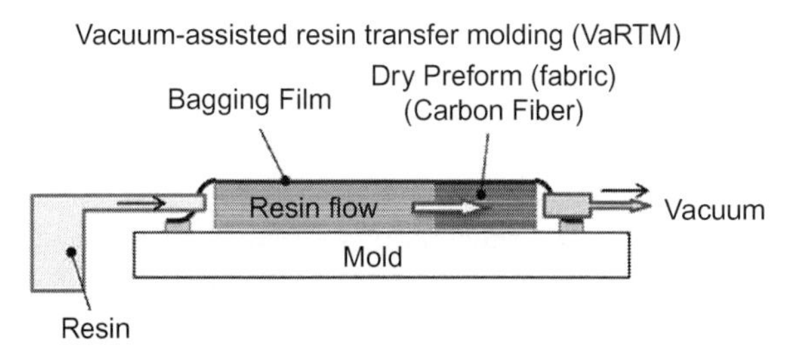

FIGURE 10 Schematic of the vacuum-assisted resin transfer molding (VaRTM) process.

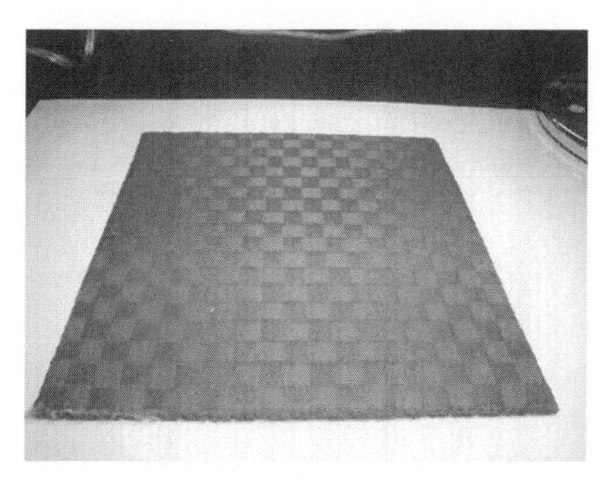

FIGURE 11 Carbon fiber-reinforced thermosetting fabricated using the vacuum-assisted resin transfer molding method.

resin is then cured by heating system. A CFRTS fabricated using the VaRTM method is represented in Figure 11.

In CFRTS manufacturing it is important that the matrix resin is cured rapidly and the cured CFRTS has high mechanical properties. Figure 12 shows an example of the flexural strength of CFRTS that were prepared using VaRTM method after conventional heating at 120°C for 3h and microwave irradiation at 120°C for 20min. The flexural strength of CFRTS cured by conventional heating at 120°C for 3h is 650MPa. On the other hand, for CFRTS cured by

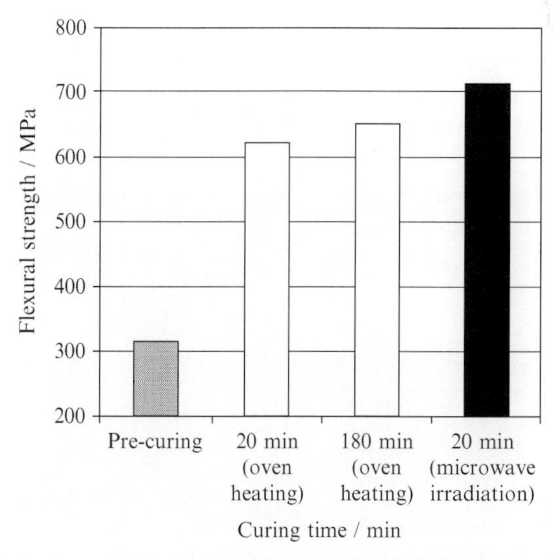

FIGURE 12 Flexural strength of carbon fiber-reinforced thermosettings that were prepared using the vacuum-assisted resin transfer molding method after conventional heating at 120°C for 20 and 180min (3h) and microwave irradiation at 120°C for 20min.

microwave irradiation, a flexural strength of 700 MPa or more was observed. Thus, adapting the microwave process for CFRTS manufacturing can lead to the rapid curing of matrix resin and improvement of mechanical properties. These results indicate that microwaves are an effective manufacturing process for CFRTPs.

4 EFFECT ON THERMAL CONDUCTIVITY OF CFRPs MATRIX RESIN ON MICROWAVE PROCESS

The previous section introduced microwave processing as effective for rapid resin-curing of CFRTS consisting of CFs and a thermosetting resin as the matrix. In this section, studies of microwave processing of CFRTP, which consists of a thermoplastic resin as the matrix, is reviewed and discussed.

Research and development for using thermoplastics as the resin matrix of CFRTPs has recently attracted attention because thermoplastics are easily able to be thermoformed [23]. The resin properties lead to excellent recycling, repairability, and reduction of manufacturing costs associated with productivity, such as a shortened manufacturing time. Therefore, the development of a novel technology for the rapid manufacture of CFRTPs with excellent mechanical properties is strongly demanded from industrial fields such as the automobile industry. However, the interface between CFs and the thermoplastic matrix in CFRTPs is easily damaged because CFs in CFRTPs are heated by an electric charge such as thunder or static electricity [24]. In fact, by applying voltage or microwave irradiation to CFRTPs, the CFs in CFRTPs are rapidly heated. As a result, the matrix resin between the CFs is melted and decomposed, as shown in Figure 13.

Thermal conductivity of the thermoplastic matrix is much low compared with that of the CFs. It is known that the thermal conductivity of general thermoplastics such as polypropylene (PP) and polyamide (PA) is 0.15~0.3 W/m·K. On the other hand, carbon fibers (CFs) have thermal conductivity of 10 W/m·K.

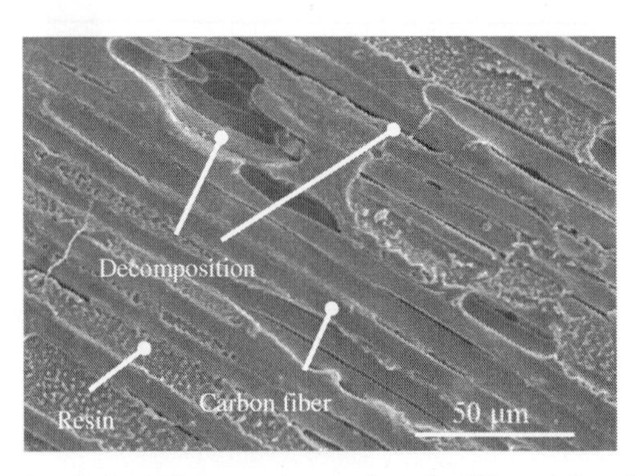

FIGURE 13 Scanning electron microscopy image of a carbon fiber-reinforced thermoplastic surface after applying voltage of 30 kV.

Thus, the heat is confined to the interfacial surface between the CFs and the thermoplastic matrix if CFs in the CFRTP are heated, indicating that thermal degradation of the thermoplastic matrix occurs at the interface between the CFs and the thermoplastic matrix in CFRTPs. Therefore, application of the microwave process for CFRTP production is difficult, and there are few applicable research reports. If the thermoplastic matrix of a CFRTP has a larger thermal conductivity, the thermal degradation at the interface between the CFs and the thermoplastic matrix may be inhibited by releasing heat to the outside or by promoting homogenous heating propagation in a composite.

Many reports of highly thermally conductive composites using hexagonal boron nitride (h-BN) have recently been published in order to improve the thermal conductive property of resins [25–27]. h-BN has a plate-like shape with flat surfaces corresponding to the basal planes of the hexagonal crystal structure. h-BN has a high thermal conductivity of $\geq$150 W/m K and excellent high-temperature reactivity as its physical properties, together with its light weight. Thus, by using a resin composite with improved thermal conductivity for a CFRTP matrix, the efficient transfer of heat to the matrix from the CFs heated by microwave irradiation can be expected, resulting in an expected suppressive effect of thermal degradation on the CFRTP. In this section the effect of a thermally conductive matrix for inhibiting the thermal degradation of a CFRTP composed of CFs, PA6, and h-BN as the thermal conductive ceramic powder is reviewed.

Figure 14 shows optical images of a conventional CFRTP with PA6 as the matrix and a CFRTP with h-BN/PA6 as the matrix after microwave irradiation for different durations at 1600 W power. Microwave irradiation was performed for 15, 30, and 60 s in the case of the conventional CFRTP, whereas the CFRTP with h-BN/PA6 was irradiated by microwaves for 60 s. Obviously, it can be seen that the conventional CFRTP without h-BN was not melted by microwave irradiation for 15 s and that the central part of the conventional CFRTP was melted via rapid heating of the CFs by microwave irradiation for 30 s. On the other hand, the CFRTP with h-BN/PA6 with a thermally conductive property was not melted by microwave irradiation for 60 s, which is a long time compared with the conventional CFRTP without h-BN. This phenomenon might be caused by

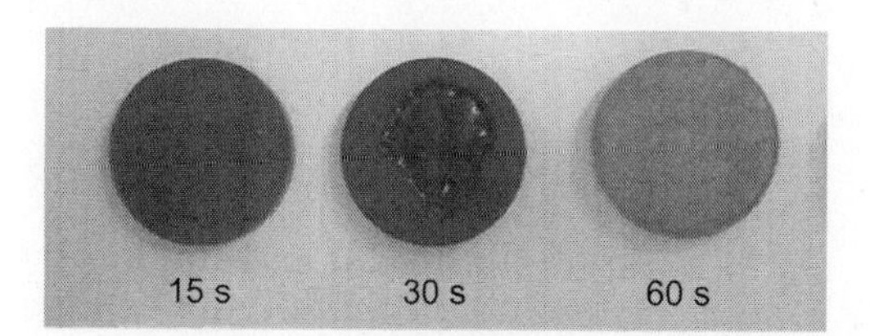

FIGURE 14 Optical images of a conventional carbon fiber-reinforced thermoplastic (CFRTP) without hexagonal boron nitride (h-BN) after microwave irradiation for 15 (left) and 30 s (center), and a CFRTP with h-BN after microwave irradiation for 60 s (right).

the heat that was conducted to the entire CFRTP through the matrix from CFs heated by microwave irradiation.

The fractural images of a conventional CFRTP irradiated by microwaves for 15 s and a CFRTP with h-BN irradiated by microwaves for 60 s at 1600 W power are shown in Figure 15a and b, respectively. In the case of the conventional CFRTP without h-BN, shown in Figure 14, the composite was not melted by microwave irradiation for 15 s. Observation of the microstructure, however, indicates that the resin degraded at the interface between the CFs and the PA6 resin, as shown in Figure 15a. On the other hand, in the CFRTP with h-BN irradiated by microwaves for 60 s, matrix degradation is not observed at the interface between the CFs and the matrix, as shown in Figure 15b, meaning that the improvement of thermal conductivity on a CFRTP matrix is one of the important factors in the development of CFRTP manufacturing using microwaves. It is effective for suppressing thermal degradation in CFRTPs to enhance the thermal conductivity of the matrix using a ceramic powder with high thermal conductivity.

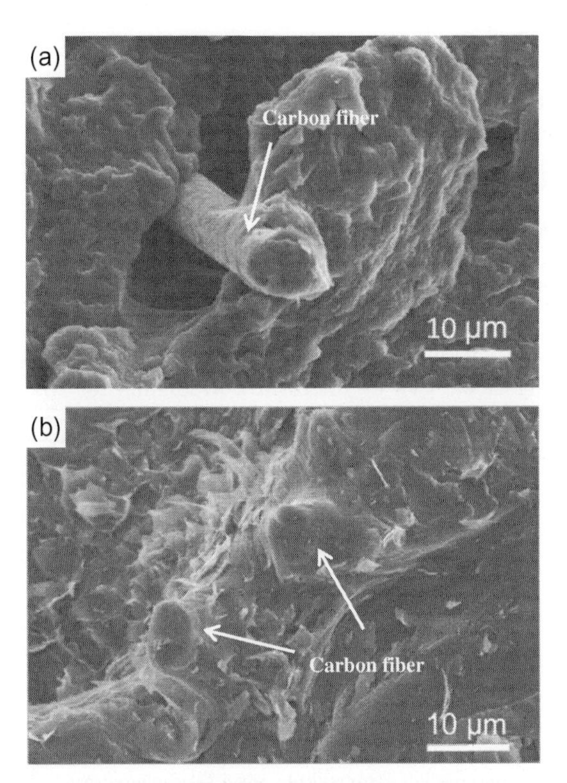

FIGURE 15 Fractural images of a conventional carbon fiber-reinforced thermoplastic (CFRTP) without hexagonal boron nitride (h-BN) that was irradiated with microwaves for 15 s (a), and a CFRTP with h-BN irradiated by microwaves for 60 s at 1600 W power (b).

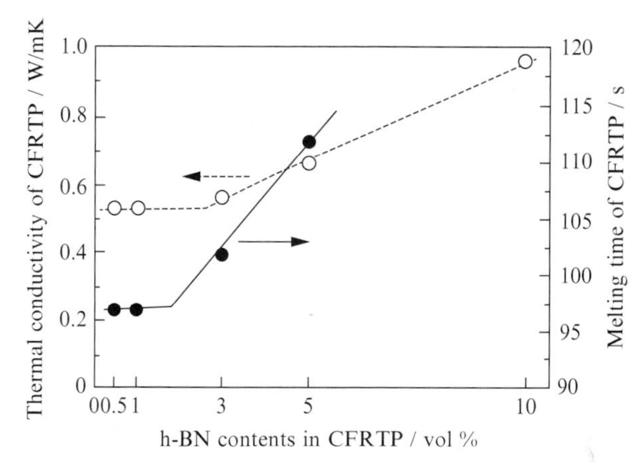

FIGURE 16 Thermal conductivity (white circles) and melting time (black circles) as a function of hexagonal boron nitride (h-BN) content in a carbon fiber-reinforced thermoplastic (CFRTP) upon microwave irradiation.

How much does thermal conductivity affect the suppression of thermal degradation at the interface between CFs and the matrix in CFRTPs? Figure 16 shows the thermal conductivity and the melting time as a function of h-BN content in a CFRTP upon microwave irradiation. It can be seen that the thermal conductivities of a conventional CFRTP without h-BN and a CFRTP with 0.12 vol% h-BN are almost identical at 0.53 W/m K. In a CFRTP with 2-10 vol% h-BN, the thermal conductivity linearly enhances from 0.53 to 0.96 W/m K because of percolation of h-BN fillers in the matrix. Moreover, the melting time of a CFRTP with h-BN increases from 97 to 112 s with an increase in the thermal conductivity of the CFRTP, meaning that the formation of thermal conduction passes between h-BN and the CFs in the CFRTP occurs when the amount of h-BN is increased. The use of thermally conductive ceramic fillers in a CFRTP matrix is very effective for suppressing the thermal degradation of the interface between the CFs and the matrix in a CFRTP.

As described above, the improvement in the thermal conductivity of a CFRTP matrix leads to the suppression of thermal degradation at the interface between the CFs and the matrix in CFRTPs manufactured using microwave processing. To efficiently produce CFRTPs, however, rapid propagation of heat to the entire CFRTP is required. In other words, the rate of heating propagated to the entire composite is important for CFRTP manufacturing. Shimamoto *et al.* [28] recently reported basic research about the possibility of rapidly heating a CFRTP using the microwave process. They prepared model CFRTP samples with one CF sandwiched between thermally conductive resin films with h-BN fillers and PP, as shown in Figure 17a, then irradiated the prepared samples with microwaves. Figure 17b and c show the microwave-irradiated CF/PP sample without h-BN and the microwave-irradiated CF/PP sample with h-BN

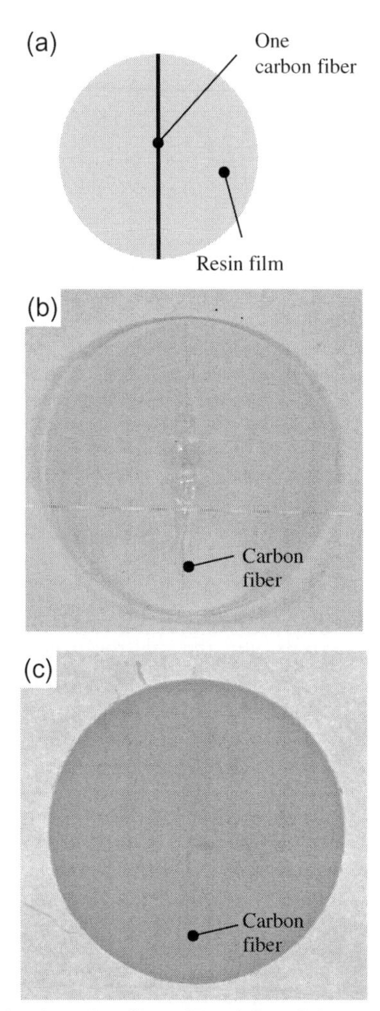

FIGURE 17 (a) Schematic of a carbon fiber (CF)-reinforced thermoplastic sample in which one CF was sandwiched between thermally conductive resin films with hexagonal boron nitride (h-BN) fillers and polypropylene (PP). (b) A Microwave-irradiated CF/PP sample without h-BN. (c) A microwave-irradiated CF/PP sample with h-BN.

(CF/PP/h-BN). The CF/PP sample, which consisted of a low thermally conductive matrix, was melted along the CFs by microwave irradiation, whereas a change of the CF/PP/h-BN sample with high thermal conductivity was not observed (Figure 17c).

Figure 18 shows photographs of thermal images on CF/PP and CF/PP/h-BN samples that were irradiated with microwaves for 30 s and observed

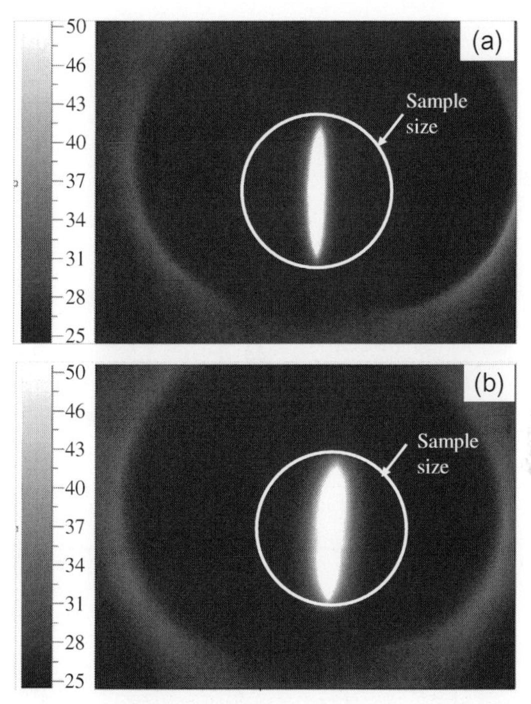

FIGURE 18 Photographs of thermal images of carbon fiber (CF)/polypropylene (PP) (a) and CF/PP/hexagonal boron nitride (b) samples undergoing microwave irradiation for 30 s. These were observed with an infrared thermographic camera.

through an infrared thermographic camera. Heat generation along the CF is observed. In particular, heat was generated over a wide area on the CF/PP/h-BN sample compared with the CF/PP sample with a matrix with low thermal conductivity, indicating that the heat of CFs that are heated by microwave irradiation is propagated strongly to the matrix. Figure 19 shows the temperature distribution in the CF/PP and CF/PP/h-BN samples according to microwave irradiation time. In Figure 19 "0 mm" represents the position of one CF. The temperature of the CF irradiated by microwaves was high compared with that of the matrix. Moreover, at a position 4 mm from the CF, while microwave irradiation for 240 s was required to increase the temperature to 50°C from 25°C in the CF/PP sample, the temperature of the matrix in the CF/PP/h-BN sample reached 50°C after only 60 s of microwave irradiation. This phenomenon indicates that the matrix with high thermal conductivity leads to rapid thermal propagation from CFs. Therefore, to manufacture CFRTPs using the microwave process, the development of matrices with thermal conductivity is one important factor.

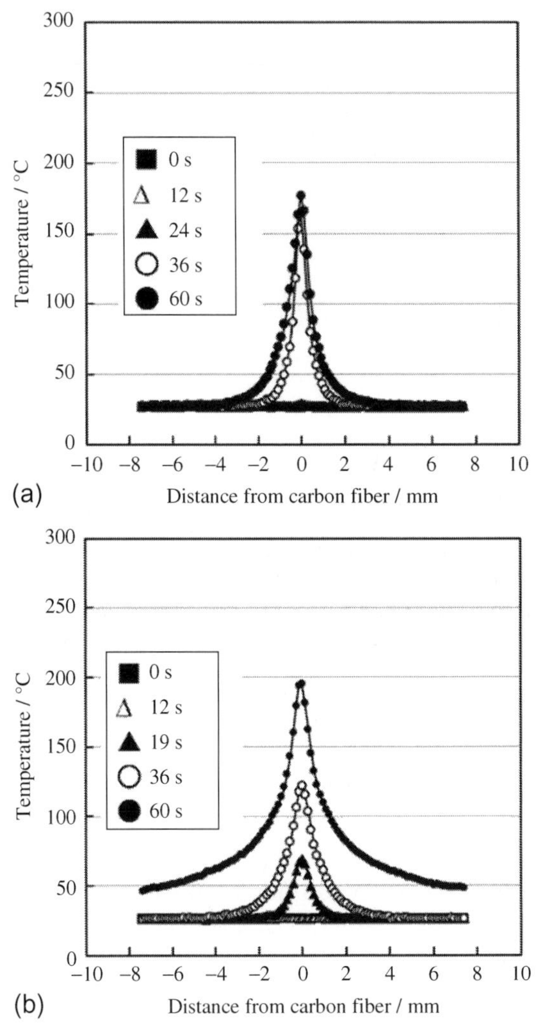

FIGURE 19 Temperature distributions of carbon fiber/polypropylene (a) and CF/PP/hexagonal boron nitride (b) samples according to microwave irradiation time.

5 SUMMARY

In this chapter the effects of microwave irradiation on CFRTS and CFRTP composites were reviewed and discussed. In the case of CFRTS manufacturing, the promotion of resin-curing by microwaves was introduced, indicating that microwave irradiation led to rapid resin-curing and enhanced mechanical properties in a short time compared with conventional heating. Microwave processing is suitable for rapid manufacturing of CFRTS composites. Moreover, in the case of CFRTP composites, to manufacture CFRTPs using the microwave process, the development of matrices with thermal conductivity is one important factor. The use of thermally conductive ceramic fillers in a CFRTP matrix is effective for suppressing the thermal degradation of the interface between CF and the matrix in CFRTPs. One important point of this chapter is that microwave processing has advantages for novel CFRTS and CFRTP manufacturing compared with conventional manufacturing.

REFERENCES

[1] E. Khosravi, O.M. Musa, Thermally degradation thermosetting materials, Eur. Polym. J. 47 (4) (2011) 465–473.

[2] A.M. Diez-Pascual, M. Naffakh, C. Marco, M. Gómez-Fatou, G.J. Ellis, Multiscale fiber-reinforced thermoplastic composite incorporating carbon nanotubes: a review, Curr. Opinion Solid State Mater. Sci. 18 (2) (2014) 62–80.

[3] S. Chand, Carbon fibers for composites, J. Mater. Sci. 35 (2000) 1303–1313.

[4] V.K. Rangari, M.S. Bhuyan, S. Jeelani, Microwave curing of CNF/EPON-862 nanocomposites and their thermal and mechanical properties, Compos. Part A 42 (7) (2011) 849–858.

[5] L. Feher, M. Thumm, Aerospace CFRP structure fabrication with the 2.45 GHz HEPHAISTOS system, in: Proceedings of MAPEES'04 Symposium, Japan, 2004, pp. 129–133, 19-22.

[6] K.D.V.P. Yarlagadda, S.-H. Hsu, Experimental studies on comparison of microwave curing and thermal curing of epoxy resin used for alternative mould materials, J. Mater. Process. Technol. 155-156 (30) (2004) 1532–1538.

[7] R.A. Witik, F. Gaille, R. Teuscher, H. Ringwald, V. Michaud, J.E. Månson, Economic and environmental assessment of alternative production methods for composite aircraft components, J. Clean. Prod. 29-30 (2012) 91–102.

[8] J.A. Menéndez, E.M. Menéndez, M.J. Iglesias, A. García, J.J. Pis, Modification of the surface chemistry of active carbons by means of microwave-induced treatments, Carbon 37 (1999) 1115–1121.

[9] J.M.V. Nabais, P.J.M. Carrott, M.M.L.R. Carrott, J.A. Menéndez, Preparation and modification of activated carbon fibres by microwave heating, Carbon 42 (2004) 1315–1320.

[10] K.R. Paton, A.H. Windle, Efficient microwave energy absorption by carbon nanotubes, Carbon 46 (2008) 1935–1941.

[11] D.A. Papargyris, R.J. Day, A. Nesbitt, D. Bakavos, Comparison of the mechanical and physical properties of a carbon fibre epoxy composite manufactured by rein transfer moulding using conventional and microwave heating, Compos. Sci. Technol. 68 (2008) 1854–1861.

[12] N. Li, Y. Li, X. Hang, J. Gao, Analysis and optimization of temperature distribution in carbon fiber reinforced composite materials during microwave curing process, J. Mater. Process. Technol. 214 (2014) 544–550.

[13] B.B. Balzer, J. McNabb, Significant effect of microwave curing on tensile strength of carbon fiber composites, J. Ind. Technol. 24 (2008) 2–9.

[14] S.Y. Fu, B. Lauke, E. Mäder, C.Y. Yue, X. Hu, Tensile properties of short-glass-fiber- and short-carbon-fiber-reinforced polypropylene composites, Composites, Part A 31 (2000) 1117–1125.

[15] D. Shimamoto, Y. Imai, Y. Hotta, Kinetic study of resin-curing on carbon fiber/epoxy resin composites by microwave irradiation, Open J. Compos. Mater. 4 (2014) 85–96.

[16] D. Wolff, K. Schlothauer, W. Tänzer, M. Fedtke, J. Spevacek, Determination of network structure in butane-1,4-diol cured bisphenol a diglycidyl ether using ^{13}C CP-MAS n.m.r. and i.r. spectroscopy, Polymer 32 (1991) 1957–1960.

[17] I.D. Maxwell, R.A. Pethrick, Low temperature rearrangement of amine cured epoxy resins, Polym. Degrad. Stab. 5 (1983) 275–301.
[18] T. Toishi, Y. Goto, Method for calculating micro-wave power, in: T. Koshijima, T. Shibata, T. Toishi, M. Norimoto, S. Yamada (Eds.), Microwave Heating Technology, NTS, Tokyo, 2004, pp. 17–26.
[19] J. Chag, G. Liang, A. Gu, S. Cai, L. Yuan, The production of carbon nanotube/epoxy composites with a very high dielectric constant and low dielectric loss by microwave curing, Carbon 50 (2) (2012) 689–698.
[20] K.J. Teoh, K.-T. Hsiao, Improved dimensional infidelity of curve-shaped VARTM composite laminates using a multi-stage curing technique—experiments and modeling, Composites, Part A 42 (7) (2011) 762–771.
[21] N. Kuentzer, P. Simacek, S.G. Advani, S. Walsh, Correlation of void distribution to VARTM manufacturing techniques, Composites, Part A 38 (3) (2007) 802–813.
[22] N.K. Nail, M. Sirisha, A. Inani, Permeability characterization of polymer matrix composites by RTM/VARTM, Prog. Aerosp. Sci. 65 (2014) 22–40.
[23] A. Jacob, Carbon fiber and cars-2013 in review, Reinf. Plast. 58 (2014) 18–19.
[24] Y. Murakami, A. Morita, J. Takahashi, K. Uzawa, in: Proc. of 18th Int. Conf. on Comps. Mater. Jeju, Korea, 2011, pp. 2126.
[25] K. Sato, H. Horibe, T. Shirai, Y. Hotta, H. Nakano, H. Nagai, K. Mitsuishi, K. Watari, Thermally conductive composite films of hexagonal boron nitride and polyimide with affinity-enhanced interfaces, J. Mater. Chem. 20 (2010) 2749–2752.
[26] J. Gu, Q. Zhang, J. Dang, C. Xie, Thermal conductivity epoxy resin composites filled with boron nitride, Polym. Adv. Technol. 23 (2012) 1025–1028.
[27] Y.K. Shin, W.S. Lee, M.J. Yoo, E.S. Kim, Effect of BN filler on thermal properties of HDPE matrix composites, Ceram. Int. 29 (2012) S569–S573.
[28] D. Shimamoto, Y. Tominaga, K. Sato, Y. Imai, Y. Hotta, Behavior of heat conduction on carbon fiber/thermal conductive thermoplastics with thermal conductive ceramic powder, in: Proc. of Spring Annual Meeting in The Society of Powder Technology, 2014, pp. 100–101.

Chapter 21

Green Manufacturing and the Application of High-Temperature Polymer-Polyphosphazenes

J. Fu and Q. Xu
School of Materials Science and Engineering, Zhengzhou University, Zhengzhou, China

1 INTRODUCTION

Because of environmental pollution and the gradual exhaustion of oil resources, using abundant and cheap inorganic minerals as materials to prepare innocuous, high temperature-resistant, aging-resistant, high-strength, and multifunctional inorganic polymer materials has been an important research field [1]. Typical examples of an organic-inorganic hybrid polymer are silicones and polyphosphazenes. The US Army Materials Research Laboratory said "It is felt that the polyphosphazenes may be the most important class of inorganic polymers since the commercialization of the silicones."

Polyphosphazenes are a class of novel organic-inorganic hybrid polymers comprising a backbone of alternating phosphorus and nitrogen atoms with two (usually organic or organometallic) side groups linked to each phosphorus. Since Allcock and Kugel [2] reported the first soluble polyphosphazene in 1965, several hundred polyphosphazenes have been synthesized, covering the following practical applications: thermally stable macromolecules with good flame resistance, low smoke evolution properties, biomaterials, photosensitive substrates, membranes, liquid crystals, ionic conductors, piezoelectricity, nonlinear optical materials, hybrid materials, elastomers, and catalysts [3–6]. The skeletal architectures of as-synthesized polyphosphazenes can extend from linear to cyclolinear, comb, star, dendritic, and block copolymeric [7]. Examples are shown in Figure 1.

The main method for synthesizing most polyphosphazenes is illustrated in Figure 2. It is a process that involves the preparation of high-molecular-weight poly(dichlorophosphazene) (PDCP), followed by replacement of all the chlorine atoms along each chain using organic or organometallic nucleophiles. In the synthesis of these polyphosphazenes a crucial role is played by PDCP, which is a polymer with a totally inorganic chemical structure known in the scientific

Green and Sustainable Manufacturing of Advanced Materials. http://dx.doi.org/10.1016/B978-0-12-411497-5.00021-7

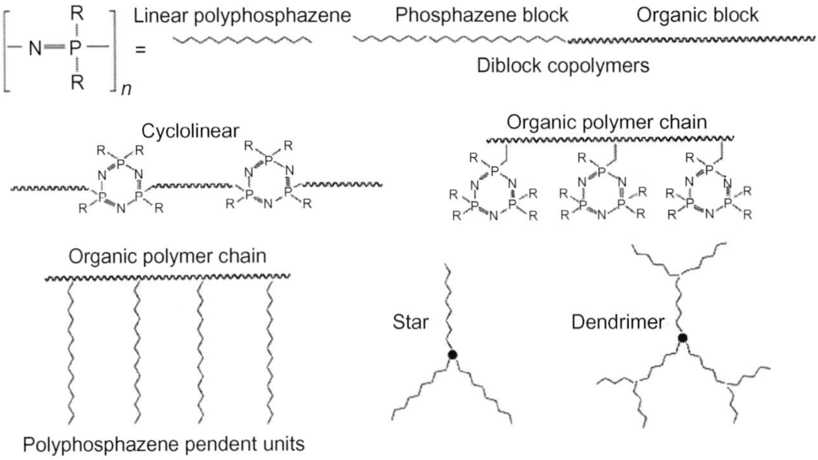

Linear polyphosphazene
Phosphazene block
Organic block
Diblock copolymers
Cyclolinear
Organic polymer chain
Organic polymer chain
Star
Dendrimer
Polyphosphazene pendent units

FIGURE 1 Different polyphosphazene architectures [7].

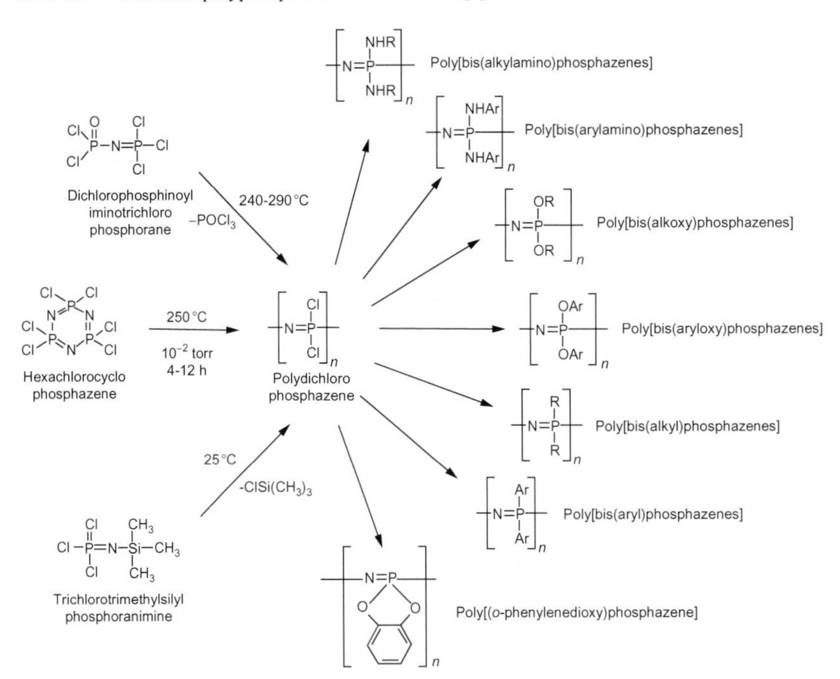

FIGURE 2 General reaction scheme for the synthesis of polyphosphazenes [3].

literature regarding phosphazene materials by the following names: polymeric intermediate, polymeric precursor, parent compound, polymer source, platform, and macromolecular reactant. All these definitions are indicative of the remarkable reactivity of the chlorine atoms bonded to the skeletal phosphorous of the phosphazene chain.

Cyclophosphazenes are cyclic molecules formed of three (or more) –P=N– repeating units, and they contain formal unsaturations in the cycle and two substituents at the phosphorus atom. Among them, hexachlorocyclotriphosphazene (HCCP) is the most common member of this series; HCCP is used not only to synthesize PDCP through ring-opening polymerization but also to construct cyclophosphazene-based polymers. For example, cyclophosphazenes with six or more reactive sites are useful in preparing star-shaped polymers, hyperbranched polymers, and dendrimers [8]. In addition, the phosphazene rings can be attached to organic polymers as side groups or linked via exocyclic groups to form cyclolinear or cross-linked polymers [9].

Many studies in the literature review the synthesis and application of polyphosphazenes, especially non-cross-linked polyphosphazenes [3–6]. In this chapter we focus on our research about cyclophosphazene-containing, cross-linked polyphosphazene micro- and nanomaterials; more precisely, a new green route for template-induced assembly was introduced to construct cross-linked poly(cyclotriphosphazene-4,4′-sulfonyldiphenol) (PZS) micro- and nanomaterials including nanofibers, nanotubes, microspheres, nanocables, and nanochains. Based on the characteristics of phosphazene chemistry, the applications of as-synthesized PZS micro- and nanomaterials when used as materials to carry noble metal nanoparticles, dye adsorbers, drug delivery carriers, or precursors for preparing porous carbon materials are reported. Also, recent research of other kinds of cyclophosphazene-containing, cross-linked polyphosphazene micro- and nanomaterials are summarized.

2 ONE-POT SYNTHESIS OF CROSS-LINKED POLYPHOSPHAZENE MICRO- AND NANOMATERIALS

In general, cyclophosphazene-containing, cross-linked polyphosphazenes are not so interesting as other members of the polyphosphazene family. The reasons may be that cross-linked polyphosphazenes are a kind of typical thermosetting material with a high degree of cross-linking and are difficult to process. They can be used only as adhesives or fillers to improve the thermal and flame-retardant properties of a matrix, which hinders the development of cross-linked polyphosphazenes [10,11].

Until 2006, the pioneering work by Zhu and coworkers [12] revealed a new route to synthesize cyclophosphazene-containing polymers that greatly pushed cross-linked polyphosphazenes forward. In this section we first describe the facile synthesis of cyclophosphazene-containing, cross-linked polyphosphazene nanofibers, together with their possible formation mechanism: *in situ* template-induced assembly. Then, based on the universality of the mechanism, we construct several types of polyphosphazene nanotubes and microspheres. The above-mentioned method to prepare cross-linked polyphosphazene micro- and nanomaterials was called the one-pot method because the polymorphic polyphosphazene materials could be formed directly during polymerization

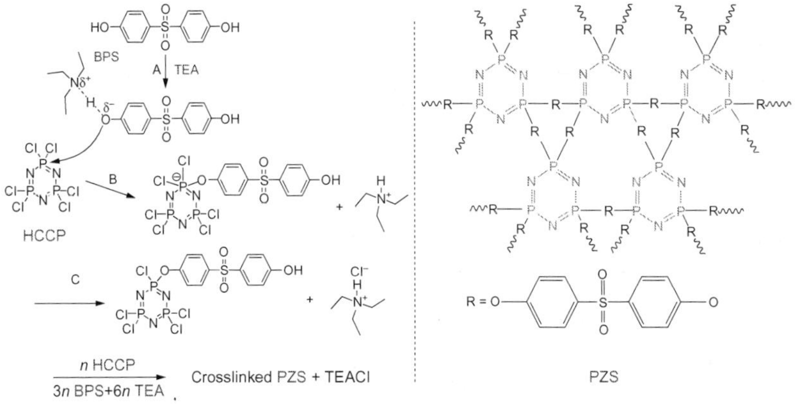

FIGURE 3 The mechanism of S_N2 reaction between hexachlorocyclotriphosphazene (HCCP) and 4,4′-sulfonyldiphenol (BPS) and the highly cross-linked chemical structure of poly(cyclotriphosphazene-4,4′-sulfonyldiphenol) (PZS). TEA, triethylamine.

without any postprocessing of the complex, which would overcome both the tedious processing and the difficulties in separating products from the templates. Therefore, the advantage of the method includes its "green-ness," simplicity, low cost, ease of scale-up, and good reproducibility. It should be noted that in the following research system, HCCP and 4,4′-sulfonyldiphenol (BPS) were used as co-monomers and triethylamine (TEA) as an acid acceptor. Figure 3 shows the mechanism of the S_N2 reaction between HCCP and BPS and the highly cross-linked chemical structure of the polycondensation product, PZS.

2.1 Synthesis of the PZS Nanofibers

PZS nanofibers were rapidly prepared as follows [13]. TEA (0.71 g, 7.02 mmol) was added to a solution of HCCP (0.27 g, 0.78 mmol) and BPS (0.59 g, 2.34 mmol) in acetone (50 mL). The reaction mixtures were stirred in an ultrasonic bath (40 kHz, 50 W) at 7°C for 10 min. The gel-like solid produced was washed four times using acetone and de-ionized water. Finally, the resulting product was dried under a vacuum to yield PZS nanofibers as a hard white agglomeration. Synthesis yield was about 75 wt%, calculated from HCCP.

Figure 4a shows scanning electron microscopy (SEM) images of bulk and uniform PZS nanofibers. As can be seen, all PZS samples obtained in the method were fibrous, with a 40- to 60-nm diameter and a length up to several thousand nanometers. A close look at the nanofibers revealed that their surface was coated by some nanoparticles with an average size of several nanometers, which appeared rather rough, as shown in Figure 4b. The coarse texture could increase the specific surface area of the PZS nanofibers, and the speculation had been proved by N_2 adsorption test. The results showed that the BET surface

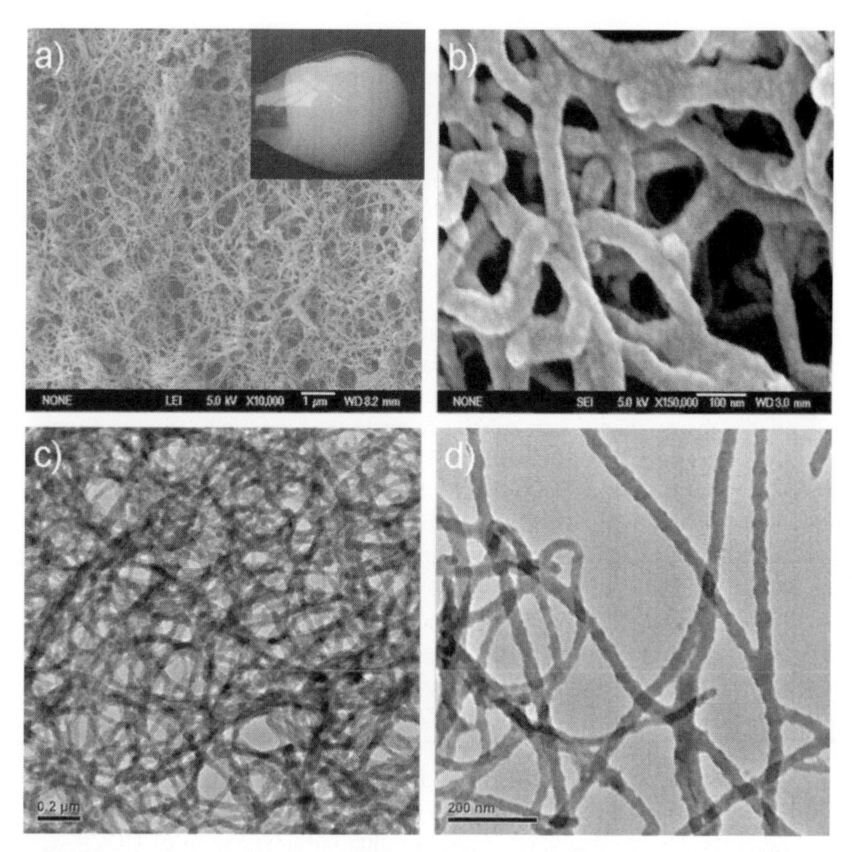

FIGURE 4 (a, b) Field-emission scanning electron microscopy images of poly(cyclotriphosphazene-4,4′-sulfonyldiphenol (PZS) nanofibers. The inset in (a) was the sample obtained after polymerization for 8 min. (c, d) High-resolution transmission electron microscopy images of the PZS nanofibers.

area of the PZS nanofibers was 79.3 $m^2 g^{-1}$, which was far larger than that of hydrogen chloride (HCl) de-doped polyaniline nanofibers (average diameter, 30 nm, 54.6 $m^2 g^{-1}$), with a relatively smooth surface reported by Huang and Kaner [14]. The rough texture could help better explain the formation mechanism of the nanofibers. Of course, from an applications point of view, the special structure also indicates that the PZS nanofibers had potential as absorbing materials. Figure 4c and d show transmission electron microscopy (TEM) images of the PZS nanofibers, which further confirmed their fibrillar morphology. Their diameters and lengths were consistent with those shown in Figure 4a and b. After carefully observing the TEM images, we found that the PZS nanofibers were solid and every nanofiber was assembled of large numbers of nanoscale particles along its axes (shown in Figure 4d).

To investigate the formation mechanism of the PZS nanofibers, we tracked the morphological evolution of PZS during condensation polymerization. The PZS products in the reaction bath were sampled periodically with a syringe for examination under an electron microscope. To avoid the formation of new *ex situ* PZS, polymerization needed to be quenched as soon as possible. Therefore, special care had to be taken by quickly depositing the reaction extract onto a glass slice or a copper net under an infrared lamp for SEM or TEM observation.

The SEM images of the product at different reaction stages are shown in Figure 5. At a very early stage of the polymerization process, a mass of irregularly shaped nanoparticles—polymer colloids (Figure 5a)—were obtained. As the reaction continued, a visible protuberance derived from a colloid aggregate (Figure 5b, inset), then the protuberance prolonged gradually (Figure 5c) and finally formed nanofibers (shown in Figure 5d and e). In addition, Figure 5f shows a high-resolution TEM image of the product at a middle stage of the reaction. A typically oriented attachment of colloid aggregates was clearly recognized, which could also help to explain the formation mechanism of nanofibers.

The SEM results show that polycondensation could finish in a rather short period, such as 8 min, which is advantageous for multiple rapid preparations of nanofibers. In addition, the results also indicate that the protuberance played an important role in the formation of PZS nanofibers. However, what was the protuberance?

To address this question, we should consider the polymerization process. Polycondensation occurred in a solution of acetone with excess TEA as an acid acceptor. Under ultrasonic irradiation, the polymerization of HCCP with equimolar BPS generated uneven, nanometer-sized polymer colloids and HCl. TEA absorbed the HCl to afford TEACl, which accelerated the polymerization. During polymerization, TEACl was precipitated out. As ion crystals, TEACl had a rather high surface energy, whereas organic polymers had relatively low surface energies ($<100\,mN\,m^{-1}$) [15,16]. As a rule, materials with low surface energies can spontaneously spread over substances with higher surface energies. One typical example based on this rule is the preparation of nanotubes by template wetting [17]. Therefore, TEACl could adhere to the polymer colloids and form large numbers of colloid aggregates. At the same time, the growing trend of TEACl along its axes guided the oriented attachment of these colloid aggregates (Figure 5f). As the reaction progressed, further cross-linking of colloids occurred. Thus, nanofibers could undergo a process from polymer colloids (Figure 5a) to colloid aggregates (Figure 5b), short nanorods (Figure 5c), and long nanofibers (Figure 5d and e). Therefore, we thought the protuberance was just the growing TEACl nanocrystal, which was covered with polymer colloids. During polymerization, the TEA nanocrystal produced *in situ* played the role of a directing template, guiding the growth and formation of PZS nanofibrous

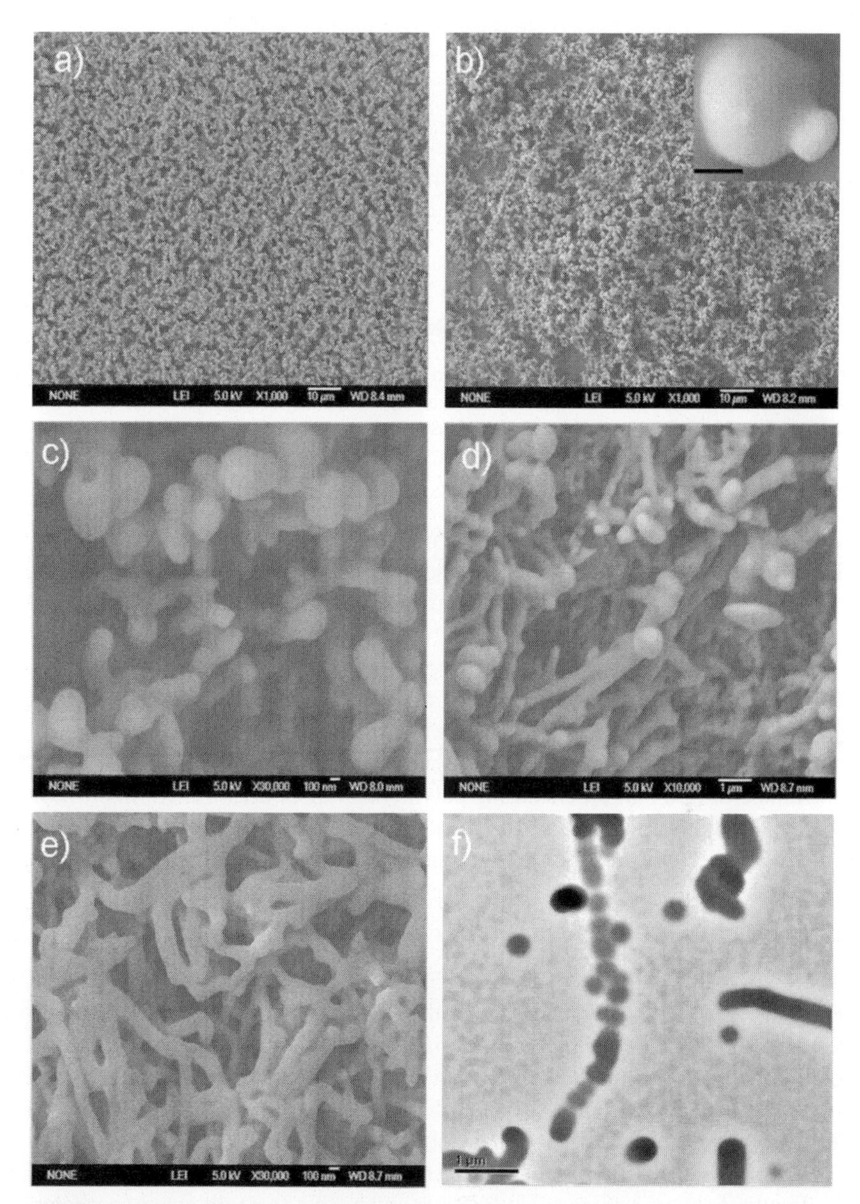

FIGURE 5 Scanning electron microscopy (a–e) and transmission electron microscopy (f) images showing the morphological evolution process of poly(cyclotriphosphazene-4,4′-sulfonyldiphenol (PZS)from sphere-like colloids to nanofibers as the reaction proceeds. The reaction times are 1 (a), 2.5 (b), 4 (c), 5 (d), 8 (e), and 4 min (f), respectively. Time zero was defined as the moment that triethylamine was added to the reaction solution. Samples (a–f) were extracted from the reaction bath and deposited immediately onto a glass slice or copper net under an infrared lamp. The inset in (b) shows the magnified image of one PZS colloid aggregate with a protuberance (scale bar = 100 nm).

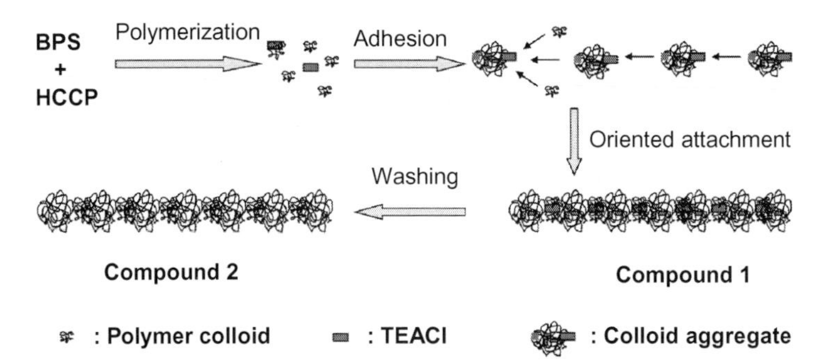

FIGURE 6 Schematic illustration of the procedure for the synthesis of the poly(cyclotriphosphazene-4,4′-sulfonyldiphenol nanofibers via an *in situ* template-induced assembly route. BPS, 4,4′-sulfonyldiphenol; HCCP, hexachlorocyclotriphosphazene.

structures. Because of the rapid polymerization process, the discontinuous TEACl template was small in bulk. As soon as it was removed, the room left by the template was easily eliminated because of the inherent flexibility of polymer colloids. So, the final nanofibers were solid.

Based on the analysis of morphological evolution, we proposed a formation mechanism of PZS nanofibers: *in situ* template-induced assembly, which is schematically expressed using Figure 6. To verify the validity of our proposed mechanism, one key issue is to prove the presence of the directing template in the middle of PZS nanofibers during polymerization. Fourier transform infrared (FTIR), X-ray diffraction, TG, solid-state nuclear magnetic resonance, and EA investigations confirmed the formation mechanism [13].

2.2 Synthesis of PZS Nanotubes

According to the above discussion, if the speed of polymerization between HCCP and BPS was slow enough and the TEACl nanocrystal, as the directing template, had time enough to grow along its axes in a special condition, a core/shell structure could predictably be formed. Zhu and coworkers [12] observed such a core/shell structure. As soon as the core was removed, a novel nanotubular structure was obtained. In fact, we obtained polymorphic PZS nanotubes by controlling the formation process of TEACl nanocrystals.

Figure 7 shows polymorphic PZS nanotubes that were obtained by changing the solvent condition or the feeding manner during the polymerization of co-monomers HCCP and BPS [18–20]. It should be noted that the selection of solvent is a key factor in the preparation of PZS nanotubes. On one hand, fibrous TEACl nanocrystals could not be solved by the selected solvent. On the other hand, the selected solvent should help to slow down the polymerization rate of co-monomers HCCP and BPS.

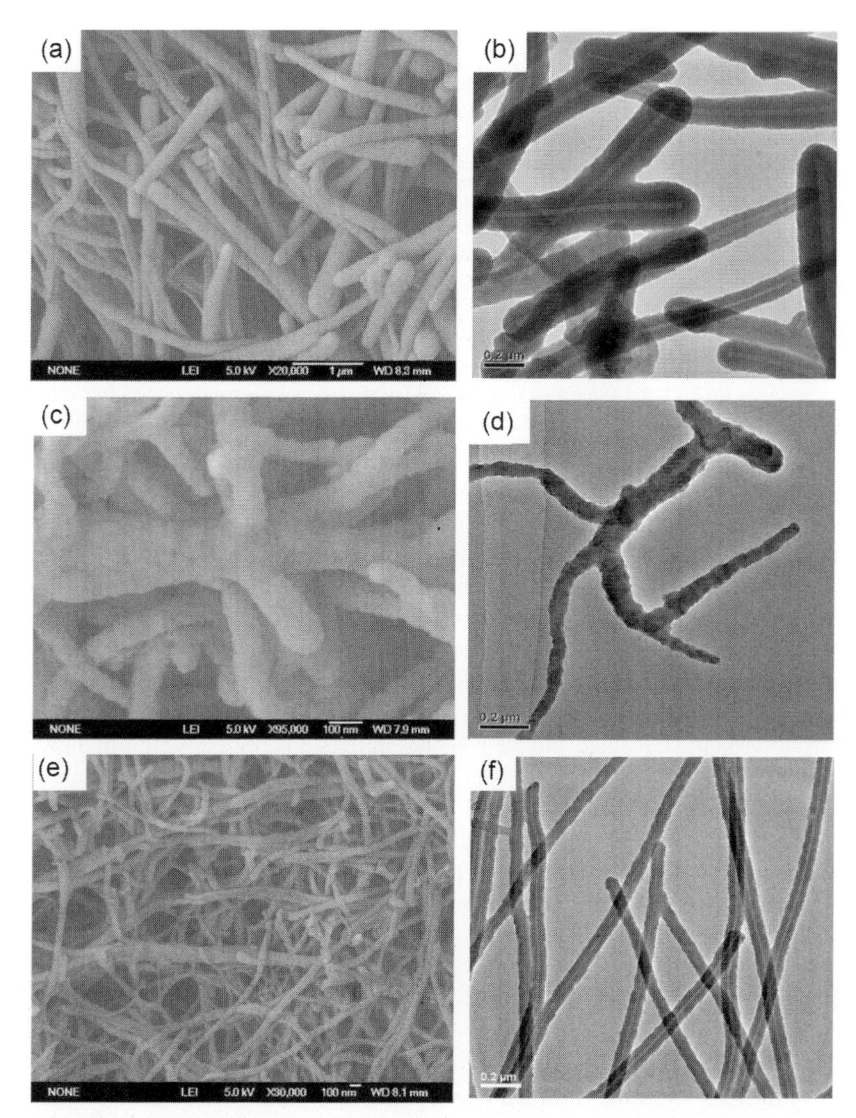

FIGURE 7 SEM and TEM images of polymorphic PZS nanotubes. (a,b) Capsicum-like nanotube, (c,d) Branched nanotubes, and (e,f) Uniform nanoutbes.

Herein we briefly describe the formation mechanism of capsicum-like PZS nanotubes. As shown in Figure 8a, under magnetic stirring, the monomers HCCP with equimolar BPS were polycondensed in a solution of acetone/toluene with excess TEA as an acid acceptor. During the initial stage, the polymerization generated nanometer-sized prepolymers and rod-like TEACl nanocrystals as *in situ* templates. Normally, organic polymers have relatively low surface energies;

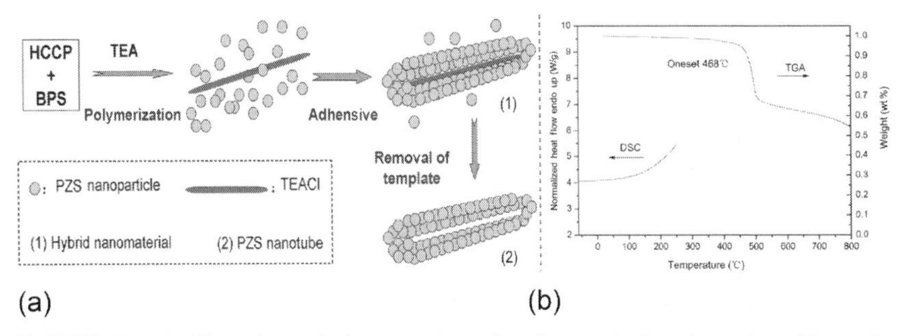

FIGURE 8 (a) Illustration of the procedure for the synthesis of capsicum-like poly (cyclotriphosphazene-4,4′-sulfonyldiphenol (PZS) nanotubes via an *in situ* template-induced assembly approach. (b) TGA and differential scanning calorimetry curves of the capsicum-like PZS nanotubes. BPS, 4,4′-sulfonyldiphenol; HCCP, hexachlorocyclotriphosphazene; TEA, triethylamine.

thus the prepolymers could easily adhere to the surfaces of the TEACl nanocrystals with relatively high surface energy. With the TEACl nanocrystals growing along their axes, further cross-linking of prepolymers occurred on the surfaces of the nanocrystals. Thus, hybrid nanocomposites with core/shell structures were obtained. Because large numbers of prepolymers were formed during the initial stage of polycondensation, their concentration was higher. Thus, more prepolymers could adhere to the surfaces of the templates, and the wall of nanotubes was thicker in this period. Comparably, the concentration of prepolymers was lower in the last stage and the wall was thinner. Finally, capsicum-like PZS nanotubes could be obtained as soon as the TEACl nanocrystals were removed by rinsing with de-ionized water.

Thermogravimetric analysis (TGA) and differential scanning calorimetry were applied to evaluate the thermal properties of the capsicum-like PZS nanotubes. As shown in Figure 8b, there was no clear glass transformation in the differential scanning calorimetry curve, which indicated the PZS nanotubes had no obvious chain relaxation; the results were consistent with the covalently cross-linked structures of PZS. TGA mass-loss curve of PZS nanotubes showed that initial decomposition of the hybrids occurred 468°C in a nitrogen atmosphere, and the ceramic residual ratio was 54% at 800°C. The enhanced thermal stability of the PZS nanotubes profited from special molecular hybrid network structures and the inherent thermal stability of cyclotriphosphazene. The superior performance and unique tubular structure make these ideal candidates for a variety of applications in chemistry, biochemistry, materials science, and medicine.

2.3 Synthesis of PZS Microspheres

In the synthesis of PZS nanofibers and nanotubes, TEACl nanocrystals played a key role as directing templates, which could grow along their axes and induce the assembly of the PZS prepolymers on their surfaces. If we destroyed the formation of TEACl nanocrystals during the polymerization of co-monomers

HCCP and BPS, the PZS microspheres could predictably be synthesized based on the classical precipitation polymerization principle. Therefore, three possible strategies can be used to synthesize PZS microspheres:

(1) Adding small amounts of a special solvent (usually ethanol), which can easily dissolve TEACl in the reaction solvent system
(2) Using strong polar solvent as the reaction solvent. According to the mechanism of the S_N2 reaction between HCCP and BPS, a strong polar solvent can help to improve the reaction rate; thus the fibrous TEACls have not enough time to form. For example, when using acetonitrile as the solvent, polymerization between HCCP and BPS can finish within 1 min, and only PZS microspheres can be obtained.
(3) Using another acid acceptor instead of TEA

Figure 9 shows the SEM images of PZS microspheres obtained through the above three strategies respectively.

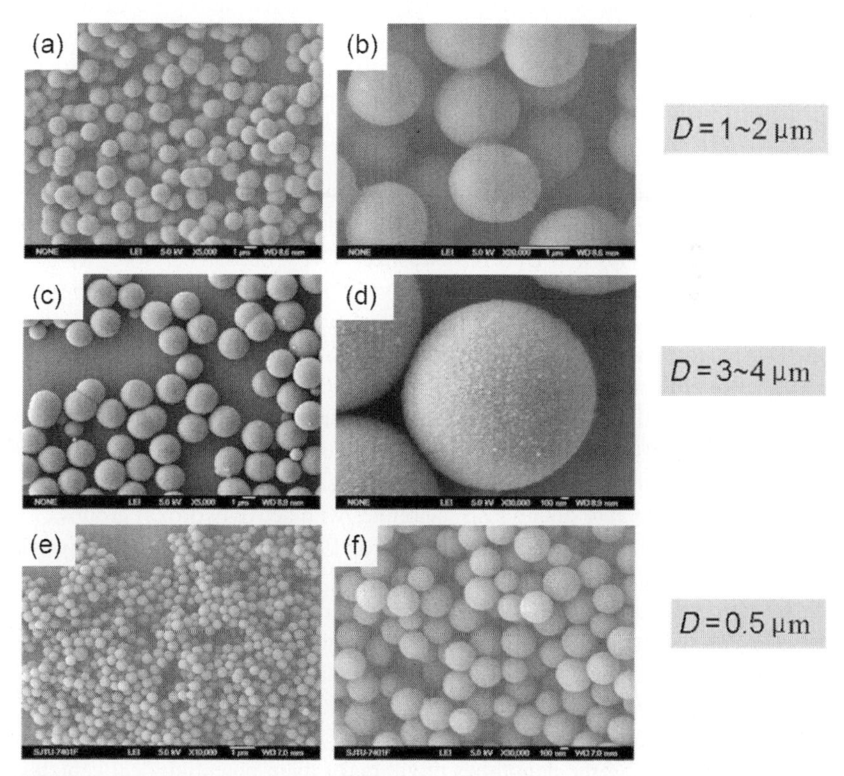

FIGURE 9 Scanning electron microscopy images of poly(cyclotriphosphazene-4,4′-sulfonyldiphenol microspheres obtained through three strategies: using the mixture of tetrahydrofuran and anhydrous alcohol (8:2 by volume) as the solvent (a); using tributylamine as the acid acceptor (b); and using acetonitrile as the solvent (c).

2.4 Synthesis of PZS Nanotubes With Active Hydroxyl Groups

Functional nanotubes have attracted much attention for their distinguishable role in fundamental studies and biological applications, including bioseparation, biolabeling, biodetection, biocatalysis, and biomolecule delivery, mainly because of their unique structure and flexible chemistry processing [21–27]. Functional nanotubes with hydroxyl, carboxyl, and amino groups on their surfaces have typically been given considerable attention because of their diverse capability to covalently bind DNA, proteins, amino acids, bioactive peptides, enzymes, and so forth [28–32]. To date, carbon nanotubes (CNTs) have widely been used for these applications because of their capability for easy chemical modification, superior stability, and remarkable mechanical strength [23,26,33]. However, biological incompatibility of CNTs restricts their range of applications. Though self-assembled nanotubes with active groups can be synthesized easily through the molecular design of precursors [27], based on common knowledge, the precursors of self-assembled nanotubes cannot be easily obtained. Since Martin [34] first used a porous anodic aluminum oxide membrane as a template, hard-template synthesis for functional nanotubes has particularly fascinated scientists [35]. However, this approach requires complicated synthetic steps, including the dissolution of the template in corrosive media. Therefore, finding a simple, cost-effective, environmentally friendly route to fabricate high-quality nanotubes with active groups is a challenge.

In the polymerization system using HCCP and BPS as co-monomers, HCCP has six free –Cl groups and BPS has two free –OH groups. Therefore, we can facilely synthesize PZS nanotubes with active hydroxyl groups by adjusting the feeding mass of co-monomers [36]. In a typical synthesis, 100 mL tetrahydrofuran (THF) with HCCP (0.8 g, 2.30 mmol) was added drop-wise into 100 mL THF with a given amount of BPS and TEA (2.08 g, 20.6 mmol) under ultrasonic irradiation (50 W, 40 Hz) at 40°C. The feeding molar ratio of BPS to HCCP equals 3:1, 3.6:1, and 4.5:1, respectively. After ultrasonic irradiation for 12 h, the solution was filtered; then the precipitates were washed with THF and deionized water three times each to obtain PZS nanotubes with active hydroxyl groups.

Figure 10a and b present SEM images of PZS nanotubes with hydroxyl groups. The products are almost uniform short nanorods with an outer diameter of ~60-80 nm and a length of several micrometers. TEM results in Figure 10c and d reveal that the nanorods possess hollow tubular structures with an inner diameter of ~20 nm. The outer diameter and length of the nanotubes are consistent with those of nanotubes shown in Figure 10a and b. In addition, it is noted that most of the nanotube ends are closed, as shown in Figure 10c and d. The unique morphology of the nanotubes might make them potential ideal vehicles for many delivery applications at controlled rates for long period of time.

Figure 11a–c depicts infrared spectra of the PZS nanotubes with hydroxyl groups prepared under different feeding molar ratios of BPS to HCCP: 4.5:1,

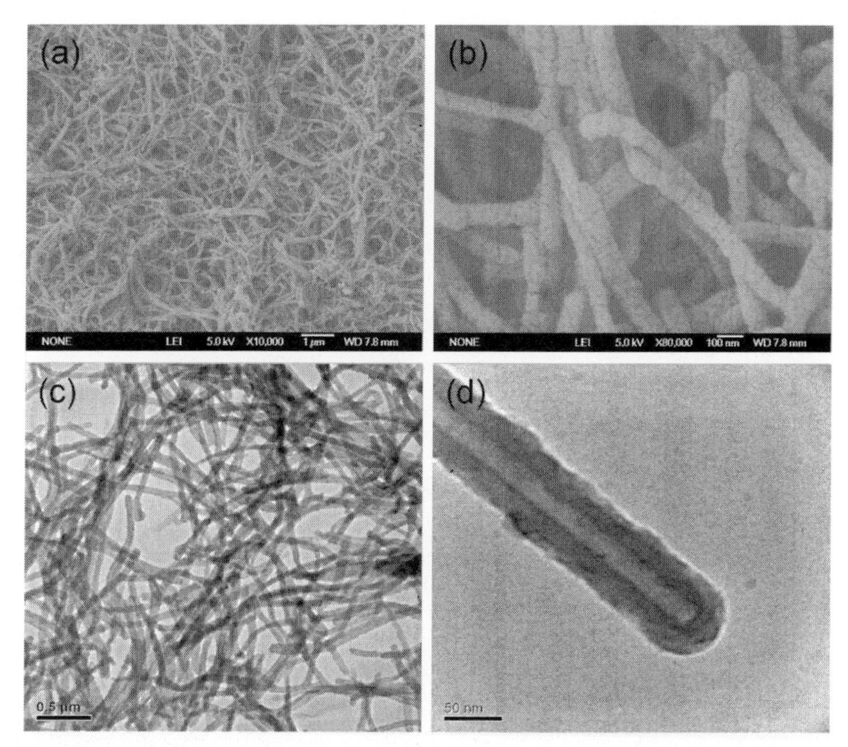

FIGURE 10 Scanning electron microscopy (a, b) and transmission electron microscopy (c, d) images of active poly(cyclotriphosphazene-4,4′-sulfonyldiphenol nanotubes prepared via an *in situ* template-induced assembly approach. The feeding molar ratio of 4,4′-sulfonyldiphenol to hexachlorocyclotriphosphazene equals 3.6:1.

3.6:1, and 3:1. Two sharp peaks at 1590 and 1490 cm^{-1} are associated with the C=C stretching vibrations in the phenylene of sulfonyldiphenol units. The strong peaks at 1187 and 880 cm^{-1} correspond to the P=N and P–N characteristic absorption of cyclotriphosphazene, respectively. The characteristic absorption of O=S=O of sulfonyldiphenol units can be seen at 1293 and 1154 cm^{-1}. The intensive absorption peak at 941 cm^{-1} is assigned to the P–O–Ar band, which is the obvious evidence proving the occurrence of polycondensation between co-monomers HCCP and BPS. It should be noted that the peaks at 3100 and 3073 cm^{-1} are assigned to the stretching vibration of the hydroxyl groups (phenolic groups). Obviously, their intensity increases significantly with an increase of the feeding molar ratio of BPS to HCCP. Therefore, the PZS nanotubes could contain hydroxyl groups with content that can be tuned by varying the feed ratio.

The reactivity of the PZS nanotube with hydroxyl groups was determined with the aid of esterification with benzoxy chloride at an ambient temperature. The successful esterification of the hydroxyl groups on the polymer nanotube surface was proved by FTIR spectra, as shown in Figure 11d. The intensive

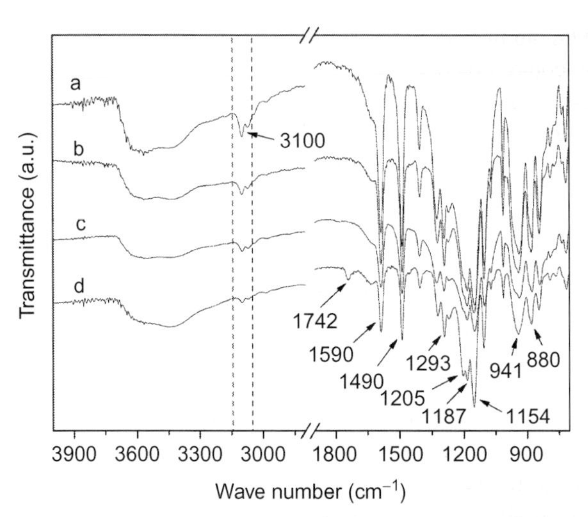

FIGURE 11 Fourier transform infrared spectra of the poly(cyclotriphosphazene-4,4′-sulfonyldiphenol)nanotubes prepared with different feeding molar ratios of 4,4′-sulfonyldiphenol to hexachlorocyclotriphosphazene: 4.5:1 (a); 3.6:1(b); 3:1 (c); and prepared at 3.6:1 and esterified later (d). a.u., arbitrary units.

absorption peaks at 1742 and 1205 cm^{-1} are assigned to the ester carbonyl group and the asymmetrical stretching vibration of C—O—Ar band, respectively, which demonstrate that the hydroxyl groups are very accessible and highly reactive.

3 SYNTHESIS OF CROSS-LINKED PZS-BASED COMPOSITE MATERIALS

From the above discussions about the formation mechanism of PZS nanofibers and nanotubes, we know that the PZS primary particles produced during the initial polymerization between HCCP and BPS have a lower surface energy and higher reactivity, which can assemble on the surface of the *in situ* template TEACl to form TEACl/PZS nanocomposites. Predictably, during the polycondensation of co-monomers HCCP and BPS, if the template effect of TEACl nanocrystals is intentionally eliminated and *ex situ* templates added, they also can predictably induce PZS primary particles to assemble on the surface of templates to construct corresponding template/PZS (core/shell) composites. Herein we give several examples.

3.1 Silver/PZS Nanocables

Metal nanowires have potential applications in the fabrication of nanoscale electronic, optoelectronic, magnetic, and sensing devices. On the nanoscale, however, metal nanowires are very sensitive to air and moisture, which degrade the performance of nanodevices [37]. One solution to the problem is to coat the nanowires with an inert polymer sheath, which has good flexibility and sealing property and could protect metal nanowires from oxidation and corrosion. Such nanocomposites of nanowires (a core) wrapped with single or multiple insulating layers (the sheath) are generally called coaxial nanocables. So far, there have been a few reports of the preparation of metal/polymer nanocables, such as cobalt/polyaniline [37], silver (Ag)/polypyrrole coaxial [38–40], gold/poly(3,4-ethylenedioxythiophene) [41], and copper/poly(vinyl alcohol) nanocables [42]. These investigations have, however, mainly been based on linear-structured polymers as the sheath materials of nanocables. As a protective sheath of metal nanowires, good mechanical and thermal stability is necessary. Compared with a linear structure, a covalently cross-linked structure is one more effective approach to achieve good stability [43–45].

On the basis of the template-induced assembly idea, using Ag nanowires as an *ex situ* template, which replaced TEACl nanocrystals formed *in situ* (TEACl can be dissolved by ethanol during polymerization), we successfully prepared uniform Ag/cross-linked PZS nanocables [46].

Figure 12a–d shows SEM images of as-synthesized Ag nanowires and Ag/PZS nanocables at different magnifications. The nanowires, characterized by

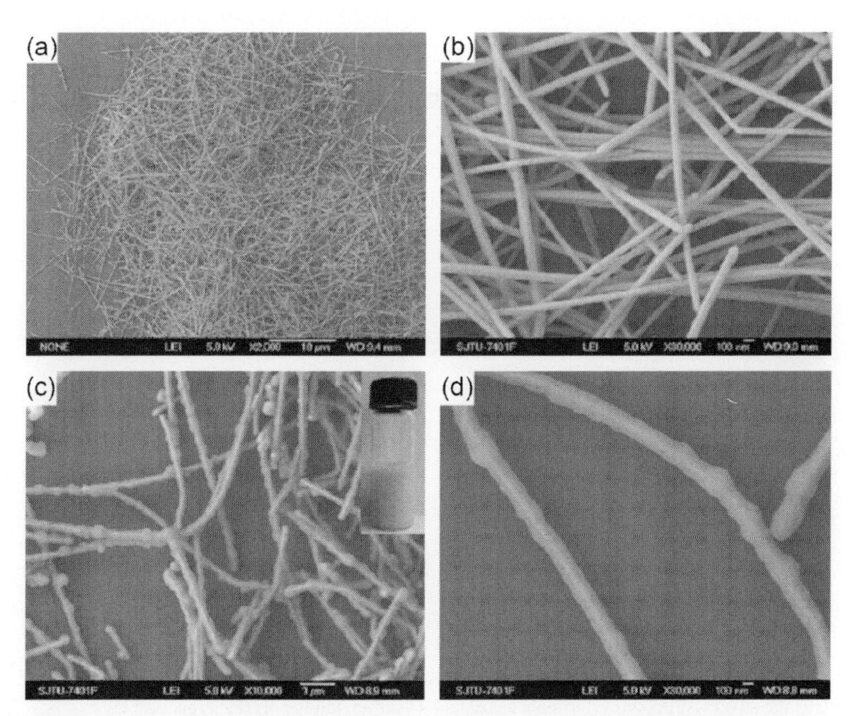

FIGURE 12 (a, b) Representative scanning electron microscopy (SEM) images of silver nanowires. (c, d) Typical SEM images of silver/poly(cyclotriphosphazene-4,4′-sulfonyldiphenol) coaxial nanocables. Inset in (c) is the sample obtained after reaction for 10 h.

diameters of 80 ± 20 nm and lengths up to tens of micrometers, display smooth outer surfaces. Almost all the Ag nanowires were coated, indicating a highly efficient preparation of nanocables. Compared with the intact Ag nanowires, nanowires wrapped with the PZS exhibit a rough surface.

Figure 13a–c are typical TEM images of the as-prepared Ag/PZS nanocables. These images demonstrate that the PZS, which exhibits a relatively lighter contrast, continuously coats the axis of the Ag nanowire to form the core/sheath nanostructure. The thickness of the sheath layer is evaluated to be about 80 nm. It should be noted that almost all the nanocables have closed ends, as shown in Figure 13c. To evaluate the environmental stability of the Ag/PZS nanocables, a thermogravimetric experiment was carried out in an air atmosphere. TGA shows that the initial decomposition of the nanocables starts at approximately 440°C, as shown in Figure 13d. The Ag/PZS nanocables possessed better thermal stability, which was interrelated with the covalently cross-linked structure of PZS layers. In addition, the sheath thickness of the Ag/PZS nanocable proved to be tunable by changing the molar ratio of Ag nanowires to co-monomers, as shown in Figure 14.

In summary, *ex situ* template-induced assembly is a simple and efficient method to construct Ag/PZS nanocables. The method can offer the following advantages when compared with previously developed methods: (1) The

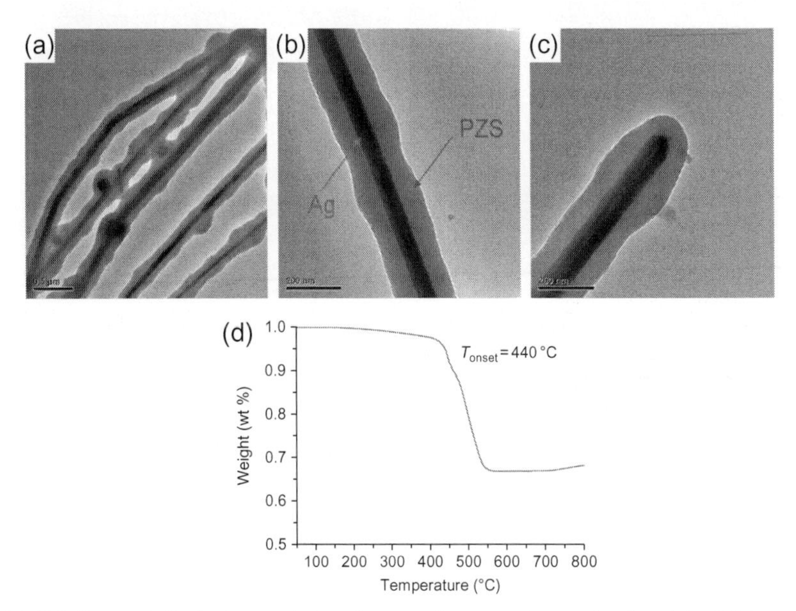

FIGURE 13 (a–c) Typical transmission electron microscopy images (a–c) and the TGA curve (d) of the silver (Ag)/poly(cyclotriphosphazene-4,4′-sulfonyldiphenol) (PZS) nanocables.

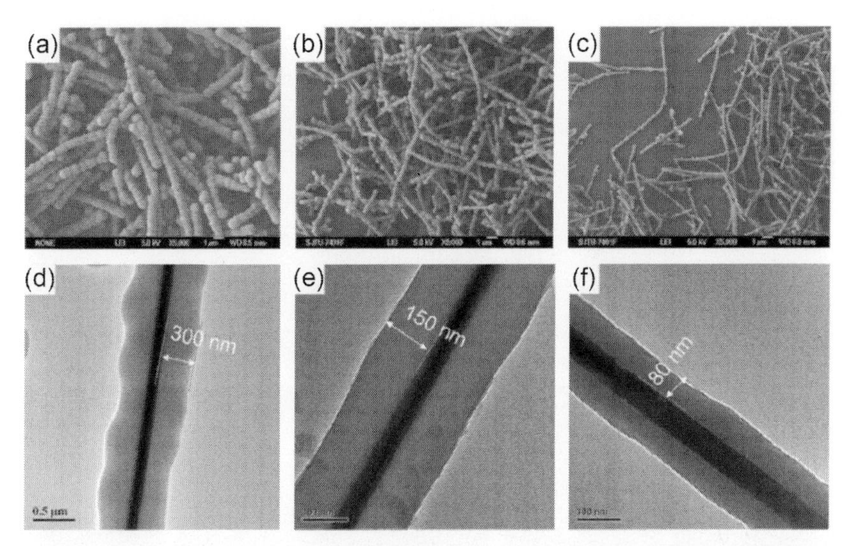

FIGURE 14 Scanning electron microscopy (a–c) and corresponding transmission electron microscopy (d–f) images of three types of silver/poly(cyclotriphosphazene-4,4′-sulfonyldiphenol) nanocables with sheath thicknesses of ~300, ~150, and ~80 nm, respectively. Reaction conditions used molar ratios of Ag to hexachlorocyclotriphosphazene to 4,4′-sulfonyldiphenol of 5:3:9 (a), 5:2:6 (b), and 5:1:3.

sheath layer of nanocables is a highly cross-linked, organic-inorganic hybrid polyphosphazene and possesses better thermal stability; (2) the sheath thickness of nanocables can be controlled by changing the molar ratio of Ag nanowires to co-monomers HCCP and BPS; (3) the process of wrapping PZS on Ag nanowires can be carried out at room temperature and does not use any surfactant or capping agent. We think the method will make it possible to prepare other high-surface-energy conductor or semiconductor/highly cross-linked PZS nanocables.

3.2 CNT/PZS Nanocomposites

CNTs have been of increasing interest because of their unique optical, electronic, mechanical, and thermal properties [47–49]. Normal untreated CNTs display poor dispersibility in liquids because of their de-wetting surface and, as a result, are difficult to process and apply in industries. One general strategy to overcome this difficulty is to coat them with a polymer through noncovalent surface modification [50]. Herein, based on the template-induced assembly mechanism, we described our success in evenly coating cross-linked PZS on multi-walled carbon nanotube (MWCNT) surfaces [51].

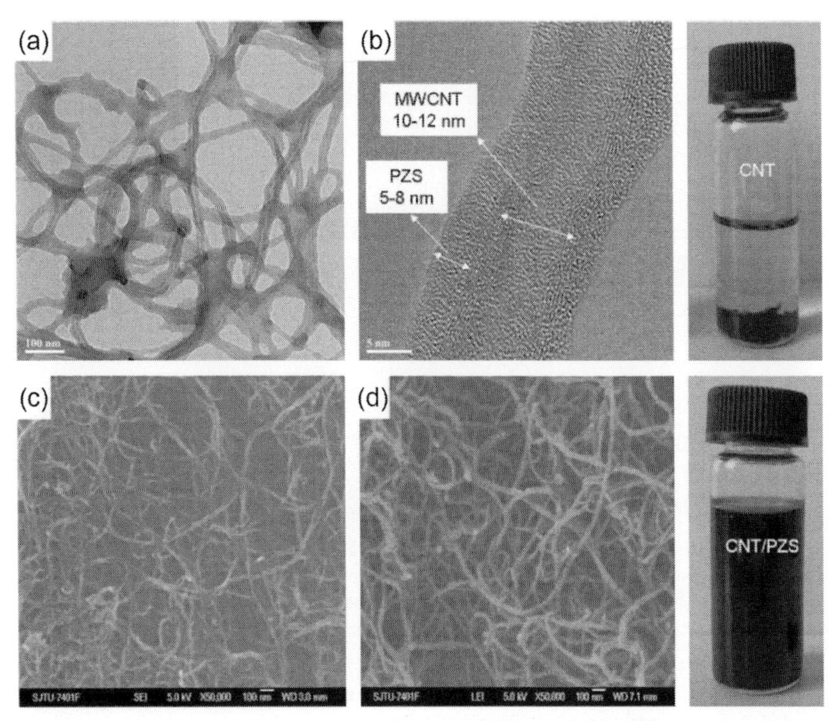

FIGURE 15 (a, b) Typical transmission electron microscopy images multiwalled carbon nanotube (MWCNT)/poly(cyclotriphosphazene-4,4′-sulfonyldiphenol) (PZS) nanocomposites. (c) Scanning electron microscopy (SEM) image of pristine carbon nanotubes. (d) SEM image of the MWCNT/PZS nanocomposites. The right photographs show tetrahydrofuran dispersions of the MWCNTs before and after modification by PZS.

The fine core/shell structures of as-prepared MWCNT/PZS nanocomposites were visibly found by high-resolution transmittance electron microscopy (HRTEM). As shown in Figure 15a and b, the PZS shell and the MWCNT graphite sheet structures are clearly observed. Typical nanocomposites were uniform, with a shell thickness in the range of 5-8 nm and a length of several micrometers (Figure 15d), and the distribution of the shell thickness (obtained by measurements from hundreds of TEM images) at different locations showed a peak centered at 7 nm. PZS-modified MWCNTs can be well dispersed in water or organic solvents including *N,N*-dimethylformamide, THF, acetone, and ethanol under a minimal ultrasonic condition, in dramatic contrast to dried non-polymer-coated CNT materials in any solvent system.

The method offers at least three merits: (1) PZS-functionalized MWCNTs have good dispersibility in water and in a variety of organic solvents; (2) the cross-linked PZS coating of CNTs is based on a template-induced assembly mechanism different from the conventional π-π stacking interactions; and (3) the process of coating PZS onto CNTs can be carried out in a one-pot process at room temperature and does not use any surfactant or other compatibilizing agent, and the thickness of the coating layer can be easily controlled.

3.3 Synthesis of Highly Magnetically Sensitive Nanochains Coated With PZS

Based on the good assembly property of PZS primary particles, Zhou and coworkers [52] presented a facile one-pot synthesis of one-dimensional (1D) magnetic nanochains coated with highly cross-linked PZS. Fe_3O_4-based colloidal nanocrystal clusters (CNCs) were selected as building blocks for 1D nanochains because of the combination of paramagnetism and a remarkable magnetic response [53–55]. The highly cross-linked PZS was chosen as the shell for the stabilization and functionalization of the 1D nanochains, and it endows the nanochains with good water dispersibility, biocompatibility, and tailored surface chemistry.

Figure 16 shows the TEM and SEM images of as-synthesized materials. Figure 16a and b demonstrate that the CNCs are very monodisperse, with an average diameter of 192 nm. Figure 16c and d show typical SEM images of as-synthesized CNCs/PZS at different magnifications. These images indicate that most of the CNCs/PZS have a chainlike morphology instead of being isolated nanoballs, and the good junction of adjacent CNCs can be clearly seen in Figure 16d. The core@shell structure of as-synthesized CNCs/PZS nanochains was further investigated by TEM. As illustrated in Figure 16e and f, the black CNCs formed 1D nanochains by head-to-tail interactions in a continuous thin PZS shell (gray). The thickness of the PZS shell is estimated to be about 20 nm (Figure 16f). It should be noted that the preorganization of linear CNC chains before being coated with PZS was crucial to the formation of nanochains. The preformed CNC spines act as a template and the PZS shell encircled the spines, leading to the formation of nanochains (Figure 16c–f). On account of the 1D assembly and the PZS coating, the nanochains display an enhanced magnetic resonance sensitivity and biocompatiblility [52].

3.4 Magnetic Fe_3O_4-PZS Hybrid Hollow Microspheres

Composites of magnetic nanoparticles with polymer hollow spheres have recently attracted significant interest on account of their bifunctionality [56–58]. In general, magnetic nanoparticles are immobilized on the outer surface of or inside the cavity of polymer hollow microspheres; thus the immobilized nanoparticles often are dropped or leaked from the polymer hollow microspheres during the application process, generally resulting in a loss of magnetism and dispersibility. Therefore, many efforts have been focused on designing one type of

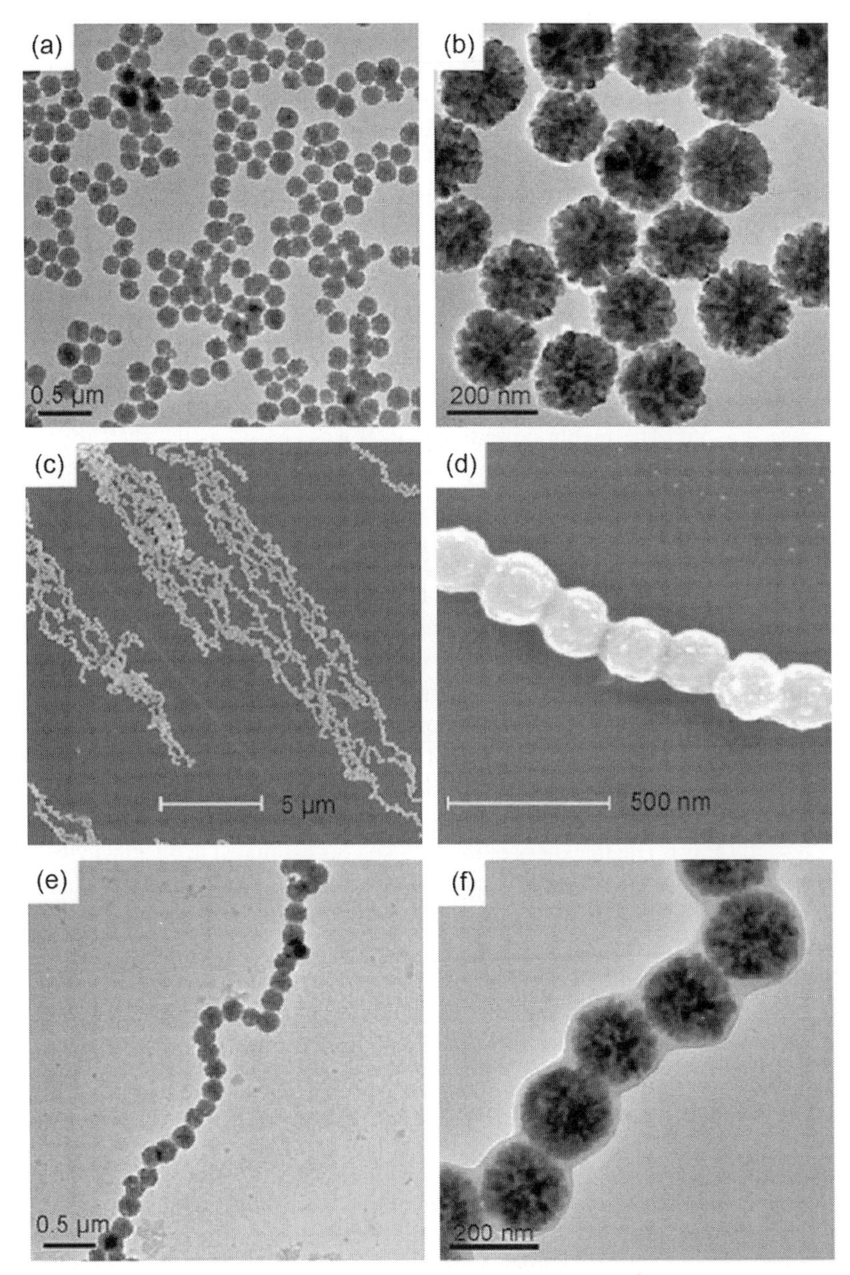

FIGURE 16 (a, b) Transmission electron microscopy (TEM) images of colloidal nanocrystal clusters (CNCs) that are about 192.0 nm in diameter. Scanning electron microscopy (c, d) and TEM images (e, f) show CNC@poly(cyclotriphosphazene-4,4′-sulfonyldiphenol) chains with an ~20-nm shell thickness.

magnetic polymer hollow sphere in which magnetic nanoparticles are incorporated into the shell layer of spheres. For example, Caruso *et al.* [59] prepared magnetic hollow microspheres through the colloid-templated, electrostatic layer-by-layer self-assembly of oppositely charged inorganic nanoparticles and polyelectrolytes, followed by removal of the core. Nonetheless, multistep processes are needed for the formation of the hollow structure, creating potential drawbacks in scaling-up the process for industrial applications. Li *et al.* [60] fabricated amphiphilic magnetic ferrite/block copolymer hollow microspheres using the solvothermal method. However, this method requires high temperature and high pressure; in addition, the preparation of the desired block copolymer is usually not a trivial task.

As a significant advance, we demonstrated a template-induced covalent assembly approach to preparing magnetic Fe_3O_4-PZS hybrid hollow microspheres, whereby Fe_3O_4 nanoparticles were firmly incorporated in a PZS shell with a highly cross-linked structure [61]. As-prepared magnetic hollow microspheres exhibit highly magnetic sensitivity (13.3 emu/g of magnetization saturation), favorable dispersion in aqueous and organic media, good thermal stability (440°C of initial decomposition temperature in a nitrogen atmosphere), and tailored surface chemistry for binding noble metal nanoparticles, which allows them to serve as ideal candidates for catalyst supports.

Figure 17 shows the procedure used to prepare magnetic Fe_3O_4-PZS hybrid hollow microspheres. First, polystyrene (PS) microspheres (1.3 μm in

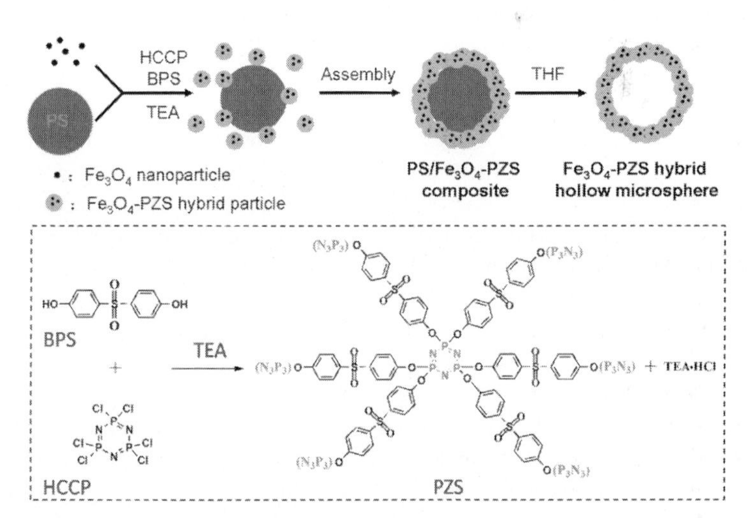

FIGURE 17 Top: Fabrication of Fe_3O_4-poly(cyclotriphosphazene-4,4′-sulfonyldiphenol) (PZS) hybrid hollow microspheres. Bottom: The polycondensation of co-monomers hexachlorocyclotriphosphazene (HCCP) and 4,4′-sulfonyldiphenol (BPS) and the cross-linked structure of the PZS shell. (P_3N_3) indicates other phosphazene cores. PS, polystyrene; TEA, triethylamine; THF, tetrahydrofuran.

diameter) and Fe_3O_4 nanoparticles with an average particle size of 15-25 nm were prepared through the common coprecipitation method and soap-free emulsion polymerization, respectively. Then Fe_3O_4 nanoparticles were embedded by active PZS generated during the polycondensation of HCCP and BPS, and the resulting active Fe_3O_4-PZS hybrid particles could self-assemble on the surface of PS micropheres and complete the cross-linking reaction between hybrid particles to produce PS/Fe_3O_4-PZS (core/shell) composites. As soon as the PS cores were removed by washing with THF, magnetic Fe_3O_4-PZS hybrid hollow micropheres were obtained. Because the Fe_3O_4 nanoparticles were firmly incorporated in the PZS shell with a highly cross-linked structure, magnetic hollow microspheres appeared to be structurally robust and well preserved after multiple rinses with an organic solvent. Furthermore, the magnetic hollow microspheres could re-disperse well in water and other polar organic solvents, such as THF, acetonitrile, acetone, dimethyl sulfoxide, and ethanol, thus allowing further potential surface modification in both aqueous and organic media.

The size and morphology of the as-prepared materials were investigated by TEM and SEM. Figure 18a shows a representative TEM image of Fe_3O_4 nanoparticles prepared using the coprecipitation method. Uniform particles with a mean diameter of 15-25 nm are clearly observed, indicating the successful preparation of Fe_3O_4 nanoparticles. Figure 18b shows an SEM image of uniform PS microspheres with a 1.3-μm diameter. It is clear that the surfaces of the PS microspheres are very smooth. When the PS microspheres are coated by Fe_3O_4-PZS hybrid particles, PS/Fe_3O_4-PZS composites clearly exhibit a rough surface (Figure 18c). The core/shell structure of as-prepared PS/Fe_3O_4-PZS composites was investigated by TEM (Figure 18c, inset). The diameter of core was about 1.3 μm, just being consistent with that of the PS microspheres. After removing the PS cores by washing with THF, magnetic Fe_3O_4-PZS hybrid hollow microspheres were formed, which can be confirmed by the broken spheres and the fragments of cracked spheres in the SEM image shown in Figure 18d. TEM characterization results in Figure 18e and f further confirm the above conclusion. Figure 18e shows these spheres with obvious contrast between the pale center and the dark edge, as reported in other hollow spheres. In particular, one big sphere shown in Figure 18f clearly exhibits a central cavity. The inner diameter of the magnetic hollow microspheres is estimated to be about 950 nm (less than the diameter of the PS templates), and the shell thickness is about 210 nm. The magnetic hollow microspheres apparently undergo significant shrinkage during the removal of colloidal templates, which may be caused by the hollow interior of the as-prepared hybrid microspheres.

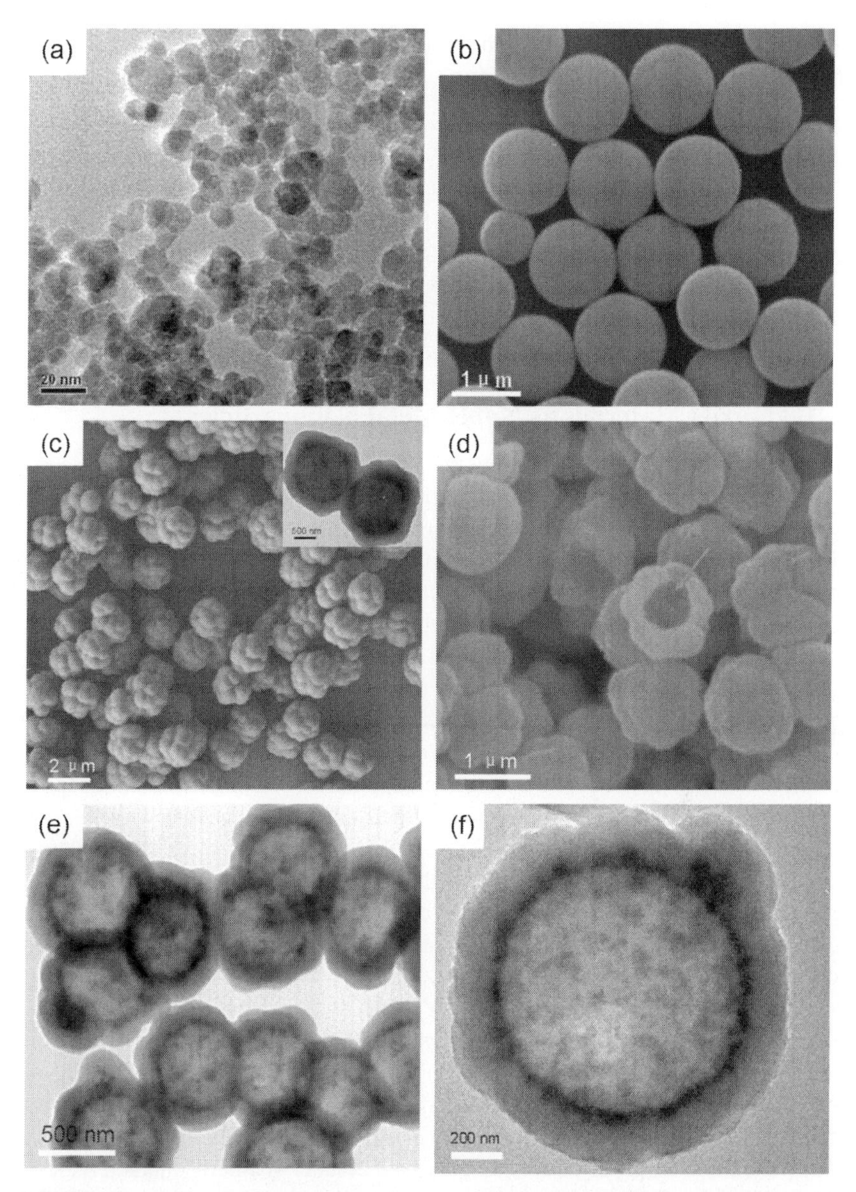

FIGURE 18 (a) Transmission electron microscopy (TEM) image of Fe_3O_4 nanoparticles. (b) Scanning electron microscopy (SEM) image of polystyrene (PS) microspheres. (c) SEM and TEM (inset) images of PS/Fe_3O_4-poly(cyclotriphosphazene-4,4′-sulfonyldiphenol) (PZS) hybrid composites. (d–f) Representative SEM and TEM images of magnetic Fe_3O_4-PZS hybrid hollow microspheres. The arrow in (d) points to a broken sphere with hollow structure, indicating the successful preparation of Fe_3O_4-PZS hybrid hollow microspheres.

4 APPLICATIONS OF PZS-BASED MICRO- AND NANOMATERIALS

As already stressed in the Introduction to this chapter, polyphosphazenes possess many applications in different domains of science and technology. Considering the chemical structure and novel micro- and nanoscale morphology of PZS materials, in this section we highlight their applications in the fields of carrier materials, adsorption materials, and drug delivery. In addition, based on the carbon-rich characteristic of PZS, we demonstrate that PZS can be used as good precursors for preparing N, P-codoped porous carbon materials.

4.1 Carrier Materials as Noble Metal Nanoparticles

In recent years, the interest in noble metal nanoparticles has grown constantly because of their unique properties and potential applications in catalysis, microelectronics, chemical sensors, data storage, and other areas [62]. Thermodynamically, naked metal nanoparticles incline to aggregate into larger particles because of van der Waals forces, resulting in a decrease in their performance. To solve this problem, different kinds of stabilizers, such as ionic liquids, dendrimers, CNTs, polymers, and ligands, have been suggested. Among them, using functionalized polymers or a coordination polymer with ligands to stabilize metal nanoparticles is a promising alternative [62,63]. In many cases, however, the facile preparation of the desired polymers is usually not a trivial task. In addition, in view of the described applications of noble metal nanoparticles, control of the size of metal nanoparticles is always necessary [64].

Cyclotriphosphazene consists of six-membered ring structures of alternating phosphorus and nitrogen atoms and has the capability to bind to metal ions via its phosphazene ring nitrogen [65]. PZS is one of the typical cyclotriphosphazene-containing polymers. Therefore, PZS materials should be good carriers to support noble metal nanoparticles.

Our research has confirmed that PZS nanospheres could be readily decorated with uniform palladium (Pd) nanoparticles through an inorganic reaction in supercritical carbon dioxide (SC CO_2)-ethanol solution using $PdCl_2$ as a metal precursor. Ethanol acts as reactant as well as a cosolvent to enhance the solubilization of $PdCl_2$ in supercritical solutions [66]. The intriguing advantages of this route lie in its lower cost, simplicity, and greener synthesis characteristics.

Figure 19a shows an SEM image of PZS nanospheres prepared through the precipitation polymerization of HCCP with BPS. Well-dispersed, uniform, spherical particles with mean diameter of 410 nm are clearly observed and no impurities are seen, indicating that we have developed a simple route to creating polymer nanospheres with narrow size distributions. A TEM image of these spherical particles reveals that the PZS nanospheres possess a solid structure and smooth surface, as shown in Figure 19b. Figure 19c displays a representative TEM image of the PZS nanospheres decorated with Pd nanoparticles with

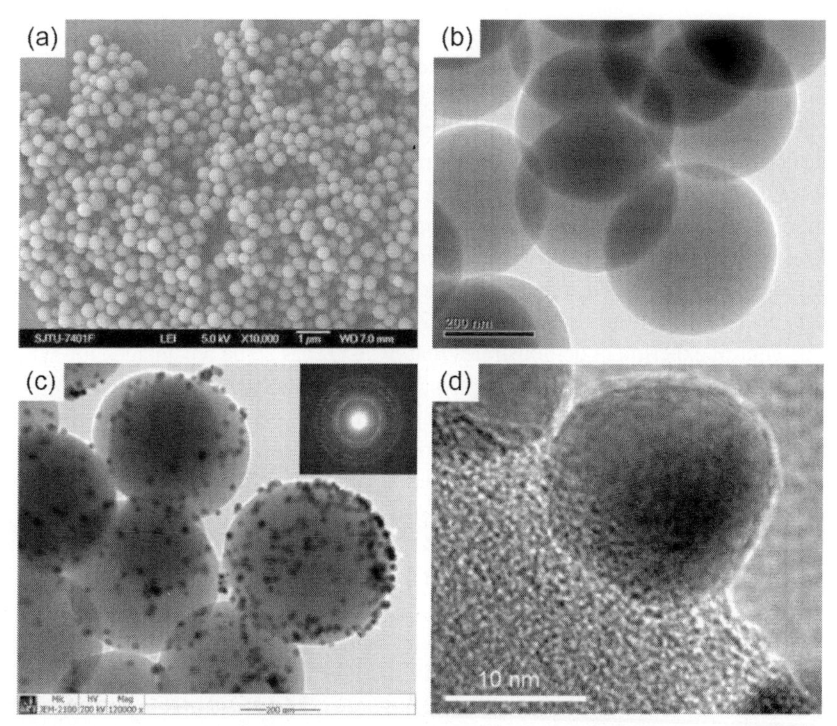

FIGURE 19 Scanning electron microscopy (a) and transmission electron microscopy (TEM) (b) images of as-synthesized poly(cyclotriphosphazene-4,4′-sulfonyldiphenol) (PZS) nanospheres. (c) TEM image of the palladium (Pd)/PZS nanocomposites. The inset displays the electron diffraction pattern of the Pd/PZS nanocomposites. (d) High-resolution TEM image of the Pd/PZS nanocomposites.

the assistance of the SC CO_2 technique. A homogeneous dispersion of spherical Pd nanoparticles on the surfaces of the PZS nanospheres can be distinguished. Digital analysis of a representative TEM image (100 particles measured) yielded an average Pd particle size of 20 nm. The Pd nanoparticles are crystalline, as indicated by the selected area electron diffraction pattern shown as the inset in Figure 19c. The HRTEM image shown in Figure 19d further verifies that the Pd nanoparticles are crystalline, with visible lattice fringes, and are well anchored onto the surfaces of the PZS nanospheres.

Usually, most of the metal nanoparticles loaded onto unfunctionalized supports could be detached from the supports after agitation or ultrasonication. In this work, the adhesion of the Pd nanoparticles loaded on the surfaces of the PZS nanospheres was investigated by ultrasonication for 30 min in ethanol. TEM observation showed that few Pd nanoparticles were detached from the surfaces of the PZS nanospheres, revealing that better adhesion had been achieved using SC CO_2 as the medium for Pd nanoparticle deposition on the PZS nanospheres. This good adhesion, as well as the homogeneous dispersion of Pd nanoparticles on the surfaces of the PZS nanospheres, should be attributed to the stabilization effect caused by the strong interaction between the phosphazene ring nitrogen and the deposited Pd nanoparticles. In addition, the SC CO_2 also played an important role during Pd deposition. To confirm this fact, two control experiments were carried out. One was performed without PZS nanospheres, yielding nonspherical Pd nanocrystals with a large particle size under similar conditions. This result illustrates that ethanol is a good reducing agent of $PdCl_2$ precursors with the aid of SC CO_2. Simultaneously, this also implies that PZS nanospheres were apparently capable of stabilizing Pd nanoparticles as well as preventing their severe growth during the process to synthesize Pd/PZS nanocomposites. The other was performed without CO_2; few Pd nanoparticles were anchored onto the surfaces of the PZS nanospheres under similar conditions. This control experiment demonstrates that SC CO_2 has some advantages for synthesizing the composites in comparison with conventional liquid solvents. With respect to liquids, the SC CO_2 has such properties as enhanced transport coefficient, low density and viscosity, and near-zero surface tension, which promote the complete wetting of PZS nanospheres and facilitate the uniform deposition of Pd nanoparticles on the surfaces of the PZS nanospheres.

In addition, our experimental results revealed that by adjusting the ethanol-reduction time of the precursor $PdCl_2$ in the condition of SC CO_2, the size of Pd nanoparticles on the surfaces of the PZS nanospheres could be easily controlled, which is crucial to their industrial applications in different fields. Figure 20 shows a set of TEM images of the Pd/PZS nanocomposites obtained after different ethanol-reduction times using $PdCl_2$ as a metal precursor with the assistance of SC CO_2. Obviously, with the increase in the reduction time, the Pd nanoparticle size on the surfaces of the PZS nanospheres increases. It should be noted that the Pd nanoparticles incline to aggregate together when the reduction time is over 3 h, as shown in Figure 20c.

Compared with PZS nanospheres, PZS nanotubes have a larger specific surface area. This characteristic endows PZS nanotubes with great potential for supporting noble metal nanoparticles. In addition, considering the chemical structure of PZS, *in situ* reduction should be a rapid route to realizing the fabrication of

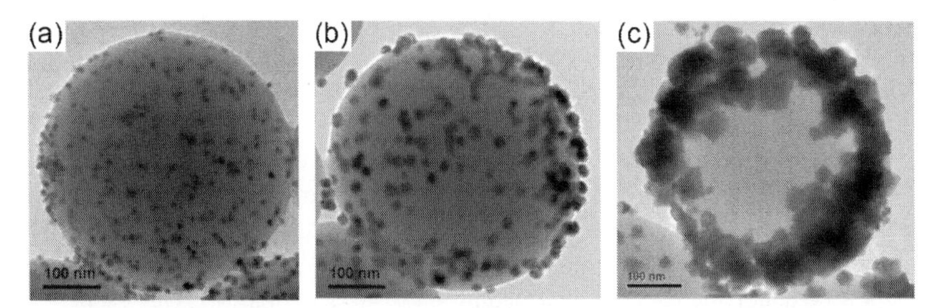

FIGURE 20 Transmission electron microscopy images of palladium (Pd)/poly(cyclotriphosphazene-4,4′-sulfonyldiphenol) (PZS) nanocomposites with a Pd particle size of ~10 nm (a), ~20 nm (b), and 35 nm (c). Reaction conditions used mass ratios of PZS nanospheres to $PdCl_2$ controlled at 10:1. The ethanol-reduction time in the condition of supercritical carbon dioxide (40°C, 13 MPa) was 30 min (a), 60 min (b), and 180 min (c).

noble metal nanoparticles loaded onto PZS materials. The experiment demonstrated that PZS nanotubes with active hydroxyl groups, a class of functional polymer nanotubes, could be used as stabilizers to synthesize *in situ* nanotube-stabilized Ag nanoparticles with sodium borohydride as the reducing reagent [67].

Figure 21a and b show TEM images of the PZS@Ag nanoparticle composites. Obviously, the PZS nanotubes were decorated with Ag NPs successfully. Ag nanoparticles with diameters in the range of 5-20 nm were clearly stabilized on the surfaces of the PZS nanotubes, and they did not aggregate together during the *in situ* reduction of sodium borohydride over the functional PZS nanotube stabilizer. Energy-dispersive X-ray spectroscopy characterization (Figure 21c) of the PZS@Ag nanoparticle composite nanotubes shows that the composite nanotubes were composed of C, O, P, S, Cl, and Ag, indicating the existence of Ag nanoparticles. These results indicated that the PZS nanotubes were highly accessible, with good ability to stabilize for Ag nanoparticles. The wide-angle X-ray diffraction patterns of the PZS@Ag nanoparticle composites are shown in Figure 22d. The broad band at $2\theta = 10\text{-}30°$ is the characteristic band of PZS [12,51]. In addition to the broad band, four distinct diffractions at about 38.1°, 44.4°, 64.3°, and 77.4°, which correspond to the (111), (200), (220), and (311) planes of Ag, respectively [68,69], indicate the presence of Ag nanoparticles.

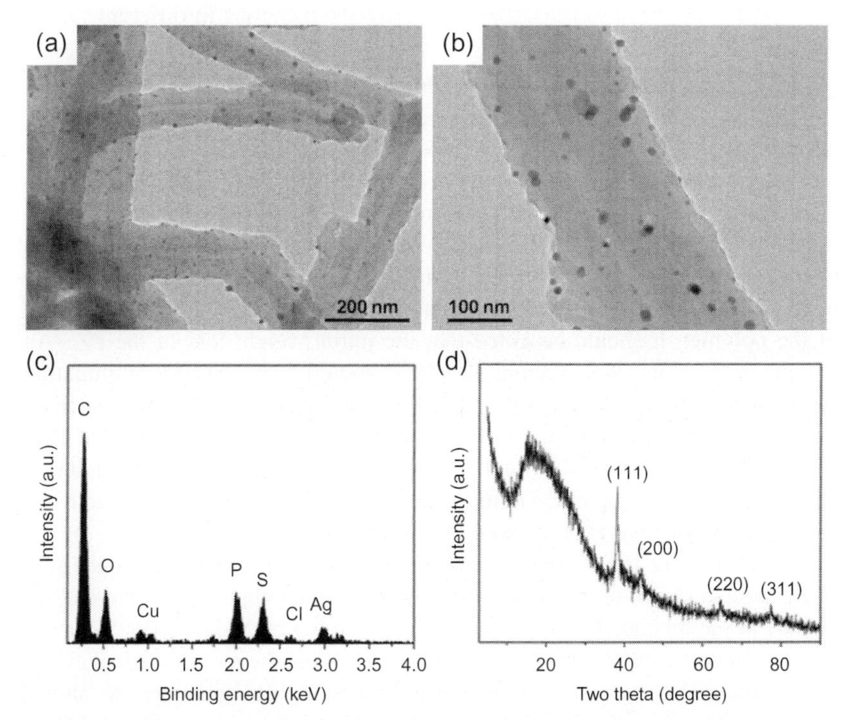

FIGURE 21 Transmission electron microscopy images (a, b), energy-dispersive X-ray spectroscopy (c), and X-ray diffraction pattern (d) of poly(cyclotriphosphazene-4,4′-sulfonyldiphenol)@silver nanoparticle composites. a.u., arbitrary units.

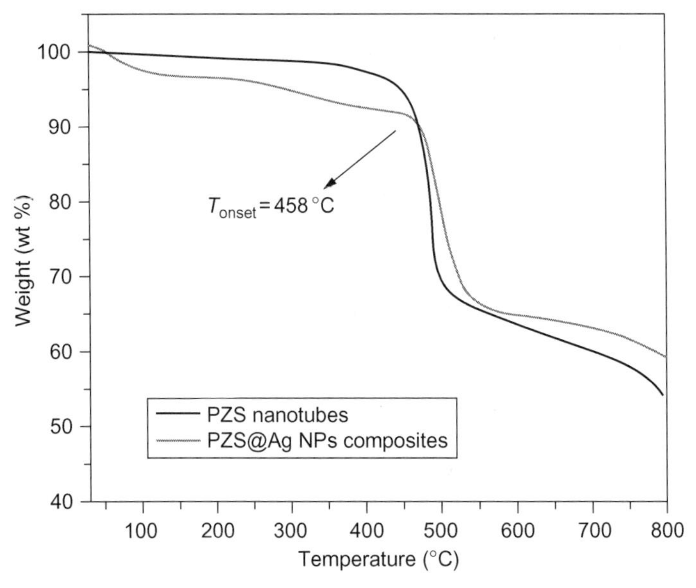

FIGURE 22 Thermogravimetric curves for pure poly(cyclotriphosphazene-4,4′-sulfonyldiphenol) (PZS) nanotubes and the PZS@silver nanoparticle (PZS@Ag NP) composites.

To characterize the behavior of the PZS nanotubes modified with Ag nanoparticles at elevated temperatures, a thermogravimetric study of the PZS@ Ag nanoparticle composites over temperatures of 30-800°C under a flowing nitrogen atmosphere was performed. The thermogravimetric analysis of pure PZS nanotubes also was added for comparison. As shown in Figure 22, the initial decomposition of the PZS nanotubes as a substrate of Ag nanoparticles starts at approximately 458°C under a nitrogen atmosphere, and the loaded Ag nanoparticles do not significantly affect the initial decomposition temperature of the polymer. It should be noted that the initial weight loss of the PZS/Ag nanoparticle composites at about 30-150°C should be assigned to elimination of the adsorbed water, and it is most likely that the presence of Ag nanoparticles enhanced the water absorbance of PZS, which is similar with that reported elsewhere [70]. Weight loss of 37 and 46 wt% occurred in the range of 150-800°C for the PZS@Ag nanoparticle composites and the PZS nanotubes, respectively, and the remaining weight increased after supporting Ag nanoparticles.

In addition, the high temperature stability of metal nanoparticles on the substrates is also important for high-temperature catalysis applications. Herein, the change in the tubular structure of the PZS nanotubes and aggregation behavior of Ag nanoparticles has been investigated by tracking the morphological evolution of the PZS@Ag nanoparticle composites at high temperature. As shown in Figure 23, when the PZS@Ag nanoparticle composites were heated from room temperature to 100°C and 250°C (Figure 23a–c), the PZS@Ag nanoparticle composites were still stable without structural collapse, and there were few

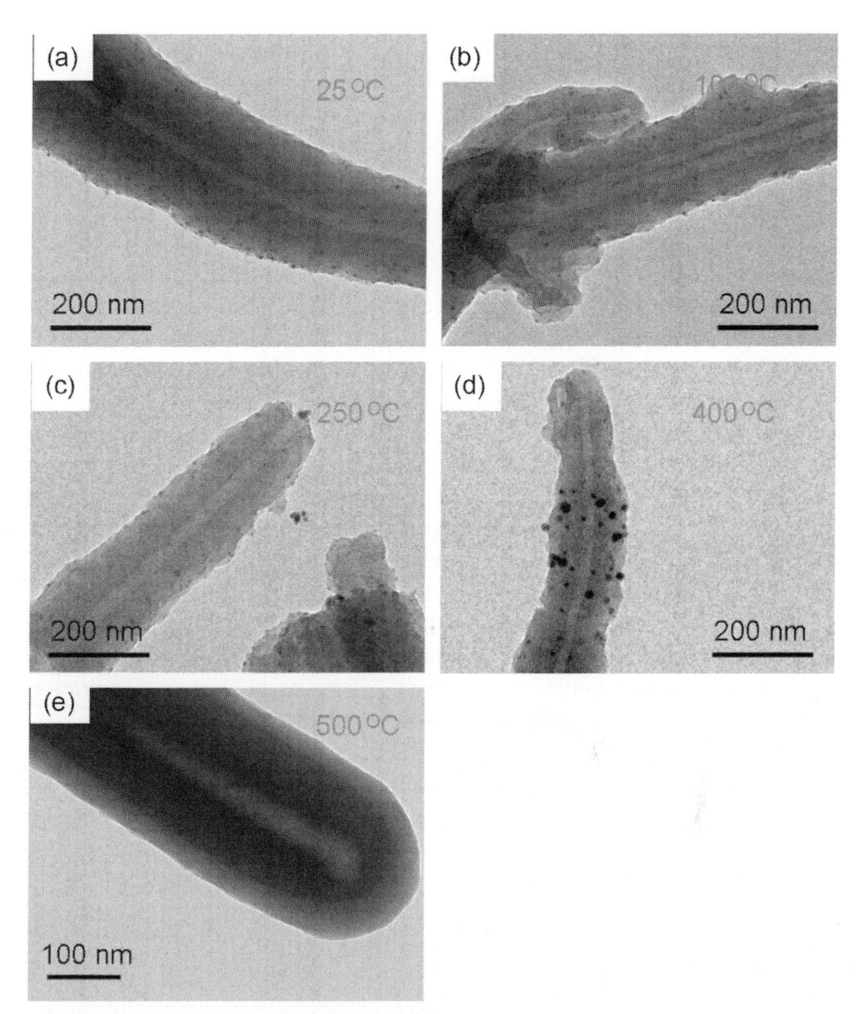

FIGURE 23 Typical transmission electron microscopy images of poly(cyclotriphosphazene-4,4′-sulfonyldiphenol)@silver nanoparticle composites after heating at different temperatures under a flowing nitrogen atmosphere: 25°C (a); 100°C (b); 250°C (c); 400°C (d); 500°C (e).

Ag nanoparticles enlarging significantly on the nanotubes. When heated up to 400°C (Figure 23d), some big, black nanoparticles appeared on the nanotubes. As reported in the literature [71,72], sintering at around 200°C is usually attributed to the reduced melting point of Ag nanoparticles compared with that of bulk metal and to surface premelting; and the melting point of Ag particles with a mean diameter of 12.6 nm prepared is about 250°C. Therefore, the big, black nanoparticles in our TEM observation should be attributed to the aggregation of Ag nanoparticles during high-temperature carbonization. When heated up to 500°C (Figure 23e), Ag nanoparticles on the carbonized samples disappeared, but the tubular structure of the carbonized samples was maintained, indicating that the PZS has a good ability to retain its original morphology.

Catalytic activities of the PZS@Ag nanoparticle composites has been examined by choosing the model catalysis reaction involving reduction of 4-NP to 4-AP with $NaBH_4$ as the reductant, as shown in Figure 24a. Such a reaction catalyzed by noble metal catalysts has been reported intensively, and can be easily and rapidly characterized [73–75]. Without the PZS@Ag nanoparticle composites, the mixture of 4-NP and $NaBH_4$ shows an absorbance band at 400 nm, corresponding to the absorbance of 4-NP in an alkaline condition, and remains unaltered with time. With the addition of the PZS@Ag nanoparticle composites to the reaction mixture, the peak changed, becoming lower, and the yellow-green color faded with the simultaneously gradual development of the new absorption peak at 297 nm, assigned to 4-AP, indicating that the reaction took place in the presence of an Ag catalyst. The catalytic reaction can be estimated to end for the complete disappearance of the ultraviolet-visible absorption at 400 nm. The turnover frequency, which is defined as the number of moles of reduced 4-NP per mole of surface M atoms per hour when conversion has reached 90%, was calculated. The turnover frequency is about 101.4 h^{-1}, comparable to that reported for Ag nanoparticles [76,77]. Furthermore, the rate constant k can be calculated from the rate equation $\ln(C_t/C_o) = -kt$, where t is the reaction time, C_o is the initial concentration of 4-NP, and C_t is the concentration of 4-NP at time t. The k was calculated to be $11.1 \times 10^{-3}\, s^{-1}$ from the linear relationship shown in Figure 24b, which is superior to that reported for smaller Ag nanoparticles as catalysts. In addition, the reusability of the PZS@Ag nanoparticle composites as a heterogeneous catalyst was investigated. Little obvious catalytic loss is evidenced within five cycles, which is attributed to the good stability of Ag nanoparticles on the PZS nanotubes. The results suggest that the PZS@Ag nanoparticles system is a better catalyst for the reactions investigated.

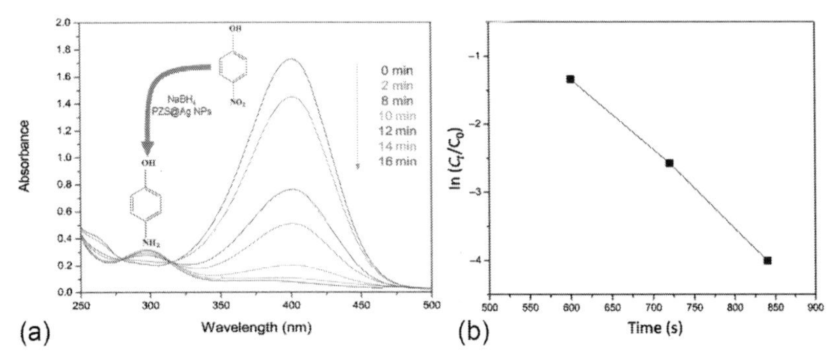

FIGURE 24 (a) Successive ultraviolet-visible absorbance spectra for the reduction of 4-NP by $NaBH_4$ in the presence of the poly(cyclotriphosphazene-4,4′-sulfonyldiphenol)@silver nanoparticle (PZS@Ag NP) composites. (b) $\ln(C_t/C_0)$ versus t for the reduction of 4-NP catalyzed by the PZS@Ag NP composite catalyst.

4.2 Adsorbent of Dyes

Dyes are colored organic compounds and are extensively used in the textile, leather, paper, food, cosmetic, and other industries [78]. They usually are present in the effluents of these industries. Without reasonable processing, they can cause serious environment pollution because they are toxic to microorganisms and can impede the photosynthesis of aqueous flora [79]. Even worse, most organic dyes are harmful to humans because of their potential mutagenic and carcinogenic effects [80,81]. Therefore, the removal of dyes from wastewater has been a seriously concern.

To date, adsorption has been an economic, effective, and easily operated process in dye removal. A wide range of adsorbent materials such as activation carbon, zeolite, silica, and natural polymeric materials, have been applied. However, these conventional adsorbents often display many defects and disadvantages, such as low adsorption efficiency and long adsorption time. Our research revealed that PZS nanotubes are a high-efficiency adsorbent for the removal of methylene blue (MB) and are useful as a model compound for removing basic dyes from aqueous solutions because of their hollow tubular structures, relatively large specific surface area, pore structures, numerous hydroxyl groups, aromatic rings, and electron-rich N and O atoms, as shown in Figure 25 [82]. The adsorption capacity at equilibrium at 25°C could reach up to 69.16 mg/g, and the corresponding contact time was only 15 min, which was shorter than the vast majority of adsorbents. Results also showed that MB adsorption onto the PZS nanotubes was highly dependent on the temperature, concentration, and pH of the MB solution. In kinetic studies the pseudo-first-order kinetic model, pseudo-second-order kinetic model, and intraparticle diffusion model were used to fit adsorption data. The pseudo-second-order kinetic model better described adsorption kinetics, and the intraparticle diffusion model also demonstrated that intraparticle diffusion was not the rate-limiting step.

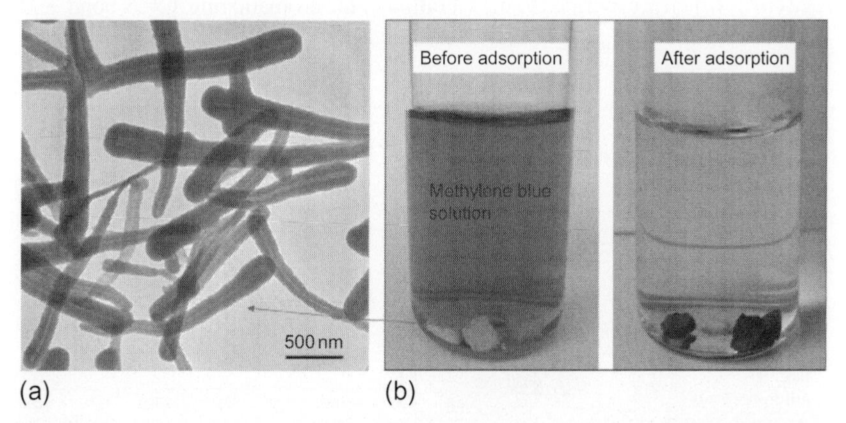

FIGURE 25 (a) Transmission electron microscopy images of as-prepared poly(cyclotriphosphazene-4,4′-sulfonyldiphenol) (PZS) nanotubes. (b) The photographs of methylene blue solutions before (left) and after (right) adsorption by PZS nanotubes.

The R^2 (0.9933) of the Langmuir isotherm and its monolayer adsorption capacity (72.83 mg/g), which highly approaches experimental data (69.16 mg/g), indicated that the adsorption of MB onto the PZS nanotubes followed the Langmuir isotherm. The values of thermodynamic parameters (ΔG^0, ΔH^0, and ΔS^0) suggested that MB adsorption onto the PZS nanotubes was endothermic and spontaneous. In addition, it was a physisorption process.

Figure 26a shows the adsorption process and mechanism for MB on the PZS nanotube. In an incubator with a shaking speed of 150 rpm, the MB molecules rapidly diffuse to the adsorbent during the film diffusion stage because of the existence of driving forces that result from the initial dye concentrations. When the cationic MB molecules are close to the external surface of the PZS nanotubes with a large number of negatively charged sites, the electrostatic attraction between MB and PZS can occur. Meanwhile, the π-π stacking interactions between them also appear because both MB molecules and PZS contain aromatic rings. Obviously, the two interactions are helpful for the quick adsorption of MB molecules by PZS nanotubes. However, the adsorption rate becomes very low and then constant in the final stage as a result of the repulsive forces on account of large amounts of MB molecules on the PZS nanotubes. FTIR analysis was carried out to gain further insight into the adsorption mechanism. Figure 26b shows the FTIR spectra of the PZS nanotubes before and after MB adsorption. Based on the aforementioned results, the surface of PZS nanotubes has abundant hydroxyl groups, confirmed by the presence of the characteristic absorption peak at 3424 cm^{-1}. After adsorption of MB, the peak at 3424 cm^{-1} appeared as an obvious red shift (3412 cm^{-1}), which reveals that the hydroxyl groups may play an important role in MB adsorption onto the PZS nanotubes. MB is a kind of cationic dye that can be adsorbed easily by electrostatic forces on negatively charged surfaces. Therefore, the red shift of the hydroxyl group peak after adsorption may be associated with the electrostatic attraction between PZS nanotubes and MB. In addition, the adsorption peaks at 1593, 1385, and 1322 cm^{-1}, which are ascribed to the vibration of the aromatic ring, C=N bond, and

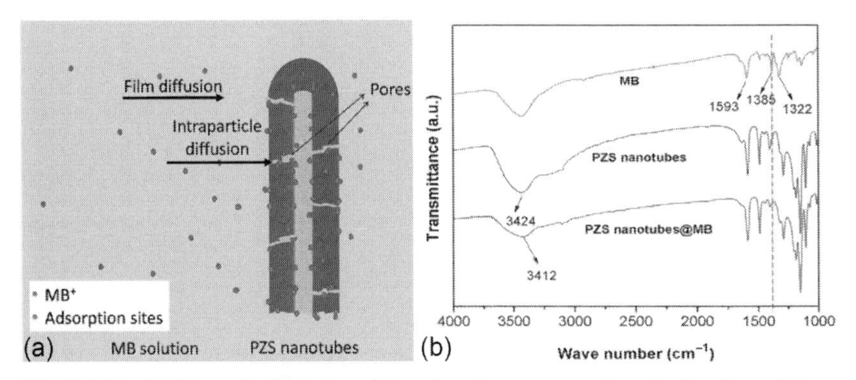

FIGURE 26 (a) Schematic illustration of the adsorption process and mechanism for methylene blue (MB) on poly(cyclotriphosphazene-4,4′-sulfonyldiphenol) (PZS) nanotubes. (b) Fourier transform infrared spectra of the PZS nanotubes before and after the MB adsorption, and for pure MB.

CH_3 group for MB, respectively [83], can be observed in the FTIR spectrum of the PZS nanotubes after MB adsorption. This indicates that the MB has been anchored on the surface of PZS nanotubes during adsorption. It should be noted that the peaks associated with the vibration of the aromatic ring and C=N bond seem to be a significant decrease in intensity, which might result from the π-π stacking interactions between the aromatic backbone of MB and the aromatic skeleton of PZS. Therefore, the electrostatic attraction and π-π stacking interactions between PZS and MB could be responsible for the high adsorption ability of the PZS nanotubes. In addition, the large BET surface area and porous structure of the PZS nanotubes are of great benefit to improving the amount of MB adsorbed on the PZS nanotubes.

4.3 Drug Delivery Carriers

Controlled drug delivery technology represents one of the most rapid advancing areas of science and involves a multidisciplinary scientific approach, contributing to human health care. The delivery systems offer many advantages compared with conventional dosage forms, which include improved efficacy, reduced toxicity, and improved patient compliance and convenience. Such systems often use hollow polymer microspheres as drug carriers because of their hollow core structure and ability to encapsulate large quantities of guest molecules. Polyphosphazene materials have excellent biocompatibility and biodegradability and have been widely used as biomaterials. Liu and coworkers [84] successfully fabricated hollow, cross-linked PZS submicrospheres through *ex situ* template-induced assembly, as shown in Figure 27. The mean diameter of the interior cavities can be well adjusted (typically in the range from 100 to 300 nm). Interestingly, there are plenty of uniform mesopores distributed in the organic-inorganic hybrid shells. The main pore size of the mesopores is about 2-4 nm. Such hollow mesoporous submicrospheres possess

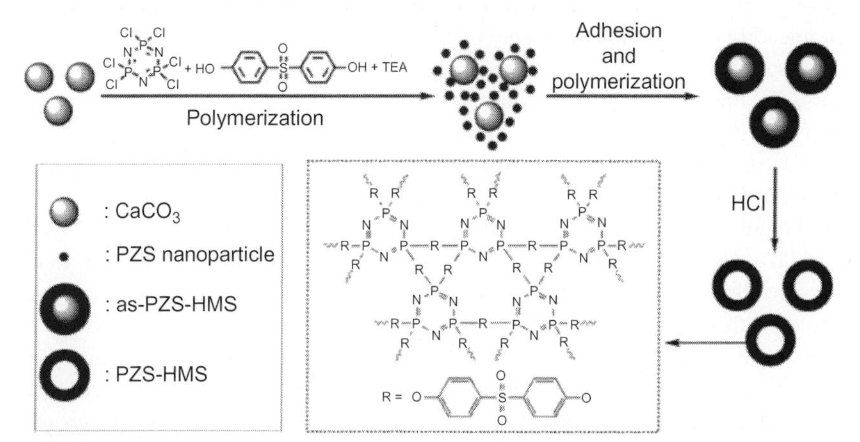

FIGURE 27 Scheme of the synthetic procedure of poly(cyclotriphosphazene-4,4′-sulfonyldiphenol) (PZS) microspheres. HMS, hollow mesoporous submicrosphere; TEA, triethylamine.

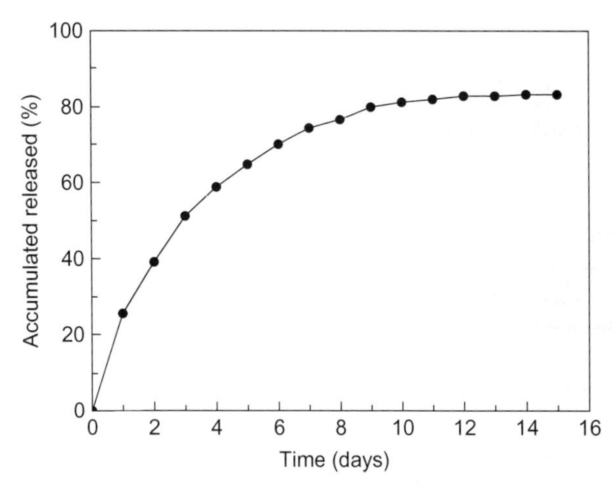

FIGURE 28 DOX release behavior of poly(cyclotriphosphazene-4,4′-sulfonyldiphenol) microspheres in phosphate-buffered saline (pH 7.2) at 37°C.

outstanding biocompatibility and dispersive ability in both aqueous and organic media. Moreover, these cross-linked polyphosphazene hollow mesoporous submicrospheres manifest a high drug storage capacity (380 mg doxorubicin hydrochloride/gram) and excellent sustained release property (up to 15 days), hence justifying their promising applications in drug delivery (Figure 28).

4.4 Precursors for Preparing Porous Carbon Materials

Carbon spheres have been a subject of considerable attention from both scientific and practical application point of views. In particular, porous carbon spheres are fascinating because of their low density, large surface area, large pore volume, chemical inertness, good mechanical stability, and good surface permeability, which makes them suitable for several potential applications including adsorbents, sensors, storage materials, catalyst supports, supercapacitors, and fuel cell electrodes. Our research showed that PZS, a common typical thermosetting resin with about 45% carbon yields, could easily form porous carbon during carbonization, providing a new and convenient way to obtain polymorphic carbon micro- and nanomaterials [85–89].

Figure 29 illustrates the preparation procedure of hollow core porous shell (HCPS) carbon spheres. First, the uniform PS colloid spheres (a hard template) were prepared through common dispersion polymerization. Second, the PS spheres were coated with a PZS layer to form PS@PZS composite spheres based on a template-induced assembly mechanism. Third, the PS@PZS composites were carbonized at 1000°C under a nitrogen atmosphere to form HCPS carbon spheres. The strategy offers at least four advantages: (1) As a hard template, the PS colloid spheres are commercially available or can be obtained through simple synthesis. (2) The process of coating PZS onto PS colloid spheres can be performed at room temperature and does not use any surfactant or other compatibilizing agent, and the thickness of the

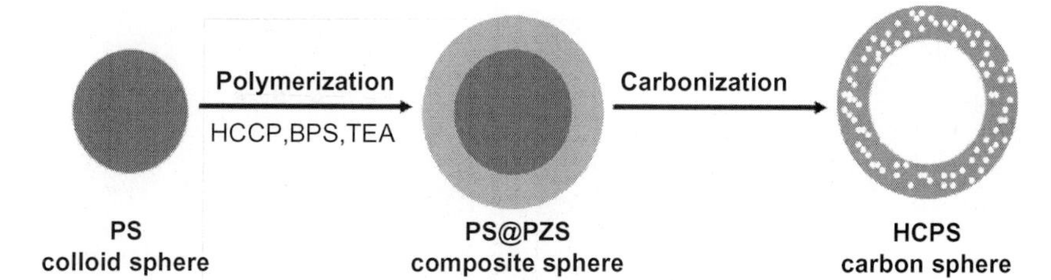

FIGURE 29 Fabrication of hollow core, porous shell (HCPS) carbon spheres. BPS, 4,4′-sulfonyldiphenol; HCCP, hexachlorocyclotriphosphazene; PS, polystyrene; PZS, poly(cyclotriphosphazene-4,4′-sulfonyldiphenol); TEA, triethylamine.

coating layer can be easily controlled. (3) The removal of the PS template and the formation of a porous carbon layer were finished during carbonization, decreasing the tedious postdisposition process and environmental pollution. (4) There were traces of oxygen, phosphorus, and sulfur in the carbon materials in spite of carbon being the main element, which modified the carbon materials.

Figure 30a and b show SEM and TEM images, respectively, of as-prepared PS colloid spheres (with an average diameter of 800 nm) through common dispersion polymerization. Every PS colloid sphere is solid and has a smooth surface. After PS colloid spheres were coated by PZS, it is clear that the surfaces of the composite spheres become relatively rough (Figure 30c). A TEM image of these composite spheres clearly reveals that almost all the PS colloid spheres were well coated by PZS, and a core/shell structure with a core size of ~800 nm and a shell thickness of ~150 nm was formed (Figure 30d).

Figure 30e shows a TEM image of the carbonized samples from the PS@PZS composite spheres after carbonization at 1000°C under a nitrogen atmosphere. A hollow sphere morphology with a hollow core size of ~560 nm and a shell thickness of ~100 nm was observed in the carbonized samples, although there was an overall reduction in size dimensions because of mass transfer flow during carbonization. The formation of a hollow structure indicates that, as a hard template, PS colloid spheres were removed successfully during carbonization, which also was confirmed by thermogravimetric analysis. Figure 30f shows the EDX pattern of the carbonized sample. Result shows that about 93 wt% carbon and a small percentage of additional elements (such as P, S, N, and O) were kept in the products, which indicated that PZS layers were well carbonized and most noncarbon elements were successfully removed during the pyrolyzing process. Meanwhile, EDX characterization indicated that heteroatom-containing porous carbon materials could be obtained by using PZS as a precursor. To investigate in depth the porous carbon microstructure, the sample was characterized by HRTEM. The results reveal that large quantities of micropores are homogeneously dispersed within the carbon shell matrix (Figure 30e, inset).

Just by changing the morphology of PZS-based precursors, we easily obtained microporous carbon nanofibers, microporous CNTs, and microporous carbon coated core/shell Si@C nanocomposite [86–88], as shown in Figures 31 through 33. Among them, the Si@C nanocomposite exhibits a stable capacity of over

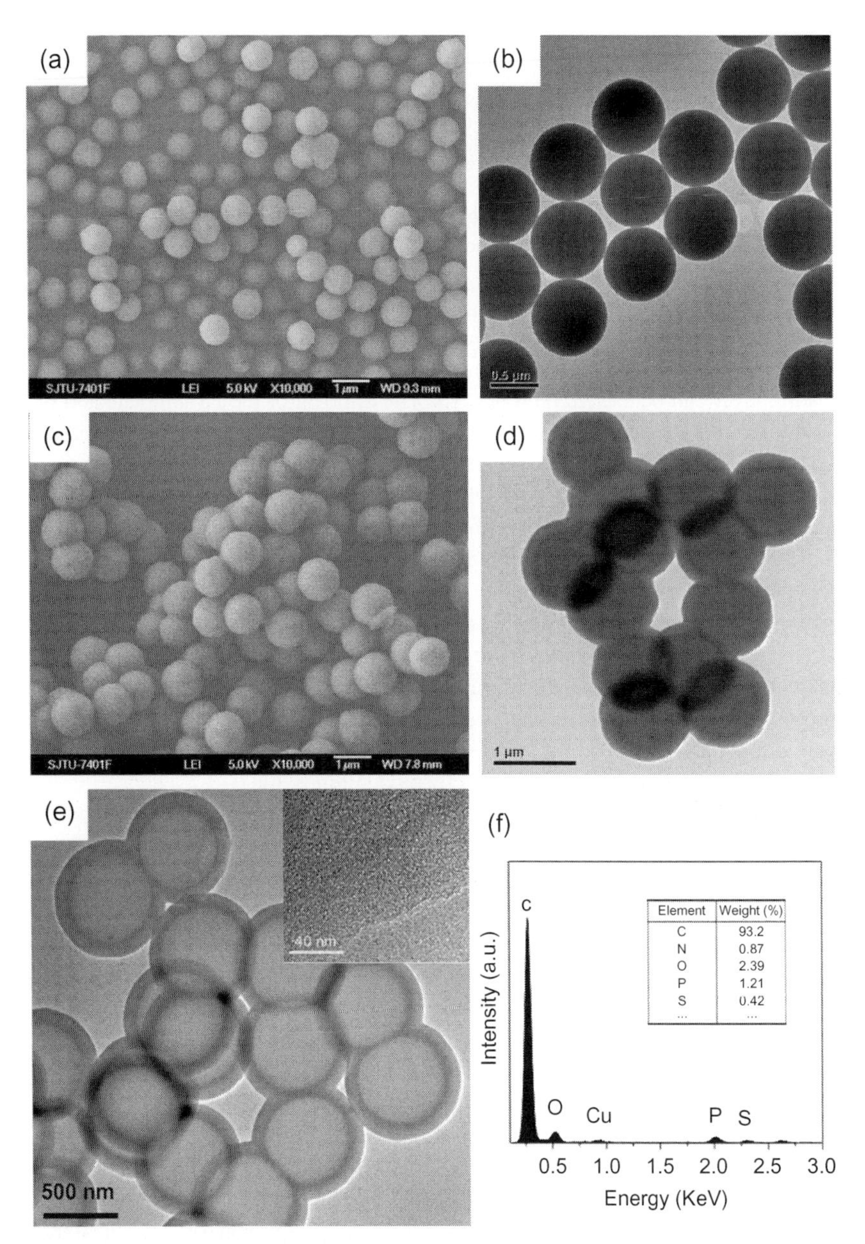

FIGURE 30 (a, b) Representative scanning electron microscopy (SEM) and transmission electron microscopy (TEM) images for polystyrene (PS) colloid spheres. (c, d) SEM and TEM images for PS/poly(cyclotriphosphazene-4,4′-sulfonyldiphenol) composite spheres. (e) TEM image of hollow core, porous shell (HCPS) carbon spheres. The inset shows a high-resolution TEM image of the carbon shell. (f) EDX spectrum of the HCPS carbon spheres.

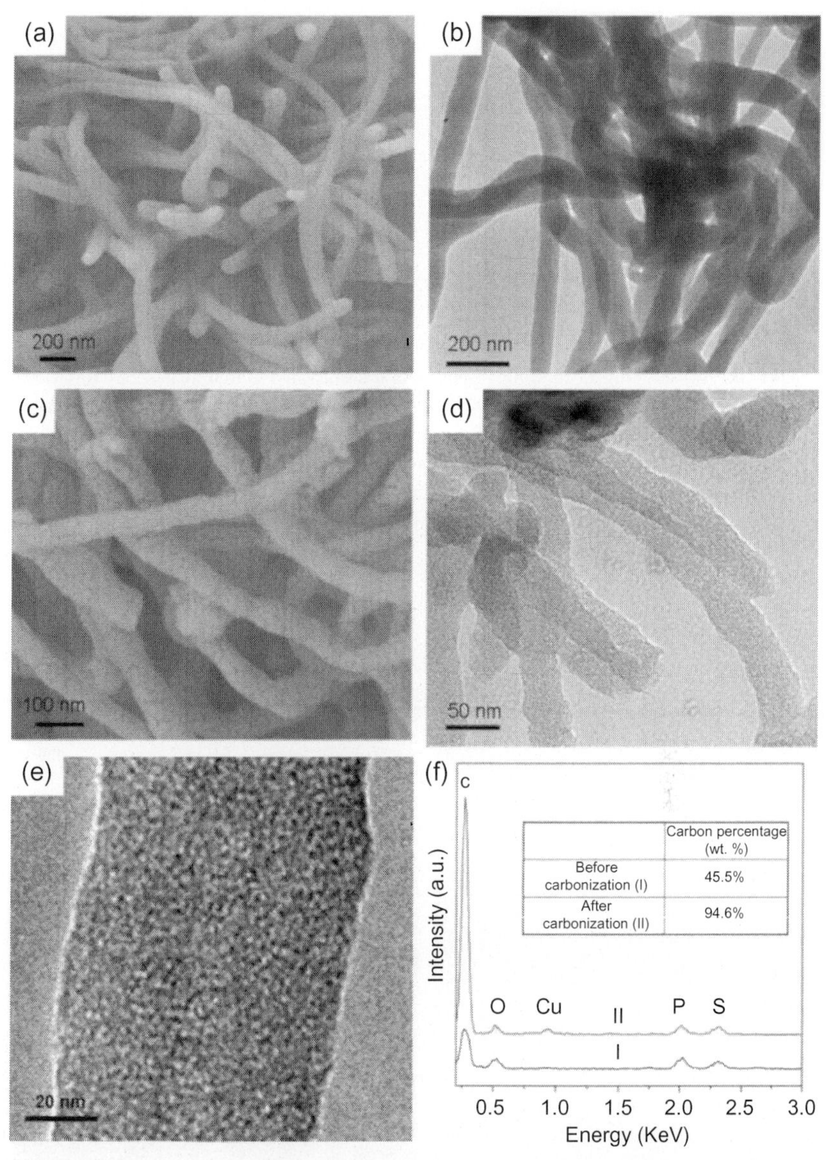

FIGURE 31 (a, b) Representative scanning electron microscopy (SEM) and transmission electron microscopy (TEM) images of poly(cyclotriphosphazene-4,4′-sulfonyldiphenol) (PZS) nanofibers. (c, d) SEM and TEM images of the as-fabricated PCNFs. (e) High-resolution TEM image of a single PCNF. (f) EDX patterns of the PZS nanofibers before and after carbonization.

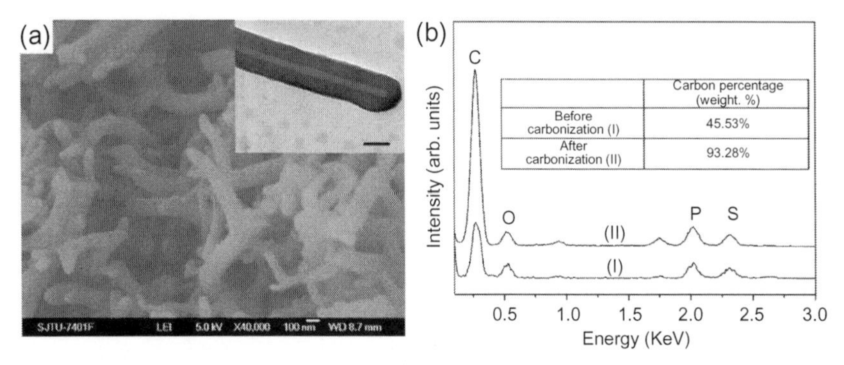

FIGURE 32 (a) scanning electron microscopy and transmission electron microscopy (inset) images of carbonized samples; scale bar of the inset = 100 nm. (b) EDX patterns of the PPZ nanotubes and the carbonized samples.

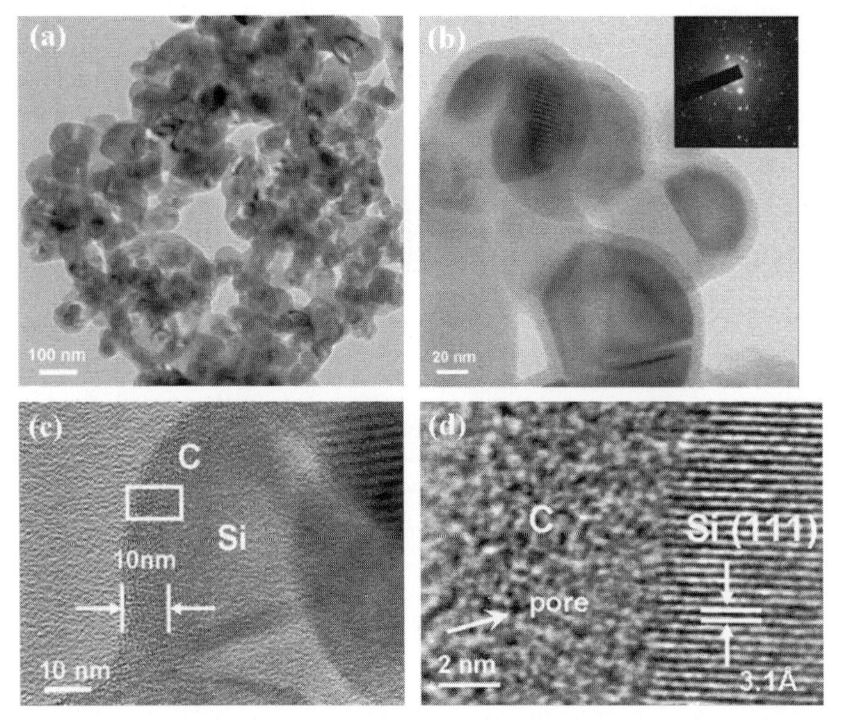

FIGURE 33 (a) Scanning electron microscopy image of the Si@C core/shell nanocomposite. (b, c) Transmission electron microscopy images of the Si@C nanocomposite. The inset in (b) is the selected area electron diffraction image of the core. (d) The Si–C interface of the Si@C nanocomposite.

1200 mAh g^{-1}, with 95.6% retention even after 40 cycles, which makes it a promising anode material for lithium ion batteries.

In addition, heteroatom-containing carbon nanospheres with micropores and mesopores also have been fabricated by forming polyphosphazene nanospheres and carbonizing them with NaOH as the activating agent [89], as shown

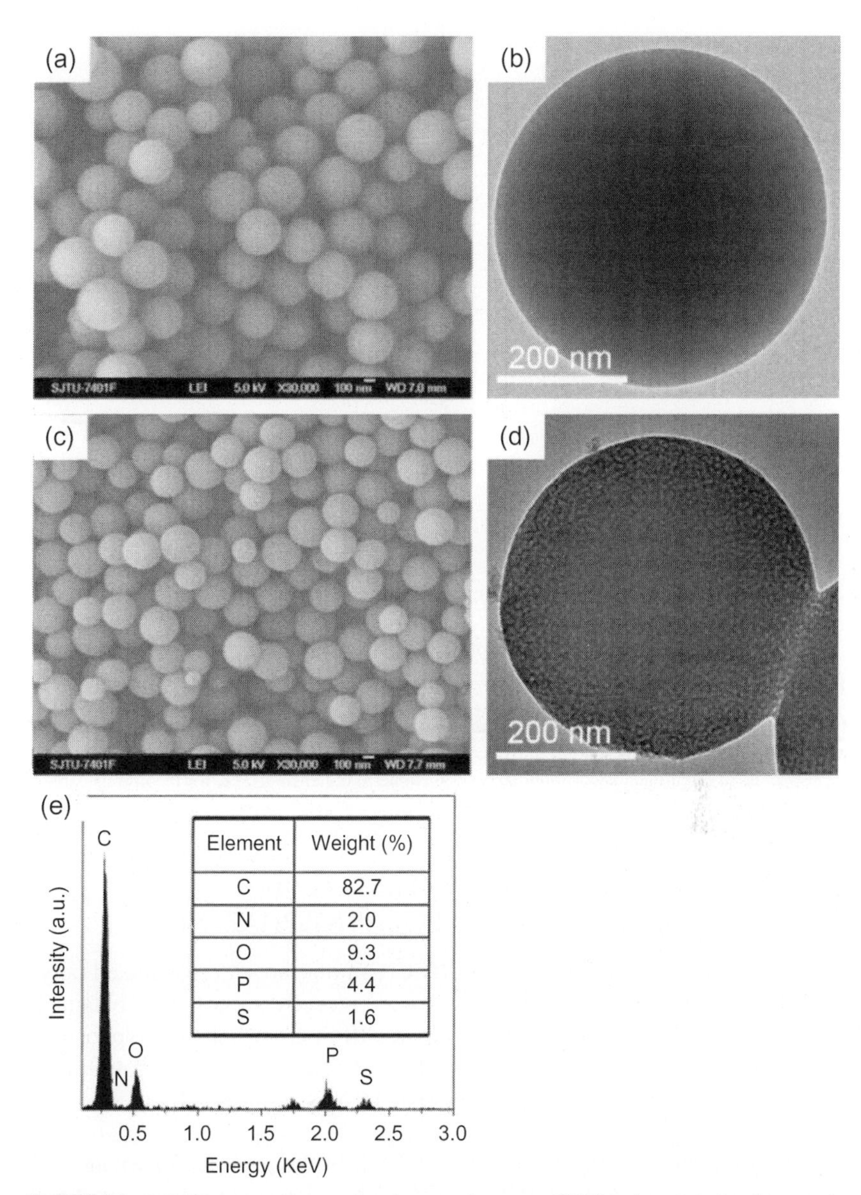

Element	Weight (%)
C	82.7
N	2.0
O	9.3
P	4.4
S	1.6

FIGURE 34 (a, b) Representative scanning electron microscopy (SEM) and transmission electron microscopy (TEM) images of poly(cyclotriphosphazene-4,4′-sulfonyldiphenol) nanospheres. (c, d) SEM and TEM images of porous carbon nanospheres. (e) The EDX pattern of porous carbon nanospheres.

in Figure 34. Moreover, N_2 and H_2 sorption measurements show that the carbon nanospheres possess a BET surface area of $1140\,m^2\,g^{-1}$, a total pore volume of $0.90\,m^3\,g^{-1}$, an ultramicropore volume of $0.30\,m^3\,g^{-1}$, a bimodal pore size distribution (3- to 5-nm and 0.6- to 0.8-nm pore diameter), and a gravimetric hydrogen uptake of 2.7% at 77 K and 1 atm (Figure 35).

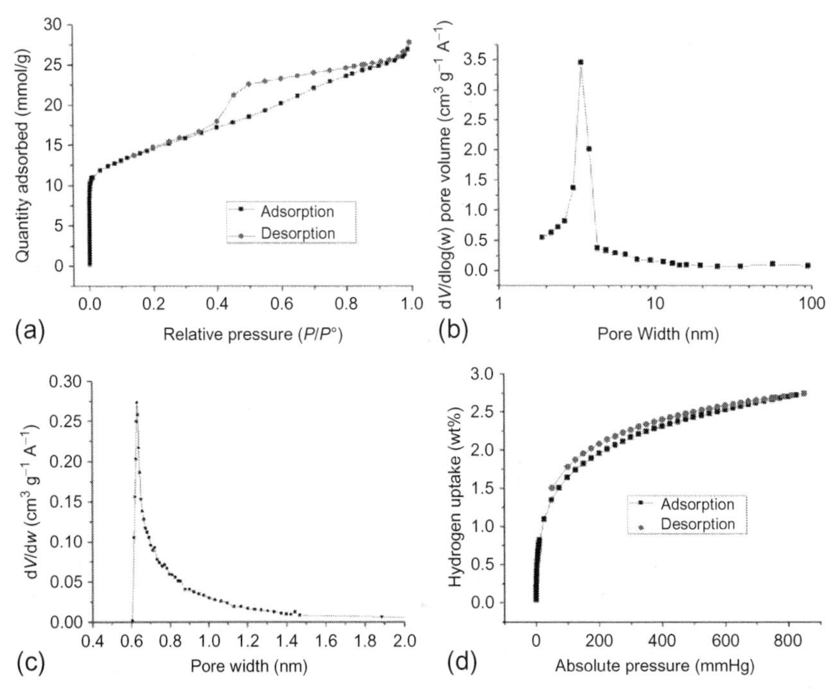

FIGURE 35 (a) Nitrogen adsorption-desorption isotherm obtained at 77 K for porous carbon nanospheres. (b, c) The mesopore and micropore size distribution curves of the porous carbon nanospheres. (d) Hydrogen adsorption-desorption isotherm obtained at 77 K for the porous carbon nanospheres.

5 OTHER KINDS OF CYCLOPHOSPHAZENE-CONTAINING, CROSS-LINKED POLYPHOSPHAZENES

By reacting different monomers with HCCP, similar novel, cross-linked polyphosphazene micro- and nanomaterials were recently successfully obtained. Using 4,4′-diaminodiphenyl ether to replace BPS, cross-linked polyphosphazene spheres with active amino groups were prepared [90]. The cross-linked polyphosphazene with active amino-modified CNTs could directly reduce gold ion in the solution [91].

Using *p*-phenylenediamine to replace BPS, in the presence of hexagonal-phase upconversion nanocrystals, NaYF4:Yb^{3+}, Er^{3+}, core/shell structured upconversion nanocrystal/*p*-phenylenediamine nanophosphors were obtained [92]. The results from this study revealed that the novel nanophosphors not only maintained efficient upconversion fluorescence but also gained good water dispersibility and the ability to chemically modify the surface because of the abundant amino groups tagged on the surface of the polyphosphazene shell.

Using 4,4′-(hexafluoroisopropylidene)diphenol to replace BPS, superhydrophobic, cross-linked polyphosphazene spheres were obtained [93]. In addition, based on the *ex situ* template-induced assembly mechanism, the PANF/CNT composites have been prepared and successfully used as good carriers for Pt and Pt–Co nanoparticles [94,95].

Using phloroglucinol to replace BPS, intrinsically fluorescent poly(cyclotriphosphazene-*co*-phloroglucinol) (PCTP) microspheres with a superior thermal stability and a broad ultraviolet-visible absorption were prepared by a facile self-assembly process [96]. Interestingly, after the PCTP microspheres were treated with acetone, a novel hollow PCTP microsphere with a unique bowl-shaped structure was successfully obtained [97].

Using resveratrol to replace BPS, highly cross-linked, porous poly (cyclotriphosphazene-*co*-resveratrol) submicrospheres with a hollow structure were prepared through an *ex situ* template-induced assembly route [98]. The hollow poly(cyclotriphosphazene-*co*-resveratrol)spheres manifested relatively high drug storage capacity (59.7 mg resveratrol/gram of spheres) and excellent controlled release property (70% (pH 7.4) and 50% (pH 5.5) (separately for up to 40 days).

Using benxidine to replace BPS, intrinsically fluorescent poly (cyclotriphosphazene-*co*-benzidine) microspheres microspheres bearing primary amine groups on the surface were successfully prepared [99]. The microspheres exhibited remarkable thermal stability, photobleaching stability, solvent resistance, and dispersibility in various solvents, including both aqueous and organic media. Further using the microspheres as a fluorescence-based nitroaromatic sensor, 2,4,6-trinitrotoluene, 2,4-dinitrotoluene, and picric acid can be effectively and sensitively detected.

6 CONCLUSIONS

This chapter deals with the synthetic strategies and application of novel cyclophosphazene-containing, cross-linked PZS materials with superior thermal stability, solvent resistance, and dispersibility in both aqueous and organic media. Based on an *in situ* template-induced assembly route, polymorphic PZS micro- and nanomaterials, including nanofibers, nanotubes, and microspheres, could be constructed facilely under mild conditions. Based on an *ex situ* template-induced assembly route, a series of core/shell structured composites using PZS as the shell layers, such as Ag/PZS nanocables, CNT/PZS nanocomposites, CNC/PZS nanochains, and PS/magnetic Fe_3O_4-PZS hybrid microspheres, could be formed easily. As-synthesized PZS micro- and nanomaterials revealed good prospects for application in the fields of carrier materials, adsorption materials, and drug delivery. Furthermore, PZS shows great potential as a good precursor in the preparation of heteroatom-containing porous carbon materials. In addition, recent research on other kinds of cyclophosphazene-containing, cross-linked polyphosphazenes also were mentioned.

Based on the chemical structure and unique micro- and nanoscale morphology of cyclophosphazene-containing, cross-linked polyphosphazene materials, we think cross-linked polymer materials have a promising future, but more scale-up and development work is needed before this becomes a practical reality.

ACKNOWLEDGMENTS

The authors acknowledge funding support from the National Natural Science Foundation of China (grant nos. 51003098 and 51173170) and the Program for New Century Excellent Talents in Universities (NCET). The authors also are grateful to X. Z. Tang and Q. H. Lu for their valuable collaboration in the work cited herein.

REFERENCES

[1] H.R. Allcock, Chemistry and Applications of Polyphosphazenes, John Wiley & Sons, Inc., Hoboken, 2003.
[2] H.R. Allcock, R.L. Kugel, J. Am. Chem. Soc. 87 (1965) 4216.
[3] M. Gleria, R. De Jaeger, J. Inorg. Organomet. Polym. 11 (2001) 1.
[4] R. De Jaeger, M. Gleria, Prog. Polym. Sci. 23 (2) (1998) 179.
[5] M. Gleria, R. De Jaeger, Top. Curr. Chem. 250 (2005) 165.
[6] M.A. Abid, L. Wang, J. Wang, H. Yu, J. Huo, J. Gao, A. Xiao, Des. Monomers Polym. 12 (2009) 357.
[7] H.R. Allcock, Soft Matter 8 (2012) 7521.
[8] K. Inoue, T. Itaya, Bull. Chem. Soc. Jpn. 74 (2001) 1.
[9] X.Z. Tang, X.B. Huang, Modern Inorganic Synthetic Chemistry, 2011 p. 295–320.
[10] H.R. Allcock, Chem. Mater. 6 (1994) 1476.
[11] H.R. Allcock, Adv. Mater. 6 (1994) 106.
[12] L. Zhu, Y. Xu, W. Yuan, J. Xi, X. Huang, X. Tang, et al., Adv. Mater. 18 (2006) 2997.
[13] J.W. Fu, X.B. Huang, Y. Zhu, Y.W. Huang, L. Zhu, X.Z. Tang, Eur. Polym. J. 44 (2008) 3466.
[14] J.X. Huang, R.B. Kaner, J. Am. Chem. Soc. 126 (2004) 851.
[15] H.W. Fox, E.F. Hare, W.A. Zisman, J. Phys. Chem. 59 (1955) 1097.
[16] S. Wu, in: D.R. Paul, S. Newman (Eds.), Polymer Blends, Academic Press, New York, 1978, pp. 244–288.
[17] M. Steinhart, R.B. Wehrspohn, U. Gosele, J.H. Wendorff, Angew. Chem. Int. Ed. 43 (2004) 1334.
[18] J.W. Fu, X.B. Huang, Y. Zhu, L. Zhu, X.Z. Tang, Macromol. Chem. Phys. 209 (2008) 1845.
[19] J.W. Fu, X.B. Huang, Y.W. Huang, L. Zhu, Y. Zhu, X.Z. Tang, Macromol. Mater. Eng. 293 (2008) 173.
[20] J.W. Fu, M.H. Wang, C. Zhang, Q. Xu, X.B. Huang, X.Z. Tang, J. Mater. Sci. 47 (2012) 1985.
[21] D.T. Mitchell, S.B. Lee, L. Trofin, N. Li, T.K. Nevanen, H. Soderlund, C.R. Martin, J. Am. Chem. Soc. 124 (2002) 11864.
[22] D. Wouters, U.S. Schubert, Angew. Chem. Int. Ed. 43 (2004) 2480.
[23] Y. Liu, D.C. Wu, W.D. Zhang, X. Jiang, C.B. He, T.S. Chung, S.H. Goh, K.W. Leong, Angew. Chem. Int. Ed. 44 (2005) 4782.
[24] J.H. Yuan, K. Wang, X.H. Xia, Adv. Funct. Mater. 15 (2005) 803.
[25] J. Barner, F. Mallwitz, L. Shu, A.D. Schluter, J.P. Rabe, Angew. Chem. Int. Ed. 42 (2003) 1932.
[26] A. Bianco, M. Prato, Adv. Mater. 15 (2003) 1765.
[27] G.E. Douberly Jr., S. Pan, D. Walters, H. Matsui, J. Phys. Chem. B 105 (2001) 7612.

[28] D.T. Bong, T.D. Clark, J.R. Granja, M.R. Ghadiri, Angew. Chem. Int. Ed. 40 (2001) 988.
[29] D. Pantarotto, R. Singh, D. McCarthy, M. Erhardt, J.P. Briand, M. Prato, K. Kostarelos, A. Bianco, Angew. Chem. Int. Ed. 43 (2004) 5242.
[30] C.C. Chen, Y.C. Liu, C.H. Wu, C.C. Yeh, M.T. Su, Y.C. Wu, Adv. Mater. 17 (2005) 404.
[31] Z.J. Guo, P.J. Sadler, S.C. Tsang, Adv. Mater. 10 (1998) 701.
[32] X.Y. Gao, H. Matsui, Adv. Mater. 17 (2005) 2037.
[33] P. Asuri, S.S. Karajanagi, E. Sellitto, D.Y. Kim, R.S. Kane, J.S. Dordick, Biotechnol. Bioeng. 95 (2006) 804.
[34] C.R. Martin, Science 266 (1994) 1961.
[35] J. Jang, S. Ko, Y. Kim, Adv. Funct. Mater. 16 (2006) 754.
[36] J.W. Fu, X.B. Huang, Y. Zhu, Y.W. Huang, X.Z. Tang, Appl. Surf. Sci. 255 (2009) 5088.
[37] H.Q. Cao, Z. Xu, H. Sang, D. Sheng, C.Y. Tie, Adv. Mater. 13 (2001) 121.
[38] A.H. Chen, K. Kamata, M. Nakagawa, T. Iyoda, H.Q. Wang, X.Y. Li, J. Phys. Chem. B 109 (2005) 18283.
[39] M.N. Nadagouda, R.S. Varma, Macromol. Rapid Commun. 28 (2007) 2106.
[40] X. Feng, H. Huang, Q. Ye, J. Zhu, W. Hou, J. Phys. Chem. C 111 (2007) 8463.
[41] Z.Y. Li, H.M. Huang, C. Wang, Macromol. Rapid Commun. 27 (2006) 152.
[42] G.W. Lu, C. Li, J.Y. Shen, Z.J. Chen, G.Q. Shi, J. Phys. Chem. C 111 (2007) 5926.
[43] V.G. Organo, A.V. Leontiev, V. Sgarlata, H.V. Rasika Dias, D.M. Rudkevich, Angew. Chem. Int. Ed. 44 (2005) 3043.
[44] Y. Kim, M.F. Mayer, S.C. Zimmerman, Angew. Chem. Int. Ed. 42 (2003) 1121.
[45] S. Hecht, A. Khan, Angew. Chem. Int. Ed. 42 (2003) 6021.
[46] J.W. Fu, X.B. Huang, Y.W. Huang, Y. Pan, Y. Zhu, X.Z. Tang, J. Phys. Chem. C 112 (2008) 16840–16844.
[47] D.M. Guldi, G.M.A. Rahman, F. Zerbetto, M. Prato, Acc. Chem. Res. 38 (2005) 871.
[48] N. Grossiord, J. Loos, O. Regev, C.E. Koning, Chem. Mater. 18 (2006) 1089.
[49] H.J. Dai, Acc. Chem. Res. 35 (2002) 1035.
[50] D. Tasis, N. Tagmatarchis, A. Bianco, M. Prato, Chem. Rev. 106 (2006) 1105.
[51] J.W. Fu, X.B. Huang, Y.W. Huang, J.W. Zhang, X.Z. Tang, Chem. Commun. 9 (2009) 1049.
[52] J.F. Zhou, L.J. Meng, X.L. Feng, X.K. Zhang, Q.H. Lu, Angew. Chem. Int. Ed. 49 (2010) 8476.
[53] J. Ge, Q. Zhang, T. Zhang, Y. Yin, Angew. Chem. 120 (2008) 9056.
[54] J. Ge, L. He, J. Goebl, Y. Yin, J. Am. Chem. Soc. 131 (2009) 3484.
[55] H. Deng, X. Li, Q. Peng, X. Wang, J. Chen, Y. Li, Angew. Chem. 117 (2005) 2842.
[56] S.C. Luo, J. Jiang, S.S. Liour, S.J. Gao, J.Y. Ying, H.H. Yu, Chem. Commun. 19 (2009) 2664.
[57] L.B. Chen, F. Zhang, C.C. Wang, Small 5 (2009) 621.
[58] R.J. Hickey, A.S. Haynes, J.M. Kikkawa, S.J. Park, J. Am. Chem. Soc. 133 (2011) 1517.
[59] F. Caruso, M. Spasova, A. Susha, M. Giersig, R.A. Caruso, Chem. Mater. 13 (2001) 109.
[60] X.H. Li, D.H. Zhang, J.S. Chen, J. Am. Chem. Soc. 128 (2006) 8382.
[61] J.W. Fu, M.H. Wang, C. Zhang, X.Z. Wang, H.F. Wang, Q. Xu, J. Mater. Sci. 48 (2013) 3557.
[62] M.C. Daniel, D. Astruc, Chem. Rev. 104 (2004) 293.
[63] S.V. Vasilyeva, M.A. Vorotyntsev, I. Bezverkhyy, E. Lesniewska, O. Heintz, R. Chassagnon, J. Phys. Chem. C 112 (2008) 19878.
[64] A.K. Manocchi, N.E. Horelik, B. Lee, H. Yi, Langmuir 26 (2010) 3670.
[65] P.I. Richards, A. Steiner, Inorg. Chem. 43 (2004) 2810.
[66] J.W. Fu, M.H. Wang, S.T. Wang, X.Z. Wang, H.F. Wang, L. Hu, Q. Xu, Appl. Surf. Sci. 257 (2011) 7129.
[67] M.H. Wang, J.W. Fu, D.D. Huang, C. Zhang, Q. Xu, Nanoscale 5 (2013) 7913.

[68] Y.H. Kim, C.W. Kim, H.G. Cha, D.K. Lee, B.K. Jo, G.W. Ahn, E.S. Hong, J.C. Kim, Y.S. Kang, J. Phys. Chem. C 113 (2009) 5105.
[69] B. Chen, X.L. Jiao, D.R. Chen, Cryst. Growth Des. 10 (2010) 3378.
[70] M. Chen, L.Y. Wang, J.T. Han, J.Y. Zhang, Z.Y. Li, D.J. Qian, J. Phys. Chem. B 110 (2006) 11224.
[71] S. Magdassi, M. Grouchko, O. Berezin, A. Kamyshny, ACS Nano 4 (2010) 1943.
[72] O.A. Yeshchenko, I.M. Dmitruk, A.A. Alexeenko, A.V. Kotko, Nanotechnology 21 (2010) 045203 1.
[73] F. Lin, R. Doong, J. Phys. Chem. C 115 (2011) 6591–6598.
[74] J. Zeng, Q. Zhang, J.Y. Chen, Y.N. Xia, Nano Lett. 10 (2009) 30.
[75] J.F. Huang, S. Vongehr, S.C. Tang, H.M. Lu, J.C. Shen, X.K. Meng, Langmuir 25 (2009) 11890.
[76] A. Leelavathi, T. Rao, T. Pradeep, Nanoscale Res. Lett. 6 (2011) 123.
[77] K. Mori, A. Kumami, M. Tomonari, H. Yamashita, J. Phys. Chem. C 113 (2009) 16850.
[78] Y.J. Yao, F.F. Xu, M. Chen, Z.X. Xu, Z.W. Zhu, Bioresour. Technol. 101 (2010) 3040.
[79] E.C. Lima, B. Royer, J.C.P. Vaghetti, N.M. Simon, B.M. da Cunha, F.A. Pavan, E.V. Benvenutti, R.C. Veses, C. Airoldi, J. Hazard. Mater. 155 (2008) 536.
[80] P.A. Carneiro, G.A. Umbuzeiro, D.P. Oliveira, M.V.B. Zanoni, J. Hazard. Mater. 174 (2010) 694.
[81] R.O.A. de Lima, A.P. Bazo, D.M.F. Salvadori, C.M. Rech, D.P. Oliveira, G.A. Umbuzeiro, Mutat. Res. Genet. Toxicol. Environ. Mutagen. 626 (2007) 53.
[82] Z.H. Chen, J.N. Zhang, J.W. Fu, M.H. Wang, X.Z. Wang, R.P. Han, Q. Xu, J. Hazard. Mater. 273 (2014) 263.
[83] L.H. Ai, C.Z. Zhang, F. Liao, Y. Wang, M. Li, L.Y. Meng, J. Jiang, J. Hazard. Mater. 198 (2011) 282.
[84] W. Liu, X.B. Huang, H. Wei, K.Y. Chen, J.X. Gao, X.Z. Tang, J. Mater. Chem. 21 (2011) 12964.
[85] J.W. Fu, Q. Xu, J.F. Chen, Z.M. Chen, X.B. Huang, X.Z. Tang, Chem. Commun. 46 (2010) 6563.
[86] J.W. Fu, Z.M. Chen, Q. Xu, J.F. Chen, X.B. Huang, X.Z. Tang, Carbon 49 (3) (2011) 1037.
[87] J.W. Fu, Y.W. Huang, Y. Pan, Y. Zhu, X.B. Huang, X.Z. Tang, Mater. Lett. 62 (2008) 4130.
[88] P.F. Gao, J.W. Fu, J. Yang, R.G. Lv, J.L. Wang, Y.N. Nuli, X.Z. Tang, Phys. Chem. Chem. Phys. 11 (2009) 11101.
[89] J.W. Fu, M.H. Wang, C. Zhang, P. Zhang, Q. Xu, Mater. Lett. 81 (2012) 215.
[90] P. Zhang, X.B. Huang, J.W. Fu, Y.W. Huang, Y. Zhu, X.Z. Tang, Macromol. Chem. Phys. 210 (2009) 792.
[91] P. Zhang, X.B. Huang, J.W. Fu, Y.W. Huang, X.Z. Tang, Macromol. Mater. Eng. 295 (2010) 437.
[92] K.R. Chen, X.B. Huang, H. Wei, X.Z. Tang, Mater. Lett. 101 (2013) 54.
[93] W. Wei, X.B. Huang, X.L. Zhao, P. Zhang, X.Z. Tang, Chem. Commun. 46 (2010) 487.
[94] X.B. Huang, W. Wei, X.L. Zhao, X.Z. Tang, Chem. Commun. 46 (2010) 8848.
[95] J.P. Qian, W. Wei, X.B. Huang, Y.M. Tao, K.Y. Chen, X.Z. Tang, J. Power Sources 210 (2012) 345.
[96] T.J. Pan, X.B. Huang, H. Wei, W. Wei, X.Z. Tang, Macromol. Chem. Phys. 213 (2012) 1590.
[97] T.J. Pan, X.B. Huang, H. Wei, X.Z. Tang, Macromol. Chem. Phys. 213 (2012) 2606.
[98] F.Q. Chang, X.B. Huang, H. Wei, K.Y. Chen, C.C. Shan, X.Z. Tang, Mater. Lett. 125 (2014) 128.
[99] W. Wei, X.B. Huang, K.Y. Chen, Y.M. Tao, X.Z. Tang, RSC Adv. 2 (2012) 3765.

Author Index

Note: Page numbers followed by *f* indicate figures and *t* indicate tables.

B

C

D

E

F

G

I

J

K

L

M

N

O

P

Q

R

S

T

X

Y

Z

Subject Index

Note: Page numbers followed by *f* indicate figures and *t* indicate tables.

D

I

J

K

L

M